BASIC ELECTRICAL ENGINEERING & ELECTRONICS

OTHER BOOKS BY THE SAME AUTHOR

- Solved Examples Electrical Calculations, 3e
- Solved Examples Electrical Calculations (Hindi)
- Viva Voce in Electrical Engineering, 5e
- Viva Voce in Electrical Engineering (Hindi), 4e
- Basic Electrical Engineering and Electronics (Hindi), 2e

BASIC ELECTRICAL ENGINEERING & ELECTRONICS

A Textbook in Electrical Engineering and Electronics for the Examination of N.C.V.T., Diploma in Engineering and A.M.I.E. Sec. A Through Questions and Answers

Fourth Edition

D.K. Sharma

A.M.I.E. (Elect. Engg.)

Ex-Lecturer Elect. Engg.,
G.B. Pant Polytechnic, Okhla, New Delhi

and

P. Sharma

M.A., B.Ed.

CBS

CBS Publishers & Distributors Pvt. Ltd.

New Delhi • Bengaluru • Chennai • Kochi • Kolkata • Mumbai
Hyderabad • Nagpur • Patna • Pune • Vijayawada

ISBN: 978-81-239-1508-1

First Edition: 1986
Second Edition: 1987
Third Edition: 1994
Fourth Edition: 2007
Reprint: 2008, 2009, 2011, 2013, 2017, 2018-2021

Published by **Satish Kumar Jain** and produced by **Varun Jain** for
CBS Publishers & Distributors Pvt. Ltd.,
4819/XI Prahlad Street, 24 Ansari Road, Daryaganj, New Delhi - 110002
delhi@cbspd.com, cbspubs@airtelmail.in • www.cbspd.com
Ph.: 23289259, 23266861, 23266867 • Fax: 011-23243014

Corporate Office: 204 FIE, Industrial Area, Patparganj, Delhi - 110 092
Ph: 49344934 • Fax: 011-49344935
E-mail: publishing@cbspd.com • publicity@cbspd.com

Branches:
• *Bengaluru:* 2975, 17th Cross, K.R. Road, Bansankari 2nd Stage,
  Bengaluru - 70 • Ph: +91-80-26771678/79 • Fax: +91-80-26771680
  E-mail: cbsbng@gmail.com, bangalore@cbspd.com
• *Chennai:* No. 7, Subbaraya Street, Shenoy Nagar, Chennai - 600030
  Ph: +91-44-26681266, 26680620 • Fax: +91-44-42032115
  E-mail: chennai@cbspd.com
• *Kochi:* Ashana House, 39/1904, A.M. Thomas Road, Valanjambalam,
  Ernakulum, Kochi • Ph: +91-484-4059061-65
  Fax: +91-484-4059065 • E-mail: cochin@cbspd.com
• *Kolkata:* 6-B, Ground Floor, Rameshwar Shaw Road, Kolkata - 700014
  Ph: +91-33-22891126/7/8 • E-mail: kolkata@cbspd.com
• *Mumbai:* 83-C, Dr. E. Moses Road, Worli, Mumbai - 400018
  Ph: +91-9833017933, 022-24902340/41 • E-mail: mumbai@cbspd.com

Representatives:

• Hyderabad: 0-9885175004	• Nagpur: 0-9021734563
• Patna: 0-9334159340	• Pune: 0-9623451994
• Jharkhand: 0-9811541605	• Uttarakhand: 0-9716462459

Printed at:
J.S. Offset Printers, Delhi (India)

Preface to the Fourth Edition

The continued popularity of this book right from 1983 when it was first published has been most encouraging to the authors. Since then a number of books by different authors have appeared in the market despite of this, this book has retained its popularity and demand amongst the students and teaching staff of various councils, boards, institutions, colleges and engineering institutions.

Now this 4th edition has been revised and enlarged. In this edition earnest endeavour has been made to include all those aspects of the subject, which the students/trainees should learn as per the syllabus and are expected to know. The power chapter is reoriented having generating power stations viz. thermal, hydroelectric, nuclear, MHD etc. Knowledge of transmission lines, poles structure, circuit breakers etc. Other chapters are also worked upon to facilitate the readers of the current knowledge and present mode of current questioning techniques.

The authors are very thankful to publisher and printer for their cooperation in bringing out the book. We are also thankful to the student/ trainees and teaching staff for their valuable cooperation. The cost which the publisher kept accessible inspite of increasing cost of paper will help it to be more conveniently accessible to the students.

Suggestions for improvement of this book will be thankfully acknowledged.

D.K. Sharma
P. Sharma

Preface to the First Edition

In the process of writing this book, we have kept in mind all the aspects of the new 'vocational and technical education' proposed by the Govt. of India in her new Education Policies likely to be implemented in the near future. This is the first book which covers the entire syllabus to be implemented by the DGE&T for NCVT courses for the sessions 1985-87 onwards. The book has been written with a view to serve as a textbook through questions and answers, as is the demand of the time. The subject matter, broadly speaking, covers sufficient ground suitable for any syllabus for National Certificate, Diploma and AMIE Sec. A examination of electrical engineering. Efforts have been made to cover all the topics and problems from the fundamentals in easy, lucid and expressive language for the convenience and better understanding of the readers.

The book comprises of 28 chapters, viz. the elementary treatment of AC and DC circuits, AC and DC machines, measuring instruments, batteries, illumination, lamps and the newly introduced electronics for the students of electrical engineering courses and so on so forth. All these have been taken up and expressed in a systematic and scientific manner in order to make them self-explanatory and more effective.

We are really thankful to Guruji Shri T.C. Sharma for his able guidance from time to time who showered his all the blessings upon us. We are also thankful to Shri O.P. Sharma, Shri S.K. Sharma and Shri Rautela for the inspiration and timely cooperation.

The success and popularity of this venture depends upon the liking of the readers, whose valuable suggestions are invariably welcomed to improve upon the book. Errors might have crept in despite utmost care and we will be thankful if these are pointed out along with the other suggestions, if any.

Karkardooma,   **D.K. SHARMA**
New Delhi - 110 092   **P. SHARMA**

Contents

PART-I

Post Office Box, Thermocouple, Maximum Demand
Indicator *KVAR* and *VARH* Measurements, Review
Questions, Numerical Problems.

PART-II

PART-III

PART-IV

SYMBOLS, ABBREVIATIONS, CONVERSION FACTORS AND CONSTANTS

The Electrical Units

	Symbol		*Symbol*
Amperes	A	Farads	F
Volts	V	Watts	W
Ohms	Ω	Kilowatts	kW
Mho	℧	Ampere hours	Ah
Coulombs	C	Watt hours	Wh
Joules	J	Kilowatt-hours	kWH
Henry	H	Kilovolt-amperes	kVA

Dimensional Prefixes

M = mega- or mega-	(1,000,000 or 10^6)
k = kilo-	(1,000 or 10^3)
m = milli-	$\left(\dfrac{1}{1,000} \ \text{or} \ 10^{-3}\right)$
μ = micro	$\left(\dfrac{1}{1,000,000} \ \text{or} \ 10^{-6}\right)$

Conversion Factors

1 cm	= 0.937 in.	1 in	= 2.54 cm
1 m	= 39.37 in.	1 in	= 0.0254 m
1 m	= 3.28 ft.	1 ft	= 0.3048 m
1 Nw	= 0.2248 lb	1 lb	= 4.45 Nw
1 Nw	= 0.1019 kg		
1 kg	= 9.81 Nw		
1 kg	= 2.205 lb	1 lb	= 0.4536 kg
1 joule	= 0.7374 ft. lb	1 ft. lb	= 1.356 joules
1 joule	= 2.388×10^{-4} kCal		
1 kCal	= 4187 joules		
1 kCal	= 3.97 B.T.U.	1 B.T.U.	= 0.252 kCal
1 kW	= 1.34 h.p.	1 h.p.	= 746 = 735.5 W
			(metric)

Constants

μ_0 = $4\pi \times 10^{-7}$ rationalised M.K.S. units (henry/m).

ϵ_0 = 8.85×10^{-12} rationalised M.K.S. units (farad/m).

c = 2.998×10^8 m/sec.

c = 2.7183

π = 3.1416

$\dfrac{1}{\pi}$ = 0.3183

$\dfrac{1}{\mu_0}$ = 0.796×10^6

Greek Alphabet

Letter		English Pronunciation	Letter		English Pronunciation
α	A	Alpha	ν	N	Nu
β	B	Beta	ζ	Ξ	Ksi
γ	Γ	Gamma	0	O	Omicron
δ	Δ	Delta	π	π	Pi
$\in$	E	Epsilon	ρ	P	Rho
ζ	Z	Zeta	σ	Σ	Sigma
η	H	(H)eta	τ	T	Tau
θ	m	Theta	ν	Y	Upsilon
i	I	Iota	ϕ	Σ	Phi
k	K	Kappa	χ	X	Chi
λ	Λ	Lamda	ψ	Ψ	Psi
μ	M	Mu	ω	Ω	Omega

Electrical Symbols

	Symbol		Symbol
Current	I		
Voltage	V	e.m.f. of each cell	e
Resistance	R	Internal resistance of each cell	r
Specific resistance	ρ	Number of cells in a row	n
Area of cross section	a	Number of rows	p or m

Length	l
Resistance Temperature coefficient	α

External resistance	R

Particulars	Symbol	Particulars	Symbol
Power mechanical	P	Power Electrical	W
Time in seconds	t	Quantity Electricity	Q
Charge in coulombs	Q	Energy consumed	kWh
Calories	Cal		
Magnetic flux	ϕ	Magnetic flux density	B
Area of cross section of		Magnetomotive force	F
magnetic field	A	Magnetising force or	
Permeability	μ	Magnetic field straight	H
Permeability in free space	μ_0	Permeability relative	μ_r
Reluctance	$Rel\ or\ S$	Steinmitz constant	η
E.m.f. induced	e	Coefficient of Self Induction	L
Coefficient of Mutual		Rate of changes in current	$\dfrac{d_i}{d_t}$
induction	M		
Permittivity	$\in$	Permittivity relative	$\in_r$
Permittivity in free space	$\in_0$	Capacitance	C
Capacity in farad	$F\ or\ \mu F$		
Electromotive force	F	Voltage at terminal	V
Armature current	I_a	Shunt field current	I_{sh}
Series field current	I_{se}	Load Current	I_L
Armature resistance	R_a	Shunt field resistance	R_{sh}
Series field resistance	R_{se}	Interpole resistance	R_i
Flux per pole	ϕ	Number of conductor	Z
Speed in r.p.m.	N	Number of parallel paths	a
Number of poles	P	Efficiency	η
Eddy current losses	W_e		
Back e.m.f.	E_b	Armature torque	T_a
Shaft torque	T_{sh}	Lost torque	T_{lost}
Frequency	f	Time period	T
Length of coil side	l	Radius	r
Angular velocity	ω	Current maximum	I_m
Current r.m.s.	$I_{r.m.s}$	Current average	I_{av}
Angular displacement	ϕ		

Power factor	$\cos \phi$	Power factor	$\cos \phi$
Inductive reactance	X_L	Capacitive reactance	X_c
Capacitance	C	Impedance	Z
Susceptance	B	Admitance	Y
Conductance	G	Reactance	X
Resonance frequency	f_r	Line current	I_L
Phase current	I_P	Form factor	K_f
Primary winding	P	Secondary winding	S
No load current	I_0	Wattful current	I_w
Magnetising current	I_μ	Phase voltage	V_P
Line voltage	V_L	Current primary winding	I_P
Current secondary winding	I_s	Primary winding Turn	N_P
Secondary winding turn	N_s	Transformation Ratio	ρ
Equivalent resistance referred to primary winding	$\overline{R}_P$	Equivalent resistance referred to secondary winding	$\overline{R}_s$
Equivalent reactance referred to primary winding	$\overline{\chi}_P$	Equivalent reactance referred to secondary winding	$\overline{X}_s$
Equivalent impedance referred to primary winding	$\overline{Z}_P$	Equivalent impedance referred to secondary winding	$\overline{Z}_s$
Distribution factor	K_d	Coil space factor	K_c
Effective resistance armature	R_e	Synchronous reactance	X_s
Synchronous impedance	Z_s	Resultant e.m.f.	E_r
Rotor e.m.f.	E_r	Rotor current	I_r
Rotor power factor	$\cos \phi_r$	Rotor frequency	f_r
Synchronous speed	N_s	Rotor speed	N_r
Stator current	I_s	Slip	S
Current in main winding	I_M	Current in starting winding	I_s
Total current taken from supply	I_T		

1

Introduction and Safety Precautions

INTRODUCTION

Electricity has comprehensively effected our daily life in all spheres. Though it has made life very comfortable and luxurious but on other side it may cause dangerous situations like electric shock, electrical accidents, casualties and even in some cases may lead to death also.

The electrical accidents may be because of damaged cable insulation, mishandling of appliances, violations of electricity rule and regulations, environmental changes and hasty habits, so it becomes necessary to be careful and observe precautions, which are laid down for the safety of human beings and properties are called safety precautions. (The details of such precautions are given in Appendix-D of I.S.-5216-1969).

1.1. SAFETY PRECAUTIONS

Q. Write down the precautions to be observed while handling or working on the electrical equipments, appliances and electrical gears.

Ans. There are the following precautions to be observed for handling the electrical equipment or appliances:

1. The electric shocks are easily received and easily avoided being careful. The risk is not always apparent, so be careful.
2. Be aware of the line conductor whether bared or insulated.
3. Never temper unnecessarily with any electrical equipment or gear which is live.
4. Before switching on, be sure that the insulation of the cable is sound and safe.

5. Before switching on, make sure that the portable equipment is properly earthed.
6. Never energise any line conductor unless you are sure that all is clear and there is no one working on the line.
7. Never temper any electrical protective, interlocking and gear unless you are specially authorised to do so, and that too after getting permission. These devices and interlockings are for the safety.
8. Never disconnect a plug point by pulling out the flexible cable.
9. The safety depends upon good earthing. Maintain good earthing connection with the appliance. The three pin plug shoe and socket should be used to enable for good earthing and earth continuity.
10. The weight of the cable should not come on the terminals of the plug shoe. It should be well supported by providing the node or by proving a screw and strip inside the plug shoe.
11. Moisture and water should be prohibited entering the current carrying portion of the appliance.
12. Joints in all the leads should be avoided if at all it is essential that should be properly insulated and mechanically strong enough.
13. While handling any electrical equipment or the appliance, merely switching off is not enough and sufficient, but it should be disconnected from the mains, leakage insulation may give shock.
14. Live wire must be controlled by the switch and there should not be any fuse on neutral wire.
15. Proper tools, in proper condition and size should be used. (Precautions, care and maintenance of tools are given in chapter "Tools" separately).

Q. **What precautions would you observe while**
    (*a*) **Working on an overhead line,**
    (*b*) **Working on a ladder,**
    (*c*) **Preparing an electrolyte and battery charging,**
    (*d*) **Replacing a blown fuse, and**
    (*e*) **In case of electric fire?**

Ans. (*a*) **Working on an overhead line:**
    (*i*) Before working on an overhead line never forget to put on your safety belt.
    (*ii*) Insulate yourself, use insulated tools and rubber gloves.
    (*iii*) While climbing over a pole, grip the pole firmly and if you don't have the enough practice don't take any risk and use ladder.
    (*iv*) Be sure whether the line conductors of the overhead line are dead or energised.

(*v*) If at all the line is off or dead, use short circuiting chain better on both side of the pole to avoid any mishappening.

(*b*) **Working on a ladder:**

(*i*) If a ladder is required to use, don't hesitate to use, it is for your safety and use.

(*ii*) Proper size of the ladder should be used. A large ladder will make a more inclination angle which will cause slipping and small ladder will leave more distance so in both the cases it is not advisable.

(*iii*) The ladder in use, should be well supported by atleast a person standing on the ground.

(*iv*) Never throw any tool from the pole, it may injure the person standing on the ground and secondly it is against the care and maintenance of the tools and will spoil it.

(*c*) **Preparing an electrolyte and battery charging.** The following precautions should be observed:

(*i*) The room should be well ventilated and try to maintain a good circulation of fresh air.

(*ii*) Always add sulphuric acid into water and not water to sulphuric acid otherwise tremendous amount of heat will develop.

(*iii*) Add sulphuric acid drop by drop from the side of the container to overcome the indothermic reaction.

(*iv*) Never bring a naked flame near the place where battery charging is done.

(*d*) **Replacing a blown fuse:**

(*i*) Always switch off the mains, before replacing a blown fuse.

(*ii*) Take out the grip and try to know the reason of blowing, (if the fuse element is just broken at a point and remaining ends are existing, it indicates the overloading, reduce the load, if the wire is just melted out and drops of molten melt exists, it shows that it is because of short circuit or earth fault, first check and remove the fault).

(*iii*) Replace the fuse according to the capacity. Don't wire it for more capacity than it is designed.

(*iv*) Wire should be tightened and housed properly because the fusing current depends upon the length and environmental conditions also.

(*v*) Put the fuse properly and close the cover. Now be at one side of the switch preferably towards handle side, to avoid any mishappening and then put the handle on.

(*e*) **In case of electric fire:**

(*i*) Put the main switch off immediately if you can do it.

(*ii*) Never throw water on the fire or the live conductor, it is danger-ous because water is a conductor of electricity.

(*iii*) Throw dry sand or dust on the fire.

(*iv*) Use fire extinguisher, but before use, make sure that it is suit-able for that purpose. Liquid CO_2 or CCl_4 extinguishers are generally suitable and use them.

1.2. IMPORTANCE OF ELECTRICITY

Q. What is the importance of electricity to an engineer and a trades man?

Ans. An engineer and the trades-man frequently come across machines which are electrically operated. A thorough knowledge of the laws governing electricity and its applications help them to carry out the assigned jobs; efficiently and effectively.

The study of electrical engineering for engineers and trades men helps in the following ways:

(*i*) It helps to acquire the complete knowledge of the machine under their charge.

(*ii*) It helps to make minor repair and remove faults if necessary.

(*iii*) It helps in selection of suitable electric drives for the machine.

(*iv*) It helps in making best use of electricity with maximum effi-ciency.

1.3. FIRST AID AND ELEMENTARY NURSING

Q. What do you understand by the first aid and elementary nursing?

Ans. "What men call accident is the doing of God providence". Acci-dents often take place at site, place and time of working, at home and elsewhere. So everyone specially the craftsman and every gen-eral people, should know what to do in such cases. If proper help is not extended in time the patient condition can get worse or if care is not taken it may even cause danger to life.

First aid means the treatment given till proper medical aid comes. The principles and practice of first aid depends on practical medi-cine and surgery. There are the following points to be kept in mind while attending a patient:

(*i*) A first aider can give skilled help, prevent death, promote re-covery and see that an injury or illness does not get worse until the doctor arrives.

(*ii*) The first aider should see such materials that are at hand.

(*iii*) The first aider does not take the doctor's place but only acts promptly.

(*iv*) A first aider should know one's limitation.

(*v*) When giving the first aid the person should see if the patient is alive, unconscious or dead, if alive the patient breathing and pulse are the first things to be noticed. Pulse rate for a child 80-90 per minute, healthy adult 72 per minute, old age 60 per minute.

Q. What are the first aid measures?

Ans. The measures can be divided into the followings:

(*a*) Serious emergencies,

(*b*) Minor cuts, abrasions and bruises,

(*c*) Extensive and deep cuts,

(*d*) Dirty wounds,

(*e*) Bleedings

(*a*) **Serious emergencies.** There are certain situations which are a matter of life and death and are classified under serious emergencies.

(*i*) When breathing stops and if help is not given at once the patient will die.

(*ii*) If there is a severe bleeding and the bleeding is not stopped the patient will die.

(*iii*) Suppose severe shock follows after an accident, unless suitable steps are taken the patient will die.

(*b*) **Minor cuts abrasions and bruises.** In this case:

(*i*) Apply a little tincture of iodine or spirit if the wound is clean.

(*ii*) Put a clean piece of gauze over it and then bandage or put a sticking plaster.

(*iii*) Instead some cotton dipped in tincture benzoin may be put on the wound.

(*c*) **Extensive and deep cut.** In this case:

(*i*) Cover the would with a sterilised dressing and bandage.

(*ii*) Send for doctor immediately.

(*d*) **Dirty wound.** In this case first cover the wound with a piece of clean cotton or gauze, clean the area of wound. Apply tincture of iodine and cover it with gauze or cotton and send for doctor.

It is very important that in case of dirty wounds the danger of tetanus is always apparent, so the doctor must be consulted.

(*e*) **Bleeding.** The bleeding may happen, it should be stopped immediately otherwise in certain cases it may lead the serious happenings.

(*i*) To check the bleeding, cover the wound with a clean piece of gauze and bandage tightly.

(*ii*) Apply the thumb pressure, if suitable or raise the injured part above the level of heart to stop bleeding or press the part above the level of heart to stop bleeding or press the part below the wound.

(*iii*) In case of bleeding from the nose, then keep the patient before a window, loose his cloths, tilt the head backward, open his mouth to take breath instead of nose, keep a pad dipped in cold water or ice bag on the patient nose.

(*iv*) In case, if internal bleeding which is not located don't delay, send for doctor or take the patient to hospital immediately.

Q. Give a brief idea of elementary equipments.

Ans. In order to look after a patient, the nursing equipments has to be considered as follows:

(*i*) Thermometer – at least two thermometers one to be hang on the wall etc. to note the room temperature and second for the patient. The second should be clinical or doctor's thermometer.

(*ii*) A hot water bottle.

(*iii*) An enema syringe or catheter and funnel.

(*iv*) A bed pan and urinal.

(*v*) A graduated glass.

(*vi*) A feeding cup.

(*vii*) Two small open bowls.

(*viii*) An air-ring cushion.

(*ix*) An ice bag.

(*x*) A jug and basin for doctors use.

1.4. TREATMENT OF ELECTRIC SHOCK

Q. Write a short note on the treatment of electric shock and different methods of artificial respiration.

Ans. When a person comes in contact with a live wire, the electric current starts flowing through the body to the earth. The value of the current depends upon the voltage, more voltage more current and less voltage less current. The current flowing through the body effects the nerves system of the body, sometimes the current is high and the heart stops its working and the person cannot breath, so when a person comes in contact with the live wire and is suffering from the electric shock, the following steps must be taken.

(*i*) **Removal from the electric shock.** If a person is in contact with the live conductor, put the main switch off; if it is near and within approach. If you cannot switch off immediately then insulate yourself on a wooden article like chair, table or any board etc. and pull the patient from the conductor. If it is not possible then use a dry coat or rope to remove the patient from the contact of electricity.

(*ii*) Extinguish the spark if any, and send for doctor immediately or inform the nearest power house or sub-station.

(*iii*) Mean while the doctor comes try for artificial respiration.

Methods of artificial respiration. The respiration can in general be defined as a cellular process in which organic food is oxidised and the energy is set free. The artificial respiration can be arranged by the following three ways:

(*a*) Schafer's method. (*b*) Neilson's method. (*c*) Mouth to nose method.

(*a*) **Schafer's method.** This method is adopted when the burns have taken place in the back position of the patient. Take the following steps:

(*i*) Lay the victim in such a way so that his chest towards the ground put his head in a rest position.

(*ii*) Kneel over the victim and keep your both hands on the back of the victim, spread your finger and thumb touching each other.

(*iii*) Now lean firmly but gently forward over the victim exerting a steady pressure downwards.

(*iv*) Now rock yourself gently backward but don't remove your hands.

(*v*) Continue both operation. Do this practice 15 times in one minute. Continue the process until doctor comes or the patient breathe as shown in Fig. 1.1.

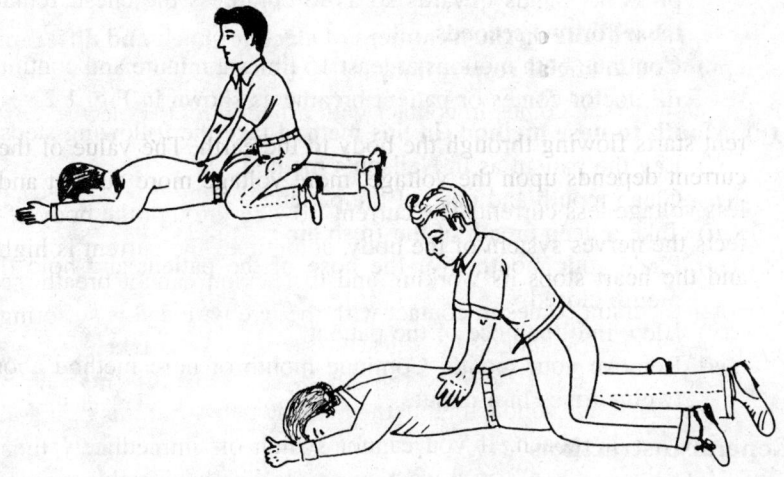

Fig. 1.1. Schafer's method

(*b*) **Neilson's method.** This method is adopted when burns have taken place in the front position of the victim, then take the following steps:

  (*i*) Lay the patient on a mat, loose his clothes around his chest and stomach. Then place a pillow or rolled up coat beneath his shoulders; so that the head falls back, open his mouth and take out his tongue.

  (*ii*) Kneel on both knees near the head of the victim.

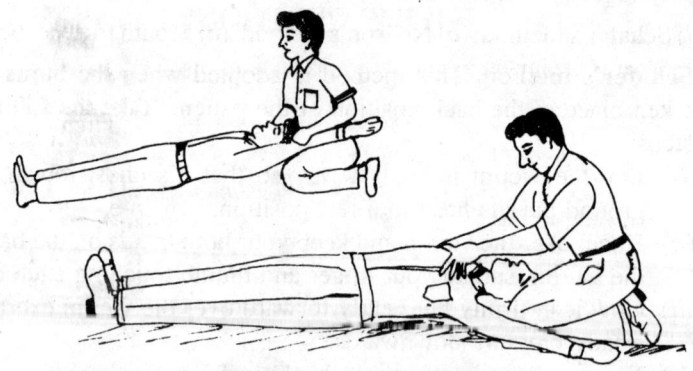

Fig. 1.2. Neilson's method

  (*iii*) Grasp the patient just below the elbow and draw his arms over his head until horizontal, retaining there for two seconds.

  (*iv*) Next bring the patient arms down on each side of his chest and press his hands inwards so as to compress the chest, remain these for two seconds.

  (*v*) Continue both motions at least 15 times a minute and continue till doctor comes or patient breathe as shown in Fig. 1.2.

(*c*) **Mouth to nose method.** In this method take the following steps:

  (*i*) Lay the patient as in Neilson's method.

  (*ii*) Clean mouth and nose of the patient.

  (*iii*) Take a deep breath of the fresh air.

  (*iv*) Place your mouth over the nose of the patient and hold the mouth closed.

  (*v*) Blow into the nose of the patient.

  (*vi*) Remove your mouth. Continue mouth of nose method about twelve times in a minute.

General Instructions

  (*i*) All methods should be continued until the patient breaths or doctor comes.

(*ii*) Never give any thing to drink or to eat to the patient before starting artificial respiration.

(*iii*) Apply proper oil on the burns etc. and save the patient from cold.

(*iv*) The flow of current through individual body depends upon individual body resistance too. The quantity of current passing through, respond in different ways, for example 1mA to 8mA may result some sensation, 8-mA-15mA painful sensation, 15mA to 20mA - burning sensation and above 20mA results burning or even death.

REVIEW QUESTIONS

1. What is the importance of safety precautions in the field of electrical engineering?
2. How the safety is ensured for electrical apparatus and installation?
3. Write down the safety precautions:
   (*a*) while handling the electrical equipments,
   (*b*) while working on a ladder,
   (*c*) while working on an over-head line.
   (*d*) while replacing a blown fuse.
   (*e*) while preparing the electrolyte for battery charging,
   (*f*) in case of electric fire.
4. Can the water be used for quenching the electric fire, if not why?
5. What do you understand by the elementary first aid?
6. Write down the importance and limitations of the first aider.
7. Name the different elementary first aid nursing equipments.
8. How will you remove a person from the electric contact?
9. What are the methods for artificial respirations? Describe any one of them.
10. Describe mouth to nose method of artificial respiration and its limitations.

Fill in the blanks

1. .... we can avoid any accident.
2. If a ladder is in use, it should be........... by another person standing on ground.
3. Use .... tool for a particular job.
4. The domestic supply voltage is ....
5. While replacing a blown fuse, do not forget to .... the main switch.
6. Use .... sand on electric fire.
7. Never .... any portable appliance just by pulling its cord.
8. Connect .... for the safety of human being.
9. Use .... gloves while working on overhead line.
10. .... mark electrical appliances be preferred.

Ans. 1. Being careful   2. supported   3. suitable   4. 230 V   5. off
6. dry   7. disconnect   8. Earth   9. rubber   10. I.S.I.

2

Electrical Engineering Signs and Symbols

The use of electrical signs and symbols make convenient and simple for clear and ease understanding of the object in use. With their use in installation and controls etc. it is very easy to represent the entire circuit on a drawing sheet.

The followings are some of the recommended basic symbols. The other additional information and details may be given on the basis of basic symbols whenever required.

I.S.I. Symbol

S.No.	Description	Sign and Symbol
1.	Main fuse board without switches, lighting	
2.	Main fuse board with switches, lighting	
3.	Main fuse board without switches, power	
4.	Main fuse board with switches, power	
5.	Distribution fuse board without switches, lighting	
6.	Distribution fuse board with switches, lighting	
7.	Distribution fuse board without switches, power	

S.No.	Description	Sign and Symbol
8.	Distribution fuse board with switches, power	
9.	Main switch lighting	
10.	Main switch power	
11.	Meter (Energy)	
12.	Single light pendent	
13.	Counter weight pendent	
14.	Rod pendent	
15.	Chain pendent	
16.	Light bracket	
17.	Batten lamp holder	
18.	Water tight light fitting	
19.	Socket out let, 2 pin 5 A	
20.	Socket outlet, 3 pin 5 A	
21.	Socket outlet and switch combined, 2 pin 5 A	
22.	Socket outlet and switch combined, 3 pin 5 A	
23.	Socket outlet, 2 pin 15 A	

S.No.	Description	Sign and Symbol
24.	Socket outlet, 3 pin 15 A	
25.	Socket outlet and switch combined, 2 pin 15 A	
26.	Socket outlet with switch, 3 pin 15 A	
27.	Bulkhead fitting	
28.	Fluorescent light double	
29.	Fluorescent light single	
30.	Power factor capacitor	
31.	Light outlet connection to an emergency system	
32.	Choke (when installed) from the lamp unit	
33.	One-way switch	
34.	Two-way switch	
35.	Intermediate switch	
36.	Pendent switch	
37.	Pull switch	
38.	Connection heater	
39.	Electric unit heater	

S.No.	Description	Sign and Symbol
40.	Immersion heater	
41.	Thermostat	
42.	Immersion heater with incorporated thermostat	
43.	Self-contained electric water heater	
44.	Humidistant	
45.	Bell push	
46.	Bell	
47.	Buzzer	
48.	Indicator (at 'N' insert number of ways)	
49.	Relay	
50.	Synchronous clock outlet	
51.	Impulse clock outlet	
52.	Master clock	
53.	Fire alarm push	
54.	Automatic contact	

S.No.	Description	Sign and Symbol
55.	Bell connected to fire alarm	
56.	Fire alarm indicator (at N-insert number of ways)	
57.	Amplifier	
58.	Control boards	
59.	Microphone outlet	
60.	Loud speaker outlet	
61.	Receiver outlet	
62.	Aerial	
63.	Ceiling fan	
64.	Bracket fan	
65.	Exhaust fan	
66.	Fan regulator	
67.	Earth point	
68.	Pilot or corridor lamp	
69.	Reset position	

S.No.	Description	Sign and Symbol
70.	Horn or hooter	
71.	Siren	
72.	Telephone instrument point public service	
73.	Telephone instrument point internal	
74.	Telephone cable distribution board public service	
75.	Telephone cable distribution board interval	
76.	Telephone private exchange or internal	

Graphical Signs and Symbols

S. No.	Description	Sign and Symbol
1.	+ve	− OR + VE
2.	−ve	− OR − VE
3.	Earth	
4.	Lamp	OR I
5.	One-way Tumbler Switch	
6.	Two-way Switch	

S.No.	Description	Sign and Symbol
7.	Intermediate Switch	
8.	I.C.D.P. Main Switch without fuses	
9.	I.C.D.P. Main Switch will fuses	
10.	I.C.D.P. Main Switch with neutral	
11.	Double pole double throw Main Switch	
12.	T.P.I.C. Main Switch without fuses	
13.	T.P.I.C. Main Switch with fuses	
14.	T.P.I.C. double throw Main Switch	
15.	Resistance fixed	

S.No.	Description	Sign and Symbol
16.	Resistance variable (rheostat)	
17.	Capacitor fixed	
18.	Capacitor variable	
19.	Inductance	
20.	Variable inductance	
21.	D.C. (direct current)	
22.	A.C. (alternating current)	
23.	D.C. voltmeter	
24.	D.C. Ammeter	
25.	A.C. voltmeter	
26.	A.C. ammeter	
27.	D.C. generator	
28.	D.C. motor	

S.No.	Description	Sign and Symbol
29.	A.C. generator (alternator)	
30.	A.C. motor	
31.	Ammeter A.C./D.C.	
32.	Choke	
33.	Auto-transformer	
34.	Transformer	
35.	Fan regulator	
36.	Cell	
37.	Battery	
38.	Isolator with fuse (H.R.C.)	
39.	Push button closing circuit	
40.	Push button opening circuit	
41.	O.C.B. (Oil Circuit Breaker)	

S.No.	Description	Sign and Symbol
42.	Lightning arrester	
43.	C.T. (current transformer)	
44.	P.T. (potential transformer)	
45.	Rectifier unit or diode	
46.	Cooker control unit	
47.	Surge-diverter	
48.	Resistance	R
49.	Inductance	L
50.	Impedance	Z
51.	Frame or chassis connections	
52.	Frame or chassis earth connections	
53.	Fault	
54.	Position of fault to frame	

S.No.	Description	Sign and Symbol
55.	Magnetic core	▅▅▅▅▅▅▅
56.	Laminated core	▬▬▬▬▬
57.	Dust core	▨▨▨▨▨▨▨

REVIEW QUESTIONS

1. What are the utilities of the electrical signs and symbols?
2. Draw the symbols for these followings: Main switch lighting, D.P.I.C. Main switch, T.P.I.C. Main switch, Double pole double throw main switch, Intermediate switch.
3. Draw the symbols for the followings:
   (a) Distribution board with fuses lighting.
   (b) Main fuse board without switch (power).
   (c) Single light pendent.
   (d) Socket outlet 3 Pin 15 A with switch.
   (e) Siren.
4. What importance of electrical sign and symbols have for reading and installation work in electrical engineering?
5. Draw the symbols for lightning arrester, rectifier, relay, transformer, fan regulation, fan point, two way switch, bell push, buzzer and aerial.

3

Electrician Hand-Tools

The workmanship of a worker does not only relate with the knowledge but very well associate with the tools he uses. The performance and proficiency is reflected by the tools, their maintenance and the use of proper tool, proper specification and size for a particular job, hence it is essential to keep the tools up to the expectations.

These are the following common tools, which are generally used by an electrician.

1. **Pliers.** The pliers are of different types:

(a) *Combination pliers.* These are used for cutting, gripping, twisting and holding the articles or wires etc. There is a provision of doing a number of operations by a single plier, so it is known as the combination plier as shown in Fig 3.1. These are also of two types:

   (i) Bare or ordinary pliers.

   (ii) Insulated pliers.

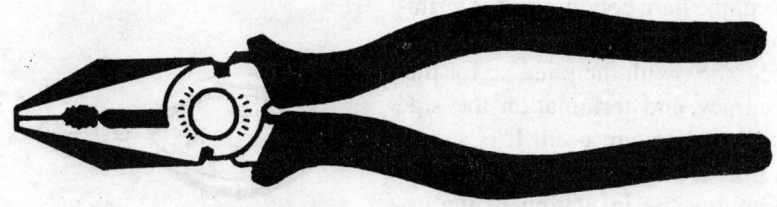

Fig. 3.1. Combination plier

The bare or ordinary plier is used for ordinary work where the danger of electricity is not there. The insulated pliers are used by an electrician or wireman because they have to work on the running

21

lines or on the live conductors. The insulation protects from the electric shock.

These are available according to lengths, *i.e.* 100 mm, 150 mm, 200 mm, 250 mm etc.

(*b*) *Flat nose pliers.* The flat nose pliers are made, as the name implies with a flat nose or jaws as shown in Fig. 3.2. These can be used for winding work, holding and drawing the windings from the stator of motor etc. and for tightening or loosening the screws and small nuts. These are also available in different sizes 100 mm, 150 mm etc.

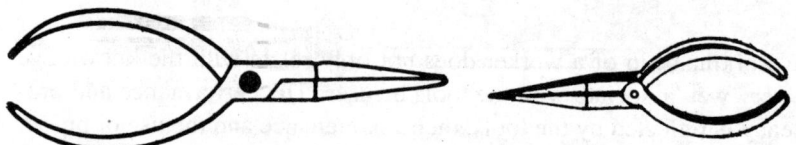

Fig. 3.2. Flat nose plier.   **Fig. 3.3.** Long nose plier.

(*c*) *Long nose pliers.* The long nose pliers are made, as the name implies, with a long nose or jaws as shown in Fig. 3.3. It is used the tightening and loosening the screws and nuts. These are also used for to make the ends of wires round, which are to be used where they are held fast under the screws. These are also used for placing and removing small items in narrow spaces.

(*d*) *Round nose pliers.* These are having their edge as round. These are also used for making the ends of wire round and also for winding work to pull out the winding from the round slots etc.

2. Side cutter or side cutting plier. These are a special type of pliers used exclusively for cutting electrical wires etc. Sometimes it is difficult to cut the bare conductors for terminating into the holders or any other accessory, with the plier, so for that accuracy and termination the side cutting pliers are used. It is shown in Fig. 3.4. It is used for cutting the removing the insulation of the cable. These are available in different sizes as 100 mm, 150 mm etc.

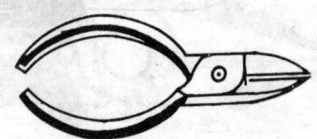

Fig. 3.4. Side cutter.

3. Screw Driver. It is an important tool. It is used for tightening and loosening the screws. These are available in different blades sizes and shapes like thin blade, heavy duty blade, square blade etc.

The screw driver has two parts, blade – generally of steel and handle

– generally wooden or plastic as shown in Fig. 3.5. A good screw driver has its edge hardened and tempered. In addition to this the screw drivers which are used for tightening and loosening the small screws of electrical accessories etc. are known as screw driver connectors. The edge of the screw drivers should not be sharp, but it should be well maintained flattened to fit slots in the head of screws or bolts. It should not be grinded to suit for different screw heads. The screw drivers are of different types, heavy duty, phillips type, helical ratchet screw driver, double ended off set screw drivers etc. These are available according to the blade sizes *i.e.* 40 mm, 100 mm, 150 mm, 200 mm, 250 mm, 300 mm, etc.

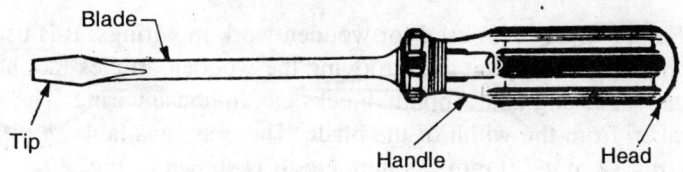

Fig. 3.5. Screw driver.

4. Electrician's knife. It is used for marking and removing the insulation of wires. It is made from high grade cutlery steel. It has two or more blades one for removing the insulation and other for cleaning the wires as shown in Fig. 3.6.

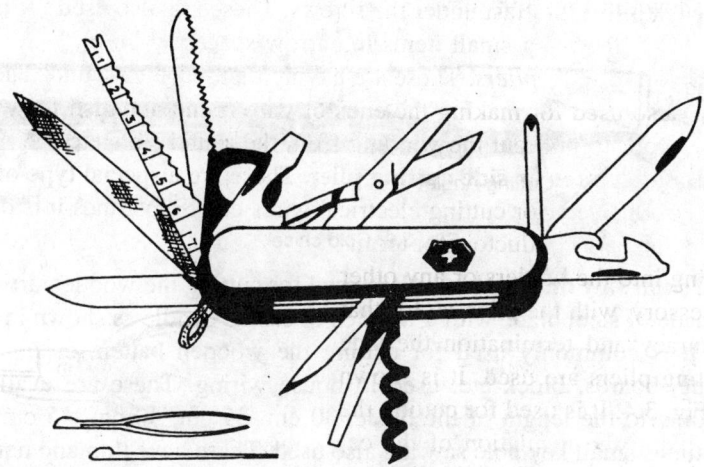

Fig. 3.6. Electrician's knife.

5. Pocker. It is used for making pilot holes for screws in wooden articles. It is a pointed tool as shown in Fig. 3.7. Generally it is made from the square blade, which is finally made with a pointed tip on the edge. It is also available in length 100 mm, 150 mm, 200 mm. etc.

Fig. 3.7. Pocker.

6. Gimlet. It is used for making holes in the wooden articles. These are of different length and different diameter.

7. Chisel. The chisels used by an electrician, are of two types:

Fig. 3.8. Firmer chisel.

(a) *Firmer Chisel* It is used for wooden work in wirings. It is used for chipping, scrapping and grooving the wooden articles like boards, batten, casing and capping, blocks etc. in house wiring. The size is taken from the width of the blade. These are available in different sizes 12 mm, 20 mm, 25 mm, etc. It is shown in Fig. 3.8.

(b) *Cold chisel.* It is used for chipping, boring and channelling in concrete and masonry work, for example making grooves for laying the conduit pipes, for digging holes for gitties, for making burrs in wall for taking out wiring from one room to another room etc. They are also available in different sizes as 100 mm, 150 mm etc. It is shown in Fig. 3.9.

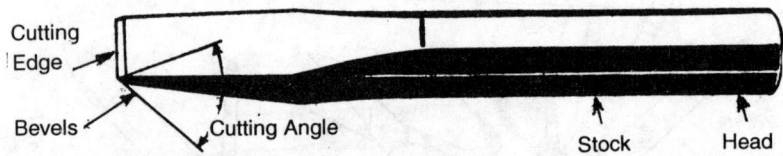

Fig. 3.9. Cold chisel.

8. Tenon saw or Hand saw. It is used for cutting the wooden articles. It is made of steel blade with a wooden or casted handle as shown in Fig. 3.10. It is commonly used for cutting the wooden batten, casing and capping, boards, block etc. used in house wiring. These are available according to the length of the blade, 30 cm, 35 cm, 40 cm, 45 cm etc. Sometimes small key hole saw are also used. These have thin and narrow

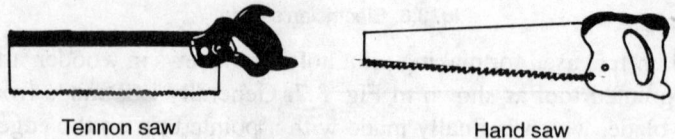

Tennon saw   Hand saw

Fig. 3.10.

blade with a hole. The blade is fixed in handle with the help of the screw and nuts. So that the blade may be replaced when it is required.

The tenon saw having a rectangular shape and a strip on the through-out of the top, as shown in Fig. 3.10 are named as tenon saw and the other as the wooden saw.

9. Hack saw. The hack saw is used for the metallic substance for cutting purposes. It is used for cutting the conduit pipe, cables etc. These are also of two types the fixed and adjustable as shown in Fig. 3.11. The fixed hack saw is fixed in size only one size of blade is used in this frame. The adjustable can be adjusted according to the size of blade available. These are available according to the blade sizes as 14 cm, 20 cm, 25 cm, 30 cm, etc.

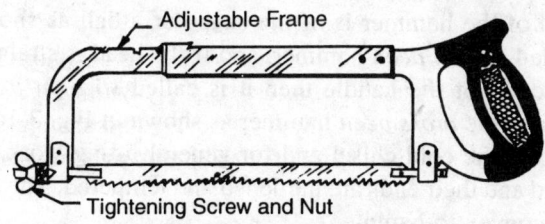

Fig. 3.11. Hack saw.

The blade is having different teeth 18, 24 and 30 teeth per inch. While fixing the blade in the hack saw, the teeth direction is adjusted in the forward direction as to cut more efficiently and fast.

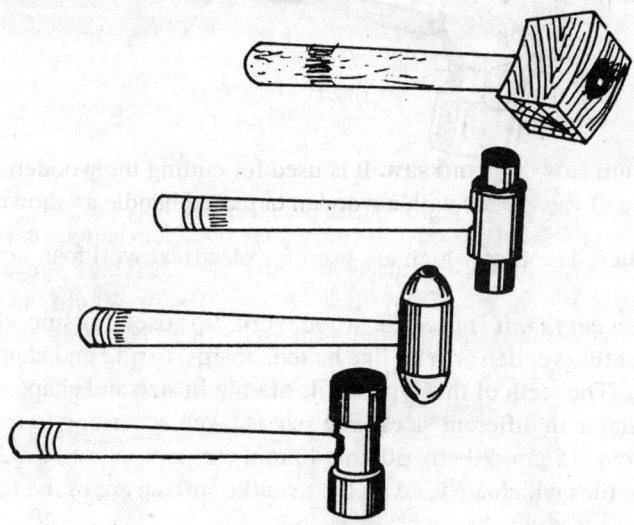

Fig. 3.12. Mallet.

10. Mallet. It is made of wood or any other soft material as shown in Fig. 3.12. It is used where the iron hammer are not advisable to use. It is used for dismantling and fixing the electric motors etc. for striking purposes, making the surface straight. It is used for making straight copper strips, sheets, thick copper or aluminium conductors, lead plates and to strike chisel etc.

11. Hammer. It is used for striking purposes. These are generally made of steel or casted by the steels. These are available according to the weight, 100 gm, 200 gm, 250 gm, 500 gm, 1 kg, 2 kg or so on. These are of the following types:

(*i*) Ball peen hammer.

(*ii*) Cross peen hammer.

(*iii*) Straight peen hammer.

If the head of the hammer is of the shape of a ball as shown in Fig. 3.13, it is called as *ball peen* hammer, and if the head is straight flat and is in the direction of the handle then it is called *straight peen* and if crossed then it is the *cross peen* hammer as shown in Fig. 3.14. Then are used for striking the cold chisel and for general fitting work. Generally they are casted and their ends are hardened and tempered. The face is also kept slightly convex to hammer.

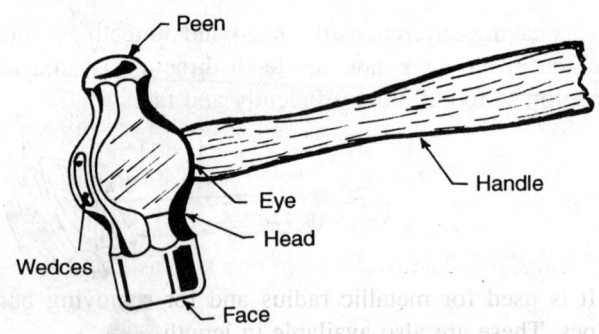

Fig. 3.13. Ball peen hammer.

12. Files. The files which are used for electrical workman are as follows:

1. *Rasp cut file.* It is used for wood work. It is used for smoothing the surface of the wooden articles like batten, boards, casing and capping and block etc. The teeth of this type of file are big in size and shape, they are also available in different sizes. The size is taken according to the length like, 12 cm, 15 cm, 20 cm, 30 cm, 40 cm, etc.

2. The files which are used on the metallic surface are of the following types:

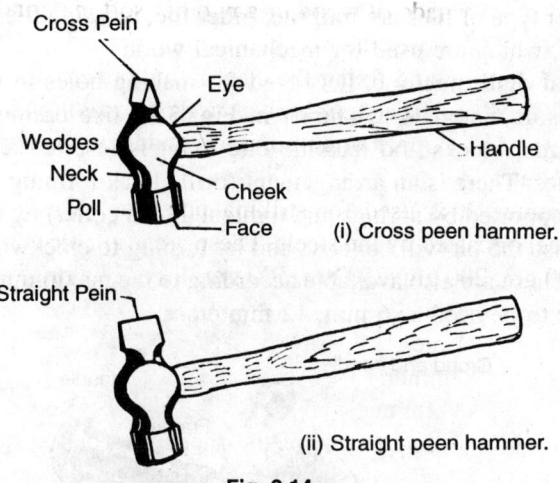

(i) Cross peen hammer.

(ii) Straight peen hammer.

Fig. 3.14.

(*a*) *Flat file.* It is parallel and about 2/3rd of its length is tapered. It is shown in Fig. 3.15. It has cuts on all the four surfaces and can be used as required. These are also fine and rough according to the requirement depending upon the construction of the teeth.

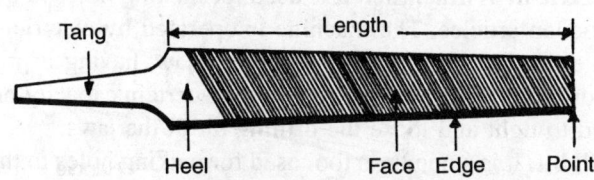

Fig. 3.15. Flat file.

(*b*) *Half round file.* These files have one side as half round and other as plane. It is used for metallic radius and for removing burrs of the conduit pipes. These are also available in length.

(*c*) *Round file.* It is round in construction. It is tapered round, more diameter towards handle side. It is used for increasing the diameter of a hole and slots. It is also used for removing the burrs of the conduit pipe. These are rough and fine, and are available in length wise.

(*d*) *Triangular file.* The triangular file has its area of cross section as a triangle. These files are used for sharpening the teeth of hand saw and tenon saw.

(*e*) *Square file.* It is used for filing corners, small squares or rectangular holes, for finishing the bottoms of narrow slots. It has the cross section as a square. All the sides are having double cut length on all four sides.

The other type of files are mill file, piller file, crossing files, knife file, cabinet file, which are used for mechanical work.

13. Hand drill machine. It is used for making holes in the wooden articles or soft materials as shown in Fig. 3.16 like casings, capping, batten, boards, blocks and bakelite etc. A bit is inserted and used for making holes. There is an arrangement (drill chuck forming a jaw, three supports, supported by springs and tightened by a coller) by turning it in anticlockwise the bit will be loose and by turning to clockwise it will be tightened. These are also available according to the maximum capacity of the drill bit to be used as 6 mm, 12 mm etc.

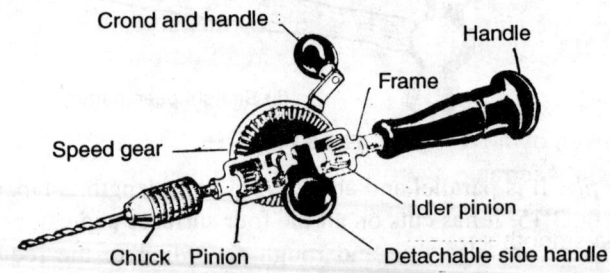

Fig. 3.16. Hand drill machine.

14. Electric drill machine. It is used for making holes in both metallic and wooden articles. The machine is operated by electricity, so it is known as electric drill machine. There is a jaw, having a provision of changing drill bits, of different size but up to certain capacity only. A key is provided to tight and loose the drilling bit in the jaws.

15. Drill bit. It is a common tool used for making holes in the metallic parts or wooden articles also. The grooves, usually called *flutes* are made in a drill and bit is called the twist drill as shown in Fig. 3.17. These are having two main parts, the body – which is the cutting unit and the shank, which is gripped by the drilling machine. These are available according to the diameter of the drill bit for example 4 mm, 6 mm, 8 mm, 10 mm, etc.

Fig. 3.17. Drill bit.

16. Ratchet bit brace. Sometimes it is required to hole in heavy materials like beams and joists, then the ratchet brace may be used. It has interlocking jaws, ball bearing head and with 25 cm sweep is best suited for electrical jobs. For drilling holes near the corner, corner angular brace is used as shown in Fig. 3.18.

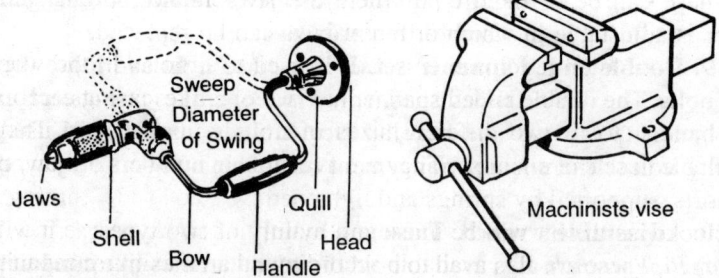

Sweep
Diameter
of Swing

Jaws

Shell

Bow

Quill

Head

Handle

Machinists vise

Fig. 3.18. Ratchet bit brace.   **Fig. 3.19.** Bench vice.

17. Vice. The vices are used for holding the things. These are used for different electrical purposes and these are of different types:

(*i*) *Bench vice.* It has parallel jaws which are opened by means of a spindle driven by a rod as shown in Fig. 3.19. The faces of the jaws are usually lightly serrated and hardened to ensure a firm grip on the work. Sometimes brass or copper caps are used to hold the soft material and protect the smooth surfaces.

(*ii*) *Pipe vice.* It is used holding and round article like conduit pipe. The jaws are of V shape which holds and round articles firmly, one jaw is fixed and other jaw is movable with handle provided on the top of the machine.

(*iii*) *Hand vice.* It is used for holding the small articles for fitting and clamping materials and temporarily drilling works etc.

(*iv*) *Draw vice.* It is commonly used by the electrician, and line man in overhead lines. It is used for drawing the wires for the overhead lines.

18. Stock and die set. It is used for making threads on the conduct pipes. It is made of casted caststeel. The dies are of two types the split die stock and the adjustable die stock. The jaws can be used of different size suitable for the particular jobs. A threading die is a flat piece of hardened steel, internally threaded with grooves or flutes intersecting the threads to form cutting edges. It is used to cut thread on the round articles. The die is split on one side, to permit turning a set screw on the side of the die. This expands the die so that the first cut may be more easily made. Both sides of the die are not same and are marked for easily and efficient threadings.

There are one set of handle provided for driving it. It is customary to reverse the movement occasionally in order to break the metal chips which might clog the die. Generally 1.25, 1.9, 2.54, 3.8 and 5 cm jaws are used for conduit pipe threadings.

These can be of fixed type where the jaws cannot be changed, the other is adjustable in which different jaws can be replaced.

19. Double ended spanner set. It is used to tight and loose the nuts and bolts. The double ended spanner has two openings one at each end of the handle, to fit two different sizes of units or bolt heads. They are available in sets of six or so. They are available in numbers or mm or inch sizes.

20. Adjustable wrench. These are mainly of two types:

(*a*) *Pipe wrench.* It is used to hold the round articles like conduit pipes by means of making the adjustment of the nut as shown in Fig. 3.20.

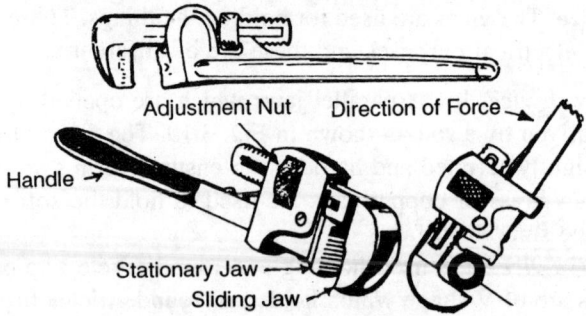

Adjustment Nut   Direction of Force

Handle

Stationary Jaw
Sliding Jaw

Fig. 3.20. (*a*) Pipe wrench.

(*b*) *Monkey wrench.* It has a movable jaw which makes it adjustable to various sizes of nuts. The heavy type of the nuts etc. are tightened by the another adjustable wrench known as monkey wrench as shown in Fig. 3.20 (*b*).

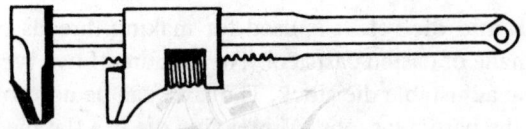

Fig. 3.20. (*b*) Monkey wrench.

21. Try square. It is used to keep the edge at right angle *i.e.* 90° as shown in Fig. 3.21. It is used in house wiring for joints to keep them at 90°, for example tee joint, right angle joint and corner joints. They are available according to length 6″, 8″ 10″ etc. or 100 mm, 150 mm, 200 mm etc.

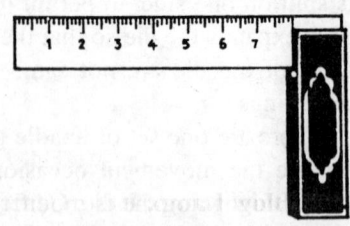

Fig. 3.21. Try square.

22. Four fold two feet rule. It is used for measuring the length of the wooden articles. It is made of having four folds to keep it accessible. The marking of cm and inch are laid down on the foot.

23. Standard wire gauge. It is used for measuring the gauge and number of wire. It is made of thin plate having a number of slots with markings on its circumstance, as shown in Fig. 3.22. The slots are made of different sizes and the numbers are marked in a systematic way. While measuring the conductor should be freely inserted in the slot.

24. Soldering iron. Electric soldering iron are used for soldering the joints in electrical circuits. Generally used in motor winding radio, T.V. and other electronic components. These are also available in different wattage as 10W, 25W, 35W, 40W, 65W and 125W etc.

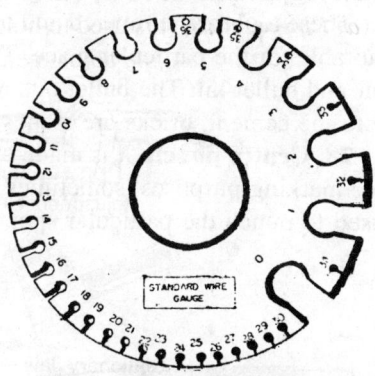

Fig. 3.22. Standard wire gauge.

The bit is used to warm up the place of jointing. The bits are of two types the pencil bit and flat bit.

25. Steel foot rule. It is made of steel and marking is done according to the inches and cm. It is used for measuring the hard substances like steel etc.

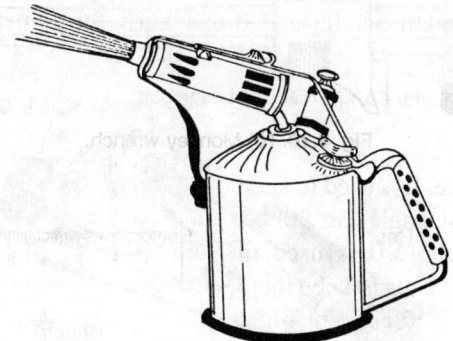

Fig. 3.23. Blow lamp.

26. Blow Lamp. It is modification of stove used for domestic works. It is used for heating the place of jointing, if it is not within the approach of soldering iron. Generally it is used to melt the lead for soldering, thumbling and cable jointing purposes. Generally kerosene oil is used for burning and they are according to the quantity of kerosene like one pint etc. It is shown in Fig. 3.23.

27. Rowl plug tool. For fixing the blocks, batten and other wiring accessories, the gitties are used but in places where only a small hole can be done and cold chisel will not do, then the rowl plug tool is used. First the drilling is done with the tool having bit, and then the wet plugs are inserted into the hole or ceiling.

These tools are numbered and only the tool of particular number is just suitable for the particular place. These bits are of two types, the drilling bit and bullet bit. The bullet bits are used in soft plaster and the drilling into the cement, bricks are stones etc.

28. Centre punch. It is made of steel as shown in Fig. 3.24. It is used for marking purposes, sometimes before drilling the holes, the punch is used to punch the particular spot, for accurate and safe drilling.

Fig. 3.24. Centre punch.   **Fig. 3.25.** Plumb bob.

29. Plumb bob. It is made of brass or sometimes other metals as shown in Fig. 3.25. It is heavy and fixed with the fish thread. There is an arrangement for fixing it with the fish thread. It is used to make the straight vertical line.

30. Putty kurfi. It is used to plaster the holes after fixing the gitties, etc. It has wooden handle and broad kurfi for cement plastering and mason work.

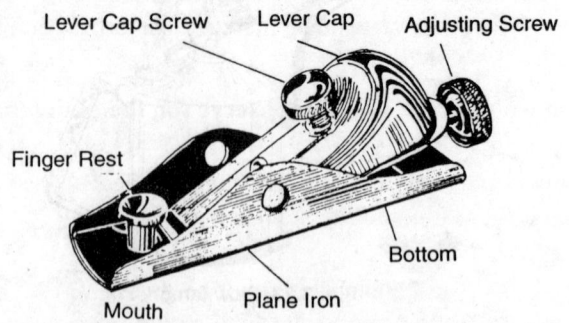

Lever Cap Screw   Lever Cap   Adjusting Screw

Finger Rest

Bottom

Mouth   Plane Iron

Fig. 3.26. Planing Smoother.

31. Planing smoother. It is used for smooth and making proper edges of the wooden articles. It is made of either wooden or cast steel.

These are of different types but only the simple planners are preferred in electrician work. There is a blade which shears the surface. These are available in different sizes and shapes.

The blade should be sharped on the oil or water stone not on the grinder. In case of chisel the cutting edge of the blade is adjust with the help of lever provided as shown in Fig. 3.26.

32. Bearing puller. It is used to pull out the bearings or pulley from the shaft of the motor or generator. It is made of cast steel. It has three legs of L shape mounted on a fixture which has a fully threaded stud to move to and fro. In order to pull out any bearing or pulley all the legs are clamped on the rigid portion of the pulley or bearing of the stud is supported on the shaft. It is tightened by means of a spanner to offer a pulling force.

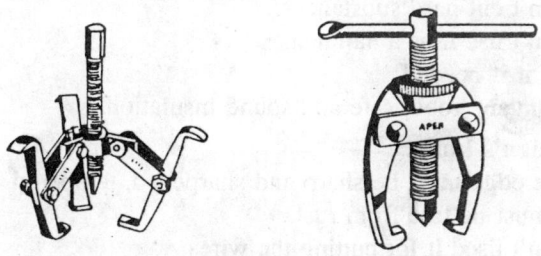

Fig. 3.27. Bearing puller.

For light work a fixture with two legs are also used to take out the bearing etc.

33. Wire cutter. It is used for cutting the wires or conductors for example G.I. wires and other wires.

34. Scriber. It is used for marking purpose, sometimes the mica, hylum, bakelite or metallic, sheets, are to be marked, this tool is used for marking line etc.

Q. Write down the care will you observe for the maintenance of the tools.

Ans. The followings are the care and maintenance to be observed for effective operation and maintenance of tools.

1. Screw driver:
  (*i*) The handle and blade should not be greasy while in use.
  (*ii*) If it is to be kept for a long time without use then apply some grease to avoid rusting.

    (*iii*) Use proper size of the bit for screw to be operated. It must fit in the slot.

    (*iv*) Don't grind the edge, to suit for different screws.

    (*v*) The edge should not be too sharp.

    (*vi*) Don't use it as a hammer or firmer chisel.

2. Pliers:

    (*i*) Don't cut steel or hard substances with the pliers.

    (*ii*) Don't hold hot substance.

    (*iii*) Don't use it as a hammer.

    (*iv*) Maintain a good insulation of the insulated plier. The insulation must be strong and sound.

    (*v*) Proper size of the pliers should be used for the particular job.

    (*vi*) Oil it if needed.

3. Side cutting plier/side cutter:

    (*i*) Don't cut hard substances.

    (*ii*) Don't use it as a hammer.

    (*iii*) Oil it if needed.

    (*iv*) Maintain good, safe and sound insulation.

4. Electrician's knife:

    (*i*) The edge must be sharp and sharpened, tempered.

    (*ii*) It must be free from rust.

    (*iii*) Don't used it for cutting the wires.

    (*iv*) Take care while in use, it is a sharpened tool may cause some mishappening.

5. Pocker:

    (*i*) Should be well pointed.

    (*ii*) Don't use it as a centre punch on metals.

    (*iii*) It is a pointed tool, use it properly, take care.

6. Gimlet:

    (*i*) It should be kept straight while drilling.

    (*ii*) Should be handled properly, it is a sharp edge tool.

7. Firmer chisel:

    (*i*) Use it on wooden articles and not on the metals.

    (*ii*) Strike it by the mallet.

    (*iii*) Should be well sharpened.

    (*iv*) It should be sharpened on the oil stone or water stone but should not be grinded.

    (*v*) Never use it as a screw driver tightening a loosening the screws.

8. Cold Chisel:

    (*i*) The edge should be well sharpened and tempered.

(*ii*) Use hammer for striking the chisel.

(*iii*) There should be no trace of oily substance.

9. Hand saw or Tennon saw:

   (*i*) It should not be used on metallic substances.

 (*ii*) The teeth should be well sharpened and well maintained.

(*iii*) Triangular file should be used for sharpening the teeth.

(*iv*) The blade should be free from rust.

 (*v*) The handle should be tight enough.

(*vi*) Apply grease if kept unused for a long time.

(*vii*) Keep straight while using.

10. Hack saw:

  (*i*) Apply water when cutting the metallic substance like conduit pipe or flat.

 (*ii*) The blade should be well tightened. It should never be loose.

(*iii*) Keep straight while cutting, a little tilt will break the blade.

(*iv*) While fixing the blade, the direction of the teeth should be forward.

11. Mallet:

  (*i*) Don't use on hard substances and pointed substances.

 (*ii*) Don't use it on nails or screws to hammer.

(*iii*) Handle should be properly fixed.

(*iv*) Striking surface should be flat and soft like tin or copper.

12. Hammer:

  (*i*) Handle should be well tightened and wedged.

 (*ii*) Should be free form oily substance.

(*iii*) Proper size of hammer should be selected.

(*iv*) The hammer handle should be gripped near the end, so that full leverage may be obtained.

 (*v*) The hammer handle must be set square with the head to ensure a proper balance.

13. Rasp cut file:

  (*i*) Should never be used on the metallic surfaces.

 (*ii*) Should always be used with handle.

(*iii*) Teeth should be cleaned and not blocked.

14. Files (used for metallic surfaces):

  (*i*) Should never be used without handle.

 (*ii*) The teeth should be properly cleaned.

(*iii*) Should never be used on oily surface.

(*iv*) Never apply oil or grease.

(*v*) Selection of the file is also to be kept in mind for the particular job.

15. Drill machines:

(*i*) The handle should be tight and properly held.

(*ii*) Moving parts should be free to move.

(*iii*) The drilling bit should be properly tightened.

(*iv*) While drilling, heavy pressure should not be applied.

(*v*) Should be kept vertical while drilling.

(*vi*) Grease the gears or apply proper lubrication.

(*vii*) In case of the big holes, apply drill of small size first and then the large one.

(*viii*) While electric drill machine is used, be sure about the earth connections and the machine should be earthed.

16. Vices:

(*i*) Don't overtight.

(*ii*) Lubricate the nuts etc.

(*iii*) Never apply oil or grease on the jaws.

(*iv*) The jaws should be cleaned.

(*v*) Don't use the vice as anvil.

(*vi*) Apply proper guards while the soft material is to be held.

17. Stock and die set:

(*i*) The jaws should be properly cleaned and sharpened.

(*ii*) Oil must be used while threading.

(*iii*) Keep the job and the handles at right angle while threading.

(*iv*) Always use forward and backward motion while cutting threads.

(*v*) The die and the conduit pipe should be of same size.

18. Spanner set:

(*i*) It should be selected according to the nuts and bolts.

(*ii*) Never apply grease or oil on the jaws.

(*iii*) Never grind the jaws to suit for other sizes.

(*iv*) Never use it as hammer.

(*v*) It should be cleaned after use.

(*vi*) Keep it safe from rust etc.

19. Try square:

(*i*) The angle should be a right angle.

(*ii*) Don't use it as hammer.

(*iii*) Keep it safe from rust etc.

(*iv*) The edge should be straight.

20. Standard wire gauge:

(*i*) It should be properly cleaned.

(*ii*) It should be oiled or greased to avoid rusting.

(*iii*) The teeth should be cleaned.

(*iv*) Never hammer it, otherwise the accuracy will be spoiled.

21. Soldering iron:

(*i*) It's bit should be cleaned.

(*ii*) Before soldering just fix the bit.

(*iii*) Don't use it on O.H. lines.

(*iv*) Solder should not be overheated.

(*v*) In case of electric soldering iron, earth continuity must be maintained.

22. Blow lamp:

(*i*) The nozzle should be cleaned.

(*ii*) Always add filtered kerosene oil.

(*iii*) Use it very carefully because the flame may cause some problem if not being careful.

(*iv*) Don't pump excess.

23. Rowl plug tool:

(*i*) Don't throw it on the ground.

(*ii*) While in use, rotate slowly.

(*iii*) Don't use it on metals.

(*iv*) Proper size of bit, handle and plugs should be used.

24. Bearing puller:

(*i*) It should not be hammered.

(*ii*) It should be periodically lubricated.

(*iii*) It should tightened with proper size of spanner.

(*iv*) It should not be tightened excentrically.

(*v*) Use properly, never use light pulley for heavy work.

Q. What precautions an electrician should observe while handling the tools?

Ans. In case of the electrician, he has to work on the running lines which may give shock if not being careful, one can injure himself in addition to the damage to the tools, so some precautions are must to observe while handling the tools.

1. The insulated tools like pliers, screw driver, side cutting plier, should have their sound and strong insulation.

2. The sharp edge tools like knife, screw driver, chisel etc. should never be put in pocket without proper shielding.

3. While using chisel the mark and cut should be away from yourself.

4. The tools which have their handles, must be properly fitted and fixed.
5. In case of sawing, a small cut should be marked first and then proceed the final cutting. For marking never use your finger near the blade, it may cause some accident.
6. After using the tools, it should be cleaned and then keep it at its proper place.
7. Only suitable tools should be used for the operation, for example the chisel should never be used in place of a screw driver.
8. In case of accident, attend immediately and send for doctor to avoid any complecacy.

REVIEW QUESTIONS

1. Name and explain the ten important hand tools generally used by an electrician.
2. Why the care and maintenance of the hand tools is essential?
3. State the different care, maintenance and precautions you will take for these following tools:
   Insulated plier, screw driver, hand drill machine, pocker and hack saw.
4. Which file is used for sharpening the teeth of a hand saw and why?
5. Which tool is used for making pilot hole in the wooden articles? Can it be used for cement or metallic bodies if not why?

Explain in brief.

1. Why should we prefer to use insulated tools, while working on live lines?
2. Why we should rescue a person who is in contact of Electricity or receiving electricshock.
3. Why you should insulate yourself before attempting a person rescuing from electric shock.
4. Why the switch be put in off position before replacing a blown fuse?
5. Why one should add acid to water while preparing electrolyte for battery charging.
6. Name the insulating materials you will use to get yourself insulated?
7. What precautions you should use while working on overhead lines.
8. Why it is important to use a chain to short circuit the off line before working on it.
9. Why should we differentiate between the distilled water and acid?
10. Should we bring a naked flame near the place of battery charging.

Protective Measures and Electrical Wiring Accessories

4.1. FUSE

The fuse was originally invented by Edison in 1880, and being considered as the weakest link in the electrical circuits.

Q. What is a fuse? State the necessity and working of a fuse.

Ans. A fuse is a protective device. It is a weakest portion in the electrical circuit. It can thus be defined as, "the fuse is a safety device which protects the electrical circuit from taking an overload current".

The fuse essentially is a small piece of metal wire connected between two terminals mounted on an insulated base. It is always kept in series with the electrical circuit.

Necessity of fuse. The fuse is required in a circuit to avoid the flow of overload current through a circuit. If overload current flows, the apparatus, instruments, machines and even the wires may be damaged. So to protect these a fuse is essentially required to limit the current in the circuit.

Working of a fuse. A fuse wire is always connected in series with a circuit so that the current of the circuit flows through the fuse. Obviously the heat is produced in a conductor, if it carries current and the heat produced is thus proportional to the square of current flowing in the circuit.

Generally, the fuse wire are of low melting temperature. These are made either of lead tin alloy or tinned copper wire. In normal conditions the fuse carries the normal current safely without heating. But when the current exceeds the normal working current, the amount of heat produced ($H\alpha\ I^2$) with increase to the melting temperature and finally the circuit will be disconnected, because the fuse element is in series of the load. Thus an open circuit will save the circuit from being overloaded.

Q. Write down a short note on fuse wires.

Ans. The fuse wires are made of such materials which have low resistance *viz,* silver, gold, lead tin alloy, tinned copper wire etc. The silver is a quite satisfactory material to be used for fusing elements, because it is not subjected to oxidation and the oxides are also even unstable. There is no deterioration of material when used in dry air and remain bright. In case of moist air and contains hydrogen sulphite, the surface is attacked and a layer is formed which protects from further attacks. But even then it is not conveniently used because of the cost, similar is the case with gold. But in very precious cases the fusing elements are also made of silver and gold.

Lead tin alloy. The lead tin fuse wire is mostly used for the low currents. In this alloy the tin is 2% to 3% which is used for low current. Generally the alloy used for normal currents contains lead 37% and tin 63%, such alloys are known as outectic alloys.

These wires are generally used in open type and semi-open type fuses for low currents. It is a homogeneous material and the heat developed is dissipated uniformly and has less tendency to spread over. But mechanically it is weak.

These are generally used up to 5A for heavy currents it is not used since the diameter of the wire used will be large and after fusing, metal release will be comparatively excess.

It is known as standard alloy.

Tinned copper wire. It is a general purpose fuse wire, which can be used for any value of currents. It is made of copper, generally tinned. The rating of current depends upon the size of the fuse wire, more thickness large current, less thickness low current. These are used for open type and semi-enclosed type fuse, up to 100 A. Some approximate sizes of fuse elements are shown in table 4.1.

Q. What are the factors which limits the rating of a fuse?

Ans. There are the following factors which effects the rating of fuse:

    (*i*) The type of material used.

    (*ii*) The cross-sectional area of the wire to be used for this purpose.

    (*iii*) The methods of placement of wire, for example the wire is open, enclosed or semi-enclosed.

    (*iv*) The surrounding atmospheric conditions.

    (*v*) The length of the fusing element as the heat developed is proportional to the length *i.e.* the resistance.

(*vi*) Number of fuse wires used in parallel, if the wires are twisted two in parallel, the current will be 1.7 times that of one element.

(*vii*) The cross-sectional area of the fusing wire.

Table 4.1. Tinned Copper Fuse Wire

Approximate Size of Fuse Elements Composed of Tinned Copper for Use in Semi-Enclosed Fuse

SWG	Current rating of fuse amperes	Approximate fusing current
40	1.5	3
39	2.5	4
38	3.0	5
37	3.5	6
36	4.5	7
35	5.5	8
34	6.0	9
33	7.0	10
32	8.0	11
31	8.0	12
30	10.0	13
29	12.0	16
28	13.0	18
27	14.0	23
26	15.0	28
25	17.0	30
24	20.0	38
23	24.0	48
22	29.0	58
21	34.0	70
20	38.0	81
19	43.0	98
18	45.0	106
17	65.0	135
16	73.0	166
15	78.0	197
14	102.0	230
13	130.0	291

Q. Define the following terms:
  (*a*) **Current rating of fuse wire** (*b*) **fusing current** (*c*) **fusing factor** (*d*) **fuse element.**

Ans. (*a*) **Current rating of fuse wire.** It is the value of current which the fuse wire carry without any melting. It is also known as the safe current of fuse wire. All the fuses are rated at this current.

  (*b*) **Fusing current.** It is the value of current at which the fuse melts. It is always more than the safe current rating of fuse.

  (*c*) **Fusing factor.** It is the ratio of the fusing current and the safe current of the wire.

$$\text{Fusing factor} = \frac{\text{Fusing current of the wire}}{\text{Current rating of fuse wire}}$$

It is always more than one.

  (*d*) **Fuse element.** It is a fuse wire connected in series with the circuit. It is a protecting device which protects the circuit from taking an overload current.

4.2. TYPES OF FUSES

Q. What are the different types of fuses? Describe any one of them with the help of a neat sketch.

Ans. The fuses are of the following types:
  (*a*) Open type fuse,
  (*b*) Semi-enclosed type, and
  (*c*) Totally enclosed type.

  Out of these three types generally for house wiring and general purpose the semi-enclosed type fuse are used.

  Semi-enclosed type fuse. In this type of fuse the fuse element is neither open nor totally enclosed. Such types of fuses are used for house wiring purpose, for example fuse cutout and kit ket fuse.

  (*i*) **Fuse cut out.** These are made of porcelain. It has two main parts known as the base and cover. The base has two terminals or plates. The fuse wire is connected between these two plates or terminals, through a

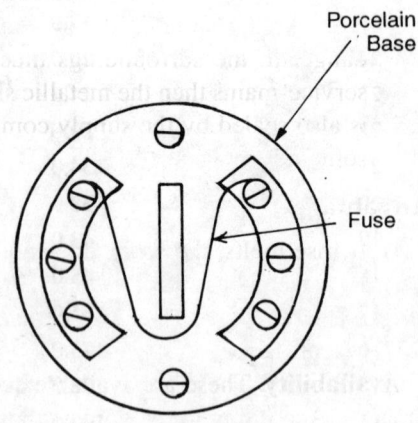

Fig. 4.1. Fuse cut-out.

porcelain channel as shown in Fig. 4.1. A cover is also provided to cover it from the external disturbances and also to avoid the sparking etc. not to come out.

(*ii*) **Kit ket fuse.** It is a common type of fuse generally used in house wiring system.

There are two main parts, the base and carrier.

1. **Fuse base.** It is made of porcelain and is a fixed part of the kit ket fuse. It has two terminals and one or two holes for fixing with the wall or board etc. with the screws as shown in Fig. 4.2. One terminal is connected with live wire and other to the load or wire connecting load. Sometimes, to avoid the earth fault, an asbestos sheet is provided over the fixing screw.

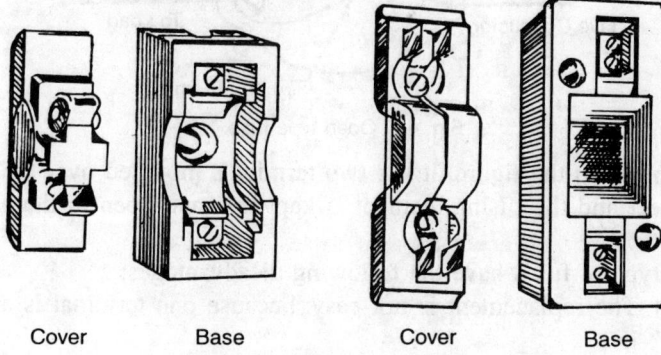

Cover   Base   Cover   Base

Fig. 4.2. Fuse base and carrier.

2. **Fuse Carrier.** It is an removable part of the fuse. It is also having two terminals generally with strips having some spring tensions, where the fusing element is connected. Sometimes the fusing element is put in a hole so that if the fuse melts, it may not cause any danger to the surroundings thus it is safe. If the fuse is used for service-mains then the metallic shielding is provided and earthed. It is also sealed by the supply company. This type of fuse are having some advantages.

Advantages:

(*i*) If fuse melts, the spark does not come out.

(*ii*) Replacement is easy.

(*iii*) All size of fuse wires can be used in this type of fuses up to some extent.

Availability. These are available according to the voltage and current capacity. According to the voltage these are 250 V and 500 V. According to current capacity these are 15 A, 30 A, 60 A, 100 A, etc.

Q. Write down a short note on the open type fuse.

Ans. The fuse which is used in open space is called open type of fuse as shown in Fig. 4.3. In this type of fuses the fusing element is kept open in the atmosphere. Generally this type of fuses are used with the overhead lines, on poles for giving the supply to the consumers. These are connected in series with the load.

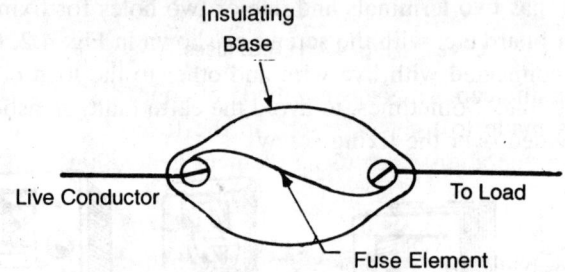

Fig. 4.3. Open type fuse.

As shown in the figure, it has two terminals mounted over the insulated sheet and the fusing element is kept free and open to the atmosphere.

This type of fuses have the following disadvantages:

(*i*) The replacement is not easy, because one terminal is always live.

(*ii*) The fusing current is not constant, as it also depends upon the environmental changes.

(*iii*) In case of fusing, the molten metal or spark may fall on such articles which may catch fire. So there is a possibility of fire also.

Q. What do you understand from cartridge fuse and H.R.C. fuse? *Or* **Describe the construction and working of the H.R.C. fuse.**

Ans. These are known as totally enclosed type fuses. In this type of fuses the fusing element is completely enclosed in a tube of glass or porcelain or fiber etc. As the fusing element does not come in contact with the atmosphere, so the fusing current remains controlled and constant.

Cartridge type fuses. In a simple construction, as shown in Fig. 4.4 the name is kept after its construction similar to the construction of a cartridge say bullet. The container is made of insulating material generally glass, fiber or porcelain. The container is filled with quartz or sand, so that the heat dissipation is maintained uniformly from the fusing element, thus maintaining the fusing current constant. The sand or quartz also helps in quenching the spark if any. Each end of the tube is covered

Fig. 4.4. H.R.C. fuses.

with the metallic cap. The caps are made of copper and then a sufficient provision is made to tight or insert, in the fuse carriers to enable the continuity in operation. The fusing element generally of silver, is fixed between these two caps. The element is made of a number of chambers, to enable the heat to dissipate.

There is a white spot on the cylindrical construction like a peep hole, which in normal condition is clear and becomes black in case of fuse blown out, so by seeing it can be judged about the condition of the fuse.

The H.R.C. fuses (high rupturing capacity) are designed with a time lag arrangement, *i.e.* for a short time, the fusing element will carry the overload current and will dissipate the heat produced. But if overload remains certainly it will blow out and can be seen from the spot already mentioned.

Operation. In case of normal working, the current flows through the fuse as it is connected in series, the heat developed is quite bearable and fuse will not blow. But, if the current exceeds the normal working limits, then the heat developed will be much more than the heat dissipated and the condition of the fuse whether blown or safe will be indicated by the spot.

These fuses are available from 1 A to 300 A. These are designed for accurate currents and are used in light and power circuits or in switch gears of high current capacity.

Q. What are the points to be kept in mind while selecting the H.R.C. fuses? Also state the advantages and disadvantages of these fuses.

Ans. There are some important points to be kept in mind before selecting the H.R.C. fuse:
  (*i*) The normal current of the circuit.
  (*ii*) The time required for the overloading to the circuit.
  (*iii*) The circuit and fuse voltage.

(*iv*) The rupturing capacity of the fuse should not be less than the current to be broken.

Advantages:

There are the following advantages:
(*i*) The fuse is very sensitive.
(*ii*) The fuse does not deteriorate with high speed (as the circuit cut off before the first peak value).
(*iii*) These are quite reliable.
(*iv*) These are cheap.
(*v*) No maintenance is required.
(*vi*) Consistent is performance.

Disadvantages:

There are the following disadvantages:
(*i*) It requires replacement after every blowouting, so they are costly in operation.
(*ii*) Inter-locking is not possible.
(*iii*) Cannot be repaired for same current capacity.
(*iv*) Produces over heating of the adjacent contacts.

Q. Can the H.R.C. fuse with tripping device be possible, if so give an idea with neat sketch?

Ans. The H.R.C. fuse with tripping devices are possible and the brief description is as under.

"The body of this type of fuses is ceramic, provided with metallic caps at the ends. The cap at one end is provided with a plunger facility, which hits the tripping contact of the circuit breaker. The plunger is electrically connected through fusible link, chemical charge and tungsten wire on the other end of the cap as shown in Fig. 4.5."

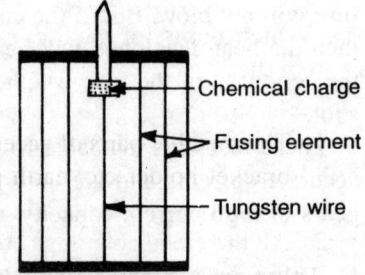

Fig. 4.5. H.R.C. fuse with tripping device.

There are a number of fusing elements made of silver, depending upon the current capacity of the fuse. These elements are connected to both caps as shown in Fig. 4.5. Whenever the fault occurs, the fuse elements are first to blow off. With the flowing of this current, the weak link in series, tungsten wire heats up and blow the chemical charge and deteriorate, thus the plung is pushed up as shown in Fig. 4.7 with a force and trip the circuit breaker.

Then with one fuse blowing, only the circuit breaker is off, as a result the single phasing is eliminated and motor etc. will not run on two phases.

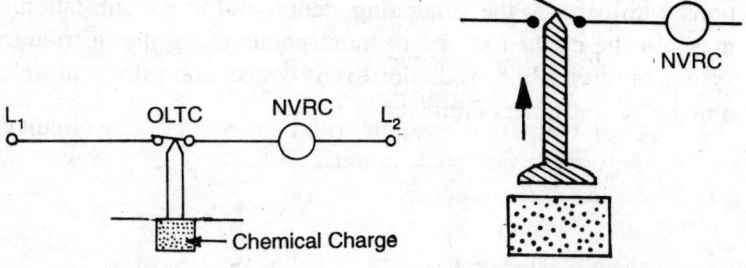

Fig. 4.6. Normal condition.   **Fig. 4.7.** Tripping operation.

4.3. EARTHING

Q. What is earthing and what is the necessity of earthing?

Ans. Earthing means the connection of all metallic parts except the current carrying conductors, with the general mass of earth, for the purpose of safety from electric shock. In the case of generating stations, the neutral point of the supply system is earthed.

The main purpose of earthing is the safety from electric shock. In case of earth fault, heavy current flows through the circuit, the circuit fuse melts out and faulty circuit will be disconnected from the supply mains.

In case of three phase supply, specially in case of unbalanced load, the unbalanced current and leakage current will flow through the earth and the voltage on each phase and neutral will be maintained.

Q. Which points of wiring, generating plants and motors should be earthed?

Ans. The following points should be connected to earth:
1. The metallic parts of accessories like batten holder, pendent holder, bracket holder etc. earth point of the 3 pin plug socket (light and power) should be efficiently earthed.
2. All the metal coverings containing or protecting and electric supply line or apparatus such as iron clad switches, iron clad distribution fuse boards, G.I. pipe and conduit pipes enclosing V.I.R. and P.V.C. cables, down rod of electric fans, should be connected to earth.
3. The frame of every generator, stationary motor and so far as is possible portable motor and metallic parts (not intended as current carrying conductors) of all transformer and any other apparatus used for regulating or controlling energy and all medium voltage energy consuming apparatus should be earthed by two separate and distinct connections with earth.

4. The neutral conductor of a 3-phase four wire system as shown in Fig. 4.8 and the middle conductor of a 2-phase three wire system should be earthed by not less than two separate and distinct connections with earth at the generating station and at the substation. It may also be earthed at one or more points along the distribution system or service line in addition to any connection with earth which may be at consumers premises.

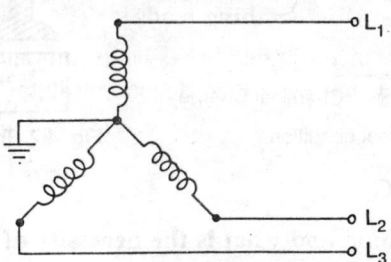

Fig. 4.8. Earthed Neutral connection

5. The metal casing of portable apparatus such as heaters, refrigerators, hand lamps, soldering irons, electric drills etc. should be connected to earth. If any one of them is fixed or installed, a separate direct connection to the earth should be provided in addition to earth within the connecting cable.

6. In case of a system comprising electric supply lines having concentric cables, the external conductor of such cables should be earthed by two separate and distinct connections with earth.

7. In case of d.c. three wire system, the middle conductor should be earthed at the generating station as shown in Fig. 4.9.

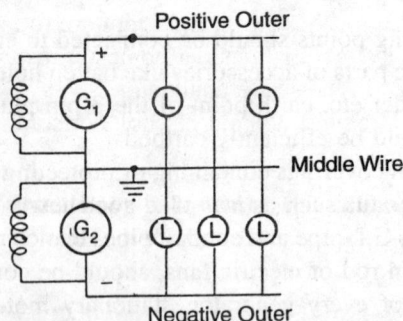

Fig. 4.9. Earthed middle wire connection.

8. Fabricated steel transmission line towers, tubler steel or rail poles carrying overhead conductors should be earthed. For this purpose a

continuous earth wire is provided and connected with earth at four points in every mile, *i.e.* 1.6 km spacing, being the points as nearly equidistant as possible, alternatively the metal work should be connected to an effective device at each individual support.

9. Stay wires provided for overhead lines should be connected to earth by connecting at least one strand to earth wire.

Q. What are the permissible value of the earth resistance and give I.S.I. specifications of earthing leads. (*N.C.V.T. 1979*)

Ans. The earth resistance should always be as minimum as it can be say tending to zero ohm, but even then depending upon the situations and places it vary as under:

1.	Large power station	0.5 ohm
2.	Major power station	1.0 ohm
3.	Small sub-station	2.0 ohm
4.	For ordinary soil	5 ohm
5.	For rocky soil	8 ohm

There are some recommended specifications as per I.S.I.

1. The earth electrode should be situated at least one and half metre away from the outside of building, whose installation is to be earthed.
2. The cross-section of the earthing conductor or lead, in general should not be less than half the section of the main supply conductor or the minimum cross-sectional area of earth lead should not be less than 0.02 sq inch or 8 S.W.G. and more than 0.1 sq inch.
3. Proper mechanical protection is to be provided to the earth continuity conductor having a shielding 12 mm G.I. pipe and below the ground level by at least 32 cm.
4. If copper plate is to be used for earthing then copper strips of 25.4 mm × 3.18 mm or 25.4 mm × 6.35 mm are to be provided.
5. The earth conductor for house, a domestic installation should not be less than 14 S.W.G. bare conductor.
6. No joint should be left without nut bolts or soldering, so that good continuity is maintained.

Q. What should be the resistance of a good earth? If it increases, what steps should be taken to reduce it? How the earth resistance is measured? (*N.C.V.T. 1979*)

Ans. The earth resistance in no case be more than 5 Ω for ordinary soil and for rocky soil should not exceed more than 8 Ω. The earth resistance in case of good soil should be tending to zero ohm or maximum to one ohm.

The resistance of earth is measured by means of the earth tester. The earth resistance tester is also basically the ohm meter. In this case two probes are fixed at 75 yards and 150 yards. The two spikes are also connected to current coil and pressure coil of the instrument. One terminal of C.C. and P.C. are short circuited and connected to earth electrode under test. The tester is on, it gives the direct reading. There are some other testers which are having hand-driven generator (For more details see instruments chapter).

Methods of reducing the earth resistance. The earth resistance can be decreased by adopting the following methods:

(a) *Maintaining moisture* In this case near the pipe or plate proper moisture is maintained by adding water to the funnel, periodically.

(b) *By increasing the area of contact.* In this case the area of contact with the earth and the plate or pipe is increased. In case of plate earthing the size of the plate is so chosen as to increase the contact area or the pipe diameter is increased to increase the surface contact with the earth.

(c) *By increasing the depth of the plate/pipe.* More the depth, more moisture and more electrical continuity or less resistivity, so by increasing the depth the resistance offered is decreased.

(d) *Parallel connections of earthing electrodes.* By connecting two resistances in parallel the equivalent resistance is reduced to half. Thus by doing double earthing we can reduce the earth resistance to a considerable extent.

This method can conveniently be adopted where the free soil is available and it is a very effective method.

(e) By increasing the proportion of salt in the mixture of salt and charcoal layers. So that the salt can preserve the water for more time.

4.4. TYPES OF EARTHING

Q. What are the types of earthing? Describe the construction and operation of the earthing, generally used for domestic installation.

Ans. The earthing is very essential for the domestic or industrial wiring. As per Indian Electricity Rules also the consumer will have one earthing facility to maintain a good earth continuity, though the supplier also provides.

There are two types of earthing:

(a) Pipe earthing, and

(b) Plate earthing.

(a) **Pipe earthing.** Generally the pipe earthing is commonly used for domestic installation owing to its accessibility and cheapness.

In this system a galvanised iron pipe of 38 mm (1.5" dia) and having the length of 2 metres (about 6.5 ft.) is used as earth electrode. This pipe has alternate holes of generally 12 mm diameter to allow the water to spread out. This pipe is buried in the moisture and is surrounded by the alternate layers of charcoal and salt. Generally the surroundings near the

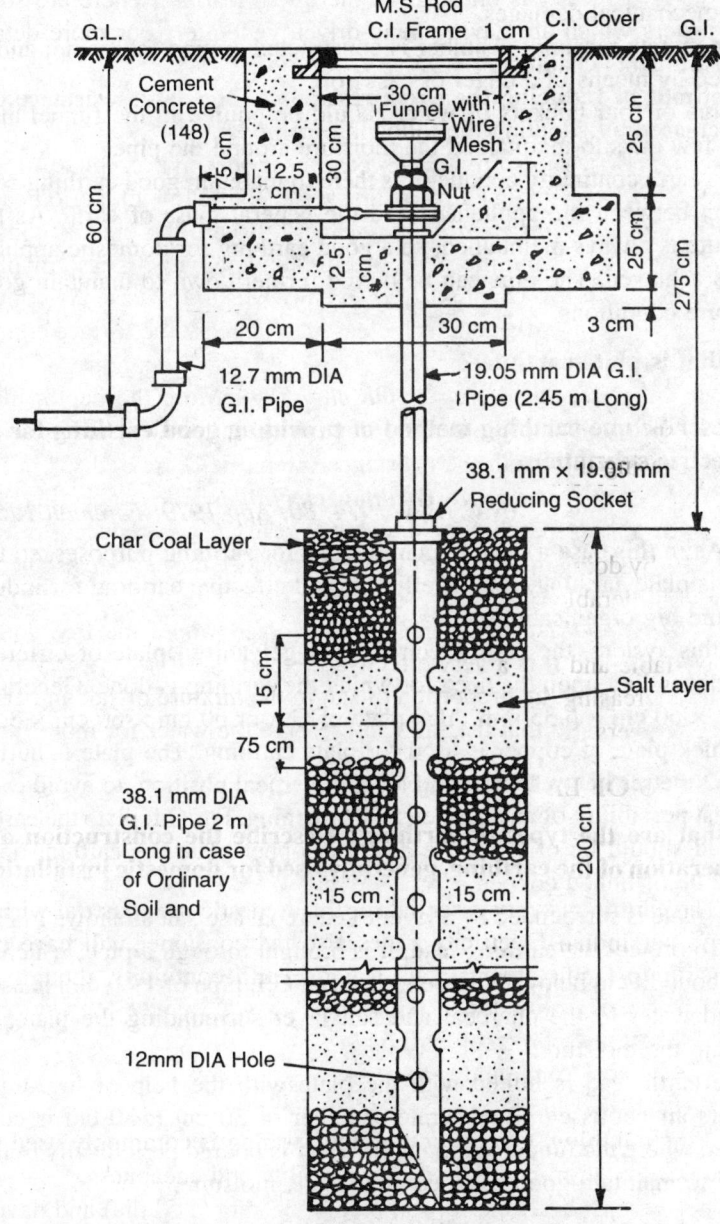

Fig. 4.10. Pipe earthing.

pipe is filled with the charcoal and salt 15 cm, alternate layers as shown in Fig. 4.10. A piece of 3/4″ diameter pipe is fixed with this pipe by means of a reducing socket. The G.I. pipe has a funnel on it's end. The earthing leads is connected to the 3/4″ diameter pipe by means of nuts, washer and bolts. The earthing lead is carried through the pipe to save it from mechanical damages.

A concrete masonry chamber is constructed around the funnel and is covered by means of a cover of cast iron.

Three or four buckets of water should be poured in the funnel after every few days to maintained the moisture around the pipe.

The earth continuity conductor is there to maintain good earthing connection between the appliances and the general mass of earth. As the moisture is always maintained, so a good earthing for domestic application is achieved. The pipe can be buried further down to maintain good moisture conditions.

Q. What is plate earthing?

OR

Describe one earthing method of providing good earthing for an electric substation.

(N.C.T.V.T. 1974, 80; App 1979; Foreman 1969)

Ans. As in this case a plate is mainly used for earthing purposes, so the name is plate earthing. It is an efficient and effective earthing for industries and big organisations.

In this system, the earth electrode is a galvanised plate of different sizes depending upon the need for which the earthing is done. Generally 60 cm × 60 cm × 6.35 mm thick plate of G.I. or 60 cm × 60 cm × 3.18 mm thick plate of copper is used for plate earthing. The plate is buried about 3 metres below the ground level in vertical position, to avoid even the least possibility of any earthen mishappening. For G.I. plate the earthing lead should be of G.I. wire and for copper plate the earthing lead should be of tinned copper.

The plate is surrounded by a layer of charcoal and salt as shown in Fig. 4.11. To protect the earthing lead, it is brought through a pipe, which is kept about 30 cm below the ground level. A G.I. pipe of 19.0 mm is used to send water to the charcoal and salt layer surrounding the plate, to maintain the moisture.

The earth lead is bolted with the plate with the help of lug, nuts, washers and bolts etc. A concrete chamber of 30 cm × 30 cm is constructed where the funnel is mounted. Water is poured periodically in the funnel to maintain good earth-continuity *i.e.* moisture.

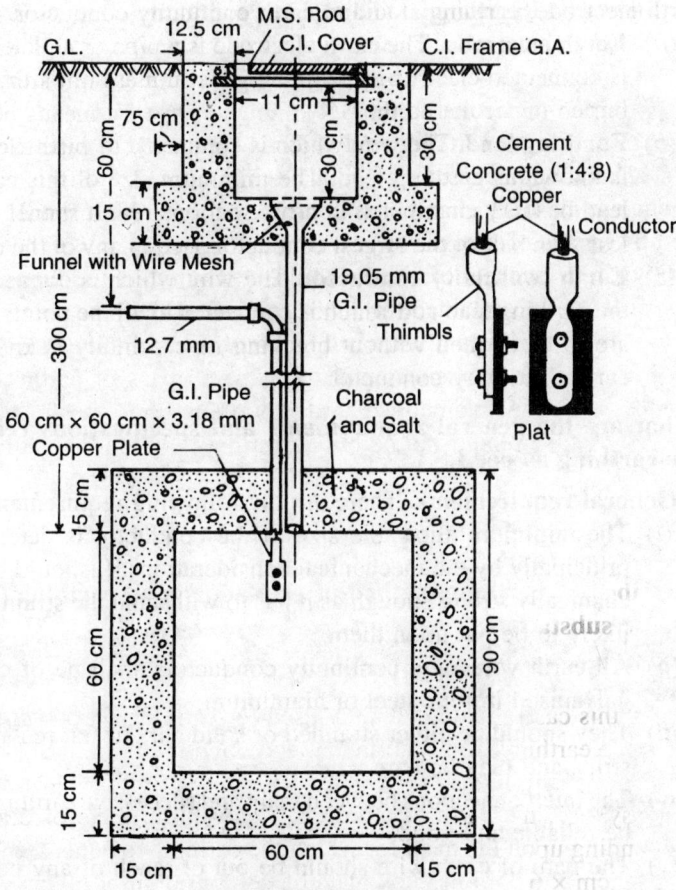

Fig. 4.11. Plate earthing.

Plate sizes

Copper plates:

30 cm × 30 cm × 3.18 mm	upto 30 H.P.
60 cm × 60 cm × 3.18 mm	upto 100 H.P.
90 cm × 90 cm × 3.18 mm	above 100 H.P.

G.I. plates:

30 cm × 30 cm × 6.36 mm	upto 30 H.P.
60 cm × 60 cm × 6.36 mm	upto 100 H.P.
90 cm × 90 cm × 6.36 mm	above 100 H.P.

Q. What are the common terms used in earthing?

Ans. There are the following very common terms used in earth engineering:

Earth electrode, earthing lead and earth continuity conductor.

(a) **Earth electrode.** The earth electrode is a pipe or a plate which is connected electrically with the general mass of earth; and is buried under the earth.

(b) **Earthing lead.** The wire which is connected to earth electrode is known as earthing lead. The minimum size of this earthing lead be 0.02 square inch in cross-section area. It should not be less than half of the largest conductor used in any of the circuit.

(c) **Earth continuity conductor.** The wire which is connected to the earthing lead and which is connected to all the points which are to be earthed without breaking its continuity is known as earth continuity conductor.

Q. What are the general requirements and specifications required for earthing as per I.S.I.?

Ans. General requirements. There are the following requirements:

(i) The minimum allowable size of the conductor is determined principally by the mechanical considerations. It should by mechanically strong enough as it has to withstand the strain that is likely to be put upon them.

(ii) All earth wires and continuity conductors shall be of copper, galvanised iron or steel or aluminium.

(iii) They should be either stranded or solid bars or flat rectangular strips and may be bare.

(iv) The interconnections between main and successive earthing shall be reliable and good.

(v) The path of earth wire should be out of reach of any person.

(vi) The neutral conductor shall not be used as earth wire.

Specifications:

(i) The earth wire should be outside at least 1.5 metre away from the building where earthing is to be done.

(ii) Possibly the earth wire should be of same material as that of earth electrode.

(iii) The minimum size should not be less than 0.02 sq inch or 8 S.W.G. wire and in general the size of earth should not be less than half the section of live wire.

(iv) The earth wire should be taken through the G.I. pipe.

(v) The earthing of the complete installation be brought to the main board where it must be firmly connected with the main earth electrode.

4.5. ELECTRICAL ACCESSORIES

All the material used in electrical installations like switches, holders, plugs and plug sockets etc. are called as electrical accessories.

Q. How the electrical accessories are classified, name them?

Ans. The electrical accessories are classified according to their uses. These are as under:

   (*i*) Controlling accessories.

   (*ii*) Holding accessories.

   (*iii*) Safety accessories.

   (*iv*) Accessories which are used to take the supply for different appliances *i.e.* outlet accessories.

   (*v*) General accessories.

Q. What do you understand by the controlling accessories? Describe the different types of switches with neat and clean diagrams.

Ans. The accessories which are used to control the circuit or a point, are called, the controlling accessories like switches. The different types of switches are as under:

   (*i*) *Single way switch.* These are used to control a point or a circuit. Only positive or phase wire is connected in the switch. These are made of bakelite or porcelain as shown in Fig. 4.12. These are rated according to the voltage and current, *e.g.*, single way switch, 6 A 250 V and 16 A 250 V, the later is generally known as single way power switch.

These are always connected in series with the circuit to be controlled. It has only two terminals which can be seen from the back of the switch without opening the cover. The base has got a mechanical lever to on and off the switch. The terminals and internal strips are designed according to the rated capacity of the switch. If more current will flow the strips will melts and switch will be out of order.

Fig. 4.12. Single way switch.

(*ii*) **Two way switch.** This switch controls the current in two ways, so it is called as two way switch. The outside construction is very much similar as that of single way switch. It has four terminals out of which two are short circuited. They are rated as 6 A 250 V. Generally these are used in stair-case wiring, where a lamp is to be controlled from two places. These are also made of bakelite or porcelain base.

(*iii*) **Intermediate switch.** It is a unique switch and is used where a lamp is to be controlled from more than two places. It has four terminals out of which two pairs of two are short circuited in one position as shown in Fig. 4.13. These are also made of bakelite and rated as 6 A 250 V.

Fig. 4.13. Position of intermediate switch.

(*iv*) **Bell push or push button.** These are made of bakelite or plastic and are used to control the bell point. It has two terminals which are short circuited when the button is pressed. It has a spring so that the short circuiting strip comes back to off position as soon as the nob is released.

(*v*) **Table lamp switch.** These are made of bakelite. These are used to control the table lamps etc. These are small on and off switch. Some times these are having the construction that in one pressing these maintain the continuity and in second pressing these release the connections and off the lamp. These are also rated as 2 A 250 V. 6 A 250 V etc.

(*vi*) **Double pole iron clad main switch.** These are also called as D.P.I.C. main switch. This used to control the live voltage going to a circuit in house wiring as shown in Fig. 4.14. In this switch both phase and neutral or positive and negative wires are connected to it. It controls both the wires simultaneously. It has either one fuse and one neutral link or both fuses as shown in Fig. 4.15.

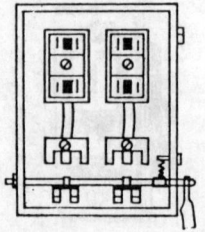

Fig. 4.14. D.P.I.C. main switch.

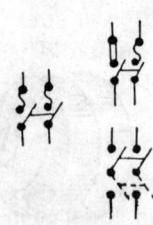

Fig. 4.15.

In house wiring the main switch having neutral link is used. These are rated according to the current and voltage ratings *e.g.*, 16 A 250 V, 32 A 250 V and 60 A 250 V etc. There is a screw connected with the body so that earth wire can be connected and the body of the main switch be earthed. In the main switch the fuse wire of proper size and proper current rating should be used. These are made of cast iron or the mild steel sheets.

In double pole main switches, there are double throw switches also. In that case, the centre position is the off position and both either positions controls the supply to the different circuits.

(*vii*) **Triple pole iron clad main switch.** This is also called T.P.I.C. main switch. This switch is used to control three phase line. It has three fuses and lines are connected to these three terminals as shown in Fig. 4.16. These are rated according to the current and voltage ratings *e.g.*, 16 A 440 V, 32 A 440 V, 60 A 500 V, 100 A 500 V etc. The fuse wire used, should be of proper size and rating. Generally in case of three phase switches two earthing provisions are provided or say double earthing is done.

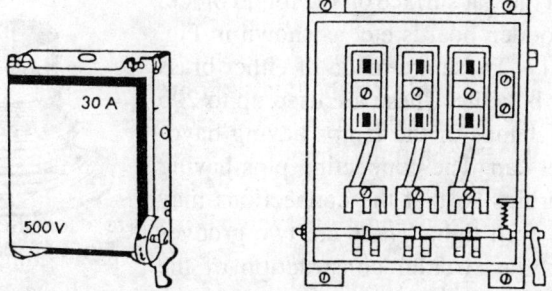

Fig. 4.16. T.P.I.C. main switch.

Sometimes these are called T.P.I.C.N. *i.e.*, the triple pole iron clad main switch with neutral link. There a separate provision for neutral wire is given. The triple pole double throw main switch controls two different circuits one by one.

(*viii*) **Knife switch.** These are used in power houses or substations to control the main lines going to a particular area. These can be of single blade, double blade, three blades. These are also rated according to their current capacity.

(*ix*) **Bed switch.** It is used to control a light point mostly in hanging position. Such switches are used to switch off the lamp while going to sleep or making the lamp 'on' while getting up at the night.

(*x*) **Pull switch.** These switches are operated with a single pull for 'on and 'off' the appliance, These are used for decoration lights, bed rooms and sometimes in bath rooms for water heater purpose etc.

(*xi*) **Toggle/Grid switch.** These are lighter. These are used with portable machines like drill machines, for lighting purpose. It is rated as 2 A 250 V, 6 A 250 V.

These are single way and with double combination also.

(*xii*) **Rotary switch.** These are rotating in nature for example the switch of the hot plate, in one position the hot plate will give less heat and in other position more heat. Similarly the switch of the stabilizer is a rotating one, so that the voltage can be boosted up or down by connecting the winding in different positions. (In that case different tappings are taken to the switch).

Q. What do you mean by the holding accessories? Describe the different types of the holders with sketches.

Ans. The accessories which are used to hold the electric lamps are known as holding accessories *i.e.*, holders. The different types of the holders are as under:

(*i*) **Batten holder.** These holders are used on the flat surface on the round block, wooden boards etc. as shown in Fig. 4.17. These are made of either brass or bakelite. These are used up to 200 W lamps or the lamps having bayonet cap. The connecting pins having springs so that the connections may be well tight. There are two grooves on the circular construction of the

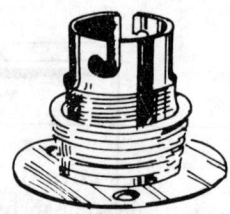

Fig. 4.17. Batten holder.

holder. The groove and pins are at right angle to each other. In this type of the holders the lamp is inserted, forced in, turned slightly and then left in position.

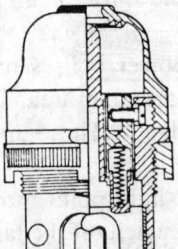
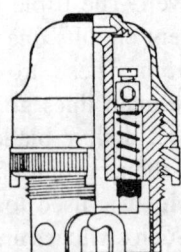

Fig. 4.18. Pendent holder.

(*ii*) **Pendent holder.** These holders are used where the lamps are required in hanging position as shown in Fig. 4.18. These are made of either brass or bakelite. These are used up to the lamps of 200 watts.

(*iii*) **Bracket holders.** These holders are used with a bracket. These are made of brass or bakelite. These are used up to the lamps of 200 W. Such holders are used to give direct light to a particular place. These are fixed on the bracket by the internal threading of the cap as shown in Fig. 4.19.

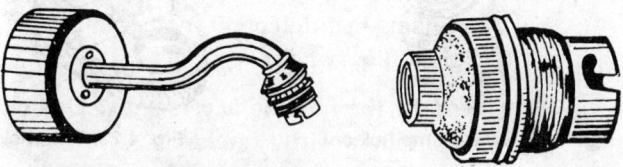

Fig. 4.19. Bracket holders.

(*iv*) **Angle holder.** These are also made of bakelite or brass. As the name explains, the lamp is kept at a particular angle as shown in Fig. 4.20. These are used up to 200 W lamps and are rated as 6 A 250 V. Generally these are used on the advertising boards, foot lights, kitchen etc.

(*v*) **Screw holder.** Generally these holders are used for the lamps of more then 200 watts. These are of two types, the Edison screw type and the Goliath screw type as shown in Fig. 4.21. It's cover is made of porcelain. Such holders are used in studios, head lights, flood lights, focussing light etc.

(*vi*) **Water tight bracket holder.** Such holders are provided with tubular glass fixed with water tight cover. These are somewhat similar to the bracket holders. such lamp holders are used outside the house and street light, where there is no shelter.

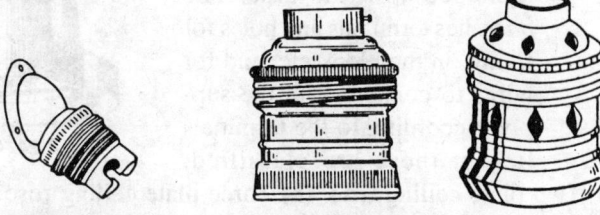

Fig. 4.20. Angle holder.   **Fig. 4.21.** Screw holder.

(*vii*) **Swivel lamp holder.** Such lamp holders are designed for controlled directional light, it beams towards a particular object, so that it can be projected. Such holders are widely used for shop window, show cases etc. There is a ball and socket joint fitted with back plate as shown in Fig. 4.22. Generally these are called focusing light.

Fig. 4.22. Swivel lamp holder.   **Fig. 4.23.** Tube holder.

(*viii*) **Tube holder.** These holders are used for tubes as shown in Fig. 4.23. These are made of plastic or bakelite. The tube is held in position by these holders. There are two strips inside the holder which actually feed the supply to the tube.

Q. What do you know about the accessories used for taking supply for the consuming appliances? Describe some electrical accessories with sketches.

Ans. The accessories by which supply is taken for the consuming appliances are covered under this type. These are of the following types:

(*i*) **Ceiling rose.** These are used to take supply for table fan, tubes, lamps, etc. from the wall or ceiling. It is shown in Fig. 4.24. These are having two parts, the base and cover. *Cover* has a hole in the centre for the connecting wires to take out. There are threading on the internal sides so that it may be fixed or tightened with the base. The *base,* has terminals and holes for fixing on the block etc. and for wires to connect with the supply. According to the terminals it has, these are classified.

Fig. 4.24. Ceiling rose.

(*a*) Two plate ceiling rose. (*b*) Three plate ceiling rose.

These are made of bakelite or porcelain. These are rated according to the current capacity as 6 A 250 V. Generally two plate ceiling rose are used. The three plate ceiling rose is used in the installation to decrease the length of the wires and can be used as the joint box also.

(*ii*) **Plug socket.** These are used to take the supply for the portable appliances, like table fan, table lamp and other portable equipments. These are made of bakelite or porcelain. These are rated as 6 A 250 V, 16 A 250 V.

These are of the following types:

(*a*) *Two pin plug socket.* These plugs are rated as 6 A 250 V and are having only two pins or terminals, the positive or phase and negative or neutral. Only two wires are connected to these plugs. Both the terminals or pins are of the same length and size.

(*b*) *Three pin plug socket.* These are also used to take supply for the portable equipments. These are rated according to the voltage and current ratings like 6 A 250 V, 16 A 250 V. The plugs 16 A 250 A are called power plugs and are used for heater, press, single phase motors etc.

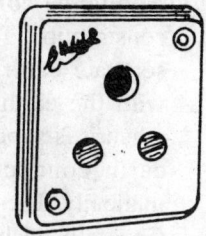

Fig. 4.24. 3 Pin plug socket.

There are three terminals marked as Live, Neutral and Earth. Live terminal is on the right hand side, neutral terminal on left hand side and on the top there is the earth terminal. The earth terminal is thick and long than the other two terminals as shown in Fig. 4.25. In every case earth wire must be connected to the earth terminal of the socket, so that the appliance must get earth wire first and then the phase and neutral.

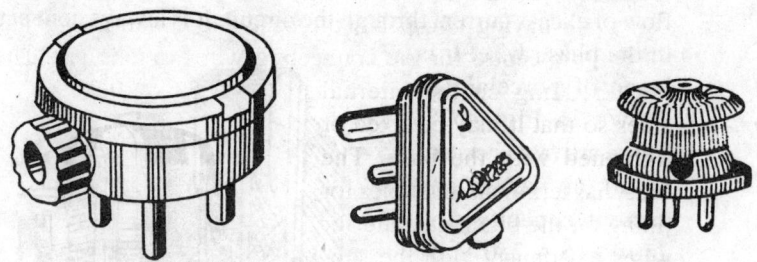

Fig. 4.25. 3 Pin plug top.

(*iii*) **Two pin plug top or shoe.** It used to take the supply from two pin plug socket for lamp etc. It is made of bakelite and rated as 6 A 250 V.

(*iv*) **Three pin plug top or shoe.** It is used to take supply from socket. It has three pins, two are similar and the third one is bigger and long for earth as shown in Fig. 4.26. The earth must be connected first of all then the phase and neutral to the appliance. These are rated according to the current i.e. 6 A 250 V, 16 A 250 V. These are made of bakelite, plastic or rubber with some plastic.

(*v*) **Adapter of plug multiplier.** These are used to take supply for a number of appliances from a single point. These are made of bakelite and can be used upto 6 A 250 V, sometimes there is a provision for two pin plug top and three pin plug tops simultaneously.

(*vi*) **Appliance connector.** It is made of bakelite and the porcelain construction. The wires are connected with the brass terminals. The separate earth connections are made with the earth terminals and in case when it is supplying the appliance the earth connections are made automatically. These are rated 16 A 250 V. Generally the heaters, hot plate, electric iron are connected across the mains through the appliance connector.

Fig. 4.27. Appliance connector.

Q. What are the safety accessories? Describe with neat and clean sketches.

Ans. The accessories which helps in checking the flow of heavy current in the circuit are known as the safety accessories. These are used in the circuit for the safety of circuit and the appliance also.

These are of the following types:

(*i*) **Porcelain kit ket fuse.** These are used in house wiring and industrial wiring for the safety of the circuit. It prevents the flow of excess current through the circuit. It is always connected in the phase wire. It is made of porcelain as shown in Fig. 4.28. These are rated according to the voltage and current capacity 6 A 250 V, 16 A 250 V, 32 A 250 A, 60 A 250 V, 16 A 500 V and 32 A 500 V etc.

It has two parts the base and the fuse carrier. The fuse is placed in the fuse carrier. The base has

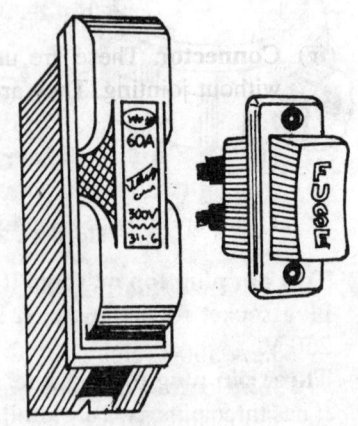

Fig. 4.28. Kit ket fuse

got the terminals, live and load wires are connected in them. The base is a fixed portion and fuse carrier can be taken out as and when needed. The fuse should be of proper capacity and size. The fuse wire should have low melting temperature. (For more details see chapter Fuse).

(*ii*) **Fuse cutout.** These are made of porcelain. It has two plates and fuse element is connected in between these plates. These are used for low voltage circuit as in railways and automobiles etc.

(*iii*) **Distribution board.** These are used where the total load is high and that is to be divided into a number of circuits. It is used where the load is more than 800 W. It has the number of fuses according to the number of circuits and a neutral link is also provided so that simultaneously the neutral wire can be taken for different circuits. All these cutouts etc. are covered in a metal box. These are two ways, three ways, four ways, six ways etc. as shown in Fig. 4.29.

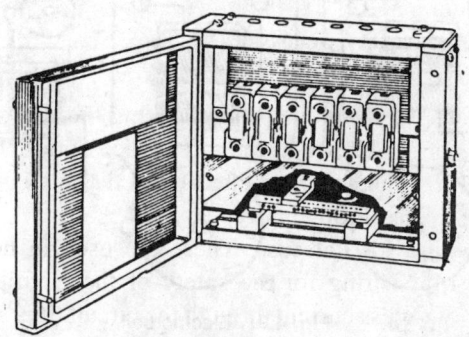

Fig. 4.29. Distribution box.

(*iv*) **Connector.** These are used to extend the length of the wire without jointing. They are made of por-celain. There is a brass sleeve with threading for small screws to tight the wire in the sleeve as shown in Fig. 4.30. These are single way, two ways, three ways etc. These are rated according to the voltage and current capacity, 6 A 250 V, 16 A 250 V, 32 A 250 V, 16 A 500 V, 32 A 500 V etc.

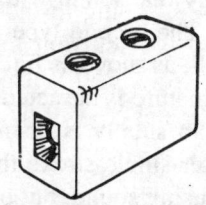

Fig. 4.30. Connector.

(*v*) **Joint cutout or joint box.** These are generally used in batten and casing wirings for jointing the wires. These are made of bakelite. These are helpful for tracing the fault.

Q. What do you understand by the general accessories? Describe with neat sketches the construction and function of the electric bell.

Ans. The accessories which are commonly used in domestic, industrial or other electrical field are called as the general accessories, *e.g.* the electric bell, electric buzzer, relays and other electro-magnetic devices etc. It is show in Fig. 4.31.

Electric bell. It is an electro-magnetic device. It is used in houses, offices and in industries also for alarming purposes.

The electric bells are of the following types,

(*a*) Simple electric bells.

(*i*) single stroke bell.

(*ii*) Trembler bell.

(*iii*) Continuous ringing bell.

(*iv*) Polarised bell.

(*b*) Electronic bell.

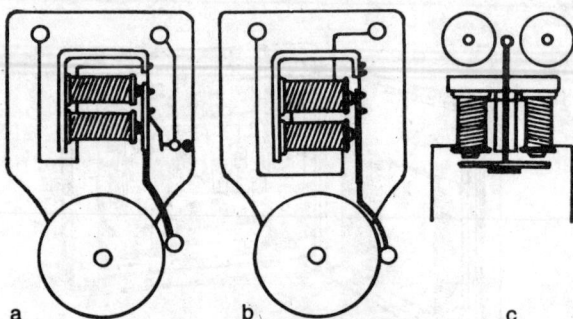

Fig. 4.31. Electric bell.

In simple construction fig. 4.31(*a*) a single stoke bell, whenever the electric current is passed through the coils the armature which·has a hammer attached to one end is attracted and rings the bell. This bell gives only one sound with one operation of bell push.

The second type of bells are generally used. It is known as trembler bell; as shown is fig. 4.31(*b*) whenever current flows through the coils the armature is attracted but as soon as it is attracted the contact of the coils from supply is opened resulting the demagnetising the armature. Thus after single-stroke the armature will again attract and hammer the gang causing sound; but because of the break in circuit and demagnetising the coils the hammer will go back. This operation will continue unless the current is switched off. Buzzer is also an example of the same operation only the gang and striker are eliminated.

The other type of bell is a continuous ringing bell. When the current flows through the coils the armature is attracted and releases the catch allowing the contact arm to move because of the auxiliary spring. This closes the contact thus producing sound independently of the bell push. It will continuously ring unless the reset cord which is installed with the bell is pulled.

There is another type of bell as shown in fig. 4.31(c) a polarised bell. It is operated on *A.C.* It is similar to that as is used in telephone circuit. When the current in one coil has one polarity pole A in strong and pole B is weak, results the striking of one bell. When the current is reversed pole 'A' is weak and pole 'B' is strong thus causing an attraction of hammer to another bell. Now the bell produces two sounds in each cycle of alternating current.

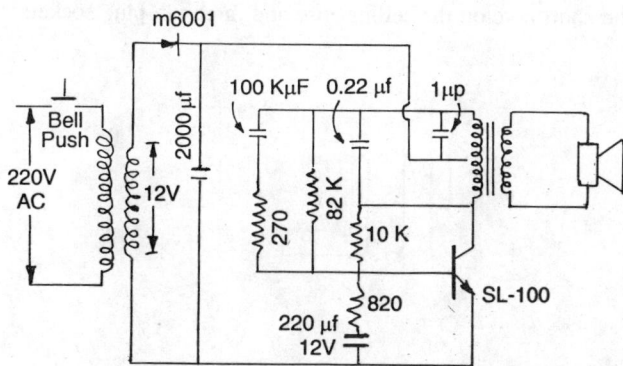

Fig. 4.32. Singing bird door bell circuit diagram.

(*b*) **Electronic bell.** In this type of bell the solid state devices are used. With the use of different components and different circuits different sounds are produced. These bell may have single, double or multitunes producing with single pressing of the bell push. A simple singing bird door circuit is shown here as under in fig. 4.32.

REVIEW QUESTIONS

1. What is the utility of a fuse in an electrical installation?
2. Why a fuse blows and how?
3. What do you understand by fusing factor?
4. What are the different types of fuses? Describe open type fuse.
5. Write a short note on semi-enclosed type fuses.
6. What are the advantages, disadvantages and uses of H.R.C. fuses?
7. Describe the construction and working of H.R.C. fuses.

8. Which types of fuses are used:
   (a) on a distribution line pole,
   (b) in houses for domestic metering,
   (c) for precious controllings,
   (d) for precious industrial work.
9. Why double earthing is preferred and essential for 3φ machines?
10. How many earths are recommended for single phase H.P. meter, 3φ 5 H.P. meter, 3φ 100 H.P. meter, ceiling fan, heater, distribution transformer etc.?
11. What is the main object of earthing?
12. What are the limitations of earthing?
13. What are the different types of earthing?
14. Why water is added periodically in pipe or plate earthing?
15. What are the classifications of accessories?
16. Describe any two controlling accessories.
17. How many types of batten holders are used? Describe the swivel type holder?
18. Write short note on the ceiling rose and three pin plug socket.

5

Fundamental Units

All the quantities used in engineering are expressed in terms of length, mass and time. These are known as fundamental units.

Q. How many system of units are generally used in engineering?

Ans. There are the following system of units generally used.

1. **C.G.S. system.** In this system, the length is expressed in centimetre, mass in grams and time in seconds. This system of unit is not commonly used nowadays.

2. **F.P.S. system.** In this system of unit, the length is expressed in foot, the mass in pounds and time in seconds. It is also not used very commonly.

3. **M.K.S. system.** In this method the length is measured in metre, the mass in kilogram and the time in seconds. It is widely used nowadays.

4. **M.K.S.A. system.** In this system of units four terms are used the length, mass, time and the current. These are expressed in metres, kilograms seconds and amperes respectively.

5. **S.I. System.** It is nothing but the rationalized M.K.S.A. system. It has six units: the length, the mass, the time, the current, the thermodynamic temperature and luminous intensity measured in metre, kilogram, seconds, amperes, kelvin and candela respectively.

   This system of units is widely used coming in the engineering field.

Q. Define

   (a) **Matter, mass, weight, speed, velocity acceleration, retardation, work and force.**

   (b) **Also define the torque, heat, temperature, calory, power and energy.**

Ans. (*a*) **Matter.** Matter is that which occupies some space, have some weight, and can be felt by one or more senses, for example wood, steel, book, water etc.

Mass. The mass can be defined as the quantity of the matter contained by a body. The mass is same at all places. It's unit is kilogram in M.K.S. system.

Weight. It is the force with which a body is attracted towards the centre of earth. As the force depends upon the distance, so the weight is different at different places. It's unit is kg. in M.K.S. system.

Speed. The speed is defined as the rate of change of motion or the rate of change of displacement of a body in any direction.

$$\text{Speed} = \frac{\text{Distance covered}}{\text{Time taken}}$$

It's unit is M.K.S. system is km/hrs or m/sec.

Velocity. The velocity is defined as the rate of change of motion or rate of change of displacement in a particular direction. It is also equal to the distance covered divided by the time taken and generally expressed in km/hrs or m/sec in M.K.S. system.

Acceleration. It is defined as the rate of change of velocity of a body. It's unit is M.K.S. system is m/sec-sec or m/sec^2.

Retardation. It is the negative acceleration. It is also the rate of change of velocity but it is taken when the body in motion is stopped or decreasing its speed. It's unit is m/sec^2 in M.K.S. system.

Work. When a force is applied on a body and the body moves through a certain distance, work is said to be done. *OR* It can otherwise be said that work is the result of applied force, so work is said to be done when a force acting on a body causes it to move.

Work done = Force × Distance covered.

The mechanical unit of work is joule.

Force The force can be defined by the effects it exhibits. Force is that which changes or tends to change the state of rest or motion of a body.

Force = Mass × Acceleration.

Its unit is M.K.S. system is Newton. One newton is that force which produces the acceleration of one metre/sec^2 in one kg weight.

(*b*) **Torque.** The torque is the turning or twisting moment of a force about its axis. The mechanical unit is joules. The torque applied is of one joule, if a force of one newton is working on a wheel of one metre radius.

Heat. Heat is the form of energy, which produces the sensation of warmness or coldness. It's unit is calory.

Calory. It is the unit of heat. One calory is that much heat which is required to raise the temperature of one gram of water through one degree centigrade. Generally calory is the small unit of heat so it is taken in kilocalory.

Temperature. It is the degree of hotness or coldness of a body. It is measured by the means of the thermometer.

Power. The rate of doing work is defined as the power. It is measured by the work done per second. It's mechanical unit is H.P. and electrical unit is Watt.

$$1 \text{ H.P.} = 746 \text{ Watts.}$$

Energy. It is defined as the capacity for doing work. The electrical unit is Kilowatt hours

$$\text{Energy} = \text{Power} \times \text{Time.}$$

Q. What do you understand by the law of conservation of energy Newton's laws of motion, and mechanical equivalent of heat?

Ans. Law of conservation of energy. This law states that, "the total energy of this universe is constant, though it may change from one form to the another".

For example, if one gram of water is heated up it will change its state of water into vapours and again if the vapours are condensed it will be water and if freezed it will be ice. So here one gram of water is changing its states but the matter exists.

Newton's laws of motion. Newton introduced three laws of motion:

First law. Every body continues in its state of rest or motion in a straight line unless it is compelled by some external force to change its state.

Second law. The rate of change of momentum of a body is directly proportional to the impressed force and takes place the same direction in which the force is applied.

Third law. To every action there is an equal and opposite reaction.

Mechanical equivalent of heat. Whenever the mechanical work is done, heat is produced, and the ratio of work done and heat developed is known as mechanical equivalent of heat.

In case of electrical engineering heat developed in a conductor is proportional to the square of current, resistance and time

i.e. $\quad H \propto I^2 \times R \times t$

or $\quad H = \dfrac{I^2 R \cdot t}{J}$

J is called the mechanical equivalent of heat.

$$J = 4.186 \text{ J/kcal} \approx 4.2 \text{ J/kcal.}$$

Q. What do you mean by the fundamental and derived units? Define S.I. system of units and their comprehensive units.

Ans. The mass, length and time are the fundamental quantities and the units which are used to represent these are known as the fundamental units. Similarly the physical quantities can be expressed in terms of fundamental units like mass, length and time, so "the units of those physical quantities which can be expressed in terms of fundamental units are called derived units".

S.I. (System International) units have been adopted by the scientists and technologist just to have a comprehensive system of measurement. In the latest S.I. system there are seven fundamental or basic units. These are as follows:

(*i*)	the metre	– the unit of length,
(*ii*)	the kilogram	– the unit of mass,
(*iii*)	the second	– the unit of time,
(*iv*)	The ampere	– the unit of current,
(*v*)	the kelvin	– the unit of temperature,
(*iv*)	candela	– the unit of luminous intensity, and
(*vii*)	mole	– the unit of amount of substance.

The S.I. system beside these seven units has two more supplementary units:

(*i*)	radian	– the unit of plane angle, and
(*ii*)	steradian	– the unit of solid angle.

6

Modern Electron Theory

6.1. ATOMIC THEORY

Q. Explain the modern theory of atomic structure.

Ans. It is commonly known as the electron theory. According to this theory all the matters whether solid, liquid or gaseous are mainly constituted by the small particles, known as molecules, which are themselves consists of further small particles known as the atom. In case of an element the smallest particles are of same nature and size and are called atoms, whereas in case of compound or mixture the smallest particle having the same properties as that of the compound or mixture has, are known as molecules and the further subdivision of the molecule where the particles do not have the same properties as that of the material, are called atoms. So far only 112 elements are discovered and the compound may be unlimited.

The atom, although extremely small particle, has a complex internal structure. The scientists are not absolutely certain about the exact structure of an atom, however the most accepted theory at present is as given.

The atom according to the modern theory, consists of a central core which is called as nucleus. The nucleus mainly consists of two particles; the protons and neutrons. The nucleus is surrounded by the number of small particles called electrons. The electrons are spinning around themselves and also revolving around the nucleus. The electron revolves in a fixed path, known as orbit. The *electrons* are negatively charged particles. The *protons* are positively charged particles. The *neutrons* are not having any charge and are neutral particles.

In an actual atom the number of electrons and protons are equal and balance each other, resulting the atom electrically neutral. The electrons are very light in weight as compared with protons and neutrons. Electrons is 1/1850th of the proton but is three times the diameter of the proton. The weight of the atom is mainly because of the protons and neutrons. The neutron is also approximately equal to the proton in weight. So the *atomic weight* of an atom is defined as the total number of protons and neutrons contained in an atom; for example the atomic weight of copper is 64, it has 35 neutrons and 29 protons which makes 35 + 29 = 64 as the atomic weight.

The electrons are firmly attracted by the nucleus. It is obvious that the electrons near the nucleus are firmly held and difficult to pull out, then those electrons which are further away from the nucleus. The electrons revolves around the nucleus in orbits. The number of electrons in any particular orbit can be calculated by the formula $2N^2$. Where N is the number of orbit. Generally the orbits are K, L, M, N or 1, 2, 3, 4 etc. The number of electrons in different orbits are as follows:

In first or K orbit $\quad 2 \times 1^2 = 2$ electrons

In second or L orbit $\quad 2 \times 2^2 = 8$ electrons

In third or M orbit $\quad 2 \times 3^2 = 18$ electrons.

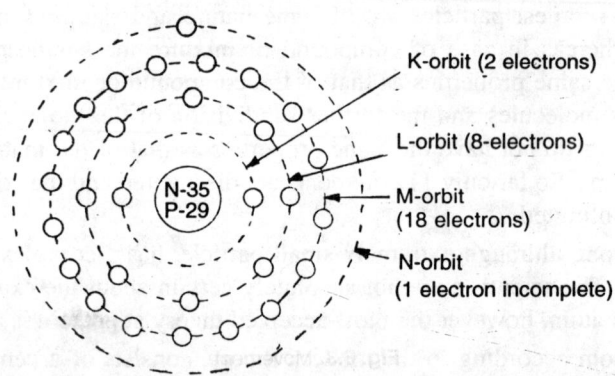

K-orbit (2 electrons)

L-orbit (8-electrons)

M-orbit (18 electrons)

N-orbit (1 electron incomplete)

Fig. 6.1. Atomic structure of copper.

The experimented researches shows that last orbit cannot have more than 8 electrons and last but one orbit cannot have more than 18 electrons. The simple arrangement of the example of helium, sodium – having 11 electrons which are arranged (2+8+1), chlorine – having 17 electrons arranged (2+8+7), and copper – 29 electrons arranged (2+8+18+1) are shown in Fig. 6.1, 6.2.

The atom is identified by its atomic number which indicates number of protons in the nucleus, for example copper has protons, so the atomic number is 29, in the periodic table, oxygen has 8 proton so the number is

8. The Table 6.1 shows the different elements with the arrangement of the electrons in their different orbits or shells.

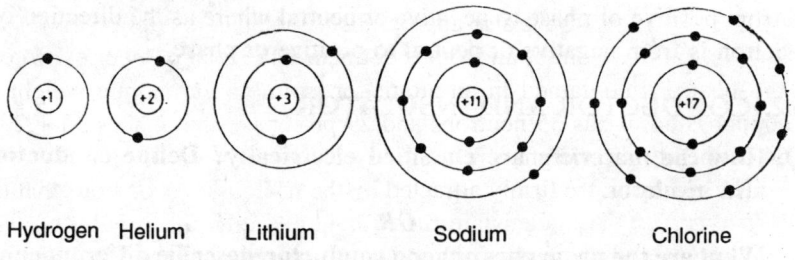

Hydrogen   Helium   Lithium   Sodium   Chlorine

Fig. 6.2. Atoms and Electrons

Q. What do you understand by the flow of electrons and the direction of current.

Ans. In an atom the number of electrons and protons are equal so the atom is electrically neutral. However if one or more free electrons (the electrons which are in the outer most orbit is not satisfied) are removed or added that atom is a charged atom or ion. If the free electrons are added, it means now that particular atom has more number of electrons than protons so the atom is negatively charged and is said anion. On the other hand if one or more free electrons are removed, the atom now has more number of proton than electrons, so it is positively charged and is known as cation.

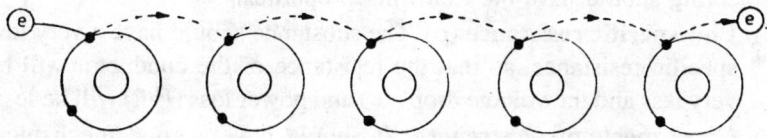

Fig. 6.3. Movement

Now let there be any conductor which has two electrons in its outer most orbit. Electrically this conductor is neutral, but now if an electric pressure is applied, the free electron is moved as shown in Fig. 6.3. The electron will keep on drifting as long as the electric potential is maintained and the drifting or movement of the electron is established as long as the energy is supplied by applying electric potential. This flow of electron is known as electric current and if 6.28×10^{18} electrons are passing from a point or junction then one ampere current is said to flow through that point.

In case of insulators the electrons are so forcibly attached with nucleus that even after applying the pressure the electron drifting is not possible, that is the reason why these are called insulator.

Direction of current. The direction of current is opposite to the direction of movement of the electrons. The conventional direction of current is from positive or phase to negative or neutral where as the direction of electron is from negative or neutral to positive or phase.

6.2. CONDUCTOR AND INSULATOR

Q. How the materials are classified electrically? Define conductor and insulator.

OR

What are the properties of good conductor, describe different conductors according to their merit of conductivity?

Ans. Electrically all the materials are divided into two classes:

(a) Conductors,

(b) Insulators.

(a) **Conductor.** The materials in which electric current can flow easily are named as conductor. *OR* The substances which offers a low resistance in the flow of electricity are known as the conductor. *OR* The substance in which from the outermost orbit, the free electrons are shifted easily, are named as conductor.

The substance whose outermost orbit is incomplete act as a good conductor *i.e.* they permit an easy detachment of their outermost electron and offer very little hindrance, such substances are called good conductor.

Properties of a good conductor. The conductor used for the electrical engineering should have the following properties:

(i) **Low specific resistance** (ρ). The substance should have a very low specific resistance, so that the resistance of the conductor will be very less and the voltage drop (IR) and power loss ($I^2 R$) will be less.

(ii) **Good mechanical strength.** If should have a good mechanical strength so that the wires drawn, may not get break in case of heavy wind etc.

(iii) **Solderability.** It should be easily soldered.

(iv) **Malleability and ductile.** The conductor should be ductile so that the wires may be drawn easily and if required the sheets can be prepared.

(v) **Easily jointed.** The conductor should be such that it may be jointed easily.

(vi) **Cost and availability.** The conductor should have a very low cost. It should not be very costly and should be easily available.

(vii) **Free from the atmospheric effects** *i.e.* **non-correcive.** It should be free from the atmospheric effects and the natural effects as the wind, the rains and chemical changes etc.

The conductivity of the conductor is inversely proportional to the resistivity. The conductor which has less resistivity will have a good conductivity and vice versa. There are some number of conductors according to their order of conductivity.

(*i*) **Silver.** It is the best conductor. It is malleable and ductile. Though it has a good properties but it is very costly so it is not used for general purposes. It is used for heavy duty electrical contacts and very precious work, *e.g.* for limiting fuse in heavy duty circuit breakers etc.

(*ii*) **Copper.** It is a good conductor second to silver. It is malleable and ductile. It is good conductor of heat. It is less costly than silver. It can be easily jointed and the mechanical strength is also good. So it is widely used in electrical work in wires and cables. But nowadays because of the world wide shortage of copper it is not so cheaply available in the market. The different values of resistances as compared with the copper is shown in Table 6.2.

Table 6.1. Details of elements and arrangement of electrons in different shells

Atomic No.	Element	Protons in the Nucleus	Arrangements of electrons in orbit					
			K	*L*	*M*	*N*	*O*	*P*
1.	Hydrogen, H	1	1					
2.	Helium, He	2	2					
3.	Lithium, Li	3	2	1				
4.	Beryllium, Be	4	2	2				
5.	Boron, B	5	2	3				
6.	Carbon, C	6	2	4				
7.	Nitrogen, N	7	2	5				
8.	Oxygen, O	8	2	6				
9.	Fluorine, F	9	2	7				
10.	Neon, Ne	10	2	8				
11.	Sodium, Na	11	2	8	1			
12.	Magnesium, Mg	12	2	8	2			
13.	Aluminium, Al	13	2	8	3			
14.	Silicon, Si	14	2	8	4			
15.	Phosphorus, P	15	2	8	5			
16.	Sulfur, S	16	2	8	6			
17.	Chlorine, Cl	17	2	8	7			

(Contd.)

Atomic No.	Element	Protons in the Nucleus	Arrangements of electrons in orbit					
			K	L	M	N	O	P
18.	Argon, A	18	2	8	8			
19.	Potassium, K	19	2	8	8	1		
20.	Calcium, Ca	20	2	8	8	2		
21.	Scandium, Sc	21	2	8	9	2		
22.	Titanium, Ti	22	2	8	10	2		
23.	Vanadium, V	23	2	8	11	2		
24.	Chromium, Cr	24	2	8	13	1		
25.	Manganese, Mn	25	2	8	13	2		
26.	Iron, Fe	26	2	8	14	2		
27.	Cobalt, Co	27	2	8	15	2		
28.	Nickel, Ni	28	2	8	16	2		
29.	Copper, Cu	29	2	8	18	1		
30.	Zinc, Zn	30	2	8	18	2		
31.	Gallium, Ga	31	2	8	18	3		
32.	Germanium, Ge	32	2	8	18	4		
33.	Arsenic, As	33	2	8	18	5		
34	Selenium, Se	34	2	8	18	6		
35.	Bromine, Br	35	2	8	18	7		
36.	Krypton, Kr	36	2	8	18	8		
37.	Rubidium, Rb	37	2	8	18	8	1	
38.	Strontium, Sr	38	2	8	18	8	2	
39.	Yttrium, Y	39	2	8	18	9	2	
40.	Zirconium, Zr	40	2	8	18	10	2	
41.	Niobium, Nb	41	2	8	18	12	1	
42.	Molybdenum, Mo	42	2	8	18	13	1	
43.	Technetium, Te	43	2	8	18	14	1	
44.	Ruthentium, Ru	44	2	8	18	15	1	
45.	Rhodium, Rh	45	2	8	18	16	1	
46.	Palladium, Pd	46	2	8	18	18	0	
47.	Silver, Ag	47	2	8	18	18	1	
48.	Cadmium, Cd	48	2	8	18	18	2	
49.	Indium, In	49	2	8	18	18	3	
50.	Tin, Sn	50	2	8	18	18	4	
51.	Antimony, Sb	51	2	8	18	18	5	
52.	Tellurium, Te	52	2	8	18	18	6	
53.	Iodine, I	53	2	8	18	18	7	
54.	Xenon, Xe	54	2	8	18	18	8	
55.	Cesium, Cs	55	2	8	18	18	8	1
56.	Barium, Ba	56	2	8	18	18	8	2
57.	Lanthanum, La	57	2	8	18	18	9	2
58.	Cerium, Ce	58	2	8	18	19	9	2
59.	Praseodymium, Pr	59	2	8	18	20	9	2

(Contd.)

Atomic No.	Element	Protons in the Nucleus	K	L	M	N	O	P

Atomic No.	Element	Protons in the Nucleus	K	L	M	N	O	P
60.	Neodymium, Nd	60	2	8	18	21	9	2
61.	Promethium, Pm	61	2	8	18	22	9	2
62.	Samarium, Sm	62	2	8	18	23	9	2
63.	Europium, Eu	63	2	8	18	24	9	2
64.	Gadolinium, Gd	64	2	8	18	25	9	2
65.	Terbium, Tb	65	2	8	18	26	9	2
66.	Dysprosium, Dy	66	2	8	18	27	9	2
67.	Holmium, Ho	67	2	8	18	28	9	2
68.	Erbium, Er	68	2	8	18	29	9	2
69.	Thulium, Tm	69	2	8	18	30	9	2
70.	Ytterbium, Yb	70	2	8	18	31	9	2
71.	Lutetium, Lu	71	2	8	18	32	9	2
72.	Hafnium, Hf	72	2	8	18	32	10	2
73.	Tantalum, Ta	73	2	8	18	32	11	2
74.	Tungsten, W	74	2	8	18	32	12	2
75.	Rhenium, Re	75	2	8	18	32	13	2
76.	Osmium, Os	76	2	8	18	32	14	2
77.	Iridium, Ir	77	2	8	18	32	15	2
78.	Platinum, Pt	78	2	8	18	32	16	2
79.	Gold, Au	79	2	8	18	32	18	1
80.	Mercury, Hg	80	2	8	18	32	18	2
81.	Thallium, Tl	81	2	8	18	32	18	3
82.	Lead, Pb	82	2	8	18	32	18	4
83.	Bismuth, Boi	83	2	8	18	32	18	5
84.	Polonium, Po	84	2	8	18	32	18	6
85.	Astatinc, At	85	2	8	18	32	18	7
86.	Radon, Rn	86	2	8	18	32	18	8

Atomic No.	Element	Protons	K	L	M	N	O	P	Q
87.	Francium, Fr	87	2	8	18	32	18	0	1
88.	Radium, Ra	88	2	8	18	32	18	0	2
89.	Actinium, Ac	89	2	8	18	32	18	8	2
90.	Thorium, Th	90	2	8	18	32	19	9	2
91.	Protactinium, Pa	91	2	8	18	32	20	9	2
92.	Uranium, U	92	2	8	18	32	21	9	2
93.	Neptunium, Np	93	2	8	18	32	22	9	2
94.	Plutonium, Pu	94	2	8	18	32	23	9	2
95.	Americium, Am	95	2	8	18	32	24	9	2
96.	Curium, Cm	96	2	8	18	32	25	9	2
97.	Berkelium, Bk	97	2	8	18	32	26	9	2
98.	Califoinium, Cf	98	2	8	18	32	27	9	2
100.	Fermium, Fm	100	2	8	18	32	29	9	2
101.	Mendelevium, Mv	101	2	8	18	32	30	9	2
102.	Noblium, No	102	2	8	18	32	31	9	2
103.	Lawrencium, Lw	103	2	8	18	32	32	9	2

(*iii*) **Aluminium.** It is next to copper. It has less weight than copper and conductivity is only 60% than that of copper. It is malleable and ductile and has a good mechanical strength. It is not much effected from the atmospheric effects. It is available in plenty, so it is widely used at present and because of the copper shortage it is widely used in the cables etc.

The aluminium conductor used in the cables etc.

The aluminium conductor used for overload line is named as A.C.S.R. (Aluminium Conductor Steel Reinforced). Aluminium conductors cannot be joined so easily.

(*iv*) **Tungsten.** It is a hard metal, it's melting point is high. It is malleable and ductile. It is widely used for making the filaments of the bulbs, tubes and electronic tube etc. It is also used in mechanical engineering for making high speed tools.

(*v*) **Zinc.** It is mostly used for preparing the brass which is the alloy of zinc and copper. It is also used for the negative platès of primary cells.

(*vi*) **Brass.** It is an alloy prepared by the mixture of (zinc and copper). It is used for preparing the terminals of the switches, plugs, plug shoes, fuses, holders and starters etc.

(*vii*) **Iron.** It has good mechanical strength. It can be easily jointed and moulded in any shape. It can be drawn into wires. It is mostly used for making the bodies of electrical machines, as motors, generator fans etc. It is also used for making galvanised wires which are used in telephone lines and for earthing purpose also.

(*viii*) **Nickel.** It is used for nickel iron alkaline cells for making the plates.

(*ix*) **Tin.** It has a low melting point. It is used for making solder wire for jointing purpose.

(*x*) **Lead.** It is used for making solder wires and for soldering the conductors and cables. It is also used for making the protective covering of underground cables.

(*xi*) **Mercury (Hg).** It is used in standard cells, gas discharge lamps, mercury vapour lamps, mercury are rectifier and d.c. energy meter etc.

(*xii*) **Eureka.** It is an alloy of Cu: Ni, (60 : 40). It's resistance is very high. It can be drawn in to wires so it is used for making resistance boxes, thermocouples and the resistance of starter and regulators.

(*xiii*) **Nicrome.** It is also an alloy of nickel and chromium (80 : 20). The resistivity is high but less than Eureka. It can be drawn into wires. It is widely used for making the heating elements in electric heaters, electric iron, kettles and furnaces etc.

(*xiv*) **Manganin.** It is also used for making resistance wires of resistance boxes and slide wire bridge etc.

(*xv*) **Carbon.** It is used for giving in or taking out the electricity from the generator, motors, fans etc. It is also used in making the carbon filament lamps and carbon are lamps etc.

Table 6.2. Resistance of the other substances as compared with the resistance of copper

S.No.	Material	Value	S.No.	Material	Value
1.	Copper	1.00	7.	Brass	4.4
2.	Silver	0.92	8.	Iron	6.67
3.	Gold	1.38	9.	Nickel	7.3
4.	Aluminium	1.6	10.	Platinum	7.8
5.	Tungsten	3.2	11.	Tin	8.2
6.	Zinc	3.62	12.	Lead	12.76

Q. What are the properties of good insulator? Describe in brief the different insulator.

Ans. The materials through which electric current can't flow easily are named as insulators. But it is not essential that if a material is insulator for a low voltage, it will be perfect for the high voltage also, which is never because every substance has its dielectric strength or say the voltage where it is punchered. So the insulators are market for voltage too, so the materials which offers a very high resistance in the flow of electric current are called insulators.

Properties of good insulator. It should have the following properties:

(*i*) **High ohmic resistance.** The value of the resistance should be as high as could be.

(*ii*) **A good dielectric strength.** As the dielectric strength is defined as the voltage with which it can withstand, so it should have high dielectric strength, it is taken in kV/mm.

(*iii*) **Non-hygroscopic.** Many conductors come in contact with atmosphere either during manufacturing or in operation or in both, so it should have the property not to be effected by the atmospheric changes *i.e.* nonhygroscopic.

(*vi*) **Ability to withstand the temperature.** An insulator with better thermal conductivity will not allow temperature rise because of better heat dissipating or transferring property. So an insulator should have better thermal conductivity and ability to withstand the temperature rise.

(v) **Ease of forming.**
It could be moulded into any shape and size.

(vi) **Not so costly and easily available.**
There are a number of insulators which are used in electrical engineering.

 (i) *Dry air.* Dry air is considered as the best insulator. In transmission and distribution lines, we are not keeping any insulator in between lines, it is only air which function as the insulator.
The dry air is the best insulator but the moisture decreases its dielectric strength.

(ii) *Vulcanised rubber.* The vulcanised rubber is prepared by mixing the pure rubber and sulphur, zinc oxide with some color. It is inflammable at 155°C. It is effected by chemicals. It is used for making the insulating coverings of copper conductors, V.I.R. wire covers, (Cu conductor in V.I.R. wire) Lead covered wire, C.T.S. wire, whether proof wire, flexible wire, and other cables etc.

(iii) *Mica.* It is a mineral substance. It can be cut into any shape. It is transparent, fire proof and does not absorb moisture. It is a good conductor of heat. It is used in electric iron kettles, soldering iron and connectors etc.

(iv) *Ebonite.* Ebonite or vulcanite is a hard substance and can be moulded into any shape. It softness by heat and burns always. It is less effected by chemicals and moisture. It is used for making container of lead acid batteries, cases of instruments and switch gears, terminal plates and low voltage pannel boards etc.

(v) *Glass.* Glass is also a good insulator. It can be moulded into any shape. It is brittle and transparent. It is fire proof and water proof. It is not effected by chemicals. It is used for making the face plates of instruments and for making the lamps and tube etc.

(vi) *Varnished paper.* It is a good insulator when dry and impregnated with oil. With heat it becomes brittle and burns away at high temperature. It absorbs moisture. Dry paper is used for low voltage, telephone wires, oiled paper is used as insulation in windings. It is also used in underground high tension cables. Paper soaked in parafinwax is used in condensers.

(vii) *Bakelite.* Bakelite is a hard substance, good insulator and good water proof. It is brittle and burns at high temperature. It can be moulded to any shape. It is used for making the bodies of switches, plugs, plug tops, holders and regulators etc.

(viii) *Leatheroid.* It is a good insulator. It is not effected by grease and oils. It burns at high rate temperature. It is used for slot insulation and coil insulation for winding works. For better performance the *glazed leatheroid* and *film leatheroids* are used.

(ix) *Insulating oil.* It is a mineral oil. It is inflammable. The moisture decreases its dielectric strength. It is used as insulating oil in transformers, starters, and switches gears of high current capacities. It quenches the spark and transfer the heat.

(x) *Porcelain.* Porcelain is made from china clay. It is a hard insulator can be moulded to any shape. It may crack when heated. The chemical effect is also very less. It is used as insulating supports for wires in overhead lines, kit ket fuse, switches fuse grips, cleats and connectors, etc.

(xi) *Slate.* It is also a good insulator of electricity. It can be cut into any shape, size, thickness. It is used for making a pannel boards.

(xii) *Empire cloths.* It is made by varnishing a cotton cloth, silk or papers. It burns away at high temperature. It is not effected by moisture. It is available in yellow and black colour in different sizes. It is used as slots insulation in winding works and for insulating the coils.

(xiii) *Press phan paper.* It used as the slot insulation is winding work for the machines of low and medium voltages.

(xiv) *Cotton tape.* It is also used for insulating the coils. It is available in different sizes depending upon the thickness. There are other tapes the silk and glass tapes which are used for taping purpose generally in case of windings. Just as 10mm, 20 mm, 25 mm, etc. The thickness of the tape is 0.006″ and width 0.75″. The half layer taping can withstand upto a voltage of 250 V and it is impregnated with any oil so that the air gaps fills up, it can withstand up to 1000 V.

(xv) *Adhesive insulating tape.* It is also known as black tape. It is used for insulating the joints or for taping the joints. It is available in size 10 mm, 20 mm, 25 mm, etc. These are also available in P.V.C. adhesive tape, cotton and bitumen tapes.

(xvi) *Bitumen.* It is used for preparing the varnish and for filling the joint boxes of underground cables. It softens when heated, burns away at high temperature.

(xvii) *Gutta percha.* It is a chemical product. The properties of which are as that of rubber. It is not effected by water. It is used as insulation for *submarines* cables.

Q. Tabulate the different insulating materials with their limiting temperatures.

OR

Classify the different insulating materials according to their class and limiting temperature.

Ans. The different insulating materials classified according to their class of insulation and limiting temperature and are shown in Table 6.3.

Table 6.3. Classification of insulations

S.No.	Class	Limiting Temperature (°C)	Insulating materials
1.	Y	90°	Cotton, silk, paper of similar organic materials neither impregnated nor immersed in any oil, rubber, polyvinyl chloride.
2.	A	105°	Impregnated paper, silk, cotton (fibrous materials), polymide resins.
3.	E	120°	Enamelled wire insulations on bases of polyvinyl formed poly-urethane and epoxy resins, moulding powder plastics materials having thermal stability 15°C higher than class A insulation.
4.	B	130°	Inorganic materials (mica, fibre glass, asbestos) impregnated with varnishes or other compounds.
5.	F	155°	Polyester epoxide varnishes and other varnishes with a high heat resistance.
6.	H	180°	Composite materials on mica, fibre glass and asbestos bases impregnated with silicons rubber etc. with suitable bonding substance such as appropriate silicon resins.
7.	C	225°	Mica, ceramics, glass, teflon porcelain.

6.3. WIRES AND CABLES

Q. What are the different classifications of the wires used in domestic/industrial wiring?

Ans. The wires used in domestic/industrial wiring are classified as under according to;

(a) **Conductors used.** Generally copper conductors and aluminium conductors are used and are named as 'copper cables' and 'aluminium cables'.

(b) **Number of core used.** These are single core, double core or twin core, three core and three and half core cables.

(c) **Voltage gradings.** According to voltage gradings these are 230/600 volts or 600/1100 volts gradings.

(d) **Type of insulation.** These are also classified according to the insulation used V.I.R. wire, P.V.C. wire, lead alloy wire and weather proof wire etc.

Q. What are the different type of wires used in domestic wiring? Describe V.I.R. wire and P.V.C. wire neat sketches.

Ans. The wires used in domestic wiring are as under:

    (a) V.I.R. wire (Vulcanised Indian Rubber wire)

    (b) P.V.C. wire (Poly Vinyl Chloride wire)

    (c) T.R.S. or C.T.S. wire (Tough Rubber Sheathed wire or Cab Taped Sheathed wire)

    (d) L.C.C. (Lead Covered Cable)

    (e) Weather proof wire.

    (f) Flexible wires.

(a) **V.I.R.** (Vulcanised Indian Rubber) wire. This type of wires are simply called as V.I.R. wire. They are always available in single core. These are available in two different grades: 250 V and 600 V grades.

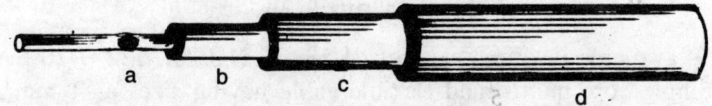

(a) Tinned copper conductor single or multistrands, or Aluminium conductor. (b) Vulcanised Indian Rubber insulation. (c) Cotton proofed tape. (d) Cotton braiding and compound.

Fig. 6.4. V.I.R. wire.

Construction. The construction is as shown in Fig. 6.4. It consist of a tinned copper conductor with a layer of Vulcanised Indian Rubber insulation, over the rubber insulation a proofed tape and a protective covering of cotton braided and compound is used. The braiding is provided with wax to protect the wires from damages while drawing through the conduit pipe. These wires may be single braided and double braided.

The sizes of wires, are as 1/18, 3/20, 3/22, 7/20, 7/22, 7/16, 19/18, 19/16 S.W.G. etc. Nowadays these wires are in stranded form having so many conductors but are as 1mm², 1.5 mm², 2 mm² or so.

Uses. The wires are used in casing and capping wiring, conduit pipe wiring, cleat wiring, and for general electrical wiring etc. These are effected by moisture and chemicals. These are available in different colours, white, black and brown.

(b) **P.V.C.** (Poly Vinyl Chloride) wire:

Construction. It consists of having a tinned copper conductor or aluminium conductor may be single or multistrands, covered with P.V.C.

covering. Nowadays aluminium conductors are widely used and these are available in both gradings 230 V/600 V. These are single core, double or twin core, three core and three and half core for different uses as shown in Fig. 6.5. Generally for domestic purpose single core cable is used. The P.V.C. insulation has more life and good appearance. The insulation has the property of resisting oils, water etc.

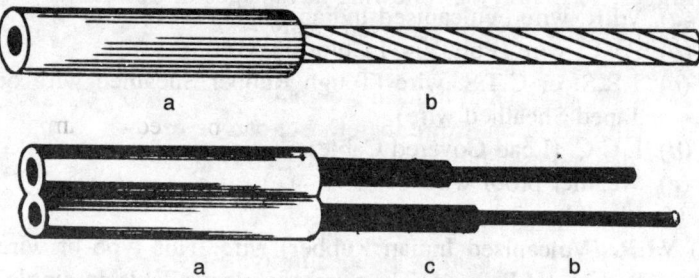

(*a*) P.V.C. insulation. (*b*) Copper or aluminium conductor single or multi strands. (*c*) P.V.C. Covering.

Fig. 6.5. P.V.C. wires.

It is available in different sizes,, 1/18, 3/22, 3/20, 7/22, 7/20 S.W.G. etc, Single core multistrand flexible cable having sizes as 1 mm^2, 1.5 mm^2, 2 mm^2 etc. It is available in different colours like, white, black, red, yellow, blue etc.

Uses. It can be used in industrial and domestic wirings, like batten wiring, casing and capping wiring, conduit pipe wiring, etc. Nowadays it is tremendously overcoming the increasing demand of electrical sphere.

Q. Write short notes on:
    (*a*) **C.T.S. wire,** (*b*) **Weather proof wire,** (*c*) **L.C.C. wires** and (*d*) **Flexible cables.**

Ans. (*a*) C.T.S. (Cab tough sheathed) wire. It is also named as T.R.S. (Tough rubber sheathed) wire.

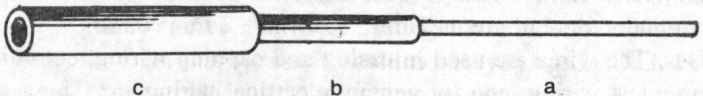

(*a*) Tinned copper conductor, (*b*) Rubber insulation and (*c*) C.T.S. covering.

Fig. 6.6. C.T.S. wire.

Construction. The conductors are protected with the C.T.S. hard rubber covering. These are single core, twin core with earth wire, the construction is as shown in Fig. 6.6. It has a tinned copper conductor or aluminium conductor covered with vulcanised rubber insulation. The covering is generally black or red. There is a tough rubber or cab tough covering over the rubber insulation. These are available in different sizes and gradings like 1/18, 3/22, 3/20, 7/22 S.W.G. etc. or 1 mm^2, 1.5 mm^2, 2 mm^2 etc.

Uses. It is generally used in batten wiring.

Advantages. There are the following advantages of this cable:

(*i*) The insulation can be removed easily.

(*ii*) It is shock proof.

(*iii*) These are cheap and light in weight.

(*iv*) Not effected by moisture therefore it can be used in damp places.

(*v*) It is also not effected by chemicals and fumes.

Disadvantage

(*i*) It cannot be used under the walls.

(*ii*) Sheathing is damaged by direct sun light.

(*b*) **Weather proof wires.** Generally it is used in service lines. It is also available in different gradings 250 V and 600 V gradings.

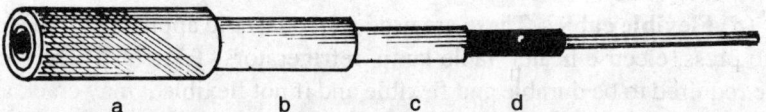

(*a*) Cotton braided and weather proof compound, (*b*) Oil dressed fabrics, (*c*) Proofed tape, (*d*) Vulcanised rubber insulation, (*e*) Tinned copper conductor single or multistrands.

Fig. 6.7. Weather proof wire.

Construction. They are either single core or twin core. It consists of the tinned copper conductor or aluminium conductor covered with a layer of vulcanised rubber insulation. This insulation is covered with the oil dressed fabrics and this all is covered with a cotton braided and weather proof compound as shown in Fig. 6.7. These are available in different sizes 1/8, 3/20, 3/22, 7/22 S.W.G. and 1mm^2, 1.5 mm^2, 2 mm^2 so on.

Uses. Generally used for weather proof wiring where the wires are placed in the sunlight and in open spaces.

These are cheaper and not effected by sun, rains or heat.

(*c*) **L.C.C. (Lead covered cables).** These are lead alloy sheathed wires. It consists of a tinned copper conductor or aluminium conductor, covered with the rubber insulation, there is a P.V.C. sheathed over the rubber insulation and then the continuous of the lead alloy as shown in Fig. 6.8. These are available in different sizes like 1/18, 3/22 S.W.G. etc. and even in single core and double or twin cores, flat three core, flat twin core with earth continuity conductor.

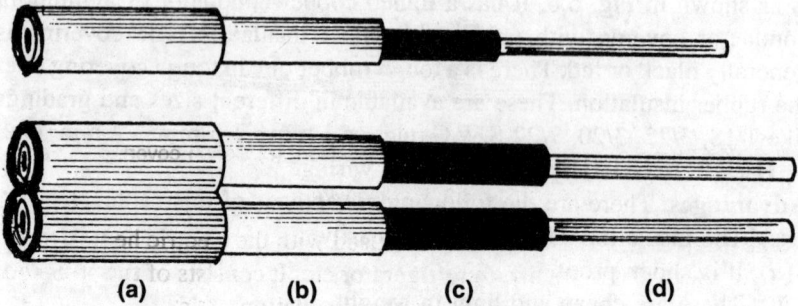

(a)   (b)   (c)   (d)

(a) Lead covering, (b) P.V.C. sheathings, (c) Rubber insulation, (d) Tinned copper
conductor single or multistrands.

Fig. 6.8. L.C.C. cables.

Uses. It is used in open spaces for taking supply like service main upto meter board from the service pole.

Advantages.

(i) It requires good skill.

(ii) It is costly.

(d) **Flexible cables.** These are used for household appliances like electric press, electric heater, table lamp, refrigerators, T.V. etc. These wires are required to be durable and flexible and if not flexible it may crack and break very soon.

Sizes. Each flexible wire consists of a number of copper conductors. The size of each conductor is 0.0076″ in diameter or 36 S.W.G. in number. The number of conductors in a single wire is 14, 23, 40, 70, 110, 140 etc. and as the size of wire is given these are 14/0.0076, 23/.0076, 40/.0076, 110/.0076, 70/.0076 and 1 mm^2, 1.5 mm^2, 2 mm^2 and so on.

Types of flexible wires. These are of the following types:

(i) Twin twisted glazed cotton covered or silk covered flexible wire.

(ii) Plastic covered flexible wire.

(iii) Twin silk covered flexible.

(iv) Three core unsinkable domestic flexible wire.

(v) Three core tough rubber sheathed flexible wire.

The *twin twisted glazed cotton covered flexible* wire is also known as cotton covered wire. Each wire has a number of conductors in stranded form, covered with rubber insulation and a covering of glazed cotton is put on as shown in Fig. 6.9. Two wires are twisted together. This wire suitable for pendent light, lamps etc. It should never be used with the appliances where heat is produced.

The plastic covered flexible wire also has a number of stranded conductors, covered with a layer of plastic covering. The plastic covering is of different colour yellow, blue, red etc. It is also used with light load, where heat is not produced.

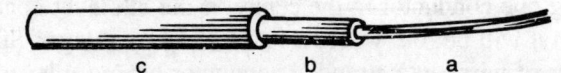

c   b   a

(a) Stranded conductor, (b) Rubber insulation, (c) Cotton covering.

Fig. 6.9. Flexible wire.

The *three core unsinkable wires* are used with the electric heater, electric iron, electric motors like in refrigerator etc. It consists of the stranded copper conductor, with a layer of rubber insulation, placed in a systematic fashion. The gap is then fitted with the cotton, then a rubber filling is also used and a cotton covering, covers the all construction as shown in figure. The colour of rubber insulation is red, for positive or phase, black for negative or neutral and green for earth. This wire is available in different sizes.

The tough rubber sheathed wire is generally known as the workshop cable. Each cable has either two core or three core or three and half core cables. These are used for electric iron, electric heater, drill machine, electric motors etc. It has red, black and green wires as positive, negative and earth respectively.

Q. Why and how the stranding of conductors is done? What is the utility of stranding the conductor?

Ans. The conductor or cable used for electrical energy, is either a solid having single conductor or stranded conductors. The stranded conductors have its own merits over the single solid conductor.

1. The stranded conductor is flexible.
2. The stranded conductor can deliver electric current if one or two strands are broken where a single solid conductor has only one conductor and can result open circuit.
3. The current carrying capacity is much in case of stranded cable than single conductor.
4. The stranded cables are easy to handle.

It is obvious that where a single conductor of large cross-section is used, it becomes rigid in construction and is liable to kinks and breaks while handling. To avoid this the conductors are made of number of thin wires, bunched together called strands. It makes the conductors flexible and eliminates to a large extent the risk of its breaking through the insulation. The stranding is done by twisting the wires together to form layers. Generally the stranding is done in opposite direction for successive layers. It means that if one layer has the stranding in the clockwise direction, the successive layer will have its strands in anticlockwise direction and so on. The general way of strands are one+six around. In a cable having one conductor in the centre, if the nth layer conductors are to be known it will be '$6n$' where n is the number of layer. Similarly the total number of wires in a stranded conductor having n layers

$$= 1 + 3n(1 + n)$$

where n = the number of layers.

So the first layer has $6n = 6$ conductors and total number of conductors = $1 + 6 = 7$. Similarly second layer has $6 \times 2 = 12$ conductors.

First layer has = $6 \times 1 = 6$ conductors.

So total number of conductors in a three layer conductor

$$= 1 + 6 + 12 = 19 \text{ conductors.}$$

So the cable can have 7, 19, 37, 61 total number of wires in the stranded conductors.

The overall diameter can also be calculated as

Overall diameter = $(1 + 2n)d$

where n = the number of layers.

d = the diameter of one wire.

for example a three layer 2.06 mm dia, wire cable has the overall diameter.

$$= (1 + 2 \times 3) \times 2.06 = 14.42 \text{ mm.}$$

The conductors can be circular stranded, compact circular stranded having one, three or four wires as the centre core.

Q. Describe some special and most popular types of the cable used for wiring.

Ans. The most popular types of cables available for home wiring:

  (*a*) **Service entrance cable.** For supply to the main distribution panel from the meter. It is usually a three-wire cable, the third wire being a bare grounding as shown in Fig. 6.10(*a*).

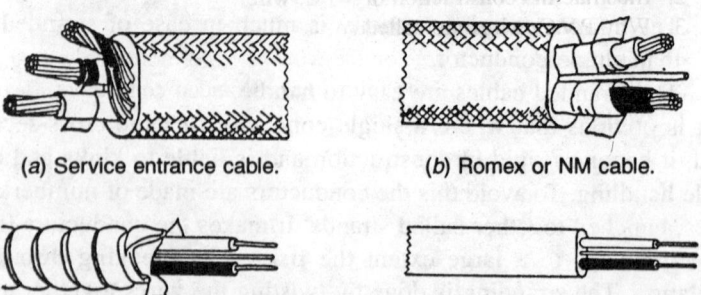

(*a*) Service entrance cable.   (*b*) Romex or NM cable.

(*c*) Underground or type UF cable.   (*d*) Armoured or BX cable.

Fig. 6.10. Cables.

(*b*) **Romex or NM cable.** For indoor use, available in three-wire and four-wire types.The three-wire cable is called two-wire with ground, and the four-wire cable is called three-wire as shown in Fig. 6.10(*b*).

(*c*) **Underground type or UF cable.** For outdoor application, the jacket is generally a molded plastic for moisture proofing. It is usually available in two-wire with ground configuration as shown in Fig. 6.10(*c*).

(*d*) **Armoured or BX cable.** It used for the same application as Romex. It is extremely durable but slightly harder to work with than Romex. It can be used only in dry, indoor circuits as shown in Fig. 6.10(*d*).

Similarly the cables having different cross-sections and cores *e.g.* the single core, double core, three core, four cores, twin concentric, ripple

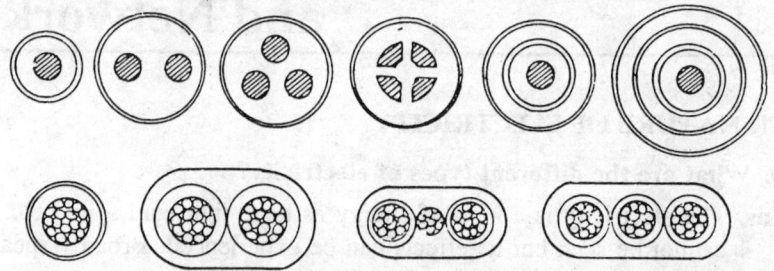

Fig. 6.11.

concentric, P.V.C. sheathed core, twin core, twin core with ball earth continuity conductor and three core cables respectively, are shown in fig. 6.11. Then cables are used in domestic and industrial purposes.

REVIEW QUESTIONS

1. How broadly the wires used in domestic and industrial applications are classified?
2. Illustrate the construction of V.I.R. wire.
3. Why P.V.C. wires are called so?
4. Write down a short note on the wires used for domestic wiring.
5. Write short notes on the followings: Stranded conductors, flexible wire, workshop cable.

Nature of Electricity, Fundamental Laws and Networks

7.1. NATURE OF ELECTRICITY

Q. What are the different types of electricity?

Ans. The electricity is a kind of energy as heat, light and sound etc. It cannot be seen but its effects can be experienced. Broadly speaking, the electricity can be classified into the followings:

(*a*) Static electricity.

(*b*) Dynamic electricity.

(*a*) **Static Electricity.** It is that branch of science which deals with the phenomenon of electricity at rest. The electricity at rest is defined as the *static electricity*. It is produced by rubbing two bodies such as glass rod with silk and ebonite rod with cat skin. This type of electricity cannot be transmitted from one place to the another, hence it is of no use for current electricity.

As basically all the substances in the universe are electrically neutral *i.e.* having no charge. Now if by any means, the electrons are increased, it means the number of electrons are more than the number of protons and the body is negatively charged.On the other hand if electrons are extracted or taken out then in that particular atom, the number of protons will be more than the number of electrons hence the body is positively charged. These two types of charges are produced:

(*i*) *Negatively charged body.* If the glass rod is rubbed with the silk it negatively charged.

90

(*ii*) *Positively charged body.* Whenever the ebonite rod is rubbed with the cat skin the body is positively charged.

(*b*) **Dynamic Electricity.** Electricity in motion is called dynamic electricity. The branch of science which deals with the phenomenon of electricity in motion is known as dynamic electricity. This type of electricity can be transmitted from one place to the another, so it has very much advantages and importance in our daily life.

This type of electricity can be produced by the cells, battery, dynamo or generator, alternator etc.

7.2. SOURCES, EFFECTS AND USES OF ELECTRICITY

Q. What are the different sources of electricity?

Ans. The different sources of electricity by which the electricity can be produced are known as the source of electricity. There are a number of sources by which the electricity can be produced, these are as follows:

(*a*) **By friction.** When two bodies or things are rubbed heat is produced, in the same way the static electricity can be produced by friction. When the glass rod is rubbed with silk or ebonite rod with cat skin in both the ways the electricity produced is of static nature and is produced by friction. This electricity cannot be transmitted from one place to the another.

(*b*) **By chemical means.** The electricity can be produced by means of chemical action for example if two dissimilar rods say of Zn and Cu are put in dilute sulphuric acid the electricity will be produced. Whenever in a jar two dis-similar rods of zinc and copper are immersed some bubbles of the nascent hydrogen over the zinc rod is seen. The nascent hydrogen is very active and negatively charged, it gives away its charge to copper. If a galvanometer is connected between copper and zinc and rod it deflects. The electricity in the cell is caused by the chemical changes.

In batteries this electrical energy is being stored in the shape of chemical energy, which can be used at any time, or in other words the battery can be said as the portable electricity. In this case the electricity flows form positive to negative.

(*c*) **By mechanical means.** In this way the mechanical energy can be covered into electrical energy by means of generators. This type of energy is the dynamic electricity. The principle is same as whenever a conductor links with the changing flux or cuts the magnetic lines of force, an e.m.f is produced in that conductor. This e.m.f.

causes the current to flow through the conductor. The current flows from positive to negative.

In this way the produced current is of two types:

(*i*) Direct current.

(*ii*) Alternating current.

The current in which the direction and value of the current does not change is known as direct current. It is produced by D.C. generator. The current which changes its value and direction after a definite interval of time is called alternating current. This current is produced by the A.C. generator called alternator.

(*c*) **By means of heat.** The electricity produced by the heat is called thermal electricity. When dis-similar metals are joint in the shape of bow and the junction is heated and the other is kept cold, the current flows from hot portion to cold portion. The current produced by the heating is called the thermal electricity. This is of very small quantity and is having a limited field.

(*e*) **By means of lighting.** When light falls on the dissimilar metals, the electricity is produced. Photoelectric cell is the example of this effect. This type of electricity is used in ratio and electronics circuits etc.

(*f*) **By means of solar cell.** Electricity can be produced by means of solar cells. When light falls over them it is converted into electricity.

Q. What are the different effects of electricity? Also states the different uses of electricity.

Ans. There are the following main effects of electricity:

(*a*) Heating effect.

(*b*) Magnetic effect.

(*c*) Chemical effect.

(*a*) **Heating effect.** When electric current passes through a conductor, heat is produced in that conductor. This effect of electricity is called the heating effect of the current. Electric heater, kettle, electric iron and electric lamps are the example of the same effect.

Heat produced in a conductor can be given by the formula

$$H \propto I^2 \cdot R \cdot t$$

H = heat, I = current in amps, R = resistance in ohms and t is the time in seconds. The unit depends upon the conversion in joule or in calories.

(*b*) **Magnetic effect.** Whenever an electric current flows through a wire, magnetic line of force are produced around the conductor. This is called the magnetic effect of the current, electric machines, generators, transformers, motors etc. works on the same effect of the current.

(c) **Chemical effect.** Whenever the current passes through an electrolyte it decomposes into its constituents or ions. All electrochemical possesses like electroplating, refinery of metals, cells, battery, works on the same effect.

Our body is also a good liquid conductor and if current passes through it, it may cause a shock. The nature of shock will depend upon the quantity of current through the body.

Uses of electric current. There are the following uses of electric current:

(a) **For lighting purposes.** Just as torch light, motor car's head lights, street and domestic lighting etc.

(b) **For heating purpose.** Whenever electric current is passed through a conductor the heat is produced, for example in electric heaters, electric iron, electric oven, electric furnaces, welding and soldering etc.

(c) **For power purpose.** For driving the machinery by electric current as electric motors, trains, electric trams, cranes, lifts, etc.

(d) For telegraphy and telephones and radio works, television, teleprinter and other signal transmitting devices.

(e) **For chemical purposes.** Just as battery charging, electroplating, refinery of metals and other electrochemical works.

(f) **For cooling purpose.** It is used for cooling as refrigerators, room coolers, water coolers etc.

(g) **For production of special rays:** In medical lines and in other defence services just as the X-ray, radar etc.

(h) **Earth satellites.** Information about weather conditions on the earth can be obtained on earth through remote controlled artificial earth satellites.

(i) **Modern industries.** Both light and heavy engineering industries can produce goods at a high rate production with utmost economy by using electricity.

(j) **Miscellaneous application.** Beside these applications there are electronics, tape recorders, players, relays, theaters and other drive devices, etc.

7.3. CIRCUIT, TERMS AND DEFINITIONS

Q. What do you understand by an electric circuit? What are the different types of simple electric circuit?

Ans. The path taken by an electric circuit is known as the electric circuit. A simple circuit consists of the following: the electric current starts from its source of supply *i.e.* battery or gen-. erator, it completes its path through a safety device *i.e.* a fuse, controlling device *i.e.* a switch, and a consuming device like lamp, heater, fan etc. The current goes to the consuming device through live wire and returns after flowing through resistance to the source of supply via return wire. The simple arrangement is shown in Fig. 7.1.

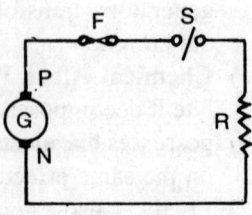

Fig. 7.1. Closed circuit.

Followings are the types of circuit:

    (*a*) Simple or closed circuit.

    (*b*) Open circuit.

    (*c*) Short circuit.

  (*a*) **Closed circuit or complete circuit.** The circuit in which the current complete its path through consuming device. In this circuit current starts from source to safety, controlling and consuming devices return to source as shown in Fig. 7.1.

  (*b*) **Open circuit.** Whenever the line or neutral wire or the resistance wire is broken then the circuit is said to be an open circuit. In this circuit the wire or accessories or the device connected, can be electrically broken as shown in Fig. 7.2.

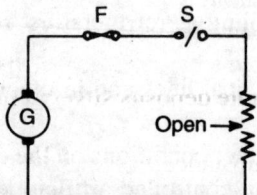

Fig. 7.2. Open circuit.

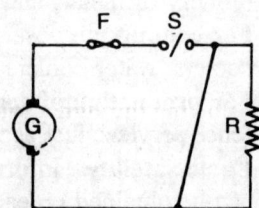

Fig. 7.3. Short circuit.

  (*c*) **Short circuit.** When the line and neutral wire meet without any resistance or the consuming device, it is called short circuit. In this case heavy current flows through the fuse wire and fuse blow out as shown in Fig. 7.3.

Q. **Define the following with their units:**

    (*a*) **Electromotive force,** (*b*) **Potential difference,** (*c*) **Current,** (*d*) **International ampere,** (*e*) **Resistance,** (*f*) **International ohm,** (*g*) **Conductance.**

Ans. (*a*) **Electromotive force.** It is also expressed by e.m.f. The electromotive force is the force which makes the electric current to flow through a circuit. This can be defined the force which compels the electrons to move. The unit of e.m.f. is volts. It cannot be measured but can be calculated or compared.

The e.m.f. can be calculated as the voltage measured plus the voltage drops. It is represented by latter E.

(b) **Potential difference.** It is also represents as P.D. It is defined as the difference of potential which exists between two points of circuit. It can otherwise be defined as the difference of potential measured between the terminals of source of supply. E.M.F. is always greater than P.D. It is represented by letter V and measured in volts.

So   P.D. = E.M.F. − Voltage drop inside the source of supply.

It is measured by the instrument known as the volt meter.

One volt. One volt is that much of electric pressure which is required to send a current of one ampere through a resistance of one Ohm.

(c) **Current.** The flow of electron is called electric current. It's units is ampere and is measured by the instrument known as ampere meter or ammeter. It is represented by the letter I.

One ampere. It is the amount of current established when one volt of electric pressure is applied to a resistance of one ohm or one ampere of current is said to flow through a wire if in any section 628×10^{16} electrons pass in one second. Since 628×10^{16} electrons are equal to one coulomb, so it can otherwise be defined as if one coulomb of charge per second flows through any section of a conductor, then one ampere current is said to flow.

(d) **International ampere.** It is that amount of current which when flowing through a solution of silver nitrate deposits silver at the rate of 0.001118 gm/sec on the cathode.

(e) **Resistance.** It is the property of the substance. It is defined as the property of the material by virtue of which it opposes the flow of current through it. The unit of resistance is ohm (Ω). It is measured by ohm meter directly. It is represented by the letter R.

One ohm. It is the resistance offered by a substance when one volt of electric pressure results one ampere current through it. The other units are as

$$\text{one megohm} = 10^6 \text{ ohm}$$
$$\text{one microohm} = 10^{-6} \text{ ohm}$$
$$\text{one milliohm} = 10^{-3} \text{ ohm}$$
$$\text{one kilo ohm} = 10^3 \text{ ohm.}$$

(f) **International Ohm.** It is the resistance offered by a conductor of mercury column at the melting ice temperature, 14.4521 gm in weight, having 106.3 cm length and the uniform cross-sectional area of 1 mm^2.

(g) **Conductance.** It is the reciprocal of resistance. It is the property of a substance by virtue of which it helps in the flow of current. It is measured by ohm meter. The unit is mho and denoted by letter G

$$G = \frac{1}{R}.$$

7.5. RESISTANCE

Q. On what factors does the resistance of a conductor depend?

Ans. There are the following factors upon which the resistance of conductor depends:

(a) **Length.** The resistance of a conductor depends upon its length. The resistance is directly proportional to the length of the conductor, *i.e.*,
$$R \propto l.$$

(b) **Area of cross section.** The resistance of the conductor is inversely proportional to the area of cross section of the conductor *i.e.*,
$$R \propto \frac{1}{a}.$$

(c) **Specific resistance.** The resistance also depends upon the nature of the material. The resistance of a conductor is directly proportional to the specific resistance of the material *i.e.*,
$$R \propto \rho.$$

(d) **Temperature.** The resistance of a conductor also depends upon the temperature, for example the resistance of pure metal increases with the increasing of the temperature.

Mathematically it can be expressed as
$$R = \rho \frac{l}{a} \ \Omega.$$

where R = Resistance of the conductor in ohm
       ρ = Specific resistance in Ω m
       l = Length of the conductor in m
      a = Area of cross-section in m^2

So the resistance of a conductor
$$= \frac{\text{Specific resistance} \times \text{Length}}{\text{Area of cross-section}} \ \Omega.$$

7.6. SPECIFIC RESISTANCE

Q. (a) **Define the specific resistance.**

    (b) **What is the effect of temperature on resistances?**

Ans. (*a*) **Specific resistance.**
The formula for resistance is

$$R = \frac{\rho l}{a} \ \Omega.$$

Let here as shown in Fig. 7.4
$$l = 1 \text{ m,}$$
$$a = 1 \text{ m}^2$$

So $\quad R = \rho$

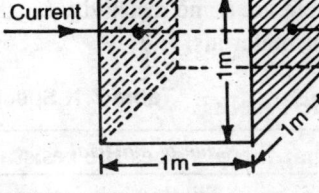

Fig. 7.4. Specific resistance.

i.e. the Resistance = the specific resistance.

It can be defined as the resistance offered by a conductor having 1 metre length and one square metre area of cross-section.

OR

So the specific resistance of the material may be defined as the resistance offered by the two opposite faces of a unit cube.

The unit of specific resistance

$$\rho = \frac{Ra}{l} \Rightarrow \frac{\text{ohm} \times \text{m}^2}{\text{m}} = \text{ohm-m.}$$

So the M.K.S. (S.I.) system of units, the unit is ohm-metre.

(*b*) **Effects of temperature.** The effect of the temperature on the resistances of various materials is as follows:

(*i*) **Most conductors.** The resistance increases with the increasing of the temperature

$$R \propto \text{Temperature}$$

for example, copper, silver, aluminium etc.

1. *With pure metals.* The resistance increases regularly with the increasing of the temperature.

2. *With alloys.* The resistance increases slightly with the increasing of temperature but not regular.

3. *With partial conductors.* The resistance decreases with the increasing of the temperature as in case of carbon. The hot resistance of carbon is less then the cold resistance.

(*ii*) **With insulators.** With the increasing of temperature, the resistance of the insulator decreases. The resistances which are perfect insulator at normal temperature, may become conductor when hot. The temperature coefficient of carbon is negative.

(*iii*) **With electrolyte.** The resistance of the electrolyte decreases with the increasing of temperature and resistance increases with the decreasing of the temperature, so an electrolyte has more resistance when cold and less when hot. It has negative temperature coefficient.

There are some resistivity *i.e.* the specific resistance of some important materials.

Table 7.1. Specific Resistances

S.No.	Name of material	Specific resistance at 20°C in $\Omega \cdot m$
1.	Silver	1.6×10^{-8}
2.	Copper	1.72×10^{-8}
3.	Aluminium	2.8×10^{-8}
4.	Nickel	7.8×10^{-8}
5.	Iron	9.8×10^{-8}
6.	Manganin	44×10^{-8}
7.	Eureka	49×10^{-8}

7.7. TEMPERATURE COEFFICIENT

Q. (a) Define the temperature coefficient of resistance.

Ans. The temperature coefficient of the resistance is defined as the change in resistance per ohm per degree rise in temperature. It is denoted by letter α and the unit is per degree centigrade. The value of temperature coefficient of resistance for copper is 0.0043 per degree. Let there by any resistance which has got the following values:

R_0 = Resistance at 0°C in ohm

R_t = Resistance at t°C in ohm

t = Rise in temperature in °C

α = The temperature coefficient.

Now when the temperature is raised from 0°C to t°C the change in resistance is from R_0 to R_t.

So the change in resistance = $(R_t - R_0)\ \Omega$.

So according to the definition, the change in resistance per ohm per degree rise of temperature *i.e.*,

$$\text{Temperature coefficient} = \frac{\text{Change in resistance}}{\text{Original value} \times \text{Temperature}}$$

$$= \frac{R_t - R_0}{R_0 \cdot t}$$

It is denoted by letter α so

$$\alpha = \frac{R_t - R_0}{R_0 t}.$$

Q. Establish a formula for temperature coefficient if the resistances at two different temperatures are given.

Ans. The temperature coefficient can be given as

$$\alpha = \frac{R_t - R_0}{R_0 \cdot t}$$

or

$$R_t - R_0 = R_0 \cdot \alpha \cdot t$$

$$R_t = R_0 + R_0 \alpha t$$

$$R_t = R_0 \left(1 + \alpha t\right)$$

When the resistance at two different temperatures are given *i.e.*

$$R_{t1} = R_0 \left(1 + \alpha t_1\right) \qquad \qquad ...(i)$$

$$R_{t2} = R_0 \left(1 + \alpha t_2\right) \qquad \qquad ...(ii)$$

where $\quad R_{t_1}$ = Resistance at $t_1 °C$ in ohm

$\qquad R_{t_2}$ = Resistance at $t_2 °C$ in ohm

$\qquad \alpha$ = Temperature coefficient

$\qquad t_1$ = Initial temperature in $°C$

$\qquad t_2$ = Final temperature in $°C$.

Now dividing (*ii*) by (*i*)

$$\frac{R_{t_2}}{R_{t_1}} = \frac{R_0 \left(1 + \alpha t_2\right)}{R_0 \left(1 + \alpha t_1\right)}$$

$$= \frac{\left(1 + \alpha t_2\right)}{\left(1 + \alpha t_1\right)} = \left(1 + \alpha t_2\right)\left(1 + \alpha t_1\right)^{-1}$$

$$= 1 + \alpha t_2 - \alpha t_1 - \alpha^2 \qquad ...\text{higher powers of } \alpha$$

Neglecting the higher powers of α, because α itself is very small.

$$= 1 + \alpha \left(t_2 - t_1\right)$$

$$\frac{R_{t_2}}{R_{t_1}} = 1 + \alpha(t_2 - t_1)$$

or
$$\alpha(t_2 - t_1) = \frac{R_{t_2}}{R_{t_1}} - 1 = \frac{R_{t_2} - R_{t_1}}{R_{t_1}}$$

$$\alpha = \frac{R_{t_2} - R_{t_1}}{R_{t_1}(t_2 - t_1)}$$

$$\text{Temp.coefficient} = \frac{\text{Final resistance} - \text{Initial resistance}}{(\text{Initial resistance} \times \text{Difference in temperature})}$$

Example. *A copper coil has a resistance of 25 Ω at 20°C and 28.225 Ω at 50°C. Find the followings:*
  (a) *Temperature coefficient at 0°C.*
  (b) *Resistance of the coil at 0°C.*
  (c) *Temperature coefficient at 20°C.*

Solution. Initial resistance of the coil at 20°C is 25 Ω. Final resistance of the coil at 50°C is 28.225 Ω. Now the temperature coefficient.

(a) $\alpha_0 = \dfrac{R_{t_2} - R_{t_1}}{R_{t_1}(t_2 - t_1)} = \dfrac{28.225 - 25}{25(50 - 20)} = \dfrac{3.225}{25 \times 30}$

$\alpha = 0.0043/°C$

(b) $R_t = R_0(1 + \alpha_0 t)$

$R_0 = \dfrac{R_t}{1 + \alpha_0 t}$

$= \dfrac{25}{1 + 0.0043 \times 20} = 23.02\,\Omega$

(c) $\alpha_t = \dfrac{\alpha_0}{1 + \alpha_0 t} = \left[\dfrac{0.0043}{1 + 0.0043 \times 20}\right]$

$\alpha_t = 0.00396/°C$ **Ans.**

Example. *A coil with 40 S.W.G. copper wire has a resistance of 420 Ω at 180°C. Calculate the resistance of the same coil at 72°C assuming temperature coefficient to be 0.0043 at 0°C.*

(N.C.V.T. 1966)

Solution. The resistance at 18°C is 420 Ω. *i.e.* R_{t_1}.

Using the formula

$$\alpha_0 = \frac{R_{t_2} - R_{t_1}}{R_{t_1} \times t - R_{t_2} \times t_1}$$

where α_0 = temperature coefficient at 0°C

R_{t_2} = resistance at 72°C

R_{t_1} = initial resistance at 18°C

t_2 = initial temperature.

t_2 = final temperature.

$$0.0043 = \frac{R_{t_2} - 420}{420 \times 70 - R_{t_2} \times 18}$$

$$R_{t_2} - 420 = 0.0043 \times 420 \times 72 - R_{t_2} \times 18 \times 0.0043$$

$$R_{t_2} + 0.0774 R_{t_2} = 420 + 130.032 = 550.032$$

$$R_{t_2} = \frac{550.032}{1.0774} = 510.52 \,\Omega$$

So the final resistance = **510.52 Ω Ans.**

Example. *Find the specific resistance of the Eureka if the 5 metre length of 0.15 cm diameter wire has 1.38 Ω resistance.*

Solution. The resistance,

$$R = \frac{\rho l}{a}$$

and, $$\rho = \frac{Ra}{l} = \frac{1.38 \times \pi \times (0.0015)^2}{5 \times 4} = 48.77 \times 10^{-9}$$

So the specific resistance = **48.77 × 10^{-8} Ω m Ans.**

Example. *If the resistance of a 4.572 metre maganin wire having 0.127 cm diameter is 1.7 ohm, find out the specific resistance.*

(*N.C.V.T. 1969*)

Solution. The resistance,

$$R = \frac{\rho l}{a} \Omega$$

Here $l = 4.572 \text{ m} = 457.2 \text{ cm}$

$$a = \frac{\pi D^2}{4} = \frac{\pi}{4} \times (0.127)^2$$

$$\therefore \qquad \rho = \frac{Ra}{l} = \frac{1.7 \times \pi \times (0.127)^2}{4.572 \times 4 \times 100} = 4.71 \times 10^{-5} \ \Omega m. \text{ Ans.}$$

Example. *A potential difference of 220 V is applied to a copper field coil of a d.c. generator at a temperature of 15°C and the current is 10 A. What will be the mean temperature of the coil when the current has reduced to 9 A; keeping the voltage same. The temperating coefficient at 15°C is 0.00187/°C.*

Solution. The resistance of the coil at 15°C.

$$= \frac{\text{voltage}}{\text{current}} = \frac{220}{10} = 22 \, \Omega$$

Let t the mean temperature of the coil, so the resistance of the coil at final temperature

$$= \frac{\text{voltage}}{\text{current}} = \frac{220}{9} = 24.44 \, \Omega$$

Now using the formula

$$R_t = R_{15} \left[1 + \alpha_{15} (t - 15) \right]$$

$$24.44 = 22 \left[1 + 0.00187 (t - 15) \right]$$

$$t - 15 = \frac{24.44 - 22}{22 \times 0.00187} = 59.3$$

Final temperature = 59.3 + 15 = **74.3°C. Ans.**

7.8. OHM'S LAW

Q. State Ohm's law and its importance.

Ans. The current flowing through a resistance and the voltage applied has a definite relationship. The relation was firstly introduced by George Simon Ohm; a great German scientist and is known by his name as Ohm's Law.

The Ohm's law is stated as, "the current flowing between the ends of a conductor is directly proportional to the applied voltage provided the temperature and physical conditions remain constant."

In the closed circuit as shown in Fig. 7.5

$$I \ \alpha \ V$$

and $$\frac{V}{I} = R, \text{ a constant.}$$

This constant is known as the resistance and is expressed by R in Ohms.

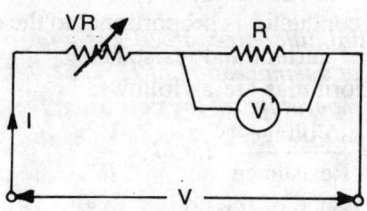

Fig. 7.5. Ohm's law.

$$\therefore \qquad \frac{V}{I} = R$$

There are the following different forms and importance of the Ohm's law. The Ohm's law for simplicity can be expressed as that in any closed circuit the current is directly proportional to the voltage impressed and inversely proportional to the resistance provided the temperature and physical conditions do not change.

i.e., $$I \alpha V$$

$\therefore \qquad I = \dfrac{V}{R}$ where R is a constant known as the resistance of the conductor.

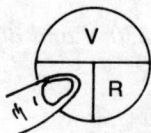  

Fig. 7.6.

These values can be obtained from Fig. 7.6 as

$$\text{the current} = \frac{\text{Voltage}}{\text{Resistance}}$$

$$\text{Voltage} = \text{Current} \times \text{Resistance}$$

$$\text{Resistance} = \frac{\text{Voltage}}{\text{Current}}$$

(a) $I = V/R$. It states that the current is directly proportional to the voltage provided the resistance is constant and thus current is equal to the voltage divided by the resistance.

(b) $I \propto 1/R$ This form states that at a given voltage the current will increase if the resistance is decreased and *vice versa*.

(c) $V = IR$. This form states that the potential difference between any two points of a conductor is proportional to the current and is given as the product of current and resistance.

So these three formulas are as follows:

$$\text{Current} = \frac{\text{Voltage}}{\text{Resistance}} \quad i.e., I = \frac{V}{R}$$

$$\text{Voltage} = \text{Current} \times \text{Resistance}, V = IR$$

$$\text{Resistance} = \frac{\text{Voltage}}{\text{Current}}, \quad R = \frac{V}{I}.$$

Example. *Calculate the value of current through a resistance of 10 Ohm, if it is connected across a battery of 1.5 V.*

Solution. The voltage = 1.5 V

The resistance = 10 Ω

Using Ohm's law

$$I = \frac{V}{R}$$

∴ $$I = \frac{1.5}{10} = 0.15$$

∴ Current = **0.15 A. Ans.**

Example. *An incandescent lamp is connected across 230 V and draws 0.26 A current. Calculate its resistance.*

Solution. The voltage = 230 V

Current drawn = 0.26A

Using Ohm's law equation for resistance

$$R = \frac{\text{Voltage}}{\text{Current}} = \frac{230}{0.26} = 844.62 \Omega$$

∴ Resistance = **884.62 ohm. Ans.**

Example. *Calculate the voltage impressed to a resistance of 15 Ohm for 2 A current in it.*

Solution. The resistance = 15 Ω

The current = 2 A

Using Ohm's law equation for voltage

Voltage $V = I \times R = 2 \times 15 = 30$ V

∴ The voltage required is **30 V. Ans.**

7.9. COMBINATION OF RESISTANCES

Q. What are the different combinations of the resistances? Describe with neat sketch the characteristics and uses of the series combination.

Ans. There are three following combinations of the resistances:

    (*i*) Series combination.

    (*ii*) Parallel combination.

    (*iii*) Series and parallel combination.

  (*a*) **Series combination.** The resistances are connected in such a way as to form a single path for the current. The end of first resistance is connected with the start of second and end of second with the start of third resistance, thus these connections are known as series connections as shown in Fig. 7.7.

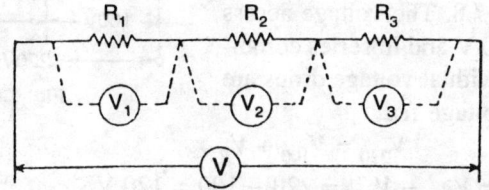

Fig. 7.7. Series circuit

The circuit has got the following characteristics:

    (*i*) There is only one path of the current to flow, hence same current will flow through all resistances.

    (*ii*) The current drawn is equal to the

$$I = \frac{\text{Voltage } (V)}{\text{Equivalent resistance } (R_{eq})}$$

hence if at a given voltage more resistances are added in series the current will decrease.

  (*iii*) The voltage drop across individual resistance depends upon the value of current and its resistance, as shown in Fig. 7.7.

$$V_1 = IR_1,\ V_2 = IR_2 \text{ and } V_3 = IR_3.$$

  (*iv*) The sum of the voltage drops across each resistance is equal to the applied voltage, *i.e.*

$$V = V_1 + V_2 + V_3 + \ldots$$

(v) Substituting for individual values in the above equation

$$V = I \cdot R_1 + I \cdot R_2 + I \cdot R_3$$
$$= I(R_1 + R_2 + R_3) .$$

Let the circuit has an equivalent resistance as $R_{eq.}$ Now the voltage is consumed by the entire circuit so,

$$V = I \cdot R_{eq.} = I(R_1 + R_2 + R_3)$$

or

$$R_{eq.} = R_1 + R_2 + R_3 \dots \Omega$$

Hence when a number of resistances are connected in series, the total resistance is equal to the sum of individual resistances.

Uses. (*i*) The series circuits are used for decoration lightings (Ladi).

(*ii*) This circuit is also used for voltage drop purposes.

(*iii*) It is also used for series testing purposes.

Example. *A lamp has a rated voltage 100 V and hot resistance 25 Ω. Find the value of the series resistance to be connected, so that it can operate on 220 V mains.*

Solution. Let R be the resistance to be connected in series with the lamp as shown in Fig. 7.8. The voltage across the lamp is 100 V and in series combination the individual voltage drops are equal to the voltage fed.

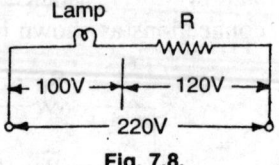
Fig. 7.8.

$$V_{220} = V_{100} + V_R$$
$$V_R = V_{220} - V_{100} = 220 - 100 = 120 \text{ V}$$

Now current drawn by the lamp

$$= \frac{100}{25} = 4 \text{ A}$$

$\therefore$ The value of external resistance $= \dfrac{V_R}{I}$

$$= \frac{120}{4} = 30 \, \Omega$$

$\therefore$ Resistance of **30 Ω. Ans.**

Example. *There resistances of 1210 Ω, 806.67 Ω and 484 Ω are connected in series across a 250 V mains. Calculate the equivalent resistance of the circuit, the current drawn and the voltage drop across each resistance.*

Solution. The resistances are connected in series so the total resistance

= Sum of all individual resistances connected

$$R_{eq} = R_1 + R_2 + R_3$$

$= 1210 + 806.67 + 484 = 2500.67 \ \Omega.$

$$\text{The current drawn} = \frac{\text{Voltage}}{\text{Equivalent resistance}}$$

$$= \frac{250}{2500.67} = 0.0999 \approx 0.1 \ \text{A}$$

Now voltage drop across each resistance

V_1 = Voltage drop a cross 1210 Ω resistance

$\quad = I \times R_1 = 0.1 \times 1210 = 121.0 \ \text{V}$

V_2 = Voltage drop across 806.67 Ω resistance

$\quad = I \times R_2 = 0.1 \times 806.67 = 80.667 \ \text{V}$

V_3 = Voltage drop across 484 Ω resistance

$\quad = I \times R_3 = 0.1 \times 484 = 48.4 \ \text{V}.$

120 V, 80.645 V, 48.38 V. Ans.

(*b*) **Parallel combination.** The circuit in which the beginning of all resistances are connected to one terminal and the ending terminals to another terminal, is known as the parallel circuit as shown in Fig. 7.9. In this circuit the branch currents are different depending upon the individual branch resistances.

This circuit has got the following characteristics:

(*i*) The voltage across each resistance is same.

(*ii*) There are as many branch currents, as many number of branches.

(*iii*) The branch currents can be calculated as the branch current

$$I_{\text{branch}} = \frac{\text{Voltage fed}}{\text{Branch resistance}}.$$

$\therefore \qquad I_1 = \dfrac{V}{R_1},$

$\qquad I_2 = \dfrac{V}{R_2},$

$\qquad I_3 = \dfrac{V}{R_3} \ \text{A}.$

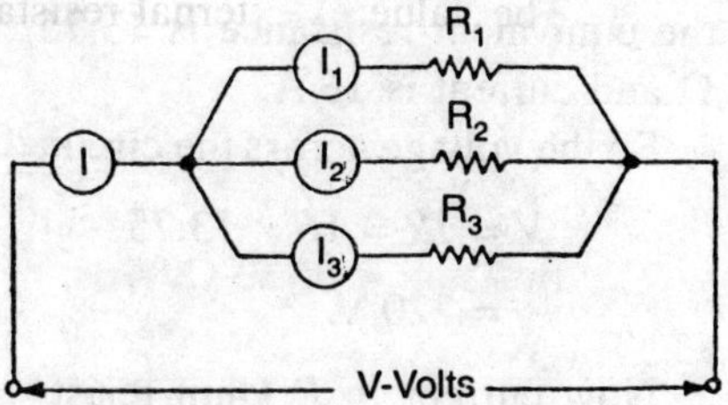

Fig. 7.9. Parallel circuit.

(*iv*) The sum of the branch currents is equal to the current drawn from the mains.

$$I_{eq.} = I_1 + I_2 + I_3.$$

(v) The equivalent or total resistance can be calculated as

$$I_{eq.} = I_1 + I_2 + I_3.$$

Let $R_{eq.}$ be the equivalent resistance of the circuit. Now substituting for the currents

$$I_{eq.} = I_1 + I_2 + I_3$$

$$\frac{V}{R_{eq}} = \frac{V}{R_1} + \frac{V}{R_2} + \frac{V}{R_3}$$

or

$$\frac{V}{R_{eq}} = V\left[\frac{1}{R_1} + \frac{1}{R_2} + \frac{1}{R_3}\right]$$

or

$$\frac{1}{R_{eq}} = \frac{1}{R_1} + \frac{1}{R_2} + \frac{1}{R_3}.$$

Hence in parallel combination, the reciprocal of the equivalent resistance is equal to the sum of reciprocals of different branch resistances.

Conductance. In parallel circuit the conductance increases with the addition of the further resistances in the parallel.

Application. It is applicable in house (domestic) and factories (industrial) wiring for connecting different appliances. Obviously all lamps, fan, kettle, motor, etc. are connected in parallel.

Note. When the equal values of resistance are connected in parallel, the equivalent resistance of the combination will be equal to the branch resistance divided by the number of branches.

Example. *In a parallel combination, the current in R Ω is one ampere the total current being 16 amp. Find the value of R ohm, if the equivalent resistance is 13.75 Ω.*

Solution. The total current, is the sum of all the branch currents. Here the equivalent resistance is 13.75 Ω and current is 16 A.

So the voltage across the circuit

$$V = IR = 16 \times 13.75$$

$$= 220 \text{ V}.$$

Now current in R Ohm resistance is one ampere, so the value of resistance

$$R = \frac{V}{I} = \frac{220}{1} = 220\,\Omega. \text{ Ans.}$$

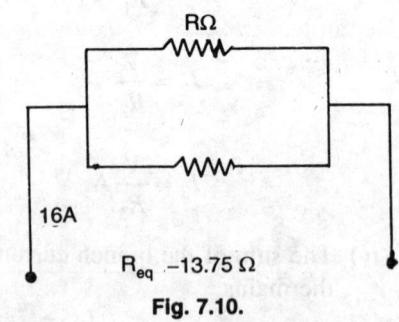

Fig. 7.10.

(c) **Series parallel combination.** In this combination, the resistances are connected in series and in parallel as shown in Fig. 7.11.

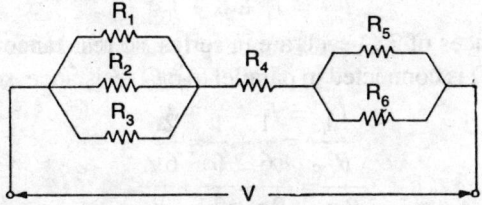

Fig. 7.11. Series-parallel circuit

This circuit is solved step by step, first the parallel branches, as shown in above figure, are solved, then the parallel circuits are resolved into the equivalent series resistances and finally the series circuit into the equivalent circuit. Thus the total resistance and current drawn are calculated.

Example. *If three resistances 1041.47 Ω, 1562.5 Ω and 625 Ω are connected in parallel, across 250 volts mains, calculate the equivalent resistance and current drawn.*

Solution. The three resistances are connected in parallel, so their equivalent or total resistance can be given by the formula,

$$\frac{1}{R_{eq}} = \frac{1}{R_1} + \frac{1}{R_2} + \frac{1}{R_3}$$

$$= \frac{1}{1041.67} + \frac{1}{1562.5} + \frac{1}{625}$$

$$= 0.00064 + 0.00096 + 0.0016$$

$$= 0.00320$$

$$R_{eq} = 312.5 \ \Omega$$

$$\text{Now current} = \frac{V}{R} = \frac{250}{312.5} = 0.736 \text{ A.}$$

So **Current = 0.736 A, R = 312.5 Ω. Ans.**

Example. *Calculate the equivalent resistance and current drawn in the given circuit.*

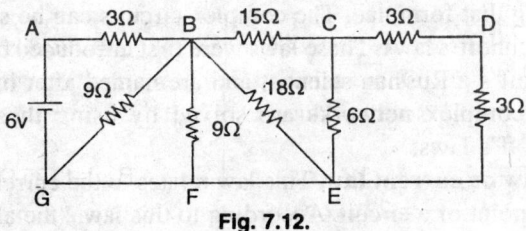

Fig. 7.12.

Solution. The circuit is a series-parallel circuit, the circuit is solved branchwise step by step.

Branch CDE.

Two resistances of 3 Ω each are in series, so resistance in $3 + 3 = 6\ \Omega$. Now this 6 Ω is connected in parallel to 6 Ω resistance, so the equivalent

$$\frac{1}{R_{CE}} = \frac{1}{6} + \frac{1}{6} = \frac{2}{6}$$

$\therefore$ $\qquad\qquad R_{CE} = 3\ \Omega.$

Branch BCE

15 Ω resistance is connected in series with 3 Ω resistance making $3 + 15 = 18\ \Omega$ resistance.

This 18 Ω resistance is in parallel of 18 Ω resistance, so

$$\frac{1}{R_{BE}} = \frac{1}{18} + \frac{1}{18}$$

$$R_{BE} = \frac{18}{2} = 9\Omega$$

Branch BEFG

There are three 9 Ω resistances connected in parallel, so

$$R_{BG} = \frac{9}{3} = 3\Omega$$

Now it is a simplified circuit having 3 Ω and 3 Ω (equivalent calculated) resistance connected in series across 6 V supply.

So the current drawn $= \dfrac{V}{R} = \dfrac{6}{6} = 1\,\text{A}.$ **Ans.**

7.10. KIRCHHOFF'S LAWS AND ITS APPLICATION

Q. Explain Kirchhoff's laws. (N.C.V.T. *1978; App. 1979*)

Ans. The simple circuits series, parallel or series and parallel circuits are solved by using Ohm's laws, but a complex circuit cannot be solved by using that formulae. The complex circuits can be solved by using Kirchhoff's laws. These laws were first introduced by the Gustav Kirchhoff's a Russian scientist and are named after his name.

The complex networks are solved by using these following Kirchhoff's laws:

(*a*) **First law or current law.** This law relates to the currents at a junction or point of a circuit. According to this law, "the algebraic sum of all the currents flowing towards a point or junction is equal to the sum of the currents going away from that point or junction".

Or

"The algebraic sum of the currents meeting at a point or junction is equal to zero."

Let I_1, I_2, I_3 are the currents flowing towards a point O and I_4, I_5 are flowing away from O as shown in Fig. 7.13. So, according to this law algebraic sum of the current flowing towards the junction is equal to the currents flowing away from the junction.

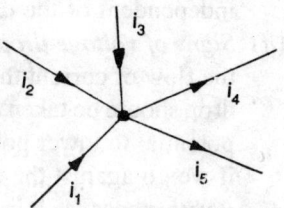

Fig. 7.13. Kirchhoff's Current law

i.e. $I_1 + I_2 + I_3 = I_4 + I_5$

or $I_1 + I_2 + I_3 - I_4 - I_5 = 0$

i.e. $\sum I = 0$

The Kirchhoff's law is true because current is the flow of electrons, which cannot be accumulated without an accumulating source.

(*b*) **Kirchhoff's 2nd law or voltage law.** This law relates to voltage fed and drops so it is called as the voltage law. According to this law, "in any closed circuit or mesh the algebraic sum of the voltage drops i.e. IR drops plus the algebraic sum of all the e.m.f.s. fed in that circuit is zero".

Or

In any closed circuit, the sum of the e.m.f. fed is equal to the sum of all voltage drops.

As shown in Fig. 7.14, this circuit diagram, V_1 and V_2 are the two e.m.f.s and drops are IR_1, IR_2 and IR_3 so according to the second law, *i.e.* voltage law.

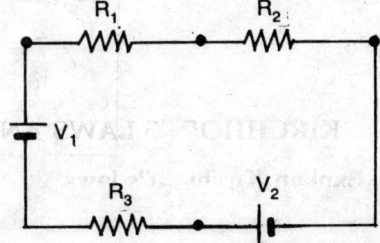

Fig. 7.14. Kirchhoff's voltage law

$V_1 + V_2 = IR_1 + IR_2 + IR_3$

or $V_1 + V_2 - IR_1 - IR_2 - IR_3 = 0$

i.e. $\sum V - \sum IR = 0$

Determination of sign of e.m.f.s. and voltage drops. In case of Kirchhoff's second law, the voltage impressed and the *IR* drops are of great significance, the mesh or closed circuit should be very carefully considered and watched.

(*i*) *Signs of e.m.f.'s.* **A rise in potential** should be considered **positive** while **fall in potential** should be considered **negative**. Thus, if we go from the −ve terminal of a battery or source to the positive terminal, there is a rise in potential and it must be considered positive. On the other hand, if we go from the +ve terminal of a battery or source to the −ve terminal, there is a fall in potential and it should be considered negative. It should be noted that sign of e.m.f. is independent of the direction of current through that branch.

(*ii*) *Signs of voltage drops.* There is a voltage drop in resistance due to the flow of current through it. If we go with the current, the voltage drop should be taken *negative,* because the current flows from higher potential to lower potential *i.e. fall in potential.* On the other hand, if we go against the current flow, the voltage drop should be taken *positive* because it is a *rise in potential.* It should be noted here that sign of voltage drop depends upon the direction of current and is independent of the polarity of e.m.f. in the circuit under consideration.

Method for solving circuits by Kirchhoff's laws

(*i*) Mark the direction of currents in various branches of the circuit in accordance with first law.

(*ii*) Choose any closed circuit and mark it.

(*iii*) Find the algebraic sum of voltage drops and e.m.f.s. in that circuit.

(*iv*) Put the algebraic sum of voltage drops *plus* algebraic sum of e.m.f.s. equal to zero.

Example. *In the circuit shown, find out the current in each resistance.*

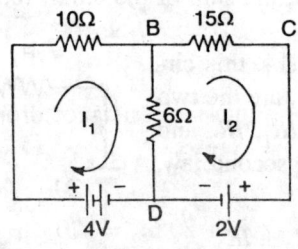

Fig. 7.15.

Solution. Let there be two currents flowing as shown in Fig. 7.15 by I_1 and I_2 amperes. Applying Kirchhoff's current law, I_1 current flowing towards B, will allow I_2 to flow from B to C and $(I_1 − I_2)$ to B to D. Now applying Kirchhoff's law to the closed network **ABDA.**

$$10i_1 + 6(I_1 − I_2) = 4$$
$$16I_1 − 6I_2 = 4 \qquad \qquad ...(i)$$

In closed mesh **BCDB**
$$15I_2 + 2 - 6\,(I_1 - I_2) = 0$$
$$-6I_1 + 21I_2 = -2 \qquad\qquad ...(ii)$$

Now solving (*i*) and (*ii*) equation for i_1 and i_2
$$i_1 = 0.32A \text{ and } i_2 = 0.183A.$$

So current in 10Ω resistance from A to B = 0.32A

current in 15Ω resistance from B to C = 0.183A

current in 6Ω resistance, (0.32 − 0.183) = 0.137A. **Ans.**

Example. *Two batteries A and B are connected to a resistance of 20 Ω as shown in Fig. 7.16. Battery A, having e.m.f. 100 V internal resistance 0.5 Ω and battery B, having 120 V e.m.f. and internal resistance 0.6 Ω. Calculate the current in each component.*

Solution. Let the current i_1 and i_2 be flowing as shown in Fig. 7.16 and $(i_1 + i_2)$ ampere in the resistance 20 Ω.

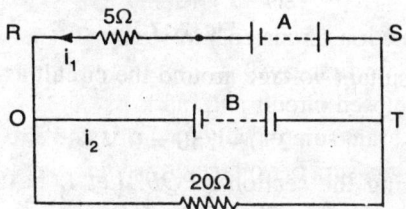

Fig. 7.16.

Now in the closed mesh RSTPQR,
$$5i_1 + 20(i_1 + i_2) + 0.5\,i_1 = 100$$
$$25.5i_1 + 20i_2 = 100 \qquad\qquad ...(i)$$
$$(0.5i_1 - \text{internal resistance drop})$$

In another closed mesh, **QTPQ**
$$0.6i_2 + 20(i_1 + i_2) = 120$$
$$20i_1 + 20.6i_2 = 120 \qquad\qquad ...(ii)$$
$$(0.6i_2 - \text{internal resistance drop})$$

Now solving equations (*i*) and (*ii*) for i_1 and i_2
$$i_1 = -2.456 \text{ A}, \ i_2 = 8.46 \text{ A}$$

It shows that current in battery *A* is the charging current i.e. 2.456 A and current by battery *B* is 8.46 A.

Current in resistance 20 Ω = 8.46 − 2.456 = 6.004 A

∴   **2.456 A, 8.46 A and 6.004 A. Ans.**

Example. *A closed ring main is fed at two points A and B at 250 V. The tapping and current drawn are shown in Fig. 7.17. Find the current in*

each section of the cable and the current supplied at each feeding point.
The resistance of each sector is given as shown.

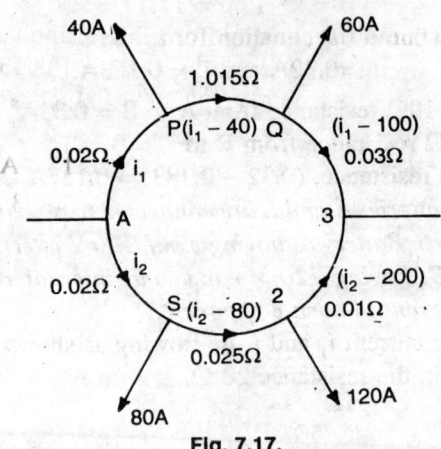

Fig. 7.17.

Solution. The resultant voltage around the circuit as shown in Fig. 7.17
is

$$250 - 250 = 0 \text{ V}$$

Now considering the section $APQB$. Let I_1 be the current flowing
through AP and rest are marked so we have

$$0.02 \times 2I_1 + 0.015 \times 2 (I_1 - 40) + 0.03 \times 2 (I_1 - 100) = 0$$

$$0.04I_1 + 0.03I_1 + 0.06I_1 - 1.2 - 6 = 0$$

$$0.13I_1 = 7.2$$

$$I_1 = 55.4 \text{ A}$$

Now current in section AP

$$= 55.4 \text{ A}$$

Drop in $\qquad AP = 55.4 \times 0.02 = 2.216 \text{ V}$

$$\text{p.d. at } P = 240 - 2.216 = 237.78 \text{ V}$$

Current in section $\qquad PQ = I_1 - 40 = 55.4 - 40 = 15.4 \text{ A}$

Current in section $\qquad QB = I_1 - 100 = 55.4 - 100 = -44.6 \text{ A}$

So current is flowing from B to $Q = 44.6$ A

Now again considering the section $ASRB$

$$0.02 \times 2I_2 + 0.025 \times 2 (I_2 - 80) + 0.01 \times (I_2 - 200) = 0$$

$$0.04I_2 + 0.05I_2 + 0.02I_2 - 4 - 4 = 0, \qquad 0.11 I_2 = 8$$

$$I_2 = \frac{8}{0.11} = 72.73 \text{ A}$$

So current in section $AS = 72.73$ A

Section $\quad\quad\quad SR = I_2 - 80 = 72.73 - 80 = -7.27$ A

It indicate the direction from R to S, value is 7.27 A

Current in section $RB = I - 200 = 72.73 - 200 = -127.27$ A

Here also the current is flowing from B to R and value $= 127.27$

Now current supplied at $A = 55.4 + 72.73 = 128.13 \simeq 128.1$ A

Current supplied at $B = 44.6 + 127.27 = 171.87 \simeq 171.9$ A

Now different currents in

section $AP = 55.4$ A $\quad\quad\quad$ section $PQ = 15.4$ A

section $BQ = 44.6$ A $\quad\quad\quad$ section $AS = 72.73$ A

section $RS = 7.27$ A $\quad\quad\quad$ section $BR = 127.27$ A

Current supplied by $A = $ **128.1 A Ans.**

Current supplied by $A = $ **171.9 A. Ans.**

7.11. WHEATSTONE BRIDGE

Q. What do you understand by the Wheatstone Bridge method of measuring resistance? Describe with neat sketch.

Ans. It is the most accurate method of determining the value of unknown resistance because it is a null or balance method. In is used to detect the null or balanced condition.

In this bridge four resistances are connected as shown in Fig. 7.18. The P and Q arms are the ratio arms and 'S' the standard variable resistor and 'R' as the unknown resistance.

A glavanometer G is connected between B and D terminals of the bridge; the cell E and a key k also connected in the circuit. The arrangement is known as the *Wheatstone bridge.*

The series circuit ABC and ADC are adjusted in such a way by adjusting the available resistance 'S' as there is no current in the connected galvanometer. When there is no current through the galvanometer the condition is said the null or balanced condition.

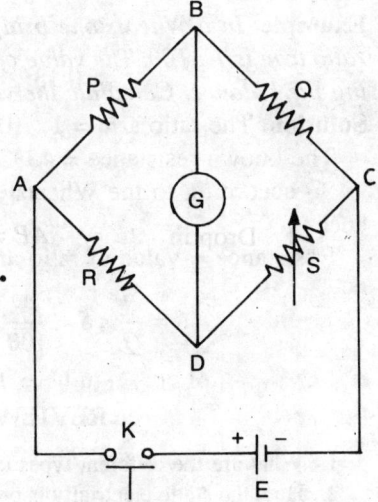

Fig. 7.18. Wheatstone Bridge

Balanced condition. In case of balanced condition there is no current through the galvanometer *i.e.*, $I_g = 0$, so the current say I_1 is flowing through P and Q and similarly if I_2 is the current in the ADC circuit than I_2 is flowing through R and S resistances. Thus if there is no current in the galvanometer the point B and D are at the same potential so voltage drop across AB is equal to the voltage drop across AD.

Similarly voltage drop across BC = voltage drop across DC *i.e.*, $I_1 \cdot P = I_2 R$ and $I_1 Q = I_2 \cdot S$

Dividing both

$$\frac{I_1 \cdot P}{I_1 \cdot Q} = \frac{I_2 \cdot R}{I_2 \cdot S}.$$

or

$$\frac{P}{Q} = \frac{R}{S}$$

and

$$R = \frac{P}{Q} \times S$$

∴ The unknown resistance = Value of Ratio arms × Known value of variable resistance.

The value of ratio arm can be varied by increasing the value of bridge, as well as to obtain greater accuracy. It may be 10,100, 1000 and 0.1, 0.01, 0.001.

Example. *In a Wheatstone bridge method of measuring resistance, the ratio arm is 1 : 100. The value of known resistance is 4532 Ω for obtaining the balance. Calculate the value of unknown resistance.*

Solution. The ratio arm = 1 : 100 *i.e.*, $P : Q$

The known resistance = 4532 Ω

So according to the Wheatstone bridge method of measuring resistance.

Resistance = Value of ratio arms × Known resistance

$$R = \frac{P}{Q} \times S = \frac{1}{100} \times 4532 = 45.32 \Omega \text{ **Ans.**}$$

REVIEW QUESTIONS

1. What are the different types of electricity, describe any one of them?
2. How the static electricity is produced? Write the advantages and disadvantages of static and dynamic electricity.
3. State the different sources of electricity.
4. State the different effects of electricity.

5. Write a short note on the uses and application of electricity.
6. What do you understand by the circuit, describe their types?
7. Define, with their units current, voltage, e.m.f. resistance, international ampere and specific resistance.
8. (*a*) What is meant by temperature coefficient of resistance?
   (*b*) Define and states the laws of resistance.
9. State Ohms law and its utility.
10. What do you understand by the combination of resistances?
11. How the load is connected in parallel combination? What are the different characteristics and uses of the parallel circuit?
12. State the Kirchohoff's laws.
13. What do you mean by the Wheatstone Bridge? How it differs from the slide wire bridge?
14. What are the effects of temperature on metal resistance and temperature coefficient?

Fill in the blanks.

1. A voltmeter is used to measure ....
2. A voltmeter is connected .... the line.
3. If a voltmeter is connected across a battery (Load switch opened) the voltage in termed as ....
4. The load voltage and without load voltage, which one will be more, ...
5. Thr e.m.f. is equal to P.D + ...
6. The resistance temperature coefficient is defined as ....
7. What do you mean by negative temperature coefficient.
8. Carbon has .... resistance temperature coefficient.
9. What is the unit of current is S.I. system of units.
10. A 5 Ω resistance is connected across a 12 V battery the current will be ....

Answers

    1. voltate   2. across   3. P.D.
    4. without load   5. drops
    6. Increase in resistance per ohm per degree
    7. that is decrease in resistance with increasing of tamperature
    8. negative   9. ampere   10. 2.4A

MCQ

1. A bulb of 200 W is connected across 200 V 50 c/s. main, the current drawn is ....
   (*a*) 1 A   (*b*) 0.2 A   (*c*) 10 A   (*d*) 0.5A
2. Three resistances 2Ω, 3Ω, 5Ω are connected in series across 10 V supply, the current drawn will be ....
   (*a*) 5 A   (*b*) 2 A   (*c*) 1 A   (*d*) 0.5A
3. Find the voltage supplied of current drown by a load of 8Ω is 1.5 A.
   (*a*) 10 V   (*b*) 15 V   (*c*) 8 V   (*d*) 12 V

4. An electric soldering iron rated 200 V, drawing 0.175 A current will have
   (a) 200 W (b) 100 W (c) 10 W (d) 35 W
5. A solding iron rated 35 W and the lamp rated 35 W, which appliance will consume more power if connected across 200 V.
   (a) Lamp (b) Soldering Iron (c) both same (d) not applicable.
6. Which is the relationship for ohm law?

   (a) $I \alpha V$ (b) $V \alpha I$ (c) $I \propto \dfrac{1}{R}$ (d) $V = IR$

7. A resistance of 12Ω is connected across a battery of 24 V the current will.
   (a) 2 A (b) 0.5 A (c) 4 A (d) 1 A
8. In any closed circuit if V = 50 V, I = 2 A, R = ?
   (a) 50Ω (b) 20Ω (c) 25Ω (d) 100Ω.
9. Fill in the blanks.
   (a) An ammeter is connected in ....
   (b) A voltmeter is connected in parallel ....
   (c) If the circuit voltage is constant, current varies in .... proportional to the resistance.
   (d) If the circuit resistance is constant, the current varies in .... proportional to the voltage.
   (e) The unit of resistance is ....
10. Which is correct.
    (i) Potential difference is bigger than EMF.
    (ii) EMF is bigger than potential.
    (iii) Resistance is directly proportional to its length.
    (iv) Resistance is inversely proportional to the area of cross-section of the conductor.
    (v) Resistance of a conductor is directly proportional to the area of cross-section.

Answers

1. (a) 2. (c) 3. (d) 4. (d)
5. (c) 6. (a) 7. (b) 8. (c)
9. (a) series (b) to main (c) inversely (d) direct
   (e) ohm
10. (i) No (ii) Yes (iii) Yes (iv) Yes
    (v) No.

State True (T) or false (F).

1. Resistance of a conductor is equal to its length multiplied by specific resistance and divided by its area of cross-section.
2. The resistance temperature coefficient is gives by expression

   $$\alpha = \frac{Rt_2 - Rt_1}{Rt_1 (t_2 - t_1)}.$$

3. The resistance of carbon increases with the increasing of temperature.
4. The tungsten has positive resistance temperature coefficient.
5. Resistance of an electrolyte decreases with the increasing of temperature.
6. If two bulbs of 60 W & 100 W are connected in series across 220 V supply, they will glow with same light.
7. The 100 W lamp will have low resistance than 60 W lamp.
8. 100 W lamp require more current than 60 W lamp to glow to full light.
9. The resistance of equivalent circuit increases with the increasing of individual resistance in series.
10. The conductance of the series circuit increases with the increasing of additional resistances.

Answers

1. T	2. T	3. F	4. T
5. T	6. F	7. T	8. T
9. T	10. F		

State True and False

1. In parallel circuit the equivalent resistance can be expressed as
   $$R_{eq} = R_1 + R_2 + R_3 + ...$$
2. In parallel combination the equivalent resistance can be given by

$$\frac{1}{R_{eq}} = \frac{1}{R_1} + \frac{1}{R_2} + \frac{1}{R_3} + ...$$

3. If four equal resistances of R Ω are connected in parallel their total resistance will be 0.25 R Ω.
4. In parallel combination the total current will be equal to the sum of individual branch currents.
5. In parallel, the conductance of the circuit increases with the increasing of resistance in parallel.
6. The equivalent resistance of the whole parallel combination decreases as more number of resistances are further added.
7. If any circuit of the parallel combination is taken out, the equivalent resistance will decrease.
8. If 15Ω resistance in connected parallel to 25Ω. The voltmeter reads 5 volt across 15Ω, the voltage across 25Ω will also be same.
9. In parallel combination if one circuit is open circuited, the remaining may work satisfactorily.
10. The power consumed will be equal to the product of voltage multiplied by total current.

Answers

1. F	2. T	3. T	4. T
5. T	6. T	7. F	8. T
9. T	10. F		

MCQ

1. The parallel circuit containing three branches carries 2 A, 5 A and 5.6 A. what is the current drawn from mains?
   - (a) 2 A
   - (b) 5 A
   - (c) 5.6 A
   - (d) 12.6 A
2. The current drawn by a (three elements) parallel combination is 16 A. If first branch carries 2 A, second 9 A what is third branch current?
   - (a) 2A
   - (b) 9A
   - (c) 16A
   - (d) 5A
3. In a three element circuit if the current drawn by all element is same and has only one path of the current, the circuit is a .... circuit.
   - (a) series,
   - (b) parallel,
   - (c) mixed,
   - (d) complex
4. Can you identify the nature of circuit if three resistances each of 100 Ω draws 6 A current from 200 V supply.
   - (a) Series,
   - (b) Parallel,
   - (c) Series-parallel
   - (d) mixed
5. Now if the same circuit (Q. 4) develops a defect and one resistance is open circuited what will be the total new current drawn.
   - (a) 2 A
   - (b) 4 A
   - (c) 6 A
   - (d) 12 A
6. A. 100 W 250 V, 200 W 250 V and 500 W 250 V lamp are connected across 250 V in parallel, but the current drawn is 2.4A, which lamp circuit is open circuited.
   - (a) 100 W 250 V
   - (b) 200 W 250 V
   - (c) 500 W 250 V
   - (d) None.
7. In Q. 6. if the equivalent current drawn is 1.2 A Which branch is opened?
   - (a) 100 W 250 V
   - (b) 200 W 250 V
   - (c) 500 W 250 V
   - (d) None.
8. Six resistance each of 60 Ω are connecting in parallel what will be the equivalent resistance.
   - (a) 10Ω
   - (b) 20Ω
   - (c) 60Ω
   - (d) 360Ω
9. Two resistance of 50Ω and XΩ are connected across 100 V supply and draws 3 A current, the valve of unknown resistance is
   - (a) 50Ω,
   - (b) 100Ω,
   - (c) 200Ω
   - (d) (Open circuit)
10. The power dissipated from a 50 Ω resistance is 200 W. What is valve of current drawn.
    - (a) 8 A
    - (b) 4 A
    - (c) 2 A
    - (d) 0.5 A

Answers

1. (d)	2. (d)	3. (a)	4. (b)
5. (b)	6. (b)	7. (c)	8. (a)
9. (b)	10. (c)		

Fill in the blanks.

1. Two resistances of 100 Ω and 50 Ω are connected in series they will carry a current of 1 A if .... voltage is applied
   - (a) 100 V
   - (b) 50 V
   - (c) 150 V
   - (d) 200 V

2. The voltage across 100 W 250 V lamp when connected in series with 60 W 250 lamp is 93.75 V. The supplied voltage is
   (a) 100 V   (b) 250 V   (c) 500 V   (d) 200 V

3. An unknown resistance is connected is series with a 100 W 250 V lamp so that it can operate on 400 V supply safely; the value will be.
   (a) 375 Ω   (b) 250 Ω   (c) 1000 Ω   (d) 400 Ω

4. In a series circuit of 100 Ω and 75 Ω if the current in 100 Ω resistance is 0.1 A what will be the total current drawn from the main
   (a) 0.1 A   (b) 0.175 A   (c) 0.75 A   (d) 1 A

5. Three 100 Ω resistance are connected in such a way that two are as a parallel combination in series with the third, can you tell the equivalent resistance offered?
   (a) 300 Ω   (b) 200 Ω   (c) 100 Ω   (d) 150 Ω

6. State True/false.
   (a) In series combination all components will carry same current.
   (b) In series the algebraic sum of all voltage drops is equal to the supply voltage.
   (c) In series circuit, the equivalent resistance of the combination is the sum of all resistances connected in series.
   (d) In series the conductance will increases and resistance will decrease with every further connected resistances.
   (e) In parallel combination the sum of all branch currents will be equal to the main line current.
   (f) The voltage across each element is same as that of supply voltage in parallel.
   (g) In house wiring all consuming devices are connected in parallel.
   (h) The power consumed is given by the product of voltage and total current drawn.
   (i) A circuit contains 10 lamps of 100 W; in the second circuit 40 lamps of 100 W are in parallel the 40 lamp circuit will have less equivalent resistance than 10 lamp circuit. Total current is equal to the voltage applied divided by the equivalent resistance.

Answers

1. (c)	2. (b)	3. (a)	4. (a)
5. (d)	6. (a) T	(b) T	(c) T
	(d) F	(e) T	(f) T	(g) T
	(h) T	(i) T		

Laws of resistance

State True (T) or False (F)

1. The resistance of a conductor is directly proportional to its length.
2. A 4 m length of a resistance will has 5 Ω resistance, 8 m length will have 10 Ω resistance.

3. The thick conductor will has less resistance than a thin conductor of the same wire.
4. The resistance of a conductor is inversely proportional to its area of cross section.
5. The conductor will have double resistance of a certain wire if the area of cross section is reduced half.
6. The resistance offered by the opposite faces of a unit cube is defined as the specific resistance.
7. The change in resistance per ohm per degree rise of temperature is defined as the resistance temperature coefficient.
8. The specific resistance is shown by ρ.
9. The resistance temperature coefficient is represented by letter α.
10. The insulation resistance of a cable is inversely proportional to its length.

Answers

1. T	2. T	3. T	4. T
5. T	6. T	7. T	8. T
9. T	10. T		

NUMERICAL PROBLEMS

1. A resistance wire of certain composition has a resistance of 500 ohm, its length being 1300 metres. What will be the resistance of 300 metres of the same wire?
2. A resistance wire of 230 metres has 15.75 ohm resistance, how much wire should be reduced to have a resistance of 9.67 ohm?
3. A piece of silver wire has a resistance of 1.5 ohm. What will be the resistance of another wire having the specific resistance 40 times the silver wire, one third the length and one third the diameter. **[Ans. 180Ω]**

8

Work, Power and Energy

Q. Differentiate between fundamental and derived units.

Ans. The units adopted to measure the fundamental quantities like mass, length and time, are called fundamental units and the units of those quantities which can be expressed in terms of fundamental units are called *derived units.*

8.1. FORCE

Q. Define force.

Ans. A force can be defined by the effects it exhibits, so the force is the physical cause which when applied to a body changes or tends to change the state of rest or motion of the body. The unit of force in M.K.S. system of units is newton.

One Newton. It is the amount of force which when applied to a mass of one kg produces the acceleration of one m/sec² in it.

$$F = m.a.$$

where F = force in Newton.

m = mass of the body in kg.

a = the acceleration in m/sec².

In the other system of units, the unit of force are poundal-in F.P.S. system, dynes in C.G.S. system.

$$1 \text{ Newton} = 10^5 \text{ dynes} = 32.2 \text{ poundals.}$$

8.2. WORK

Q. Define work.

Ans. The work is said to be done when a force acting on a body causes it to move. The work done can be calculated by the product of force and the displacement of the point of application in the direction of force.

So work done = force × distance.

The unit of work in M.K.S. system is joule.

One joule. It is defined as the work done by a force of one Newton which acts through a distance of one metre in the direction of force, so if a force F acts on a body through a distance of d-metre then

$$W = F \times d \text{ joule.}$$

The other units are foot poundal in F.P.S. system, ergs-in C.G.S. system.

$$\text{One joule} = 1 \text{ Nm} = 10^5 \text{ dynes} \times 100 \text{ cm}$$
$$= 10^7 \text{ dynes or ergs.}$$

The electrical unit of work is watt sec.

One watt sec. It is the amount of work done in an electric circuit, when two points in a circuit having one volt electric pressure has one ampere electric current flowing through it for one second.

One watt sec = 1 volt × 1 amp. × 1 sec.

8.3. POWER AND ENERGY

Q. Define power and Energy

Ans. Power. The power is defined as the rate of doing work. Mathematically it can be said as

$$\text{Power} = \frac{\text{Work done}}{\text{Time}}$$

The unit of power is joule/sec or watt in M.K.S. or S.I. system of unit.

The unit of power in various other systems of units are erg/sec, kgf-m/sec and gm-cm/sec and ft-lb/sec.

In electrical engineering the unit of power is watt; which can be given as

$$\text{Power} = \text{Voltage} \times \text{Current Watts}$$
$$= (\text{Current})^2 \times \text{Resistance Watts}$$
$$= \frac{(\text{Voltage})^2}{\text{Resistance}} \text{ Watts}$$

The bigger unit is kilowatt.

$$1 \text{ kW} = 1000 \text{ W}$$
$$1 \text{ Meg. Watt} = 10^6 \text{ W.}$$

The mechanical unit of power is horse power, if an agent does the work at the rate of 550 ft-1b/sec. or 33000 ft-lb/min or 75 kg-m/sec. then the power is said to be one horse power.

Relationship

(i) *H.P. (British) and watt.*

1 H.P. = 550 ft lb/sec

= 550 × 30.48 × 453.6 × 9.81 erg/sec

= 746 × 10^7 erg/sec

1 H.P. = 746 joule/sec or watts

(ii) *H.P. (metric) and watts*

1 H.P. = 75 kg m/sec

= 75 × 9.81 kg m/sec

∴ 1 H.P. = 735.5 joules/sec or watt.

Energy. The energy is defined as the capacity for doing work. Its units in electrical engineering is Watt sec.

The unit in M.K.S. system is joule and can be defined as the work done when the point of application of the force of one Newton, moves the body one metre in the direction of force applied.

The energy of this universe is constant but it may change its state and forms. It can be divided into the followings:

(a) *Potential energy.* It is the energy gained or released by a body when it is moving vertically and is given as under.

Potential energy = $m \times g \times h$.

Where m = mass in kg

g = gravitational pull 9.81

h = height in m.

(b) *Kinetic energy.* It is the energy due to movement and is equal to the product of half the mass multiplied by the square of velocity

$$= \frac{1}{2} mV^2$$

where m = mass in kg, V = velocity in m/sec.

Relationship between electrical and mechanical energy:

(i) kWh and joule

1 kWh = 1000 W × 3600 sec

= 3600000 W-sec or

∴ 1 kWh = 36 × 10^5 joules.

(ii) kWh and ft lb

kWh = 1 kW × 1 hr = 1000 W × 3600 sec

= 3600000 watt sec = 36 × 10^5 joule

$$= \frac{36 \times 10^5 \times 550}{746} \text{ ft lb}$$

1 kWh = 2654 × 10^3 ft lb

8.4. HEATING EFFECT

Q. What do you understand by heat? Define joule's law of heating.

Ans. Heat is the form of energy which produces the sensation of warmness or coldness. The unit is calories and bigger unit is kcal.

$$1 \text{ kcal} = 1000 \text{ calorie} = 10^3 \text{ cal}$$

Mathematical it can be expressed as

$$H = m \cdot s \cdot \theta$$

where H = heat in calorie

m = mass in gm

s = specific heat

θ = change in temperature *i.e.*, $(t_2 - t_1)^0$

One calorie. It is that amount of heat which is required to raise the temperature of one gm of water through one degree rise in temperature in °C.

Joules law. Joules law relates between the electrical energy expended and the amount of heat produced. According to that law, "the amount of the heat produced is directly proportional to the electrical energy expended", *i.e.*

Heat developed (H) α electrical energy (W)

$$\frac{W}{H} = \text{constant}, J. \qquad \qquad ...(i)$$

where J is known as the mechanical equivalent of heat. Its value is 4.18 joules per calories.

Let there be a resistance R ohm connected across V volts and causes I amp. current for t sec. then the work done during that period is

$$= \text{Voltage} \times \text{current} \times \text{time joule}$$

$$= V \times I \times t \text{ joule}$$

$$= I^2 \cdot R \cdot t \text{ joule} (\because V = IR)$$

So electrically work done $W = I^2 \cdot R \cdot t$ joule substituting in (i) equation,

$$\frac{I^2 \cdot R \cdot t}{H} = J$$

$$H = \frac{I^2 \cdot R \cdot t}{J}$$

$$H = \frac{I^2 \cdot R \cdot t}{4.18} \text{ cal} \qquad (\because J = 4.18)$$

Thus the heat developed in a conductor is
(*i*) directly proportional to the square of current,
(*ii*) directly proportional to the resistance,
(*iii*) directly proportional to the time taken for which current is flowing.
So the heat can be given as

$$H = \frac{I^2 Rt}{4.18} = \frac{VIt}{4.18} = \frac{V^2 t}{R \times 4.18} \text{ calories}$$

$$= 0.24 I^2 Rt = 0.24 V \cdot I \cdot t = \frac{0.24 V^2 \cdot t}{R} \text{ cal.}$$

Relationships:

(*a*) Heat and mechanical energy
  (*i*)   1 cal = 4.18 joule
  (*ii*) C.H.U. and joule
    1 C.H.U. = 1 lb × 1°C
      = 453.6 gm × 1°C = 453.6 cal.
      = 453.6 × 4.18 = 1896 joule
  ∴ **1 C.H.U. = 1896 joule.**
  (*iii*) B.Th.U. and joule
    1 B.Th.U. = 1 lb × 1°F

$$= 453.6 \text{ gm} \times \frac{5}{9}°C = 453.6 \times \frac{5}{9} \times 4.18$$

$$= 1053 \text{ joule}$$

  ∴ **1 B.Th.U = 1053 joule.**
(*b*) Electrical energy and heat
  (*i*) 1 kWh and calorie
    1 kWh = 1 kW × hr

$$= 1000 \text{ W} \times 3600 \text{ sec} = \frac{36 \times 10^5}{4.18} \text{ cal.}$$

  ∴ **1 kWh = 860 × 10³ cal or 860 Kcal**
  (*ii*) kWh and B.Th.U.
    1 kWh = 36 × 10⁵ joule

$$= \frac{36 \times 10^5}{1053} = 3400 \text{ B.Th.U.}$$

∴   **1 kWh = 3400 B.Th.U.**

Q. Define the thermal efficiency of an electrical heating appliance.

Ans. The electrical appliance which is producing heat is known as heating appliance like heater, kettle etc. The thermal efficiency can be defined as the ratio of useful heat to the total heat developed electrically.

$$\text{Thermal efficiency, } \eta_{th} = \frac{\text{Useful heat}}{\text{Total heat developed electrically}}.$$

Example. *A lamp which is connected on 230 V draws 0.261 A. Calculate the power and hot resistance of the lamp.*

Solution. The lamp is working on 230 V and draws 0.261 A, so according to the formula

$$W = V \times I = 230 \times 0.261 = 60 \text{ W}$$

$$\text{Now the resistance} = \frac{V}{I} \text{ or } \frac{V^2}{W}$$

$$\therefore \qquad R = \frac{230 \times 230}{60} = 881.23 \text{ Ohms}$$

$$\simeq \textbf{881 Ohms.}$$

Example. *A lamp of 200 W 230 V is connected across the supply. Calculate the current drawn and the resistance of the lamp.*

Solution. The lamp wattage is 200 W at 230 V. When the lamp is connected, the current drawn can be calculated using the formula

$$W = V \times I$$

or

$$I = \frac{W}{V} = \frac{200}{230} = 0.86956$$

$$\simeq \textbf{0.87 A.}$$

The resistance R can be calculated using the formula

$$I^2 R = W$$

$$R = \frac{W}{I^2} = \frac{200}{(0.86956)^2} = 264.5 \text{ ohms}$$

$$\textbf{0.87 A 264.5 } \Omega. \textbf{ Ans.}$$

Example. *An electric kettle is rated 1.5 kW 230 V takes 5 minutes to bring 1 kg of water to boiling point from 15°C. Find the efficiency of the kettle.*

Solution. The electric kettle is rated 1.5 kW = 1500 W

Time taken is min. $= 5 \times 60 = 300$ sec

So the input $= V.I. \ t = Wt$

$$= 1500 \times 300 = 450000 \text{ Watts-sec}$$

The heat required to boil water

$$= m \cdot s \cdot \theta = 1000 \times 1 \times (100 - 15)$$
$$= 85000 \text{ or } 85 \text{ Kcal}$$

The efficiency $\quad = \dfrac{\text{Useful heat}}{\text{Input power}} = \dfrac{85000 \times 4.18}{450000} \times 100$

$$= \textbf{78.9\% Ans.}$$

Example. *An electric kettle having an efficiency 90% is required to raise the temperature of one litre of water 15°C to boiling point in 10 minutes. Calculate:*

(i) *The resistance of the heating element if the working voltage is 230 V.*

(ii) *The energy consumption and cost of energy at the rate of Rs. 4.50 per unit.*

Solution. The heat required $= m \cdot s \cdot \theta$ where m = mass in kg, s = specific heat, θ = change in temperature.

$$\therefore \quad H = 1000 \times 1 \times (100 - 15) = 85000 \text{ calories}$$

The efficiency of the kettle 90%. So the input to the kettle,

$$\text{Input} = \frac{\text{Output}}{\eta} = \frac{85000}{0.9 \times 1000} = 94.44 \text{ kcal}$$

The kWh equivalent $= \dfrac{94.444}{860}$ $\qquad$ ($\because$ 1 kWh = 860 cal)

$$= 0.1098 \text{ kWh}$$

Now the energy input $= \dfrac{V^2}{R} \times$ times in hrs.

$$\therefore \quad 0.1098 \times 10^3 = \frac{230 \times 230}{R} \times \frac{10}{60}$$

$$R = \frac{230 \times 230 \times 10}{0.1098 \times 1000 \times 60} = 80.297$$

$$\simeq \textbf{80.3 } \Omega \textbf{ Ans.}$$

Cost of energy $\quad$ = Energy consumed $\times$ cost/kWh

$$= 0.1098 \times 4.5 = 0.4941 \text{ P.}$$

$$= \textbf{0.494} \simeq \textbf{Re. 0.50 Ans.}$$

Example. *An electrical Installation consists of 15 light points of 60 W each, 8 light points of 40 W. Lamp, 4 fans of 60 W capacity and a pump* *motor of 0.5, h.p. Assuming that 50% of light and fans are used for 4 hours per day and that the water pump works for 3 hours daily. Find out the monthly consumption and cost of electricity bill, based on tariff of 33 paise per kWh.* (*N.C.V.T., 1966*)

Solution. The lighting load consists of the lamp, fans, etc. Total electrical wattage:

15 lamps of 60 W each *i.e.*, $15 \times 60 = 900$ W

8 lamps of 40 W each *i.e.*, $8 \times 40 = 320$ W

4 fans of 60 W each *i.e.*, $4 \times 60 = 240$ W

The total wattage of light and fans $= 900 + 320 + 240 = 1460$ W

The 50% of these lights are used for 4 hours a day

So the energy consumption $= 1460 \times 4 \times 0.5 = 2920$ W hours

Now the power load of the motor of 1/2 h.p. and working for 3 hours a day

$\therefore$ Wh $= 0.5 \times 746 \times 3 = 1119$ Wh

Now total Wh per day $=$ light $+$ power

$= 2920 + 1119 = 4039$ Wh

The power consumed in the 30 working days

$= 4039 \times 30 = 121170$ Wh

The amount at the rate of 33 paise/unit

$$= \frac{121170 \times 33}{1000 \times 100} = \text{Rs. } 39.99$$

$\approx$ **Rs. 40. Ans.**

Example. *An office of electrical installation comprises the following loads. Calculate the energy charge paid to the supply authority for the month of November 1967. Energy consumptions 10 paise per unit for power load and 20 paise per unit for lighting load. Air-conditioner, pump, heater, calculators are connected to the power circuit. The office worked for 25 days of that month except security section which worked on all days.*

60 Nos. 4 ft 40 W fluorescent tubes 8 hr/day, 4 Nos. 60 W lamps, 12 hrs/day for security purposes, 16 Nos. 60 W ceiling fans 6 hrs/day.

1 No. 1,000 W air-conditioner for 5 hrs/day.

1 No. 750 W pump 2 hrs/day.

2 Nos. electric calculator 500 W 1 hr/day.

One electric heater 1,500 W for 2 hrs per day.

Meter rent at Rs. 2.00 per meter and 10% surcharge on charges are also made by supply authorities. (*A.I. Compt. 1968*)

Solution. The electrical energy consumption in Wh is calculated first as under:

60 tubes of 40 W each working 8 hrs/day

$$= 60 \times 40 \times 8 = 19200 \text{ Wh}$$

4 lamps of 60 W each working 12 hrs/day

$$= 4 \times 60 \times 12 = 2880 \text{ Wh}$$

16 fans of 60 W each working 6 hrs/day

$$= 16 \times 60 \times 6 = 5760 \text{ Wh.}$$

Now total light consumption during that month will be as the office is working for 25 days so the lights, tube and fan worked for those days only and remains the lights for 30 days because Nov. 1967 is of 30 days. So the load $19200 \times 25 = 480000$ Wh

$$2880 \times 30 = 86400 \text{ Wh}$$

$$5760 \times 25 = 144000 \text{ Wh}$$

and total consumption = 710400 Wh

Total amount during that month =

Energy consumption in kWh × rate/kWh.

$$= \frac{710400 \times 200}{1000 \times 100} = 142.08$$

= Rs. 142.08. Ans.

The power load comprises of the followings:
One air conditioner working 5 hrs/day

$$= 1 \times 1000 \times 5 = 5000 \text{ Wh}$$

One 750 W pump working 2 hrs/day

$$= 1 \times 750 \times 2 = 1500 \text{ Wh}$$

Two calculator 500 W each working 1 hr/day

$$= 2 \times 500 \times 1 = 1000 \text{ Wh}$$

One heater 1500 W each working 2 hrs/day

$$= 1 \times 1500 \times 2 = 3000 \text{ W}$$

Total power consumption = 5000 + 1500 + 1000 + 3000 = 10500 W

The office worked only for 25 days so the energy consumption during that month at the rate of 10 paise/unit

$$= \frac{10500 \times 25 \times 10}{1000 \times 100} = \text{Rs. } 26.25$$

Total charges   = lighting + power

$$= 142.08 + 26.25 = \text{Rs. } 168.33$$

Surcharge @ 10%   = Rs. 16.883

∴   Total charges   = 168.32 + 16.833 = 185.163

There are two separate meters one for power and one for light so the meter charges are Rs. 4.00.

∴   Bill for that month = 185.163 + 4 = 189.163

≈ **Rs. 189.20. Ans.**

Example. *Calculate the bill of electricity charges for the following load fitted in an electrical installation:*

1. *20 lamps 100 W each working 6 hrs a day.*
2. *10 ceiling fans 120 W each working 12 hrs a day.*
3. *2 kilowatt heater working 3 hrs a day.*
4. *2 B.H.P. motor having 85% efficiency working 4 hrs a day.*

Rate of charges for light and fans is 20 paise per unit and heater and motor is 15 paise per unit.   (*N.C.V.T. 1973*)

Solution. The rate of energy consumptions are different for lighting and power. So first the *lighting load* consumption.

20 lamps 100 W each working 6 hrs/day

$$= 20 \times 100 \times 6 = 12000 \text{ Wh}$$

10 Ceiling fans 120 W each working 12 hrs/day

$$= 10 \times 120 \times 12 = 14400 \text{ Wh}$$

So total lighting load = 12000 + 14400 = 26400 Wh

$$= 26.4 \text{ kWh or units.}$$

The power load

One heater 2 kW working for 3 hrs/day

$$= 2000 \times 3 = 6000 \text{ Wh}$$

2 B.H.P. motor (η = 85%) working for 4 hrs daily

$$\frac{2 \times 746 \times 4}{0.85} = 7021 \text{ Wh}$$

Total power load = 6000 + 7021 = 13021 Wh

$$= \frac{13021}{1000} = 13.021 \text{ kWh or units}$$

The lighting expenditure @ Rs. 0.20/unit

$$= 26.4 \times 0.20 = \text{Rs. } 5.28$$

Power expenditure @ Rs. 0.15/unit

$$= 13.021 \times 0.15 = \text{Rs. } 1.95$$

So total expenditure during that period

= **Rs. 5.28 + Rs. 1.95 = Rs. 7.23. Ans.**

Example. *An electric installation consists of the following:*

(*i*) *20 lamp of 100 W each working 6 hrs per day.*

(*ii*) *A 1,500 W heater, working 8 hrs per day.*

(*iii*) *A 2 H.P. motor (efficiency 85%) working four hours a day.*

Calculate the total number of units consumed during one month of 30 days if all these operate as shown against each day.

(*N.C.V.T. 1970*)

Solution. The total number of units consumed during the month of 30 days can be calculated by calculating the kWh consumed by individual load per day and then multiplied by 30. So

20 lamps 100 W each working 6 hrs/day

$$= 20 \times 100 \times 6 = 12000 \text{ Wh}$$

One heater 1500 W working 8 hrs/day

$$= 1 \times 1508 \times 86 = 12000 \text{ Wh}$$

One 2 H.P. motor with 85% efficiency working for 4 hrs/day

$$= \frac{2 \times 746 \times 4}{0.85} = 7021 \text{ Wh}$$

So total watt hours = 12000 + 12000 + 7021 = 31021

$$= \frac{31021}{1000} = 31.021 \text{ kWh or units.}$$

Total consumption during the month

$$= 31.021 \times 30 = \textbf{930.63 units. Ans.}$$

Example. *A 230 V dynamo supplies load consisting of 100 lamps of 60 W each and 40 lamps of 100 W each at a distance of 250 metres. The voltage across the lamps is 220 V. Calculate the size of the wires of the specific resistance 17 microhm/cm³.* (*A.I. Compt., 1969*)

Solution. The load supplied consists of lamps. So the total wattage

100 lamps 60 W each = 100 × 60 = 6000 W

40 lamps 100 W each = 40 × 100 = 4000 W

So total load   = 10000 W

Now voltage at the lamps end = 220 V

∴   Current $= \dfrac{\text{Wattage}}{\text{Voltage}} = \dfrac{10000}{220} = 45.45 \text{ A}$

∴   drops in the line = 230 − 220 = 10 V the current is 45.45 A.

$$\text{So the resistance } = \frac{10}{45.45} = 0.22 \ \Omega$$

$$R = \frac{\rho l}{a} \ \Omega \text{ here } R = 0.22 \ \Omega, \ \rho = 1.7 \ \mu\Omega \text{ cm., } l = 250 \text{ cm}$$

here
$$a = \frac{\rho l}{R} = \frac{1.7 \times 10^{-6} \times 250 \times 100 \times 2}{0.22} = 0.3862 \text{ cm}^2$$

the diameter of the wire
$$d = \sqrt{\frac{4a}{\pi}} \qquad \left(\because a = \frac{\pi d^2}{4} \right)$$

$$= \sqrt{\frac{4 \times 0.3862}{3.14}} = 0.70123 \text{ cm}$$

Diameter of the wire = 0.7 cm. Ans.

Example. (*a*) *Two lamps of 100 W and 150 W 220 V are connected in series across 220 V supply. Calculate (i) the respective voltage and across each lamp (ii) power consumption.*

(*b*) *What would be the power consumption of the two lamps are connected in parallel?*   (*N.C.V.T. Compt., 1970*)

Solution. The lamps are rated as 100 W and 150 W 220 V. So the resistance of individual lamp

$$\text{Resistance of 100 W lamp} = \frac{E^2}{W} = \frac{220 \times 220}{100} = 484 \text{ Ohm}$$

$$\text{Resistance of 150 W lamp} = \frac{220 \times 220}{150} = 322.67 \text{ Ohm}$$

When both lamps are in series the total resistance will be
$$R_1 + R_2 = 484 + 322.67$$
$$= 806.67 \ \Omega$$

$$\text{Now current} = \frac{\text{Voltage}}{\text{Resistance}} = \frac{220}{806.67} = 0.273$$

$$= 0.273 \text{ A}$$

Now the voltage drops across 100 W lamp
$$= 484 \times 0.273 = 132.132 \text{ V}$$

So the voltage drop across 100 W lamp
$$= \textbf{132.13 V Ans.}$$

and voltage drop across 150 W lamp = **87.87 V Ans.**
The power consumption = V.I = 220 × 0.273 = 60.06
$$\approx \textbf{60 W. Ans.}$$

(b) Current taken by 100 W lamp

$$= \frac{100}{220} = 0.4545 \text{ A}$$

Current taken by 150 W lamp

$$= \frac{150}{220} = 0.6818 \text{ A}$$

Total current = 1.1364 A
∴ Power = $V \times I$ = 220 × 1.1364 = **250 W. Ans.**

Example. *An electric lift is required to raise a load of 5 tons through a height of 100 ft. One quarter of electrical energy supplied to the lift is lost in motor and gearing. Calculate (a) energy consumed in kWh, (b) if the time required to raise the load is 27 minutes. Find the B.H.P. of the motor and current taken by the motor, the supply voltage being 230 V. D.C.*

(*N.C.V.T., 1968*)

Solution. The weight required to lift is 5 ton = 5 × 2240 = 11200 lb
The weight is to be lifted to 100 ft so the work done for lifting the weight

$$= 100 \times 11200 = 1120000 \text{ ft lb}$$

The 25% energy is lost in the gears etc. So the work performed is nearly 75% *i.e.* the work done or output of the motor is 1120000 ft lb.
Now actual output of the motor

$$= \frac{1120000}{0.75} = 1493300 \text{ ft lb}$$

$$= 149.33 \times 10^4 \text{ ft. lb}$$

Now the equivalent is

$$1 \text{ kWh} = 2654 \times 10^3 \text{ ft lb}$$

or $$1 \text{ ft lb} = \frac{1}{2654 \times 10^3} \text{ kWh}$$

$$= 3.7679 \times 10^{-7} \text{ kWh}$$

So the output in kWh = 149.3333 × 10^4 × 3.7679 × 10^{-7}
$$= 0.56267 \text{ kWh}$$
$$= \textbf{0.563 kWh. Ans.}$$

The energy supplied by the motor in 27 min,
So energy supplied = Voltage × Current × Time in hrs.

$$0.56267 = \frac{230 \times I}{1000} \times \frac{27}{60}$$

$$I = \frac{0.56267 \times 60 \times 1000}{230 \times 27}$$

$$= 5.4364 \text{ A} = 5.44 \text{ A. Ans.}$$

Example. *A 50 K.V.A. 3 phase 400 V alternator has an efficiency of 85% and a power factor of 0.8. Calculate the B.H.P. of the engine driven at full load.*

Calculate also the full load line current of the alternator.

(All India Compt., 1970)

Solution. The output of the alternator is 50 kV A at 0.8 p.f. So the power output

$$= 50 \times 0.8 = 40 \text{ kW.}$$

The input to the alternator

$$= \frac{\text{output}}{\eta} = \frac{40}{0.85} \text{ kW}$$

$$= \frac{40 \times 1000}{0.85 \times 746} = 64 \text{ H.P.}$$

So the B.H.P. of the driving engine
= 64 H.P. Ans.

The output of the alternator

$$50000 = \sqrt{3} \times 400 \times I_L$$

$$I_L = \frac{50000}{\sqrt{3} \times 400} = 72.1688 \approx \textbf{72.2 Ans.}$$

Example. *A 5 km long overhead line having a resistance of 0.025 Ohm per kilometer supplies a load of 50 H.P. at 440 V from a generating station. Find the P.D. at sending end.* (All India Compt., 1970)

Solution. The generating station is 5 km away so the conductor will have the total length

$$= 5 \times 2 = 10 \text{ km.}$$

The load current $= \dfrac{50 \times 746}{440} = 84.77 \text{ A}$

The resistance of the line
$$= 0.025 \times 5 \times 2 = 0.25 \ \Omega.$$
$\therefore$ Voltage drop $= IR = 84.77 \times 0.25 = 21.193$ V.
So the P.D. at the sending end
$$= \text{Voltage at load} + \text{drops}$$
$$= 440 + 21.193 = \textbf{461.193 V Ans.}$$

Example. *Calculate the B.H.P. of an Engine to drive an alternator supplying a load of 60 A on 3 phase 400 V supply at a power factor of 0.85, efficiency of the alternator 90%.* (*Inter I.T.I., 1970*)
Solution. The alternator is 3ϕ 400 V having 90% efficiency and load power factor as 0.85.

So the power
$$= \sqrt{3} \cdot V_L \cdot I_L \cdot \cos\phi$$

$$= \sqrt{3} \times 400 \times 60 \times 0.85 = 35333.84 \text{ W}$$

Now input the alternator

$$= \frac{\text{Output}}{\eta} = \frac{35333.84}{0.90} \text{ W}$$

The input to the alternator is the B.H.P. of the engine

$$= \frac{35333.85}{0.90 \times 746} = 52.63 \text{ H.P.}$$

$$= \textbf{52.63 H.P. Ans.}$$

Example. *Two lamps of 100 W and 60 W rated for 220 V supply are connected to 440 V supply mains in series. How will it be possible with the help of additional resistance to get 220 V on each lamp. Find out the value of this resistance and show how will it be connected?*
(*N.C.V.T., 1973 W/C*)
Solution. The 100 W 220 V lamp the quite sufficient to meet the voltage independently but 60 W has to bear the same 220 V but current should also be of the same value.

Current taken by 100 W lamp $\dfrac{100}{220} = 0.4545$ A.

Current in 60 W lamp $= \dfrac{60}{220} = 0.2727$ A.

The difference $0.4545 - 0.2727 = 0.1818$ A has to bypass through an external resistance shunted to 60 W lamp. So the resistance when carrying 0.1818 A at 200 V

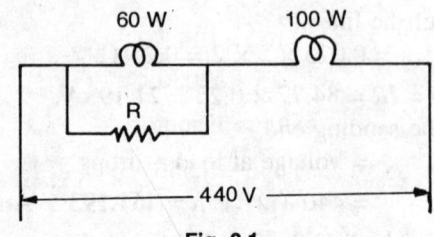

Fig. 8.1.

$$R = \frac{V}{I} = \frac{220}{0.1818} = 1210.12 \ \Omega.$$

So a resistance of **1210.12 ohm** is to be placed in parallel with the 60 W lamp. **Ans.**

Example. *An electrically driven pump is required to the pump 10,000 gallons of water per hour to a height of 80 ft. Taking the motor efficiency 0.85 and pump efficiency as 0.70, calculate the B.H.P. of the motor required to drive the pump.* (*N.C.V.T., 1968*)

Solution. The 10000 gallons of water is to be raised to 80 ft in one hour, so the work done per minute

$$= \frac{100000 \times 10 \times 80}{60} \ \text{ft lb} = 133333.33 \ \text{ft lb}$$

The H.P. of the pump $= \dfrac{10000 \times 10 \times 80}{60 \times 33000} = 4.0404 \ \text{H.P.}$

The input to pump = Output of the motor
So the output of the motor

$$= \frac{4.0404}{0.70} = 5.772 \ \text{H.P.}$$

So the B.H.P. of the motor
$$= 5.772 \ \text{H.P. Ans.}$$

Example. *A pump driven by 3 phase 440 V A.C. motor is required to pump 1,100 gallons of water every 15 minutes to a height of 70 feet. If efficiency of pump is 70% find B.H.P. of motor and current taken by it at p.f. 0.8. Take efficiency of motor as 85%.*

(*A.I. Compt., 1968; Inter I.T.I. Compt., 1969*)

Solution. The work done $= \dfrac{\text{Water to be lifted} \times \text{Height}}{\text{Minutes}}$

$$= \frac{1100 \times 10 \times 70}{15} \ \text{ft lb}$$

$\therefore$ Output in H.P. $= \dfrac{1100 \times 10 \times 70}{15 \times 33000}$ H.P. $= 1.556$

Now input of pump = Output of motor

$$= \dfrac{1.556}{\eta_a} = \dfrac{1.556}{0.70} = 2.22 \text{ H.P.}$$

The output of motor = Input of pump = **2.22 H.P.**

$\therefore$ Input of motor $= \dfrac{2.22}{\eta_m} = \dfrac{2.22 \times 746}{0.85}$ W

So $I_L = \dfrac{W}{\sqrt{3} \cdot V_L \times \cos\phi} = \dfrac{2.22 \times 746}{0.85 \times \sqrt{3} \times 440 \times 0.8}$

$$\simeq 3.2 \text{ A.} \quad \textbf{Ans.}$$

Example. *An electric oven has three heating elements of resistance 20 Ohm each. They can be switched either all in series or all in parallel. Calculate the rate of production of heat in calories per second in both cases when the supply voltage is 230 V.* (*N.C.V.T. 1970*)

Solution. The three elements consisting of 20 Ω each may either be connected in parallel or in series when these are connected in *series*, the total resistance will be the sum of all these three so

$$R_{eq} = 20 + 20 + 20 = 60 \ \Omega$$

Heat developed per second $= \dfrac{V \cdot I}{4 \cdot 2}$ or $\dfrac{V^2}{R \times 4 \cdot 2}$ cal

$$= \dfrac{220 \times 230}{60 \times 4 \cdot 2} = \textbf{209.92 cal. Ans.}$$

When these are connected in *parallel*, the total resistance will be

$$= \dfrac{20}{3} \Omega$$

heat developed $= \dfrac{230 \times 230 \times 3}{20 \times 4 \cdot 2}$ cal.

$$= \textbf{1889.3 cal. Ans.}$$

Example. *An electric heater is required to dissipate 1 kW connected to 230 V supply. Calculate the resistance of the wire. Find also the time taken to raise the temperature of 1.5 litres of water at 15°C to boiling* *point by means of this heater, if the efficiency of the heater is 80%.*

(*N.C.V.T., 1971*)

Solution. The heater has 1 kW capacity working on 230 V so the heater resistance

$$= \frac{V^2}{W} = \frac{230 \times 230}{1000} = 52.9 \, \Omega \quad \textbf{Ans.}$$

It is required to heat 1.5 litre *i.e.* 1.5 kg of water from 15°C to boiling point. So heat required.

$$H = m \cdot s \cdot \theta = 1.5 \times 1000 \times 1 \times (100 - 15)$$
$$= 127500 \text{ cal}$$

Heater has the efficiency 80%.

So heat actually generated $= \dfrac{127500}{0.8} = 159375 \text{ cal} = \textbf{159.375 kcal}$

Now heat developed by the heater

$$= \frac{W \cdot t}{4 \cdot 2} \text{ cal.}$$

$$\frac{1000 \times t}{4 \cdot 2} = 159375$$

$$t = \frac{159375 \times 4 \cdot 2}{1000} = 669.375 \text{ sec.}$$

$$= \textbf{11 min 9.4 sec. Ans.}$$

Example. *An electric crane raises a load of 5 tonnes to a height of 30 metres in one minute. Calculate the H.P. (metric) of the motor and current taken from a 230 V d.c. supply if the efficiency of the crane is 75% and that of the motor 85%.* (*B.T.E. Raj., Nov. 1974*)

Solution. The work done is to lift the 5 tonnes of weight to a height of 30 metres.

$\therefore$ Work done $= 5 \times 1000 \times 30 \times 9.81 = 1471500$ joules.

The efficiency of the crane is 75% so the input to the crane

$$= \text{Output of the motor}$$

$$= \frac{1471500}{0.75} \text{ joules.}$$

This work is preformed in 60 sec, so work done per sec *i.e.*, joule/sec or watts

$$= \frac{1471500}{0.75 \times 60} = \textbf{32700 joule/sec or watts.}$$

Now the output of the motor (metric)

$$= \frac{32700}{735.5} = 44.46 \text{ H.P.}$$

The motor efficiency is 85% so the input to motor

$$= \frac{\text{motor output}}{\eta} = \frac{327500}{0.85} = 38470 \text{ W}$$

The working voltage is 230 V, so the line current is given as

$$I = \frac{38470}{230} = 167.26 \text{ A}$$

Now the **H.P. = 44.46 H.P. (metric)**

and **I_L = 167.26 A. Ans.**

Example. *A pump driven by a d.c. motor lifts 250000 kg of water per hour in a height of 50 metres. The efficiency of the pump is 80% and that of the motor is 90% of the supply voltage is 500 V. Calculate the current drawn by the motor.* (*B.T.E. Haryana, 1974*)

Solution. The mass of the water to be lifted is 250000 kg in one hour, so the water lifted/sec.

$$= \frac{250000}{60 \times 60} = 69.44 \text{ kg}$$

The height of the lift is 50 metre

∴ Work done/sec = 69.44 × 50 × 9.81

$$= 34062.5 \text{ J/sec or watts.}$$

Now the efficiency of the pump is 80% so the input to pump

$$= \frac{\text{Output}}{\eta} = \frac{34062.5}{0.8} = 42579.375 \text{ W}$$

The input to the motor $= \dfrac{\text{Output of motor}}{\eta} = \dfrac{\text{Input to pump}}{\eta}$

$$= \frac{42579.375}{0.9} = 47310 \text{ W}$$

Now current drawn from 500 V mains

$$\frac{47310}{500} = \textbf{94.62 A.} \quad \textbf{Ans.}$$

Example. *An hydro electric station is supplied from a catchment area of 150 km² with an annual rainfall of 200 cm and effective height of 300 metres. Assuming a yield factor of 60%. Calculate*

(i) *the available continuous power.*

(ii) *the ratings of generators installed, and*

(iii) *net energy available in kWh.*

(A.M.I.E. Sec. B. May, 1975)

Solution. The generating station is a hydroelectric station where the water is the main source for generation. So the quantity of the water available per annum

$$= \text{Catchment area} \times \text{Rainfall}$$

$$= 150 \times 10^6 \times \frac{200}{100} = 300 \times 10^6 \text{ m}^2$$

The volume of water utilized

$$= \text{Volume of water available} \times \text{Yield factor}$$

$$= 300 \times 10^6 \times 0.6 = 180 \times 10^6 \text{ } m^3/\text{annum}$$

Since the 1 m³ of water = 1000 kg, so the volume of utilized is kg.

$$= (180 \times 10^6 \times 1000) \text{ kg.}$$

The effective head is 400 metre. So the energy stored in catchment area = mgh joules

$$= 180 \times 10^6 \times 1000 \times 9.81 \times 300 \text{ joules}$$

Now the net energy available

$$= \text{Energy stored} \times \text{Over all efficiency of the plant}$$

Let the efficiency be 80%, then

$$= 180 \times 10^6 \times 1000 \times 9.81 \times 300 \times 0.8$$

$$= 423792 \times 10^9 \text{ watt sec or joules}$$

$$\text{or} \quad = \frac{423792 \times 10^9}{60 \times 60 \times 1000} = 117.7 \times 10^6 \text{ kWh}$$

The energy available = **117.7 × 10⁶ kWh. Ans.**

The available continuous power

$$= \frac{\text{Net energy available}}{8760} = \frac{117.7 \times 10^6}{8760}$$

$$= \textbf{13425 kWh. Ans.}$$

The rating of the generating station

$$= \frac{\text{Available continuous power}}{\text{Load factor}}$$

Let the load factor be 0.5

So $\qquad = \dfrac{13425}{0.5} = 26850 \text{ kW}$

13425 kW, 26850 kW, 117.7 × 10^7 kWh. Ans.

8.5. HEATING, SOLDERING AND WELDING

Q. What do you understand by electric heating? Write down the advantages of electric heating.

Ans. Electric heating. The process of producing heat by passing the electric current through a conductor or resistance is known as electric heating. It is of great importance and use in industries and houses.

Advantages of electric heating. There are the following advantages:

(a) *Cleanliness.* There is no dust, ash, fumes or smoke in electric heating so it makes the system clean.

(b) *Economical.* It is cheap than other types of heating because electricity is cheap, electric furnaces have low initial cost and maintenance cost, hence requires less attention and there is no necessity of storage of fuel.

(c) *Response.* It is quick to respond and heat can be obtained easily and quickly.

(d) *Ease to control.* Quick, simple, accurate and reliable, temperature control can be obtained very easily. Such temperature controlling is not possible in other types of heating.

(e) *Pollution.* This heating do not spoil or pollute the atmosphere.

(f) *Special heating.* In certain cases heating to a particular portion or uniform heating or heating of non-conducting material, is required which is only possible in case of electric heating.

(g) *Upper limit of temperature.* There is no upper limit to the temperature obtainable except the ability of material to withstand the heat.

(h) *Efficiency.* The electric losses only a little losses therefore the efficiency is more and heating is efficient.

(i) *Better working condition.* It does not produce irritating noise, the radiation losses are less and the working on the furnace etc. is convenient and cool.

(j) *Safety.* it is quite safe.

Q. What are the different methods of electric heating? Describe core less induction type furnace.

Ans. There are the following methods of electric heating:

(a) Resistance heating.
  (i) Direct heating.   (ii) Indirect heating.
(b) Induction heating.
  (i) Direct heating.   (ii) Indirect heating.
(c) Dielectric heating.
(d) Electric arc heating.

Core less induction type furnace. This type of furnace is suitable on high frequencies 500 – 3000 Hz, mains. In this case the body (charge) to be melted is made the secondary of the transformer. The primary winding is wound over a crucible 'A' itself and separated from it by a thin layer of heat insulating material.

The coil 'C' which acts as primary winding of a two winding transformer, is energised by a high frequency supply. It produces the eddy currents in the charge which is made as the secondary as shown in Fig. 8.2. The eddy currents induced in the charge, heat it and melts. An automatic stirring due to electro-dynamic force is obtained in this process.

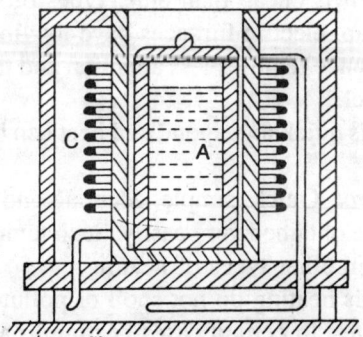

Fig. 8.2. Core less insulation type

The frequency employed usually depends upon the size of the furnaces. Even 3000 Hz supply is supplied for lower size of machine and 500 Hz for higher sizes *i.e.* 100 kW. The melting capacity range from 50 kg to 20 tonnes.

Q. Write short note on the followings:

(a) **Soldering and use of flux,** (b) **Difference between soldering, brazing and welding,** (c) **Types of electric welding.**

Ans. (a) **Soldering.** Soldering is a process of jointing two piece of metal with the help of a third metal called solder. It is done at a low temperature not at higher temperature. The solder used is an alloy of lead and tin. The solder forms a surface uniform, and increases the mechanical strength of the joint. It increases the conductivity of the joint, fill up the gap in the joint.

For different purpose different solders are used as shown in Table 8.1.

It is important that the melting temperature of the solder must be less than the metal to be soldered.

Flux. For soldering the jobs, the flux is needed, these are the main functions of the flux;

(*i*) It cleans the job.

(*ii*) It helps the solder to spread uniformly over the job.

(*iii*) It maintain the temperature of the solder and the solder does not cool soon.

(*iv*) It prevents the oxidation of the metal or jobs.

Table 8.1. Solders

S.No.	Purpose	Tin	Lead	Melting point
1.	Electrical purpose	95%	5%	220°C
2.	Soldering joints	60%	40%	192°C
3.	Cable joint soldering	67%	33%	190°C
4.	Ordinary soldering	50%	50%	205°C

Resin, the flux which is known as fluxite, is used for soldering when lead tin alloy is used as the solder.

Procedure of soldering. Take the following steps:

(*i*) Clean the part or job to be soldered.

(*ii*) Heat the part to be soldered.

(*iii*) Carry the flux on the spot and then apply solder.

(*iv*) When the job has been soldered properly remove the soldering iron and clean the part which is soldered.

(*b*) **Difference between soldering, brazing and welding.** There are there following processes which have their own characteristics:

(*i*) *Soldering.* It is a process used for jointing two metallic pieces with the help of third metal known as solder. It is done at low temperature. Generally lead tin alloy is used for soldering purpose.

(*ii*) *Brazing.* Soldering at higher temperature using brass as solder is called brazing. Brazing is used where the strong joint is needed and welding is not possible.

In soldering and brazing the metal to be soldered does not melt and fill up the gap.

(*iii*) *Welding.* Welding is a process of jointing two metallic pieces with the help of third metal called as electrode. In welding the metallic piece to be soldered are melted and filled up the gap. It is done at a high temperature.

(c) **Types of electric welding.** The heat is developed electrically in case of welding so it is known as electric welding. It is of the following types:

(i) Resistance welding.

(ii) Arc welding.

(i) *Resistance welding.* Resistance welding is a process in which heat is developed by passing current through the points to be welded. It is again classified as:

1. Butt welding.
2. Spot welding.
3. Seam welding.

In case of butt welding it is done for joining rods, wires and tubes etc. A heavy current at low voltage is send across the joint. The spot welding, is used for joining the thin metal sheets. The metal sheets to be joined are placed one over the other between a pair of pointed copper electrode and heavy current at low voltage is passed through the joint. Seam welding is used where a continuous point is required between two overlapping metal sheets. In this case also low voltage heavy current intermittently is passed through the spot.

(ii) *Arc welding.* The process of welding two metal pieces by striking an electric arc is known as arc welding. It is again subdivided as

1. Carbon arc welding
2. Metal arc welding
3. Atomic hydrogen arc welding
4. Helium or argon arc welding.

In carbon arc welding this method is suitable on d.c. carbon rod is used are soldering. The voltage for arc varies from 35 to 50 V with 200 to 1000 A. In metal arc welding the metal electrode is used which itself feeds additional material to fill up the gap. In case of hydrogen or atomic arc welding – the arc is struck between the tungsten electrode and hydrogen gas is passed through the arc. In case of helium or argon arc welding, the arc is struck between the tungsten electrodes and in an atmosphere of inert gas like argon or helium.

REVIEW QUESTIONS

1. Define power, energy, work and force.
2. What do you understand by the derived units?
3. What is difference between mechanical and electrical powers?
4. Describe with neat sketch an induction furnace.

5. What is soldering? Why flux is used for soldering?
6. What are the different types of welding?

Heat

1. The heat produced according to Joule's law can be given as
   (a) Proportional to square of voltage applied.
   (b) Proportional to the square of current.
   (c) Proportional to the container.
   (d) Proportional to the square of current, resistance and time.
2. The temperature of a furnace is measured by
   (a) Ammeter　　　　　　　　(b) Wattmeter
   (c) Pyrometer　　　　　　　(d) Potentiometer.
3. In case of pyrometer the heat energy is first converted into
   (a) Electrical energy　　　　(b) mechanical energy
   (c) chemical energy　　　　(d) sound energy.
4. The arcing between two opposite polarity conductors spaced by few mm is because of
   (a) more voltage (b) more current (c) thin conductors (d) ioning the space.
5. The heat transfer can be possible by
   (a) Conduction　　　　　　(b) Convection
   (c) radiation　　　　　　　(d) Conduction, convection and radiation.
6. The thermometer used for measuring temperature actually measures.
   (a) the degree of hotness or coldness.
   (b) the quantity of heat developed.
   (c) the temperature w.r.t. the surroundings.
   (d) None of the above.
7. The temperature range of degree celsius thermometer is
   (a) $0 - 80°$　　　　　　　(b) $0 - 100°$
   (c) $32° - 212°$　　　　　(d) $0 - 273°$
8. The temperature from one scale to another scale can be transferred by using the formula.
   (a) $\dfrac{C}{5} = \dfrac{F}{4}$　　　　　　(b) $\dfrac{C}{5} = \dfrac{F - 23}{9}$

   (c) $\dfrac{C}{9} = \dfrac{F}{32}$　　　　　(d) $\dfrac{C}{5} = \dfrac{F - 32}{9}$
9. In degree kelvin scale the starting point in taken as
   (a) $0°$　　　　(b) $32°$　　　　(c) $-273°$　　　　(d) $100°C$
10. Mica is a good conductor of .... and bad conductor of ....
    (a) heat, electricity　　　　(b) electricity, heat
    (c) heat, heat　　　　　　　(d) electricity, electricity

NUMERICAL PROBLEMS

1. A lamp is marked 100 W 250 V, determine the current drawn by the lamp

when connected to rated voltage. What is its hot resistance?

(B.T.E. Haryana, Nov. 1977)

[Ans. 0.4 A, 625 Ω]

2. An electric kettle of resistance 30 Ω takes 5 A current. Calculate the heat developed in joules in one minute. How much energy would be consumed in 6 hours? *(B.T.E.U.P., 1980)*

[Ans. 45000J. 4.5 kWh]

3. Ten lamps of 100 W each, one lamp of 200 W are illuminated for five hours a day. The working voltage is 230 V. Calculate the energy consumed, the current drawn and resistance of 200 W lamp.

[Ans. 6 kWh, 5.22 A, 264.5 Ω]

4. A resistance R is connected in series with a paralled combination of 12 Ω and 8 Ω. Calculate R, if the total power dissipated in the circuit is 80 W and voltage applied is 25 V. **[Ans. 3.0125 Ω]**

5. An electric kettle is marked 1.5 kW 230 V takes 7.5 minutes to raise the temperature of 1.5 kg of water from 15°C to boiling point. Find the efficiency of the kettle. (a joule = 0.24 cal)

[Ans. 78.7%]

6. An electric motor driven pump lifts 2.4×10^5 kg water per hour to a height of 30 metres. The efficiency of the pump and motor are 80% and 90% respectively. Calculate the motor input and current taken by it at 110 V.

[Ans. 27.25 kW, 62 A]

7. An electric installation consists of the following:
   (*i*) ten lamps of 200 W each working 8 hours a day.
   (*ii*) five ceiling fans of 120 W each working 12 hours a day.
   (*iii*) ten lamps of 60 W each working 12 hours a day.
   (*iv*) an electric exhaust fan of 0.5 h.p. efficiency 85% working three hours a day.
   (*v*) An electric pump motor 1.5 h.p. efficiency 85% working three hours a day.

   Calculate the connected load in kW. Also calculate the monthly bill for Sep. 1978 when energy is supplied at 20 paise/unit for light and 15 P/unit for power. **[Ans. 4.955 kW Rs. 206.10]**

8. A pump which is gear driven by a d.c. motor delivers 1000 litres of water per minute to a tank 25 metres above the level of the pump. If the efficiency of the pump is 80%, gears 90% and that of motor 85%. Calculate the current drawn from the mains, if the supply voltage is 400 V.

[Ans. 16.68A]

9. 20 lamps of 60 W and ten lamps of 100 W are run for 5 hours from 230 V supply. Calculate the units consumed and current through the mains. Also calculate the resistance of 60 W lamp. **[Ans. 11 units, 881.67 Ω]**

10. A 2 kW motor, a 500 W heater and 750 W electric iron, are used at the same time on 230 V line. Determine the cost of operating all the three for 4 hours if the energy cost is 30 paised per unit. *(B.T.E.U.P., Sep. 1978)*

[Ans. Rs. 3.90]

11. It is required to heat the 120 litres of water in a copper container of 20 kg by an immersion heater of 3 kW. Calculate the time required to raise the temperature from 10°C to 60°C if 20% of the energy supplied is wasted in heat losses. (Assume specific heat of copper 0.095 and J = 4.2 J/cal)

    [**Ans. 2 hours, 58 min**]

12. An electric kettle contains 1.5 kg of water at 15°C. It takes 15 minutes to raise the temperature to 95°C. Assuming the heat losses due to radiation and heating the kettle to be 14 kcal. Find the current taken, the voltage is 100 V. [**Ans. 6.23 A**]

Chemical Effects of Electric Current

9.1. ELECTROLYSIS

Q. Define and explain the electrolysis.

Ans. Whenever an electric current is passed through the electrolyte *i.e.* the liquid conductor, it decomposes into its ions. This phenomenon is used in electrolysis and electroplating etc. So the electrolysis is defined as, "it is the process in which the electrolyte is decomposed into its ions." Figure 9.1 shows the example of electrolysis of NaCl.

Two electrodes are immersed in the solution, or the electrolyte of NaCl. The electrodes which is connected to the positive of the battery is called the anode and that which is connected with the negative is called cathode.

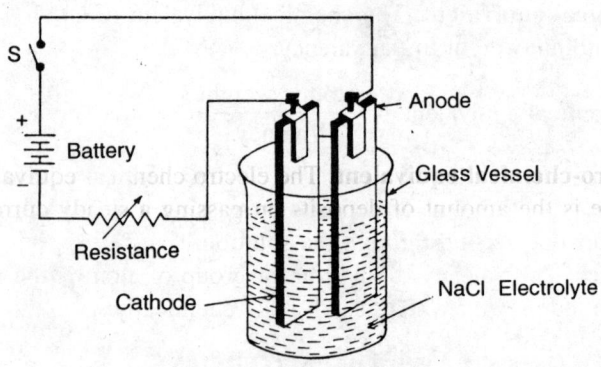

Fig. 9.1. Electrolysis.

Whenever the switch is on, the electric current starts flowing. Now NaCl will decompose into its constituents Na^+ and Cl^-. Here Na^+ being positively charged will be attracted towards negative *i.e.* cathode and Cl⁻ ions being negatively charged will be attracted towards anode. so

$$NaCl \rightarrow Na^+ + Cl^-$$

The liberation of these ions and decomposition is known as the electrolysis. It is used in chemical industries for electroplating, electrotyping etc.

Q. Define the followings: electrolyte, cathode, anode, ion, anion, cation, chemical equivalent, electro-chemical equivalent, atomic weight and valency.

Ans. Electrolyte. It is a liquid conductor or a solution which undergoes the chemical changes when an electric current is passed through it, for example dil. H_2SO_4, solution of $CuSO_4$ and $AgNO_3$ etc.

Cathode. It is a metallic plate or electrode which is immersed in electrolyte and is connected to the negative terminal of the battery or the source of supply.

Anode. It is a metallic plate or electrode which is immersed in the electrolyte and is connected to the positive terminal of the battery or source of supply.

Ions. It is a group of the atoms of the same size or molecules of the decomposed electrolyte. These are the charged particles of a solution.

Cation. The charged particles which are having positive charge, are known as cation.

Anion. The charged particles which are having negative charge are known as anion.

Chemical equivalent. The chemical equivalent of a substance is the ratio of atomic weight to the valency.

$$\text{So chemical equivalent} = \frac{\text{Atomic weight}}{\text{Valency}}$$

Electro-chemical equivalent. The electro chemical equivalent of the substance is the amount of deposits on passing a steady current of one ampere for one second through the solution.

Valency. The valency of an atom or group of atom is the number of hydrogen atoms with which it will react chemically.

9.2. FARADAY'S LAWS OF ELECTROLYSIS

Q. Explain the Faraday's laws of electrolysis.

Ans. Faradays established two well known laws regarding the electrolysis, which are named as the Faraday's laws of electrolysis.

Ist law. The weight of the liberated or deposited ions during electrolysis is directly proportional to the quantity of electricity which is passed through it. So

$$m \propto q$$

where

m = Mass of the liberated ions

q = Quantity of electricity *i.e. It*

I = Current in amperes

t = Time in seconds

$$m \propto It$$

$$= Z. \ It.$$

Z is the electro-chemical equivalent of the substance. If one ampere current is passed for one second then

$$m = Z$$

i.e., the electro-chemical equivalent is defined as the mass of the liberated ions when one ampere current is passed through the electrolyte for one second.

Second law. Whenever the same quantity of electricity is passed through different electrolytes the amount of the different substances liberated is directly proportional to their chemical equivalents

$\therefore$ $$m_1 \propto k_1$$

and $$m_2 \propto k_2 \text{ etc.}$$

Q. What are the main applications of electrolysis, describe any one of them?

Ans. The phenomenon of electrolysis is of great importance and has so many applications. Some of these applications are as under:

(*a*) Electroplating,

(*b*) Extraction and purification of metals,

(*c*) Electrotyping.

(*a*) **Electroplating.** Electroplating is a process by which a metal is deposited on any other metal by passing current in the electrolyte. The basic principle is the electrolysis, *i.e.* the decomposition of the electrolyte used. The material to be deposited is made as anode and where deposited as cathode.

For electroplating the d.c. low voltage is used upto 25 V or so. The d.c. can be taken from the d.c. generator or battery or rectifier. The current rating depends upon the area to be plated. The current is also different for different metal and articles.

The electrolyte is a solution of the metal element which is do be deposited or plated; for example, for silver plating the solution is $AgNO_3$;

for copper plating the solution of $CuSO_4$ and for gold plating K $[Au(CN)_2]$ solution.

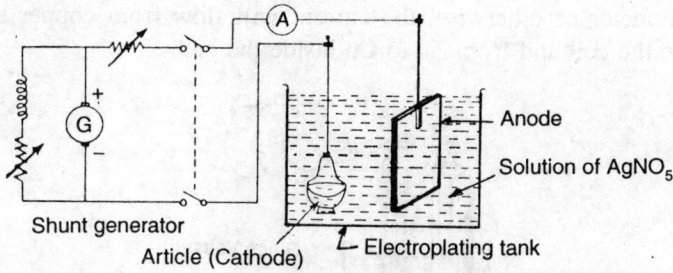

Fig. 9.2. Electroplating.

Process of electroplating. Figure 9.2 shows a simple arrangement of silver plating. The solution is the silver nitrate ($AgNO_3$). There are two electrodes, one Ag plate which is anode and the article to be plated as cathode. A variable resistance is connected in the circuit to control the voltage and to limit the current. As the current is passed $AgNO_3$ decomposes into its ions

$$AgNO_3 \rightarrow Ag^+ + NO_3^-$$

The Ag^+ being positively charged is attracted towards the cathode where the object to be plated is kept. So the Ag particles are collected layerwise forming a uniform layer of silver particles. The anode is the Ag electrode so that the gravity of the solution is maintained.

9.3. CELLS

Q. What is a cell? What are their different types? Explain simple voltaic cell.

Ans. Cell is a device which converts chemical energy into electrical energy.

The cells are of two types:

    (a) Primary cells,
    (b) Secondary cells.

In both the types the electrical energy is covered from chemical energy.

The *primary cell* is that which is able to deliver the current as soon as its components are assembled. It is the cell which cannot be recharged again. The *secondary cell* is the cell which have to be charged before putting into service, more over there are having completely reversible reaction and can be recharged after discharge.

9.3.1. Primary cell (Voltaic cell)

It is a primary cell. It consists of a glass vessel containing dil. H_2SO_4, and two dissimilar metals used as the electrodes. The copper and zinc plates are used. Both the electrodes are immersed in dil. H_2SO_4 solution as shown in Fig. 9.3. If both the electrodes are connected through glavanometer or otherwise, the current will flow from copper to zinc outside the cell and from Zn to Cu inside the cell.

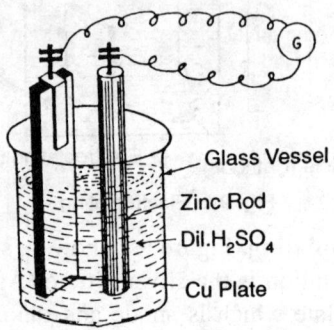

Fig. 9.3. Voltaic cell.

The Zn rod reacts with the dil. H_2SO_4 and forms hydrogen.

$$Zn + H_2SO_4 \rightarrow ZnSO4 + 2H$$

$$2H + 2e \rightarrow H_2$$

The hydrogen being positively charged travels towards the copper plate and after gaining two electrons liberates in the form of atomic hydrogen around the copper plate in the shape of bubbles. Thus the copper plate behaves like the anode and Zn plate as cathode.

Defects. The cell suffers two major defects the local action and polarization.

Local Action. The Zn rod used in this cell is not a pure Zn rod but has some impurities like Sn, Fe, Cu, etc. In that case, even if the load is not connected even than small cells are formed. The hydrogen gives away its charge to the dissimilar metal particles. As a result due to this action there is a wastage of Zn, and life is also decreased. This defect is known as the local action as shown in Fig. 9.4.

This defects can be removed by amalgamating the Zn rod. In this method the mercury layer is deposited over the zinc rod. Thus the mercury will not permit the other impurity to react and will only allow the Zn particles to react with dil. H_2SO_4.

Polarization. The nascent hydrogen after giving its charge to copper and finally gaining two electrons, liberates in the form of atomic hydrogen which is neutral and behaves as an insulator for the other charged particles of hydrogen. It does not allow other hydrogen particles to give its charge to copper plate and hence the current is reduced as shown in Fig. 9.5. The rate of deposition will increase with the time and finally the cell will stop its working. This action is known as the polarization.

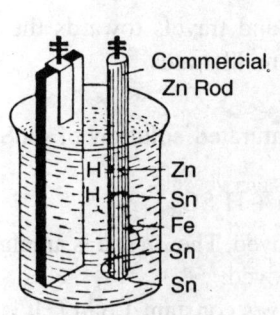

Fig. 9.4. Local Action

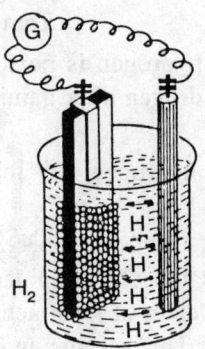

Fig. 9.5. Polarisation

The defect can be removed by two ways:

(a) **By mechanical means.** In this method the copper plate is taken out, brushed and again used.

(b) **By using depolarizer.** In this case a chemical agent is used, which converts the hydrogen into other forms say water etc. The chemical is known as the depolarizer as MnO_2 etc.

Q. Describe the construction and working of Daniel cell.

Ans. Construction. It is a double fluid cell. Figure 9.6 shows the construction of the Daniel cell. It consists of an outer copper vessel, which works as positive pole of the cell. It contains saturated solution of $CuSO_4$ crystals placed in the perforated shelves as shown in Fig. 9.6.

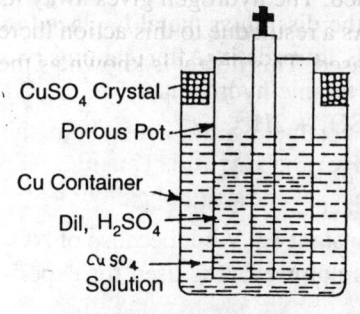

Fig. 9.6. Daniel cell.

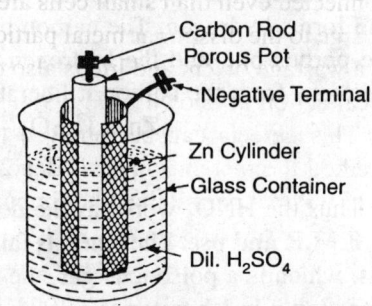

Fig. 9.7. Bunsen cell.

A porous pot is placed inside the vessel. It contains the electrolyte dil.

H_2SO_4 and an amalgamated Zn rod. The Zn rod acts as the negative terminal of the cell.

Working. On completing the circuit, the Zn rod reacts with dil. H_2SO_4 and forms nascent hydrogen.

$$Zn + H_2SO_4 \rightarrow ZnSO_4 + 2H$$

The hydrogen is positively charged and travels towards the copper. This hydrogen after gaining two electrons liberates.

$$2H + 2e \rightarrow H_2$$

This hydrogen will react with the saturated solution of $CuSO_4$ and forms

$$CuSO_4 + H_2 \rightarrow Cu + H_2SO_4$$

Thus the defect of polarization is removed. The Zn rod is amalgamated so the defect of local action is also removed.

Uses. This has an e.m.f. of 1.1 V. It gives constant 1 e.m.f. It is mostly used in laboratories etc.

Note. Sometimes it is also called the Daniel gravity cell. In this cell both the electrolytes are placed in the same container. There is no porous pot but because of the gravity of the liquids, the dil. H_2SO_4 has less gravity so it is on the upper portion and $CuSO_4$ solution in the bottom. The Zn rod reacts with the dil. H_2SO_4 and $CuSO_4$ there works as the depolarizer.

Q. Describe the construction and working of Bunsen's cell.

Ans. It is a double fluid cell. Figure 9.7 shows the construction of the Bunsen cell. The container is made of glass. It has cylindrical shape. The jar contains the electrolyte the dil. H_2SO_4. Here HNO_3, which is filled inside the porous pot works as the depolarizer. The carbon rod works as positive pole.

Working. On completing the circuit, the Zn react with the dil H_2SO_4 and forms hydrogen. The carbon being the dissimilar metal is placed in the porous pot, and the hydrogen travels towards it. After gaining two electrons from the carbon it liberates as atomic hydrogen.

$$Zn + H_2SO_4 \rightarrow ZnSO_4 + 2H$$

$$2H + 2e \rightarrow H_2$$

Thus the HNO_3 works as the depolarizer.

E.M.F. and use. The e.m.f. is fairly constant 1.9 V but because of NO_2 gas, which is a poisonous gas, the cell is not used. It is used for experiments only in laboratories.

Q. Draw the diagrams of the Bichromate cell and Weston standard cell.

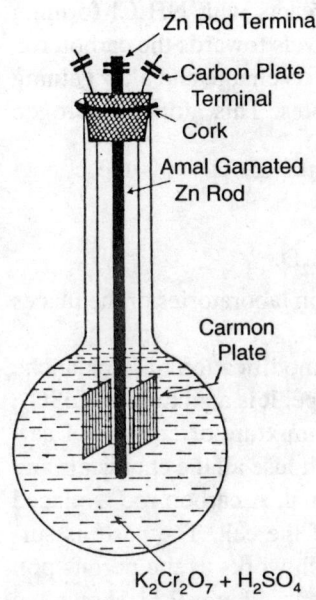

Fig. 9.8. Bichromate cell

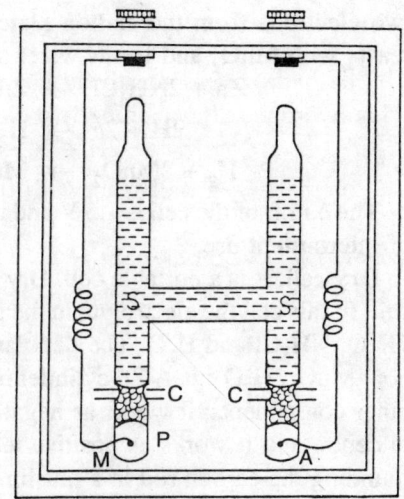

Fig. 9.9. Weston cadmium cell.

M – Pure distilled Hg. (+ve)
A – 12.5% cadmium amalgamated obtained by mixing 1 gm of Cd + 7 gm Hg. (–ve)
P – Obtained by grinding pure mercurus sulphate cadmium sulphate and Hg. in the ratio of 8 : 4 : 1 and dissolved in the solution of $CdSO_4$.
S – Saturated solution of $CdSO_4$ electrolyte.
C – Crystals of $CdSO_4$.

External cover made by double covering of wood and is also perfectly insulated thermally and electrically.

Q. Describe the lachlanche cell and dry cell.

Ans. Lachlanche Cell Construction. It consists of glass vessel. The electrolyte is a solution of NH_4Cl. A amalgamated Zn rod is also used in the container as shown in Fig. 9.10, it works as the negative terminal. A porous pot containing a carbon rod and a mixture of MnO_2 and carbon crystals etc. is placed is the electrolyte. The carbon rod works as the positive plate of the cell. The binding screws are provided over the carbon and Zn rods for connections.

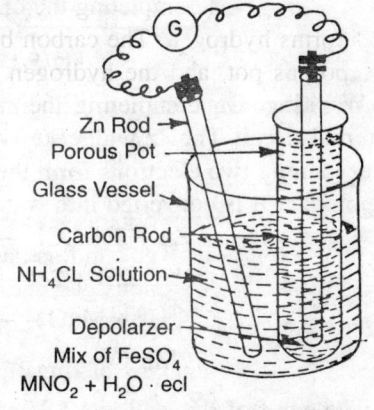

Fig. 9.9. Lachlanche Cell.

Working. On completing the circuit Zn reacts with NH_4Cl forming hydrogen and $ZnCl_2$. These hydrogen ions travels towards the carbon rod through porous pot. The hydrogen gives away its charge and after gaining two electrons from the carbon plate it liberates. This atomic hydrogen reacts with MnO_2 and forms water and Mn_2O_3.

$$Zn + 2NH_4Cl \rightarrow ZnCl_2 + 2NH_3 + 2H$$

$$2H + 2e \rightarrow H_2$$

$$H_2 + 2MnO_2 \rightarrow Mn_2O_3 + H_2O.$$

The e.m.f. of the cell is 1.5 V and it is used in laboratories or the places of intermittent use.

Dry cell. It is a portable cell. Dry cell is a modification of Lachlanche cell. In this cell the electrolyte in the paste shape. It is a mixture of $FeSO_4$, $7H_2O$, NH_4Cl and H_2O. The depolarizer is a mixture of crystals of carbon, MnO_2, H_2O etc. A Zn cylinder is used to house all the chemicals and other components; it works as negative terminal. A carbon rod is placed in centre and it works as positive terminal of the cell. The paste is surrounding the carbon rod in a muslin bag which works as the porous pot. A vent plug is provided to escape the NH_3 gas. Figure 9.11 shows the simple construction of the dry cell.

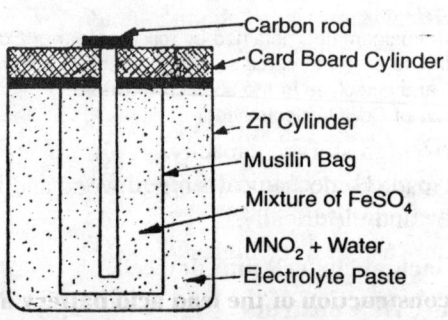

Fig. 9.11. Dry cell.

Working. On completing the circuit the current flows through the external circuit. The Zn will react with NH_4Cl and forms hydrogen; which after gaining two electrons form the carbon liberates and because of the depolarizer it is converted into water.

$$Zn + 2NH_4Cl \rightarrow ZnCl_2 + 2NH_3 + 2H$$

$$2H + 2e \rightarrow H_2$$

$$H_2 + MnO_2 \rightarrow Mn_2O_3 + H_2O$$

$$2Mn_2O_3 + O_2 \rightarrow 4MnO_2$$

The e.m.f of this cell is 1.5 V and has got numerous uses *e.g.* torch, toys, transistors, bells, automatic guns, laboratories and other so many places.

9.3.2. Secondary Cells

Q. What do you understand by the secondary cell? What are the characteristics of the secondary cells?

Ans. A cell in which the chemical and physical states of the electrodes and electrolyte may be restored by charging is called the secondary cell or storage cell. It can otherwise be stated that the secondary cell is one in which the random material can be brought back to original state by passing the current in reverse direction *i.e.*, charging. These cells are of two types:
1. Lead acid cell.
2. Alkaline cell. These are of two types:
   (*a*) NiFe cell.
   (*b*) NiCd cell.

Characteristics. The characteristics of the secondary cells are as follows.
1. Completely reversible reaction – It should be able to convert complete chemical energy into electrical energy in case of discharging and the electrical energy into chemical energy while charging.
2. It should have low internal resistance.
3. It should have higher efficiency.
4. The e.m.f. per cell should be fairly constant.
5. It should have high storage capacity.
6. It should have good mechanical strength.
7. It should have long life.
8. The cost in no case be very high.

Q. Describe the construction of the lead acid battery and explain the chemical reactions during charging and discharging.

Ans. The combination of two or more than two cells suitably connected together is known as a battery. In case of lead acid cell, the cell has got the following parts.

Parts of lead acid battery. The different parts are studied independently:
(*a*) **Container.** It is used to accumulate all the parts of the cell or battery *viz.* plates, separators, electrolyte etc. The container is divided into a number of chambers or compartments equal to the number of cells used for that battery. There is a mud space provided in the bottom where the fallen active material may be collected. The container is made of hard rubber compound, **celluloid and glass** etc. It

is open from the top. After the assembly the cell and top are sealed by a sealing compound.

(b) **Plates.** There are two types of the plates the positive plate and negative plate. The active material of the positive plate is PbO_2 (lead peroxide) and spongy lead for negative plate.

According to the construction the plates are divided into the followings:

   (i) Plante type. (ii) Faure type.

   (i) *Plante type plates.* These are prepared from the pure lead by repeated charge and discharge. The grid of pure lead is immersed in dil. H_2SO_4 and d.c. is passed through them, after several charge and discharge the plate is covered with a layer of PbO_2, *i.e.* positive plate. The other plate is covered with spongy lead and is known as negative plate. This type of plates have comparatively longer life.

   (ii) *Faure type plate.* In this case the active material is filled in the pockets of lead orbits. The paste of the active material is filled in the pockets of a grid which is prepared by the lead antimony alloy. The paste is a mixture of oxides of lead and dil. H_2SO_4. After filling them properly, the plates are put on charge. The plates are prepared, the active material of positive plate is lead per oxide and of negative plate the spongy lead.

   Such plates are heavier then the Plante type plates. The colour of positive plate is of chocolate brown colour and the negative plate is of grey colour.

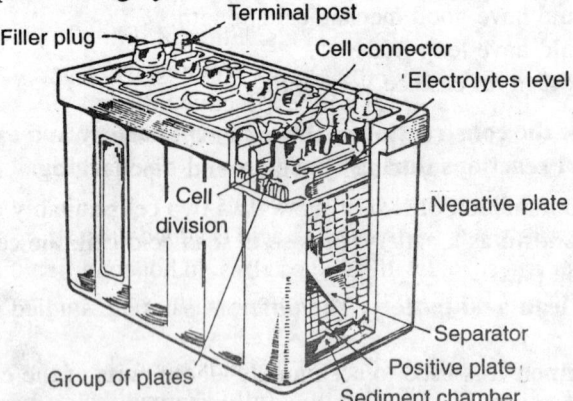

Fig. 9.12. Parts of lead acid battery.

The positive plates are joined at one terminal which is known as positive terminal and the negative plates which another terminal which is known as negative terminal. The batteries are categorised according to the number of plates i.e. 15 plates, 17 plates and 19 plates, etc.

(c) **Separators.** The separators are used between the positive and negative plates, to avoid the short circuit. These are made of perforated wood, perforated rubber, perforated celluloid or the glass wood. It has one side plane and other side grooved to enable the electrolyte to come in contact with the positive plates. The number of separators are always double than the number of positive plates or one less than the total number of plates (positive and negative). These should have adequate mechanical strength.

(d) **Plate connector or terminal.** These are made of lead. All the positive plates are connected to one positive pole. The plates are welded or soldered with the connector. There are two such connectors per cell, one positive and other as negative.

(e) **Cell connector.** These are used to connect the cell in series. These are made of lead and antimony. The plates connector is welded with cell connector. The cell connector must be so thick to withstand the current capacity of the load without heating.

(f) **Cell cover.** A cell cover is used to cover the cell. It is made of moulded hard rubber compound and are sealed with the container to cover the cell. There are two holes for connections and one for filling the electrolyte or distilled water etc.

(g) **Vent plug.** A vent plug is used to cover the filling hole. It contains so many small holes to escape the gas.

(h) **Sealed compound.** It is used to seal the cell cover and also to acid tight joints.

(i) **Electrolyte.** The electrolyte is a solution of dil. H_2SO_4. It is a carrier of current between the positive and negative plates. The active material of the plate will not be effective unless it is covered with electrolyte. The distilled water is added in the H_2SO_4 to make it dilute. The ratio of acid and water depends upon the specific gravity required or as per makers instructions.

While preparing the electrolyte the acid is added drop by drop into the distilled water and stirred with a rod of glass. In hot condition the specific gravity is between 1280-1300 and while cold the specific gravity should be 1250-1280.

The internal resistance of the cell depends upon the electrolyte.

Chemical action. While the cell is fully charged the positive plate is of dark chocolate brown and negative of grey colour. In case of discharging the H_2SO_4 decomposes into its ions the hydrogen and sulphate.

$$\overset{\text{+plate}}{PbO_2} + H_2 + H_2SO_4 \longrightarrow PbSO_4 + 2H_2O$$

$$\overset{-\text{plate}}{Pb} + SO_4 \longrightarrow PbSO_4$$

and when it is recharged

$$PbSO_4 + H_2 \longrightarrow Pb + H_2SO_4$$

and $\qquad PbSO_4 + SO_4 + 2H_2O \longrightarrow PbO_2 + 2H_2SO_4.$

Thus the complete equation can be given in one reversible reaction

$$\overset{+\text{plate}}{PbO_2} + 2H_2SO_4 + \overset{-\text{plate}}{Pb} \underset{\text{charging}}{\overset{\text{discharging}}{\rightleftharpoons}} \overset{+\text{plate}}{PbSO_4} + H_2O + \overset{-\text{plate}}{PbSO}$$

The charging and discharging is concluded in the table form as under:

Table 9.1. Charging and discharging

S.No.	Discharging	Charging
1.	It supplies the load.	It works as the load.
2.	Both the positive and negative plates becomes whitish in colour ($PbSO_4$).	Both $PbSO_4$, changes to PbO_2 for anode and Pb for cathode.
3.	The specific gravity decreases.	The specific gravity increases.
4.	The voltage/cell decreases.	The voltage/cell increases.

Q. What are the differences between the primary and secondary cells?

Ans. The differences between the primary and secondary cells can be tabulated as under:

Table 9.2. Primary and secondary cells

S.No.	Primary cell	Secondary cell
1.	The e.m.f. is induced as soon as its components are assembled.	The e.m.f. and current is available if it is charged.
2.	If discharged cannot be recharged again.	If discharged can be recharged again with its old components.
3.	It cannot work continuously for long time.	It can work continuously for a long time.
4.	These are light in weight.	These are heavy.
5.	The life is short.	It has long life.
6.	It has higher internal resistance.	The internal resistance is comparatively low.
7.	These are Voltaic, Lachlanche, **Dry, Bunsen,** Danial, Bichromate, Standard cell.	These are lead acid cells, NiFe cells, NiCd cells.

Q. How can you judge the condition of the battery by checking the specific gravity of the electrolyte?

Ans. The specific gravity of the cell gives away the idea of the condition of the battery as shown in table 9.3.

It should be clearly understood that the specific gravity is written and spoken as 1280 but actual value is 1.28. The electrolyte is prepared by adding sulphuric acid into distilled water and not *vice-versa*. The indothermic reaction causes a temperature rise rapidly so the stirring is essential and add slowly.

Table 9.3. Specific gravity

S.No.	Specific gravity	Battery condition
1.	1280-1300	Fully charged
2.	1230-1280	About 75% charged
3.	1200-1230	About 50% charged
4.	1170-1200	About 25% charged
5.	1110-1170	Discharged

9.4. EFFICIENCY

Q. Define the efficiency of the secondary cell. How these are classified?

Ans. The efficiency is the ratio of output to input. The efficiency is generally expressed in percentage. The efficiency of the secondary cell are classified as under:

(*a*) **Ampere hour efficiency.** It is defined as the ratio of ampere hour discharging to the ampere hour charging, shown in percentage,

$$\% \text{ ampere hour efficiency} = \frac{\text{Ampere hour discharge}}{\text{Ampere hour charge}} \times 100$$

$$\% \eta_{AH} = \frac{I_d \cdot T_d}{I_c \cdot T_c} \times 1000$$

The ampere hour efficiency of the lead acid cell varies from 90-95%.

(*b*) **Watt hour efficiency.** It is defined as the ratio of output in watt-hours to input in watt-hours shown in percentage. It is also known as the energy or watt-hour efficiency.

So $$\% \eta_{wh} = \frac{\text{Output in watt hours}}{\text{Input in watt hours}} \times 100$$

$$= \frac{V_d \cdot I_d \cdot T_d}{V_c \cdot I_c \cdot T_c} \times 100$$

$$= \frac{V_d}{V_c} \times \% \, \eta_{AH}$$

$$= \frac{\text{Discharging voltage}}{\text{Charging voltage}} \times \% \text{ ampere hour efficiency}$$

Here V_c = Charging voltage in volts

V_d = Discharging voltage in volts

I_c = Charging current in amperes

I_d = Discharging current in amperes

T_c = Time of charging

T_d = Time of discharging.

9.5 CHARGING OF SECONDARY CELLS

Q. Define the following terms used for secondary cells:

    (*a*) **Charging,** (*b*) **charging p.d.** (*c*) **fully charged p.d.,** (*d*) **fully discharged p.d.,** (*e*) **specific gravity,** (*f*) **charging rate,** (*g*) **topping up,** and (*h*) **gassing.**

Ans. (*a*) **Charging.** When a cell is receiving electrical energy from external d.c. source it is said to be the charging.

  (*b*) **Charging p.d.** The potential difference at which the cell is charged is known as charging p.d. The charging p.d. is 2.5 volt to 2.75 volts per cell, for example for charging a 6 volt battery the potential difference should be 7.5 volts.

  (*c*) **Fully charged p.d.** The potential difference of the cell at full charged condition is known as fully charged p.d. It varies from 2.2 V to 2.5 V per cell.

    Discharging. When a cell is delivering power or current to the electrical load it is said to be discharging.

  (*d*) **Fully discharged p.d.** The potential difference at fully discharged condition is called fully discharged potential difference. It is 1.7 V per cell.

  (*e*) **Specific gravity.** The ratio of the weight of any liquid and water of equal volume is called the specific gravity of that liquid. The specific gravity of the electrolyte in full charging condition is 1250 to 1280 and on discharging condition 1150-1200.

  (*f*) **Charging rate.** The rate in amperes at which the cell is charged is called the charging rate. The charging rate depends upon the method of charging and makers instructions. The normal charging rate is 4 to 6 A but new battery is charged at low rate that is known as initial charging.

(g) **Topping up.** To maintain the level of electrolyte up to 10 mm to 20 mm above the plates is called topping up. For recharging a battery the distilled water is added to maintain the level of electrolyte.

(h) **Gassing.** When a cell is put on charge and it reaches to full charged condition gasses comes out of the cell. It is known as gassing. The electrical energy now is not being converted into chemical shape but it decomposes the electrolyte so the gasses are released.

Q. What are the different types of charging according to the charging rate?

Ans. According to the charging rate, the charging is of the following types:

(a) **Initial charging.** It is the first charge given to the new battery after purchasing. In this charge, the battery is charged at a low rate, generally 2 A. While putting on charge the makers instructions and battery conditions must be strictly followed.

(b) **Normal charging.** In this type of charging the battery is charged at normal rate generally 4 to 6A. This charging is suitable for the batteries working in normal conditions and having no defect.

(c) **Boosting charging.** Charging at a high rate is known as boosting charging. The maximum limit of the charge is six times the normal charging rate. It is suitable for the batteries which are normal and are required in short time.

(d) **Trickle charging.** This charging is done at a very low rate, *i.e.* even 1/50th of the normal charging rate. It is applied when the battery has not been used and lying idle.

(e) **Equilizing charging.** This charging used to remove the sulphation from the plates. In this method the low charging rate is used, which is generally $1/4^{th}$ of the normal charging rate.

Q. What are the different methods of charging a battery?

Ans. There are two main methods of charging a battery:

(a) **Constant current method.** In this charging method the batteries are charged at a constant current. The charging current is set by introducing some resistance in the circuit. This method has its own drawbacks because the state of charge of the battery is not taken into account. Initially the charging rate may be high but when the battery is charged up to some extent the charging rate will be less.

(*b*) **Constant voltage method.** In this method the batteries are charged at a constant voltage. The voltage is given to the battery by means of the d.c. shunt generator or rectifier. With this charging method the time of charging is reduced considerably.

Q. How the batteries are charged
   (*a*) **when D.C. is available?**
   (*b*) **when A.C. is available?**

Ans. (*a*) **When DC is available.** The D.C. voltage may be available in two conditions the low voltage and high voltage.

   (*i*) *D.C. low voltage.* In this case of low voltage the battery is connected across the supply through suitable controllings. As in case of cars, trucks etc. the shunt generator in used. In this case the positive of the source is connected with the positive of the battery and negative with the negative of the battery.
   If more than one batteries are to be charged then the batteries are connected in parallel.

   (*ii*) *D.C. high voltage.* If more voltage D.C. is available then the constant current method is used and a battery charging board is used. The number of carbon filament lamps are used to control the current. The lamps are so connected that they are in parallel to each other and in series with the battery as shown in Fig. 9.13. The number of lamps of a given wattage to be connected can be calculated as:

$$N = \frac{IE^2}{(E - Eb) \times W}$$

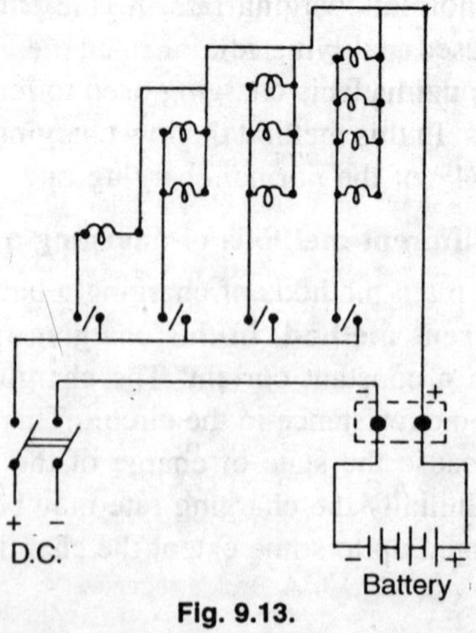

Fig. 9.13.

where N = The number of lamps.

 I = The charging current in ampere.

 E = The supply voltage in volts.

 E_b = The battery voltage in volts.

and W = The wattage of the lamp in watts.

Disadvantage. This method is not useful if the batteries are less in number because the power loss in the lamps will be merely a wastage.

(*b*) **When A.C. is available.** In this method the batteries are charged by the following ways:

(*i*) *By means of M.G. set.* In this set the a.c. motor is coupled with d.c. generator. The motor is energised by a.c. supply and it drives the d.c. generator hence d.c. is produced. The d.c. voltage is used for charging the battery at a constant voltage method. It is costly and requires more maintenance.

(*ii*) *By means of battery charger.* This method is mostly and commonly used nowadays. The battery charger has a step down transformer so a low voltage is obtained and converted into d.c. by means of the rectifier. The rectifier can be the metal rectifier or bulb rectifier.

Metal rectifier. In this method a selenium or copper oxide rectifier is used. The property of the oxide layer is to allow the current in one direction and check in the reverse direction. Figure 9.14 shows the connections of a bridge rectifier used for battery charging.

In bulb rectifier the A.C. is converted into D.C. by means of a tunger bulb rectifier. Figure 9.15 shows a full wave tunger bulb rectifier. There are two anodes and one cathode which emits the electrons. The electrons are attracted towards the positive plate (anode) during one complete cycle. The anodes are alternatively attracts the electrons and full wave is rectified. The voltmeter and ammeter connected indicates the voltage and current for charging the battery.

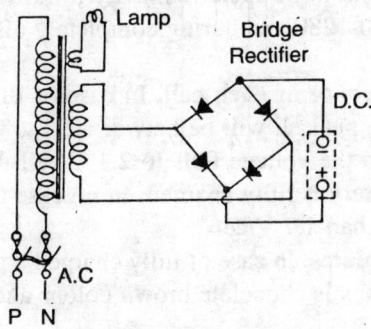

Fig. 9.14. Bridge rectifier.

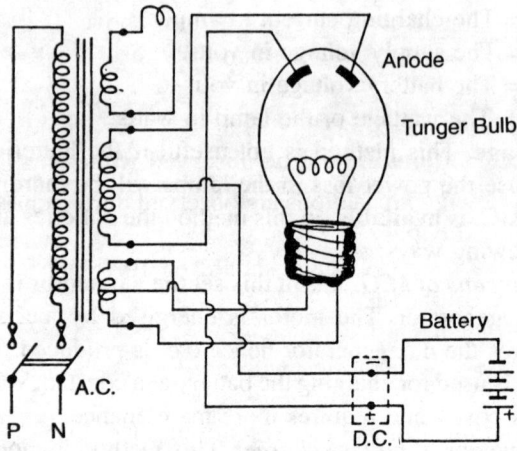

Fig. 9.15. Tunger bulb rectifier.

Advantages of battery charger. There are the following advantages:
1. Power loss is less.
2. It requires less maintenance.
3. Requires less attention during charging.
4. Easy to handle and connections are also easy.

Q. What are the indications of fully charged and fully discharged battery?

Ans. There are the following indications:

(a) **Gassing.** When a battery is put on charge and it is fully charged, the gassing takes place. The gassing is because the current is not used for battery charging but decomposes the water and gases are released.

(b) **Specific gravity.** Measure the specific gravity of the electrolyte three or four times. The specific gravity in fully charged condition is between 1250-1280 and during completely discharged condition 1150-1180.

(c) **Potential difference of each cell.** In case of fully charged condition the voltage per cell will be between 2.2 to 2.5 V per cell. But after sometimes the voltage falls to 2.1 V/cell. When this stage is reached the battery is fully charged. In no case the discharged p.d. should be less than 1.7 V/cell.

(d) **Colour of the plates.** In case of fully charged battery, the colour of the positive plates is chocolate brown colour and of negative plate is grey colour.

(*e*) **Level of electrolyte.** In case of fully charged cell the level of electrolyte should be 10 mm to 20 mm above the plates. But it decreases from the initial level in fully discharged condition.

9.6. DEFECTS, CARE AND MAINTENANCE

Q. What precautions should be observed while charging a battery?

Ans. The following precautions are observed during charging and before charging a battery:

(*i*) The charging voltage should be 10% higher than the battery full charged p.d.

(*ii*) The cell must be charged on D.C. only. The positive of the battery is connected with positive of the source and negative with negative of the source.

(*iii*) The level of the electrolyte must be 10 to 20 mm above the plates. To maintain the level of the electrolyte distilled water is added.

(*iv*) The electrolyte should not contain any impurity.

(*v*) The vent plugs must be opened.

(*vi*) Charge the cells as per rate and instructions supplied.

(*vii*) The connections must be tight, right and cleaned.

(*viii*) The battery charging room must be well ventilated.

(*ix*) Don't bring any naked flame near the place where the battery is being charged because hydrogen gas is flammable.

(*x*) The jars of acid and distilled water must be properly marked.

(*xi*) A battery must be recharged as soon as it is discharged otherwise the sulphation will take place and the battery would not be in a position to be recharged.

Q. What are the main defects and their remedies of the lead acid battery?

Ans. There may be the following main defects in a lead acid battery.

(*a*) **Sulphation.** Formation of the lead sulphate layer on positive and negative plate is known as the sulphation.

Effects. The capacity, life and the efficiency of the cell is decreased.

Reasons. There are the following reasons:

(*i*) Keeping the battery in charging and discharging conditions for a long time.

(*ii*) Higher rate of charging and discharging.

(*iii*) Improper level of electrolyte.

(*iv*) Over discharging condition.

Remedy. The defect can be overcome by the following methods:

(*i*) By charging the battery at a low rate *i.e.* trickle charging for a long time.

(*ii*) If the layer is thick then take out the plate and remove the sulphate layer by mechanical means. If the plates are in working condition then reassemble the battery and charge again.

(*b*) **Buckling.** Deformation of plates is known as buckling.

Effects. Life is reduced.

Reasons. There are the following reasons:

(*i*) High rate charging and discharging.

(*ii*) Short circuit inside or outside the cell.

(*iii*) Over charging and discharging.

Remedy. (*i*) Take out the plate, make them straight and if the plates are in a condition to be used, reassemble and charge.

(*ii*) Change the battery plates.

(*c*) **Sedimentation.** Falling of active material is known as the sedimentation. The active material may be positive plate (PbO_2) or negative plate (Pb).

Effects. Capacity and life is reduced.

Reasons. (*i*) Overheating due to flow of overload current .

(*ii*) Vibrations.

Remedy. Dismantle the battery, take out the fallen material from the mud space, clean the container with distilled water, check the plates, if these are straight, reassemble the battery and charge it.

(*d*) **Internal short circuit.** The plates, positive and negative, are coming in contact with each other without any external resistance, the defect is known as internal short circuit.

Effect.

(*i*) Low voltage.

(*ii*) Low discharged potential difference.

(*iii*) Fully charged cell will be discharged very soon even at low current.

(*iv*) Overheating in idle condition also.

Remedy. Check and clean the plates, separates the mud space and electrolyte and if these are in working condition then assemble and recharge the battery.

(*c*) **Loss of electrolyte.** If the proper level is not maintained then it is said the loss of electrolyte.

Effect. The voltage and capacity of the cell is reduced.

Reason. Crack in container or hole in the container or in cell cover etc.

Remedy. Change the container.

Q. What are the care and maintenance of the lead acid cell?

Ans. There are the following precautions to be observed for the care and maintenance of the lead acid battery:

1. Strictly observe the makers instructions and rate of charging and discharging.
2. Do not keep the battery idle for a long time.
3. For storage, charge the battery, remove the electrolyte, dry the battery and keep it in cool dry place after reassembling.
4. Keep the terminals clean and dry and apply grease or Vaseline.
5. Check the level of electrolyte, it should be at least 10 mm above the top of the plates.
6. Do not fully discharge the battery, send for charging after half discharge. In no case the voltage per cell be less than 1.7 volt/cell.
7. Add distilled water to the battery for recharging.
8. The charging rate should be proper.
9. Do not use high specific gravity electrolyte.
10. Do not short circuit the battery.
11. In mobile vanes the battery must be kept carefully tightened and fixed.
12. The temperature must not exceed 50°C and below the freezing point of water.
13. Keep the plugs tight and vent holes cleaned.
14. Never bring naked flame near the battery while charging the battery and room should be well ventilated.
15. Never charge the lead acid and NiFe batteries together.

9.7 ALKALINE CELL

Q. Describe in brief the construction and working of NiFe cell. What are the advantages and disadvantages of nickel iron cell over the lead acid cell?

Ans. It is a secondary cell. The active material in case of charged and discharged conditions is NiO_2 for positive plate and Fe for negative plates. The electrolyte is a mixture of KOH + H_2O(21% 79%).

Construction. It has the following parts.

(a) **Container.** It is made of nickel plated steel box. The joints are welded, so that electrolyte may not come out. This box is coated with a thick layer of varnish to avoid short circuit in side the cell.

(b) **Plates.** The positive and negative plates are used in the cell. The active material is filled in perforated tubes and flats pockets of nickel plated steel.

(*i*) *Positive plate.* The active material of the positive plates is NiO_2 or its hydrates $Ni(OH)_2$. The mixture is filled in the perforated nickel plated steel tubes, having a length of 11.5 cm and dia 6 mm. A number of these tubes (15, 30 or 45) are assembled in a frame of nickel plated steel.

(*ii*) *Negative plates.* The active material of the negative plate is iron oxide. It is filled in perforated pockets of nickel plated steel. These pockets are assembled in a frame of nickel plated steel. A little amount of mercury is added to improve its conductivity.

All the positive plates are welded together with one terminal and negative plates with other terminal known as negative terminal or negative pole. The number of negative plate are one more than the positive plates. A cell has 9, 13 or 15 plates. The type of the cell is determined by the size of the tubes and number of tubes for positive plate.

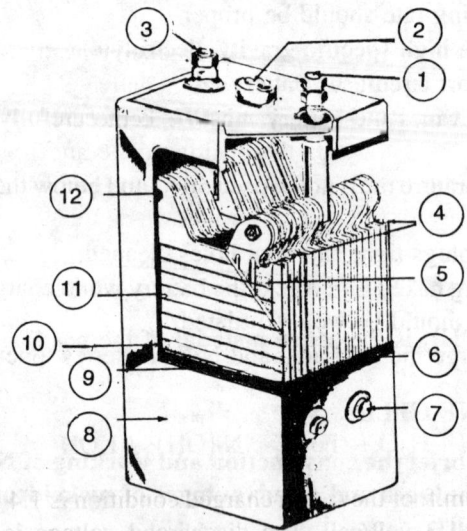

1. Positive terminal 2. Vent 3. Negative terminal 4. Frame and plate separators (polystyrene) 5. Steel negative plate containing cadmium oxide 6. Hard rubber and steel 7. Suspension less 8. Steel Tank top and bottom welded to sides 9. Plate frame (steel) 10. Sided sheet (hard rubber) 11. Positive plate nickel-plated steel containing nickel hydroxide 12. Plate bus bar (steel).

Fig. 9.16. Alkaline battery.

A – type contains two rows of 15 tubes each.

B – type contains one row of 15 tubes each.

C – type contains three rows of 15 tubes each.

D – type contains two rows of 20 tubes each.

The diameter and length of tubes in A B C tubes is 0.25″ and 4.5″ respectively and discharging rate is, 0.25 A per tube for five hours, similarly for D-types plates the diameter 0.1875″ and length 4.5″, the discharging rate is 0.1875 amp for five hours.

  (*c*) **Separators.** To avoid the short circuit between the positive and negative plates, the ebonite separators are used.

  (*d*) **Electrolyte.** The electrolyte is a mixture of 21% KOH and 79% water.

    (*i*) **Specific gravity.** The specific gravity of the electrolyte is 1230 which remains constant in charged and discharged conditions of the cell.

    (*ii*) **Level of Electrolyte.** It should be 15 mm to 75 mm above the surface of the plates or as per makers instructions.

The NiFe battery is of one cell but in some cases if more than one cell are required then these should be joined by means of heavy conductors and proper insulations.

Chemical changes inside the cell. In charged condition, the active material of the positive plate is NiO_2 and negative plate Fe. The electrolyte is a solution of KOH and water. While the cell is discharged the hydrogen ions travels towards the positive plate and forms water molecules. The OH ions travels towards the negative plate and forms Fe_3O_4.

The complete reversible reaction is shown as under

$$\overset{+\text{plate}}{6NiO_2} + 8KOH + H_2O + \overset{+\text{plate}}{3Fe} \underset{\text{charging}}{\overset{\text{discharging}}{\rightleftharpoons}} \overset{-\text{plate}}{2Ni_3O_4} + 8KOH + H_2O + \overset{-\text{plate}}{Fe_3O_4}$$

Note. If $Ni(OH)_4$ is the active material of the positive plate then the reaction is

$$\overset{+\text{plate}}{Ni(OH)_4} + KOH + H_2O + \overset{-\text{plate}}{Fe} \rightleftharpoons \overset{+\text{plate}}{Ni(OH)_2} + KOH + H_2O + \overset{-\text{plate}}{Fe(OH)_2}$$

E.M.F. The e.m.f. of the cell in charged condition is 1.4 volt/cell which drops to 1.2 or 1.3 volt/cell. The discharged voltage is 1 V/cell. The charging voltage is 1.7 volt/cell.

Q. Prepare a table for comparing different cells.

Ans. (See Table 9.4 on page 152).

Q. Compare the lead acid cell and NiFe cell.

Ans. The comparison of the L.A. and NiFe cells is shown in Table 9.5 (see on page 153).

Example. *A battery has an e.m.f. 100 V. It is to be charged with 200 V d.c. mains. Find the number of 100 W lamps to be connected in series if the charging current is 4 A.*

Table 9.4. Cells in a glance

Name of the Cell	Positive Plate	Negative Plate	Electrolyte	Depolarizer	Voltage	Uses
Primary cells						
1. Simple voltaic cell	Cu	Zn	dil H_2SO_4	–	1.1 V	Rarely used, only in laboratories.
2. Lachlanche cell	Carbon	Zn	NH_4Cl	MnO_2	1.5 V	Used for intermittent use and in laboratories.
3. Dry cell	Carbon	Zn	NH_4Cl	MnO_2	1.5 V	General purpose cell, used in torch, transistor lab, instruments etc.
4. Daniel cell or gravity cell	Cu	Zn	dil H_2SO_4	$CuSO_4$ solution	1.4 V	Generally used in laboratories.
5. Bunsen cell	Carbon	Zn	dil. H_2SO_4	HNO_3	1.9 V	Voltage is fairly constant but not in use.
6. Bichromate cell	Carbon	Zn	dil. H_2SO_4	$K_2Cr_2O_7$	2.0 V	Used in laboratorites.
7. Standard cell	Hg	Cd	Cd SO_4	Mercury sulphate	1.01864 V	It is used to compare the e.m.f. of the other cells.
Secondary cells						
1. Lead acid cell	PbO_2	Pb	dil. H_2SO_4	–	2.0 V/cell	It is widely used in automobile, ontrol pannels emergency light, locomotives, etc.
2. NiFe cell	$Ni(OH)_2$ or NiO_2	Iron oxide	KOH solution	–	1.2 V/cell	-do-

Solution. The d.c. main voltage,

$$V = 200 \text{ V}$$

The battery voltage $E_b = 100 \text{ V}$

The charging current $I = 4 \text{ A}$

and wattage of the lamp $= 100 \text{ W}$

Now the formula to be used, while the battery is charged and lamps are required to be connected in series.

$$N = \frac{I \cdot E^2}{(E - E_b)W}$$

So $= \dfrac{4 \times 200^2}{(200 - 100) \times 100}$

$$= \frac{4 \times 200 \times 200}{100 \times 10} = 16 \text{ lamps. Ans.}$$

Example. *A battery has an e.m.f. of 150 V with internal resistance 2.5 ohm. It is to be charged on 200 V d.c. mains. Find the resistance to be added in series to have a charging current.*

5A. *Also calculate the amount of energy consumed, if the cost of energy is Rs. 4.50 per unit and battery takes hours to charge.*

Sol. The main voltage is 200 V d.c.

The battery voltage is 150 V

The internal resistance $= 2.5 \ \Omega$

and charging current $= 5A$

Now the voltage to be dropped

$$= 200 - 150 = 50 \text{ V}$$

The resistance required $= \dfrac{V}{I} = \dfrac{50}{5} = 10 \Omega$

But the internal resistance of the cell is $2.5 \ \Omega$

So the resistance to be added

$$= 10 - 2.5 = 7.5 \ \Omega. \text{ Ans.}$$

Now the energy consumption

$$= V.I.T.$$

$$= 200 \times 5 \times 5$$

$$= 5000 \text{ W or 5 units}$$

The amount of energy consumption @ Rs. 4.50/unit

$$= 5 \times 4.5 = \textbf{Rs. 22.5 Ans.}$$

Table 9.5. Comparison between lead acid and nickel iron cells

Particulars	L.A. cell	NiFe cell
1. Container	Made of hard rubber compound	Made of nickel plated steel
2. Positive plate	Lead per oxide, PbO_2	Nickel hydride $Ni(OH)_2$ or NiO_2
3. Negative plate	Spongy lead, Pb	Iron oxide
4. Electrolyte	Dil. H_2SO_4	KOH-solution
5. Specific gravity		
(a) charged	1280	1230
(b) discharged	1180	1230
6. Average e.m.f.	2 volt/cell	1.2 volt/cell
7. Internal resistance	Low	High
8. Cost	Less than NiFe cell	High than L.A. cell
9. Maintenance cost	High	Low
10. Life	Can be charged and discharged about 1200 times	At least five years
11. Efficiency		
(a) Ampere hour	90-95%	80%
(b) Watt hour	72-80%	60%
12. Care and maintenance	Needs much care and maintenance	Robust, strong, and can stand heavy charge and discharging currents free from corrosive liquids and fumes
13. Weight	Heavier	Lighter

Example. *A 12V discharged storage cell is charged at 5A for 3.5 hours. If it is discharged through a resistance of R Ohm, the duration of discharge is 12 hours and the terminal voltage remains constant. The ampere hour efficiency of the cell is 85%. Calculate the value of resistance R.*

Solution. The charging current is $I_c = 5$ A

The charging time is 3.5 hours

and voltage is 12 V

Time taken for discharge is 12 hours, the $\eta AH = 85\%$

Now the ampere hour efficiency

$$= \frac{\text{Total amp hour output}}{\text{Total A-h input}}$$

$$\frac{85}{100} = \frac{I_d \times T_d}{I_c \times T_c}$$

$$\therefore \qquad I_d = \frac{85 \, I_c \cdot T_c}{T_d \times 100}$$

$$= \frac{85 \times 5 \times 3.5}{12 \times 100} = 1.24 \text{ A}$$

Now the voltage is 12 V and current is 1.24 A

$$\text{Resistance} = \frac{\text{Voltage at discharge}}{\text{Discharging current}}$$

$$= \frac{12}{1.24} = 9.676 \, \Omega \text{ Ans.}$$

Example. *A nickel iron alkaline cell is discharged at a steady current of 5 A for 12 hrs. The average terminal voltage being 1.2 V. To restore it to its original stage of charge a steady current of 3.8 A is required for 20 hrs. The average voltage is 1.44 V. Calculate the ampere hour and wattful efficiencies.*

Solution. The A.H. efficiency

$$= \frac{\text{Total output A-h}}{\text{Total input A-h}}$$

The ampere hour output

$$= \text{Current} \times \text{Time in hours}$$
$$= 5 \times 12 = 60 \text{ A-hr}$$

The ampere hour input $= \text{Current} \times \text{Time in hours}$

$$= 3.8 \times 20 \text{ A-hr}$$

$$\%\eta_{AH} = \frac{\text{Output}}{\text{Input}}$$

$$= \frac{5 \times 12}{3.8 \times 20} \times 100 = \textbf{78.95\% Ans.}$$

Now watt hour efficiency $= \dfrac{\text{Watt hour output}}{\text{Watt hour input}}$

$$= \eta_{AH} \times \frac{\text{Output voltage}}{\text{Input voltage}}$$

$$= 78.95 \times \frac{1.2}{1.44} = 65.79\% \text{ Ans.}$$

Example. *A 6 V 124 A-h, lead acid battery is to be charged from 220 V d.c. mains, at a rate of 8 A. Find out the number of 150W lamps to be connected in the circuit.*

Solution.
$$N = \frac{I \times E^2}{(E - E_b)W}$$

$$= \frac{8 \times 220 \times 220}{(220 - 6) \times 150} = \textbf{12 lamps. Ans.}$$

Example. *A rectangular disc of 15 cm long and 12 cm wide is to be plated with nickel coating of 0.1 mm thick. If a steady current of 8 A is passed through the electrolyte, how long must the disc be immersed. Density of nickel 8.9 gm/cc and E.C.E. is 0.000304 gm/coul.*

Solution. The area to be coated is the area of the rectangular disc. Let the disc be a lamina, having negligible thickness, so the area to be coated.

$$= 2 \times \text{length} \times \text{breadth}$$

$$= 2 \times 15 \times 12 = 360 \text{ cm}^2$$

The thickness of coating
$$= 0.01 \text{ cm}$$

∴   Volume required $= 360 \times 0.01 = 3.6 \text{ cm}^3$

Now the mass of the liberated ions = density × volume
$$= 8.9 \times 3.6 = 32.04 \text{ gm}$$

Obviously $\qquad m = Z.i.t$

or $\qquad t = \dfrac{m}{Zi} = \dfrac{32.04}{0.000304 \times 8} = \textbf{13174.34 sec}$

$$= \textbf{3 hrs 39 min 34.3 sec. Ans.}$$

9.8. GROUPING OF THE CELLS

Q. Define and explain the grouping of the cells.

Ans. The cells can be grouped into the following ways:

  (*a*) Series combination
  (*b*) Parallel combination

(c) Series-parallel combination

(a) **Series combination.** The combination of cells in which the cells forms a chain shape is known as the series combination. In this system the negative of one cell is connected with the positive of the second cell and so on as shown in Fig. 9.17. This type of the connections are used when the voltage required is more than the voltage of a cell. So to obtain more voltage the cells are grouped in series. The combination has got the following characteristics:

(i) The voltage of the circuit increases as the number of cells connected in series are increased.

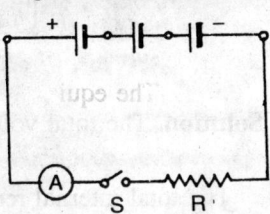

Fig. 9.17. Series combination.

Let e = the voltage per cell

r = the internal resistance per cell

E = voltage of the combination

R = the external resistance.

Let there are n cells connected in series, so the e.m.f. will be

$$= e_1 + e_2 + e_3 + \ldots e_n$$

There are n cells of e volt e.m.f. per cell

So $\quad E = ne$ V $\quad (e_1 = e_2 = e_3 \ldots = e_n)$

(ii) The voltage of the combination is equal to the e.m.f per cell multiplied by the number of cell connected in series.

(iii) The total internal resistance increases as more number of cells are added.

The total internal resistance $= r_1 + r_2 + r_3 + \ldots r_n$

$\quad$ = No. of cells connected × Internal resistance per cell

$\quad = n \times r$ ohm.

(iv) The total current of the combination is same as there is only one path of the current and

$$= \frac{\text{Total voltage}}{\text{Total resistance}}$$

$$= \frac{\text{No. of cells} \times \text{e.m.f. per cell}}{\text{External resistance} + \text{Total internal resistance}}$$

$$I = \frac{ne}{R + nr} \text{ amp.}$$

Example. *Twenty cells each of 1.5 V e.m.f. and internal resistance per cell 0.3 ohm are connected in series. Calculate the current through the* *external resistance of 20 ohm and also find out the potential difference across it.*

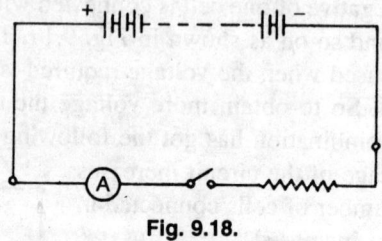

Fig. 9.18.

Solution. The total voltage obtained

$$= ne = 20 \times 1.5 = 30 \text{ V}$$

The total internal resistance

$$= nr = 20 \times 0.3 = 6 \ \Omega$$

Now total resistance $= R + nr = 20 + 6 = 26 \ \Omega$

The current, $I \qquad = \dfrac{ne}{R + nr}$

$$= \dfrac{30}{26} = \textbf{1.154 A. Ans.}$$

The potential difference

$$= I \times R = 1.154 \times 20$$

$$= 23.08 \simeq 23 \text{ V. Ans.}$$

(b) **Parallel combination.** The system in which the positive of each cell are connected at one terminal and negative on the other terminal, is known as parallel combi-nation as shown in Fig. 9.19. This system is used when more current is required.

It has the following character-istics:

(i) The voltage of the combi-nation is the same as that of the voltage per cell.

(ii) The equivalent internal re-sistance being connected

in parallel will be $\dfrac{r}{n} \Omega$

(iii) The total current is equal to the sum of the current delivered by each cell.

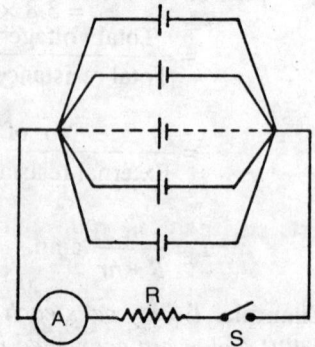

Fig. 9.19. Parallel combination.

$$I = i_1 + i_2 + i_3 + \dots + i_n \text{ amp}$$

$$I = \frac{\text{Net e.m.f. of the combination}}{\text{Total (internal and external) resistance}}$$

$$I = \frac{e}{R + \dfrac{r}{n}} \text{ amp} \qquad I = \frac{ne}{nR + r} \text{ amp}$$

Example. *Twenty cells, each of 1.5 V e.m.f and 0.3 ohm internal resistance are connected in parallel. Find the current through an external resistance of 2 ohm.*

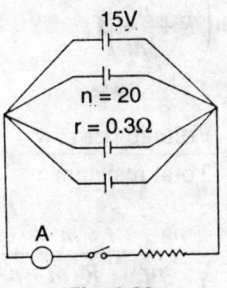

Fig. 9.20.

Solution. The e.m.f. cell

$$e = 1.5 \text{ V}$$

The total internal resistance $= \dfrac{r}{n} = \dfrac{0.3}{2} = 0.015 \, \Omega$

The total resistance $= R + \dfrac{r}{n} = 2 + 0.015 = 2.015 \, \Omega$

Now current $= \dfrac{e}{R + \dfrac{r}{n}} = \dfrac{1.5}{2 + 0.015}$

$$= 0.744 \text{ A. Ans.}$$

(c) **Series-parallel combination.** The combination is used when the voltage and current are required more than the values per cell. In this combination the cells are connected in both series and parallel as shown in Fig. 9.21. This system is preferred when both series and parallel combinations properties are required.

Let the combination has

$$n = \text{cells per row}$$
$$m = \text{number of rows}$$

(*i*) The voltage per row
$$= ne \text{ V}$$

(*ii*) The internal resistance per row
$$= nr \ \Omega$$

and the internal resistance of the combination
$$= \frac{nr}{m} \Omega$$

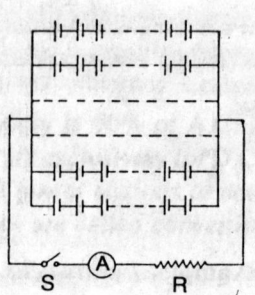

Fig. 9.21. Series-parallel combination.

(*iii*) The equivalent resistance
$$= \left(R + \frac{nr}{m} \right) \Omega$$

(*iv*) The total current
$$= \frac{\text{Voltage per row}}{\text{Total resistance}}$$

$$I = \frac{ne}{R + \dfrac{nr}{m}} = \frac{m \cdot n \cdot e}{R \cdot m + r \cdot n} \text{ amp}$$

The maximum current can be obtained if the denominator is minimum.

$$R = \frac{nr}{m}$$

or   The external resistance = Internal resistance of the combination.
But it is not used because of more power wastage.

Example. *Twenty cells of 1.5 V and 0.3 ohm internal resistance are connected in five rows, four cells per row. Calculate the current through the external resistance of 10 Ω.*

Solution The voltage per row
$$= n \times e = 5 \times 1.5 = 7.5 \text{ V}$$

The internal resistance per row
$$= nr = 5 \times 0.3 = 1.5 \ \Omega$$

The internal resistance of the battery
$$= \frac{nr}{m} = \frac{5 \times 0.3}{4} = 0.375 \Omega$$

The total resistance $= 10 + 0.375 = 10.375 \ \Omega$

The current $= \dfrac{7.5}{10.375} = \mathbf{0.723\,A. \ Ans.}$

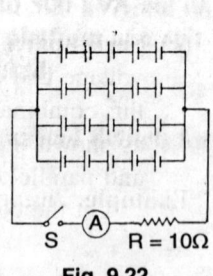

Fig. 9.22.

Example. *Find the smallest number of cells required to operate an electrical apparatus having 40 Ω resistance and current rating one amp. The voltage per cell is 2 V and internal resistance 2 ohm per cell.*

Solution. It is very much obvious that the combination is a series parallel combination. So for maximum current, the internal resistance must be equal to external resistance.

$\therefore$ $nr = mR$

$\therefore$ $2n = 40m (\because r = 2\ \Omega, R = 40\ \Omega)$

$\therefore$ $n = 20\ m$

The current as given is one amp. So

$$I = \frac{m \cdot n \cdot r}{nr + mR}$$

Substituting for n in this

$$1 = \frac{m \times 20m \times 2.0}{20m \times 2 + m \times 40}$$

or

$$= \frac{40m^2}{40m + 40m} = \frac{40m^2}{80m}$$

$$I = \frac{m}{2}$$

or $m = 2$

and $n = 20\ m = 20 \times 2 = 40$

So there are two rows, each row has 40 cells.

So total cells are **80. Ans.**

Example. *Calculate the amount of a current in a resistance of 5 ohm, if 8 cells each of 1.5 V and 0.5 ohm internal resistance are connected in (i) Series (ii) Parallel and (iii) Series parallel of two rows.*

Solution. (i) When the cells are connected in series the current

$$I = \frac{ne}{R + nr}$$

where n = number of cells in a row

              e = e.m.f. per cell in volts

              r = internal resistance in ohm

              R = external resistance in ohm

$\therefore$ $I = \dfrac{8 \times 1.5}{5 + 8 \times 0.5} = 1.33$ **A. Ans.**

(*ii*) When the cells are grouped in parallel

$$I = \frac{e}{R + \dfrac{r}{n}} \text{ amp}$$

$$= \frac{ne}{nR + r} = \frac{8 \times 1.5}{5 \times 8 + 0.5}$$

$$= 0.296 \approx \mathbf{0.3\ A. Ans.}$$

(*c*) When the cells are grouped in series and parallel
$m = 2$, $n = 4$, $e = 1.5$ V, $r = 0.5\ \Omega$ $R = 5\ \Omega$

$$I = \frac{m \cdot n \cdot r}{Rm + nr} = \frac{2 \times 4 \times 1.5}{5 \times 2 + 4 \times 0.5}$$

$$= \frac{12}{10 + 2} = \mathbf{1A.\ Ans.}$$

Q. Draw a sketch of hydrometer. How it is used to check the specific gravity of the electrolyte?

Ans. It is used to measure the specific gravity of the electrolyte. Fig. 9.23 shows a view of the hydrometer and the method of use for checking the specific gravity of the electrolyte.

Note: For gravity details of L-A Cell see Page 165, Table 9.3.

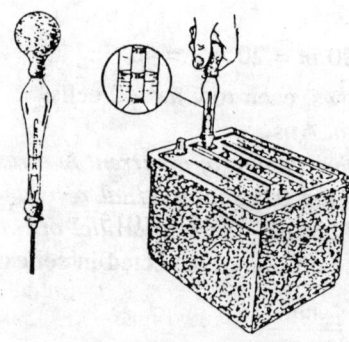

Fig. 9.23. Hydrometer.

Objective Type Questions

Electrolysis

1. The process by which the electrolyte decomposes into its constituents is known as
   (*a*) Electromagnetism   (*b*) Electrolysis
   (*c*) Filtration   (*d*) Sedimentation

2. The weight of the liberated ions during electrolysis is directly proportional to the quantity of electricity, is the .... law of Electrolysis.
   (*i*) First,   (*ii*) Second   (*iii*) Third   (*iv*) None.

2. If the same quantity of Electricity is passed through different electrolytes, the mass of thus liberated ions is proportional to their chemical equivalents. It is the .... law of Electrolysis.
   (*i*) First   (*ii*) Second   (*iii*) third   (*iv*) None

3. According to the first law of electrolysis
   (*a*) $m \propto It$ (*b*) $m \propto \dfrac{I}{t}$ (*c*) $m \propto \dfrac{t}{I}$ (*d*) $m \propto \dfrac{1}{It}$

4. If one ampere current is passed through a solution of $AgNo_3$ the weight of the liberated Ag-ions will be equal to
   (*a*) 0.001118 gm/sec   (*b*) 0.01118 gm/sec
   (*c*) 0.1118 gm/s   (*d*) 10118 gm/sec.

5. Name the type of cell which starts delivering current as soon as its components are assembled.
   (*a*) Primary cell   (*b*) Secondary cell
   (*c*) Plant cell   (*d*) None of the above.

Cells

1. A cell is a Chemical device which converts the Chemical energy into .... energy.
   (*a*) Electrical   (*b*) mechanical
   (*c*) magnetic   (*d*) sound.

2. How many electrodes does a cell have?
   (*a*) Two   (*b*) one
   (*c*) four   (*d*) three

3. Name the cell having container of Zn; having Dil H_2So_4 and Zn & Cu as electrodes.
   (*a*) Bensen,   (*b*) Lachlanche
   (*c*) Voltaic   (*d*) dry

4. Collection of hydrogen particles around the positive electrode of the cell is called.
   (*a*) polarisation   (*b*) amalgamation
   (*c*) decomposition   (*d*) neutralisation

5. The simple voltaic cell has two main defects these are local action and ....
   (*a*) ionisation   (*b*) polarisation
   (*c*) amalgamation   (*d*) neutralization

6. The cells are connected in series to
   (*a*) increase the current output   (*b*) decrease the current output.
   (*c*) increase the voltage output   (*d*) decrease the voltage output.

7. While connected cells in series the total internal resistance will
   (*a*) Increase   (*b*) decrease
   (*c*) remains same   (*d*) Not applicable

8. Cells are connected in parallel to increase–
   (a) the current output of the combination.
   (b) the voltage output of the combination.
   (c) the internal resistance of the system.
   (d) None of the above.
9. If the voltage and current of the combination is to be increased which combination is preferred
   (a) Series, (b) parallel, (c) series-parallel, (d) none
10. The total equivalent internal resistance of the Combination in parallel will be .... mode cells are included.
    (a) same, (b) increased, (d) decreased, (d) unaffected.

Electroplating

Q. Calculate the electrochemical equivalent of silver if a current of 2A deposits a layer of 2.01214 gm. in 15 min. over an article during electroplating.

Sol.:   Mass of deposit i.e., $m = 2.0124$ gm

Time = 15 min. or $15 \times 60 = 900$ sec.

Current = 2 Amp

Obviously $m = Zit$

or $$Z = \frac{m.}{it.} = \frac{2.0124}{2 \times 900}$$

$$= 0.001118 \text{ gm/sec.} \quad \textbf{Ans.}$$

Q. A rectangular disc of 15 cm long and 12 cm wide is to be plated completely nickel plated with a coating of 0.01 cm thick. If a steady current of 8A is passed through the electrolyte, how long must the disc remain immersed? Density of nickel 8.9 gm/cc and ECE is 0.0000304 gm/cc.

Sol.: The plate is to be plated completely that is the plated area will be twice.

$= 2 \times$ one side area.

The plate coating is L = 15 cm, B = 12 cm and thickness is 0.01 cm.

∴   Area of plate $= 15 \times 12$ cm^2
and total area plated $= 15 \times 12 \times 2$ cm^2
      Mass $=$ Volume $\times$ Density
∴   Volume $=$ area of plating $\times$ thickness
                         $= 15 \times 12 \times 2 \times 0.01$ cm^3
and mass $= 15 \times 12 \times 2 \times 0.01 \times 8.9$
                         $= 32.04$ gm

According to Faraday's Laws of electrolysis

$$m = Zit.$$

$$32.04 = 0.000304 \times 8 \times t$$

$$\therefore \quad t = \frac{32.04}{0.000304 \times 8} \text{ Sec.}$$

$$= \frac{32.04}{0.000304 \times 8 \times 3600} \text{ Hrs.}$$

= 3 Hours 39 min and 34.34 sec. Ans.

Q. It is required to electroplate a copper sheet of 20 × 20 cm. with silver coating of 0.1 mm on both side of it. If a current of 10 A is used how long will it take to electroplate (ECE of silver is 0.001118 gm/sec and density of silver is 10.5 gm/cc).

Sol. According to the Faradays Laws of electrolysis

$$m = Zit$$

Here as given $Z = 0.001118$ gm/cc. $I = 10$ A; and time = ?

Now for calculating, mass of silver the volume of silver to be deposited is calculated and is multiplied by the density of silver.

$$\text{Volume of silver} = \text{Area of plating} \times \text{thickness}$$

$$= \text{length} \times \text{breadth} \times \text{thickness}$$

$$= 20 \times 20 \times 0.01$$

$$\therefore \text{ mass} = VD = 20 \times 20 \times 0.01 \times 10.5$$

As the plating is to be coated on both sides so

$$\text{mass actual} = (20 \times 20 \times 0.01 \times 105 \times 2) \text{ gm}$$

$$m = Zit.$$

$$t = \frac{m}{Zt} = \frac{20 \times 20 \times 0.01 \times 10.5 \times 2}{0.001118 \times 10}$$

$$= 7513.4 \text{ sec or}$$

= 2 hrs. 5 min. 13.4 sec. Ans.

10

Electrostatic and Capacitance

10.1. ELECTROSTATIC

Q. Define the followings:

(*a*) **Electric field,** (*b*) **electrostatic induction,** (*c*) **laws of electrostatics,** (*d*) **electric flux,** (*e*) **electric flux density and** (*f*) **permittivity.**

Ans. (*a*) **Electric field.** The space around a charge in which an electric charge experience a mechanical force and stress is called an electric field. The direction of the field is that of the force on a positive charge placed in that field.

(*b*) **Electrostatic induction.** It is observed that whenever an uncharged body is placed near an electrostatically charged body, it acquire some charge, this phenomenon of acquiring the charge is defined as the electrostatic induction. The positive charge induces a negative charge and *vice versa*. The important point to note is that the charge acquired or induced is equal to the inducing charge.

(*c*) **Laws of electrostatics.** There are these following laws:

(*i*) *First law.* Like and similar charges of electricity repel each other, whereas the dissimilar charges attracts each other.

(*ii*) *Second law.* According to this law the force exerted between two charges is directly proportional to the product of their strength and inversely proportional to the square of the distance between them. It is also inversely proportional to the absolute permittivity of the surrounding medium.

This law is known as the "Coulomb's law".

Let F : the force between two charges in Newton

Q_1, Q_2 : the charges in Coulombs

d : the distance between them in metre

$\epsilon = \epsilon_0 \epsilon_r$: the absolute permittivity

ϵ_0 : permittivity in free space or air

ϵ_r : relative permittivity in farad/metre.

Now the force exerted can be, according the laws of electrostatics.

$$F \propto Q_1 Q_2$$

$$\propto \frac{1}{d^2}$$

$$\propto \frac{1}{\epsilon} \text{ or } \frac{1}{\epsilon_0 \epsilon_r}$$

$$\therefore \qquad F = K \cdot \frac{Q_1 \cdot Q_2}{\epsilon_0 \epsilon_r \cdot d^2}$$

where K = the constant and is $\dfrac{1}{4\pi}$

$$\therefore \qquad F = K \cdot \frac{Q_1 \cdot Q_2}{4\pi \cdot \epsilon_0 \cdot \epsilon_r \cdot d^2} \text{Nw}$$

So $\qquad F = 9 \times 10^9 \times \dfrac{Q_1 \cdot Q_2}{\epsilon_r \cdot d^2} \text{Nw}$ in any medium

and $\qquad F = 9 \times 10^9 \dfrac{Q_1 \cdot Q_2}{d^2} \text{Nw}$ is air

Now let $\qquad Q_1 = Q_2 = Q, d = 1$ m, $F = 9 \times 10^9$ Nw

then $\qquad 1 = Q^2$

or $\qquad Q = \pm 1$.

Thus the unit charge can be defined as the charge which when placed in air at a distance of one metre from an equal charge (similar or dissimilar) repel or attracts it with a force of 9×10^9 Newton force.

(d) **Electric flux.** The electric flux can be represented by lines or sometimes called tubes of force. It may be defined as the region or space enclosed within the tubular surface formed by drawing lines of force through every point of a small closed curve in the electric field. The unit tube of flux is known as the Faradays Tube. The electric flux is shown by Ψ. Since the electric flux is equal to the charge

$$\Psi = Q \text{ Coulomb}$$

(e) **Electric flux density.** It is defined as the number of lines of force or electric flux per unit area emerging normally from the surface. It's unit is Coulombs per metre square and represented by the letter *D*.

$$\therefore \quad D = \frac{\psi}{4 \cdot \pi \cdot r^2} = \frac{Q}{4\pi r^2} \text{ Coulombs / m}^2$$

It is related with the electric field intensity in this manner
$$D = \epsilon_0 \cdot \epsilon_r E. \quad \text{in any medium}$$
$$= \epsilon \cdot E \quad \text{in air or free space.}$$

(f) **Permittivity.** The ratio of electric field density and electric field intensity is known as the permittivity of free space or the electric space constant. It is denoted by letter ϵ_0 and is given by the value 9×10^9 Farad/metre.

The relative permittivity or the dielectric constant ϵ_r is like μ_r and has got different values for different mediums. It is unity in air or vacuum, 2-3 paper, 3-7 for mica and 5 to 10 for glass etc.

10.2. CAPACITOR OR CONDENSER

Q. (a) **Define a condenser and how it works?**
    (b) **Define coulomb, charging of capacitor, dielectric strength, discharging of capacitor, capacitance, and farad.**

Ans. (a) A device which stores the electricity in the shape of electrostatic charge is known as the condenser or capacitor.

Any two conductors between which an electric field can be maintained, forms a capacitor or condenser.

Let there be two metallic plates *A* and *B*, separated from each other by means of air. Both plates are very near to each other and are connected to a battery as shown in Fig. 10.1. Plate *A* is connected to the positive terminal of the battery and *B* plate with negative terminal of the battery. On closing the switch a current starts flowing through the battery from positive to negative terminal. Thus there is a flow of electrons from positive plate *A* to *B* plate via battery, but there is no current through the air between plates *A* and *B*. The direction of flow of electron is shown in Fig. 9.1. Now both the plates are charged and an electric field will be set up between these two plates. Thus

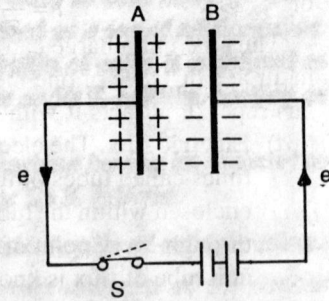

Fig. 10.1. Simple condenser or capacitor.

the number of charges on plate A are equal to the number of charge on B-plate.

Now when switch is put off the charge will be static on the plate and an electric field will continue to establish between the plates. Thus a capacitor is formed.

(b) **Coulomb.** It is the unit of quantity of electricity. The charge of a capacitor is measured in coulomb. A capacitor has a charge of one coulomb when one ampere of current flows for one second through the capacitor

One coulomb = One ampere   sec

The small unit of charge is micro coulomb

∴ $1\mu c = 10^{-6}$ coulomb.

The charge is proportional to the voltage applied and the capacity of the capacitor.

$$Q \, \alpha \, V \text{ and } Q \, \alpha \, C$$

∴ $$Q = CV \text{ farad.}$$

Charging of a capacitor. Whenever a condenser or capacitor takes current from a source of supply, it is known as charging. In this process the capacitor acquire charge and stores electricity.

Dielectric constant. It is also known as the permittivity or specific inductive capacitance. The dielectric constant of a material is the ratio of the capacitance of a capacitor with the given material as dielectric to the capacitance of the same capacitor with air as dielectric.

So dielectric constant,

$$K = \frac{\text{Capacitance of a capacitor with given material as dielectric}}{\text{Capacitance of the same capacitor air as dielectric}}$$

It is denoted by letter K

Discharging of a capacitor. Whenever a charged capacitor is connected across the resistance or load, it delivers the stored energy to the resistance or load, thus electrons flow from negative to positive plate, so when the potential of both the plates becomes equal and zero, it is said to be discharged. In this condition there will not be any charge.

Capacitance. It is the ratio of the charge and voltage of a capacitor. It is represented as

$$C = \frac{Q}{V} \text{ farad.}$$

The unit of the capacitance is farad. The other units are micro and pico farad

$$1 \, \mu F = 10^{-6} \, F$$

$$1 \text{ mmF} = 10^{-12} \text{ F}$$

$\mu\mu$F is called pico farad.

It is defined as the property of a capacitor to store electricity.

Farad. It is the unit of the capacity of a capacitor. The capacity is said to be one farad when one coulomb of charge is required to maintain a potential difference of one volt

$$\text{One farad} = \frac{\text{one coulomb}}{\text{one volt}}.$$

Factors effecting the capacitance of a capacitor

There are the following factors which actually effects the capacitance of a capacitor.

(a) **Area of the plates.** The capacitance is directly proportional to the area of the plates i.e.

$$C \propto a$$

(b) **Distance between plates.** The capacitance is inversely proportional to the distance between the plates. In other words it is clear that this distance is nothing but the thickness of the dielectric used i.e.

$$C \propto \frac{1}{d}$$

(c) **Dielectric constant.** The capacitance is directly propostional to the dielectric constant of the dielectric used i.e.

$$C \propto \in$$

Here this $\in$-permittivity i.e. dielectric constant is equal to $\in = \in_0 \in_r$, where $\in_0$ is the permittivity in vacuum or free space and $\in_r$ is the relative permittivity of the dielectric medium.

$$\therefore \qquad C = \frac{\in_0 \in_r \cdot A}{d}$$

If it is a multiplate capacitor it has to be multiplied by the actual capacitor formed. Let there be n plates than $(n-1)$ are the number of capacitor actually formed so the capacity of a multiplate capacitor is

$$\therefore \qquad C = \frac{\in_0 \in_r \cdot A (n-1)}{d} \text{ Farads}$$

$$\boxed{C = \frac{\in A \cdot (n-1)}{d}}$$

where C = Capacitance in farad.

$\in$ = Permittivity

$\in_0$ = Permittivity in free space or vacuum = 8.854×10^{-2} F/m

$\in_r$ = relative permittivity.

n = the number of plates.

d = Thickness of dielectric or distance between plates.

A = Area of the plates.

Table: Showing $\in_r$ for different mediums

S.No.	Name of the medium	Permittivity
1.	Air	1
2.	Paper Press Board	2
3.	Paper oiled	2
4.	Empire Cloth	2
5.	Ebonite	2.5
6.	Rubber	2.5
7.	Polythene	2.3
8.	Paraffine wax	3
9.	Gatta Parcha	4
10.	Bakalite	6
11.	Mica	7
12.	Porcelian	7
13.	Glass	7

Farad is a bigger unit so we have small too

∴ $$1 \, \mu F = 10^{-6} F$$

$$1 \, PF = 10^{-12} F = 1 \, \mu\mu F.$$

Q.1.Find the capacitance of a capacitor formed by two parallel plates each 100 cm² in area. The thickness of the air as dielectric is 2 mm. The $\in_0 = 8.854 \times 10^{-12}$ F/m.

Solution.

$$\in_0 = 8.854 \times 10^{-12} \text{ F/m.}$$

$$A = 100 \text{ cm}^2 = 100 \text{ cm}^2 = 100 \times 10^{-4} \text{ m}^2$$

$$d = 2 \text{ mm} = 2 \times 10^{-3} \text{ m.}$$

$$C = \in_0 \frac{A}{d}$$

$$= \frac{8.854 \times 10^{-12} \times 100 \times 10^{-4}}{2 \times 10^{-3}}$$

$$= 4.427 \times 10^{-10} = 4.427 \times 10^{-4} \ \mu F \ \textbf{Ans.}$$

or we can say 442.7 P.F.

Q.2.Find out the capacitance of a multiplate capacitance whose area of each plate is 5 cm², seperated by 0.1 mm air as dielectric. There are 21 plates. Find also the charge on it if it is connected across 200 volts mains.

Solution. Given $A = 5 \ cm^2 = 5 \times 10^{-4} \ m^2$

$$d = 1 \ mm = 1 \times 10^{-3} \ m.$$

$$\epsilon_0 = 8.854 \times 10^{-12} \ F/m.$$

So substituting in the expression,

$$C = \frac{\epsilon_0 \cdot A \cdot (n-1)}{d}$$

$$= \frac{8.854 \times 10^{-12} \times 20 \times 5 \times 10^{-4}}{10^{-4}}$$

$$= 8.854 \times 10^{-10} = 8.854 \times 10^{-4} \ \mu F. \ \textbf{Ans.}$$

Charge $\quad Q = CV.$

$$= 8.854 \times 10^{-10} \times 200$$

$$= 0.17708 \times 10^{-6} \ \text{or} \ 0.17708 \ \mu C \ \textbf{Ans.}$$

Also find out the capacitance if the dielectric with $\epsilon_r = 5$ is used.

In that case $\quad C = \dfrac{\epsilon_0 \epsilon_r \cdot A(n-1)}{d}$

$$= \frac{8.854 \times 10^{-12} \times 5 \times 5 \times 10^{-4} \times 20}{0.01 \times 10^{-2}}$$

$$= 0.4427 \ pf \ \text{or} \ 44.27 \times 10^{-4} \ \mu F \ \textbf{Ans.}$$

Q. Find the capacitance of a parallel plate capacitor of plate area 0.2 m² with dielectric 2.0 cm thick between them of relative permittivity as 5.

Solution. Surface area as given = 0.2 m²

The thickness of the dielectric d = 2 cm = 2×10^{-2} m

Relative permittivity $\epsilon_r = 5$.

Permittivity in free space $\epsilon_0 = 8.854 \times 10^{-12}$ F/m.

Now capacitance $= \dfrac{\epsilon_0 \, \epsilon_r \cdot A}{d}$

$$= \dfrac{8.854 \times 10^{-12} \times 5 \times 0.2}{2 \times 10^{-2}}$$

$$= 442.7 \; \mu\mu\text{F} \quad \text{Ans.}$$

Example. *Two similar and equal charges are placed 2 metre apart in air. Calculate the force exerted if charge is 8×10^{-6} coulombs.*

Solution. Since the charges are similar and equal so the force will be a repulsive force which can be given as

$$F = \dfrac{Q_1 Q_2}{4\pi \, \epsilon_0 \, E_r \cdot d^2} \; \text{N}$$

Since the medium is air so

$$E_r = 1$$

$$F = \dfrac{Q_1 \cdot Q_2}{4\pi \, \epsilon_0 \, d^2} \; \text{N}$$

Here $Q_1 = Q_2 = 8 \times 10^{-6}$ C, $d = 2$m, $\epsilon_0 = 8.854 \times 10^{-12}$

$$\therefore \quad F = \dfrac{8 \times 10^{-6} \times 8 \times 10^{-6}}{4\pi \times 8.854 \times 10^{-12} \times 2^2} = 0.1438 \; \text{N. Ans.}$$

Example. *Two similar charges placed at 12 cm apart in air repel each other with a force of 95 Newton. Determine the quantity of charge on each in coulomb and the nature also.*

Solution. The force is a repulsive force of 95 N so the charges are of same nature. The force exerted when the charges are placed 12 cm apart in air.

$$F = \dfrac{Q_1 \cdot Q_2}{4\pi \, \epsilon_0 \cdot d^2} \text{N}$$

Since the charges are same, let be Q.

$$F = \dfrac{Q^2}{4\pi \, \epsilon_0 \cdot d^2}$$

$$Q^2 = F \times 4\pi \cdot \epsilon_0 \cdot d^2$$

$$= 95 \times 4\pi \times 8.854 \times 10^{-12} \times (0.12)^2$$

$$= 1522073 \times 10^{-16}$$
$$Q = 1.2337 \times 10^{-6} \text{ coulomb.}$$
$$= 12.337 \ \mu C. \textbf{ Ans.}$$

Q. Derive an expression for finding out the energy stored by a capacitor.

Ans. Whenever a capacitor is charged it stores electrical energy in the shape of static charge and delivers if it is connected to the circuit for discharge.The electrical energy is stored in the capacitor and an electric field is set up in the dielectric medium.

Let there be a capacitor of C farad, which is connected across a potential difference of V volt as shown in Fig. 10.2. During charging work is done for shifting the electrons from positive plate to negative plate via battery. Let at any particular stage the charge being 'q' coulomb and voltage being 'V' volt.

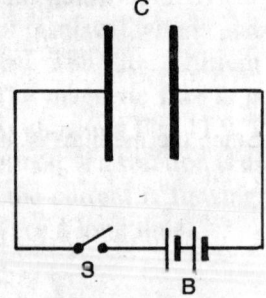

Then $C = \dfrac{q}{v}$ coul.

Fig. 10.2. Capacitor.

Now some more charge is transferred then the amount of the work done is given by

$$dw = v \cdot d_q$$

or $\qquad = Cv \cdot dv \qquad\qquad\qquad [\because dq = d(cv) = c \cdot dv]$

Then the total work done in raising the potential of the uncharged capacitor to 'V' volts is given by

$$W = \int_0^v C_v \cdot dq = C \int_0^v v \cdot dv$$

$$= C \left[\frac{V^2}{2} \right]_0^V = \frac{1}{2} \cdot CV^2$$

So the energy stored is given by the formula $\dfrac{1}{2}CV^2$. This formula can other wise be re-written as

Energy stored $= \dfrac{1}{2}CV^2$ joules.

$$= \frac{1}{2} Q \cdot V \text{ joules} \qquad (\because Q = CV)$$

$$= \frac{1}{2} \frac{Q^2}{C} \text{ joules} \qquad \left(\because V = \frac{Q}{C} \right)$$

10.3. COMBINATION OF CAPACITORS

Q. State the different groupings of the capacitors, also describe their characteristics.

Ans. The capacitors can be connected in the following groups:
  (*a*) Series combinations.
  (*b*) Parallel combination.
  (*c*) Series-parallel combinations.

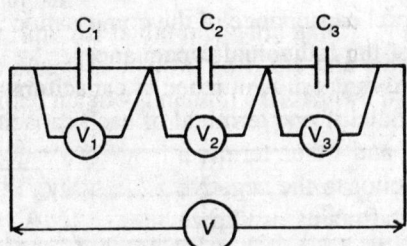

Fig. 10.3. Series combination.

(*a*) **Series combination.** Whenever a number of capacitors are connected so as to form a chain, are said to be connected in series, as shown in Fig. 10.3. Let there are three capacitors C_1, C_2 and C_3 connected in series across a voltage V volt. The voltage across each capacitor being v_1, v_2 and v_3 respectively. This circuit has got the following characteristics:
  1. The charge on all capacitors connected in series, is same.
  2. The voltage across each capacitor is different and depends upon

  their capacity. So $v_1 = \dfrac{Q}{C_1}$, $v_2 = \dfrac{Q}{C_2}$ and $v_3 = \dfrac{Q}{C_3}$.

  3. The voltage supplied to the combination is equal to the sum of all these voltages, *i.e.*, $V = v_1 + v_2 + v_3$ volts.
  4. The total or equivalent capacitance can be determined, let C_{eq} – the equivalent capacitance and charge being Q coulomb. Now

$$V = v_1 + v_2 + v_3$$

$$= \frac{Q}{C_1} + \frac{Q}{C_2} + \frac{Q}{C_3}$$

$$= Q\left[\frac{1}{C_1} + \frac{1}{C_2} + \frac{1}{C_3}\right]$$

$$V = \frac{Q}{C_{eq}} = Q \cdot \left[\frac{1}{C_1} + \frac{1}{C_2} + \frac{1}{C_3}\right]$$

or

$$\frac{1}{C_{eq}} = \frac{1}{C_1} + \frac{1}{C_2} + \frac{1}{C_3}$$

So whenever the different capacitors are connected in series then the reciprocal of the total capacitance of the combination is equal to the sum of the reciprocal of the individual capacitances.

(b) **Parallel combination.** A number of capacitors are said to be connected in parallel if one terminal of each capacitor is connected to positive end and other terminal of each capacitor to the negative terminal of the mains as shown in Fig. 10.4. These connections have got the following characteristics:

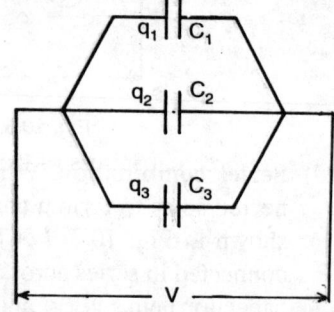

Fig. 10.4. Parallel combination.

1. The voltage across each capacitor is same.
2. The charge on each capacitor is different and depends upon the individual capacitances:

$$q_1 = C_1 V, \quad q_2 = C_2 V, \quad q_3 = C_2 V.$$

3. The total charge of the combination is equal to the sum of the individual charges:

$$Q = q_1 + q_2 + q_3 \text{ coulombs.}$$

4. The total or equivalent capacitance C_{eq}, can be seen as

$$Q = q_1 + q_2 + q_3 = C_1 V + C_2 V + C_3 V = V(C_1 + C_2 + C_3)$$

$$Q = C_{eq} V = V(C_1 + C_2 + C_3)$$

∴

$$C_{eq} = C_1 + C_2 + C_3.$$

Hence when a number of capacitors are connected in parallel the equivalent capacitance will be equal to the sum of the individual capacitances connected in parallel.

(c) **Series-parallel Combination.** Whenever capacitors are so connected as to have series and parallel combinations like C_1 and C_2, C_3 and C_4 in series, C_3, C_4 and C_5 are in parallel as shown in Fig. 10.5. the combination is said to be the series-parallel combination. In this case the circuit is solved step by step.

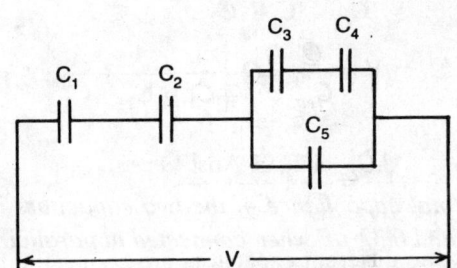

Fig. 10.5. Series-parallel combinations.

Example. *What will be the total capacitance if three capacitors of equal capacitance are put in series?*

Solution. Whenever the capacitors are connected in parallel, the equivalent capacitance C_{eq} can be obtained as

$$\frac{1}{C_{eq}} = \frac{1}{C_1} + \frac{1}{C_2} + \frac{1}{C_3}$$

here are three capacitors of same value, let it be C.

$$\frac{1}{C_{eq}} = \frac{1}{C} + \frac{1}{C} + \frac{1}{C} = \frac{3}{C}$$

and

$$C_{eq} = \frac{C}{3}.$$

So the equivalent value of capacitance will be

$$= \frac{\text{Capacity of similar capacity capacitors}}{\text{Number of capacitors connected in series}}$$

$$= \frac{C}{3}. \text{ Ans.}$$

Example. *A capacitor of 6 μF capacity is joined in (a) parallel (b) series, with another capacitor of 3 μF. Calculate the total capacitance.*

Solution. In case of parallel combination the equivalent capacity

$$C_{eq} = C_1 + C_2 + C_3$$

Here $C_1 = 6\mu F$ and $C_2 = 3\mu F$

So $C_{eq} = 6 + 3 = 9\ \mu F.$ **Ans.**

In case of series equivalent capacity C_{eq},

$$\frac{1}{C_{eq}} = \frac{1}{C_1} + \frac{1}{C_2}$$

$$= \frac{1}{6} + \frac{1}{3} = \frac{3}{6}$$

or $C_{eq} = 2\ \mu F.$ **Ans.**

Example. *The total capacitance of the two capacitors is 0.06 μF when joined in series and 0.32 μF when connected in parallel. Find the capacity of each capacitor.*

Solution. Let the different capacity be C_1 and C_2, when these are connected in parallel

$$C_{eq} = C_1 + C_2 = 0.32\ \mu F \qquad \qquad \dots(i)$$

and when in series

$$\frac{1}{C_{eq}} = \frac{1}{C_1} + \frac{1}{C_2}$$

or $$C_{eq} = \frac{C_1 C_2}{C_1 + C_2} = 0.06\ \mu F$$

or $$C_1 C_2 = 0.06 \times 0.32$$

$$C_1 C_2 = 0.0192 \qquad \qquad \dots(ii)$$

Simplifying, (*i*) and (*ii*) equations, we have

$$C_1 = 0.24\ \mu F\ \text{or}\ 0.08\ \mu F$$

$$C_2 = 0.08\ \mu F\ \text{or}\ 0.24\ \mu F.\ \textbf{Ans.}$$

Example. *Two capacitors 8 μF and 16 μF are connected in parallel and the combination is connected in series with 48 μF capacitor. Find the total capacitance, total charge of capacitors if the combination is fed by 100 V mains.*

Solution. It is an example of mixed or series-parallel combination. The 8μF and 16 μF capacitors are connected in parallel, so the group will have a capacity

$$C_{ea} = 8 + 16 = 24\ \mu F.$$

Now both 24 μF and 48 μF. capacitances are connected in series so the equivalent capacitance is equal to

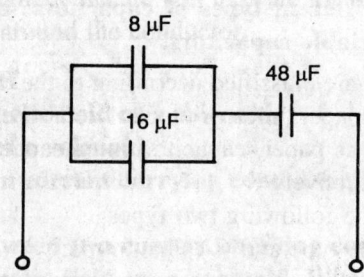

Fig. 10.6

$$\frac{1}{C_{eq}} = \frac{1}{24} + \frac{1}{48} = \frac{3}{48} \text{ or } C_{eq} = 16 \text{ μ F. Ans.}$$

The charge, $Q = CV$
$$= 16 \times 10^{-6} \times 100 = 1600 \times 10^{-6}$$
$$= \textbf{1600 μC. Ans.}$$

Example. *A parallel plate capacitor with five plates has each plate one metre square area at a distance of 1.5 mm and a dielectric of relative permittivity 3.5. If the capacitor is charged to 1000 V, how much energy will be stored in it.*

Solution. The energy stored by a capacitor can be understood by the formula $\frac{1}{2}CV^2$ joules.

The capacitance, $$C = \frac{(n-1)\,\epsilon_0 \cdot \epsilon_r, \, A}{d} \text{ F}$$

Here $n = 5$, $\epsilon_0 = 8.854 \times 10^{-12}$, $\epsilon_r = 3.5$, $A = 1 \text{ m}^2$ and $d = 1.5$ mm or 0.015 m.

So $$C = \frac{(5-1)\,4 \times 8.854 \times 10^{-12} \times 3.5 \times 1}{0.015} = 8.263733 \times 10^{-8} \text{ F}$$

Now energy stored

$$= \frac{1}{2}CV^2$$

$$= \frac{1}{2} \times 8.263733 \times 10^{-8} \times 1000 \times 1000$$

$$= \textbf{4.132} \times \textbf{10}^{-2} \textbf{ joules. Ans.}$$

10.4. TYPES

Q. What are the different types of capacitors? Describe with neat sketches the variable capacitors.

Ans. The capacitors are classified according to the dielectric used. If the capacitors has paper, mica or electrolyte as the dielectric, these are named after it as paper, capacitor, mica-capacitor and electrolytic capacitors respectively.

These are of the following two types:

(*a*) Variable type and (*b*) Fixed type:

(*a*) **Variable type capacitors.** The capacitors which have variable capacitances are known as the variable type capacitors. There are the following two methods of obtaining the variable capacitance.

(*i*) *By using a number of capacitors.* In this system a number of capacitors of different values are used, one wire is common and directly taken out from the common terminal of the selecting switch as shown in Fig. 10.7. The different value can be obtained just by rotating the selector switch in the required position. In this case the capacitors used are generally the paper or electrolyte type capacitors.

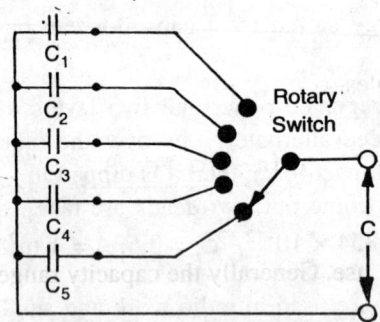

Fig. 10.7. Variable capacitors.

(*ii*) *By using variable air capacitor.* In this capacitor, air is used as the dielectric, so it is known as air capacitor, secondly the capacitance can be changed or varied so this a variable capacitor also. It is a multiplate capacitor the metallic plates are of two types one is fixed and other is variable. Plates are semicircular made of aluminium and attached on a metallic shaft. The fixed plates are joined together by means of metallic strip. The shaft is rotated to obtain the required capacitance.

The capacitor has maximum capacitance when the rotating and fixed plates meshed together; and has got minimum value when these are separated out completely a shown in Fig. 10.8 and Fig. 10.9.

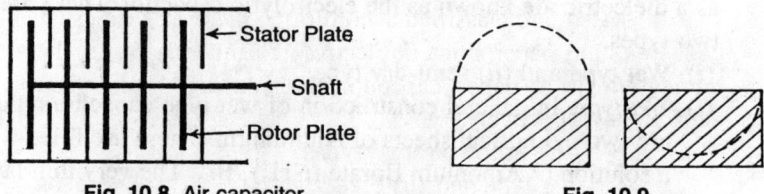

Fig. 10.8. Air capacitor.   Fig. 10.9.

Capacity and use. These types of the capacitors have low value capacitances *i.e.*, ranging from 10 pico farad to 500 pico farad. These are used in radio and T.V. works.

Q. Describe the different types of fixed capacitors used in engineering.

Ans. The capacitors which have got a fixed value, are known as the fixed capacitors and their capacitance cannot be varied. These are of the following types:

(*a*) Paper capacitors. (*b*) Mica capacitors. (*c*) Electrolytic capacitors.

(*a*) **Paper capacitors.** In this capacitor, the paper is used as a dielectric, so it is known as the paper condenser or capacitor. The paper is usually impregnated with wax or oil. Generally there are two layers of impregnated paper and two layers of tinned foil. These plates are placed alternately one over the other and rolled in the cylindrical compact form; then it is dipped in wax so as the air and moisture may come out; two leads are taken out from the tinned foil.

Capacity and use. Generally the capacity ranges from 0.001 μF to 0.1 μF. These are used in radio work and small electrical work to minimise the sparking.

(*b*) **Mica capacitor.** In this capacitor mica is used as the dielectric. The alternate layers of mica and metallic foils are placed closely together. These alternate layers of foil are joined together at one end and one terminal is taken out, similarly on the other side second alternate plates or foils are joined and the terminal is taken out. This type of construction is generally used in multiplate condensers. Then the assembly is kept in a case of bakelite or other covering. These capacitors are very accurate and can withstand the temperature variations.

Capacity and use. The capacity ranges from 50 PF to 500 PF. These are used in radio and telecommunication work, for high frequency work.

(c) **Electrolytic capacitors.** The capacitor in which electrolyte is used as a dielectric are known as the electrolytic capacitor. These are of two types.

(i) Wet type and (ii) semi-dry type.

(i) *Wet type.* In general construction of wet type capacitors, there are two cylindrical sheets of Aluminium. These are filled with a solution of Amonium Borate $(NH_2)_2 Br_2$. The very thin layer of amonium oxide deposits on the positive plate, this layer works as a dielectric in between electrolyte and positive plate. As this layer is very thin so the capacity of this capacitor is very high as compared with the other capacitor of the same size.

The indication for the polarity is marked over the capacitor. The positive wire must be connected with the positive terminal and negative with negative plate.

(ii) *Semi-dry type.* In the semi-dry type capacitors the plate of Amonium Borate is deposited on a paper. It is now used in place of electrolyte in between two layers of the metallic foils. These are rolled up in cylindrical form and placed in the container. Mostly negative wire is connected with the body of the container and positive with the centre terminal.

Capacity and use. These are available in high capacities. The polarised capacitors (only for A.C.) is used in the automobile and D.C. circuits etc.

The electrolytic capacitors are also designed to work on A.C. and D.C. both. These are known as non-polarised electrolytic capacitor. These are also used in A.C. single phase motors like, fans etc.

10.5. TESTING OF CAPACITOR

Q. How will you test a capacitor?

Ans. A capacitor can be tested by the following methods:

(a) **By means of screw driver.** It is also known as the charging and discharging method. In this method the capacitor is connected across the supply of proper voltage, as rated on the capacitor. After some interval the supply is disconnected and the two terminals of the capacitor are short circuited by means of screw driver or a piece of wire as shown in Fig. 10.10, on joining both the terminals if the capacitor gives away the sparking, it indicates that the capacitor is okay.

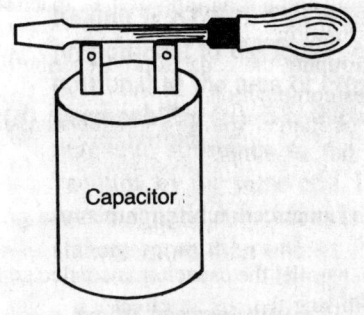

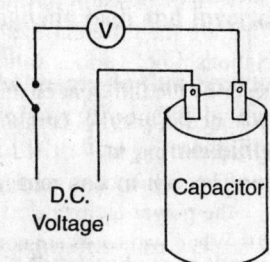

Fig. 10.10. Testing of capacitor.   **Fig. 10.11.** Testing of Capacitor.

If spark does not occur, it indicates that the capacitor is not in proper order there is an open circuit. Secondly if just by connecting the capacitor on supply fuse blow out, it indicates there is a dead short circuit.

This test can be performed on A.C. or D.C.

(*b*) **By means of high resistance meter.** In this method a high resistance meter say voltmeter is connected in series with the capacitor across the d.c. supply of suitable voltage as shown in Fig. 10.11. When ever switch is put on, the readings are observed as:

(*i*) The pointer remains on the zero position and does not deflect – it means there is an open circuit in the circuit.

(*ii*) The meter gives away a continuous reading, it indicates there is a short circuit in the capacitor.

(*iii*) If the meter is giving some deflection less than the supply voltage, continuously – it indicates a leakage in the capacitor.

(*iv*) When deflection is noticed in the first stage and slowing down to zero position it indicates the capacitor is okay.
Note. If this test is performed on A.C. then the continuous reading will be obtained. If the lamp of suitable wattage, is inserted in series, then the full light means the short circuit, no light means open circuit, and dim light means okay.

(*v*) By measuring the capacity of the capacitor. In this method the capacity of the capacitor is measured by means of a Capacity Bridge meter.

REVIEW QUESTIONS

1. Define farad, coulomb, electric flux, unit charge and permittivity.
2. State the different laws of electrostatics.
3. What do you understand by the dielectric strength, also state the factors which effects the dielectric strength of a dielectric?

4. What do you understand by the charging and discharging of the capacitor?

5. Derive as expression for the energy stored in a capacitor.

6. What are the different methods of grouping the capacitors? Describe the relationship of various values in series combination.

7. Describe the different characteristics of charge, voltage and equivalent capacitance in parallel combination of the capacitors.

8. Fill up the blanks:

   (a) The capacitor is connected in the fluorescent tube circuit is to............ the power factor.

   (b) When two tubes are connected in parallel the capacitor is connected in parallel of one set helps in minimising the............ effect?

   (c) To improve the power factor of the line or system a............ of capacitors is used in parallel.

   (d) In single phase motors, the capacitor is connected in series with starting winding which helps in............ the angular displacement between the running and starting winding currents.

   (e) In series combination the total capacitance............ with the increasing of the capacitor in series.

9. (a) Three capacitors of 10, 20 and 40 µF are connected in parallel across 50 V mains. Find out the equivalent value of the capacitance.

   (b) Now if these are connected in series calculate the equivalent capacitance and charge on the capacitors.

   [Ans. (a) 70 µF; (b) 5.71 µF, 288.5 µC]

10. A capacitor of 80 µF is connected across 100 V mains, calculate the charge contained.   **[Ans. 8000 µC]**

11. What is the necessity of grouping the capacitors? Calculate the energy stored by a 100 µF capacitor if it is connected across 230 V supply.

    [Ans. 132.25 joule]

Magnetism and Electromagnetism

11.1. MAGNET

Q. What is a magnet? What are the different properties of a magnet?

Ans. A magnet is a substance which has the property of attracting the iron and its ores or the magnetic substances. It has the following properties:

1. A magnet attracts iron, its alloys and other magnetic substances.
2. It has the property of direction, if a magnet is freely suspended, it always indicates towards north and south.
3. It has two poles, the pole which indicates towards geographical north is known as north pole and the one which indicates towards south is known as south pole.
4. Like poles repel each other and unlike poles attract each other.
5. The pole always exists in pair, one isolated pole cannot exist.
6. It induces the opposite polarity as soon as the magnetic material is brought near it, *i.e.,* it induces the magnetic property into the other.

11.2. TYPES OF MAGNET

Q. What are the different types of magnets?

Ans. These are the following types of the magnet:
   (*a*) Natural magnet (*b*) Artificial magnet
 (*a*) **Natural magnet.** A substance is found in the nature which exhibits the property of attracting the iron and its filings, simultaneously the property of directing north and south. It is in the shape of a stone as shown in Fig. 11.1. In the ancient era it was used by the navigators

for direction purposes. It is known as leading stone or lode stone.

(*b*) **Artificial magnet.** The magnets which are prepared by the human beings by artificial means like touch method etc. are known as artificial magnets. These are prepared by iron or steel.

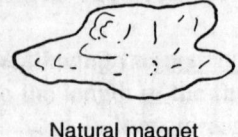

Natural magnet

Fig. 11.1.

The artificial magnets are of the following types:

(*i*) Temporary magnet

(*ii*) Permanent magnet.

(*i*) *Temporary magnets.* The magnets which loses their magnetism as soon as the magnetising means or force is removed. All the electro-magnets are the temporary magnets as shown in Fig. 11.2. The metal used in this case is the soft iron. The temporary magnets are used in electric bell, buzzers, bell indicators, etc.

(*ii*) *Permanent magnets.* The permanent magnets are those which retains their magnetism for a long time. It is observed that if a piece of hard steel is magnetised, it acquire a substantial magnetism which it retains for an indefinite time. Such magnets are permanent magnets. The material used for permanent magnets are tungsten steel, high carbon steel and other hardened steel etc.

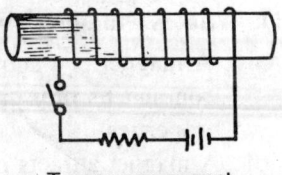

Temporary magnet

Fig. 11.2.

Shapes. There are the different shapes like bar magnet, horse shoe magnet, U-shape magnet and magnetic needle, as shown in Fig. 11.3.

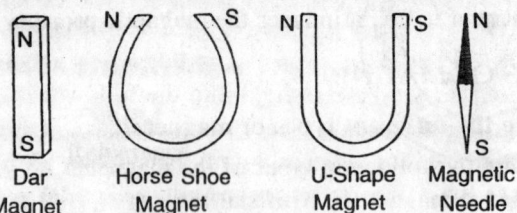

| Dar Magnet | Horse Shoe Magnet | U-Shape Magnet | Magnetic Needle |

Fig. 11.3. Shape of Permanent magnets.

Q. Describe the molecular theory of a magnet.

Ans. It is also known as the Webber's theory of magnetism. If a permanent magnet is broken into pieces, every individual piece will behave like a separate and perfect magnet. There will be two poles north and south as shown in Fig. 11.4 (*a*).

Every substance is made of a number of molecules. Each molecule is an independent magnet; having north and south. In an ordinary stage these molecules are not arranged in a proper sequence as shown Fig. 11.4(*b*) these are placed haphazardly. Thus the molecules neutralizes the effect of magnetism and as a result the piece do not behave like a magnet.

If a magnetising force is applied, the molecules starts arranging in a proper sequence and the substance starts exhibiting the property of a magnet as shown in Fig. 11.4 (*c*). Otherwise also if the molecules are dearranged and if a bar magnet is brought as one end the molecules starts arranging in a straight line, thus two poles are formed. The end near the north of a magnet, will magnetise to be south of the piece and so on.

The magnetic power of the piece to be magnetised will be proportional to the number of molecules arranged if less molecules are arranged the magnetism will be less and if more molecules then the magnetic power will be more. In case where all the molecules are arranged the maximum power will be exhibited, after this if the magnetising power is increased even then the magnetism will not increase, this point is known as saturation point.

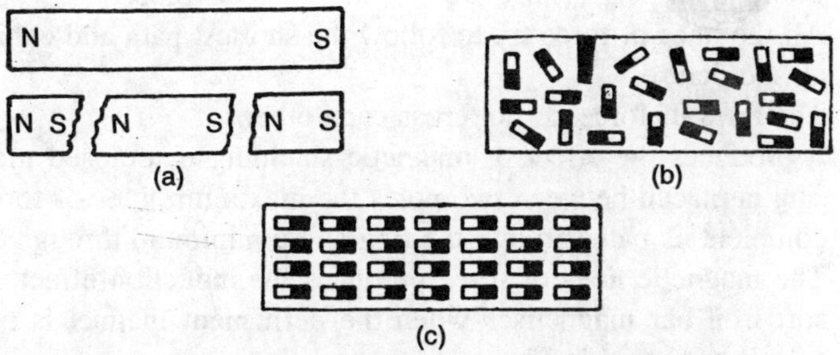

Fig. 11.4. Molecular structure of magnet.

Q. What are the classifications of the substances according to the magnetic properties?

Ans. The substances can be classified into the following three types:

(*a*) **Paramagnetic materials.** These materials have small but positive relative permeability *i.e.* nearly unity but greater than unity. These are ordinarily the non-magnetic materials like Cu, Al, etc. In such materials the individual atomic dipoles are oriented in the random fashion. On the application of the magnetising force the dipoles orient themself but not appreciably.

(*b*) **Diamagnetic material.** The materials which have small and negative relative permeability but very nearly equal to unity. In this material the permanent magnetic dipoles are absent or even nearly negligible. If the magnetising force is applied the magnetism is in opposite direction, thus the relative permeability is negative. This makes diamagnetic materials unimportant for electrical engineering applications.

(c) **Ferro-magnetic material.** The magnetic materials in which the permeability varies considerably high and positive. It varies from 200 to 1000. The substance shows magnetic properties strongly high when subjected to a magnetic force. These materials are generally crystalline solid. The permanent magnetic atomic dipoles are aligned parallel to each other and even a small amount of magnetising force causes a considerable change in the magnetism.

Q. What are the properties of the lines of force, also state the care and maintenance of the permanent magnet?

Ans. Following are the properties of the magnetic lines of force:

1. All the lines of force starts from the north pole and terminates to south pole outside the magnet and from south to north inside the magnet.
2. There is no perfect insulator for the lines of force.
3. All the lines of force try to follow the shortest path and exhibit the elastic nature.
4. The lines of force do not cross each other.
5. It produces the effect of magnetic shielding if a closed magnetic ring is placed between two poles the maximum lines of force will complete its path through the ring and a minimum through the air.
6. The magnetic lines of force produces the induction effect. As the soft iron bar magnetises when the permanent magnet is brought near it as shown in Fig. 11.5.

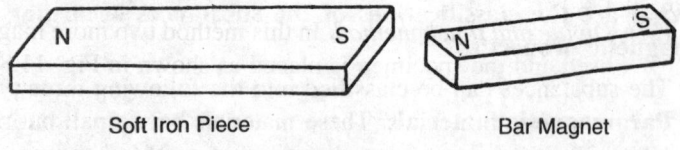

Soft Iron Piece   Bar Magnet

Fig. 11.5.

Care and maintenance. If the molecules of the permanent magnet are de-arranged then the magnet is demagnetised so the following precautions should be taken:

(i) It should be used properly and should not be thrown.
(ii) It should not be heated up.
(iii) It should not be hammered

(*iv*) If a pair of magnets is kept, the keepers should be used and magnets should be kept facing the opposite polarity.

11.2. METHODS OF MAGNETISATION

Q. What are the different methods of magnetisation? Describe with neat sketches.

Ans. There are the following different methods.
  (*a*) Touch method,
  (*b*) Electrical method,
  (*c*) Other methods.

(*a*) **Touch method.** The touch method can be further categorised into the following methods:

(*i*) *Single touch method.* In this method a permanent magnet is rubbed with the soft iron or steel specimen, to be magnetised. The direction of the movement is shown in the Fig. 11.6. The direction is uniform. The process is continued for long time, so this method is not frequently used.

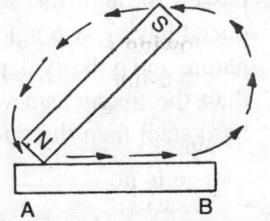

Fig. 11.6. Single touch method.

(*ii*) *Double touch method.* In this method two magnets are taken and are rubbed over the specimen to be magnetised as shown in Fig. 11.7.

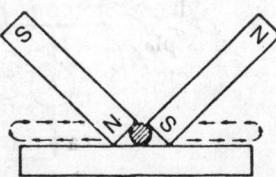

Fig. 11.7. Double touch method.

(*iii*) *Divide and touch method.* In this method two more magnets are used and the specimen is placed as shown in Fig. 11.8. In this

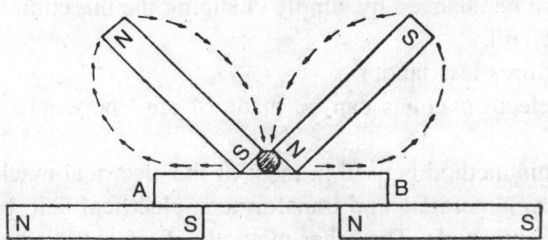

Fig. 11.8. Divide and touch method.

case the specimen is magnetised in less time. Both magnets are moved as shown by dotted lines. In this method a non-magnetic piece is kept between the magnets to avoid the possibility of attraction and sticking the magnetic poles.

Disadvantages. There are the following disadvantages of the touch method:

1. It requires more time.
2. The magnet cannot be of the required strength.
3. The specimen cannot have more strength than the magnet used for magnetisation.

Uses. This method is not used for practical purposes.

(b) **Electrical method.** In this case the coil of insulated wire is placed over the soft iron piece to be magnetised. The direct current is passed through the coil to energise it. The magnetic flux is produced having two poles north and south as shown in Fig. 11.9. The nature and polarity depends upon the specimen, if a soft iron is used than the magnetism will last as long as the current is flowing, if hard steel then the magnetism will retain for a long time.

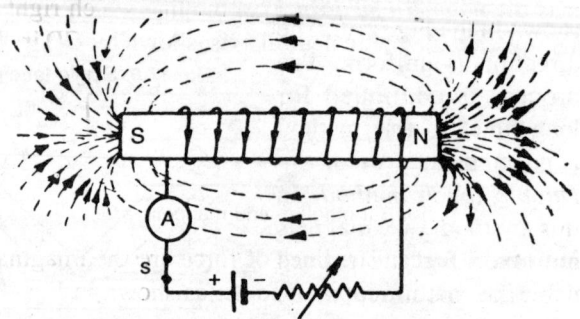

Fig. 11.9. Electrical method.

Advantages. There are the following advantages of electromagnet:

1. It requires less time.
2. The specimen can be magnetised in any of the polarities. The polarity can be changed by simply changing the direction of the current in the coil.
3. It requires less labour.
4. The electromagnets can be made of any shape, size and polarity also.

Uses. This method is used in most of the electrical machines and appliances *viz.*, generators and transformers, electrical bell, buzzer etc.

(c) **Other methods.** The other methods of magnetisation are:

(*i*) If you pound a higher retentivity substance with a hammer while the substance is placed in the direction of earth's magnetic field, the substance becomes a permanent magnet.

(*ii*) A high retentivity substance can be heated and subsequently cooled while it is placed in the direction of the earth magnetic field, the substance thereby formed into a permanent magnet.

11.3. TERMS AND DEFINITIONS

Q. Define the followings:
Magnetic pole, magnetic axis, magnetic lines of force, magnetic flux, magnetic flux density and magnetic field strength.

Ans. Magnetic pole. The two ends of the magnet where the magnetic power is maximum or concentrated, are known as the magnetic poles. There are two poles North and South.

Magnetic axis. It is an imaginary straight line joining the north and south poles of a magnet.

Here the joining line *AB* is the magnetic axis.

The other is the *magnetic neutral axis,* the line which right bisect the magnetic axis is called magnetic neutral axis, shown by *CD* in Fig. 11.10.

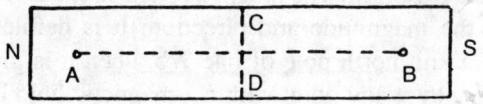

Fig. 11.10. Bar magnet, and magnetic axis.

Magnetic lines of force. The lines of force are the imaginary curves through which a magnet is supposed to act as shown in Fig. 11.11.

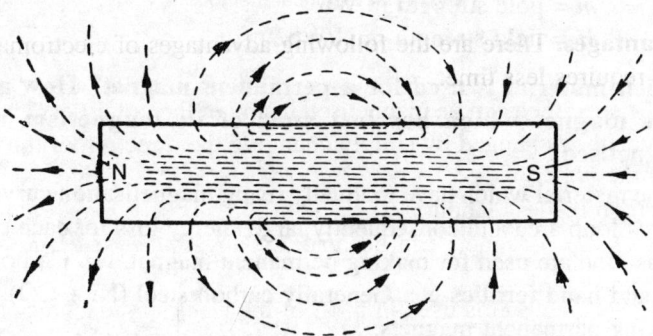

Fig. 11.11. Magnetic field

It can otherwise be defined as the path along which a **unit north** pole would move if it is free to move.

OR

It is that imaginary curve, the tangent on which indicates the direction of field. It's unit is maxwell and lines.

Magnetic field. The space or region occupied by the lines of force around a magnet is called magnetic field.

Magnetic flux. The total number of lines of force in a magnetic field is called magnetic flux. The unit is line and in MKS system webber.

$$1 \text{ Wb} = 10^8 \text{ lines or maxwell}$$
$$1 \text{ mega lines} = 10^6 \text{ lines}$$

The symbol is ϕ.

Magnetic flux density. It is defined as the magnetic flux per unit area. It is represented by letter B and unit is Wb/m^2 or max/cm^2 or lines/cm^2.

It is $$B = \frac{\phi}{A}$$

where B = flux density in Wb/m^2
ϕ = flux in Wb
A = area of cross-section in m^2.

Magnetic field strength. The magnetic field strength is a vector quantity possessing the magnitude and direction. It is defined as the force experienced by a unit north pole of one Wb when it is placed there.

The field strength at any point within a magnetic field is measured by the force experienced by a north pole of one webber placed at that point.

It is denoted by letter H and the unit is N/Wb or ampere Turns/meter.

It is given as $F = mH$

where F = force in newton
m = pole strength in Wb
H = field strength in N/Wb.

Q. Which material is used for a permanent magnet? How a horse shoe magnet which has lost most of its magnetism will be magnetised?

Ans. The material which have gradually rising magnetisation curve, large hysteresis loop area and consequently large energy loss for each cycle of magnetisation are used for making permanent magnet, *viz.* Carbon steel, Alnico and hand ferrities etc. Generally carbon steel (Ni + CO) is used for making permanent magnets.

The horse shoe magnet can be brought back in magnetised state by using an electromagnet method. Wrap the super enamelled wire over the magnet as shown in Fig. 11.12 and connect to d.c. low voltage supply, for sufficient time and intermittently. In this case the direction of current should be strictly observed.

It can otherwise be magnetised on the pole charger the polarities must be observed carefully.

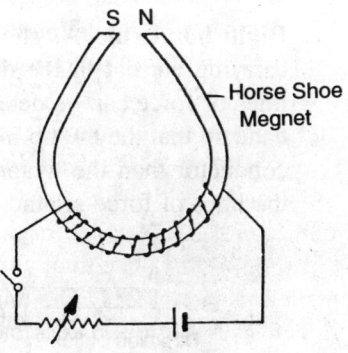

Fig. 11.12.

Q. What do you understand by the soft and hard magnetic materials, give example?

Ans. The ferromagnetic materials may be divided into the two groups:

(*a*) **Soft magnetic materials.** The materials which have a steeply rising magnetisation curve, relatively small and narrow hysteresis loop and small energy loss per cycle of magnetisation are called soft magnetic material. These are generally used for cores of transformer, alternator etc. These are soft iron, silicone steel, nickel iron alloy and soft ferrites.

(*b*) **Hard magnetic materials.** The materials which have gradually rising magnetisation curve, large hysteresis loop area and consequently large energy losses for each cycle of magnetisation are called hard magnetic materials. Such materials are used for making permanent magnets. These are carbon steel, tungsten steel, cobalt steel, Alnico, hard ferrites [$BaO(Fe_2O_3)_6$], etc.

11.4. ELECTROMAGNETISM

Q. What do you understand by the magnetic effect of current and define electromagnet?

Ans. Whenever the current is passed through a conductor magnetic lines of force are set up around the conductor, this effect of current is known as the magnetic effect of current.

The magnet which works on the principle of magnetic effect of current is known as the electromagnet.

Q. State the different methods of determining the direction of magnetic flux around a current carrying conductor.

Ans. The direction of the magnetic lines of force around a current carrying conductor can be determined by these following methods:

(*a*) Right hand rule.

(*b*) Maxwell's cork screw rule.

(*c*) End rule or arrow rule.

(*d*) By means of the magnetic compass.

(a) **Right hand rule.** As shown in Fig. 11.13 a long conductor *AB* is carrying current in the direction as marked. The direction of the lines of force can be determined, hold the conductor in your right hand so that the thumb indicates the direction of the current in the conductor than the wrapped fingers will indicate the direction of the lines of force around the conductor.

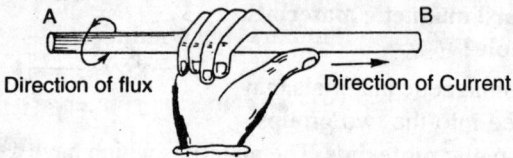

Direction of flux   Direction of Current

Fig. 11.13. Right hand rule.

(b) **Maxwell's cork screw rule.** Generally it is known as the screw rule. Imagine a right handed wood screw as shown in Fig. 11.14. Now rotate the screw so that the tip or point should indicate the direction of current then the direction of rotation will indicate the direction of magnetic lines of force around the conductor.

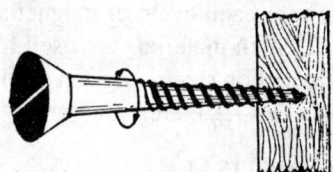

Fig. 11.14. Cork's screw rule.

(c) **End rule or arrow rule.** This method is commonly employed for finding out the direction of lines of force around a current carrying conductor. The direction is observed from one end. If the current is going away from the observer this is marked as ⊕ and if coming towards the observer it is marked as ⊙ sign.

Now the direction of magnetic lines of force will be anticlockwise if the current has ⊙ direction *i.e.,* coming towards the observer and it will have clockwise direction, if the current is going away from the observer *i.e.* ⊕ sign as shown in Fig. 11.15.

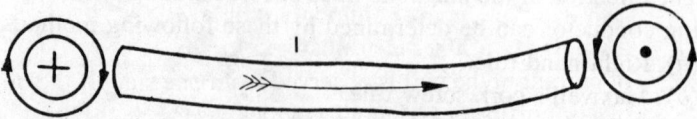

Fig. 11.15. End rule.

(*d*) **By means of the magnetic compass.** A magnetic compass is placed around a conductor which is live or carrying current, then the north pole of the magnetic needle will indicate the direction of magnetic lines of force around the conductor.

Q. Describe in brief
   (*a*) **The magnetic field of a solenoid.**
   (*b*) **Polarity of a solenoid.**
   (*c*) **Force on a current carrying conductor, lying in a magnetic field.**
   (*d*) **Force between two current carrying conductors.**

Ans. (*a*) **The magnetic field or a solenoid.** Whenever the current is flowing in the solenoid, the magnetic lines of force of different turns strengthened and finally forming a magnetic belt constitutes the strong magnetic flux. The direction can be determined by the right hand rule: thus the north and south poles are shown in Fig. 11.16. The magnitude of the flux depends upon the quantity of current and the number of turns in the solenoid.

   (*b*) **Polarity of a solenoid.** The direction of the north pole or south pole or the polarity can be determined by these following methods.

   (*i*) *Right hand rule.* In this case hold the solenoid in your right hand so as the tips of the fingers indicate the direction of current in the solenoid then the thumb will indicate the direction of the north pole so produced.

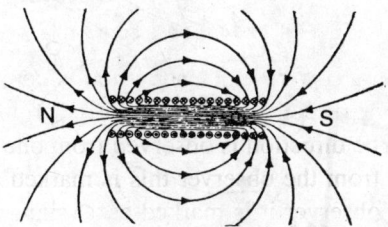

Fig. 11.16. Solenoid.

   (*ii*) *Screw rule.* In this method have a right handed screw, rotates it in the direction of current in the coil then the position of point or tip of the screw will indicate the north pole.

   (*iii*) *End rule.* It is most commonly employed rule for finding out the polarity of the electromagnet. In this method the direction of current in the coil is observed from one side. If the direction of current in the solenoid is clockwise than that end will be south pole and if the direction is anticlockwise the end will be north pole.

(iv) *Magnetic compass.* Bring a magnetic compass near the one end of the solenoid, the attraction or repulsion will take place between the formed pole and the compass, so if the south pole is attracted then the north pole is formed, similarly south pole is determined.

(c) **Force on a current carrying conductor lying in the magnetic field.** It is observed that whenever a current carrying conductor is placed in the magnetic field, a force tends to move the conductor at right angle to the main magnetic field. Here let the current in the conductor be downward inducing the magnetic field as shown say clockwise. The direction of the flux is also given, the flux will be crowded at one end and spreaded on the other end, as shown in Fig. 11.17. The net result will be the development of a torque in the direction indicated. This force can be given as

$$F = BIl \text{ Newton}$$

where F = force in the Newton

B = Flux density in Wb/m^2

I = Current in the coil/conductor in amp.

l = Length of conductor in metre; lying in the magnetic field.

So $F = \mu_0\mu_r\lambda \, NIl\cdot$ Newton $(\because B = \mu_0\mu_r\cdot N)$

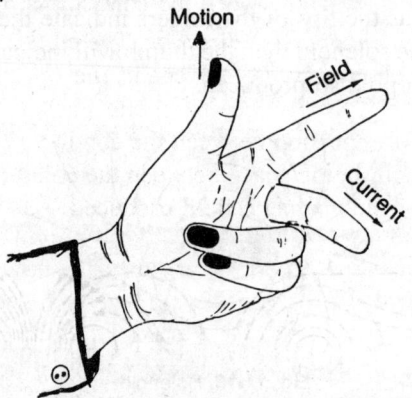

Fig. 11.17. Flaming's Left hand rule.

The direction of torque developed can be determined by the Fleming's left hand rule, "stretch your left hand in such a way so that the thumb, forefinger and middle finger are at right angle to each other. Now if the forefinger indicates the direction of the magnetic field and the middle finger the current in the conductor then the thumb will indicate the direction of the torque developed" as shown in Fig. 11.17.

(*d*) **Force between two current carrying conductors.** If two parallel conductors are laid down the current can be in these two different directions:

(*i*) *Current in the same direction.* Let there be two conductors *A* and *B* carrying currents I_1 and I_2 amp in the same direction. The magnetic lines of force are induced around the conductors as shown in the Fig. 11.18.

Each one of the pair of parallel conductors experience a mechanical force because of the magnetic field produced by them. Each conductor sets up its own magnetic field consisting of circular lines of force and resultant field is the resultant of both the magnetic fields and a magnetic belt results which exhibit a tendency of magnetic attraction.

Hence the two conductors carrying current in the same direction will experience a force of attraction between them.

$$F = \frac{\mu_0 I_1 I_2}{2\pi d} l \cdot N$$

where F = force in newton

            I_1, I_2 = currents in the conductors

            l = length of the conductor

            d = the distance between the conductors.

(*ii*) *Currents in the opposite direction.* The magnetic field set up by the conductors carrying current in the opposite directions is shown in Fig. 11.19. The magnetic lines of force exhibits a tendency of repulsion between the conductors. In this case the field strength in the space between the conductors is increased because of the magnetic field produced.

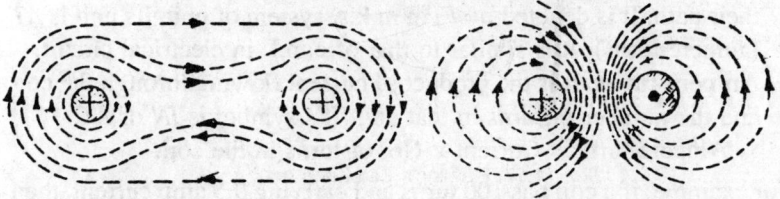

Fig. 11.18. Current in same direction.   **Fig. 11.19.** Current in Opposite direction.

Q. What are the differences between the permanent and electromagnets?

Ans. There are the following differences and similarities between the permanent and electromagnets as shown in Table 11.1.

Table 11.1. Difference between Permanent and Electromagnet

Electromagnet	*Permanent magnet*
1. Easy to prepare.	Difficult and takes more time.
2. Can be made of any strength.	Strength cannot be more the magnet used to prepare.
3. The magnetic strength can be varied.	Magnetic strength cannot be varied.
4. The polarity can be changed.	Polarity cannot be changed.
5. It is costly.	It is not so costly.
6. It is suitably used in D.C. motor, generators for large size.	It cannot be used, only suitable for less output.
7. Efficiently used in electric bell, signal indicator etc.	Not possible.
8. Efficiently used as lifting magnets in steel mills etc.	Cannot be used.
9. Cannot be used as the magnetic compass for navigators.	It is very efficiently used.
10. It can be made of any size, shape and polarity.	In cannot.

Q. Define the following:
   (a) Magnetomotive force, (b) Ampere turns, (c) Reluctance, (d) Permeability, (e) Magnetic leakage, (f) Residual magnetism, (g) Permeance, (h) Intensity of magnetisation, (i) Susceptibility.

(a) **Magetomotive force (m.m.f.).** The magnetomotive force is defined as the force which compels the magnetic lines of force to complete their path. It is denoted by F, in m.k.s. system of unit, its unit is AT (ampere turns). It is similar to that of e.m.f. in electrical circuit.

(b) **Ampere turns.** It is the product of current flowing through the coil and the number of turns in that coil. It's symbol is IN or AT.

   Ampere turns = Current × No. of turns in the coil.

For example, if a coil has 100 turns and carrying 0.5 amp current, then the ampere turns are

$$= 0.5 \times 100 = 50 \text{ AT}$$

(c) **Reluctance.** It is the opposition offered by the magnetic path to establish the magnetic flux in the magnetic circuit. It is represented as Rel. It is similar to that of the resistance in the electrical circuits.

$$\text{Reluctance} = \frac{\text{Magnetomotive force}}{\text{Magnetic flux}}$$

Its unit is AT/Wb. The reluctance of the magnetic path is directly proportional to the length of the magnetic path and inversely proportional to the area of cross-section.

(d) **Permeability (μ).** It is the ratio of the magnetic flux produced in a magnetic substance to the magnetic flux produced in the air or vacuum, by the same coil. Its symbol is μ. The permeability of the non-magnetic substances is less than one and of the magnetic substances more than one.

$$\text{So the permeability} = \frac{\text{The magnetic flux produce in the medium}}{\left(\begin{array}{c}\text{The flux produced in the vacuum}\\ \text{or air by the same coil}\end{array}\right)}$$

In terms of magnetic flux density and magnetising force it is given as

$$\mu = \frac{B}{H} = \frac{\text{Magnetic flux density}}{\text{Magnetising force}}$$

(e) **Magnetic leakage.** Whenever the magnetic lines of force takes some other path then the desired one, it is called the magnetic leakage. The sum of all the lines of force which cannot be utilized for doing useful work, because their path takes them away from the desired one, are called leakage flux.

Leakage factor. As shown in Fig. 11.20 the flux passing through the air gap between two poles is the useful flux and this flux is used in electric machine, so flux which does not take the required path and cannot be utilized is known as leakage flux. The total flux is the sum of the leakage flux and the useful flux.

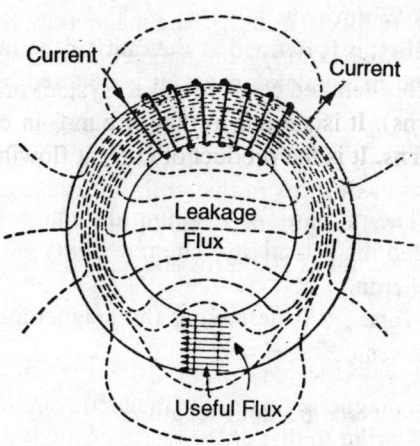

Fig. 11.20. Magnetic flux.

$$\therefore \qquad \phi_T = \phi_U + \phi_L$$

The leakage factor is defined as the ratio of total flux to the useful flux.

$$\text{Leakage factor} = \frac{\text{Total flux produced}}{\text{Useful flux}} = \frac{\phi_U + \Phi_L}{\phi_U} = 1 + \frac{\phi_L}{\phi_U}$$

The leakage factor is always more than one. It is generally 1.1 or 1.15.

(*f*) **Residual magnetism.** It is the magnetism left in a magnetic substance after removing the magnetomotive force or the lagging of the magnetism behind the magnetising force is called the residual magnetism. It is used in the generator for producing the initial voltages etc.

(*g*) **Permeance.** It is the property of the magnetic circuit which helps in establishing the magnetic flux. It is the reciprocal of the magnetic reluctance and similar to that of the conductance in electric circuits. It is denoted by letter ρ.

$$\rho = \frac{1}{\text{Rel.}} = \frac{\phi}{\text{m.m.f.}} \, \text{Wb/AT}$$

The unit is Wb/AT.

(*h*) **Intensity of magnetisation.** It is defined as the pole strength developed per unit area of the magnet. Let a magnet has the pole strength as *m*, pole face area as *A*. Then the intensity of magnetisation is

$$I = \frac{m}{A} \, \text{Wb}/\text{m}^2$$

The unit is Wb/m^2.

(*ii*) **Susceptibility.** It is defined as the ratio of the intensity of magnetisation to the magnetising force. It is denoted by letter *K*.

$$K = \frac{I}{H} = \frac{\text{Intensity of magnetisation}}{\text{Magnetising force}} \, \text{A/m}$$

Retentivity. The property of retaining the magnetism by a magnetic substance is called the retentivity. The retentivity of steel is higher than the retentivity of iron.

Magnetising force. It is defined as the magnetomotive force per unit length. It is represented by *H* and

$$\text{Magnetomotive force per unit length} = \frac{IN}{l}.$$

The unit is AT/m.

11.5 LAWS OF MAGNETISM

There are the following laws:
 (*i*) *First law.* Like poles repel each other; unlike poles attract each other.
 (*ii*) *Second law.* The force exerted between two magnetic poles in proportional to the product of their pole strengths and is inversely proportional to the square of distance between them.
 These can be expressed by the formula,

$$F \propto m_1 \times m_2$$

$$\propto \frac{1}{d^2}$$

$$F = K \frac{m_1 \times m_2}{d_2}$$

Where m_1 and m_2 are their pole strengths and d as the distance between them and K is a constant the value of which depends upon the nature of the medium in which the poles are placed.

In S.I. system of units, $K = \dfrac{1}{4\pi\mu_0\mu_r}$

Where μ_0 permeability of the vacuum and μ_r is the relative permeability of the medium with respect to the permeability in free space or vacuum.

Unit pole: Let $m_1 = m_2 = m \cdot$ pole strength, d the distance as 1m., and force exerted as $\dfrac{1}{4\pi\mu_0}$N

$$1 = \frac{m \times m}{1}$$

or $$m^2 = 1$$

or $$m = \pm 1$$

Hence a unit magnetic pole is defined as that pole which when placed at one metre apart from a similar pole exert a attractive or repulsive force of 1/4 $\pi\mu_0 N$

Or

A unit pole is also defined (in CGS system of units) if, when placed at a distance of one cm. from a similar pole, causes a repulsive or attractive force of one dyne.

Q. (*a*) **What are the factors which effects the magnetic reluctance.**

(*b*) **Derive an expression for the ampere-turns required for a single and composite circuit.**

Ans. The magnetic reluctance depends upon the following factors:

(*i*) The reluctance is directly proportional to the length of the flux path.

(*ii*) The reluctance is inversely proportional to the area of cross-section of the flux path.

So $$R \propto l$$

$$\propto \frac{1}{\alpha}$$

$$\therefore \qquad R = \frac{1}{\mu A} \text{AT/Wb}$$

Here μ = Constant of proportionality, *i.e.*, $\mu = \mu_0 \cdot \mu_r$

μ_0 = Permeability for free space and

μ_r = relative permeability of any medium.

$$\therefore \qquad R = \frac{1}{\mu_0 \mu_r \cdot A} \text{AT/Wb}$$

(*b*) Let in any magnetic circuit,

I = Current to energise the coil in amp.

N = Number of turns in the coil.

ϕ = The flux established in Wb.

l = The length of the path.

μ_0 = Permeability in free space or vacuum.

μ_r = Relative permeability.

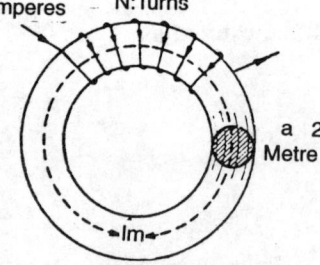

I-Amperes   N:Turns

a 2 Metre

Fig. 11.21.

$$\therefore \qquad \text{Flux} = \frac{\text{m.m.f}}{\text{Reluctance}}$$

The m.m.f. = IN

$$\text{Rel.} = \frac{l}{\mu_0 \mu_r \cdot A}$$

$$\therefore \qquad \phi = \frac{IN}{\dfrac{l}{\mu_0 \mu_r A}} = \frac{IN \cdot \mu_0 \mu_r \cdot A}{l}$$

or $\quad IN = \dfrac{\phi I}{\mu_0 \mu_r \cdot A} = \dfrac{\phi}{A} \cdot \dfrac{l}{\mu_0 \mu_r} = \dfrac{B \cdot l}{\mu_0 \mu_r}$ $\qquad \left(\because \dfrac{\phi}{A} = B \right)$

Obviously $\mu = B/H$. or $H = B/\mu$, so $\qquad \left(\mu = \mu_0\, \mu_r \right)$

$$IN = Hl \cdot \text{AT}$$

Similarly for a composite circuit, let A_1, A_2 and A_3 be the area of cross section and l_1, l_2 and l_3 the lengths respectively, the total ampere turns can be determined,

Total $\qquad \text{Rel} = \text{Rel}_1 + \text{Rel}_2 + \text{Rel}_3 + \dots\dots$

and $\qquad \phi = \dfrac{\text{m.m.f.}}{\text{Total reluctance}}$

$\therefore \qquad \phi = \dfrac{IN}{\dfrac{l_1}{\mu_0 \mu_r A_1} + \dfrac{l_1}{\mu_0 \mu_2 A_2} + \dfrac{l_1}{\mu_0 \mu_3 A_3}} + \dots\dots$

or $\qquad IN = \phi \left[\dfrac{l_1}{\mu_0 \mu_1 A_1} + \dfrac{l_2}{\mu_0 \mu_2 A_2} + \dfrac{l_3}{\mu_0 \mu_3 A_3} + \dots\dots \right]$

$\qquad\qquad = \phi \left[\dfrac{\phi l_1}{A_1 \mu_0 \mu_1} + \dfrac{\phi l_2}{A_2 \mu_0 \mu_2} + \dfrac{\phi l_3}{A_3 \mu_0 \mu_3} + \dots\dots \right]$

$\qquad\qquad = \left[\dfrac{B_1 l_1}{\mu_0 \mu_1} + \dfrac{B_2 l_2}{\mu_0 \mu_2} + \dfrac{B_3 l_3}{\mu_0 \mu_3} + \dots\dots \right]$

$\qquad\qquad\qquad [\because B_1 = \phi/A_1,\ B_2 = \phi/A_2 \text{ and } B_3 = \phi/A_3]$

Obviously $\qquad H = \dfrac{B}{\mu}$

So $\qquad IN = H_1 l_1 + H_2 l_2 + H_3 l_3 + \dots\dots$

Q. Define magnetic circuit and compare the electric and magnetic circuits.

Ans. The magnetic circuit consists of the magnetic path followed by the magnetic flux.

The magnetic flux in the magnetic circuit is given as

$$\text{Flux} = \dfrac{\text{m.m.f.}}{\text{Reluctance}}$$

It is similar to that of the current in electric circuit so it is called as the Ohm's law in magnetic circuit.

Table 11.2. Comparison between Electric Circuit and Magnetic Circuit

Electric circuit	*Magnetic circuit*
1. E.m.f. is the force which causes the current to flow.	m.m.f. is the force which causes the flux to establish.
2. $\text{Current} = \dfrac{\text{e.m.f.}}{\text{Resistance}}$	$\text{Flux} = \dfrac{\text{m.m.f.}}{\text{Reluctance}}$
3. $\text{Resistance} = \dfrac{\rho l}{a} \Omega$	$\text{Reluctance} = \dfrac{1}{\mu \cdot a} \text{AT/Wb}$
4. Resistance $\propto$ length	Reluctance $\propto$ length
$\propto \dfrac{1}{\text{Area of cross-section}}$	$\propto \dfrac{1}{\text{Area of cross-section}}$
5. $R = R_1 + R_2 + R_3 + \ldots\ldots$ (for series combination)	$\text{Rel} = \text{Rel}_1 + \text{Rel}_2 + \text{Rel}_3 + \ldots\ldots$ (for series combination)
6. For parallel combination $\dfrac{1}{R} = \dfrac{1}{R_1} + \dfrac{1}{R_2} + \dfrac{1}{R_3} + \ldots$	For parallel combination, $\dfrac{1}{\text{Rel}} = \dfrac{1}{\text{Rel}_1} + \dfrac{1}{\text{Rel}_2} + \dfrac{1}{\text{Rel}_3} + \ldots$
7. $\text{Current density} = \dfrac{\text{Amp}}{\text{m}^2}$.	$\text{Flux density} = \dfrac{\text{Flux}}{\text{m}^2}$.
8. In an electric circuit, current actually flow.	In the magnetic circuit flux does not actually flow.
9. Energy is needed as long as the current flows.	Energy is needed only to create the flux not to maintain it.
10. Current reduces to zero after removing the applied electromagnetomotive force.	The flux persists even after removing the magnetomotive Force.

Example. *A magnet has a pole strength of 1.5 × 10⁻³ Wb and the rectangular cross-section 0.8 × 2.3 cm. Calculate the field strength at a distance of 15 cm from the magnetic pole in air.*

Solution. The pole strength

$$m = 1.5 \times 10^{-3} \text{ Wb}$$

The distance = 15 cm = 0.15 m

The field strength $= \dfrac{m}{4\pi \cdot \mu \cdot \mu_r \cdot d^2}$ $(\because \ \mu_r = 1 \text{ m air})$

$$= \frac{1.5 \times 10^{-3}}{4\pi \times 4\pi \times 10^{-7} \times 1 \times (0.15)^2}$$

$$= 4221.71 \text{ N/Wb. Ans.}$$

Example. *A permanent magnet of a circular shape has a circular area of cross-section of 2 cm². If the flux produced is 1.55 m Wb calculate the flux density in the air gap.*

Solution. The area of cross-section

$$= 2 \text{ cm}^2$$

The flux produced = 1.55 m Wb

∴ Flux density $= \dfrac{\phi}{A} = \dfrac{1.55 \times 10^{-3}}{2 \times 10^{-4}}$

$$= 7.75 \text{ Wb/m}^2. \quad \text{Ans.}$$

Example. *A solenoid 10 cm long consists of 1000 turns of a super-enamelled wire uniformly wound over a hollow cylindrical bobin. The outside diameter being 4 cm calculate the flux density with in solenoid and total flux produced, if the current is 5 A.*

Solution. The length of the solenoid = 10 cm = 0.1 m.

Diameter is 4 cm so the area of the cross-section.

$$= \frac{\pi D^2}{4} = \frac{\pi (0.04)^2}{4}$$

$$= 0.001257 \text{ m}^2$$

The number of turns = 1000 turns

The flux density $= B = \dfrac{\mu_0 \cdot N \cdot I}{l}$

$$= \frac{4\pi \times 10^{-7} \times 1000 \times 5}{0.1}$$

$$= 0.0628 \text{ Wb/m}^2$$

The total flux = flux density × Area

= 0.0628 ×12.57 × 10⁻⁴ **Ans.**

Example. *A solenoid 1.5 m long consists of 5000 turns of wire uniformly wound over an insulated bobin having outside diameter 0.038 m, if the flux produced is 56.4 × 10⁻⁶ Wb. Calculate the current supplied and the flux density in the centre of the solenoid.*

Solution. The solenoid length = 1.5 m

The number of turns = 5000 turns

The diameter = 0.038 m

So the area of cross-section $= \dfrac{\pi D^2}{4}$

$$= \frac{\pi \cdot (0.038)^2}{4} = 1.1341 \times 10^{-3} \text{ m}^2$$

The total flux $\phi = B \cdot A$

$$56.4 \times 10^{-6} = B \times 1.1341 \times 10^{-3}$$

$$B = \frac{56.4 \times 10^{-6}}{1.1341 \times 10^{-3}} = 0.04973 \text{ Wb}/\text{m}^2$$

Obviously $B = \dfrac{\mu_0 N I}{l}$

Here $N = 5000$, $l = 1.5$, $\mu_0 = 4\pi \times 10^{-7}$

and $I = \dfrac{Bl}{\mu_0 N} = \dfrac{0.04973 \times 1.5}{4\pi \times 10^{-7} \times 5000} = \textbf{11.87 A Ans.}$

Example. *A conductor lying perpendicular to a magnetic field of 0.65 Wb/m² flux density is carrying a current of 10 amperes. Find the force acting on the conductor.*

Solution. Force exerted $= B \cdot I \cdot l$ N

$$= 0.65 \times 10 \times 1 = 6.5 \text{ N/m}$$

Here $l = 1$ m, so the force per unit length is **6.5 N/m. Ans.**

Example. *The air gap of an electromagnetic relay is 6 mm long. It is desired to operate with a flux density of 0.5 Wb/m² in the air gap. Calculate the ampere-turns to produce the same flux density.*

Solution. The flux density

$$= 0.5 \text{ Wb/m}^2.$$

and $B = \mu_0 H$

The total ampere-turns required

$$= Hl$$

$$= 0.7958 \times 10^{-6} \times B \times l$$

$$= 0.7958 \times 10^{-6} \times 0.5 \times 6 \times 10^{-2}$$

$$= 2387.4 \text{ AT}$$

$$\cong \textbf{2388 AT. Ans.}$$

Example. *Find the ampere-turns required to produce a flux of 0.4 m-Wb in the air gap of the magnetic circuit as shown in Fig. 11.22. The* μ_r = *1800 and leakage coefficient 1.15.*

Solution. The flux density

$$= \frac{\text{flux}}{\text{area}} = \frac{0.4 \times 10^{-2}}{0.0004}$$

$$= 1 \text{ Wb/m}^2$$

The length of the air gap $= 0.0005$ m

$$\therefore \quad AT_g = \frac{Bl}{\mu_0} = 1 \times 0.005 \times 0.7958 \times 10^{-6}$$

$$= 3979 \text{ AT}$$

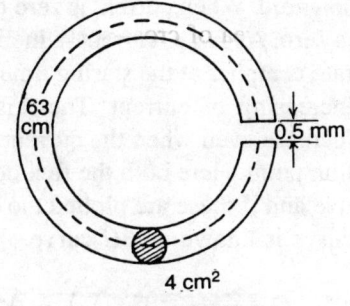

Fig. 11.22.

The flux density in iron cores

$$= \text{Flux density in air} \times \text{Leakage coefficient}$$
$$= 1 \times 1.15 = 1.15 \text{ Wb/m}^2$$

The length $= 63 \text{ cm} = 0.63 \text{ m.}$

$$\therefore \qquad AT_i = \frac{0.7958 \times 10^{-6} \times 1.15 \times 63 \times 10^{-2}}{1800}$$

$$= 321 \text{ Turns}$$

Now total ampere-turns

$$= AT_g + AT_i = 3979 + 321 = 4300 \text{ AT}$$

So total $\qquad AT = \textbf{4300 AT. Ans.}$

11.6. HYSTERESIS

Q. What do you understand by the magnetic hysteresis? Describe the BH curve and hysteresis loop.

Ans. The phenomenon of lagging of magnetisation or induction density behind the magnetising force when a specimen of ferromagnetic material is taken through a cycle of magnetisation is known as the magnetic hysteresis.

BH Curve. The curve which shows the relationship between the magnetising force and magnetic flux density of a magnetic substance is known as the BH curve, where B – the flux density *i.e.* ϕ/A and H – the magnetising force *i.e.* IN/l.

Let the coil has N turns and l, the length which is constant in this particular case. So $H \propto I$ and the current shown by ammeter will itself indicate the magnetising force.

Let the specimen AB be an unmagnetised soft iron piece, which is wound as shown in Fig. 11.23. The magnetising force can be increased or decreased by increasing or decreasing the current through the wound solenoid. When current is zero the *H* is also zero and flux density initially is zero. As *H i.e.* current is increased the flux density also increases. The rate is higher at the staring time and the rate gradually decreases with the increasing of current. There is one stage when flux density does not increase even when the current is increased, that is known as the saturation point. Here both the flux density and the magnetising force are positive and if these are plotted the curve will be as shown in Fig. 11.24. The curve is known as *BH* curve.

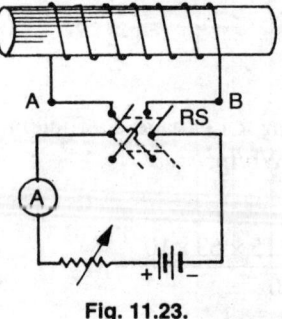

Fig. 11.23.   **Fig. 11.24.** B.H. curve.

Uses. 1. It is used to find out the saturation point of the magnetic materials, so it is useful for designing purpose.

2. It is used to find out the permeability *i.e. B/H* of the material.

Hysteresis loop. The specimen *AB* is subjected to the different magnetising force, more or less, positive or negative. Similarly the flux density s produced will also be effected and changed. Let the magnetising force *H*, here *I*, is increased in positive direction, then flux density *B* is also increased up to the saturation point. When gradually the magnetising force is decreased, the flux density does not follow the same curve and even if the magnetising force is zero, the flux density is not zero but follow *ab* as shown in Fig. 11.25. At 'b' though the magnetising force is zero but *Ob* is the magnetic flux density left behind the magnetising force and is known as the *residual magnetism.*

To demagnetise the specimen, the reversed magnetising force is to be used, which is '*Oc*' and is known as *coercive force.* Now if the current is further increased, the piece will be magnetised in the reversed order till the saturation. Again when the force is decreased the curve takes it new position as '*de*' and here *Oe* is the residual magnetism and *Of* is the coercive force which is applied to demagnetise the residual magnetism. If this magnetising force is increased the curve finally comes to point *'a'* making a closed loop *'a, b, c, d, e, f, a'* and this loop is known as the *hysteresis loop.*

In this loop *'Ob'* and *'Oe'* represent the residual magnetism and *'Oc'* and *'Of'* represent the coercive force.

The hysteresis results due to the fact that the molecules of a magnetic material are not perfectly elastic, once arranged in an orderly fashion by the magnetising force, the molecules, molecular magnetic do not return exactly to their original position, when the magnetising force is removed, but retains some residual magnetism. This results in lagging of *B* behind the *H* and called magnetic hysteresis.

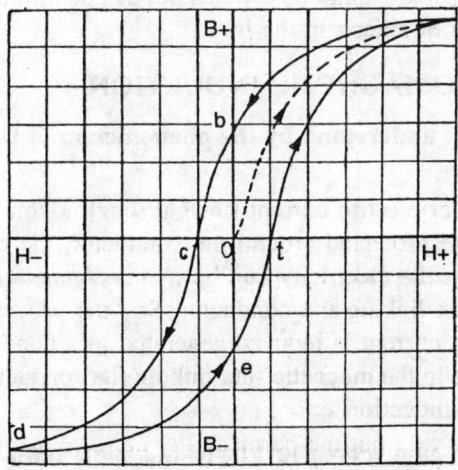

Fig. 11.25. Hysteressis loop.

Hysteresis loss. The coercive force does not preform any useful work. It is a kind of loss and is known as hysteresis loss. It produces heat.

The hysteresis loss depends upon:

(a) *Area of the hysteresis loop.* The larger the area of the loop greater will be the hysteresis loss.

(b) *Frequency of reversal of magnetisation.* The loss is directly proportional to the frequency of reversal.

(c) *Volume of the material.* Greater will be the volume of the material larger will be the hysteresis loss.

(d) *The flux density.* This loss is directly proportional to the flux density $(B_m^{1.6})$.

(e) *Type of the material.* It also depends upon the type of the material.

Now the hysteresis loss can be given as

$$W_h = \eta \cdot B_m^{1.6} \cdot f \cdot V \text{ watts}$$

where W_h = Hysteresis loss in watts

η = Steinmitz constant or hysteresis constant

B_m = The maximum flux density in Wb/m^2

f = Frequency

V = Volume of the material.

Effects. It has the following effects:

(i) More hysteresis loss produces more heat.

(ii) Loss of electrical energy is increased.

(iii) As heat is increased so the insulation of the machine may be damaged or weakened.

The hysteresis loss cannot be avoided but can be minimised by adding some percentage of silicon in the iron.

11.7. ELECTROMAGNETIC INDUCTION

Q. What do you understand by the phenomenon of Electromagnetic Induction?

Ans.   Whenever electric current flows through a conductor magnetic lines of force are produced around the conductor, the converse is also equally true, that the electricity can be induced/generated by changing the magnetic flux linking the conductor. So "the process by which an e.m.f. and hence current is induced/generated in a conductor whenever there is a change in the magnetic flux linking the conductor, is called the electromagnetic induction".

Q. Describe Faraday's laws of Electromagnetic induction.

(N.C.V.T. 1972, 79, 80)

Ans.   These are the following laws, which are known as the Faraday's laws of electromagnetic induction.

1st law. According to the first law of electromagnetic induction, "whenever the flux linking with a coil or circuit changes, an e.m.f. is induced in it".

This linkage can be obtained either by rotating the conductor in the magnetic field or by rotating the magnetic field keeping the conductor stationary.

2nd Law. According to the second law of electromagnetic induction, "the magnitude of the induced e.m.f. in a coil is directly proportional to the rate of change of flux linkage".

$\therefore$ $e \propto$ rate of change of flux linkage.

Let there be a coil having 'N' turns, the initial flux being ϕ_1 and the final flux value after time 't' is ϕ_2.

The net change in flux
$$= \phi_2 - \phi_1$$
and the rate of change of flux linkage
$$= \frac{N\phi_2 - N\phi_1}{t}$$

Now according to Faradays laws of Electromagnetic induction. The e.m.f. $\propto$ rate of change of flux linkage

$$\propto \frac{N\phi_2 - N\phi_1}{t} = k \frac{(\phi_2 - \phi_1)}{t} \times N \text{ volts}$$

where k is the constant of proportionality and here being unity

So
$$e = \frac{(\phi_2 - \phi_1)}{t} \times N \text{ Volts}$$

Induced e.m.f. = rate of change of flux × No. of conductors and it can be otherwise stated as

$$e = \frac{d\phi}{dt} \times N \text{ Volts}$$

Example. *A magnetic flux of 0.004 Wb is made to link a coil having 500 turns in 0.1 sec find the average induced e.m.f.*
Solution. The induced e.m.f. = rate of change of flux linkage.

Here $N = 500$, $t = 0.1$ sec and flux 0.004 Wb

$$\therefore \quad e = \frac{d\phi}{dt} \times N$$

$$= \frac{0.004}{0.1} \times 500$$

$$= 20 \text{ V. Ans.}$$

Example. *A coil of 100 ohm is placed in a magnetic flux of 0.1 m Wb. The coil has 500 turns and a galvanometer of 400 ohm resistance is connected in series with it. The coil is moved in 0.1 sec from the given field to 0.3 Wb find,*

(a) *An average induced e.m.f.*

(b) *Average current through the coil.*

Solution. The change of flux is from 0.3 m-Wb to 0.1 m-Wb in 0.1 sec and coil has 500 turns.

So the induced e.m.f. = rate of change of flux linkage

$$= \frac{(0.3 - 0.1) \times 10^{-3}}{0.1} \times 500$$

$$= \frac{0.2 \times 10^{-3}}{0.1} \times 500 = 1 \text{ V}$$

The induced e.m.f.   **= 1 V. Ans.**

The current $= \dfrac{\text{Volts}}{(R_m + R_{ex})} = \dfrac{1}{100 + 400} = \dfrac{1}{500}$

= 0.002 A or 2 mA.   Ans.

Q. Define the followings:
   (*a*) **Lenz's law.** (*b*) **Fleming's right hand rule.** (*c*) **dynamically induced e.m.f., self-induction and its uses, mutual induction and eddy currents:**

Ans. (*a*) **Lenz's law.** The e.m.f. induced, according to the Faradays laws of electromagnetic induction, has got only the strength of the magnitude and is silent over the direction. The direction was stated by Lenz's laws in 1835.

According to Lenz's law, the direction of the induced e.m.f. because of the electromagnetic induction is such a way as to oppose the cause which is responsible for the production of this e.m.f.

The equation can be given as

$$e = -\frac{d\phi}{dt} \times N \text{ volts}$$

The minus sign indicates the direction of the induced e.m.f. *i.e.* opposing.

   (*b*) **Fleming's right hand rule.** This rule is applied to find out the direction of the induced e.m.f. in a conductor. According to this law, stretch your right hand in such a way as to keep the thumb, middle finger, and forefinger at right angle to each other. If the thumb indicates the direction of motion and the forefinger indicates direction of the

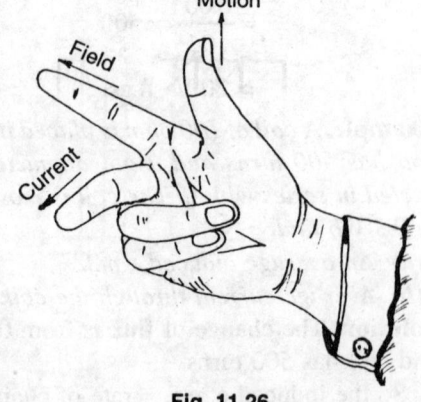

Fig. 11.26.

magnetic flux, then the middle finger will indicate the direction of the induced e.m.f. in the conductor as shown in Fig. 11.26.
The e.m.f. can be induced by the following ways:

(*i*) Dynamically induced e.m.f. (*ii*) Statically induced e.m.f.

(*i*) **Dynamically induced e.m.f.** The induced e.m.f. can be given as the rate of change of flux multiplied by the number of conductors. The production of the e.m.f. can be illustrated as shown in Fig. 11.27. The coil is connected across the galvanometer. The bar magnet has certain magnetic pole strength. If the coil is kept stationary and the bar magnet is moved as indicated, the flux linking the conductors will induce the voltage as long as the change of flux is continued. The e.m.f. induced is proportional to the rate of change of flux linking. When the magnet is brought back, the flux decreases and the e.m.f. is produced in the reverse direction. But if the magnet is made stationary at any stage, there will not be any change of flux, resulting no induction and hence no e.m.f.

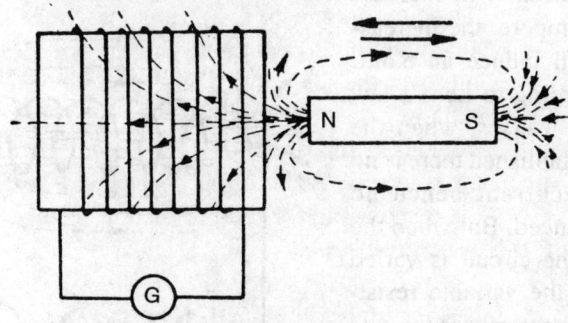

Fig. 11.27.

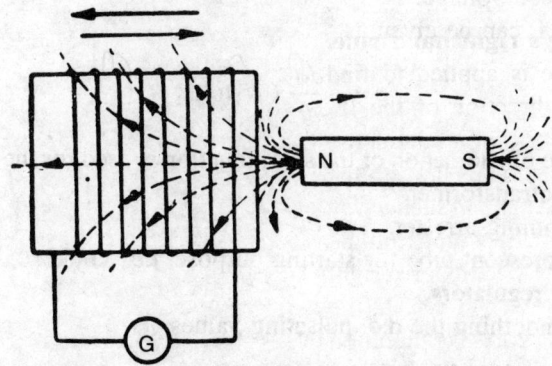

Fig. 11.28.

It is also observed as shown in Fig. 11.28. that if the magnet or the magnetic flux is made stationary and the conductor (coil) is moved than the e.m.f. is induced depending upon the flux linkage. If there are more number of turns the e.m.f. induced will be more. The d.c. generator is an example of dynamically induced e.m.f.

Statically induced e.m.f. Whenever a conductor links with the changing flux an e.m.f. is induced in that conductor. In this system there is no moving part. The static induction can be of two types, the self induction and the mutual induction.

Self Induction. The property of the circuit by virtue of which an e.m.f. is induced in the same circuit in which the current is changed is known as self inductance. This induced e.m.f. is known as self induced e.m.f. and the phenomenon is known as self induction.

The Fig. 11.29 Shows a coil connected to a battery through an ammeter, variable resistance so that the current can be varied. When the switch is on the current will increase from zero ampere, the increasing flux will induce an e.m.f. which according to Lenz's law opposes the causes, when the current is established there is no change of current hence no e.m.f. is induced. But when the current in the circuit is varied by varying the variable resistance the current causes the rate of change of flux and induction of self-induced voltage.

This e.m.f. can be given as

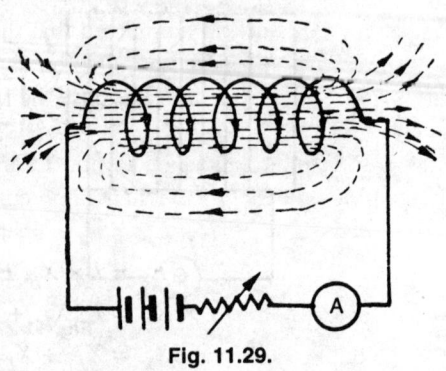

Fig. 11.29.

$$e = -\frac{d\phi}{dt} \times N \text{ volts}$$

Uses. The phenomenon of the self induction is used for the followings:
1. In autotransformer.
2. In lightning arrester.
3. In fluorescent tube for starting purpose, *i.e.*, choke.
4. In fan regulators.
5. For smoothing the d.c. pulsating values.

Coefficient of self inductance (L)

It is represented by letter L. The coefficient of self inductance of a coil is defined as the weber turns per ampere in the coil *i.e.*

$$L = \frac{N\phi}{I}$$

The unit of self induction is henry. If in the above expression $I = 1A$, and $N\phi = 1$ Wb. Turn than $L = 1$ Henry.

Hence a coil will have a self inductance of one henry if a current of one ampere when flowing through it produces a flux linkage of 1 Wb, Turn in it.

It can otherwise be defined as "a coil has a self inductance of one henry if a change of one ampere/sec causes an induction of one volt in it."

Combination of inductances

The combination of inductances can be as under;
- (*a*) Inductances in series.
- (*b*) Inductances in parallel.
- (*c*) Inductances is series & parallel.

(*a*) *Inductances in series*

When the coils are so connected that the end of coil one is connected to the starting of second coil and so on as to form a shape of a chain, these are connected in series. In this case the flux of first coil does not effect the second coil at all.

Let their inductances be L_1 and L_2 and are connected across E volt of f-frequency. In that case when same current is flowing in both the coils.

$$E_R = E_1 + E_2$$
$$E_r = I \times X_L, E_1 = I \times X_{L1} \qquad E_2 = I \times X_{L2},$$

or $\qquad I \times X_L = I \cdot X_{L1} + I X_{L2}$

or $\qquad X_L = X_{L1} + X_{L2} \qquad\qquad (X_L = 2\pi f L)$

So $\qquad 2\pi f L = 2\pi f L_1 + 2\pi f L_2$

or $\qquad L = L_1 + L_2$

It is a very ideal case. But in actual practice the mutual induction takes place and let M be the Coefficient of mutual induction. Than equivalent inductance,

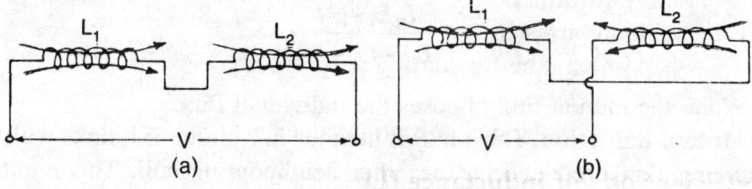

(a)   (b)

Fig. 11.30.

(*i*) If both the coils are connection in such a way as to result a additive *mmfs* than equivalent inductance is

$$L = L_1 + L_2 + 2M.$$

(*ii*) If both the coils are connected in such a way as to result a subtractive *mmfs* than equivalent inductance is

$$L = L_1 + L_2 - 2M$$

(b) Inductance is parallel

When the coils are so connected as to have a common voltage i.e. their startings are connected at one end and their ends are connected at another end across a voltage, these are said to be connected in parallel, here

$$I = I_1 + I_2$$

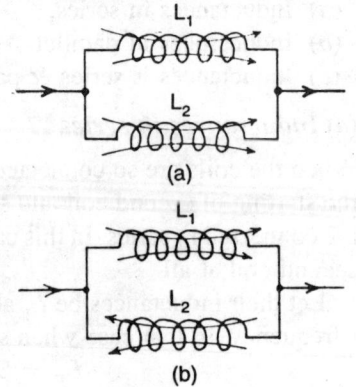

$$\frac{E}{X_L} = \frac{E}{X_{L_1}} = \frac{E}{X_{L_2}}$$

or

$$\frac{E}{2\pi f L} = \frac{E}{2\pi f L_1} + \frac{E}{2\pi f L_2}$$

$$\frac{1}{L} = \frac{1}{L_1} + \frac{1}{L_2}$$

(a)

(b)

Fig. 11.31.

Thus the reciprocal of total inductance is equal to the sum of reciprocal of individual inductances connected in parallel.

In actual practice the flux links or effects the other coil and the coefficient of mutual inductance *M*, comes into existence,

Then

$$L = \frac{L_1 L_2 - M^2}{L_1 L_2 - 2M}$$

When the mutual flux helps the individual flux.

and

$$L = \frac{L_1 L_1 - M^2}{L_1 + L_2 + 2M}$$

When the mutual flux opposes the individual flux.

Mutual induction. The e.m.f. is induced in a circuit if it links with the changing flux produced by the other neighbouring coil. This e.m.f. is known as mutually induced e.m.f. and the phenomenon is known as the mutual induction.

The figure shows two coils *A* and *B* placed side by side. A galvanometer is connected across the coil *B* and coil *A* is connected to a battery through an ammeter and variable resistance to vary the current. When the switch is on, the current changes producing a changing magnetic field which links the coil *B* and according to the Faradays laws of electromagnetic induction the e.m.f. is induced in coil *B*. The induction of voltage will last only as long as the change of flux is there. This e.m.f. is known as mutually induced e.m.f. and the phenomenon of induction as the mutual induction.

Uses. This principle is used in two winding transformers on a.c. where the direction and magnitude of the current always keeps on changing.

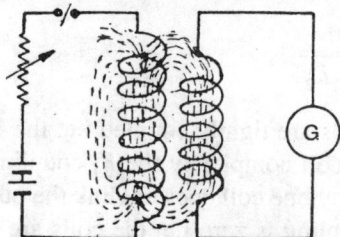

Fig. 11.32. Mutual induction.

Coefficient of mutual induction (M)

It is denoted by the letter M.

If there are two coils having N_1 and N_2 number of turns; than the coefficient of mutual inductance between the two coils is defined as the weber-turns in one coil due to one ampere current in the other neighbouring coil.

Let there be I, ampere current flowing in one coil produces ϕ_1 Wb. flux. Now this flux links with the second coil without loss of any flux.

Now the flux linkage (Weber turns) in the second coil for unit current in the first coil are

$$= \frac{N_2 \cdot \phi_1}{I_1}$$

Now by definition we can say that

$$M = \frac{N_2 \phi_1}{I_1} \text{ Henry}$$

if $N_2\phi_1 = 1$ and $I = 1A$ in that case $M = 1$, which can be defined as the two coils are said to have a mutual inductance of 1H if one ampere current when flowing in one coil produces a flux linkage of one Wb-turn in the other neighbouring coil.

It can otherwise be defined as two coils will have a mutual inductance f one henry if a current changing at the rate of one ampere per second in one coil induces an e.m.f. of one volt in the other coil.

Coefficient of coupling

Let L_1 = Coefficient of self inductance of coil 1

        L_2 = Coefficient of self inductance of second coil.

        M = Coefficient of mutual inductance,

Than $M = K\sqrt{L_1 L_2}$

Or $K = \dfrac{M}{\sqrt{L_1 L_2}}$

When the two coils are tightly coupled i.e. the flux produced by coil links with the other coil completely the K, coupling coefficient is unity. If the flux produced by one coil does not link the other coil than the value of coefficient of coupling is zero i.e. the coils are magnetically isolated from each other.

Eddy currents. The eddy currents are known as the Faucault currents. Whenever a material is placed in an alternating magnetic field, the eddy currents are induced in it. It is because the material is subjected to the rate of change flux linkage and in accordance with the Faradays laws of electromagnetic induction. The direction of these currents is always opposing the cause to produce them. These currents causes loss of energy (I^2R) in the material where I is the induced eddy current, R the resistance of the material. It results in the heating of the material.

The eddy current loss is proportional to the square of frequency, square of the thickness of the material and inversely proportional to the resistivity of the material. This is the main consideration that the magnetic cores to be used in an alternating magnetic field are made of thin sheets, called lamination, instead of a solid block. An insulation is also used in between the laminations, to avoid the short circuiting. The insulation may be of thin layer of insulating varnish, a paper or sometimes a sheet of paper. An efficient insulation for silicone steel sheet is a film of phosphate chemically deposited on the surface.

The eddy current losses are proportional to the square of the flux density.

i.e. $W_e \propto B^2 \cdot f^2 \lambda$

REVIEW QUESTIONS

A. State whether true or false:

(*i*) Magnetite is a natural magnet.

(*ii*) The northern pole of a freely suspended magnet points exactly towards the geographic north pole.

(*iii*) Attraction of unlike poles in the sure test for magnets.

(*iv*) If a magnet is broken into two pieces, one piece is a north pole and the other piece is a south pole.

(*v*) The molecular magnets within a domain point in random direction is an unmagnetised piece of iron.

(*vi*) A south pole brought near one end of an unmagnetised iron bar induces a south pole at that end.

(*vii*) Electric current flowing through a wire develops a magnetic field around itself.

(*viii*) If a current flowing through the coil of an electromagnet is increased, the strength of the electromagnet increases.

(*ix*) Heating a magnet can make it weaker.

(*x*) Keepers, used in storage of magnets, are made of woods.

(*xi*) Electromagnets are temporary magnets.

(*xii*) Two bar magnets should never be stored with their unlike poles side by side.

Fill in the blanks

(*i*) Magnetism was first discovered in an iron ore called ....

(*ii*) Man made magnets are called .... magnets.

(*iii*) The points of a magnet where the magnetic force is the strongest is called its ....

(*iv*) .... is used for making permanent magnets.

(*v*) .... is suitable for use as electro magnets.

(*vi*) .... is the sure test for magnets.

(*vii*) A .... magnet is dumb-bell shaped.

(*viii*) The region around a magnet where its magnetic inference can be experienced is known as its ....

(*ix*) The soft iron pieces placed across the poles of a horse shoe magnet is called ....

(*x*) The instrument having a magnetic needle which is used for indicating the direction of magnetic field is known as ....

Electric Circuit

Fill in the blanks:

1. The heat produced in a conductor depends upon the .... and the .... for which the current is passed.

2. A energised solenoid behaves like a .... magnet.

3. An .... behaves like a magnet when electric current is passed through it.

4. The strength of the magnetic field produced by an electromagnet depends upon the .... and .... in the electromagnet.
5. The polarity of an electromagnet can be found out by .... rule.

MCQ

1. An ammeter is used to measure ....
   (a) current   (b) voltage   (c) power   (d) energy
2. A voltmeter is used to measure ....
   (a) current   (b) voltage   (c) power   (d) energy
3. A charged body with excess electrons is
   (a) negatively charged body   (b) positively charged body
   (c) neutral   (d) None of the above.
4. The rechargeable cell used in torch is
   (a) Dry cell   (b) Lachlanche cell
   (c) Lead acid cell   (d) Nickle cadmium cell

Fill in the blanks:

1. The electrons in motion constitutes ....
2. The cells in which the chemical reaction is not irreversible are ....
3. In dry cell the positive terminal is ....
4. In Bunsen cell the positive terminal is made up of .... and negative is of ....
5. In Lachlanche cell the positive terminal is of .... negative of .... and the electrolyte is ....
6. The direction of current is .... to that of the direction of electrons.

NUMERICAL PROBLEMS

1. A solenoid of one metre long and 1 cm in diameter uniformly wound with 500 turns of super enamelled copper wire. Find the magnetic flux density at the centre of the solenoid, when it is carrying a current of 4 amperes.
   (**Hint.** $H = NI/l$)   [**Ans. 2000 AT/m**]
2. An air cored coil has a flux density of 2.5 Wb/m². When the iron cores is inserted in the coil, the flux density increases to 2500 Wb/m². What is the μ of the iron? (**Hint.** $\mu = B/H$)   [**Ans. 1000**]
3. A conductor of 25 cm long on a periphery of an armature of diameter 50 cm rotates at 1000 r.p.m. If the field strength under the pole is 0.6 Wb/m², find the average e.m.f. induced in the conductor.   [**Ans. 3.927 V**]
4. A straight long conductor of 3 m length carrying a current of 50 A is placed at right angle to uniform magnetic field strength of 1.2 Wb/m². Determine the mechanical force acting on the of 1.2 Wb/m². Determine the mechanical force acting on the conductor.
   (**Hint.** $F = B \cdot I \cdot l \sin \theta$)   [**Ans. 180 N**]
5. Calculate the reluctance, m.m.f., magnetising force (H) necessary to produce flux density of 0.75 Wb/m² in a transformer core of mean magnetic length 50 cm and cross-section 40 cm². Assume the relative permeability to be 1000 at the above flux density.
   [**Ans. 9.95 × 10⁴ AT/Wb, 298.5 AT, 596.5 AT/M**]

12

D.C. Generator

The d.c. generator is a machine which converts the mechanical energy into d.c. electrical energy. The mechanical energy can be obtained from the different sources *viz.* steam turbine, hydro turbine and diesel engine etc.

12.1. CONSTRUCTION

Q. What is a dynamo?

Ans. The machine which converts mechanical energy into electrical energy or the electrical energy into mechanical energy, is known as the dynamo.

If the mechanical energy is converted into d.c. electrical energy the machine is said to be the d.c. generator; and if the d.c. electrical energy is converted into mechanical energy then the machine is said to be the d.c. motor.

Q. Describe with neat sketches, the different parts of the d.c. generator.

Ans. The parts of d.c. generator can be broadly divided into the followings:

    (*a*) Stationary parts and (*b*) Rotating parts.

Stationary parts. The parts of the d.c. generator which remains stationary during the working of the generator, are known as stationary parts. These are as follows:

Eye bolt, body or yoke, poles, side covers, brushes and rocker, bearings and legs and bed sheet.

Rotating parts. The parts of d.c. generator which rotates during the working of d.c. generator, are known as rotating parts. These are as follows:

Armature, commutator, fan and shaft.

The brief description of these parts is as follows:

(a) Stationary Parts

(i) **Eye bolt.** It is a stationary part and fixed on the top of the body or yoke. It is used for lifting the machine. These are one of two depending upon the frame size of the machine.

(ii) **Body or yoke.** It is the outer frame of the machine. It accommodates all parts of the generator. It is made of forged steel or cast steel or cast iron. The cast steel is used for the machines of large capacity because of the good magnetic properties. The cast iron frames are used for the machines of low capacity: because the magnetic properties are not so good.

It is experienced that the frame size of cast iron and cast steel differ for the same capacity, cast steel requires half the size. Nowadays the yokes of large machines are almost invariably fabricated steel because this material has good magnetic properties, for example the permeability is twice that of cast iron; and has hence the weight.

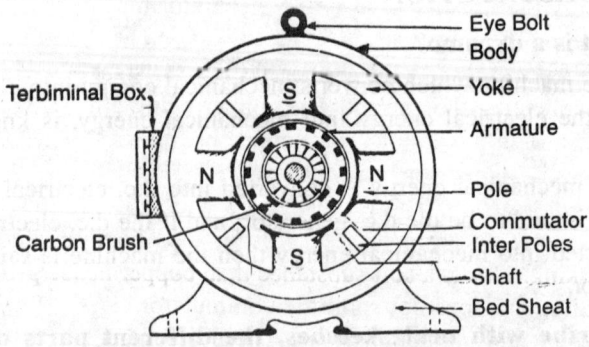

Fig. 12.1. Parts of D.C. Generator.

The yoke provides the magnetic path for the magnetic flux. The flux in yoke is half than that of the flux per pole, because the flux passing diverts in two paths causing half the flux through the yoke. Sometimes the body and yoke of the machine are different but in most of the cases these are same.

(iii) **Poles.** The main magnetic field is produced by the poles excited by the field coils. The poles are made either of the cast iron or soft steel or the laminations of silicone steel. In small machines the poles are casted with the body. In some construction the pole faces are separate and attached to the mountings. Nowadays the complete poles are made by the laminations of silicone steel which are pressed hydraulically and riveted together. The poles are attached with the yoke or body by means of bolts etc.

The field coils are wound with different number of conductors (shunt field coil – having more number of turns of thin conductors series field coil – less number of turns of thick conductor) are placed over the poles, in such a way so that it may not come out. The pole faces are made circular as to provide uniform air gap around the armature and uniform flux density also. The poles are always in pair; *i.e.* 2, 4, 6 and 8 etc.

(*iv*) **Brushes and brush gears.** The main function of brushes is to collect the current from the commutator and supply to the external load circuit. These are housed in rectangular chamber. The brushes are provided with a spring as to offer some pressure which could easily be adjusted by the spring loading finger. The assembly is called brush holder, generally made of brass. These brush holders are mounted over the round construction known as the 'Rocker'. By the displacement of rocker over the commutator, the brush position can be changed.

Generally the brushes used in d.c. generator machines are of copper, carbon and copper plus carbon.

(*i*) *Copper brushes:* As the copper has low resistance; so it can be used for collecting the current. The copper being a hard substance may spoil the commutator. The copper brushes are used for the machines designed for heavy current.

(*ii*) *Carbon brushes.* These brushes are widely used at present. These are made of carbon. These brushes also help in minimising the sparking. This is a soft substance than copper hence producing less friction and does not spoil the commutator.

(*iii*) *Carbon and copper brushes.* These are made of the mixture of copper and carbon. These are having less resistance and less friction. These are used for large capacity generators and for heavy currents also for example in electroplating works etc.

(*iv*) *Terminal box.* It is attached with the body. All main terminals from armature, field windings and interpoles are brought here. The terminals are insulated from each other and fixed on the insulated plate. The terminals are insulated from the body of the machine also. The terminals are marked as $A_1 A_2$ (armature), $Z_1 Z_2$ (shunt field), and Se_1 Se_2 (series field).

The bed sheet is also attached with the body. There the holes are made in the bed sheet so that the machine may be fixed with foundation.

(*v*) *Side covers.* The side covers are used to cover the machine from the sides. There are a number of grooves provided in the side cover, some are used for the air circulation and the central one used for the bearing so that the armature shaft can be in the very much centre of the body. These are attached with the body with the help of the screws or stud and nuts. These covers are the two types, the *front end cover* which is on the commutator side and the *rear end cover*, which is on the other side that is on pulley side. Both the covers are made of cast iron and after proper machining these are fixed.

(vi) *Bearings.* The bearings are fitted in the covers, so as to have minimum friction between the rotating and stationary portions (shaft and side covers). The bearings also help in keeping the armature in the centre for smooth running.

Mainly the bearings are classified into the followings:

(i) *Sleeve bearing.* These are made of gun metal. These are filled in the cover with a reasonable tolerance, so that the movement is smooth. Generally oil is used for periodic lubrication purpose. These are used for low capacity machines. Either the continuous grooves or holes are provided in these bearings for continuous lubrication; otherwise production of due to friction will damage the bearing.

(ii) *Ball bearing.* The bearings are filled in the side covers with the side flanges. The covers are provided to keep the bearings intact and in position. In these bearings the friction is reduced further by reducing the area of contact between both the surfaces of bearing. For housing the steel balls the provision is provided. These are available in different sizes according to the diameter of the numbers. Grease is generally used for lubrication purpose in these bearings. These are widely used nowadays.

(iii) *Roller bearings.* In these bearings steel rollers and used in place of steel balls. These are used for big machines. These bearings can bear more mechanical thrust and life is also good. Grease is generally used as a lubricant. For continuous lubrication the oil or grease should be of good quality and clean and should be changed periodically. The bearings should be cleaned with petrol or kerosene oil. These should never be hammered and there should not be any play.

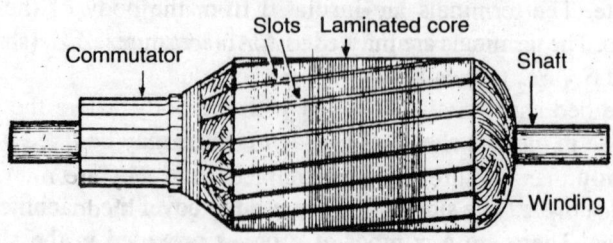

Fig. 12.2. Armature.

(b) Rotating or rotary parts

The parts which revolves during operation is called the rotating parts.

(i) **Armature.** The armature rotates in the magnetic field. The conductors in which the e.m.f. is induced, are housed in the slots of the armature. The armature is made of laminations of silicone steel to reduce the eddy current and hysteresis loss. The laminations are assembled and riveted under hydraulic pressure to avoid any air gap between the laminations. The laminations are insulated from each other by means of the varnish or sometimes thin insulating paper. By adding the silicon in steel, the resistance is increased and thus decreases the eddy currents which in other words decreases the eddy current losses, the lamination instead of a solid block, causes the further reduction in current and ultimately reducing both the eddy current and hysteresis losses. Generally the thickness of each lamination is 0.4 to 0.6 mm.

In order to dissipate the heat, thus produced, some ventilating ducts are provided. The air circulation through these ducts increases the heat dissipation and keeps the machine temperature under specified limits.

Function of armature. It solves two purposes:

(i) It accommodates the armature conductors.

(ii) It provides the low reluctance path for the magnetic flux.

The conductors are placed/housed in the slots. The slots are generally of the following types:

(i) *Closed type.* These types of the slots are used for large machines. These are totally closed. The conductors are inserted in the slots one by one.

(ii) *Open type.* In this type of slots are having the same width from top to bottom. Readymade coils are used for these type of slots. The bamboo or fibre wedges are used on the top of the slot to protect the coils from coming out.

(iii) *Semi-enclosed type:* These types of slots are not symmetrical. The size of the slot is less on the top and more at bottom. In this type of slot the coils are inserted turn by turn. The wedges are placed on the top of the slot to stop the possibility of coming out of the coil. This type of slots are commonly used for medium size machines.

Generally the super enamelled copper conductors are used for winding purpose.

In case of large diameters of stampings where the laminations are not conveniently manufactured in one complete unit, they are made in segments. The spider is used to house these segments with dovetailed joints. The spider is made of cast iron and is *keyed* to the shaft.

Commutator. It is made of the hard drawn copper segments which are insulated from each other and from shaft by means of mica or micanite. It is mounted on the shaft of the machine. The copper segments are tapered and there is a riser on one side of each segment. There is a space provided to solder the conductor with the riser or segment as shown in Fig. 12.3. It is round in shape to facilitate the collection of current (in case of generator) from the armature.

The construction of the segment is such, that it has 'V' grooves on both the sides to protect the segment from coming out because of the centrifugal forces. Every segment is properly insulated from every side *i.e.* from segment to segment, from segment to shaft and from segment to the sleeve and 'V' check nuts by means of mica or micanite.

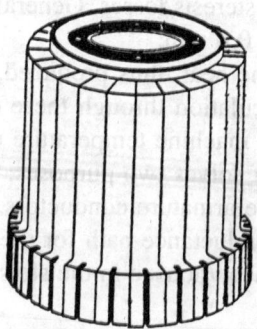

Fig. 12.3. Commutator.

Fan and shaft. A fan is mounted over the shaft in opposite direction of the commutator. It is made of cast iron or thick mild steel sheets. It circulates the air through armature, armature winding etc. to keep the temperature down.

A shaft generally of mild steel is used, which carries the armature, commutator, fan and bearings. The pulley is also mounted on the shaft after the side covers to enable the mechanical energy to the load.

12.2. WORKING PRINCIPLE AND GENERATED E.M.F.

Q. What is the working principle of d.c. generator?

Ans. The d.c. generator works on the principle of "Faraday's laws of electromagnetic induction and that too the dynamic induction". The e.m.f. thus induced or produced is known as the dynamically induced e.m.f.

According to that "principle", whenever a conductor cuts the magnetic lines of force, an e.m.f. is generated in that conductor. The e.m.f. generated is directly proportional to the rate of change of flux *i.e.* the angle of flux linkage. The total e.m.f. generated in the armature is also proportional to the number of conductors.

Q. How many things are essential for the production of dynamically induced e.m.f.?

Ans. The following things are essential for the production of the e.m.f.:

(i) **Magnetic flux.** As the e.m.f. induced is directly proportional to the rate of change of flux, so the magnetic flux is essential. In case of d.c. generator there are magnetic poles which are stationary. These are always in pairs, 2, 4, 6 or 8 etc. The poles may be permanent a electromagnets. In case of electromagnetic type d.c. is used to energise the poles.

(ii) **Conductors.** The total generated e.m.f. is directly proportional to the number of conductors hence a number of conductors are arranged systematically on the armature of d.c. generator. These conductors are insulated from each other. Super enamelled copper conductors are used for winding purpose.

(iii) **Source of prime mover.** The rate of change of flux causes the induction of voltage, hence rotation is must, which is given by means of a prime mover generally the external source. The mechanical energy for motion is obtained from the steam turbine, hydroturbine, atomic energy and diesel engine set also, these turbines are called the prime movers.

Q. How an e.m.f. is induced in a single loop d.c. generator?

Ans. Whenever a conductor links the magnetic flux or cuts the magnetic flux an e.m.f. is induced in the conductor, depending upon the rate of change of flux. Let there be a single loop *ABCD* mounted on the shaft and placed in a uniform magnetic field of definite flux density.

The conductor is moved by means of a mechanical power (not shown here) so that the flux linkage changes continuously, resulting an induction for dynamically induced e.m.f. This e.m.f. at any particular instant is proportional to the rate of change of flux. The direction of this e.m.f. can be found out by means of Fleming's right hand rule.

Different positions. Now let the conductor be rotated in clockwise direction at a uniform speed, the rate of flux linkage is changing, as

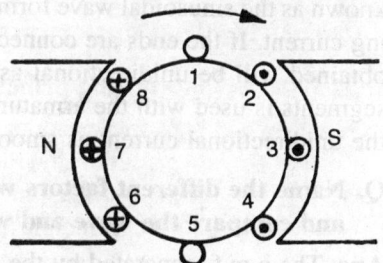

Fig. 12.4. Generation of e.m.f.

a result the e.m.f. is induced. This e.m.f. has got different instantaneous values at different instants as shown in Fig. 12.4.

When the conductor *AB* is on the top or say at position no. 1 and conductor *CD* on the bottom, the direction of movement of the conductor and flux linking are parallel. Here the linkage is zero hence no e.m.f. is induced (zero e.m.f. at *MNP*). As the coil moves ahead, the conductor start cutting the magnetic flux and as there is rate of change of flux experienced by the conductor, there is a production of the e.m.f. According to the Fleming's right hand rule, the direction of induced e.m.f. is in inward direction in *CD* and outwards in *AB* conductor.

When the conductor *AB* is moved further ahead *i.e.* just at right angle to the main magnetic field, it will experience the maximum rate of change of flux, resulting the production of maximum e.m.f. In the next quarter the rate of change of flux reduces to zero at *MNP* and reducing this e.m.f. to zero value. All these positions of conductor *AB* are shown in Fig. 12.5.

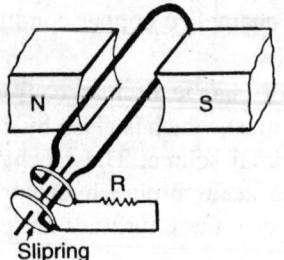

Fig. 12.5. Slipring method.

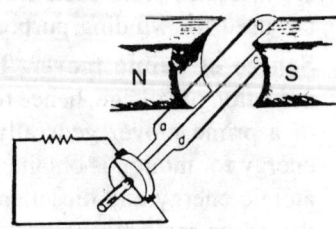

Fig. 12.6. Spltring method.

When the conductor moves ahead from *MNP* under north pole, the direction of movement is upward and according to Fleming's right hand rule the direction of induced e.m.f. in conductor *AB* changes from outward to inwards. This e.m.f. will be maximum at magnetic axis and minimum (zero) at magnetic neutral axis or plane.

Now the two ends of the coil are connected to two slip-rings shown in Fig. 12.5, the e.m.f. thus obtained if plotted, it will have a wave as shown, known as the sinusoidal wave form. The nature of this current is Alternating current. If the ends are connected to the split rings then the wave so obtained will be unidirectional as shown in Fig. 12.6. An assembly of segments is used with the armature, known as commutator just to make the unidirectional current as smooth or steady direct current.

Q. Name the different factors which the e.m.f. depends. Also define and compare the wave and windings.

Ans. The e.m.f. generated by the generator depends upon the following main factors:

(*i*) **Flux.** The e.m.f. generated depends upon the flux; more flux more e.m.f. and less flux less e.m.f. *i.e.*

$$E \propto \phi$$

The flux is also a depending factor upon the flux per pole and the number of poles.

(*ii*) **Number of conductors.** The e.m.f. generated is directly proportional to the number of conductors, more conductors more e.m.f. less conductor less e.m.f.

(*iii*) **Speed or rate of change of flux.** The e.m.f. induced is directly proportional to the rate of change of flux and that is otherwise related with the speed in r.p.m. which actually constitutes the rate of change of flux, more speed more rate of change of flux and *vice versa*.

(*iv*) **Number of parallel paths.** It is an important factor for the e.m.f. or voltage used for external circuit. It depends upon the type of winding, the lap winding and wave winding. An armature winding will have the parallel paths of the winding depending upon the type of winding and the voltage per parallel path will be the voltage used for external circuit.

Wave winding. The armature winding which has only two parallel paths what so ever be the number of poles of the machine may be, is known as the wave winding as shown in Fig. 12.7.

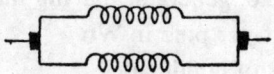

Fig. 12.7. Wave winding.

Lap winding. The armature winding in which the number of parallel paths are same as that of the number of poles in the machine is known as the lap winding as shown in Fig. 12.8.

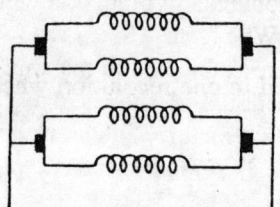

Fig. 12.8. Lap winding.

Table 12.1. Comparison between wave winding and Lap winding

Wave winding	Lap winding
1. There are only two parallel paths and are independent of the number of poles of the machine.	The parallel paths are as many as the number of poles in the machine and depends upon the poles.
2. The total load current is equally contributed by the parallel paths hence current per parallel path will be half the load current.	The load current is contributed by each parallel hence current per parallel path will be less.
3. It is used for high voltage and low current.	It is used for low voltage and high current.
4. The set of brushes are equal to two only.	The set of brushes are equal to the sets of winding.

Q. State the e.m.f. equation of d.c. generator.

Ans. According to Faradays laws of electromagnetic induction, if a conductor cuts the magnetic flux or it experience a rate of change of flux, an e.m.f. is generated in that conductor, this e.m.f. can be given as

$$e = \frac{d\phi}{dt} \text{ V}$$

Now here in case of d.c. generator, having the following datas:

ϕ = The flux per pole in Wb

Z = Number of conductors

N = Speed in r.p.m.

P = Number of poles

a = Number of parallel paths

and E = the induced e.m.f. in volts.

Now the total flux produced

$$= (P \times \phi) \text{ Wb}$$

Now the time required in one revolution when running at N r.p.m.

$$= \frac{1}{N} \text{ min.}$$

$$= \frac{60}{N} \text{ sec}$$

Now e.m.f. induced = rate of change of flux

$$= \frac{\phi P}{\frac{60}{N}} \text{ V}$$

$$= \frac{\phi \times P \times N}{60} \text{ V}$$

It is the voltage induced per conductor. But there are Z conductors so the e.m.f.

$$= \frac{\phi \cdot Z \cdot N \cdot P}{60} \text{ V}$$

In d.c. armature winding the closed type winding is done so the total armature conductors (Z) are divided into the number of parallel paths depending upon the type of winding. The voltage taken out will be the voltage induced per parallel path. Here let 'a' be the parallel paths, so the e.m.f. generated can be given as

$$E = \frac{\phi \times Z \times N \times P}{60 \times a} \text{ V}$$

The above equation is said the e.m.f. equation of the generator.

12.3. TYPE AND CHARACTERISTICS OF GENERATORS

Q. Name the different types of d.c. generators. Describe the permanent magnet type d.c. generator.

Ans. The d.c. generators are classified as under.

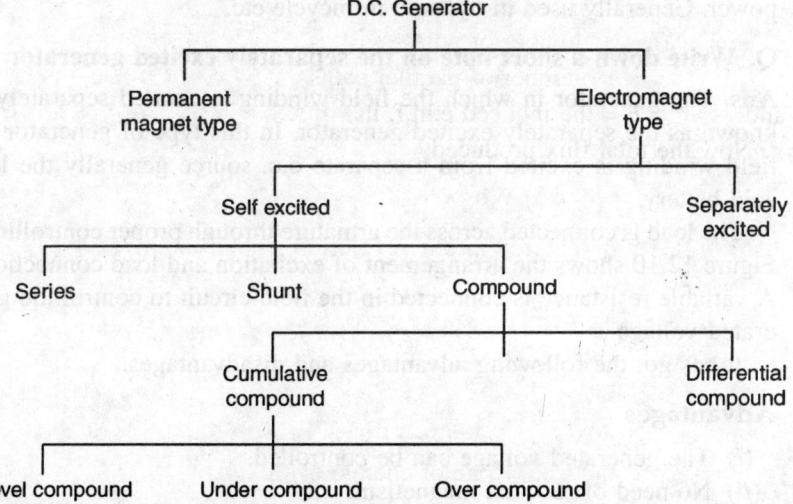

Permanent magnet type d.c. generator. In this type of d.c., generator the flux is produced by the permanent magnets as shown in Fig. 12.9. The permanent magnets are fixed with the yoke or body. The armature is rotated by means of the prime mover in the magnetic poles.

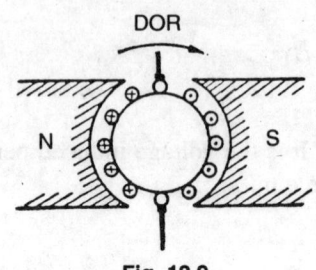

Fig. 12.9.

Operation. Whenever a conductor rotates in the magnetic field and it experience the flux linkage, an e.m.f. is induced in it.

Here in permanent magnet type generator the field is produced because of the permanent magnets. The armature is rotated in the field, hence the e.m.f. is produced in the armature depending upon the speed of the armature. This e.m.f. is taken for external load through the brushes placed on the commutator.

This generator has got the following advantages and disadvantages:

Advantages. No need of any field excitor.

Disadvantages. These are the followings:

(i) The voltage cannot be controlled and regulated as the field strength is constant.

(ii) Because of the aging factor, the field strength decreases and the generated voltage also decreases.

(iii) Polarity of induced e.m.f. cannot be changed.

Characteristics. The voltage falls down with the increasing of load.

Uses. This generator is used for low voltage, low current, and low power. Generally used in dynamo in bicycle etc.

Q. Write down a short note on the separately excited generator.

Ans. The generator in which the field winding is excited separately, is known as the separately excited generator. In this type of generator the field winding is excited from a separate d.c. source generally the lead acid battery.

The load is connected across the armature through proper controllings. Figure 12.10 shows the arrangement of excitation and load connections. A variable resistance is connected in the field circuit to control the generated voltage.

It has got the following advantages and disadvantages.

Advantages

(i) The generated voltage can be controlled.

(ii) No need of residual magnetism.

(*iii*) The load can be connected at the time of starting.

(*iv*) Polarity of induced e.m.f. can be changed.

Disadvantages

(*i*) External source of supply is required for field excitation.

(*ii*) More space is required

(*iii*) Cost of the complete set is more.

Characteristics. (*i*) The voltage induced increases with the increasing of field excitation up to saturation as shown in Fig. 12.11(*a*).

(*ii*) The voltage falls down with the increasing of load as shown in Fig. 12.12(*b*).

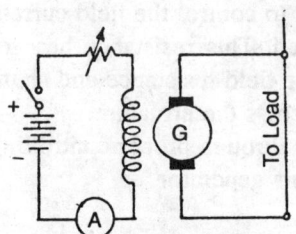

Fig. 12.10. Separately excited Generator.

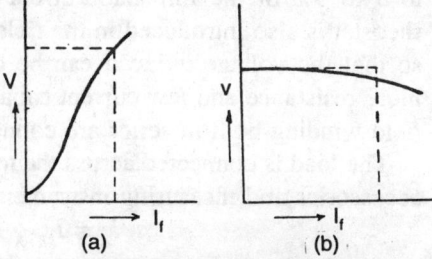

Fig. 12.11(a) O.C.C. curve (b) L.C.C. curve – Characteristics curve

The drop is because of the armature drop and armature reaction.

Uses. This type of generators are suitable for low voltage and low current in electroplating and refinery of metals etc.

Q. What do you understand by the self-excited generators? Describe with neat sketches the shunt field generator.

Ans. The generator in which the exciter is not required but the field winding is excited by the same generator, is known as the self-excited generator. In this case the field winding is also energised by the armature of the same machine.

These are of the following types:

(*i*) Shunt generator

(*ii*) Series generator

(*iii*) Compound generator.

Shunt generator. The self-excited generator in which the field winding is connected across the armature of the same machine is known as the shunt field generator as shown in Fig. 12.12.

The field winding is designed to be connected across the armature. In other words this field winding has to withstand the generated voltage. Therefore the resistance of the winding is more. It is wound with thin

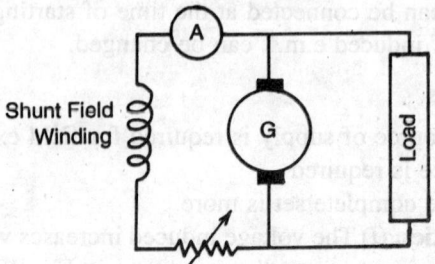

Fig. 12.12. Shunt Generator.

wire having more number of turns. Generally the field current is limited to 3 to 5% of the full load current capacity of the generator. A field rheostat is also introduced in the field circuit to control the field current so that the voltage induced can be controlled. This resistance has got more resistance and low current capacity. The field resistance and shunt field winding both in series are connected across the armature.

The load is connected across the terminals through proper controlling accessories and measuring instruments. In this generator

$$I_a = I_{sh} + I_L$$

where I_a = armature current, I_{sh} = shunt field winding current and

$$I_L = \text{the load current.}$$

Building up of the voltage. In this generator the residual magnetism is the fundamental requirement for the production of the e.m.f. The load switch is put off before starting the generator. The field regulator is also kept out of circuit. Whenever the prime mover is started, the conductor moves in the magnetic field produced by the residual magnetism and generating some voltage. Now this e.m.f. will circulate some current depending upon the field winding resistance. This current will again strengthen the magnetic field produced. This increased magnetic field will again generate more voltage and this more voltage will again cause more current and generating more voltage. This will continue until the saturation point comes and hence establishing the rated voltage. Now this generated voltage can be regulated by regulating the field rheostat.

Now the load switch can be on.

Advantages. These are the following advantages:

  (i) No need of external field excitor.
  (ii) Less space is required.
  (iii) Cost is less.
  (iv) Maintenance cost is also less.
  (v) Voltage regulation is easy.

Disadvantages. There are the following disadvantages:

(*i*) The generator cannot be started with load.

(*ii*) If the direction of current in the field winding is reversed the residual magnetism will be destroyed and then no e.m.f. will be induced.

Characteristics.

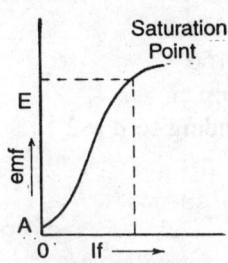

Fig. 12.13. O.C.C. curve.

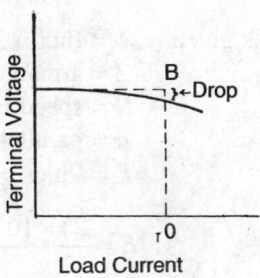

Fig. 12.14. L.C.C. Curve

Uses. It is a general purpose generator. Generally it is used for excitation purpose in alternator, lighting load, battery charging and electroplating etc.

Example. *A shunt generator supplies a load of 50 kW at 220 V through a pair of feeders of total resistance 0.04Ω armature resistance is 0.15Ω, shunt field resistance 50Ω. Find the terminal voltage and e.m.f. generated.*

Solution. The load current

$$= \frac{\text{Power}}{\text{Voltage}} = \frac{50 \times 1000}{220} = 227.3 \text{ A}$$

Now voltage drop in feeders

$$= I \times R = 227.3 \times 0.04 = 9.09 \text{ V}$$

The voltage across armature,

$$V_{ter} = V_L + V_{feeder} = 220 + 9.09$$
$$= 229.09 \text{ V. Ans.}$$

Shunt field current $= \dfrac{V_{sh}}{R_{sh}} = \dfrac{229.09}{50} = 4.58 \text{ A}$

Armature current $I_L + I_f = 227.3 + 4.58 = 231.88 \text{ A}$

Armature drop, $V_a = I_a \cdot R_a = 231.88 \times 0.15 = 34.78 \text{ V}$

E.m.f. generated, $E = V_L + V_{feeder} + V_a$
$$= 220 + 9.09 + 34.78$$
$$= 263.87 \text{ V. Ans.}$$

Example. *A four pole wave wound armature has 160 conductors and runs at 1200 r.p.m. If the flux per pole is 25 m-Wb. Find the e.m.f. generated.*
Solution. The formula for calculating the e.m.f generated by a d.c. generator is

$$E = \frac{\phi \cdot Z \cdot N \cdot P}{60 \cdot a} \text{ V}$$

here as given is, ϕ = flux is Wb = 25×10^{-3} Wb
$\quad\quad$ Z = armature conductors = 160
$\quad\quad$ N = speed in r.p.m. = 1200 r.p.m.
$\quad\quad$ a = parallel paths, wave winding so a = 2
$\quad\quad$ P = Poles = 4

$$\therefore \quad E = \frac{25 \times 10^{-3} \times 160 \times 1200 \times 4}{60 \times 2}$$

$$= \textbf{160 V. Ans.}$$

Example. *A 4 pole, lap wound armature when driven at 900 r.p.m. generates 200 V. If the flux per pole is 0.03 Wb. Find the number of armature conductors.*
Solution. The e.m.f. generated,

$$E = \frac{\phi \cdot Z \cdot N \cdot P}{60 \cdot a} \text{ V}$$

$$Z = \frac{E \times a \times 60}{\phi \cdot N \cdot P}$$

$$Z = \frac{200 \times 4 \times 60}{0.03 \times 900 \times 4}$$

$$= 444.4 \cong 445 \text{ conductors. } \textbf{Ans.}$$

Q. Write short note on the D.C. series generator. Also states how it can be used as a booster?

Ans. It is a self excited generator. In this case the field winding is connected in series with the armature. The load, the armature and field windings are connected in series, therefore same current flows through the armature, series field and load. The field winding has low resistance and less number of turns of thick wire. The armature also has less resistance: as show in Fig. 12.15.

It is very essential to keep the load switch on for generating the voltage. The induced voltage inherently depends upon the field current *i.e.* the load current up to the saturation point. Therefore the generator has the rising characteristics. The voltage increases with the increasing of load; and after saturation point the voltage falls down with the increasing of load.

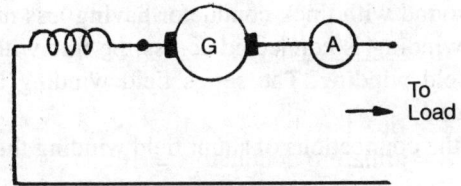

Fig. 12.15. Series generator.

Disadvantages.

(*i*) As the circuit complete via load, so it is very essential for generation to keep the load switch on.

(*ii*) Constant voltage can only be possible, in case of constant load.

Uses.

(*i*) It is used where the load is constant.

(*ii*) It is used as a booster in d.c. distribution lines.

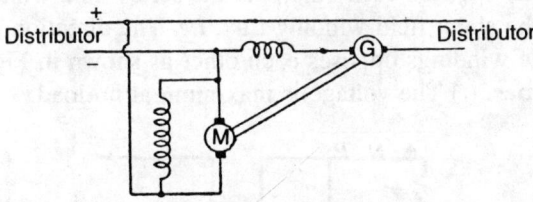

Fig. 12.16. Booster.

Booster. As shown in Fig. 12.16. in the long d.c. distribution lines the d.c. series generator is used as a booster. We know the voltage produced is proportional to the current in the field winding *i.e.* the load current in case of d.c. series generator.

The voltage drop in the distribution line is proportional to the load current $(I_L \times R_f)$ and the consumer will not get the constant voltage. Just to keep the voltage constant at the consumers end the series generator is put in the line to maintain the constant voltage. The series generator is driven by the d.c. shunt motor. The generator will induce the voltage just proportional to the load current and that voltage is injected in the distribution line to boost the voltage after checking the proper polarity.

Q. What do you understand by the d.c. compound generators; explain with the help of neat sketches?

Ans. The generator, in which both the series and shunt field windings are used is known as compound generator. Both the windings are placed on the same pole. The shunt field winding has more resistance wound with thin conductor having more number of turns. The series field winding has low resistance wound with thick conductor having less number of turns. The shunt field winding is connected across the line with or sometimes without series field winding. The series field winding is connected in series with the load.

According to the connections of shunt field winding the generators are classified as

Short shunt compound generator. The generator in which the shunt field winding is directly connected across the armature is known as short shunt compound generator as shown in Fig. 12.18(a).

Long shunt compound generator. The generator in which the shunt field winding is connected across the armature through the series field winding, is known as long shunt compound generator.

Here the voltage drop in series field = $(I_L + I_{sh}) \times R_{se}$.

Types of compound generators. These are mainly two types depending upon the direction of current in series and shunt field windings.

(a) **Differential compound generator.** In this type of generator both the fields opposes each other, *i.e.* the series field winding flux opposes the shunt field winding flux, *i.e.* The direction of current in both the windings opposes each other as shown in Fig. 12.17.

Characteristics. (i) The voltage is maximum at no load.

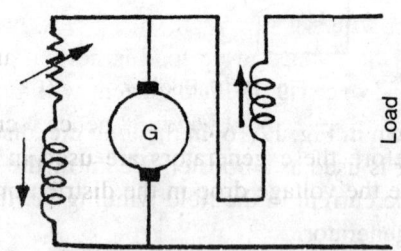

Fig. 12.17. Differential compound generator.

(ii) Voltage falls down with the increasing of load, very rapidly.

Uses. Generally it is used for the welding purpose, because at the time of starting it gives maximum voltage which drops down rapidly in short circuiting condition: which helps in producing and maintaining the are between the plate and electrode. The voltage depends upon the length of the arc, long arc more voltage and short arc low voltage.

(b) **Cumulative compound generator.** In this type of generator the series field winding helps the shunt field winding flux *i.e.* the direction of current in both the series and shunt field winding flux *i.e.* the direction of current in both the series and shunt field windings is same, helping each other as shown in Fig. 12.8. The compound generators are further classified depending upon the number of turns in the series field winding. These are:

(*i*) Over compounded generator;

(*ii*) Level compounded generator;

(*iii*) Under compounded generator.

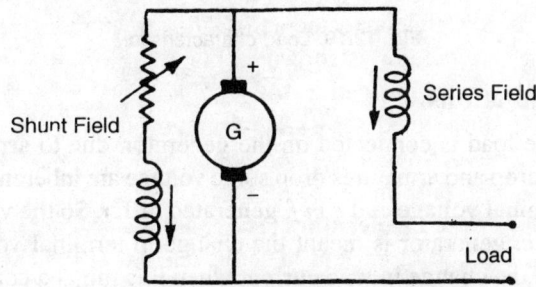

Fig. 12.18. Cumulative compound generator.

(*i*) **Over compounded generator.** This generator has more number of series field winding turns; as compared with other types of compound generators. The series field winding is connected in series with the armature winding as to allow the load current to flow through the series field winding.

In this generator the voltage at no load is normal and increases with the increasing of load because the load current will increase the amount of flux produced by the series field winding, hence increasing the generated voltage. Therefore these generators are used in long distribution lines, to compensate the voltage drop in the distributors and to keep the voltage constant at the consumer's end.

Level compounded generator. In this generator the voltage remains constant from no load to full load. In series field winding the number of turns are less than the over compounded generator. In this case the series field winding generates the extra voltage to compensate the voltage drops due to armature drops and armature reaction also.

These are used for the distribution purpose in factories, houses etc. which are not at a far distance from the generator.

Under compounded generator. In this generator it has less number of turns in series field winding than even the level compounded generator. The voltage at no load is maximum and decreases with the increasing of load. These are used for low voltage and high currents near them, *i.e.* for factories etc. The figure for all over level and under compound generators is shown in Fig. 12.19.

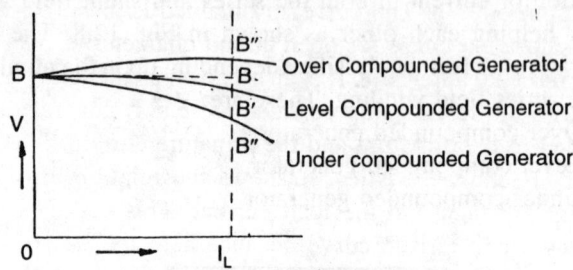

Fig. 12.19. Load characteristic.

Voltage regulations

Whenever the load is connected on the generator, due to armature reaction, contact drop and armatures drop some voltage are inherently dropped. Thus the terminal voltage and *e.m.f.* generated differ. So the voltage regulation of a *d.c.* generator is meant the change in terminal voltage (rated voltage) with the change in load current when it is run at a constant speed i.e. let the *e.m.f.* generated be E, and terminal voltage be V, than,

$$\% \text{ Voltage regulation} = \frac{E - V}{V} \times 100$$

So the voltage regulation is defined as the change in rated voltage when the load is reduced to zero, expressed in percentage of rated load voltage. If the load current is taken the full load current then the regulation is full load voltage regulation and so on.

If the change in voltage from no load be full load is small the regulation a good and if large the voltage regulation is poor.

For example as shown in fig, 12-20.

The voltage OB' or OB'' or DB''' are considered as load voltage, depending upon the type of generator.

$$\% \text{ regulation} = \frac{OB - OB'}{OB'} \times 100$$

(OB' can be replaced by OB'' or OB''' depending upon the type of generator)

Q. Name the different characteristics of d.c. generators, also draw and explain the external characteristic curve of d.c. shunt generator.

Ans. There the following characteristics:

(*i*) **Open circuit characteristics (O.C.C.).** It is also known as the magnetising characteristic of the d.c. generator. It shows the relationship between the no load generated voltage (or e.m.f.) and the exciting current (I_f). The curve so obtained is also known as the no load saturation curve for open circuit characteristic curve *n*.

(*ii*) **Internal characteristic.** It is also known as the total characteristic. It shows the relationship between the e.m.f. generated or actually induced in the armature and the armature current.

(*iii*) **External characteristics.** It shows the relationship between the terminal voltage and the load current. This curve lies below the internal characteristic curve because here the terminal voltage is taken after considering the armature drops.

External characteristic curve of d.c. shunt generator. As shown in Fig. 12.20 the field regulator is so adjusted as to give the normal voltage. The speed and field excitation are strictly watched and kept constant.

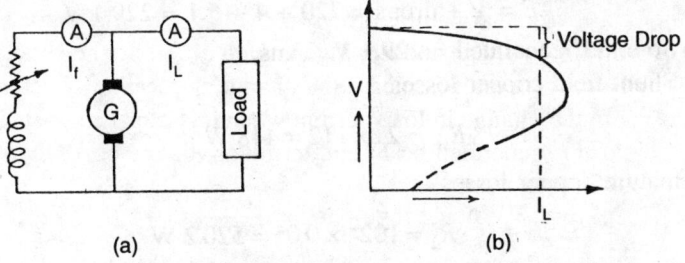

(a) (b)

Fig. 12.20. Shunt generator with load.

Now the load is put on the machine and increased gradually. Both readings for current and voltage are taken. The voltage will drop as shown in Fig. 12.20, it is because of the armature drops ($I_a R_a$) and the armature reaction drops, brushes contact drop and the cumulative drop of the field current because of less voltage.

As the load is increased beyond the full load capacity of the machine, then after a certain value the curve start turning back as shown and voltage falls down very rapidly. The fall is also because the operation is now on the unsaturated (knee) portion of the magnetisation curve of d.c. shunt generator. The drop in voltage can be compensated by adjusting the field regulator.

Example. *A short shunt compound generator supplies 100 A at 220 V. Armature resistance 0.05Ω, shunt field resistance is 112 Ω, series field resistance 0.04 Ω. Iron and friction losses amount to 960 W. Find*

(*a*) *e.m.f. generated*

(*b*) *copper losses; (individual)*

(*c*) *overall efficiency.*

Solution. (*a*) The load current is 100 A at 220 V so the drop in the series field is

$$= I_{se} \cdot R_{se} = 100 \times 0.04 = 4 \text{ V}$$

So voltage across armature

$$= V_L + \text{drop} = 220 + 4 = 224 \text{ V}$$

The shunt field current

$$= \frac{V_{sh}}{R_{sh}} = \frac{224}{112} = 2 \text{ A}$$

Now armature current

$$= I_L + I_f = 100 + 2 = 102 \text{ A}$$

Now armature drop,

$$V_a = I_a R_a = 102 \times 0.05 = 5.1 \text{ V}$$

So the e.m.f. generated,

$$E = V + \text{drops} = 220 + 4 + 5.1 = 229.1 \text{ V}$$

So e.m.f. generated = **229.1 V.** **Ans.**

(*b*) Shunt field copper losses

$$= I_{sh}^2 R_{sh} = 2^2 \times 112 = 448 \text{ W}$$

Armature copper losses

$$= I_a^2 \cdot R_a = 102^2 \times 0.05 = 520.2 \text{ W}$$

Series field copper losses

$$= I_{se}^2 \cdot R_{se} = 100^2 \times 0.04 = 400 \text{ W}$$

Total copper losses = 448 + 520.2 + 400 = **1368.2 W. Ans.**

(*c*) Over all efficiency

$$= \frac{\text{Output}}{\text{Input}} \times 100$$

Input = output + losses

$$= 100 \times 220 + 1368.2 + 960 = 24328.2 \text{ W}$$

So $\quad \% \eta = \dfrac{100 \times 220}{24328.2} \times 100 = \mathbf{90.4\%}$ **Ans.**

12.4. ARMATURE REACTION

Q. What do you understand by the armature reaction in d.c. genera-tor? Explain with the help of suitable sketches.

Ans. The armature reaction is basically the effect of the flux produced by the current carrying armature conductors on the main magnetic flux. So, "the armature reaction is defined as the effect of the magnetic field produced by the armature conductors on the distribution of the flux under the main poles".

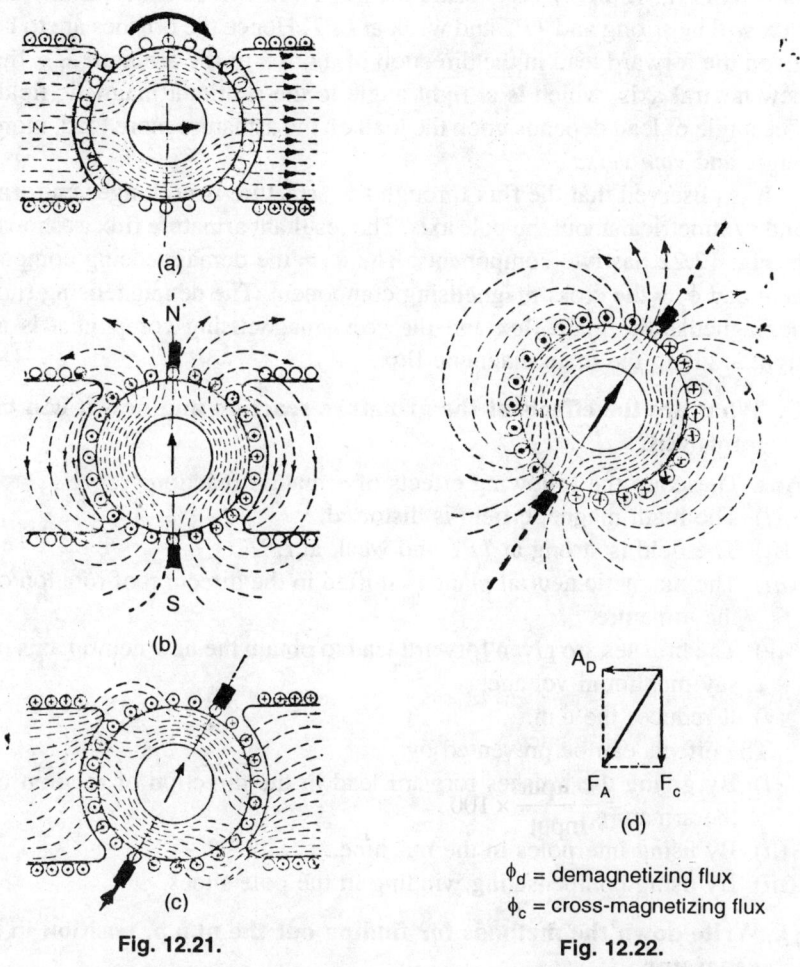

(a)

(b)

(c)

Fig. 12.21.

ϕ_d = demagnetizing flux
ϕ_c = cross-magnetizing flux

Fig. 12.22.

Now consider an armature rotating in the magnetic field produced by the two pole machine, whenever load is not connected on the armature so no current is flowing through the armature conductors, the magnetic field produced is merely because of the main magnetic field as shown in Fig. 12.21(*a*).

According to Fleming's right hand rule the conductors under north pole are carrying current in such a direction as shown. Now the magnetic field produced by the armature conductors is shown in Fig. 12.21(b). The magnetic flux will contribute the magnetic belt in both sides of *MNA* and hence the direction of armature flux is upwards. Now two magnetic fluxes, because of main magnetic field and because of the armature conductors are working at right angle to each other when these are energised as shown in Fig. 12.21(c). As a result, the main flux will be distorted and the flux will be strong and *TPT* and weak at *LPT*. Hence the brushes are to be given the forward lead in the direction of motion of the armature, *i.e.* the new neutral axis, which is at right angle to the resultant magnetic field. The angle of lead depends upon the load on the armature more load, more angle and *vice versa*.

It is observed that the flux through the armature is no longer uniform and symmetrical about the pole axis. The resultant armature flux as shown in Fig. 12.22 has two components: The ϕ_d = the demagnetising component and ϕ_c = the cross-magnetising component. The demagnetising flux demagnetise the main flux and the cross-magnetising component is at right angle to the main magnetic flux.

Q. What are the effects of the armature reaction and how it can be improved?

Ans. These are the following effects of armature reaction:

(i) The main magnetic field is distorted.

(ii) The field is strong at *TPT* and weak at *LPT*.

(iii) The magnetic neutral plane is shifted in the direction of rotation of the armature.

(iv) The brushes are given forward lead to obtain the new neutral axis or say maximum voltage.

(v) It reduces the e.m.f.

The effects can be prevented by:

(i) By giving the brushes forward lead in the direction of rotation of the armature.

(ii) By using interpoles in the machine.

(iii) By using compensating winding in the pole shoes.

Q. Write down the methods for finding out the m.n.p. position in a generator.

Ans. The magnetic neutral plane is the plane in which the conductors do not have any e.m.f. induced in them. This position can be determined by the following methods:

(a) By means of voltmeter

(b) By galvanometer.

(c) With the help of an ammeter.

(*a*) **By means of voltmeter.** In this method connect a suitable voltmeter across brushes, run the generator and adjust the position of brushes. The position where the maximum voltage is obtained will be the m.n.p. and the generator gives maximum voltage with minimum sparking, as shown in Fig. 12.23.

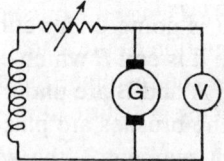   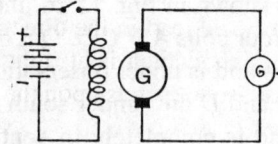

Fig. 12.23. Voltmeter method.   **Fig. 12.24.** Galvanometer-method.

(*b*) **By galvanometer.** In this case as shown in Fig. 12.24 the small current is passed in the field winding of the generator. A centre zero galvanometer is connected across the brushes. At m.n.p. the galvanometer will show zero and by changing the position on the commutator the direction of deflection will also change when the switch is put on or off.

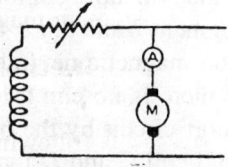

Fig. 12.25.

(*c*) **With the help of an ammeter.** This method is generally adopted in case of d.c. motor. Connect one ammeter in series with the armature as shown in Fig. 12.25. At m.n.p. the ammeter reading will be minimum.

12.5. COMMUTATION

Q. What do you understand by commutation in D.C. generator? What are its effects and states the methods of improving the commutation?

OR

Explain with the help of neat sketches, what is commutation and how to improve it? What do you understand by the compensating winding?

Ans. Commutation means, the changes takes place in an armature coil during the period of short circuiting by the brush. Obviously the e.m.f. induced in a armature coil is of alternating nature. The e.m.f. changes its direction when it passes from one pole to the another pole. If in a coil the direction of current *I*, is in clockwise direction under north pole, it will change or reverse under south pole *i.e.,* in anticlockwise direction. Thus a current of +*I* to −*I* will change causing the total change of 2*I* A current.

The current does not change to −*I* ampere in reverse direction in the same specific time, but takes some more time. The failure of current to reach in full value in reverse direction is the main reason of sparking at commutator.

Here as shown in Fig. 12.26, the coil *B* is going under commutation. There are four coils *A B C D*. Out of which it is coil *B* which experiences the changes and is under observation. Coil A and B are under north pole and coil *C* and *D* are under south pole. The brushes are placed at *MNP* position and is completely in contact with segment 3, the width of the brush is equal to one segment and one mica. The direction of current *I*, coil *A* and *B* is in anti-clockwise and in coil *C* and *D* current *I* ampere in clockwise direction. Thus resulting 2*I* ampere current flowing out from brushes.

Now the armature, commutator is moving ahead resulting the brush to be on segment 2 and 3, *i.e.* the B which is connected on segment 2 and 3 is now on magnetic neutral plane between north and south poles. Now actually there is no e.m.f. in this coil being on *MNA* and secondly it is being short circuit by the brush. Now the current will be flowing from coil *A* and coil *C* and *D* ignoring coil *B*.

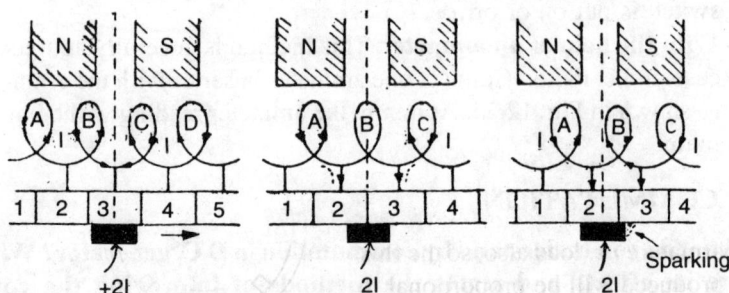

Fig. 12.26. Commutation

In the third position coil *B* is proceeding towards south pole, here the direction of current will be reversed. The current which is zero will now try to reach *I* ampere in opposite direction. But the current is opposed by the voltage induced in the coil because of the inductance of the coil *B* or say the reactance voltage of the coil *B*. The current is coil *C* is of full value *I* ampere but does not have full value in coil *B*. So partially some current (the opposed amount) will jump directly from the coil *C* to brush. The current in air is known as the sparking at commutator. After some-times when the reactance voltage dies the amount of current start flowing in full value *I* amperes and there will not be any jumping of current from coil *C* to brush so there will not be any sparking.

Reason of sparking. The main reason of sparking at the brush is the failure of current to reach to its full value in opposite direction in the coil which is going under commutation. The failure is due to the reactance voltage, when the current starts from 0 to *I* amperes in opposite direction, this rise in current induces voltage in the coil which is due to self-induction. According to Lenz's law the e.m.f. will oppose the cause which is responsible for the production of this e.m.f. so the reactance voltage will oppose the rising current and will cause the sparking on commutator.

Effects. There are the following effects on the commutation:

(*i*) Sparking on the commutator and sparking is proportional to the load current.

(*ii*) Due to sparking life of brushes and commutator is less.

(*iii*) The commutator will be blackish causing the decreasing in efficiency of the commutator.

(*iv*) More temperature rise in the armature.

Method of improving. The commutation can be improved by the following methods:

(*i*) *By means of inter-poles or commutating poles.* These are also known as the auxiliary poles. These are placed in between the main poles. The inter-poles are wound with thick having few turns and are connected in series with the armature poles are small in size than the main poles about half or one third of the main poles as shown in Fig. 12.27. In case of d.c. generator the polarity of the poles is the same as that of the polarity of the forward main pole. As it is connected in series with the armature *i.e.* load also, so the flux produced will be proportional to the load current.

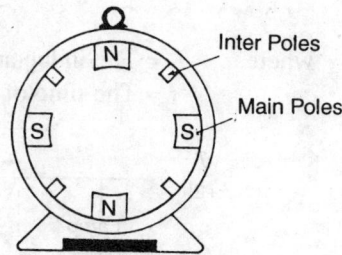

Fig. 12.27. Interpoles

There are the following advantages:

Advantages.

(*i*) The flux of inter-poles helps the current to rise in reverse direction, hence it helps to neutralize the reactance voltage.

(*ii*) Reduces the armature reaction.

(*iii*) The machine can be overloaded upto approximately 30%.

(*iv*) Life of commutator and brushes increases.

(*ii*) **By using high resistance brushes.** The second method of improving the commutation is by the use of high resistance brushes. If the contact resistance of the brushes is high then some circulating current will flow through the coil in reverse direction to that of reactance voltage in the coil undergoing commutation. Thus the reactance voltage is reduced and the commutation is improved.

Compensating winding. In case of large capacity machines for examples in rolling mills etc. the compensating winding is placed in the poles face of the poles. The winding is placed in series with the armature. Thus the magnetic flux produced by this winding is proportional to the load current. The polarity is so produced that the flux is in opposite direction than that of the cross-magnetising field which actually improves the Armature reaction.

Q. What do you understand by the term reactance voltage during commutation?

Ans. Whenever the current during commutation changes from $+I$ ampere to $-I$ ampere, the total change being $2I$ amperes. This rate of change of current causes a flux which is proportional to this current. Now the coil undergoing commutation is linking with this flux, this linkage will induce the voltage in it. This induced voltage according to Lenz's law opposes the cause responsible for the production of this voltage, so it opposes the cause *i.e.* the current to reach to $-I$ ampere in reverse direction. This voltage is known as the reactance voltage and can be given by the formula as

$$= L \times \frac{2I}{t} \text{ V}$$

where L = The inductance of the coil, $2I$ = the current and
t = The time of commutation as shown in Fig. 12.28.

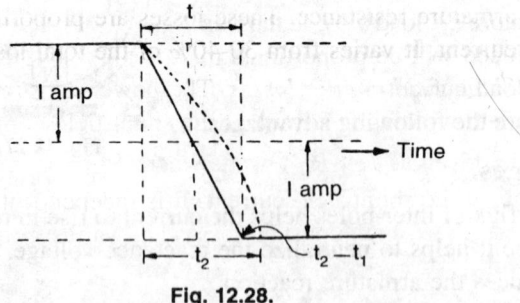

Fig. 12.28.

12.6. LOSSES AND EFFICIENCY

Q. Define the efficiency of a d.c. generator.

Ans. The efficiency of d.c. generator is defined as the ratio of output to input, generally expressed in terms of percentage.

i.e.
$$\%\eta = \frac{\text{Output}}{\text{Input}} \times 100$$

The *output* of the d.c. generator is electrical and is the product of the voltage and current known as wattage or kilowattage. Therefore the generators are rated in kilowatts. This rating shows the maximum amount of current at this rated voltage, so the rating of d.c. generator in

$$kW = \frac{\text{Voltage} \times \text{Full load current}}{1000}$$

The *input* of d.c. generator is the mechanical energy. The generator is driven by means of prime mover. The input is expressed in H.P. or B.H.P. This is the power which actually compensates the output of the generator and the losses.

Q. What are the different losses in the d.c. generator?

Ans. Whenever the mechanical energy is being converted into electrical energy by means of d.c. generator, the losses are inherent. These losses are divided into the followings:
Copper losses, iron losses and mechanical losses.

(*a*) **Copper losses.** These losses are related with the current carrying conductors of the shunt field, armature winding, series field winding and the inter-pole winding.

(*i*) *Armature copper losses.* The power consumed in the armature of the d.c. generator is called the armature copper loss. These are $I_a^2 \cdot R_a$ watts where I_a = the armature current and R_a = the armature resistance. These losses are proportional to the load current. It varies from 30–40% of the total losses.

(*ii*) *Shunt field copper losses.* The power consumed by the shunt field winding is known as the shunt field copper losses. It can be given as $(I_{sh}^2 R_{sh})$ watts where R_{sh} = shunt field resistance, I_{sh} = the shunt field current. It is independent of the load current. These are constant losses and varies from 20–25% of the full load losses.

(*iii*) *Series field copper losses.* It is the $I_{se}^2 \times R_{se}$ watts where I_{se} = current in series field, R_{se} = resistance of series field. It is the total wattage consumed in series field.

(iv) *Inter pole copper losses.* These are also calculated $I_p^2 \times R_{sp}$
where I_p = current in inter-pole and R_p = the resistance of the interpole.

(b) **Iron losses.** The losses which are in the armature cores are known as the iron losses. These are constant losses, because of the flux and speed of the generator. These are hysteresis and eddy current losses.

(i) *Hysteresis.* These losses are due to hysteresis in the armature and other cores if any.

$$W_h \propto B^{1.6} \times \text{speed}$$

It produces heat in the armature.

(ii) *Eddy current losses.* The armature is made of laminations of silicone steel and rotates in the magnetic field. Therefore some currents are inherently induced in the cores, known as the eddy current. These currents causes heat production in the armature, so the energy consumed in these currents and power loss is known as eddy current loses.

$$W_c \propto B^2 \times (\text{speed})^2$$

These losses are 20–30% of the total full losses.

(c) **Mechanical losses.** The power wasted in the mechanical parts that is the friction at bearings and commutator and windage losses due to weight of the armature etc. is known as mechanical losses.
These losses are 10–20% of the full load losses.

12.7. PARALLEL OPERATION OF D.C. GENERATORS

Q. (a) **What is the necessity of parallel operation?**
  (b) **What are the conditions for parallel operation?**
  (c) **Describe the method of parallel operation of the shunt generators.**

Ans. The electrical power is generated at the power stations by means of the generators. Instead of having a single generator of huge capacity, it is always preferable to use more than one generator in parallel, because of the following reasons:

(i) Generators can be put into operation according to the load required, when load is less, one generator may feed and if more others can be started to share the load.

(ii) Efficiency will be good.

(iii) If by chance single unit fails there will be break down and no power, but if more generators are in use they can share the load and faulty generator can be shut down thus the continuity of the supply is maintained.

(*iv*) Generators can be shut down for periodic overhauling. Thus the life of the set is increased.

So it is always better to install more than single unit in the generating station.

(*b*) **Conditions.** There are the following conditions for parallel operation.

    (*i*) The voltage of the incoming machine should be equal to the bus bar voltage.

    (*ii*) The polarity of the incoming machine and bus bar should be identical. The positive of the generator should be connected with the positive of the bus bar and negative with the negative of the bus bar. For checking the polarity a centre zero moving coil voltmeter is generally preferred.

The *bus bars* are two copper strips in this case. These are flat in shape and run behind the switch board. These are marked as positive and negative bus bar. In case of compound generator one extra bus bar known as the equalizing bus bar or equalizer is used. All generators are connected with the bus bar through their circuit breaker and their protections.

(*c*) **Parallel operation of shunt generator.** Figure 12.29 shows as arrangement of two shunt generators with their controlling and measuring devices. Generator No. 1 is already supplying the load and Generator No. 2 is to be brought into operation to share the load.

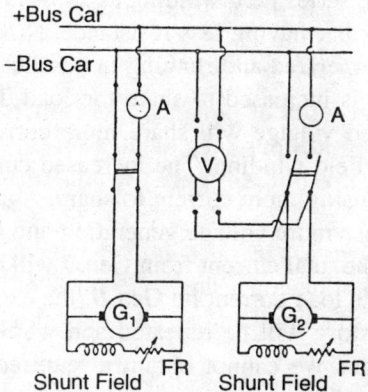

Fig. 12.29. Parallel operation of D.C. shunt Generator.

The main switch of the incoming machine must be in off position and rheostate out of circuit then, the prime mover of generator no. 2. is started. Now as it generates the voltage, the polarity is checked by means of M.C. type instrument. The voltage is then adjusted and regulated by means of the field regulator. It is kept approximately 1% higher than the bus bar voltage. When both the polarity and voltage are checked the main switch of the incoming machine is put in on position. In this condition the incoming machine will neither deliver any power not draw any power from the bus bar, this situation is said 'the *floating* on the bus bar'.

To share the current the voltage of the incoming machine is increased by adjusting the field regulator. Now current will start increasing on this machine and will decrease on the first generator. When the required sharing is achieved the regulator is adjusted and stopped. Now both machines will share the further load automatically. When generator is to be shut down the load is brought to zero ampere by adjusting the field regulator. Then the main switch is put off and then only the prime mover of the set is stopped.

Q. What do you understand by the equalizer or equilizing bus bar in case of compound generators? Describe the method how the d.c. compound generators are run in parallel.

Ans. A number of d.c. generators are run in parallel. In case of compound generators generally the flat loop or slightly over compounded generators are used. Three bus bars are generally used in this case, these are positive bus bar, negative bus bar and equilizing bus bar.

Equalising bus bar. It is generally known as the equalizer. It is used to equalize the load current on the machines. It is connected with the inner terminals of the series field winding as shown in Fig. 12.30. It is a conductor or say bus bar having less resistance. Now if two compound generators are to be operated and running in parallel of the field excitation of any machine is increased to share the load, the voltage will increase. This increased voltage will share more current, resulting more current in the series field winding. The increased current will boost the generated voltage causing more current to share. Again this more series field current will cause more voltage generation and hence more current to share and finally the total current from Gen. *I* will come on to Gen. *II*.

Now if to bring the load current on Gen *II*, the excitation of Gen *II* is increased the same story will be repeated and whole load current will come on Gen *II*. Thus we cannot obtain a required shearing in these circumstances.

To equalize the current on both machines the inner terminal of series fields are connected to a bus bar called equalizer whenever the load increased that will be divided automatically in the series fields causing no instability and the current will be shared.

Utility. If two compound generators Gen. I and Gen. II are to be connected in parallel, they are connected through the main switches having a provision of equalizing bus bar connections too. Let us say the generator no. I is already supplying the bus bar. Now the second generator – II is to be connected in parallel. Open the main switch or say the circuit breaker is off and field regulator is also out of circuit. Now start the prime mover of the generator and adjust the voltage 1 to 2% higher the bus bar voltage check the polarity and bus bar connections also. After being satisfied 'on' the circuit breaker.

Now in this stage the generator will be floating on the bus bar and will not take or supply any power. Increase the field excitation of the incoming machine to share the load. The current start flowing towards machine no. II, when proper sharing is done, stop the field regulation.

Now any further load will be automatically shared. The equalizing bus bar will protect any possibility of instability.

To shut down the generator, decrease the field excitation of the generator and reduce the load current to zero amperes. Now 'off' the circuit breaker and stop the prime mover. It should be clearly understood that in no case the prime mover be stopped before disconnecting the respective generator from the mains.

Q. What is the back torque and states its effects?

Ans. The back torque may otherwise be defined as the magnetic drag. "whenever a current carrying conductor is placed in the magnetic field, a torque is developed over the conductor which tends to move it at right angle to the main magnetic field".

Similarly whenever the load current flows through the armature conductors a magnetic field is set up by the current known as the armature flux. This armature flux is proportional to the load current. There are, thus two magnetic fields, one because of the main poles and other because of armature current. In this case, the effect of the magnetic field when the current carrying (i.e., loaded) armature conductor are placed in it is to produce a torque which will have the tendency to oppose the reason responsible for the production of this *i.e.* the rotation of the armature (Lenz's Law). Thus a torque which is in opposite direction will be produced which is known as the back torque. The strength of this back torque is proportional to the armature current or in other words the load current.

Effects of back torque. The main effect is that the power of the prime mover is always kept more than the maximum back torque developed in the armature so the back torque governs the power of the prime mover.

12.8. DEFECTS AND REMEDIES

Q. Name the different defects, symptoms and their remedies in case of d.c. generator.

Ans. There are the following defects, symptoms and their remedies.

(a) The generator fails to build up the voltage.

S.No.	Defects	Symptoms	Remarks
1.	No residual magnetism.	(i) No reading in voltmeter. (ii) Even if the soft iron piece is placed on the poles it does not exhibits any magnetic property.	Re-excite poles by means of d.c. source.
2.	Reverse direction of rotation.	Sound will be slightly unusual and voltmeter meter reading is zero	Check it and change the d.o.r. of the prime mover.
3.	Open circuit in (i) Field coil (ii) Armature circuit (iii) Field regulator.	Test these separately bymeans of ohmmeter or test lamp.	Check the continuity and connect the broken wire.
4.	Condition of load.	No reading of voltmeter	Check the proper starting (i) Shunt generator should not be started with load. (ii) Series generator must be started with load.
5.	Brushes not in proper contact with the commutator	No reading of voltmeter.	Check the spring tension and make good arrangement for tension so that the brushes are in good contact with commutator.

(b) Heavy sparking at commutator

	Defects	Symptoms	Remarks
1.	Brushes are not in *MNP* position.	Continuous sparking and reduced voltmeter reading.	Check and set the brushes in *MNP* position.
2.	Lack in armature shaft.	Armature will not appear rotating smoothly.	Get it rectified on lathe machine.
3.	Commutator not properly round.	Surface of commutator will not be smooth and will not appear rolling smooth.	Check it and get rectified on lathe machine.

(Contd.)

	Defects	Symptoms	Remarks
4.	Short circuit in segments due to carbon fillings.	Intermittant sparking on the commutator surface.	Clean and under cut the mica of the commutator.
5.	Low spring tension on the brush.	Continuous slow sparking on commutator.	Increase the spring tension.
6.	Short circuit in coil.	Production of heat.	Test the coil by means of ohm meter and remove the fault.
7.	Dirty commutator due to grease or oil or carbon.	Black colour appear on the commutator.	Clean the commutator by means of sand paper, cotton waste or petrol.
8.	Leakage between rocker and brushes holder of opposite polarity.	–	Increase the insulation resistance between the brushes and rocker.
9.	Wrong interpole polarity	–	Test and check the polarity and if wrong change the polarity.
10.	Reverse connection of armature coil.	Irregular sparking on the commutator surface.	Test the armature coil and remove the fault.
11.	More load on the armature.	Ammeter reading exceeds.	Reduce the load on the machine.

(c) Heat in the generator.

	Defects	Symptoms	Remarks
1.	Sparking on commutator.	–	Remove the sparking.
2.	Defective bearing.	Sound is also improper and heat will be produced.	Check and change the faulty bearings.
3.	Lubrication not proper.	Sound and heat will result.	Lubricate the generator properly.
4.	Short circuit in field coils and armature coils.	–	Test the winding and removing the fault.
5.	Weak insulation between the laminations.	Core will be heated up.	Get the laminations properly insulated.

(Contd.)

(d) **Sound in generator.**

Defects	Symptoms	Remarks
1. Defective bearings.	Heat developed.	Check and change the bearings if required.
2. Loose fitting of foundation.	Vibration in the machine.	Tight the foundation bolt.
3. Improper fitting of the side covers.	Sound improper and low speed.	Fit the cover and bearings properly.
4. Poles field coils not properly fitted.	–	Check and tight them.
5. Resistance of armature and field coils unequal.	Jerk in shaft at a particular point.	Test the coils and rectify the fault.

REVIEW QUESTIONS

1. What do you understand by a D.C. generator, explain its working principle with sketches?
2. Give the constructional features and the principle of working of a d.c. generator.
3. Name the different parts of d.c. generator. Why is a commutator and brushes arrangement necessary for the operation of a d.c. machine?
4. What do you understand by a commutator? Explain how the e.m.f. induced in the armature is changed as d.c. by the commutator?
5. Derive an expression for induced e.m.f. in the armature of d.c. generator.
6. What are the classification of the generators?
7. What do you understand by the self excited generator? Describe the shunt field generator.
8. What are the separately excited generators describe with sketches?
9. Differentiate between the long shunt and short shunt compound generators.
10. What do you understand by the differential and cumulative compound generators?
11. What are the classifications of the d.c. compound generators?
12. Explain the armature reaction in the d.c. generators, explain its effects and the methods of improving.
13. What are the interpoles? Why their windings are connected in series with the armature circuit?
14. Write short notes on the following:
    (a) Armature reaction of d.c. generator.
    (b) Commutation.
    (c) Reactance voltage.
    (d) Parallel operation of shunt generator.
15. What do you mean by the Commutation? Explain clearly the methods adopted for minimising the sparking at the brushes?

16. Why parallel operation is needed? What are the utilities of the equilizing bus bar?

17. What are variable and constant losses, Define the efficiency and how to calculate it?

NUMERICAL PROBLEMS

1. A 4 pole, 1250 r.p.m. d.c. generator with lap wound armature has 72 slots and 12 conductors per slot. The flux per pole is 0.02 Wb. Calculate the e.m.f. generated in the armature. **[Ans. 360 V]**

2. A 4 pole d.c. generator with a lap wound armature has 960 conductors. The flux per pole is 0.05 Wb and the armature is driven at 600 r.p.m. Calculate the generated e.m.f. **[Ans. 480 V]**

3. A 4 pole, 600 r.p.m. generator with a wave wound armature has 65 slots and 12 conductors per slot. The flux per pole is 0.02 Wb. Determine the generated e.m.f. in the armature. **[Ans. 312 V]**

4. A 8 pole lap wound d.c. generator armature has 960 conductors, flux 40 mWb and speed at 400 r.p.m. Calculate the e.m.f. generated on open circuit. If the same armature is rewound in wave winding at what speed must it be driven to generate 400 V. **[Ans. 256 V, 156.25 r.p.m.]**

5. A 30 kW 300 V d.c. shunt generator has armature and field resistance 0.05 Ω and 100 Ω respectively. Calculate the total power developed by the armature when it delivers full load output power. **[Ans. 31.43 kW]**

6. A short shunt cumulative compound generator supplies 7.5 kW at 230 V. The shunt field, series field and armature resistance are 100 Ω, 0.3 Ω and 0.4 Ω respectively. Calculate the generated e.m.f. **[Ans. 253.78 V]**

7. Calculate the number of conductors required on the armature of a d.c. generator having simple lap winding, 12 poles, flux per pole 0.094 Wb and running at 250 r.p.m. The resistance of the main circuit of the dynamo is 0.01 Ω and full load current 1400 A at 550 V. **[Ans. 1440]**

8. A 4 pole shunt generator with lap winding supplies a load of 200 A at 200 V. Shunt field resistance 50 Ω and armature resistance 0.05 Ω calculate.

    (a) The total armature current.

    (b) Current per parallel path.

    (c) E.m.f. generated.

    Allow a brush contact drop as 2 V. **[Ans. 204 A, 51 A, 212.20 V]**

13

D.C. Motors

Q. What is a D.C. motor?

Ans. D.C. motor is defined as the machine which converts D.C. electrical energy into mechanical energy. The construction of D.C. generator and motor is almost similar but their functions are entirely different, one converts the mechanical energy into electrical energy, whereas the other converts the electrical energy into mechanical energy.

13.1 WORKING PRINCIPLE

Q. Explain the working principle of D.C. motor.

Ans. A.D.C. motor works on the principle of the magnetic drag. Whenever the current carrying conductor is placed in the magnetic field, a torque is developed over the conductor which tends to move it at right angle to the main magnetic field.

The developed torque can be illustrated as shown in Fig. 13.1, let there is a current carrying conductor placed in between the magnetic poles. Now there are two magnetic field, one because of the current in the conductor and the second because of the main magnets. Because of the characteristics of the magnetic lines of force, the magnetic field will be

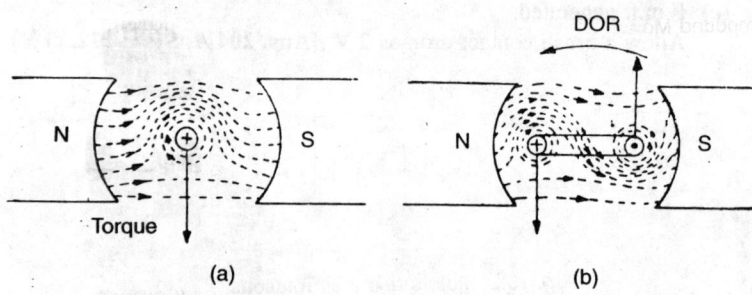

(a)   (b)

Fig. 13.1. Production of Torque.

strong on one side and weak on the other side as shown in Fig. 13.1. Thus a magnetic drag will be experienced by the conductor from strong to weaker side and the conductor will tend to move.

The direction of the torque developed can be determined by the Fleming's left hand rule, "Stretch your left hand in such a way so that the forefinger, middle finger and thumb are at right angle to each other. If the forefinger indicates the direction of magnetic field, middle finger the direction of current in the conductor, then the thumb will indicate the direction of motion, so developed".

As there are number of conductors systematically placed on the armature, so collectively the force developed will rotate the armature and a smooth torque is obtained.

Q. Illustrate the different methods of changing the direction of rotation of the D.C. motors.

Ans. The direction of rotation can be changed as shown in Fig. 13.2.

(a) Series Motor

(b) Shunt Motor

(c) Compound Motor

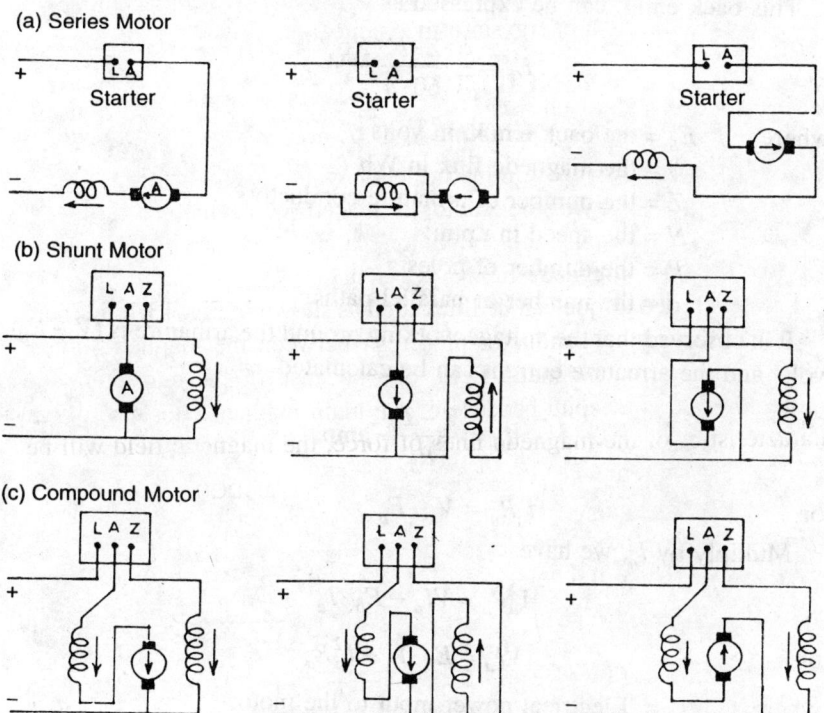

Fig. 13.2. Motors and their rotations.

(*i*) by changing the direction of main magnetic field,

(*ii*) by changing the direction of current in the armature.

If the direction of current is changed in field and armature both, the direction of rotation will not be changed and will remain the same.

It is customary to change the direction of current in armature not in field winding. It is because the change of current may spoil the residual magnetism and in case of compound motor the type of motor from commutative to differential compound or *vice versa*.

Q. Define and express the back e.m.f. and its importance.

Ans. Whenever the D.C. supply is connected to the motor its armature rotates due to the torque produced. Let us consider that the conductors are rotating in the magnetic field and cuts the magnetic field, according to the Fleming's right hand rule, there will be a production of e.m.f. in the armature conductors also. This induced e.m.f. according to Lenz's law, opposes the cause responsible for the production of this *i.e.*, the voltage, hence it will always oppose the applied voltage and so it is called the back e.m.f. (E_b).

This back e.m.f. can be expressed as

$$E_b = \frac{\phi \cdot Z \cdot N \cdot P}{60 \ a} \text{V}$$

where E_b = the back e.m.k. in volts
         ϕ = the magnetic flux in Wb
         Z = the number of armature conductors
         N = the speed in r.p.m.
         P = the number of poles
         a = the number of parallel paths.

It is observed that the voltage working around the armature is $(V - E_b)$ volts and the armature current can be calculated as

$$I_a = \frac{V - E_b}{R_a} \text{ amp.}$$

or
$$I_a R_a = V - E_b$$

Multiply by I_a, we have

$$I_a^2 R_a = VI_a - E_b \cdot I_a$$

or
$$VI_a = E_b \cdot I_a + I_a^2 R_a$$

where VI_a = Electrical power input to the motor
         $E_b I_a$ = Electrical power developed in armature for moving or say equivalent mechanical power developed in armature.

         $I_a^2 R_a$ = Copper losses in the armature.

Importance. The back e.m.f. is the self-governing factor for the armature current. Due to back e.m.f. the armature takes more current from the supply when it is loaded, so armature current automatically changes with the increase of the load on the motor.

If there is no back e.m.f. the armature may take very high current and winding may be damaged. At the time of starting there is no back e.m.f., therefore the motor takes a high current. To avoid the high starting current the starters are used with the motor for its safety.

The back e.m.f. is always less than the supply voltage.

Q. Define (*a*) **torque,** (*b*) **armature torque,** (*c*) **shaft torque and** (*d*) **lost torque.**

Ans. (*a*) **Torque.** The torque is the turning and twisting moment of a force about its axis. It is the product of the force applied and the perpendicular distance from the line of action to the force.

∴ Torque = Force × Perpendicular distance.

(*b*) **Armature torque** (T_a). Let T_a be the torque developed in the armature of a motor running at N r.p.m.

$$T_a = 0.159\left(\frac{\phi \cdot Z \cdot P}{a}\right) \cdot I_a$$

$$= 0.0162 \cdot \frac{\phi Z P}{a} I_a \text{ kgm}$$

or

$$= 0.0162 \cdot \frac{E_b \cdot I_a}{N} \text{ kgm}$$

It is the torque developed in the armature for turning it.

(*c*) **Shaft torque** (T_{sh}). The torque available at shaft for doing the useful work, is known as the shaft torque. It is also called as the b.h.p. of the machine.

$$\text{B.H.P.(metric)} = \frac{T_{sh} \times 2\pi N}{735.5 \times 60}$$

or

$$T_{sh} = \frac{735.5 \times 60 \times \text{B.H.P(metric)}}{2\pi N}$$

or

$$T_{sh} = \frac{\text{Output in watts}}{2\pi N} \text{ Nm.}$$

(*d*) **Lost torque.** The difference between the armature torque and shaft torque is known as the lost torque, *i.e.*, $(T_a - T_{sh})$.

So $\quad (T_a - T_{sh}) = 0.159 \cdot \dfrac{\text{Iron and friction loss in watts}}{N}$ Nm

$$= 0.162 \times \dfrac{\text{Iron and friction loss in watts}}{N} \text{ kgm.}$$

13.2. TYPES AND CHARACTERISTICS

Q. What are the different types of D.C. motors? Explain with the help of diagram. (*N.C.V.T. 1984; Delhi Class I 1964, 69, 80*)

Ans. The D.C. motors are classified into the following categories:

(*a*) **Series motor.** The motor in which the field winding is connected in series with the armature is known as series motor. The series field winding has low resistance, wound with thick conductor having less number of turns as shown in Fig. 13.3.

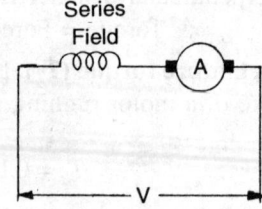

Fig. 13.3. Series motor

In this motor the same current flows through the field winding and armature. At no load the current is minimum and increases with the increasing of load. At no load the speed of motor is very high and dangerous. The speed decreases with the increasing of load. The motor has high starting torque.

Uses. The motor is used for the following purposes:

(*i*) Whenever the load is directly coupled with motor.

(*ii*) Where high starting torque is required.

(*iii*) In case of cranes, trains, trams, D.C. fans, trolleys and conveyors etc.

(*b*) **Shunt motor.** The motor, is which the field winding is connected across the armature is known as shunt motor. The resistance of the field winding is designed as to withstand the supply voltage as shown in Fig. 13.4. The field winding is done with thin conductor having more number of turns. The field winding has high resistance. Generally the shunt field current is allowed up to 5% of the full load current capacity of the motor.

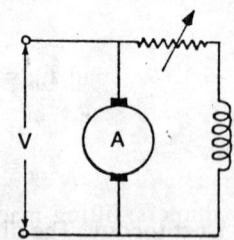

Fig. 13.4. Shunt motor.

In this motor

$$I_L = I_{sh} + I_a$$

Therefore $\qquad I_a = I_L - I_{sh}$

Therefore $I_a = I_L - I_{sh}$ in case of shunt motor the field current is almost constant, but the armature current depends upon the load, *i.e.* it increases with the increasing of load and decreases with the decreasing of load. The speed of the motor is approximately constant, but it decreases with the increasing of load. The speed is maximum without load. Starting torque developed is almost 1.5 to 2 times the full load torque.

Uses. It is a general purpose motor. It is used where almost a constant speed is required for the load as in common shaft or line shaft etc. It is used for lathe machine, wood working machine and other machine tools.

(*c*) **Compound motor.** The compound motor has both windings, the shunt field winding and series field winding. The shunt field winding has more resistance wound with thin conductor having more number of turns. The series field winding has less resistance wound with thick conductor having less number of turns. The current in the shunt field winding is approximately constant, but in series field winding the current is proportional to the load, more load more current, and less load less current in series field winding.

These are of two types:

(*i*) *Cummulative compound motor.*
In this compound motor both series and shunt field windings helps each other as shown in Fig. 13.5. The direction of current *i.e.*, poles, because of both windings remains the same.

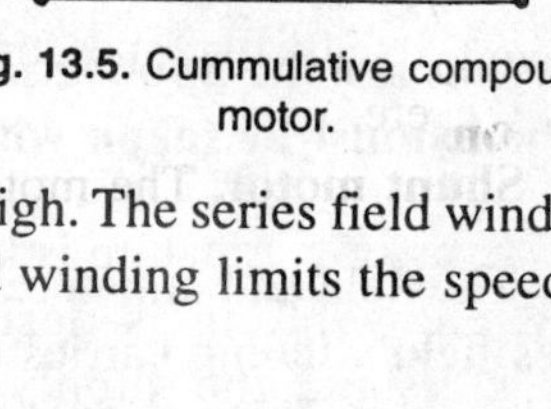

The speed is maximum at no load and falls with the increasing of load.

Fig. 13.5. Cummulative compound motor.

The starting torque is more and sufficiently high. The series field winding helps for high starting torque and shunt field winding limits the speed at light load.

Uses. It is used.

(*i*) where the load suddenly comes and goes after sometimes,

(*ii*) where the constant speed is not essential but torque is the main consideration, and

(*iii*) generally for shearing machines, punching machine, presses, power hammers, lifting machine, rolling mills etc.

Generally in such machines a flywheel is used. This wheel stores the energy when load is not on the shaft and gives away the stored energy when load comes.

(*ii*) *Differential compound motor.* In this compound motor, the direction of current in both, the series and shunt field windings is opposite to each other. When load is increased the resultant flux decreases and if load is reduced the resultant magnetic flux increases. The current in shunt field is approximately constant, but in series field winding it is proportional to the load. The speed of the motor at no load in normal and increases with the increasing of load. The starting torque is poor because the series field winding carries heavy current neutralises the shunt field flux.

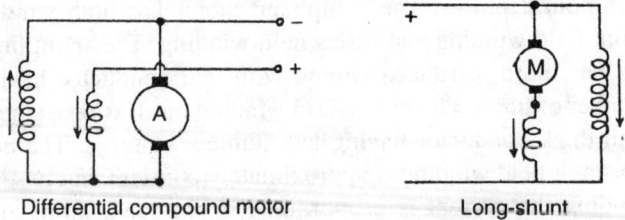

Differential compound motor   Long-shunt

Fig. 13.6. Compound motors

At the time of starting the series field winding needs to be short circuited, otherwise the series field would rise to its full value before the shunt field and motor may start in reverse direction with high speed. The differential and commulative motors are shown in Fig. 13.6.

Uses. It is not a general purpose motor and can only be used where a constant speed is required for a particular load as in battery charging set or booster etc.

Q. What are the applications of D.C. series motor and what precautions must be taken with respect to the load on a series motor?

(*N.C.V.T. 1959; App 1981, Oct.*)

Ans. The field and armature of this motor are connected in series. So the series field winding carries the load current. The speed of the D.C. motor is inversely proportional to the magnetic field. Hence at neg ligible load, the current taken is less, causing less magnetic field and high speed. At this time the resultant torque is also low; because the

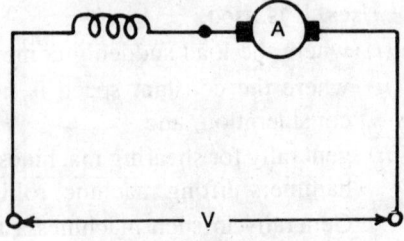

Fig. 13.7. Series motor.

flux produced and current in the armature both the low. But as load is put the current increases, increasing the flux produced. The speed decreases and torque so developed is also increased (see Fig. 13.7).

At the time of switching on, the motor draws heavy current resulting high starting torque. So at the time of starting the motor should never be starting without load, otherwise the speed will be so much that it can break even the foundation also.

Uses. It is used where high starting torque is required for driving trains, trams, trolleys, fan, blowers, hoists etc.

Q. Draw the characteristic curves of series, shunt and compound motors. Give their brief description also.

Ans. There are the following types of characteristics:

(*a*) **Speed characteristic** (*N/I*$_a$). The speed characteristic shows the relationship between the load on the motor and the speed of the motor. It shows whether the speed falls down with the increasing of load or increases with the increasing of load.

(*b*) **Torque characteristic** (*T$_d$/I$_a$*). It shows the effects of load current on the torque produced.

(*c*) **Speed torque characteristic** (*N/T$_a$*). It shows the relationship between the speed and torque of the motor. It is also known as the mechanical characteristic of the motor.

Now let us study the characteristic separately for individual motor.

Series motor

(*a*) **Speed characteristic.** As shown in Fig. 13.8, the current in the field winding at no load is less hence the field thus produced is also less *i.e.* weak enough, therefore the motor runs at very high speed without load. The speed falls down with the increasing of load.

(*b*) **Torque characteristic.** The torque is proportional to the flux and the armature current. So in series motor the flux is also proportional to the armature current ($\phi \propto I_a$). Thus

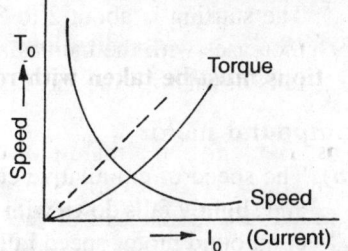

Fig. 13.8. Speed-Torque Characteristic of Series motor.

the torque is proportional to the square of armature current $\left(T_a \propto I_a^2\right)$. Hence the torque changes parabolically with the change in armature current as shown in Fig. 13.8. After saturation point the flux becomes constant and torque $(T_a \mu I_a)$ characteristic behaves like a straight line.

(c) The torque increases with the increasing of load. The starting current is high so the starting torque is about 5 times the full load torque of the motor. The torque is minimum when speed is maximum and speed decreases with the increasing of torque as shown in the characteristic curve.

Shunt motor

(a) **Speed characteristic.** The speed si maximum at no load and slightly decreases with the increasing of load as shown in Fig. 13.9.

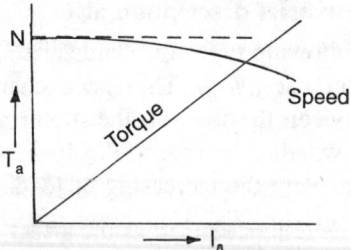

Fig. 13.9. Speed Torque characteristic of D.C. shunt motor.

(b) **Torque characteristic.** The torque increases with the increasing of load. The torque

$$T \propto \phi \cdot I_a$$

here in shunt motor the flux is almost constant so we can say that the torque is proportional to the armature current. So torque increases with the increasing of load and decreases with the decreasing of load as shown in Fig. 13.9.

(c) The starting is about 2 to 3 times the full load torque. The torque increases with the increasing of load and simultaneously the speed falls down with the increasing of load as shown in Fig. 13.9.

Compound motor

(a) The speed of cumulative compound motor is maximum at no load and slightly falls down with the increasing of load. But in differential compound motor speed falls down very rapidly with the increasing of load as shown.

(b) The torque increases with the increasing of load in case of cumulative compound motor. But in case of differential compound motor it is not effected so much.

(c) The speed is cumulative compound motor slightly falls down with the increasing of load and in case of differential compound motor the speed increases with the increasing of load as shown in Fig. 13.10.

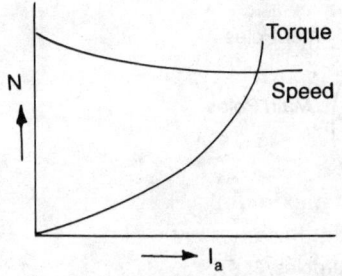

Fig. 13.10. (*a*) Cumulative compound motor.

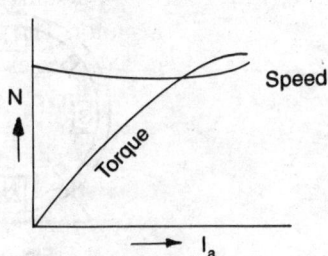

Fig. 13.10. (*b*) Differential compound motor.

13.3. ARMATURE REACTION

Q. Define the armature reaction in D.C. motor.

Ans. The armature reaction is defined as the effect of the magnetic field produced by the armature conductors (when current is flowing), on the distribution of flux under the main poles. The different independent and combined effects are shown in Fig. 12.20. When the field is alone, secondly when armature is loaded, when the current carrying armature is placed in the field and the altered situation of the armature field is illustrated in figures.

In case of D.C. motor the effects of the armature reaction are as follows:

(*i*) The main field is distorted.

(*ii*) The main field becomes strong under L.T.P. and weak at T.P.T.

(*iii*) The main field is reduced.

(*iv*) M.N.P. is shifted backwards to the direction of rotation of the armature, therefore the brushes are given the backward lead to bring the brushes in M.N.P. position.

The angle of shifting the brushes position depends upon the load *i.e.* the armature current.

Q. What are the interpoles? How they are mounted on D.C. motor? How their polarity is maintained?

Ans. These are the poles mounded in between the main poles. These are small in size than the main poles. Generally these are 1/3 to 3/4 th of the main poles as shown in Fig. 13.11. The winding is done with thick conductor. It is connected in series with the armature. The magnetic field produced by these poles is proportional to the load current.

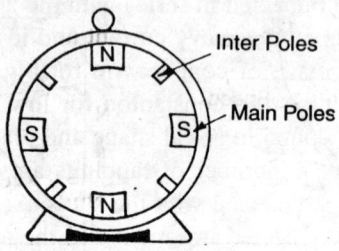

Fig. 13.11. Interpoles.

In case of D.C. motor, the polarity of interpoles will be opposite to the main poles ahead in the direction of rotation of the armature.

13.4. STARTERS

Q. Why starters are used in D.C. motor?

Ans. The D.C. motors should never be connected across the line directly. Initially the motor is at standstill and there is no back e.m.f. in the armature. Therefore by connecting directly to the mains, the motor will take very high current as the armature resistance is very low ($I = V/R_a$). The heavy current may damage the armature winding, commutator and brushes etc. so to reduce the high starting current the D.C. motor are always started with the starters.

Q. What are the functions of D.C. motor starters ? What are their types and describe the construction and working of a D.C. three point starter?   *(N.C.V.T. 1965, 69, 79)*

Ans. The starter is a device which is used for the safety of motor. It compensates the back e.m.f. at the time of starting by introducing external armature resistance and protects the motor from taking an over load current.

It helps in starting and stopping of motor.

It protects the motor from taking over load current and it protects the motor in case of supply failure also.

Types. There are the following types of the D.C. motor starters:

    (*a*) Three point D.C. motor starter.

    (*b*) Four point D.C. motor starter.

    (*c*) Drum controller starter.

(*a*) **Three point D.C. motor starter.** The starter as shown in Fig. 13.12 has got only three terminals marked *L*, *A* and *Z*, so it is called as three point D.C. motor starter.

Construction. It has got the following main parts:

(*i*) *Variable resistance.* There is a variable resistance provided in the starter. It is to be connected in series with the armature circuit. The resistance is designed for heavy current and low resistance values. The resistance is made of copper wire for big machines, iron for medium size machine and constanton for low capacity machines. The resistance is wound in spiral shape and placed on the asbestos sheet or porcelain. A number of tappings are taken out from the resistance and are connected with the studs. At the time of starting full resistance is connected in series with the armature and gradually reduced to zero value as the motor attains the speed.

(*ii*) *No volt release coil* (*N.V.R.C.*). It is a coil mounted on the 'ON' side of the starter. It keeps the handle in 'ON' position when the motor is running and releases the handle in case of supply failure. It is wound with a number of turns of fine wire. It has high resistance. It is connected in series with the shunt field winding.

(*iii*) *Over load release coil* (*O.L.R.C.*). This coil protects the motor from taking an over load current; so it is called over load release coil. It is connected in series with the line. It has low resistance. It is wound with thick conductor having less number of turns.

A soft iron strip, is placed near the over load release coil. The distance between the cores and strip can be adjusted, according to the current carrying capacity of the motor by means of over load current setting. Whenever normal current flows, the strip will not be attracted and if any over loading comes, the strip will be attracted and finally short circuit the no volt release coil, as a result the no volt release coil is demagnetised and releases the handle.

(*iv*) *Handle.* The handle is insulated from the starter. A copper strip is placed under the handle to make contact with the studs. It has an armature of soft iron core, which is attracted by no volt release coil and the handle remains in 'ON' position as long as the current flows through the no volt release coil. A spring is also attached with the handle which brings the handle in 'OFF' position when the no volt release coil is de-energised or demagnetised.

(*v*) *Box.* All the parts of the starter are enclosed in a metallic box, which is mounted on the panel etc.

Working. As shown in Fig. 13.12, the line goes to the handle through the over load release coil. Whenever handle is moved the current starts to the armature through resistance and to the field winding through no volt release coil. As the armature resistance is in series, so the current is reduced considerably and motor starts. As the motor speeds up the armature current reading comes down, the handle is then moved further. The handle should be moved slowly as not to allow the current to exceed the limit, finally the resistance is out of circuit and the motor will be working on the supply voltage.

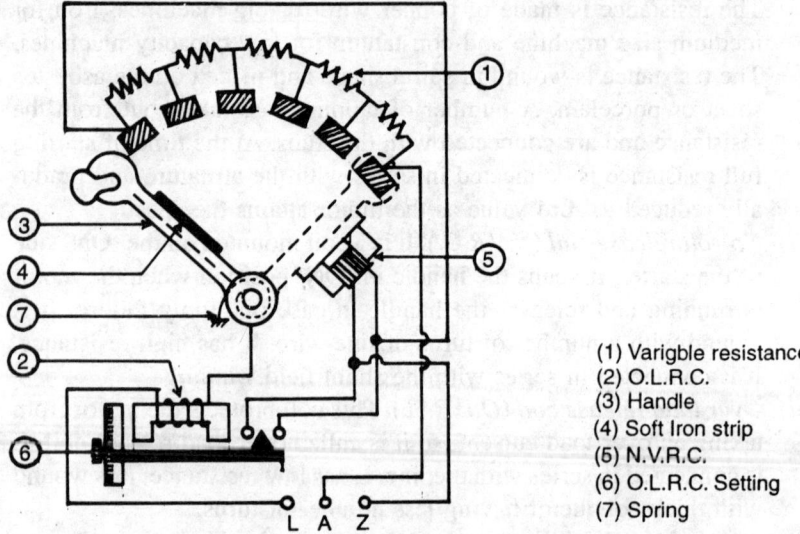

(1) Varigble resistance
(2) O.L.R.C.
(3) Handle
(4) Soft Iron strip
(5) N.V.R.C.
(6) O.L.R.C. Setting
(7) Spring

Fig. 13.12. D.C. three point starter

In some starters there is a strip provided on the face plate and on the handle as to make contact and complete the field circuit.

As the complete resistance is out of circuit, the no volt release coil produces magnetic flux and attracts the soft iron strip attached with the handle. Now the handle is in 'ON' position and motor works on full supply voltage.

Now if at any time the voltage fails, the no volt release coil, which is otherwise known as holding magnet is demagnetised, then the spring pulls back the handle to 'OFF' position and motor stops.

In case of overloading, the current through the over load release coil increases producing more magnetic field, which attracts the strip and short circuit both the contacts of no volt release coil. Thus the NVRC is demagnetised and releases the handle which is pulled back by the spring to 'OFF' position.

Q. Draw the connection diagram of a three point d.c. motor starter with shunt and compound motor.

Ans. The main supply is controlled by the main switch. The positive terminal is connected to the *L* terminal of the starter. The *A* terminal of starter is connected with the armature of the motor and Z with the field winding of the motor. The negative terminal of shunt field and armature winding as shown in Fig. 13.13.

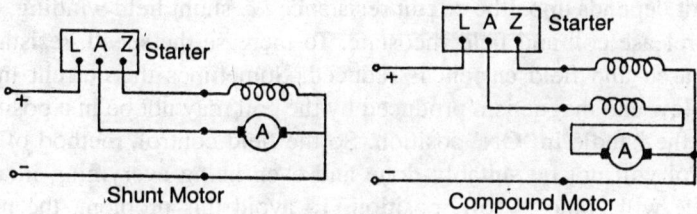

Fig. 13.13. Three point starter with motors.

Q. Draw the internal connection diagram of a four point d.c. motor starter and state how it differs from a three point starter.

(*N.C.V.T. 1967*)

Ans. This starter can be used with d.c. shunt and compound motors. It has four terminals marked *L, N, A* and *Z. L* – for positive wire, *N* – for negative wire, *A* – for armature terminal and *Z* = for shunt field terminal. In this starter the no volt release coil is not connected through the field winding but independently connected across the line *i.e.* positive and negative.

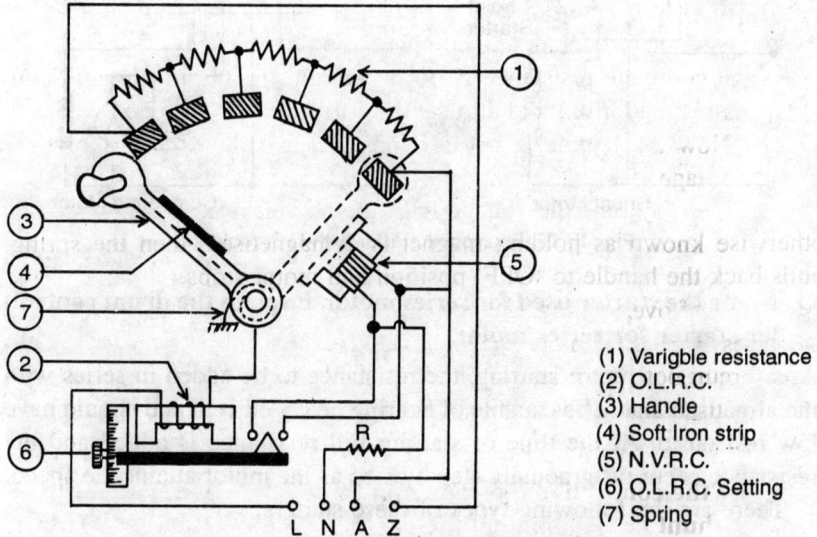

(1) Varigble resistance
(2) O.L.R.C.
(3) Handle
(4) Soft Iron strip
(5) N.V.R.C.
(6) O.L.R.C. Setting
(7) Spring

Fig. 13.14. Four Point D.C. motor starter.

To avoid the short circuit in case of protection against the over loading, a limiting resistance is introduced in the over load protection circuit as shown in Fig. 13.14.

The basic difference is that in case of three point starter, whenever the field control method of speed control is adopted, the current in field circuit depends upon the circuit resistance *i.e.* shunt field winding, the no volt release coil and field rheostate. To increase the speed, resistance is increased and field current is reduced. Sometimes the current may be very low and magnetism produced by the coil may not be in a position to hold the handle in 'ON' position. So the field control, method of speed control will not be suitably done and even being everything intact the handle will come to 'off' position. To avoid this problem, the no volt release coil is designed to have an independent circuit. Thus the field circuit and *NVRC* circuit are independent and the field control can be suitably performed.

In this starter to avoid direct short circuit because of the over loading, a limiting resistance *R* is placed in the *NVRC* circuit as shown '*R*' in Fig. 13.14.

Q. Draw the connection diagram showing a four point starter with a compound and shunt motors.

Ans. The connections diagrams are shown in Fig. 13.15.

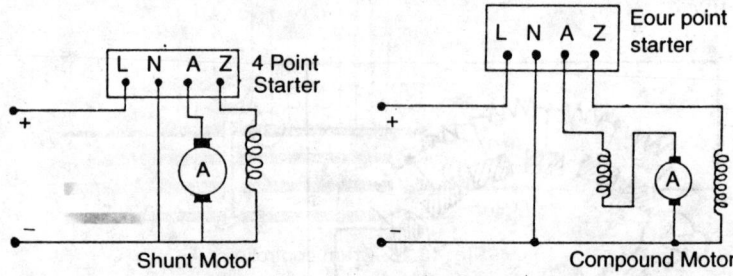

Fig. 13.15. Four point starter with D.C. motors.

Q. Name the starter used for series motor. Explain the drum controller starter for series motor.

Ans. For series motor starting, the resistance to be added in series with the armature should be capable of bearing heavy current and should have low resistance. At the time of starting full resistance is added and the resistance is cut off gradually step by step as the motor attains the speed.

There are the following types of these starters:

    (*i*) Starting with simple resistance.

    (*ii*) Liquid type resistance starter.

Drum controller starter. The series motors of large capacity are started by means of drum controller starter, generally called as drum type d.c. motor starter. The starter is provided with the reversing arrangement. It has a drum generally made of wood or thin sheet with insulating coating. A number of copper strips are placed on the drum. These strips are mounted on the insulated drum and tightened with counter shunk screws etc. When the drum rotates these strips makes contact with the contacts, called fingers. The fingers are mounted on the insulated rod and are having fixed contacts where the connections are made.

The diagram shows the schematic view of a drum controller with reversing arrangement. There are fourteen fixed contacts. These contacts are known as figure contacts. The connections for supply, resistance and motor are connected with these contacts. The resistance is connected between the contacts and armature and supply is connected as shown in Fig. 13.16.

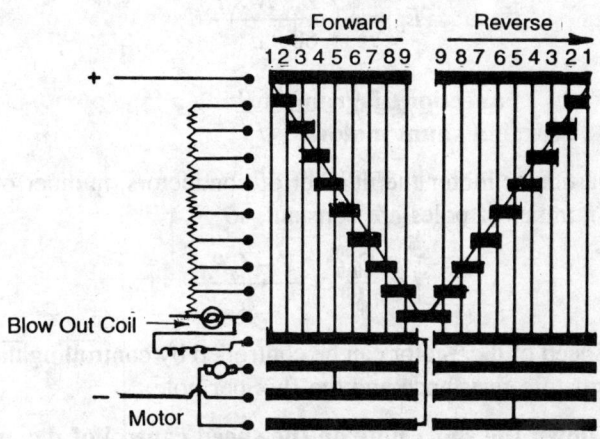

Fig. 13.16. Drum controller

At the time of starting when the handle is moved, say in forward direction, the supply goes to the armature and field of the motor, thus the motor starts because the current carrying conductors are placed in the magnetic field and the torque resulted causes the movement of the rotor. Now the resistance is reduced gradually as the motor is accelerated.

If the direction is to be reversed the handle is moved in reverse direction. The speed of the motor can also be controlled by introducing the resistance in motor circuit.

There is a blow out coil provided in the starter to avoid the possibility of sparking. It is connected in series. It quenches the sparking if any between the drum and fixed contacts.

In case of big motors the starters are dipped in oil.

13.5. SPEED CONTROL

Q. Name the different procedures to obtain the speed control in D.C. motors.

Ans. There are some industrial and domestic applications where the constant speed is not always appreciable, but for the operation, we require the variable speed. The speed can be controlled by the following methods.

 (*i*) **By mechanical means.** In this system the speed is controlled by mechanical gears, using the gears of different diameters and teeth. For example in lathe machine and other machine tools.

 (*ii*) **By electrical means.** In this method the speed of the motor is controlled electrically and we do not require any gear.
Obviously

$$E_b = \frac{\phi \times Z \times N \times P}{60 \times a}$$

or

$$N = \frac{E_b \times 60 \times a}{\phi \times Z \times P}$$

For a particular motor the number of conductors, number of parallel paths and number of poles are constant, so

$$N \propto \frac{E_b}{\phi} \text{ or } \frac{V - I_a R_a}{\phi}$$

So the speed of d.c. motor can be controlled by controlling the voltage fed, the armature resistance and the flux per pole.

Q. Write down the short note on the speed control of d.c. motors.

Ans. There are the following methods of controlling the speed in case of d.c. motors:

 (*a*) The field control,
 (*b*) The armature control,
 (*c*) The voltage control, and
 (*d*) Ward Leonard system of speed control.

 (*a*) **Field control** $\left(N \propto \dfrac{1}{\phi} \right)$. In this method, the speed is inversely proportional to the flux. If the flux is more the speed will be less and *vice versa*. In this method a resistance of high value and low current capacity is connected in series with the field winding of the motor with the introduction of the resistance in field winding the current is controlled and hence the flux produced. In this method the speed can be controlled above normal.

(*b*) **Armature control.** In this method the voltage across the armature is controlled, by connecting a suitable resistance of low value and more current capacity in series with the armature. The speed of the motor is directly proportional to the impressed voltage. In this method the speed is controlled below normal.

(*c*) **Voltage control.** In this method the voltage across the motor is controlled. The speed is also proportional to the voltage supplied, if less voltage the speed is less and more if the voltage supplied is more. In this method also the speed is controlled below normal.

(*d*) **Ward Leonard control.** In this method, the voltage given to the armature of the motor to be controlled is regulated. The field winding is directly connected across the line. In this system a M.G. set is used to control the voltage to be fed to the main motor. It was introduced by Ward Leonard so it is named as Ward Leonard method of speed control.

Q. Explain with diagrams how the speed can be varied in d.c. shunt motor?

Ans. The shunt motor speed can be controlled by the following methods:

(*a*) **Field control method.** To control the speed above normal this method is adopted. A variable resistance of more value and low To avoid the short circuit in case of protection against the over loading, a limiting resistance is introduced in the over load protection circuit as shown in Fig. 13.14. current ca-pacity is connected in series with the shunt field circuit. When full resist-ance is out of the circuit, the flux produced by the field winding is maximum and the speed, say, is nor-mal. As soon as the resistance is added, the current will decrease and flux produced by the winding also. This will increase the speed of the motor as the flux and speed are in-versely proportional to each other.

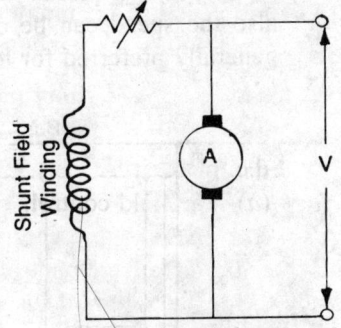

Fig. 13.17. Field control.

It is an economic method of speed controlling. Initially if the field rheostat is in the centre and speed in normal, the speed in that case can be controlled in either way, when the motor is driving some load this operation of speed control should be carried over very carefully. The torque is the product of current (I_a) and flux. So when flux is less current will be more. In the motor having interpoles the ratio of speed can be up to 1 : 4.

(*b*) **Armature control.** This method is used where the speed is required below normal. A resistance of low value and more current capacity is added in series with the armature circuit. The resistance decreases as shown in Fig. 13.18 the voltage given to the armature and hence speed is changed. This method is not so economical. There are the following disadvantages:

(*i*) More power loss in rheostat (I^2R).

(*ii*) When load increases voltage drop in the resistance increases and voltage across the armature further decreases, decreasing the speed.

(*iii*) Because of losses the efficiency will be less.

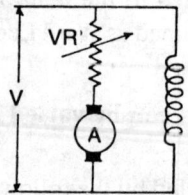 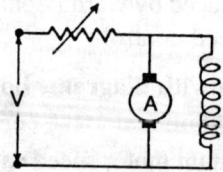

Fig. 13.18. Armature control.   **Fig. 13.19.** Voltage control.

(*c*) **Voltage control.** In this method the rheostate is connected is series with the line. In this case a resistance of low value with more current capacity is placed in series as shown in Fig. 13.19. In this case also the speed can be controlled below normal. This method is generally preferred for low capacity motors.

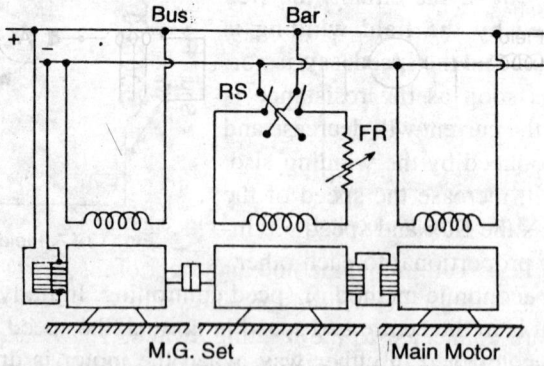

Fig. 13.20. Ward Leonard control.

(d) **Ward Leonard control.** In this method as shown in Fig. 13.20. a variable voltage is applied to the armature in obtain the variable speed. The field is connected across the line, thus the field is produced of normal strength.

Q. Explain the speed control of D.C. series motor.

(N.C.V.T. 1963, 67, 77)

Ans. The speed of D.C. series motor can be controlled by the following methods:

(i) **Field control.** In this method the speed is controlled above normal. A resistance of low value and more current is connected across the field winding as shown in Fig. 13.21 and is known as the field diverter.

Here the current in the field winding is bifurcated and some current flows through the field winding to armature and some, a part of the load is bypassed through the diverted. Thus the flux which is proportional to the current is controlled and the speed is changed. The speed is controlled above normal.

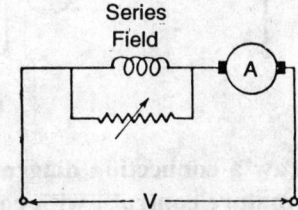

Fig. 13.21. Field control series motor.

(ii) **Armature control.** In this method the diverter is placed across the armature. Here the current in armature is bypassed by connecting the diverter across the armature as shown in Fig. 13.22. Here the speed is controlled below normal. This method is usually adopted for traction purpose like in cranes, hoist motors etc.

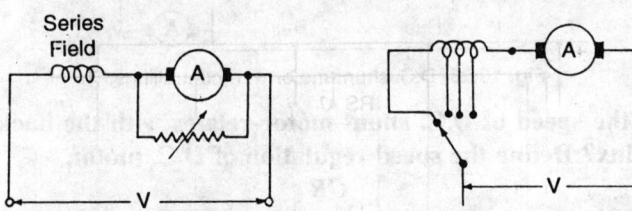

Fig. 13.22. Armature control.   Fig. 3.23. Tapped field control.

(iii) **Tapped control.** In this method a number of tappings are taken out from the series field winding as shown in Fig. 13.23. These tappings are connected to the rotating switch. The flux produced by the series field winding depends upon the effective number of turns of the series winding. The flux is proportional to the current and number of turns also, so by changing the number of turns the flux is changed and hence the speed.

(*iv*) **Voltage control.** In this case the voltage across the motor terminals is controlled. A rheostat is connected in series with the line or the motor. The voltage is dropped in the resistance and less voltage is fed to motor and the speed is controlled. In this method speed is controlled below normal. This system is used with small motors like fan, sewing machines etc.

(*v*) **Paralleling field coil.** In this case the field coils are connected in series, parallel and in series and parallel as shown in Fig. 13.24, thus the field strength is changed, resulting the change in speed.

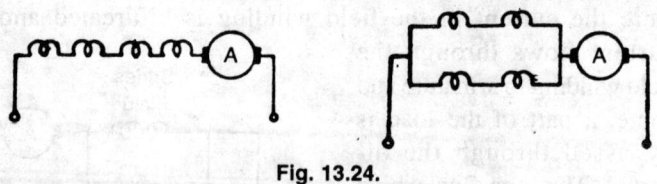

Fig. 13.24.

Q. Draw a connection diagram for shunt motor, showing field and armature controls, with controlling devices.

Ans. The connection diagram is shown in Fig. 13.25.

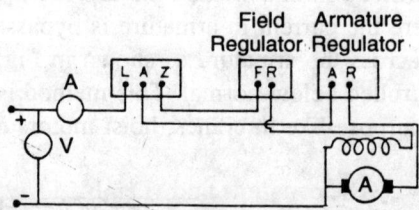

Fig. 13.25. D.C. shunt motor with controllings.

Q. How the speed of D.C. shunt motor relates with the back e.m.f. and flux? Define the speed regulation of D.C. motor.

OR

Define the speed regulation of D.C. motor. What relationship exits between the flux, back e.m.f. and the speed?

Ans. The speed regulation is defined as the rise in speed when the full load from the shaft of the motor is reduced to zero, shown in the percentage of the rated load speed. Let us consider N_0 = the speed at no load, N_f = the full load speed.

$$\therefore \qquad \% \text{ Regulation} = \frac{N_0 - N_f}{N_f} \times 100$$

Obviously

$$E_b = \frac{\phi ZNP}{60 \cdot a} \text{ volts}$$

and being Z, P, a constant for a particular motor, the back e.m.f. can be given as

$$E_b \propto \phi N \quad \text{or} \quad N \propto \frac{E_b}{\phi}.$$

If a motor has N_1 speed at E_{b1} and ϕ_1, the same motor if rotated at N_2 speed the other values will be E_{b2} and ϕ_2 then,

$$\frac{N_2}{N_1} = \frac{E_{b2}}{E_{b1}} \times \frac{\phi_1}{\phi_2}$$

In case of D.C. shunt motor $\phi_1 = \phi_2 = \phi$ say, because the field current approximately is constant.

$$\frac{N_2}{N_1} = \frac{E_{b2}}{E_{b1}} \qquad\qquad \text{(shunt motor)}$$

In case of series motor $\phi_{se} \propto I_{se}$, so

$$\frac{N_2}{N_1} = \frac{E_{b2}}{E_{b1}} \times \frac{I_{a1}}{Ia_2} \quad \text{or} \quad \frac{E_{b2}}{E_{b1}} \times \frac{I_{se1}}{I_{se2}}.$$

13.6. LOSSES AND EFFICIENCY

Q. What is meant by the constant and variable losses in D.C. motor?

OR

Describe the losses in D.C. motors.

Ans. The losses in D.C. motor can be divided in the followings:

(a) **Constant losses.** The losses which are constant from no load to full load and do not change with the loading conditions are known as constant losses. The mechanical losses (friction losses at bearings, brushes etc.) windage losses and iron or core losses are the constant losses.

(b) **Variable losses.** The losses which are not constant from no load to full load, are known as the variable loses. These are proportional to the load current in a definite proportion. These are copper losses in fields and armature windings.

Regarding the *constant losses* the frictional and windage losses, these are depending upon the speed of the motor. Generally, the speed of the motor is not effected appreciably with the loading conditions, but only a very small percentage, so these losses are somewhat constant. Regarding the eddy current and hysteresis losses, these are also constant, because these are depending upon the flux density, frequency and the volume of the material (cores) used.

$$\therefore \qquad W_e \propto B_m^2 \times f^2$$

where W_e = the eddy losses, B_m = maximum flux density is the cores, and f = the frequency in c/s.

and
$$W_h \propto B_m^{1-6} \times f$$

where W_e = the hysteresis losses, B_m = maximum flux density and f = the frequency of supply in c/s.

The iron and friction losses are also known as the stray losses.

Regarding the *variable losses*, these losses are the shunt field, armature, series field, inter-pole losses.

(*i*) *Armature copper loss.* Power lost in the armature because of the current in the armature, is known as the armature copper losses, *i.e.*,

$$W_a = I_a^2 R_a \text{ watts.}$$

(*ii*) *Shunt field copper loss.* Power lost in the shunt field winding is the shunt field copper loss, *i.e.*,

$$W_{sh} = I_{sh}^2 R_{sh} \text{ watts.}$$

(*iii*) *Series field copper loss.* Power lost in the series field of the motor is called series field copper loss, *i.e.*,

$$W_{se} = I_{se}^2 R_{se} \text{ watts.}$$

(*iv*) *Interpole copper loss.* Power lost in the interpoles of the machine is called interpole copper loss, *i.e.*,

$$W_c = I_i^2 R_i \text{ watts.}$$

and the total copper losses are

$$W_e = \left(W_a + W_{sh} + W_i + W_{se}\right) \text{ watts}$$

where W_c = total copper loss.

13.7. TESTING OF MOTORS

Q. What do you understand by the efficiency of D.C. motor? Describe any one method of determining the efficiency of D.C. motor.

Ans. The efficiency of the motor is defined as

$$\% \, \eta = \frac{\text{Output}}{\text{Input}} \times 100.$$

The *output* of the motor is the mechanical power taken from the motor's pulley or shaft. Generally, it is expressed in terms of B.H.P. The *input* of the motor is the electrical power supplied to the motor. Whenever the losses of a motor are known, then

$$\text{Output} = \text{Input} - \text{Losses}$$

or $$\text{Input} = \text{Output} + \text{Losses}.$$

It has practically observed that a machine will have is maximum efficiency, if its variable losses are equal to the constant losses.

The motor is tested for checking the efficiency at a particular load, for example: half load, full load and quarter load or any other desired load. The temperature rise, speed, winding resistances, insulation performance etc. are also tested.

There are the following methods of testing:

(*a*) Direct testing.

(*b*) Indirect Testing.

(*c*) Hopkinson or Regenerative test.

(*a*) **Direct testing.** This is a direct testing method, the motor is subjected to actual loading conditions. This is also known as the brake test. Figure 13.26 shows the arrangement for this test. It consists of a water pulley placed on the shaft of the D.C. motor. The brake belt

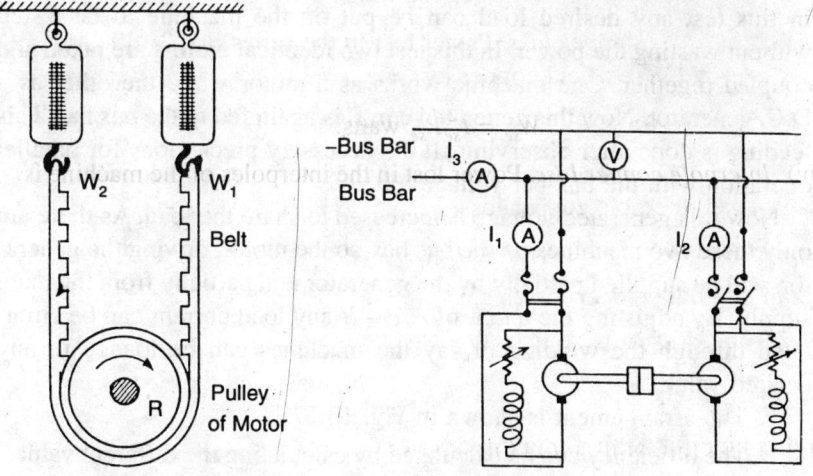

Fig. 13.26. Brake Test

Fig. 13.27. Regenerative Test or Hopkinson's Test.

is fixed with the help of wooden blocks gripping the pulley. The two ends of the belt are connected with two spring balances attached with a stand. A spring balance is provided with the adjustable arrangement, where the other is fixed with wall etc. The motor is then, with proper controllings. started.

By adjusting the pressure with the adjustable arrangement, the motor can be loaded for the desired load. Now

The input to motor = $V \times I$ watts

The output of the motor = $2\pi \cdot N \cdot T_{sh}$

The shaft torque developed = $(W_1 - W_2) R \times 9.81$ watts

Here the efficiency $= \dfrac{\text{Output}}{\text{Input}}$

$$= \frac{2\pi N \times (W_1 - W_2) R \times 9.81}{V \cdot I}$$

$$\eta = \frac{61.68\, N \times R \times (W_1 - W_2)}{V \cdot I}$$

The test is useful for small motors. In case of series motors the belt should be properly tightened, otherwise the motor will run at a fast speed.

Q. What do you understand by the Hopkinson's test for testing the D.C. motors? Describe with the help of neat sketch.

Ans. This test is also known as the Regenerative test or back to back test. In this test any desired load can be put on the machine to be tested without wasting the power. In this test two identical motors are taken and coupled together. One machine works as a motor while the other as a D.C. generator. Now this generated e.m.f. is again fed to the bus bar. This feeding is done after observing all the necessary precautions for parallel operation with the bus bar voltage.

Now this generated voltage is increased to share the load. As there are only these two machines on the bus bar, so the motor, driving the generator, will be supplied partially by the generator and partially from the main supply. By adjusting the e.m.f. of *m/c – II* any load current can be circulated through the winding or say the machines can be loaded to any desired value.

The arrangement is shown in Fig. 13.27.

The efficiency can be calculated by calculating the different values.

The current taken by the motor $I_1 = (I_1 + I_3)$ ampere

Now input to $\qquad m/c - I = V \cdot I_1$ watts

$$= V(I_2 + I_3) \text{ watts}$$

The output of this motor, when the efficiency is η_1

$$\text{Output} = V \cdot (I_2 + I_3) \cdot \eta_1 \text{ watts}$$

Now output of $\qquad m/c - II = V \cdot I_2$ watts

The input of $m/c - II$, when it has the efficiency η_2

$$\text{Input} = \frac{VI_2}{\eta_2}$$

But both the machines are mechanically coupled so the output of $m/c - I$ is the input of $m/c - II$, so it can be said,

$$\frac{VI_2}{\eta_2} = V(I_2 + I_3) \times \eta_1$$

or

$$VI_2 = V(I_2 + I_3) \times \eta_1 \eta_2$$

or

$$\eta_1 \eta_2 = \frac{I_2}{I_2 + I_3}$$

As both the machines are identical, so

$$\eta_1 = \eta_2 = \eta$$

So

$$\eta^2 = \frac{I_2}{I_2 + I_3}$$

or

$$\eta = \sqrt{\frac{I_2}{I_2 + I_3}}.$$

Thus the efficiency can be calculated. In this case the machines can be put on load for any desired load and time without consuming much energy. The power is required just to feed the losses of the machines. The performance can be tested for the temperature rise, stray losses and resistances etc.

Q. What do you understand by the indirect testing of a d.c. motor? Write its merits and demerits.

Ans. Generally this test is also known as the *Swinburn's test*. In this test the motor is not actually loaded but assumed to be loaded and the efficiency is calculated at different loads by calculation.

At the time of staring the motor is started without load and the armature current, line current and shunt field current are measured. In this test the resistances of the armature, shunt field, series field or inter-poles should be known.

Let V is the supply voltage and I_0 = the current without load. At no load the power consumed is nearly the wastage, it includes the copper losses and iron and friction losses.

∴   Input to the motor = VI_0 watts

The copper losses can be calculated,

Armature copper loss, $W_a = I_a^2 \times R_a$ watts

Shunt field copper loss $W_{sh} = I_{sh}^2 \times R_{sh}$ watts

Series field copper loss $W_{se} = I_{se}^2 R_{se}$ watts

and Interpole copper losses $W_i = I_i^2 R_i$ watts

In copper losses add W_{se} and W_i, if the series field and interpole windings are there.

∴ $W_c = W_a + W_{sh} + W_{se} + W_i$ watts.

The iron and friction losses (constant losses) can be calculated, because the copper losses and input without load is given, so

$$W_s = VI_0 - W_c \text{ watts}$$

where W_s = stray losses or iron and friction losses; there losses are constant from no load to full load.

Let us take the example of the motor as shown in Fig. 13.28. If it is loaded and full load current is I_L, then.

The input to the motor = VI_L watts.

The copper and iron and friction loss are calculated as discussed earlier here, armature copper losses are $I_a^2 R_a$ or $(I_L - I_{sh})^2 \times R_a$ and shunt field copper losses are $I_{sh}^2 R_{sh}$.

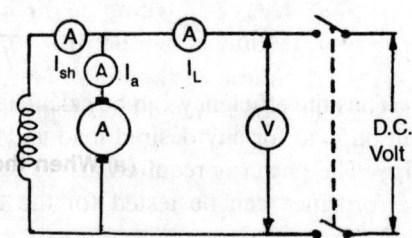

Fig. 13.28. Indirect Testing of motor

Let W_c = the copper losses at that particular load and W_i = the iron and friction losses.

∴ $W_T = W_c + W_i$

Now output of the machine

$$= (V \times I_L - W_T) \text{ watts}$$

$$\% \eta = \frac{\text{Output}}{\text{Input}} \times 100 = \left(\frac{V \times I_L - W_T}{V \times I_L} \right) \times 100$$

Hence we can calculate the efficiency at any load, but it has got its *merits* and *demerits*.

Merits:

(*i*) Less energy is required as the machine is not actually loaded so it is economical.

(*ii*) The efficiency at any load is predetermined which helps for the better performance.

Demerits:

(*i*) Effect of temperature rise is not taken into consideration, which will certainly effect the stray and copper losses of the machine.

(*ii*) It is not suitable and rather possible to use this test on D.C. series motors.

(*iii*) It does not indicate whether the commutation on full load or at any desired load will be satisfactory.

13.8. FAULTS AND THEIR REMEDIES

Q. What may be the trouble and how to rectify them when

　　(*a*) **D.C. motor fails to start.**

　　(*b*) **The motor is running at high speed at the time of starting.**

　　(*c*) **Fuses blows out at the time of starting.**

　　(*d*) **Heavy sparking at the brushes.**

　　(*e*) **Motor is heated up.**

　　(*f*) **Sound in the motor?**

Ans. The troubles and their remedies are as follows:

(a) When the motor fails to start:

Causes	Remedies
1. Failure of supply.	Check the supply and arrange, if not coming.
2. No fuse in the main switch.	Check and replace.
3. Brushes not in proper contact	Check and refix them also increase the tension on the brushes if needed.
4. Break in armature circuit.	Test the armature and replace the coil or rewind it.
5. Break in field circuit.	Check and repair.
6. Break in starter resistance.	Check the starter and repair the resistance.

(b) Motor running at high speed at the time of starting:

1. Open circuit in the field coils of the motor.	Test the continuity of the field and remove the fault.
2. Open circuit in no volt coil circuit.	Test and remove the fault.
3. Series motor might be	Never start series motor without load.
4. In case of compound motor the series field would be connected in differential type.	Check and reverse the connection of series field.

(c) Fuses blow out at the time of sparking:

1. Bad starting, moving handle very quickly.	Start the motor properly move the handle slowly.
2. More load.	Reduce the load.
3. An open circuit either in field or in no volt release oil.	Test the continuity and remove the fault.
4. Short circuit in the armature.	Test the continuity and remove the fault.
5. Short circuit in the rheostat of armature circuit.	Check the rheostat and remove the fault.
6. Short circuit in the brushes at rocker.	Test the insulation resistance between brushes and rocker.

(d) Badly sparking on the commutator:

Causes	Remedies
1. Brushes loose contact with the commutator.	Adjust the brushes tension.
2. Brushes not at *MNP.*	Adjust the *MNP* position.
3. Commutator surface is not smooth and round.	Turn it on lathe and also under cut mica.
4. Dirty commutator.	Clean the commutator with sand paper petrol and cotton waste.
5. Short circuit in armature coil.	Test the armature coil and remove the fault.
6. Reverse connections in the armature coil.	Check and remove the fault.
7. More load on the armature *i.e.* over loading.	Reduce the load.
8. Improper bedding of brushes.	Properly bed the brushes.
9. Wrong inter-pole polarity.	Check and correct it.

(e) Motor is heating up:

1. Due to sparking at the commutator.	Remove the cause and reduce sparking.
2. Short circuit in the field winding or armature coil.	Test the winding and remove the fault.
3. Improper lubrication.	Lubricate the motor.

(f) Sound in the motor:

1. Wrong fitting.	Tight the nuts etc.
2. Defective bearings.	Check and replace if needed.
3. Lack in shaft.	Turn on the lathe and get the fault rectified.

Full Load Current – Current capacity of D.C. motor at 230 volts

Motor H.P.	Approximate Full load current in Amp.	Motor H.P.	Approximate Full load current in Amp.
0.5	2.4	20	72
1.0	.3	30	108
2	8.5	40	143
3	12.5	50	176
4	16.5	60	212
5	20.5	70	247
6	24.3	80	270
7.5	30	90	314
10	38	100	349
15	55		

REVIEW QUESTIONS

1. What is the D.C. motor? Describe its construction.
2. Discuss the construction of D.C. motor and explain its working principle.
3. What is the back e.m.f. in D.C. motor, describe its importance in starting and running of motor?
4. What are the different types of D.C. motors? Describe the shunt motor.
5. What are the classifications of D.C. compound motors?
6. How is torque produced in D.C. motor?
7. Explain the characteristics of D.C. motor. Draw the torque-speed curve of D.C. shunt and series motors.

8. Describe the applications of D.C. series, shunt and compound motors.
9. Explain armature reaction in D.C. motors, indicating few remedies.
10. What do you understand by the starter? Why it is necessary for D.C. motors?
11. Why are starter used for D.C. motors? Draw a diagram of a 3 point starter and explain the function of each component.
12. What are the differences of 3 point and 4 point starters? Describe the construction of 4 point starter with sketch.
13. Fill up the followings:
    (*i*) The d.o.r. of a D.C. motor can be determined by ....
    (*ii*) When the line terminals of D.C. motor are inter changed the speed will ....
    (*iii*) The d.o.r. of a D.C. motor can be changed by changing the direction of current in .... or ....
    (*iv*) The field control method is adopted to ....the speed of the motor.
    (*v*) D.C. shunt motor is a .... speed motor.
14. Describe the various methods of speed control employed for D.C. series and shunt motors.
15. What are the constant and variable losses in D.C. motor?
16. Explain the different methods for determining the efficiency of a D.C. machine.
17. What is the Hopkinson test and why it is called regenerative test? Describe with neat sketch.
18. Write short notes on the followings:
    (*i*) Rating of D.C. motors.   (*ii*) Speed control of D.C. series motors.
    (*iii*) Working principle of D.C. motor.   (*iv*) D.C. shunt motor starter.
    (*v*) Production of torque in D.C. motor.

NUMERICAL PROBLEMS

1. Find the speed of a shunt motor having four pole wave winding and taking 42 A at 410 V. The flux per pole is 0.03 Wb, armature having 39 slots, 12 conductor per slot. Armature resistance is 0.25 Ω and shunt field resistance 210 Ω.   **[Ans. 855 r.p.m]**
2. Calculate the value of back e.m.f. of a motor operating on 440 V when the armature current is 50 A and armature circuit resistance is 0.75 Ω.
        [Ans. 402.5 V]
3. Calculate the armature conductors for a wave wound D.C. 4 pole motor running at 900 r.p.m. on full load at 440 V supply. The full load armature current is 25A flux per pole is 0.0117 Wb. and the armature circuit resistance is 1.2 Ω.   **[Ans. 1168 conductors]**
4. The armature resistance of a 220 V.D.C. shunt motor is 0.5 Ω and no load armature is 2 A. If the armature current is 50 A, the speed is 1200 r.p.m. Find the no load speed.   **[Ans. 1348 r.p.m.]**
5. A D.C. motor is running at 1000 r.p.m. and the torque exerted on pulley is 1200 Nw-m. Calculate the b.h.p. of the motor.   **[Ans. 170.85 H.P.]**

6. A.D.C. series motor operating on 230 V draws 50 A the armature resistance and field resistance are 0.2 Ω and 0.1 Ω respectively. Calculate the back e.m.f. of the motor. **[Ans. 215 V]**

7. The armature of a four pole D.C. motor carries 314 conductors which are lap wound. For an armature current of 20 A, the torque produced is 50 N-m. Determine the required flux per pole. **[Ans. 0.05 Wb]**

8. A 200 V shunt motor is taking a current of 30 A. Armature resistance 0.2 Ω, shunt field resistance 100 Ω, iron and friction losses 600 W. Calculate the B.H.P. of the motor and commercial efficiency.
    [Ans. 6.49 H.P., 80.7%]

14

Wiring System

14.1. WIRING

Q. What is meant by the wiring? Illustrate the simple arrangement of commencing the supply to the consumers.

Ans. Wiring means to connect the electric load to the supply mains, according to the Indian Electricity rules. The electrical energy is used in houses for lighting and other purposes whereas in industries it is used for power and lighting purposes. So the wiring is either single phase or three phase.

A.C. single phase or D.C. wiring. This wiring is used in houses for lighting and power purposes. It can be A.C. single phase or D.C. wiring. In case of D.C. wiring there are two wires positive and negative. The load is connected across these wires through proper controllings and safety devices. In A.C. single phase the phase and neutral are connected to the load through proper controlling and safely accessories. The third wire is an important wire *i.e.* earth wire in both A.C. or D.C. systems. The earth wire is connected to the all metallic portions of the wiring and appliances except the current carrying conductors.

Three phase wiring. The three phase wiring is used for three phase power load and for single phase loads also. In A.C. single phase loads to be connected on three phase system, every individual phase is connected across three independent circuits, the total load is divided into three balanced circuits, generally three phase four wire system is used. In case of three phase power loads the line voltage is 400/440 volts. The three phase load may be in star or in delta. The single phase circuit is connected across 220/230 volts, across one phase and neutral.

A simple arrangement is shown in Fig. 14.1. The supply is taken from the distribution line through service mains. A pole fuse is provided, the function of which is to protect the service line against over loading. The supply is directly fed to the energy meter and after the energy meter a service fuse, which is generally an iron clad fuse. is provided for the safety purpose. The iron clad cutout and energy meter are the property of the supplier. So both are sealed by the supplier.

The mains are taken from the energy meter and are controlled by means of neutral linked I.C.D.P. main switch. If the load is more and has to be divided into a number of circuits, the distribution boards and boxes, depending upon the circuits required are used *e.g.* two ways, three ways, four ways etc. One phase and one neutral is taken for independent circuits. If it is required, depending upon the load, the sub-distribution boxes may be used. Thus in this arrangement the pole fuses, distribution boards, earth wire are the essential elements which are illustrated in Fig. 14.1. The earth continuity is maintained throughout the installation.

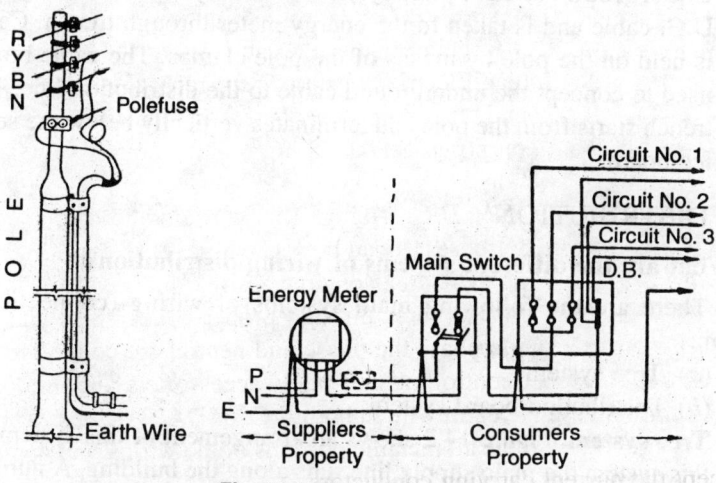

Fig. 14.1. Method of supply.

Q. What are the methods of installing the service mains?

Ans. There are two methods of installing the service mains:

    (*a*) Over head service line system.

    (*b*) Under ground service line system.

(*a*) **Over head service line system.** In this case the service mains runs over the height. Depending upon the prevailing conditions the over head lines are:

    (*i*) *For double and multistoried buildings.* In this case a bracket is embedded into the wall on a suitable height and then the line is installed and taken to the energy meter. Generally weather proof wire or P.V.C. cable is used for this installation, this cable is known as service cable. The G.I. pipe is bend back to prevent the unwanted water entry into the pipe.

(ii) *Single storeyed building.* In this case a G.I. pipe is raised above the roof to a suitable height and fixed with wall with clamps. Similarly the pipe is bend back to prevent the water entry into the pipe. From the service bracket the service cable is taken to the meter.

(iii) *Weather proof cable and P.V.C. cable method.* In this method the service main is not of the bare conductors as described in method I and II, but are insulated weather proof or P.V.C. cables. It is used for high buildings, road crossing etc. In this case an 8 S.W.G. G.I. wire is stretched from the service pole to an eye bolt fixed to the wall. The insulated cable is clipped to this G.I. wire by means of clips or binding wires.

(b) **Underground cable.** In this system the supply is taken through the U.G. cable and is taken to the energy meter through trench. Cable is held on the pole by means of the pole clamps. The cable box is used to concept the underground cable to the distribution line. The trench starts from the pole and terminates vertically below the service board.

14.2. DISTRIBUTION

Q. What are the different systems of wiring distribution?

Ans. There are the following main systems of wiring (connections) system:

(a) Tree system.

(b) Distribution board system.

(a) **Tree system.** Figure 14.2 shows an arrangement of this system. In this system the main supply line runs along the building. A number of tappings are taken from the main line for different circuits.

There are the following advantages and disadvantages:

Advantages: It is less costly.

Disadvantages:

(i) The voltage across the appliances does not remain constant. The voltage near the supply main circuit will have more voltage than the voltage at a far circuit.

(ii) The fuses of the different circuits are at different places.

(iii) The tappings can be loose and fault location is not easy.

(iv) The main supply wire are bigger size. This system is not used generally.

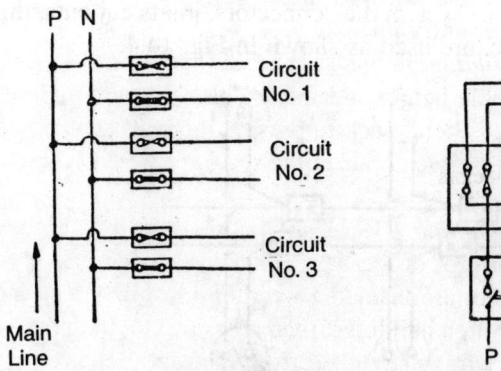
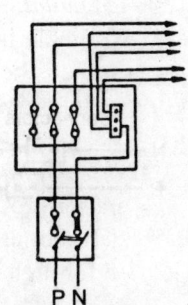

Fig. 14.2. Tree system.　　**Fig. 14.3.** Distribution system.

(*b*) **Distribution board system.** This system is mostly used at present. In this system all the appliances etc. are connected across the same voltage. The leads from the main switch are taken to the copper strip in a main distribution board. There are a number of fuse provided in the box, from where the different circuits can be drawn. If the load is more, then the sub-distribution boxes are also used. The line goes from the main to sub-distribution box and then to the individual load circuits as shown in Fig. 14.3.

Nowadays, instead of fuse cutouts the MCB are preferred. The fuse is being replaced by MCB and neutral is connected with the neutral link. Thus separate circuits one drawn and all are controlled independently by the individual MCB.

There are the following advantages and disadvantages of this method:

Advantages:

(*i*) All the loads are connected across the same voltage.
(*ii*) Fault location is easy.
(*iii*) The fuses are at one places.
(*iv*) Appearance is good.
(*v*) Extension is easy.

Disadvantages:

(*i*) Costly than the tree system.
(*ii*) Skilled labour is required.

Q. What are the different methods of connections to the accessories?

Ans. There are the following two methods of connections:

(*a*) T-connections.
(*b*) Looping in system.

(a) **T-Connections.** In this system the connectors, joints cut outs, three plate ceiling rose etc. are used as shown in Fig. 14.4.

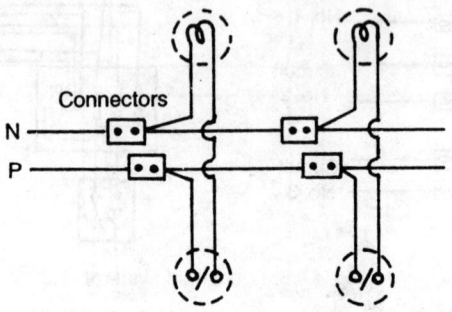

Fig. 14.4. T-Connections.

There are the following advantages and disadvantages:

Advantages:

(i) Less wire is required.

(ii) Less cost.

(iii) Less voltage drop in the wires and less power loss in the wiring.

Disadvantages:

(i) A number of joints are is the wiring. The joint is the weakest point of the installation.

(ii) The joint may be loose and so the faults are frequent and the fault location is not so easy.

(b) **Looping is system.** In this system the connector etc. are not used but the connections are made as shown in Fig. 14.5 from the switches or batten holders.

There are the following advantages and disadvantages:

Advantages:

(i) There are less number of joints.

(ii) Good insulation resistance of the wiring.

(iii) Appearance is good.

Disadvantages:

(i) A greater length is required.

(ii) It is costly than the tree system.

(iii) More voltage drop and power loss in the installation.

(iv) Because of more wires in the conduit or casing and capping etc. The fault location is not so easy.

(v) Looping in switches and holders terminals is usually difficult.

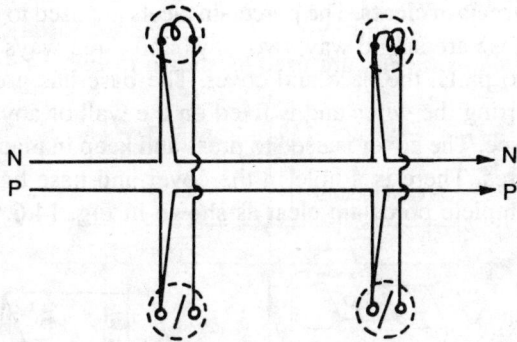

Fig. 14.5. Looping in system.

14.3. TYPES OF WIRING

Q. Name the different types of wiring. What are the main points to be considered for selecting the types of wiring for a particular place?

Ans. There are the following types of the wiring:

    (*a*) Cleat wiring.

    (*b*) Casing and capping wiring.

    (*c*) C.T.S. or T.R.S. wiring.

    (*d*) Lead sheathed wiring.

    (*e*) Conduit pipe wiring.

        (*i*) Surface conduit pipe wiring.

        (*ii*) Concealed conduit or Underground pipe wiring.

There are the following points to be considered before selecting the types of wiring for any particular place.

  (*i*) **Durability.** The wires used in that wiring must be durable and should be safe from the fire and weather changes etc.

  (*ii*) **Safety.** The wiring should not be risky to any human being.

  (*iii*) **Cost.** It should be within the approach of the consumer and should not be costly.

  (*iv*) **Appearance.** The outlook of the wiring should be good enough and should be according to the construction and design of the building.

  (*v*) **Accessibility.** It should be easily accessible and easy to extend.

Q. Describe the material required, main points for the installation and uses of the cleat wiring.

Ans. The cleat wiring is not a permanent type of wiring, it is a temporary type of wiring. The different materials required for this wiring are as under:

(*i*) **Porcelain cleats.** The porcelain cleats are used to hold the wires. These are single way, two ways and three ways etc. There are two parts, the base and cover. The base has grooves for supporting the wires and is fixed on the wall or any other suitable place. The cover is used to press and keep in place and tight the wires. There is a hole in the cover and base both to tight the complete porcelain cleat as shown in Fig. 14.6.

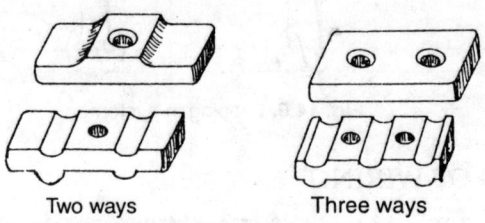

Two ways   Three ways

Fig. 14.6. Porcelain cleats.

Sometimes wooden cleats are also used.

Wires. Generally P.V.C. or V.I.R. wires are run in the grooves of the cleats. These wires are held intact by the wooden or porcelain cleats.

Gitties. These are used to fix the cleats on the wall or ceiling and the screws are used to fix cleats. The gitties are made of wood.

Screws. Different sizes of screws are used to fix the cleats, blocks, accessories and board etc.

Round blocks and board. Wooden or P.V.C. round or hexagonal or rectangular blocks are used to tight and fix the accessories.

Accessories. Different accessories like switches, holders, wall sockets are used depending upon the requirement. The complete diagram is shown in Fig. 14.7.

There are the following points to be kept in mind during installation.

(*i*) This wiring should be done above two metres from the floor. If it is required to install then use conduit pipe.

(*ii*) The maximum distance between the cleats should not be more than 60 cm and the safe distance is 30 cm.

(*iii*) The wires should be in stretched condition. The wires should not come in contact with the wall or with each other.

(*iv*) The bend should not be sharp and conduit pipe must be used with a wooden bushing at both the ends which it is required to cross a wall. The cleats must be used near the bend, turning and accessories.

(*v*) Wires should not cross each other. Use bridge in case if it has to cross.

(*vi*) The wires should not be run near the water or gas pipes.

(*vii*) This wiring should not be used for high voltage.

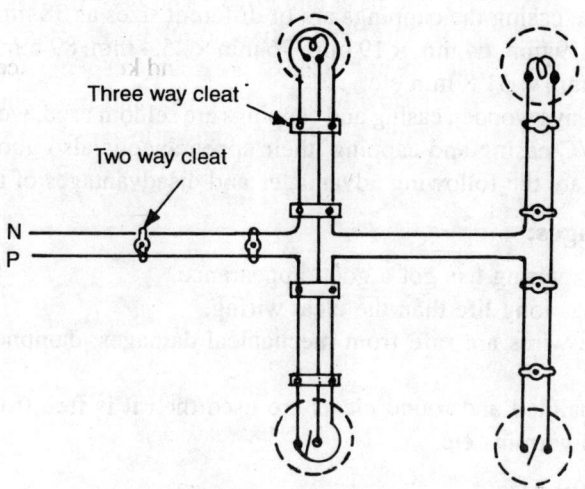

Three way cleat

Two way cleat

N

P

Fig. 14.7. Cleat wiring.

There are the following advantages and disadvantages.

Advantages:

(*i*) It requires less labour and workmanship.

(*ii*) It requires less time and less cost.

(*iii*) It can be easily removed when not required.

(*iv*) Extension and fault location is easy.

Disadvantages:

(*i*) Less life.

(*ii*) Less efficiency.

(*iii*) Dust and dirt spoil the appearance.

(*iv*) The white wash, oil, dust and smoke injure the wires.

(*v*) No safety from fire, dampness, mechanical damages and rats etc.

Uses. This type of wiring is only used for temporary purposes. It must not be done in the damp places, blacksmith shop, moulding shop etc. It is only used in the camps for temporary purpose. In no case it should be used for more than six months. If it is required to use for more than three months it should be got inspected and tested again.

Q. Explain the casing and capping wiring.

Ans. In this wiring system V.I.R. wire or P.V.C. wires run in the grooves of the teak wood casing and capping that is why this system is known as casing and capping system of wiring. The casing is first fixed on the surface of wall or ceiling by means of wooden gitties and wood screws. Generally there are two grooves in the casing in which wires are laid down. The casing the cappings are of different sizes as 38 mm × 16 mm, 51 mm × 19 mm, 64 mm × 19 mm, 76 mm × 25.4 mm, 89 mm × 31.8 mm and 102 mm × 31.8 mm etc.

Nowadays wooden casing and cappings are seldom used, we are switching to PVC casing and capping, their appearance is also good.

There are the following advantages and disadvantages of this system:

Advantages:

(*i*) This wiring has got a good appearance.

(*ii*) It has long life than the cleat wiring.

(*iii*) The wires are safe from mechanical damages, dampness and rats etc.

(*iv*) If painted and round cleats are used then it is free from moisture and vermins etc.

Disadvantages:

(*i*) It requires better workmanship so labour is costly.

(*ii*) There is no safety from fire.

(*iii*) The two wires of opposite polarity cannot run in the same groove but should run in different grooves.

(*iv*) No protection from dampness.

(*v*) If not painted and varnished, the vermins may eat the wood and life is considerably reduced.

(*vi*) Extension is not so easy.

(*vii*) Fault location is not easy as the wires are covered.

General rules for wiring:

(*i*) Round porcelain cleats should be used under the surface of casing and capping to keep the wiring safe from moisture.

(*ii*) The wiring should be painted and varnished to save it from vermins and dampness.

(*iii*) Where there is a a crossing in the wires of opposite polarity bridge must be provided.

(*iv*) There should not be any gap between the casing and capping.

(*v*) The width of both casing and capping should be same.

(*vi*) In no case the casing be capping be burried under the plaster etc.

(*vii*) When the wiring is to be passed through the wall, conduit pipe with bushings should be used.

(*viii*) At bends the grooves of casing and capping should be well rounded to avoid the damages to the insulation of the wiring.

(*ix*) The wiring should not be done near the water pipes and gas pipes.

(*x*) The wooden gitties must be used near the bend, accessories and joints etc.

(*xi*) 15 mm screws should be used with capping and 35 mm screws for casing. The distance between the two gitties, in no case be more than 70-90 cm.

Uses. It is a common type of house wiring. It should never be used where there is a risk of fire (blacksmith and moulding shops) and in damp places.

Q. (*a*) Describe the C.T.S. or T.R.S. batten wiring.

(*b*) What do you understand by the lead covered wiring?

Ans. In this system teak wood batten of different sizes are used. The battens are available in different sizes according to the width and breadth. The width are 13, 19, 25, 31, 38, 44, 50, 63, 69 and 75 mm, the breadth 13 mm and 19 mm. The wiring is also of two types:

1. C.T.S. or T.R.S. batten wiring.

2. Lead alloy sheathed batten wiring or lead covered cable wiring.

C.T.S./T.R.S. batten wiring. In this system C.T.S. or T.R.S. wires are used. The wires may be single core or twin cores. In most of the cases single core cable is used. The wires are fixed on the wooden batten with the help of teak wood batten by means of clips. The clips are also of two types the link clips and joint clips as shown in Fig. 14.8. These clips are made of iron or lead or sometimes fibre. These are available length-wise 16 mm, 25 mm, 30 mm, 40 mm, 50 mm, 80 mm etc.

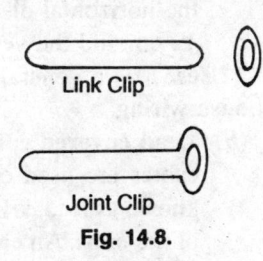

Link Clip

Joint Clip

Fig. 14.8.

There are the following advantages and disadvantages:

Advantages:

(*i*) Semi-skilled worker can do it, so the labour cost is less.

(*ii*) It is less costly than casing and capping.

(*iii*) It has good appearance.

(*iv*) It has long life.

(*v*) The C.T.S. wire is not so effected by the moisture, therefore, this wiring can be done in damp places also.

(*vi*) If round cleats are used and batten is painted and varnished then the wiring is free from moisture vermins etc.

(*vii*) If single core wire is used then the extension is very easy, fault location is very easy and removal of the wiring is also easy.

Disadvantages:

(*i*) There is no safety from fire and shock.

(*ii*) It cannot be done in the places where it is exposed to sun and rains.

General rules for wiring:

(*i*) In damp places round porcelain cleats must be used and the batten be painted and varnished.

(*ii*) In damp places the metallic batten holders and other metallic accessories be earthed.

(*iii*) There should not be any tension on the wires.

(*iv*) The joints should be in joint cutouts.

(*v*) The maximum distance between two round cleats in no case be more than 75 cm and the safe distance is 30 cm.

(*vi*) All the plugs and metallic installations should be earthed.

(*vii*) When the wires of opposite polarity are crossing the bridges should be used.

(*viii*) If the wiring is to cross a wall or ceiling the conduit pipe should be used.

(*ix*) While fixing the clips on the batten, it should be kept in mind that the horizontal distance between the clips should not be more than 12 cm and the vertical distance 18 cm.

Uses. It is a general purpose domestic wiring and generally used in house wiring.

(*b*) **Lead covered cable wiring.** In this system of wiring lead covered cables are used on a batten or sometimes without batten, so it is known as lead covered cable wiring. The wire is supported by means of the clips. An earth wire runs along to maintain the earth continuity. The wires can be single or twin core.

There are the following advantages and disadvantages:

Advantages:

(*i*) These wires are safe from fire, moisture and mechanical damages.

(*ii*) It has long life.

(*iii*) It can be done in damp places, open areas even which are exposed to sun and rains also.

(*iv*) It has good appearance.

Disadvantages:

(*i*) Skilled worker required so the labour is costly.

(*ii*) It is costly because of lead covering.

(*iii*) Extension is not so easy.

(*iv*) Fault location is not easy.

The power wiring should be carried out separately from the lighting circuit. In this wiring all the joint boxes and ends of the wires are sealed with compound, so that moisture may not enter in the cable or wires.

Uses. It is not a common type of wiring. It is used in the places where dampness, exposing to run and rain etc. are there. It should not be used the places, where in chemical corrosion may occur.

Q. What are the common accessories used in conduit pipe wiring?

Ans. These are the following conduit accessories which are commonly used:

(*a*) **Conduit pipe.** The conduit pipe which is used for wiring purpose is of two types:

   (*i*) *Light gauge.* These are made from thin sheets of alloyed iron or aluminium or P.V.C. The two ends of these pipes are opened. Threading cannot be done over these pipes, because of less thickness. These are cheap but not suitable for good wiring.

   (*ii*) *Heavy gauge.* These pipes are either solid drawn or the ends are welded together. It is made of mild steel sheets. In this pipe threading can be done. It is painted to prevent the rusting. These pipes are commonly used. In case of P.V.C. pipe the threading is not required our the housing is so moulded.

(*b*) **Flexible conduit pipe.** It is usually called as flexible pipe. This pipe is used, where a rigid pipe cannot be used, *e.g.,* near the terminal box of motors, generators, transformers etc. It is made of galvanised steel strips specially wound upon each other. These are available in different sizes as 20 mm, 25 mm, 31 mm, 38 mm, 50 mm, 56 mm, 69 mm, 75 mm etc. in diameter. P.V.C. flexible pipes are also available of different diameter.

(*c*) **Couplings.** These are used to join two conduit pipes in a straight line or this is used to couple two pipes together. Couplings are threaded in the inner portion to which conduit pipes are tightened.

(*d*) **Flexible coupling.** It is used to join the flexible pipe where the wire leaves the rigid pipe. By using the flexible coupling, the insulation resistance of the wire is not damaged, while drawing the wires. These are made of wood, bakelite, cast iron. The iron bushings are threaded which serves as the lock nut.

(*e*) **Bushings.** These are used at the mouth of the pipe, where the wires leave the pipe. It has two parts which can be joint together by means of four screws.

(f) **Lock nuts or Check nuts.** These are used at the end of conduit pipe entering the box. These are made of generally cast iron. One end of it clamps the flexible pipe, while the other end is threaded for check nut to fix with the box. These are available in different sizes according to the diameter of the pipe.

(h) **Conduit nipple.** These are same as that of the conduit bushings. These are used to increase the length of the conduit pipe and sometimes used to join elbow or two tees etc.

(i) **Conduit reducers.** These are used at the places where a number of conduit specially two conduits of different diameters are joined.

(j) **Conduit boxes.** These are used where a number of conduit pipes comes from different directions. These boxes are provided with covers, so it can be opened to inspect the wires. These are of different shapes: square, round, hexagonal or rectangular etc.

(k) **Junction boxes.** These are used to join two conduit pipes. It has a cover to opening of which the wires can be drawn and inspected also as shown in Fig. 14.9. These are made of casted cast iron. These are two ways, three ways, four ways. Sometimes the switch and other accessories are also fixed over that as shown in Fig. 14.10.

   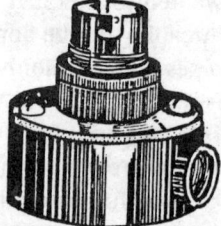

Fig. 14.9. Junction box.   **Fig. 14.10.** Junction box with holder.

(l) **T-solid or Inspection.** These are used where the wires are to be brought down from the running line. The solid T is used in concealed conduit wiring and the inspection T, in the surface conduit pipe wiring, so that the wires can be inspected and drawn easily. These are available according to the diameter of the pipe.

(m) **Elbow or bend.** These are used in the wiring, where the pipes are bend at 90°. These are made of cast iron or sometimes by simply bending the pipes. By the use of the inspection bend, the wires can be drawn and inspected also as shown in Fig. 14.11. These are available in according to the sizes of the pipes. P.V.C. elbow and bends of different diamters are available and used for wiring purpose.

(*n*) **Conduit shaddles.** The conduit pipe is fixed with the help of the conduit shaddles. Sometimes the shaddles are named as clamps, conduit straps. These are available in different sizes according to the pipe diameter.

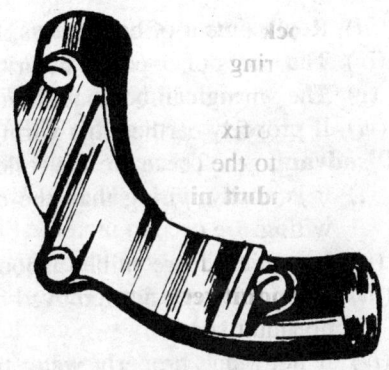

(*o*) **Earthing clips.** These are made of copper. These are fixed over the conduit pipe

Fig. 14.11. Conduit bend.

having earth wire. These clips help in maintaining earth continuity throughout the installation.

Q. Explain the conduit pipe wiring. What are the advantages, disadvantages and uses of this type of wiring?

Ans. In this wiring system the V.I.R. wire or P.V.C. wire run in the conduit pipe of different size as $\frac{3}{4}''$, $1''$, $1\frac{1}{4}''$, $1\frac{1}{2}''$, $2''$ and $2\frac{1}{2}''$ etc. The conduits are electrically and mechanically continuous and are connected to earth at some suitable points to maintain the earth continuity throughout the installation. This wiring is of two types:

(*a*) **Surface conduit pipe wiring.** In this system the conduit pipe is fixed on the wall or ceiling by means of the shaddles or clips. First the conduit is fixed and then the wires are drawn. The earth wire is fixed by means of the earth clips. This type of wiring is suitable for factory or workshops etc.

(*b*) **Concealed conduit pipe wiring.** In this system the conduit pipe (nowadays P.V.C. pipe is also used) is fixed under the wall or ceiling. In this case first the channels are made then the conduit pipe is fixed in the channel by means of hooks and clamps. Then the wires are drawn into the pipes.

As the wiring is done under the plaster so the whole of the system is made water tight to prevent the entering of moisture. The wires are drawn in the conduit pipe by means of either the drawing system or pushing or fish wire system.

There are the following advantages and disadvantages:

Advantages. There are the following advantages:

(*i*) The wires are safe from mechanical damages, moisture and fire etc.

(*ii*) The life is sufficiently long.

(*iii*) Replacement of conductors, fault location and extension is easy.
(*iv*) The wires of opposite polarity can run together in the same pipe.
(*v*) The wiring can be done in any places.
(*vi*) If properly earthed this wiring is shock proof.

Disadvantages. These are the following disadvantages:

(*i*) It is costly wiring than the cleat, batten and casing and capping wiring.
(*ii*) It requires more skilled labour so labour cost is more.
(*iii*) If the burrs are not removed these may spoil the conductor insulation.
(*iv*) If not made properly water tight the moisture may enter and the insulation may damage.

General rules for conduit pipe wiring:

1. The conduit pipe cutting should be done very carefully and the burrs should be removed with round file or reamer.
2. The threading should be accurate for satisfactory joints. Apply oil while cutting threadings.
3. In order to maintain good electrical continuity the threading should be painted with aluminium paint.
4. The conduit pipe should be cleaned and there should not be any moisture or water drops etc. inside the pipe.
5. First the pipe should be fixed on the wall or ceiling then the wires should be drawn with the help of draw wires.
6. The pipe should be fixed with shaddles and the distance in no case, should exceed one metre. The wooden plugs should be used to separate pipe from a wall or ceiling.
7. The conduit should be painted to save the pipe from rust etc.
8. Use proper bushings at the end of the pipe where the wire leaves the pipe.
9. Conduit should be electrically continuous and properly earthed.
10. The conduit should be errected away from the water and gas pipe.
11. The pipe should be water tight.
12. The number of wires should be according to the diameter of the pipe. The over crowding should be avoided.
13. The wires should not be in twisting form inside the pipe, avoid the possibility also.
14. If pipe bending is essential use pipe bending machine.
15. Use proper conduit pipe accessories for the wiring.
16. The solid accessories should be used in case of concealed conduit pipe wiring and inspection accessories for surface conduit pipe wiring.

Table 14.1

Size of cable (No. and dia. in mm)	Material of conductor	No of cables that can be accommodated in the conduit of sized											
		19 mm		25.4 mm		31.8 mm		38.4 mm		50.8 mm		63.5 mm	
		250 V	660 V	250 V	660 V	250 V	660 V	250 V	660 V	250 V	660 V	250 V	660 V
1/1.12	Copper	6	4	10	9	14	12	—	—	—	—	—	—
3/0.736	Copper	6	4	10	9	14	10	—	—	—	—	—	—
1/1.40	Aluminium	6	4	10	9	14	10	—	—	—	—	—	—
0/0.915	Copper	5	4	10	8	14	9	—	—	—	—	—	—
1/1.80	Aluminium	5	3	10	6	14	8	—	—	—	—	—	—
7/0.738	Copper	5	3	10	6	14	8	—	—	—	—	—	—
1/2.24	Aluminium	4	2	6	5	10	7	—	—	—	—	—	—
7/0.915	Copper	4	2	6	5	10	7	—	—	—	—	—	—
1/2.80	Aluminium	4	—	6	4	10	6	—	—	—	—	—	—
7/1.12	Copper	4	—	6	4	10	6	—	—	—	—	—	—
7/1.32	Copper	—	—	4	3	8	5	8	6	—	—	—	—
1/3.55	Aluminium	2	1	4	3	5	5	7	5	—	—	—	—
7/1.626	Copper	—	—	4	2	5	4	7	5	—	—	—	—
7/1.70	Aluminium	—	—	2	2	4	3	6	4	—	—	—	—
19/1.12	Copper	—	—	—	—	4	3	6	4	7	7	8	—
19/1.32	Copper	—	—	—	—	3	2	4	3	6	6	7	—
7/2.24	Aluminium	—	—	—	—	2	—	3	3	6	6	7	—
7/2.50	Aluminium	—	—	—	—	2	—	3	2	5	5	6	—
19/1.626	Copper	—	—	—	—	—	—	2	2	5	5	6	—
7/3.00	Aluminium	—	—	—	—	—	—	2	—	3	3	5	—
19/1.80	Aluminium	—	—	—	—	—	—	—	—	3	3	5	—

Table 14.2. Different types of wiring in a glance

Particulars	Cleat wiring	Casing and capping wiring	Batten wiring		Conduit pipe wiring	Remarks
			T.R.S.	L.C.C.		
1. Material	Cleat, V.I.R. or P.V.C. wire, screws, gitties, board & block	T.W. or P.V.C. casing & capping, V.I.R., P.V.C. wires, wooden gitties, screws	T.W. batten, T.R.S., C.T.S. or P.V.C. wires, gitties, screws, nails, clips	Batten, lead covered wire, gitties, screws clips, board and blocks	Conduit pipe, shaddles, hooks, wooden gitties, I.C. Bend and socket and other accessories, screws, block and board V.I.R. or P.V.C. wire.	
2. Cost	Cheap	Fairly expensive	Cheap	Expensive	Expensive	
3. Life	Short	Fairly long	Long	Long	Very long	
4. Mechanical protection	Nil	Fair	Fair	Good	Very good	
5. Safety from dampness	Nil	Fair	Good	Good	Very good	
6. Labour	Semi-skilled	Highly skilled	Skilled	Skilled	Highly skilled	
7. Extension and removal	Very easy	Difficult	Easy	Difficult	Not so easy and is costly	
8. Time	Short	Fairly long	Short	Fairly long	Very much	
9. General reliability	Poor	Good	Good	Fairly good	Very good	
10. Appearance	Not so good	Good	Good	Good	Very good	

14.4. TESTING OF INSTALLATION

Q. What is the necessity of testing the new installation? What apparatus would you like to use for testing the domestic installation?

Ans. All types of electrical installations, domestic and industrial, should be tested and inspected before energising. It must be made sure that there is no short circuit, earth leakage or discontinuity etc., moreover the wiring is done according to the Indian Electricity Rules. According to the Indian Electricity Rule No. 47, the new installation or extension of the existing installation shall not be energised by any supplier unless the supplier makes sure that the wiring has been done according to the Indian Electricity Rule and the wiring will not give shock and prove harmful to any human being or supplier.

So before connecting the supply the supplier will test the wiring. The instrument generally used for this testing is the megger. An authorised contractor will state and certify that the wiring has been done according to the Indian Electricity Rule and leakage current will not exceed 1/5000th of the total current between either two conductors of opposite polarity or between any live conductor and the earth *i.e.*

$$\text{Leakage current} = \frac{250}{\text{Insulation resistance in mega ohm}} \text{ Amp}$$

Megger. The wiring is tested with this instrument. It is used to measure the insulation resistance of the wires in mega ohm. The megger consists of a hand driven d.c. permanent magnet generator. Generally for the domestic wiring purpose the megger of 500 V is used and for power wiring testing 1000 V megger is used. There are two terminals marked as *L* and *E i.e.* live and earth. A pointer deflects over a calibrated scale which is divided from 0 to ∞. The handle should be rotated at 160 r.p.m.

Precautions for using the megger. These following precautions should be carefully observed:

(*i*) Generally the voltage of the megger should be double then the voltage rating, for domestic installation 500 V and for industrial installation etc. 1000 V megger is used.

(*ii*) In no case it should be connecting on the line.

(*iii*) Before testing the installation, the megger should be tested itself and make sure whether it is in working order or not. For that testing short circuit both *L* and *E* terminals of the megger, on rotating the handle it should show zero.

(*iv*) In case of short circuit the megger should show zero and in case of open circuit the megger should show infinity.

(v) The handle should be rotated at 160 r.p.m. or as mentioned in the instructions supplied by the manufacturer.

(vi) While testing, the connections should be well tight.

(vii) The megger should be handled carefully.

Q. Describe the various tests to be performed before energising the new domestic installation.

Ans. The new installation should never be energised unless it has the insulation resistance at least of one mega ohm. The insulation resistance can be found out as

$$IR = \frac{50}{\text{No. of outlets (Points + Switches)}} \, M\Omega$$

There are the following different tests to be carried over before energising the new installation.

(i) Continuity test.

(ii) Insulation resistance between the conductors and earth.

(iii) Insulation resistance between conductors of opposite polarity.

(iv) Polarity test.

(v) Earth continuity test.

(i) **Continuity test.** It is also known as the open circuit test. The main object of this test is to test the continuity of the wires and of each point.

Put the main switch off and connect both L and E terminals of the megger to positive and negative or phase and neural respectively of the main switch. Put all the lamps in the lamp holder, short circuit the plug socket and for easy and effective testing put the switches 'off' before starting the test. Rotate the handle and turn the switch 'on' turn by turn. The megger should show zero with every individual switch and if it does not, then there is a break in the circuit, check and remove the fault.

(ii) **Insulation resistance between the conductors and earth.** This test is performed to test the insulation resistance between the conductors (phase and neutral) and earth. If the insulation resistance is less there will be the leakage current from conductors to earth, take the following steps for this test:

(i) Insert all lamps in the holders and short circuit or put load in the sockets, outlets and 'ON' the switch.

(ii) Now short P and N, both the terminals of the installation and connect to 'L' terminal of the megger. Connect E terminal to earth wire as shown in Fig. 14.12.

(iii) On rotating the handle, it should show infinity or at least one megaohm.

The insulation resistance should be $= \dfrac{50}{\text{No. of outlets (Points + Switches)}} \text{M}\Omega$

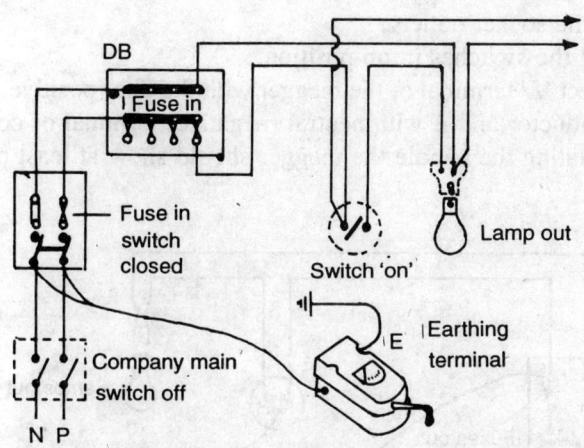

Fig. 14.12. Insulation resistance test between conductors and earth.

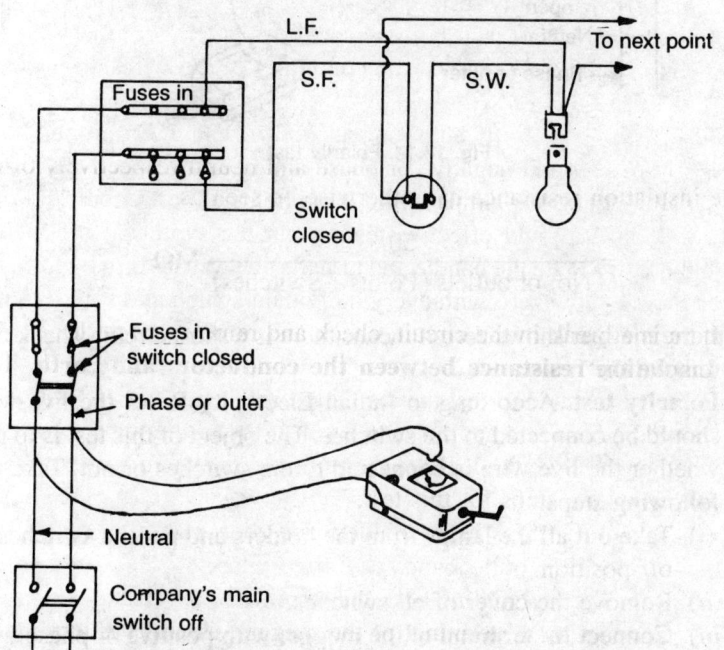

Fig. 14.13. Insulation resistance between conductors of opposite polarity.

3. **Insulation resistance between the conductors of opposite polarity.** The object of this test is to test the insulation resistance between the conductors of opposite polarity. Take the following steps:

(i) Take out all the lamps from the holders and short circuiting wire from the socket outlets.

(ii) Put all the switches in on position.

(iii) Connect 'L' terminal of the megger with the line (positive) terminal or conductor and E with neutral (negative) terminal or conductor.

(iv) On rotating the handle the megger should show at least one mega ohm.

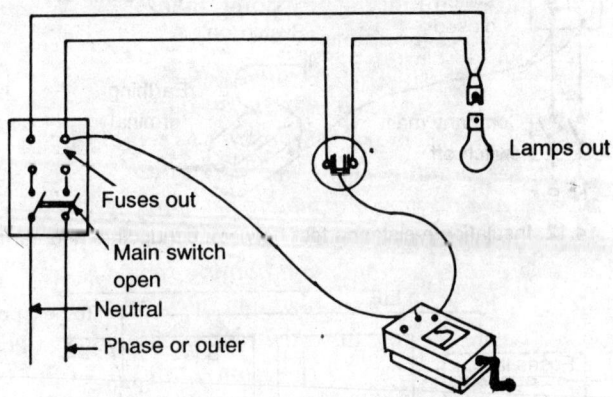

Lamps out

Fuses out

Main switch
open

Neutral

Phase or outer

Fig. 14.14. Polarity test.

The insulation resistance can otherwise be seen as

$$= \frac{50}{\text{No. of outlets (Points + Switches)}} \ \text{M}\Omega$$

If the megger shows zero it means there is a short circuit, check and remove the fault.

4. **Polarity test.** According to Indian Electricity Rules the live wire should be connected to the switches. The object of this test is to test whether the live wire is connected to the switches or not. Take the following steps:

(i) Take out all the lamps from the holders and put the switches in off position.

(ii) Remove the cover of all switches.

(iii) Connect 'L' terminal of the megger with positive or phase wire in the main switch and E terminal to the individual switch turn by turn.

On rotating the handle the megger should indicate zero with every individual switch. If it shows infinity it means that connection are not according to I.E. rules, check it and remove the fault.

5. **Earth continuity test.** The earth continuity throughout the installation should be maintained and all the metallic parts except the current carrying conductor be properly earthed.

To perform this test take the following steps:

Connect '*E*' terminal of the megger with earth wire at the main switch. The '*L*' terminal of the megger should be connected to all the metallic parts of the installation except current carrying conductors; earth terminal of the three pin plug sockets. In every case the megger should show zero and if it indicates infinity at any point, there is a break in the earth wire, check and remove the fault.

Example. *A room 5 × 4 × 3.5 m is required to be provided with one fan, one tube light, one lamp point and a socket outlet. Each point is to be controlled independently. Mark the positions of lights, fan and socket outlet and switches suitably. Draw the installation plan. Calculate the length of the wire and prepare a complete list of the material required for wiring the room in concealed conduit pipe system. No main switch is to be provided as the entry is from the adjoining room.*

Solution. Keeping in mind certain laws and rules, which are laid down as under, let us calculate the length of the phase and neutral wires:

(*i*) The height of the horizontal run from the floor is 3 m.

(*ii*) The fan point in this particular room should be kept in the centre of the ceiling and the light point should be aside adjoining walls 0.5 m below the ceiling.

(*iii*) The height of the switch board from the floor is 1.5 m.

(*iv*) The H.R. and V.R. are horizontal run and vertical run respectively.

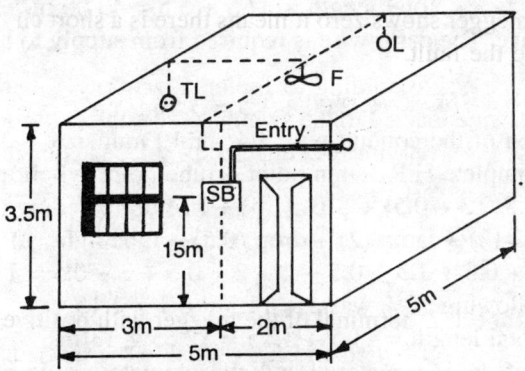

Fig. 14.15. Domestic installation.

(v) S.B. stands for switch board.
Calculation of the length of phase wire of 1/1.4 mm^2.
   (i) From the point of entry of circuit into the room upto switch board
      = 2 m.
      Now horizontal + vertical run = 2 + 1.5 = 3.5 m.
   (ii) From switch board to fan.
      S.B. to H.R. = H.R. + top point of vertical run to start of rise
      +H.R. to ceiling + along the ceiling to fan
         = 1.5 + 0.5 + 0.5 + 2 = 4.5 m.
   (iii) From switch board to light point (lamp)
         = S.B. to fan point + fan point to the wall or
         corner of ceiling + drop
         = 4.5 + 2 + 0.5 = 7.00 m.
   (iv) From switch board to tube light
         = S.B. to tube point + tube to ceiling corner + drop
         = 4.5 + 2 + 0.5 = 7 m.
      Now total length = light + fan + tube + entry
         = 7 + 4.5 + 7 + 3.5 = 22 m.

Neutral wire:

   (i) From point of entry to switch board = 2 + 1.5 = 3.5 m.
   (ii) From S.B. to fan point   = 1.5 + 0.5 + 0.5 + 2 = 4.5 m.
   (iii) From fan to lamp point   = 2 + 0.5 = 2.5 m.
   (iv) From fan to tube light   = 0.5 + 2 + 0.5 + 2 = 5 m.
   (Fan to lamp and lamp to fan and then to tube = 5 m.)
   Now total length = 3.5 + 4.5 + 2.5 + 5 = 15.5 m
   Total length of phase and neutral wire
         = 22 + 15.5 = 37.5 m
   Now allowing 15% wastage = 37.5 × 0.15 = 5.5 m
∴   Total length = 37.5 + 5.5 = 43.00 m.

   Earth wire. The earth wire is required from supply to board for 3 pin
socket
         = 2.00 + 1.5 = 3.5 m
   Estimation of the conduit pipe 3/4″ or 19 mm size
         = complete H.R. + fan point + tube point + light point
         = H.R. (2 + 0.5) + S.B. (1.5) + rise up to ceiling (0.5) + ceiling
         fan (2) + lamp (2) + drop (0.5) + tube light (2) + drop (0.5)
         = 2 + 0.5 + 1.5 + 0.5 + 2 + 2 + 0.5 + 2 + 0.5 = 11.5 m
         Allowing 10% wastage 11.5 × 0.1 = 1.15
∴   Total length = 11.5 + 1.15 = 12.65 ≅ 13 m.
   The list of the raw material and approximate cost is shown in Table
14.3.

Table 14.3

Specifications	Qty.	Rate	Per	Amount	Remarks
1. 1/1.4 mm 250 V grade single core P.V.C. Al. wire	43 m	250/-	100/m	125	The
2. Earth wire 14 S.W.G.G.I. (in one kg = 30 m app)	35 m	50/-	kg	55	labour
3. Teak wood board underground (25 × 20) cm	1 No.	15/-	each	15	and other
4. Bakelite cover @ Rs. 2/-cm²	1 No.	10/-	each	10	charges
5. Flush switch 5 A	3 No.	12/-	each	36	are not
6. Flush socket with switch 5 A 250 V	1 No.	24/-	each	24	included.
7. Celling rose 2 plate (for fan and tube)	2 No.	12/-	each	24	
8. Brass bracket with holder for lamp	1 No.	15/-	each	15	
9. Conduit pipe 19 mm dia, light gauge	13 m	10/-	m	130	
10. Single way junction box	2 No.	5/-	each	10	
11. Two way junction box at take off point	1 No.	5/-	each	5	
12. Three way Junction box	2 No.	8/-	each	16	
13. Conduit bends 19 mm dia.	3 No.	4/-	each	12	
14. Conduit couplers	6 No.	4/-	each	24	
15. Crampets	12 No.	2/-	each	24	
16. Screws of all sizes	20 No.	5/-	doz.	10	
Total cost				Rs. 535/-	

REVIEW QUESTIONS

1. What do you understand by wiring?
2. How supply is fed to a consumer from the distribution pole?
3. What is service main? What are their classifications?
4. Describe the different systems of distribution of supply to the consumers.
5. What are the different systems of wiring used for domestic installations? what are the test to be performed under the Indian Electricity Rules before energising domestic installation?　　　　　　　　　(*N.C.V.T. 1964*)
6. What are the advantages and disadvantages of the batten and casing and capping wirings?
7. How conduit wiring is done? What are the different classifications?
8. Which rules and regulations will you observe while performing the surface conduit pipe wiring?
9. Differentiate the different types of wiring.
10. Estimate the basic requirements, wiring and expenditure for a domestic living room of 15′ × 12′ × 10′ size.
11. Fill up the blanks:
    (*i*) The system of wiring distribution is tree and ....
    (*ii*) The allowable leakage current is ....
    (*iii*) The insulation resistance of an installation should never be less than ....
    (*iv*) The .... is used to measure the insulation resistance of the installation.
12. Answer the following questions:
    (*i*) Why pole fuse is used for service mains?
    (*ii*) Why the iron clad service fuse is sealed?
    (*iii*) Why distribution system is popular?
    (*iv*) Which system of wirings are used for wet place, Blacksmith shop and cold storage?
    (*v*) Why tests are essential before energising the new installation?

15

A.C. Fundamentals and Circuit

15.1. ALTERNATING CURRENT

Q. What do you understand by A.C.? Write down the advantages and disadvantages of A.C. over D.C. keeping in view the generation, transmission and distribution.

Ans. At present about 90% of the generation of electricity is of A.C. nature. The A.C. can be defined as the quantity of electricity which changes its magnitude and direction after a definite interval of time or say periodically.

There are the following advantages and disadvantages.

Advantages. Following are the advantages in:

(*a*) **Generation.**

  (*i*) The D.C. can be generated only up to 650 V but A.C. can be generated at higher voltage say 11 kV or even more or even more.

 (*ii*) For same frame size of machine, A.C. generator has more output than D.C. generator.

(*iii*) The construction of A.C. generator is much easy than D.C. generator.

(*iv*) Due to easy construction, the cost is less and the maintenance cost is also less than D.C. generator.

 (*v*) Less possibility of faults in A.C. generator *i.e.* the alternator than D.C. generator.

(*b*) **Transmission.** To send the electricity at a far distance is known as transmission. The transmission is always done at a higher voltage. These are the following advantages:

  (*i*) A.C. can be stepped up and stepped down with the help of transformer but D.C. cannot.

(*ii*) By increasing the voltage, the current is reduced for the same power.

(*iii*) Due to less current, the size of conductor as well as the cost is reduced.

(*iv*) Because of low current, the voltage drop ($I \times R$) and the power loss ($I^2 \times R$) in the line are less.

(*v*) With the reduction in size of the conductor, the poles can be placed at a far distance, the strength of the poles is also reduced, the labour cost is also reduced, the hence the cost of transmission is less, thus the transmission is economical.

(*vi*) The voltage regulation is also good in case of A.C. than D.C.

(*c*) **Distribution.** To distribute the electrical energy to the consumers is known as distribution. These are the following advantages of A.C. distribution over D.C. distribution:

(*i*) The light and power can be taken from the same distribution line. All three phases for power and one phase and neutral for single phase power and lighting loads.

(*ii*) Less number of conductors are required for lighting and power loads than D.C. distribution.

(*d*) **Utilization.** To use the electrical energy by the consumers is known as utilization. There are the following advantages:

(*i*) All kind of lamps and tubes can work as A.C. where the neon lamps, H.P.M.V. lamp and sodium vapour lamps cannot work on D.C.

(*ii*) A.C. motors can be build up for higher output.

(*iii*) The construction of A.C. motor is much easy, simple and robust than D.C. motor.

(*iv*) The maintenance cost is also less and possibility of faults is also less than D.C. motors.

(*v*) A.C. motors of small power say upto 5 H.P. can be started directly, but the D.C. motor cannot be started without starter.

(*vi*) Efficiency of A.C. motors is very good.

(*vii*) A.C. can be stepped down for the lamps of low voltage.

(*viii*) A.C. can easily be converted to D.C. but D.C. conversion is not so cheap and economical.

Disadvantages. There are the following disadvantages of A.C. over D.C.

(*i*) A.C. cannot be used for electro-chemical works; as in electroplating, battery charging, electrolysis and refinery of metals etc.

(*ii*) A.C. calculations are difficult than D.C. calculations because of power factor, inductance, capacitance and other complex quantities.

(*iii*) Speed control is not so accurate and fine in A.C. as in case of D.C. motors.

(iv) For traction purpose D.C. motor (series) are much useful. They are widely used in trains, trolley, cranes etc.

(v) Arc lamps are much economical and effective on D.C. than A.C.

(vi) Relays and other controllings are much effective on D.C. than A.C.

15.2 PRODUCTION

Q. Describe with the help of a suitable sketch how A.C. is induced?

Ans. Whenever a conductor links or associates the magnetic flux or cuts the magnetic flux at an angle, an e.m.f. is induced in the conductor. The induced e.m.f. is proportional to the number of conductors and the angle at which it cuts the magnetic flux.

As shown in Fig. 15.1. whenever the conductors are rotated in the magnetic poles *i.e.* North and South, the conductor cuts the magnetic flux at different angles on different positions as shown by the different positions from No. 1 to 8 as shown in Fig. 15.2.

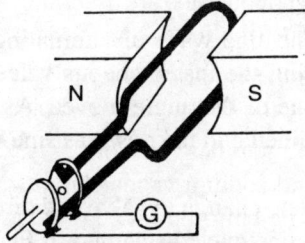

Fig. 15.1. Simple alternator.

At positions number 4 and 8 the direction of flux and direction of movement *i.e.* velocity both are parallel, hence no cutting is achieved. Thus e.m.f. induced will be zero. Now if we plot the different angular positions and the e.m.f. induced, the curve will have the shape as shown in Fig. 15.2.

The e.m.f. induced can be given as,

$$e = \text{Rate of change of flux} \times \text{No. of conductors}$$

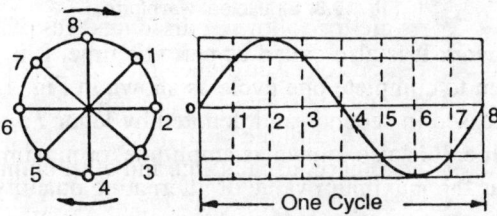

Fig. 15.2. Induced E.m.f.

In generalised form the e.m.f. can be represented as

$$e = E_{max} \sin\theta$$

where e = E.m.f. induced at any instant

E_{max} = Maximum value of induced e.m.f.

θ = Angle covered = ωt

where ω = The angular velocity

and t = The time taken.

Similarly current equation can also be generalised as

$$i = I_{max} \sin\theta.$$

15.3 TERMS AND DEFINITIONS

Q. Define the followings:

(*a*) **Sine wave,** (*b*) **Cycle,** (*c*) **Time period,** (*d*) **Peak valve,** (*e*) **Instantaneous value,** (*f*) **Average value,** (*g*) **Effective or r.m.s. value,** (*h*) **Frequency,** (*i*) **Form factor,** (*j*) **Peak Factor** and (*k*) **Phase difference.**

Ans. (*a*) **Sine wave.** The sine wave of alternating quantity is the curve obtained by plotting the instantaneous values of alternating quantity against the sine of the angle moved. As shown in Fig. 15.3. The curve so obtained and is known as sine wave of the alternating quantity.

(*b*) **Cycle.** The complete change in valve and direction of an alternating quantity is called one cycle as shown in Fig. 15.3. It is completed during 360° electrical.

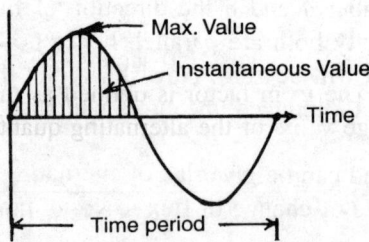

Fig. 15.3. Sinusoidal waveform.

(*c*) **Time period.** It is also called as periodic time. It is defined as the time taken to complete one cycle as shown in Fig. 15.3. Generally it is expressed in seconds and denoted by letter *T*.

(*d*) **Peak value.** It is also known as amplitude, or maximum value. It is defined as the maximum value of alternating quantity attained during positive or negative half cycle, as shown in Fig. 15.3.

(e) **Instantaneous value.** It is defined as the value of alternating quantity at any particular instant, as shown by i_1, i_2 in Fig. 15.3.

(f) **Average value.** The average value of alternating quantity is defined as the average of all instantaneous values attained during half cycle either positive or negative.

It can otherwise be given as

$$I_{av} = 0.637\, I_{max}$$
or $$V_{av} = 0.637\, V_{max}$$

(g) **Effective or r.m.s. value.** It is also known as virtual value of alternating quantity. The r.m.s. value of alternating quantity is defined as that value which when applied to a given circuit for a given time produces the same amount of heat as when D.C. is applied to the same circuit for the same interval of time.

$$I_{r.m.s.} = 0.707\, I_{max}$$
or $$V_{r.m.s.} = 0.707\, V_{max}$$

(h) **Frequency.** It is defined as the number of cycles per seconds. It is denoted by f or Hz. It is given by the formula

$$f = \frac{P \cdot N}{120} c/s$$

where f is the frequency is c/s

and P = The number of poles

N = The speed in r.p.m.

The time period and frequency are related as

$$T = \frac{1}{f} \text{ sec}$$

For example the 50 c/s has 1/50 sec *i.e.* 0.02 sec time period

(i) **Form factor.** The form factor is defined as the ratio of the r.m.s. value to average value of the alternating quantity,

i.e. Form factor = $\dfrac{\text{r.m.s. value of alternating quantity}}{\text{Average valve of alternating quantity}}$

$$= \frac{0.707 \text{ Max. value}}{0.637 \text{ Max. value}} = 1.11.$$

(j) **Peak factor.** The peak factor is defined as the ratio of maximum value to r.m.s. value of an alternating quantity.

i.e. Peak factor = $\dfrac{\text{Max. value of alternating quantity}}{\text{R.M.S. valve of same alternating quantity}}$

$$= \frac{\text{Max. value}}{0.707 \, \text{Max.value}} = 1.414.$$

(k) **Phase difference.** The phase difference in alternating quantity is defined as the angular displacement between two alternating quantities *i.e.* voltage and current, current and current, voltage and voltage.

Q. Define the followings: (*a*) **Power factor;** (*b*) **Inphase,** (*c*) **Out of phase,** (*d*) **Inductance,** (*e*) **Capacitive reactance,** (*f*) **Inductive reactance,** (*g*) **Impedance,** (*h*) **Admittance,** and (*i*) **Susceptance.**

Ans. (*a*) **Power factor.** It is a ratio. It is defined as the ratio of true power to apparent power in A.C. circuit.

i.e. $\text{power factor} = \dfrac{\text{True power}}{\text{Apparent power}}.$

It can otherwise be defined as the cosine of the angle between current and voltage or the cosine of the angular displacement between two alternating quantities

In A.C. series circuit the ratio of resistance to impedance also gives away the power factor of the circuit.

(*b*) **In-phase.** Because of the phase difference or say angular displacement, the alternating quantities do not attain their maximum and minimum values at the same time but they may differ also.

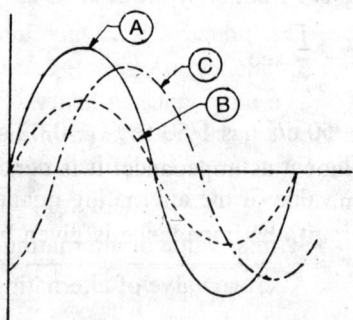

Fig. 15.4. Phase displacement.

So whenever two alternating quantities attains their minimum and maximum value at the same time they are called in-phase. For example quantity *A* and *B* in Fig. 15.4. attains their maximum and minimum values at the same time so they are in-phase.

(*c*) **Out of phase.** If two alternating quantities do not attain their maximum and minimum values at the same time, these are termed as out of phase. For example quantity A and C in Fig. 15.4 are out of phase.

Here quantity A is attaining its maximum and minimum value before quantity C so quantity A is leading the quantity C or quantity C is lagging behind quantity A.

(d) **Inductance.** It is the property of the circuit by virtue of which an e.m.f. is induced in the same circuit. This induced e.m.f. opposes the cause responsible for the production of this e.m.f. The unit of inductance in Henry and represented by L.

Henry is the unit of inductance and *one henry* can be defined as the circuit will have an inductance of one henry if a change of one ampere/sec causes an induction of one volt in it.

(e) **Capacitive reactance.** It is the opposition offered by the capacitance in the flow of alternating current circuit. It is denoted by X_c and measured in ohm.

$$X_C = \frac{1}{2\pi f C}$$

where X_c = Capacitive reactance in ohm.
f = Frequency in c/s.
C = Capacitance in farad.

(f) **Inductive reactance.** It is the opposition offered by an inductance in the flow of current in A.C. It is denoted by X_L and measured in ohms.

$$X_L = 2\pi f \cdot L \ \Omega.$$

where X_L = The inductive reactance in ohm.
f = Frequency in c/s.
and L = The inductance in henry.

(g) **Impedance.** The total opposition offered in the flow of current in A.C. circuit, is known as impedance. It is denoted by letter Z and measured in ohm.

In A.C. series circuit, the impedance is given by the formula

$$Z = \sqrt{R^2 + \left(X_L \sim X_C\right)^2}$$

and in A.C. parallel circuit is can be given as

$$\frac{1}{Z} = \sqrt{\frac{1}{R^2} + \left(\frac{1}{X_L} \sim \frac{1}{X_C}\right)^2}$$

(h) **Admittance.** It is the reciprocal of the impedance, and can be defined as the property of the circuit which helps in the flow of current. It is denoted by letter Y and unit is mho.

(*i*) **Susceptance.** It is the reciprocal of the reactance. It is denoted by letter *s* or *b* and unit is mho.

Q. Define the vector and scalar quantities. How the alternating quantities and added?

Ans. Vector. The quantity which has got the magnitude and direction both, are known as the vector quantity, *e.g.* the velocity, current and voltage etc. Such quantities are recognised by their magnitude, direction and the sense of direction in which actually these are drawn. Generally these are represented by a straight line with an arrow head. The length of the arrow indicates the magnitude of the quantity, the position vertical or horizontal or inclined represent the direction and the arrow head, the sense of direction.

An alternating quantity is also a vector quantity and that can be represented by a vector.

Scalar. The quantity which has got the magnitude or direction is known as the scalar quantity *e.g.* the speed, weight etc.

Addition of two vectors. The addition can be divided into these forms:

(*a*) *Both vectors are at zero degree to each other.* If the quantities are revolving with the same speed or both quantities are inphase, they are not having any angular displacement, then they can be added up as

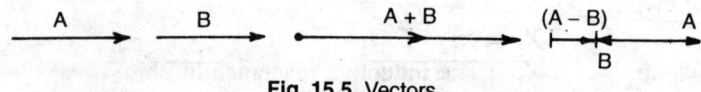

Fig. 15.5. Vectors.

when these are added, then $R = A + B$
and in case of subtraction, then $R = A - B$.

(*b*) *Both are at right angle to each other.* Their resultant will be

$$R = \sqrt{A^2 + B^2}.$$

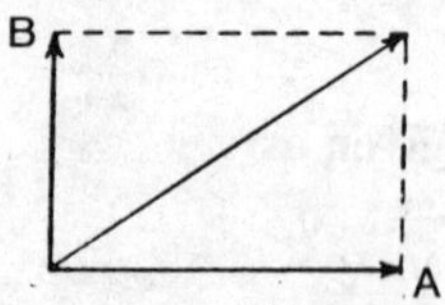

Fig. 15.6. 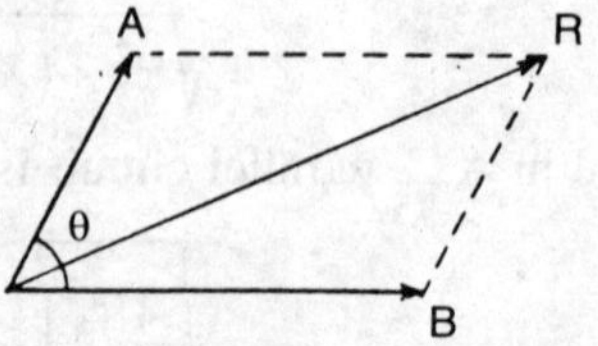   **Fig. 15.7.**

(*c*) *Both are at an angle 'θ' to each other.* Whenever these are working at an angle 'θ' in that case their resultant is

$$R = \sqrt{A^2 + B^2 + 2AB\cos\theta}.$$

(d) If there are *more vectors* working on a body *at different angles* then, the vectors are resolved horizontally and vertically and then their resultant is calculated.

The similar components are added up ΣH and ΣV. Their resultant will be

$$R = \sqrt{(\Sigma H)^2 + (\Sigma V)^2}.$$

Example. *Find out the current drawn by a parallel circuit if one branch is having 6 A at 45° to voltage vector and other having 8 A at unity power factor.*

Solution. *Method I.* Now the current having unity power factor, will be zero degree displacement from the voltage vector and the other current at 45°.

Now the resultant will be

$$I = \sqrt{6^2 + 8^2 + 2 \times 6 \times 8 \times \cos 45°}$$

$$= \sqrt{36 + 64 + 96 \times 0.707} = \textbf{12.956 A. Ans.}$$

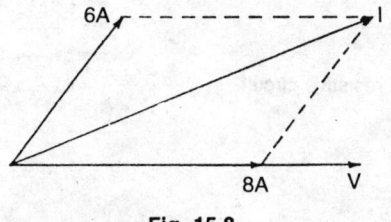

Fig. 15.8.   **Fig. 15.9.**

Method II. In this case the currents are resolved horizontally and vertically.

Horizontal components

$$= OB + OA \cos 45°$$

$$= 8 + 6 \cos 45 = 12.242 \text{ A}$$

Vertical components

$$= OA \sin 45 = 6 \sin 45° = 4.242 \text{ A}$$

Now the resultant current will be

$$I = \sqrt{(\Sigma I_H)^2 + (\Sigma I_V)^2}$$

$$= \sqrt{(12.242)^2 + (4.242)^2}$$

$$= \textbf{12.956 A. Ans.}$$

15.4 A.C. PURE CIRCUITS

Q. Name the different obstructions in A.C. circuit. State the behaviour of current, voltage, power factor and power in case of purely resistive circuit.

Ans. In case of A.C. circuits the opposition is mainly because of resistance, inductance and capacitance.

A purely resistive circuit is that which neither offer the inductive reactance nor capacitive reactance. It has only the resistive component as shown in Fig. 15.10.

Followings are the characteristics of this circuit:

(*i*) The voltage and current are in phase, *i.e.* the current is maximum when the voltage is maximum and zero when voltage is zero. The current can be given as

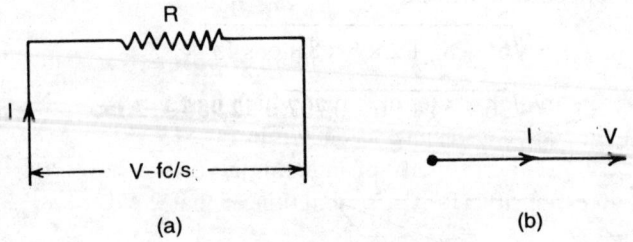

Fig. 15.10. Pure resistive circuit.

$$i = \frac{V}{R}$$

$$= \frac{V_{max} \cdot \sin \omega t}{R} \qquad (\because v = V_{max} \cdot \sin \omega t)$$

or $\qquad i = I_{max} \cdot \sin \omega t \ A.$

(*ii*) The phase difference between current and voltage is zero as shown in Fig. 15.10(*b*). So the power factor is unity.

(*iii*) The power in a.c. resistive circuit is the product of current and voltage

i.e. $\qquad W = V \cdot I \cdot$

Obviously $\qquad w = v_i$

$$= V_m \cdot \sin \omega t \cdot I_{max} \cdot \sin \omega t$$

$$= V_m \cdot I_m \cdot \sin^2 \omega t$$

$$= \frac{1}{2} \cdot V_m \cdot I_m \cdot (1 - \cos 2\omega t)$$

$$= \frac{V_m \cdot I_m}{2} - \frac{V_m \cdot I_m}{2} (\cos 2\omega t)$$

The later portion is having double the frequency and net amount of the product will be zero for the entire cycle.

So

$$P = \frac{V_{max} \cdot I_{max}}{2} = \frac{V_m}{\sqrt{2}} \cdot \frac{I_m}{\sqrt{2}}.$$

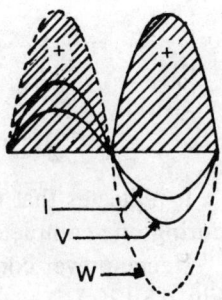

Fig. 15.11. Power resistive.

$$= V_{r.m.s} \cdot I_{r.m.s}.$$

So the power is or the product of voltage and current as shown in Fig. 15.11.

Q. Describe the characteristics of the a.c. purely inductive circuit.

Ans. A purely inductive circuit is that which has only inductance not the capacitance or resistance as shown in Fig. 15.12. An inductive circuit has got the property of inducing a voltage which opposes the cause responsible for the production of that voltage.

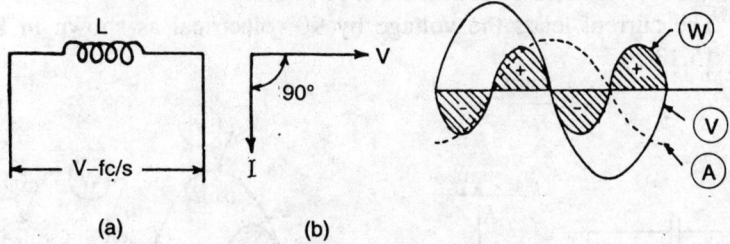

Fig. 15.12. Pure inductance.

It has got the following properties:

(*i*) The voltage and current are at 90° elect. to each other. The current lags behind the voltage.

(*ii*) The current and voltage are out of phase.

(*iii*) The power consumed by the circuit is

$$w = v \cdot i$$

$$= V_m \cdot \sin \omega t \cdot I_m \cdot \sin (\omega t - 90)$$

$$= V_m \cdot I_m \cdot \sin \omega t \cdot \sin (\omega t - 90)$$

$$= \frac{1}{2} \cdot V_m \cdot I_m \cdot \left[2\sin \omega t \cdot \sin(\omega t - 90) \right]$$

$$= -\frac{1}{2} \cdot V_m \cdot I_m \cdot \sin 2\omega t.$$

It indicates that this expression do not have any constant factor and during one complete cycle this factor comes zero.

Hence power consumed by purely inductive load is zero as shown in Fig. 15.12(c).

(iv) The net opposition offered is known as the inductive reactance and is equal to

$$X_L = 2\pi f \cdot L \ \Omega$$

where X_L = inductive reactance in ohm, f-supply frequency in c/s and L—the inductance in henry.

Q. How current, voltage and power behaves in purely capacitive circuit?

Ans. The purely capacitive circuit has only the capacitance neither the inductance nor resistance as shown in Fig. 15.13.

It has the following properties:

(i) The voltage and current are out of phase.

(ii) The current leads the voltage by 90° electrical as shown in Fig. 15.13(b).

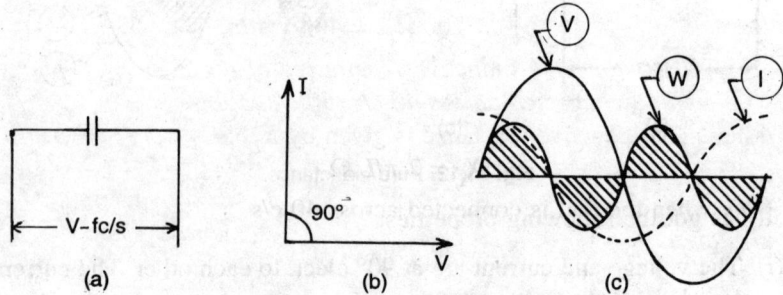

Fig. 15.13. Pure capacitance.

(iii) The power consumed
$$= \text{Voltage} \times \text{Current} \times \text{Power factor}$$
$$= V \times I \times \cos 90° = 0$$

The waveform is shown in Fig. 15.13 (c).

(iv) The opposition offered is X_c and can be given as

$$X_c = \frac{1}{2\pi f C} \text{ ohm}$$

where X_c = capacitive reactance in ohm, C = the capacitance in farad.

Example. *Calculate the amount of current taken by a pure resistance of 100 Ω, when connected across 200 V A.C. mains. Also give the r.m.s. and maximum value of the line current and the power consumed.*

Solution. It is a purely resistance so the current taken will be

$$= \frac{\text{Voltage}}{\text{Resistance}}$$

$$= \frac{200}{100} = 2 \text{A}$$

Now generally this value is the r.m.s. value because the voltage 200 V is also the r.m.s. value of the voltage.

Now   Peak factor $= \dfrac{I_{max}}{I_{rms}} = 1.414$

∴ $I_{max} = I_{rms} \times 1.414 = 2 \times 1.414$

$$= 2.828 \text{ A}$$

∴ I_{rms} = **2 A**, I_{max} = **2.828 A. Ans.**

The power $= V \cdot I$ watts

$$= 200 \times 2 = \textbf{400 watts. Ans.}$$

Example. *Calculate the inductive reactances of a coil having 0.05 H inductance and connected across 40 c/s, 50 c/s and 60 c/s mains.*

Solution. The inductive reactance is given by X_L i.e.

$$X_L = 2\pi f L \ \Omega$$

Now when the coil is connected across 40 c/s

$$X_{L1} = 2\pi \times 40 \times 0.05\,\Omega$$

$$= 12.566\,\Omega$$

$$X_{L2} = 2\pi \times 50 \times 0.05\,\Omega \qquad \text{(when } f = 50 \ c/s\text{)}$$

$$= 15.708\,\Omega$$

$$X_{L3} = 2\pi \times 60 \times 0.05\,\Omega \qquad \text{(when } f = 60 \ c/s\text{)}$$

$$= 18.85\,\Omega$$

∴   **12.566 Ω, 15.708 Ω and 18.85 Ω.   Ans.**

Example. *A purely inductive load of 0.2 H is connected across 200 V 50 c/s mains. Calculate the inductive reactance the current drawn and power absorbed.*

Solution. The inductive reactance X_L is

$$X_L = 2 \; \pi f L = 2 \times \pi \times 50 \times 0.2 = 62.8 \; \Omega$$

$$\text{The current} \quad = \frac{\text{Voltage}}{\text{Inductive reactance}}$$

$$= \frac{200}{62.8} = 3.185 \, A$$

The power absorbed $= V \cdot I \cos \phi$

$$= 200 \times 3.185 \times \cos 90° = 0 \; W.$$

$\therefore$ **62.8 Ω, 3.185 A and zero W.   Ans.**

Example. *Calculate the inductance of a coil (purely inductive) taking 2.1 A at 250 V 60 c/s. Find also the current taken by the same coil if connected across 100 V 40 c/s mains.*

Solution. The current drawn

$$= \frac{V}{X_L} \text{ or } X_L = \frac{V}{I} \Omega$$

So $$X_L = \frac{250}{2 \cdot 1} = 119.05 \, \Omega$$

Obviously $$X_L = 2\pi f L$$

$\therefore$ $$L = \frac{X_L}{2\pi f} = \frac{119.05}{2 \cdot \pi \cdot 60} = 0.316 \; H$$

Now the coil is connected across 100 V 40 c/s. So the current drawn

$$= \frac{V}{X_L} = \frac{100}{2 \times \pi \times 40 \times 0.316} = 1.26 \, A$$

0.316 H and 1.26 A. Ans.

Example. *Calculate the capacitive reactance of a 100 μF capacitor, if it is connected to (a) 25 c/s (b) 40 c/s (c) 50 c/s.*

Solution. The capacitive reactance

$$X_c = \frac{1}{2\pi f C} \Omega$$

$$X_{c_1} = \frac{1}{2 \times \pi \times 25 \times 100 \times 10^{-6}} = 63.66\,\Omega$$

$$X_{c_2} = \frac{1}{2 \times \pi \times 40 \times 100 \times 10^{-6}} = 39.79\,\Omega$$

and $$X_{c_3} = \frac{1}{2 \times \pi \times 50 \times 100 \times 10^{-6}} = 31.83\,\Omega$$

$$= 63.66\ \Omega,\ 39.79\ \Omega,\ 31.83\ \Omega \quad \textbf{Ans.}$$

Example. *A purely capacitive load of 150 μF is connected across 200 V 50 c/s mains. Find the capacitive reactance, current drawn and power absorbed.*

Solution. The capacitive reactance $= \dfrac{1}{2\pi fC}\,\Omega$

$$= \frac{1}{2 \times \pi \times 50 \times 150 \times 10^{-6}} = 21.22\,\Omega$$

$\therefore$ $$X_L = 21.22\,\Omega$$

$$I = \frac{V}{X_L} = \frac{200}{21.22} = 9.425\,\text{A}$$

The power $$= 220 \times 9.425 \times \cos 90° = 0\ \text{W}$$

$\therefore$ **21.22 Ω 9.425 A and zero W. Ans.**

15.5 A.C. SIMPLE SERIES CIRCUIT

Q. What do you understand by R.L series circuit? Derive the expressions for impedance, current and power for the circuit containing resistance and inductance in series.

Ans. Whenever the purely resistance and purely inductance are connected in series the circuit is known as the *R-L* series circuit as shown in Fig. 15.14(*a*). This circuit has got the following properties:

   (*i*) The current is same in both resistance and inductance.
  (*ii*) The voltage given constitutes the V_R and V_L, V_R = The voltage drop across resistance and V_L = Voltage drop across the inductance.
 (*iii*) The current is inphase with the voltage in resistance and out of phase in inductance, lagging behind the voltage by 90° as shown in Fig. 15.14 (*b*).

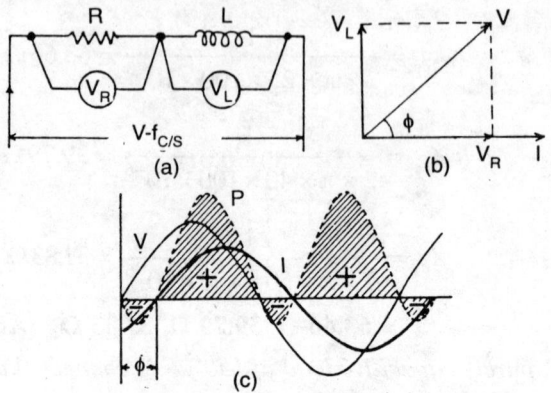

Fig. 15.14. *R-L* series circuit.

(*iv*) The voltage given in the vector sum of V_R and V_L

i.e. $\qquad V = \sqrt{V_R^2 + V_L^2}$ volts.

$\qquad V_R = I \cdot R$ and $V_L = IXX_L$,

(*v*) $V_R = I \cdot R$ and $V_L = I X_L$,
Let the impedance of the circuit be Z
then $\qquad V = I Z$ volts

Now $\qquad V = \sqrt{(I \cdot R)^2 + (I \cdot X_L)^2} = I\sqrt{R^2 + X_L^2}$

or $I Z = I\sqrt{R^2 + X_L^2}$ or $Z = \sqrt{R^2 + X_L^2}\ \Omega$.

So the impedance of the circuit is the square root of the sum of squares of resistance and inductive reactance, connected in the circuit.

(*vi*) The current $= \dfrac{\text{Voltage}}{\text{Impedance}} = \dfrac{V}{\sqrt{R^2 + X_L^2}}$ A

(*vii*) The power factor is the ratio of true power to apparent power

$$\cos\phi = \frac{\text{True power}}{\text{Apparent power}} = \frac{V \cdot I \cos\phi}{V \cdot I} = \frac{I^2 R}{I^2 Z} = \frac{R}{Z}$$

(*viii*) The power consumed
$\qquad\qquad\qquad = \text{Voltage} \times \text{Current} \times \text{Power factor}$
$\qquad\qquad\qquad = V \cdot I \cos\phi$

The wave form is shown in Fig. 15.14(*c*).

Example. *An inductive circuit has a resistance of 2 Ω in series with an inductance of 0.015 H. Find (i) Current, (ii) Power factor (iii) Power absorbed, when connected across 20 V 50 c/s mains.*

Solution. The inductive reactance

$$= 2 \pi f L \ \Omega$$

$$= 2 \cdot \pi \ 50 \cdot 0.015 = 4.71 \ \Omega$$

The impedance
$$= \sqrt{R^2 + X_L^2}$$

$$= \sqrt{2^2 + 4.71^2} = 5.11 \Omega$$

(*i*)Current
$$= \frac{\text{Voltage}}{\text{Impedance}} = \frac{20}{5.11} = 3.91$$

$$= \textbf{3.91 A. \quad Ans.}$$

(*ii*) Power factor
$$= \frac{R}{Z} = \frac{2}{5.11} = \textbf{0.391 Ans.}$$

(*iii*) Power absorbed
$$= V \cdot I \cos \phi$$

$$= \textbf{20} \times \textbf{3.91} \times \textbf{0.391} = \textbf{30.57 W Ans.}$$

Q. Derive an expression for current, voltage, impedance and power consumed for an A.C. series circuit containing resistance and capacitance.

Ans. Whenever the resistance and capacitance are connected in series across the A.C. voltage, the circuit is said to be the *R-C* series circuit as shown in Fig. 15.15. It has the following characteristics.

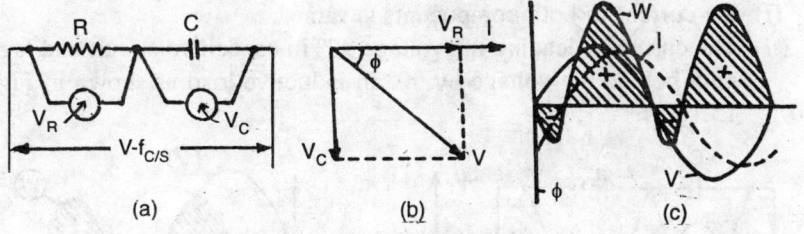

(a)   (b)   (c)

Fig. 15.15. *R-C* series circuit.

(*i*) The current in both components is same.

(*ii*) Current and voltage in resistance are in phase with the voltage and in capacitance current leading the voltage by 90° elect. as shown in Fig. 15.15(*b*).

(*iii*) The voltage drop

$$V_R = I \cdot R \text{ volts}$$

and
$$V_c = I \cdot X_c \text{ volts}$$

The supply voltage is the vector sum of the V_R and V_c.

$$\therefore \quad V = \sqrt{V_R^2 + V_c^2}$$

(*iv*) Now substituting for V_R and V_c

$$V = \sqrt{(IR)^2 + (IX_c)^2} = I\sqrt{R^2 + X_L^2}$$

Let the impedance of the circuit be Z, then

$$V = IZ = \sqrt{R^2 + X_c^2}$$

or $\quad Z = \sqrt{R^2 + X_c^2} \ \Omega.$

(*v*) The current drawn is $\dfrac{\text{Voltage}}{\text{Impedance}} = \dfrac{R}{Z}$

(*vi*) The power factor

$$= \frac{\text{Resistance}}{\text{Impedance}}$$

(*vii*) The power taken

$$= V \cdot I \cos \phi \text{ watts}$$

the sinusoidal wave form as shown in Fig. 15.15(*c*).

Q. Explain the characteristics of a L-C series circuit.

Ans. If the purely inductance and capacitance are connected in series the circuit is known as *L-C* series circuit as shown in Fig. 15.16 (*a*).

It has the following characteristics:

(*i*) the current in both components is same.

(*ii*) The current is leading the voltage 90° in capacitive circuit and lagging behind the voltage by 90° in inductive load as shown in Fig. 15.16(*b*).

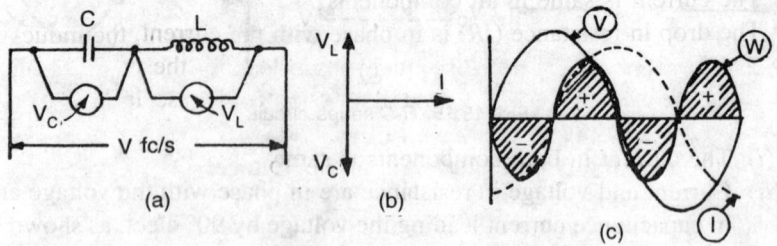

(a)   (b)   (c)

Fig. 15.16. *L-C* series circuit.

(*iii*) Here $\qquad\qquad V_L = I X_L \text{ volts}$

and $\qquad\qquad V_c = I \cdot X_c \text{ volts, both are } 180° \text{ apart}$

So their resultant will be the voltage fed

i.e. $\qquad V = V_L \sim V_c$ volts.

(*iv*) Their impedance will be the total reactance

i.e. $\qquad Z = X_L \sim X_c = X \, \Omega.$

(*v*) Current $\qquad = \dfrac{V}{Z} = \dfrac{\text{Voltage}}{\text{Reactance}}$ amp. $\left(\text{in this case}\right)$

(*vi*) The power factor is the cosine of the angle between the current and voltage *i.e.* cos 90 = 0, so the power factor is zero.

(*vii*) The power consumed which is the product of current voltage and power factor is zero, as shown in Fig. 15.16 (*c*).

Q. Derive the expressions for current, voltage, impedance and power for the A.C. general series circuit.

OR

Explain the characteristics of A.C. circuit containing resistance, inductance and capacitance in series.

Ans. As shown in Fig. 15.17(*a*) if the pure resistance, pure inductance and pure capacitance are connected in series, the circuit is known as *RLC* or general series circuit.

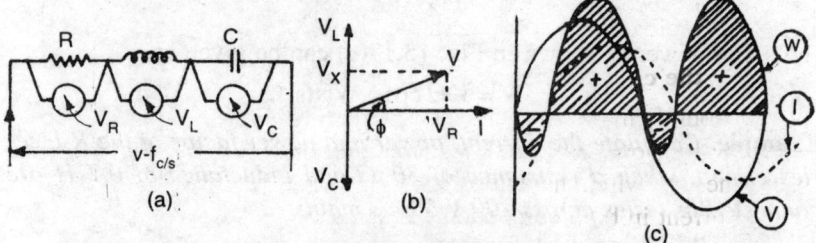

Fig. 15.17. RLC series circuit.

This circuit has got the following characteristics:

(*i*) The current is same in all components.

(*ii*) The drop in resistance (*IR*) is in phase with the current, the inductive reactance (IX_L) drop is out of phase leading the current and capacitive reactance (IX_c) drop is also out of phase lagging the current by 90° elect. as shown in Fig. 15.17(*b*).

(*iii*) The voltage fed is the vector sum of the voltages

i.e. $\qquad V = \sqrt{V_R^2 + V_X^2}$

(*iv*) For impedance, substitute the values

$$V = \sqrt{(IR)^2 + (IX)^2}$$

$$= \sqrt{(IR)^2 + (IX_L \sim IX_c)^2}$$

$$= I\sqrt{R^2 + (X_L \sim X_c)^2}$$

Let Z be the total impedance of the circuit so

$$IZ = I\sqrt{R^2 + (X_L \sim X_c)^2}$$

or
$$Z = \sqrt{R^2 + (X_L \sim X_c)^2} \; \Omega$$

(v) The current drawn is equal to

$$I = \frac{V}{\sqrt{R^2 + (X_L \sim X_c)^2}} \; \text{Amp.}$$

(vi)
$$\text{Power factor} = \frac{\text{True power}}{\text{Apparent power}}$$

$$= \frac{I^2 R}{I^2 Z} = \frac{R}{Z}.$$

(vii) The power as shown in Fig. 15.17(c) can be given as,

$$W = V \cdot I \cos \phi \; \text{Watts.}$$

Example. *Calculate the current, power and power factor of the R-L series circuit, when a resistance of 30 Ω and inductance of 0.2 H are connected in series across 200 V 50 c/s mains.*

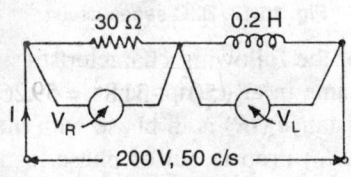

Fig. 15.18.

Solution. The inductive reactance

$$X_L = 2\pi f L \; \Omega$$

$$= 2 \times \pi \times 50 \times 0.2 = 62.8 \; \Omega$$

The impedance of a series circuit

$$= \sqrt{R^2 + X_L^2}$$

$$Z = \sqrt{30^2 + 62.8^2} = 69.6 \ \Omega$$

Current,
$$I = \frac{200}{69.6} = 2.87 \, \text{A}$$

Power factor
$$= \frac{R}{Z} = \frac{30}{62.8} = 0.478$$

Now power consumed

$$= V \cdot I \cdot \cos\phi$$
$$= 200 \times 2.87 \times 0.478 \, \text{W}$$
$$W = 274.4 \, \text{W}$$

2.87 A, 274.4 W and 0.478. Ans.

Example. *Determine the value of current, impedance, power factor and power consumed by a R-C series circuit if a 50 Ω resistance and a 100 μF capacitance are connected in series across 250 V 50 c/s mains.*
Solution. The capacitive reactance

$$X_c = \frac{1}{2\pi f C} \ \Omega$$

$$X_c = \frac{1 \times 10^6}{2 \times \pi \times 50 \times 100} = 31.8 \ \Omega$$

The impedance in R-C series circuit is

$$Z = \sqrt{R^2 + X_c^2}$$

$$= \sqrt{50^2 + 31.8^2} = 59.26 \Omega$$

Now current
$$= \frac{V}{Z} = \frac{250}{59.26} = 4.22 \, \text{A}$$

The power factor
$$= \frac{R}{Z} = \frac{50}{59.26} = 0.844$$

The power
$$= V \cdot I \cos\phi$$
$$= 250 \times 4.22 \times 0.844 = 890.42 \, \text{W}$$

4.22 A 59.26 Ω, 0.844, 890.42 W. Ans.

Example. *An alternating voltage of v = 100 sin 314 t is applied to a circuit consisting of a 5 Ω pure resistance and a pure inductance of 0.02*

H, connected in series. Show that the current drawn be given by the following expression,

$$i = 12.45 \sin(314t - 51.4°) \text{ A.}$$

Solution. The inductive reactance

$$= 2\pi f L \ \Omega$$

$$= 2 \cdot \pi \cdot \left(\frac{314}{2\pi}\right) \cdot 0.02 \qquad \left(\because f = \frac{\omega}{2\pi}\right)$$

$$= 6.28 \ \Omega$$

$$Z = \sqrt{R^2 + X_L^2} = \sqrt{5^2 + 6.28^2} = 8.03 \ \Omega$$

Now the maximum current

$$= \frac{V_{max}}{R} \text{ and } V_m = 100 \text{ V}$$

as given in the fundamental equation of voltage, so

$$I_{max} = \frac{100}{8.03} = 12.45 \text{ A}$$

The power factor

$$= \frac{R}{Z} = \frac{5}{8.03} = 0.6226$$

$$\therefore \qquad \phi = \cos^{-1}(0.6226) = 51.4°$$

Now as the equation for current,

$$i = I_{max} \sin(\omega t - \phi) \text{ A}$$

So $\qquad$ **i = 12.45 sin (314t − 51.4°) A. Ans.**

Example. *A series A.C. circuit consisting of 100 Ω; inductance 0.2 H and capacitance of 120 μF is connected across 200 V 50 c/s. Calculate the impedance, current, power factor and power absorbed.*

Solution. The inductive reactance

$$= 2\pi f L$$

$$X_L = 2 \times \pi \times 50 \times 0.2 = 62.83 \ \Omega$$

The capacitive reactance $= \dfrac{1}{2\pi f C} \Omega$

$$\therefore \qquad X_c = \frac{1 \times 10^6}{2 \times \pi \times 50 \times 120} = 26.53 \Omega$$

Here $\qquad X_L \sim X_c = 62.83 - 26.53 = 36.30 \ \Omega$

The impedance $\qquad = \sqrt{R^2 + (X_L - X_c)^2}$

$$= \sqrt{100^2 + (36.3)^2} = 106.4 \ \Omega$$

The current $\qquad = \dfrac{\text{Voltage}}{\text{Impedance}} = \dfrac{200}{106.4} = 1.88 \text{ A}$

The power factor $\qquad = \dfrac{R}{Z} = \dfrac{100}{106.4} = 0.94$

The power absorbed $\quad = V \cdot I \cos \phi$

$$= 200 \times 1.88 \times 0.94 = 353.4 \text{ W}$$

$\therefore \qquad$ **106.4 Ω, 1.88 A, 0.94, 353.4 W.   Ans.**

Example. *An inductive coil having 25 Ω resistance and 0.4 H inductance is connected in series across 200 V 50 c/s main. Find the current, power factor and power absorbed. Also find out the value of capacitor to be connected in series to improve the power factor to unity.*

Solution. The inductive reactance

$$= 2 \pi f L$$

$$= 2 \times \pi \times 50 \times 0.4 = 125.66 \ \Omega$$

$$Z = \sqrt{R^2 + X_L^2} = \sqrt{25^2 + 125.66^2} = 128.12 \Omega$$

So impedance is 128.12 Ω

Current drawn $\qquad = \dfrac{V}{Z} = \dfrac{200}{128.12} = 1.56 \text{ A}$

The power factor $\qquad = \dfrac{R}{Z} = \dfrac{25}{128.12} = 0.195$

Now this power factor is to be brought to unity. It means the current will be only taken by the resistance, *i.e.*

$$= \frac{200}{25} = 8 \text{ A}.$$

For that the inductive component is to be brought to equal to the capacitive components.

$\therefore \qquad\qquad X_L = X_c$

$$125.66 = \frac{1 \times 10^6}{2\pi \cdot 50 \cdot C}$$

or
$$C = \frac{125.66 \times 2 \times \pi \times 50}{10^6} = 0.0395\,\mu F$$

The condenser should be of **0.0395 μF. Ans.**

15.6. POWER AND POWER FACTOR

Q. What do you understand by (a) apparent power, (b) true power and (c) reactive power? Name the instrument with which the power in A.C. circuit is measured.

Ans. The power consumed by the electrical circuit is measured by means of wattmeter. A wattmeter has two coils: the current coil and pressure coil. The current coil has less resistance, less turns wound with thick conductor and connected in series with the load. The pressure coil is having more resistance wound with the conductor having more number of turns and always connected across the supply. The torque produced depends upon the flux produced by the current coil and pressure coil.

(a) **Apparent power.** The apparent power is the product of the voltage and current taken by the load.

∴   Apparent power = Voltage × Current = $V \cdot I$.

All the A.C. machines are rated in kVA.

(b) **True power.** It is the power which is actually consumed in the circuit to do the useful work or it is the power actually shown by wattmeter or power consumed in a resistance. It is also known as actual power, useful power, real power or wattful power. It is measured in watts or kW by means of wattmeter.

True power = $V \cdot I \cos \phi = I^2 R$ watts.

(c) **Reactive power:** The power which is consumed in reactance of the circuit. This is also known as wattless power

Reactive power = $VI \sin \phi = I^2 x$ watts.

Q. Define power factor. What are the causes and effects of low power factor?

Ans. The power factor is defined as the ratio of true power to apparent power.

i.e. $$\text{Power factor} = \frac{\text{True power}}{\text{Apparent power}}.$$

It is the cosine of the angular displacement between current and voltage. It is denoted by cos ϕ.

Depending upon the angular position the power factor it is of two types: the leading power factor and lagging power factor. If the current is ahead in the direction of rotation then the voltage, the power factor is said to be leading power factor. If the current is trailing behind the voltage, the power factor is lagging.

The power factor is always less than unity.

Causes of low power factor. There are the following reasons:

(*i*) In industrial and domestic fields, the induction motors are widely used. The induction motors are always taking lagging current which results low power factor.

(*ii*) The industrial furnaces have low power factor.

(*iii*) The transformers at substations have lagging power factor because of load and their magnetising currents.

(*iv*) Because of inductive load in houses like tubes, fans, motors etc.

Effects of low power factor. There are the following effects of low power factor:

Disadvantages

(*i*) For the same power to be transmitted over a distance, it will have to carry more current at low power factor. As a result the area of cross-section of the conductors is more, causing more cost, labour cost, line losses, line drops and lowering the efficiency of transmission.

(*ii*) The transformer, switch and switch gears, generators has to carry more current to meet the same power with low power factor. It will again effect the size, cost, efficiency and life of the machine and apparatus.

(*iii*) The prime mover capacity will also be effected with low power factor, as the alternator has to induce more power to meet the demand at low power factor.

(*iv*) Low power factor will effect the voltage regulation of generator, transformer, transmission line also.

Advantages of power factor improvement. There are certain advantages of improving the power factor:

(*i*) The output capacity of the prime mover is better utilized.

(*ii*) The output of alternator is increased.

(*iii*) The kW output of transformer and line are increased.

(*iv*) Efficiency of plant is increased.

(v) The voltage regulation is also improved.

(vi) The machines generator, transformer, lines etc. will carry less cross-sectional area for the same power to be transmitted and utilized for improved power factor than low power factor.

Q. What do you understand by the impedance triangle?

Ans. In the R.L. series circuit it has already been seen that

$$Z = \sqrt{R^2 + X_L^2}.$$

So to find out the impedance the resistance and reactance must be added vectorically. Thus forming a right angle triangle. In fact the impedance, resistance and reactance can be represented by these three sides of the right angled triangle. This triangle is known as the impedance triangle as shown in Fig. 15.19.

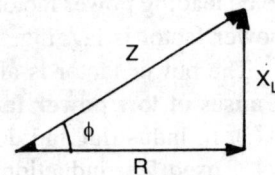

Fig. 15.19. Impedance triangle

With this triangle it is found that the resistance reactance and impedance can be noticed.

Hence, an *impedance triangle* is a right angled triangle, the base of which represents resistance, perpendicular the reactance and the hypotenous, the impedance of the circuit.

The impedance triangle can give:

(i) The impedance of the circuit i.e., $Z = \sqrt{R^2 + X^2}$.

(ii) The phase angle and power factor $\tan\phi = \dfrac{X}{R}$, here as shown in Fig. 15.19.

(iii) It can also show the nature of the power factor whether leading or lagging.

15.7 A.C. SIMPLE PARALLEL CIRCUIT

Q. Derive an expression for current, impedance, power factor and power consumed for an R-L parallel circuit.

Ans. Whenever the resistance and inductance are connected in parallel the combination is said *R-L* parallel circuit as shown in Fig. 15.20. It has the following characteristics:

(i) The voltage is same across each component.

(ii) The current and voltage are in phase in resistive component and in inductance, current lags behind the voltage by 90° electrical, as shown in Fig. 15.20(b).

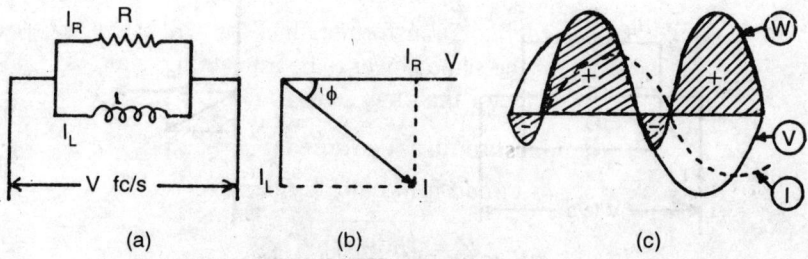

Fig. 15.20. R-L parallel circuit.

(*iii*) The current taken by resistance is V/R and in inductance V/X_L amp. The line current is the vector sum of both I_R and I_L.

$$\therefore \qquad I = \sqrt{I_R^2 + I_L^2}$$

(*iv*) Substituting for I_R and I_L, we have

$$I = \sqrt{\left(\frac{V}{R}\right)^2 + \left(\frac{V}{X_L}\right)^2}$$

Let Z be the impedance, then

$$I = \frac{V}{Z} = \sqrt{\left(\frac{V}{R}\right)^2 + \left(\frac{V}{X_L}\right)^2} \text{ or } \frac{1}{Z} = \sqrt{\frac{1}{R^2} + \frac{1}{X_L^2}}.$$

(*v*) The power factor is the ratio of true power to apparent power.

$$\therefore \qquad \text{Power factor} \qquad = \frac{\dfrac{V^2}{R}}{\dfrac{V^2}{Z}} = \frac{Z}{R}.$$

(*vi*) The power consumed $= V \cdot I \cos \phi$ as shown in Fig. 15.20.

Q. Derive the expressions for current, voltage, impedance and power etc. for the A.C. general parallel circuit.

OR

Explain the characteristics of A.C. circuit having resistance, inductance and capacitance connected in parallel.

Ans. When the resistance, inductance and capacitances are connected in parallel across the A.C. voltage the circuit is known as general parallel circuit as shown in Fig. 15.21.

It has the following characteristics:

(*i*) The voltage is same across each element.

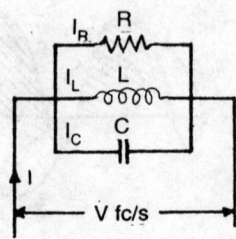 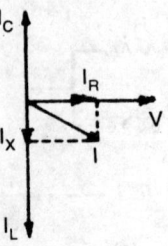

Fig. 15.21. RLC parallel circuit.

(*ii*) The current $I_R = V/R$, is in-phase with the voltage, $I_L = V/X_L$, the inductive current lagging behind the voltage by 90° and $I_c = V/X_c$ leading the voltage by 90°.

(*iii*) The line current is the vector sum of all the branch currents

$$I = \sqrt{I_R^2 + (I_L \sim I_c)^2}$$

(*iv*) Substituting for the values,

$$I = \sqrt{\left(\frac{V}{R}\right)^2 + \left(\frac{V}{X_L} \sim \frac{V}{X_c}\right)^2}$$

$$= V \sqrt{\frac{1}{R^2} + \left(\frac{1}{X_L} \sim \frac{1}{X_c}\right)^2}$$

Let Z be the impedance than

$$\frac{V}{Z} = V \sqrt{\frac{1}{R^2} + \left(\frac{1}{X_L} \sim \frac{1}{X_c}\right)^2}$$

or $\quad \dfrac{1}{Z} = \sqrt{\dfrac{1}{R^2} + \left(\dfrac{1}{X_L} \sim \dfrac{1}{X_c}\right)^2}$

(*v*) The power factor is the ratio of impedance to resistance.

(*vi*) The power consumed is

$$= V \cdot I \cos \phi \text{ Watts}$$

Example. *Name the different methods of solving the A.C. parallel circuits.*

An inductive coil is connected in parallel across a non-inductive resistance of 30 Ω and this parallel circuit is connected across 150 V 50 c/s. The total current taken from the mains is 9 A. When the current is non-inductive resistance is 5 A and in inductive coil 6.5 A. Calculate

(a) *Inductance of the coil.*

(b) *Power factor of the circuit.*

(c) *Power absorbed by the parallel circuit.*

Solution. Generally the following methods are used the solving the parallel A.C. circuits.

(i) Vector method.

(ii) Admittance method.

(iii) J-method or symbolic method.

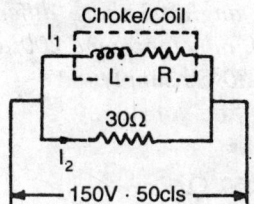 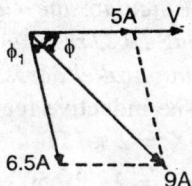

Fig. 15.22.

The resistive component carries 5 A current as shown in Fig. 15.22 so the voltage applied = $I_2 \times R = 5 \times 30 = 150$ V.

Now current drawn by the coil is 6.5 A, so he impedance of the branch

$$= \frac{V}{I_1} = \frac{150}{6.5} = 23 \ \Omega$$

Now the main current is 9 A, so the angular displacement between the currents can be determined as

$$I^2 = I_1^2 + I_2^2 + 2I_1 I_2 \cos\phi_1$$

∴ $$9^2 = 5^2 + 6.5^2 + 2 \times 5 \times 6.5 \cos \phi_1$$

$$81 = 67.25 + 65 \cos \phi_1$$

∴ $$\cos\phi_1 = \frac{81 - 67.25}{65} = 0.2115$$

and $$\sin \phi_1 = 0.9774$$

The inductive reactance of the coil $X_L = Z_1 \sin \phi_1$
$$= 23 \times 0.9774 = 22.48 \ \Omega$$

So inductance of the coil $= \dfrac{X_L}{2\pi f} = \dfrac{22.48}{2 \times 3.14 \times 50} = 0.07156 \, \text{H}$

So the inductance of the coil = 71.6 mH.

As shown in fig. 15.22, $9 \cos \phi = I_2 + I_1 \cos \phi_1$

$$\cos\phi = \frac{5 + 6.5 \times 0.2115}{9} = 0.708$$

$$\phi = \cos^{-1}(0.708) = 44.9°$$

The power absorbed $= V \cdot I \cdot \cos\phi = 150 \times 9 \times 0.708$

$$= 955.8 \text{ W}.$$

71.6 mH, 0.708, 955.8 W.   Ans.

Example. *Calculate the current and angle of phase difference of the circuit having 10 Ω resistance, 0.04 H inductance and 200 μF capacitor connected in parallel across a 200 V 50 c/s supply.*

Solution. The inductive reactance

$$X_L = 2\pi f L\,\Omega$$

$$= 2 \cdot \pi \cdot 50 \cdot 0.04 = 12.57 \ \Omega$$

The capacitive reactance

$$X_c = \frac{1}{2\pi f C}$$

$$= \frac{10^6}{2 \times \pi \times 50 \times 200} = 15.9 \ \Omega$$

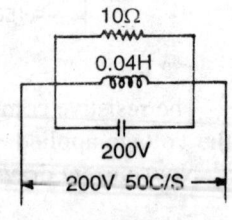

Fig. 15.23.

Now the impedance

$$\frac{1}{Z} = \sqrt{\frac{1}{R^2} + \left(\frac{1}{X_L} - \frac{1}{X_c}\right)^2}$$

$$= \sqrt{\frac{1}{10^2} + \left(\frac{1}{12.57} - \frac{1}{15.9}\right)^2}$$

$$= \sqrt{(0.1)^2 + (0.0166)^2}$$

$$= \sqrt{0.010277} = 0.10137$$

and $\quad Z = 9.86 \ \Omega$

Current $\quad = \dfrac{200}{9.86} = 20.28\,\text{A}$

The reactive component $= 15.9 - 12.57 = 2.33 \ \Omega$

It is a capacitive reactance so the power factor is of leading in nature.

$$\tan \phi = \frac{X_c}{R} = \frac{2.33}{10} = 0.233$$

$$\phi = \tan^{-1}(0.233) = 13.12°$$

and $\qquad \cos \phi = 0.974$

20.28 A, 13.12° Lagging Ans.

REVIEW QUESTIONS

1. Explain why a large percentage of electrical energy used for commercial purpose is generated as A.C.?
2. Define the followings:
   Alternating quantities, periodic time, frequency, inphase, out of phase and cycle.
3. Define and explain the following terms used in A.C. quantities: (*a*) R.M.S. value, (*b*) average value, (*c*) peak factor, (*d*) power factor and (*e*) form factor.
4. What do you understand by the phase difference, explain the terms lag and lead?
5. How A.C. behaves in purely resistive circuit, draw vector diagram.
6. How current, voltage and power relates in case of purely inductive and capacitive circuit in alternating current circuits?
7. What are the different combinations of A.C. components? Draw the vector diagram and relationship between current, voltage, power and power factor of general RLC series circuit.
8. Explain the followings:
   (*i*) What do you understand by 'ω'?
   (*ii*) How A.C. quantities are added?
   (*iii*) How two A.C. quantities, one lagging are represented vectorically?
   (*iv*) How R.M.S. value of A.C. is calculated?
   (*v*) How average value of A.C. is taken and why?
9. Explain the terms, true power, apparent power, reactive power and power factor.
10. Why series resonance is called voltage resonance? Deduce the formula for series resonance.
11. Define general parallel circuit, deduce expressions for impedance, current power factors and power.
12. Draw the vector diagram for a parallel resonance circuit and calculate the resonance frequency.
13. Answers the followings:
    (*i*) What is the phase relationship between current and voltage of a purely inductive circuit?
    (*ii*) What is the effect of frequency on the capacitive reactance?
    (*iii*) What is the relationship between true, apparent and reactive powers?
    (*iv*) What is series resonance?

(v) What is the power factor?
14. Fill up the blanks:
    (i) The power factor in pure capacitive circuit is ....
    (ii) The power in A.C. is given by the formula ....
    (iii) The product of voltage, current and power factor in A.C. circuit is known as .... power.
    (iv) The power factor of D.C. circuit is always ....
    (v) The power factor of a resonance circuit is ....

NUMERICAL PROBLEMS

1. Find the maximum, r.m.s. and average value and frequency of the voltage $e = 300 \sin 628t$.   **[Ans. 300 V, 212 V, 191 V, 100 c/s]**
2. Calculate the r.m.s. value of an a.c. having the instantaneous values of equal interval as 0, 10, 20, 30, 20, 10, 0, −10, −20, −30, −20, −10, 0.
        [Ans. 4.7 A]
3. A 100 Ω pure resistance is connected on 220 V 50 c/s mains. Calculate the impedance, power factor and current taken from the mains.
        [Ans. 100 Ω, Unity, 2.2 A]
4. A coil has resistance 5 Ω and inductance 0.0398 H. It is connected to 100 V 60 Hz. Calculate (a) the impedance of the coil, (b) current in the coil, (c) power factor of the coil. **[Ans. (a) 15.8 Ω, (b) 6.33 A, (c) 0.316 lagging]**
5. A 70 Ω resistance is connected in series with an inductance of 0.22 H across a certain voltage of 50 c/s. If the current in the circuit is 2A, find the applied voltage and voltage across each element.
        [Ans. 196.75 V, 140 V, 138.23 V]
6. An inductive coil takes a current of 32.24 A from 230 V 50 c/s mains. The current lags behind the voltage by 30° electrical. If the resistance of the coil is 6 Ω, calculate the power taken by the coil.   **[Ans. 6.26 KW]**
7. Calculate the current and power factor of a series circuit consisting of 10 Ω resistance, 0.04 H inductance and 200 μF capacitor, connected to 200 V 50 c/s mains.   **[Ans. 19 A, 0.95]**
8. A circuit containing a resistance of 40 Ω in parallel with an inductive reactance of 30 Ω is connected across 240 V 50 c/s mains. Calculate the current drawn by each element total current, the total power factor and power taken from the mains.
        [Ans. 6 A, 8 A, 10 A, 0.6 lagging, 1440 W]
9. A choking coil when connected to 230 V 50 c/s mains draws 15 A and consumes 1300 W, calculate the power factor.   **[Ans. 0.3768]**
10. A coil having resistance 45 Ω and inductance 0.4 H is connected in parallel to 20 μF capacitor across 230 V 50 c/s mains, calculate the current taken, power factor and power taken from the mains.
         [Ans. 0.615 A, 0.951, 0.134 kW]
11. Calculate the inductive reactances of a 0.5 H inductor when it is connected to 50 c/s, 60 c/s and 100 c/s supply frequencies.
         (Ans. 157 Ω, 188.5 Ω, 628.32 Ω)

12. An inductor having 2 H inductance when connected to 100 V takes 0.318 A current find out the frequency to which it is energised.

    (**Ans. 25 Hz**)

13. An inductor connected to 25 V 60 Hz frequency takes 0.2 current. If it is connected to 12 V 50 c/s. How much current will it draw? What is the inductance of it? (**Ans. 0.1157 A, 0.33 H**)

14. An inductive circuit of negligible resistance limits to 50 mA current at 125 volts a.c. If the working voltage is 12.5 V 200 Hz, estimate its inductance?

    (**Ans. 0.1989 H**)

15. It is required in an electronics network and inductance to limit a current to 0.01 A at 0.5 V 350 KHz. Calculate the inductances to be connected.

    (**Ans. 0.227 μH**)

16. A choke of a fluorescent tube carries 0.43 A when the tube is connected to 230 V 50 Hz. The drop across choke is 166 V. The d.c. resistance is 85 Ω calculate the inductance and inductive reactance of the choke.

    (**Ans. 1.199 H, 376.57 Ω**)

17. An inductive circuit having a resistance of 150 Ω and inductance 0.5 H when connected to 230 V results an impedance of 216.2 Ω. Calculate the supply frequency and power absorbed. (**Ans. 50 c/s, 168.2 Watts**)

18. An inductive circuit draws a current of 40 Amp from 200 V 50 c/s mains at a power factor of 0.75 lagging. Calculate the power consumed, also calculate the active and reactive components of the current.

    (**Ans. 6 kW, 30 A, 26.5 A**)

19. A coil having a resistance of 5 Ω and inductance of 0.02 H is connected across 25 V 50 Hz mains. Calculate the impedance of the coil, the current drawn, the power factor and power consumed.

    (**Ans. 8 Ω, 3.12 A, 0.625, 48.75 W**)

20. Calculate the impedance, current drawn, angle of lagging, power factor, and power consumed by a coil having inductance 0.1 H and resistance as 5 Ω when connected to 230 V 50 Hz mains.

    (**Ans. 31.8 Ω, 7.23 A. 81°, 0.157, 261 W**)

21. Calculate the capacitive reactance offered by a capacitor of 100 μF at 100 V 50 Hz supply. (**Ans. 31.83 Ω**)

22. A 50 μF capacitor is connected to 50 V of certain frequency supply if the current drawn is 0.78 A calculate the supply frequency. (**Ans. 50 Hz**)

23. A coil has a resistance of 4 Ω and an inductance of 6 Ω, find out the current taken by the coil if connected across 250 V 50 Hz. Also calculated the kVA power taken and actual power consumed.

    (**Ans. 34.7 A, 8.667 kVA, 4.815 kW**).

24. A 15 Ω resistance and 0.5 H inductance coil is connected to 230 V 50 Hz. AC supply find the impedance, current taken and power consumed.

    (**Ans. 157.8 Ω, 1.458 A, 31.87 W**)

25. A choking coil of 0.02 H inductance and 4 Ω resistance is connected across 50 V 50 Hz supply, calculate the current taken, power consumed and power factor of the coil. (**Ans. 6.71 A, 180.2 W, 0.537**)

16

Polyphase Supply System

16.1. POLYPHASE—TWO PHASE AND THREE PHASE

Q. What is meant by the polyphase supply system? Write the advantages and disadvantages of polyphase system over single phase supply system. (N.C.V.T. 1977, 84, 85)

Ans. The supply system which contains more than one phase, is known as polyphase system, *e.g.* two phase, three phase supply systems. There are the following advantages and disadvantages of polyphase system over single phase supply system:

 (*i*) The polyphase alternator has more output than a single phase alternator for the same frame size of machine. A two phase alternator has approximately 40% more output and a three phase alternator has 70% more output than a single phase alternator.

 (*ii*) To transmit the required amount of power over a given distance a polyphase system *i.e.* polyphase transmission line requires less copper than a single phase transmission line.

 (*iii*) In case of three phase system the lighting and power circuits can be drawn from the same line but in single phase separate lines are required.

 (*iv*) Single phase motors are not self-starting where the three phase motors are self-starting.

 (*v*) Three phase motors have smooth torque where single phase motors have the pulsating torque.

 (*vi*) Polyphase motors can be build in large capacity while single phase motors cannot.

 (*vii*) For the same size of machine the output of single phase motors is always less than the output of three phase motors.

(*viii*) The power factor and efficiency of three phase motor is always better than the single phase motor.

(*ix*) Fault location is always easy in three phase system.

(*x*) Starting of three phase motor is always easy than single phase motors.

Q. (*a*) **Describe two phase supply system; state the different connections.**

(*b*) **A two phase 50 c/s alternator is inducing a phase voltage as 100 V. Calculate the possible line voltage and load current if the load is 10 kW.**

Ans. The supply system which has only two phases is known as two phase supply system. In this case there are two sets of windings which are placed at 90° to each other. The four ends are taken out and connected to the slip rings. The ends or terminals of the windings are connected in the following manner:

(*a*) **Two phase four wire supply system.** In this case four wires are taken out from the four slip rings. Two wires from each phase are connected with two separate circuits as shown in Fig. 16.1.

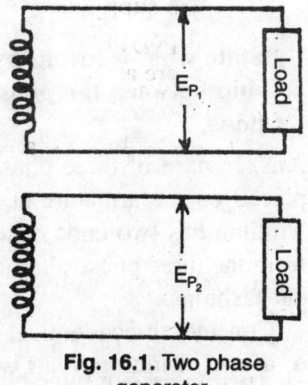

In case of equal and balanced load on both the circuits, the different values are as, let the phase voltages be E_{P1} and E_{P2} and currents I_{P1} and I_{P2} respectively.

Thus the power of circuit No 1

$$= E_{P1}I_{P1}\cos\phi$$

and of the second circuit $= E_{P2}I_{P2}\cos\phi$

Let both $\qquad E_{P1} = E_{P2} = E_P$

and $\qquad\qquad I_{P1} = I_{P2} = I_P$

∴ The total power $= E_P \cdot I_P \cdot \cos\phi$

$$+ E_P \cdot I_P \cos\phi$$

Fig. 16.1. Two phase generator.

or $\qquad = 2 \cdot E_P I_p \cos\phi$

So total power = 2 × Phase voltage × Phase current × Power factor

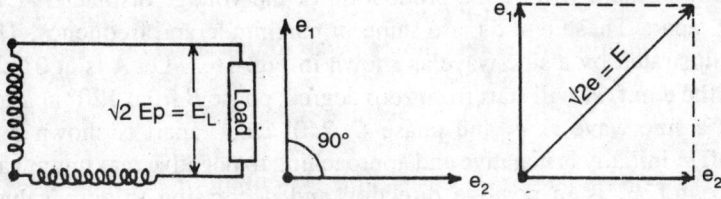

Fig. 16.2. Two phase connections.

(b) **Two phase three wire system.** In this case two ends are connected together and only three wires are taken out as shown in Fig. 16.2. The line and phase voltages are taken from the different terminals. The line voltage is obtained from the two outer ends and the phase voltage between the middle wire and either of the outers. These both voltages are at right angle to each other.

So
$$E_L = \sqrt{E_{P1}^2 + E_{P2}^2}$$

Let
$$E_{P1} = E_{P2} = E_P$$

∴
$$E_L = \sqrt{2E_P^2} = \sqrt{2}\,E_P$$

and the power = Line voltage × Line current × Power factor

$$W = \sqrt{2}\,E_L \cdot I_L \cos\phi \text{ Watts.}$$

Q. Describe the three phase supply system also state which type of polyphase supply system is used nowadays.

OR

State what is meant by star and delta connections. Give relationship between the phase and line values in star and delta connections.

Ans. In case of three phase supply system three different windings are placed on the armature at an angular displacement of 120° elect. Each winding has two ends or terminals, so there are six terminals taken out from the three phase alternators. These terminals are connected in different fashions.

Consider three identical windings, A, B and C winding. A has two ends a_1 and a_2, winding B has two b_1 and b_2, winding C has again two c_1 and c_2. These are placed at 120° apart. Let these be rotated in the magnetic field produced by a bipolar machine in anticlockwise direction with a speed equal to ω radian/sec. Now depending upon the Fleming's right hand rule, there will be the production of the voltage displaced at 120° elect. apart. These e.m.f.s. are same in magnitude and frequency. These are illustrated by a sine wave as shown in Fig. 16.3. Let A is at 0° elect. then the e.m.f. e_1 will start from zero degree, phase B from 120° apart will have a sine wave as e_2 and phase C, 240° elect. apart is shown by e_3. E.m.f. e_2 initially is negative and approaching to negative maximum value. The e.m.f. e_3 is in positive direction and decreasing to zero value as shown.

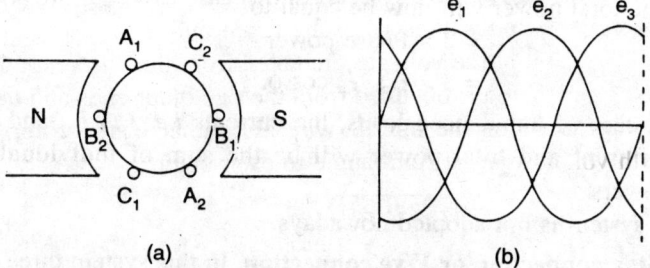

Fig. 16.3. Three phase generation.

These e.m.f.s. can be said as

$$e_1 = E_m \sin \omega t$$
$$e_2 = E_m \sin (\omega t - 120)$$
$$e_3 = E_m \sin (\omega t - 240)$$

Similarly the current can be drawn keeping in mind the angle of lag or lead.

16.2. INTER CONNECTIONS OF THREE PHASES

There are the following different ways of connecting these three windings for obtaining three phase supply for the load.

(*a*) **Three phase six wire system.** In this case the load is divided into three equal circuit and every load circuit is connected across individual circuits; as shown in Fig. 16.4.

Let E_{P_1}, E_{P_2} and E_{P_3} be the phase voltages and I_{P_1}, I_{P_2} and I_{P_3} be the currents respectively. So the power in different phase are:

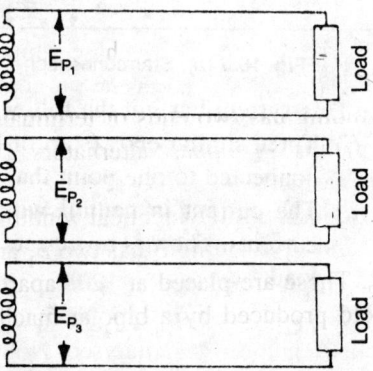

Phase $A = E_{P1} I_{P1} \cos \phi$ Watts.

Phase $B = E_{P1} I_{P2} \cos \phi$ Watts.

Phase $C = E_{P3} I_{P3} \cos \phi$ Watts.

Fig. 16.4. Three phase windings.

In a particular machine when phase windings are identical so it will have the characteristic as given.

  (*i*) The voltage across each phase is same because the windings are identical.

  (*ii*) In case when load is equal, the same current will flow through each phase.

(*iii*) The total power will now be equal to

$$= 3 \times \text{Phase power}$$

$$= 3 \times E_P \cdot I_P \cos \phi.$$

(*iv*) In case of unbalanced loads, the currents *i.e.* I_{P_1}, I_{P_2} and I_{P_3} are different and total power will be the sum of individual phase powers.

This system is not adopted nowadays.

(*b*) **Star connection or Wye connection.** In this system three similar ends of the phase windings are connected at one point which is known as the star point or neutral point. The remaining three ends are brought out for supply this supply system is known as three phase four wires supply system as shown in Fig. 16.5.

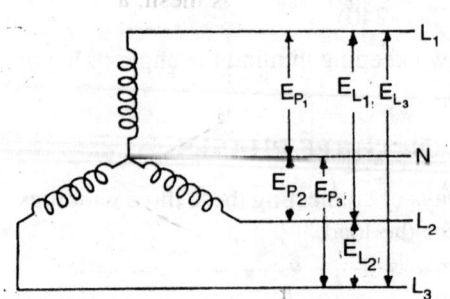

Fig. 16.5 (*a*). Star connection

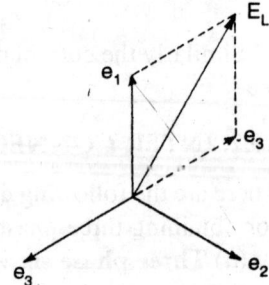

Fig. 16.5 (*b*). Vector diagram

This system has got the following characteristics:

(*i*) Three similar ends (start or finish) of the three phase windings are connected to one point that is known the star or neutral point.

(*ii*) The current in neutral wire will be the vector sum of all phase currents. There is no current in the neutral wire in case of balanced load.

i.e. $\qquad \dot{I}_{P_1} + \dot{I}_{P_2} + \dot{I}_3 = 0.$

(*iii*) The phase voltage $= \dfrac{\text{Line voltage}}{\sqrt{3}}.$

(*iv*) The line current is equal to phase current.

(*v*) The line voltage leads the phase voltage by 30° elect.

(*vi*) The power is $W = 3 \times \text{Phase power}$

$$= 3 \times E_P \cdot I_P \cos \phi = 3 E_P \cdot I_P \cos \phi.$$

where E_P is the phase voltage and I_P the phase current.

$$W = 3 \times \frac{E_L}{\sqrt{3}} \cdot I_L \cos\phi \qquad \left(\because E_p = \frac{E_L}{\sqrt{3}}, \text{ and } I_L = I_P \right)$$

$$= \sqrt{3} \cdot E_L \cdot I_L \cos\phi.$$

so the power= $\sqrt{3}$ × line voltage × line current × power factor.

In case of unbalanced loads the total power can be given by the algebraic sum of the individual phase powers.

This system is generally used because for domestic and industrial purposes both voltages light and power are possible from the same line.

Delta connection or Mesh connections. In this system two dissimilar ends *i.e.* the starting end of one winding is connected to the finishing end of the other winding and so on. Thus forming a mesh or closed circuit. The three lines are taken from the three junctions of this mesh; as shown in Fig. 16.6. So this system is known as delta or mesh connection.

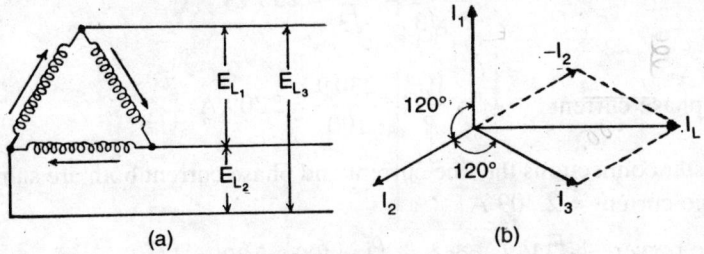

(a)   (b)

Fig. 16.6. Delta connections and vector diagrams.

These connections have got the following characteristics:

(*i*) The dissimilar ends are connected and leads are taken from the junctions of the closed circuit.

(*ii*) The line voltage = phase voltage.

(*iii*) The line current = $\sqrt{3}$ phase current.

(*iv*) The power in the circuit.

$$W = 3 \times \text{Phase power}$$

$$= 3 \times E_P \cdot I_P \cos\phi$$

$$= 3 \times E_L \times \frac{I_L}{\sqrt{3}} \cos\phi$$

$$= \sqrt{3} \ E_L \cdot I_L \cos\phi$$

so the power = $\sqrt{3}$ × Line voltage × Line current × Power factor.

In most of the cases these connections are used for the primary winding connections of the distribution transformer and secondary winding of the transmission transformers.

Example. *Calculate the phase and line currents in the given circuit Fig. 16.7. Also calculate the power taken from 400 V 50 c/s mains.*

L1
400V
50c/s
100 Ω
100 Ω
100 Ω
L2
L3

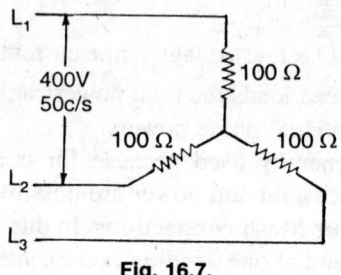

Fig. 16.7.

Solution. The phase voltage

$$= \frac{V_L}{\sqrt{3}} = \frac{400}{\sqrt{3}} = 230.9\,\text{V}$$

Now phase current $$= \frac{V_P}{R} = \frac{230.9}{100} = 2.309\text{ A}$$

In star connections the line current and phase current both are same so the line current = 2.309 A

The power $= \sqrt{3}\, V_L I_L \cos\phi = \sqrt{3} \times 400 \times 2.309 \times 1$

$= 1599.999 \approx$ **1600 W. Ans.**

Example. *Calculate the current taken and power consumed by the given load as shown in Fig. 16.8.*

Solution. The connections, as shown, having 20 Ω resistors connected in delta. The phase voltage is equal to line voltage.

So phase voltage = 100 V

The phase current $= \frac{100}{20} = 5\,\text{A}$

$= \sqrt{3}\, I_P = \sqrt{3} \times 5$

= 8.66 A. Ans.

20 Ω
20 Ω
20 Ω
100V
50c/s

Fig. 16.8.

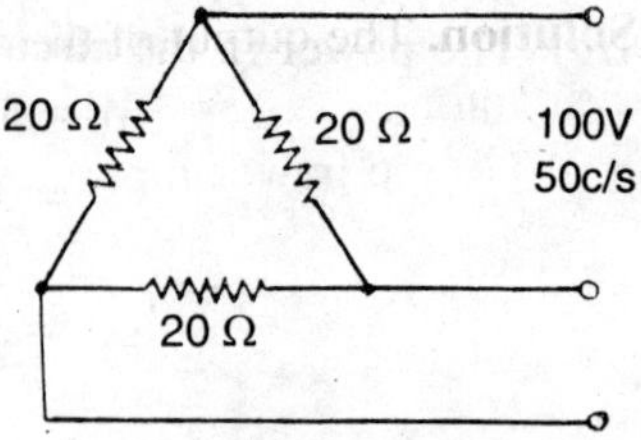

The line current

$= \sqrt{3}\, V_L I_L \cos\phi = \sqrt{3} \times 100 \times 8.66 \times 1$

= 1499.95 ≅ 1500 W. Ans.

Example. *Calculate the line and phase values of current of A.C. 3 φ) 400 V 10 B.H.P. motor operating at 0.8 power factor and 90% efficiency, when the motor is connected in (a) star (b) delta.*

Solution. The 10 B.H.P. motor is operating at 0.8 P.F. and 90% efficiency. So the input to the motor

$$= \frac{\text{Output}}{\eta} = \frac{10}{0.9} = 11.11 \text{ H.P.}$$

$$\text{Power} = \frac{10 \times 746}{0.9} = 8288.89 \text{ W}$$

The power in 3φ circuit

$$= \sqrt{3} \, V_L I_L \cos \phi$$

∴

$$I_L = \frac{W}{\sqrt{3} \, V_L \cos \phi}$$

$$= \frac{8288.89}{\sqrt{3} \times 400 \times 0.8} = 14.955 \approx 15 \text{ A}$$

(a) When the motor is connected in star

The line current = Phase current = **14.955** ≈ **15 A. Ans.**

(b) When the motor is connected in delta

Phase current = line current ÷ $\sqrt{3}$

$$= \frac{15}{\sqrt{3}} = 8.63 \text{ A} = \textbf{8.63 A. Ans.}$$

Example. *A 3 φ 400 V A.C. motor requires a current of 85 A, calculate its B.H.P. at a power factor of 0.85 and efficiency 90%.* (*N.C.V.T., 1966*)

Solution. The output of the motor is efficiency multiplied by the input to the motor.

In 3φ the power input

$$= \sqrt{3} \, V_L I_L \cos \phi$$

$$= \sqrt{3} \times 400 \times 85 \times 0.85 = 50056.27 \text{ W}$$

The output $= \eta \times \text{Input}$

$$= \frac{90}{100} \times 50056.27 = 45050.64 \text{ W}$$

The output in H.P. $= \dfrac{45050.64}{764} = $ **60.4 H. P. Ans.**

Example. *Find the normal full load current of motor of 20 H.P. on a 3 ϕ, 400 V A C. supply. What capacity of fuse wire should be put on the main switch for this purpose?* (N.C.V.T., 1963)

Solution. The power factor is not given, let us assume 0.8 and the efficiency of the motor as 85%.

The output of the motor
$$= 20 \text{ H.P.} = 20 \times 746 = 14920 \text{ W}$$

Now the input $= \dfrac{\text{Output}}{\eta} = \dfrac{20 \times 746}{0.85} = 17552.9 \text{ W}$

or $$\text{Input} = \sqrt{3}\, V_L I_L \cos\phi$$

$\therefore$ $$I_L = \dfrac{\text{Input}}{\sqrt{3} \cdot V_L \cos\phi}$$

$$= \dfrac{20 \times 746}{\sqrt{3} \times 400 \times 0.8 \times 0.85} = 31.67 \text{ A}$$

The motor carries 31.67 A as full load current if the motor is connected in star. If the motor is in star the rating should be
$$= 31.67 \times 1.5 = 47.5 \text{ A}$$

Secondly if the motor is in delta then the fuse rating should be

$$= 31.67 \times \sqrt{3} \times 1.5 = 82.3 \text{ A}$$

Thus is star, the fuse of 50 A and in delta 80 A which are very near to the calculated values, should be used. **Ans.**

Example. *Find the power factor of 420 V 3 phase motor giving out 25 H.P., while taking a line current of 35 A. At this load the efficiency of the motor is 87%.*

Solution. The output of the motor is 25 H.P.

i.e. $$25 \times 746 = 18650 \text{ W}$$

The 3 ϕ power, input $= \sqrt{3}\, V_L I_L \cos\phi = \dfrac{\text{Output}}{\eta}$

and power factor, $\cos\phi$ $= \dfrac{\text{Output}}{\sqrt{3}\, V_L I_L \cdot \eta}$

$$= \frac{25 \times 746}{\sqrt{3} \times 420 \times 35 \times 0.87} = 0.842 \text{ Ans.}$$

16.3 POWER MEASUREMENT

Q. What are the different methods of power measurement in three phase supply system? Describe one wattmeter method.

Ans. The power in three phase, which is equal to $\sqrt{3}\, V_L I_L \cos \phi$, can be measured by these following methods:
   (a) By single wattmeter of single phase,
      (i) when neutral is available,
      (ii) when neutral is not available.
   (b) By two wattmeters method.
   (c) By three wattmeters of single phase method.
   (d) By one wattmeter of three phase method.
   (a) **By single wattmeter method.** This method is adopted in case of balanced load. The total power consumed will be equal to the three multiplied by one wattmeter reading.

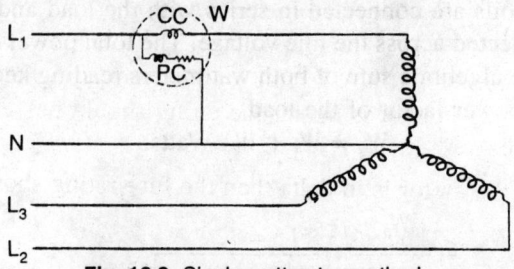

Fig. 16.9. Single wattmeter method.

   (i) *When neutral is available.* In that case the current coil of the wattmeter is connected in series with load and pressure coil across one phase and neutral as shown in Fig. 16.9. The watt

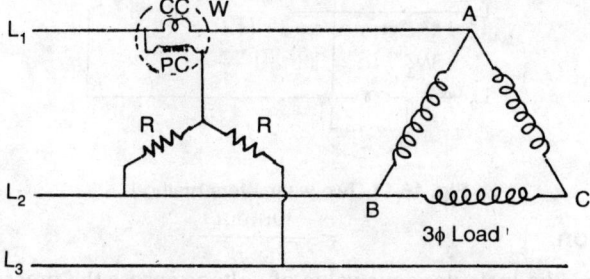

Fig. 16.10. Single wattmeter method making artificial neutral.

meter reading is taken and multiplied by three. Thus total power $W = 3 \times$ Wattmeter reading.

(*ii*) *When neutral is not available.* In that case the artificial neutral is made. Generally in case of delta connections the neutral is not available. The artificial neutral is made just by connecting the two similar resistances having equivalent value of the same value as that of the resistance of pressure coil of the meter, as shown in Fig. 16.10.

The total power is equal to

$$W_T = 3 \times \text{Wattmeter reading.}$$

Q. Draw the circuit diagram for measuring the power of a three phase delta connected induction motor with two wattmeter method.

OR

Explain with the help of calculation and diagram how three phase power in A.C. circuit can be measured by two wattmeters method.

(*N.C.V.T. 1964, 75, 76, 77, 71*)

Ans. This method of power measurement can be used for both the balanced or unbalanced loads of 3 ϕ 4 wires or 3 ϕ 3 wire systems.

In this method two wattmeters are connected as shown in Fig. 16.11. The current coils are connected in series with the load and the pressure coils are connected across the line voltage. The total power consumed by the load is the algebraic sum of both wattmeters reading keeping in view the value of power factor of the load

i.e. $\qquad W_T = W_1 + W_2$ Watts.

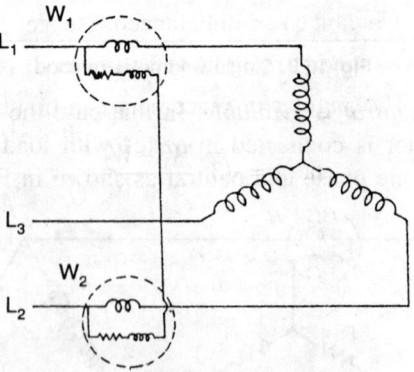

Fig. 16.11. Two wattmeters method.

Calculation:

Let $\quad {}_1e_2 =$ the instantaneous value of voltage across the pressure coil of wattmeter W_1

$_3e_2 =$ The instantaneous voltage across the pressure coil of wattmeter W_2

v_1, v_2 and v_3 are the instantaneous voltages of each phase.

i_1, i_2 and i_3 are the instantaneous value of currents in the respective
lines.

W_T, the total power consumed by the load.

In polyphase system, the vector sum of all the currents is equal to zero
i.e.,

$$i_1 + i_2 + i_3 = 0$$

or $$i_2 = -(i_1 + i_3)\ \text{A}$$

Now total power consumed W_T

$$W_T = e_1 i_1 + e_2 i_2 + e_3 i_3$$

Substituting for $\quad i_2 = -(i_1 + i_3)$

$$= e_1 i_1 - e_2(i_1 + i_3) + e_3 i_3$$

$$= e_1 i_1 - e_2 i_1 - e_2 i_3 + e_3 i_3$$

$$= i_1(e_1 - e_2) + i_3(e_3 - e_2)$$

Here $\quad e_1 - e_2 =$ the voltage impressed across the P.C. of
wattmeter W_1

$e_3 - e_2 =$ Voltage impressed across the P.C. of
wattmeter W_2

So $$W_T = i_1 \cdot {}_1e_2 + i_3 \cdot {}_3e_2$$

$$= W_1 + W_2\ \text{Watts.}$$

This proves that the total power is the sum of the wattmeters reading
whether the load is balanced or unbalanced. Hence, total power

$$W_T = \text{Wattmeter-I reading} + \text{Wattmeter-II reading.}$$

Power factor. The angular displacement can be obtained as

$$\tan\phi = \sqrt{3} \cdot \frac{W_1 - W_2}{W_1 + W_2}$$

and $$\phi = \tan^{-1} \sqrt{3}\left(\frac{W_1 - W_2}{W_1 + W_2}\right)$$

Thus $$\cos\phi = \cos\left[\tan^{-1}\left(\sqrt{3}\ \frac{W_1 - W_2}{W_1 + W_2}\right)\right]$$

and thus the power factor of the load can be calculated.

Depending upon the power factor, the readings of the wattmeters are effected. There are the following different cases.

(i) *When power factor is unity.* In this case both the wattmeters will give positive readings and equal also

i.e. $W_1 = W_2$ Watts.

(ii) *When Power factor ≥ 0.5.* Both wattmeters will read positive but different readings. In case if the power factor is 0.5, only one wattmeter will give reading and total power will be equal the this reading.

(iii) *When power factor < 0.5.* In this case the one wattmeter will read in opposite direction, change the connection of the current coil of this wattmeter and get positive reading.

16.4. POWER FACTOR CORRECTION

Q. What do you understand by the power factor correction equipments?

Ans. In a.c. circuits the power factor has its own significance. There is a lowest limit beyond which it cannot be permitted. In big industries and organisations, special arrangements are made to indicate and limit the lowest power factor for example the power factor demand indicators etc.

There are the following equipments generally employed for the power factor correction:

(a) **Synchronous capacitors.** The over excited synchronous motor works as the power factor correction equipment, and in this case it is known as synchronous capacitor. The motor draws leading current and thus leading kVAR which neutralizes or balances the lagging kVAR of the system, thus increasing the power factor. The biggest advantages of this system is to have required power factor improvement just by varying the field excitation. Secondly the mechanical energy is also available on the shaft which can be utilized.

(b) **Capacitor bank or static capacitors.** A group or bank of the capacitors are generally installed in the substation or at the site to improve the power factor of the load. These are frequently used near the group of a.c. motors. The capacitor draws leading current which meets the lagging reactive component of the current hence improving the power factor. These are employed with an automatic switching off arrangement to avoid the over compensation.

(c) **Phase advancer.** These are also used to improve the power factor. These are installed with industrial machines.

Q. Show how will you connect three single phase wattmeters and one three phase wattmeter for measuring the power in three phase power load.

Ans. Three phase power can also be measured by these following methods:

(a) **By using three wattmeter of single phase.** In this case three single phase wattmeters are used. This method can be used for balanced or unbalanced loads. The c.c. of each wattmeter is connected in series with the load and the pressure coil across each phase in star as shown in Fig. 16.12. In this case neutral may or may not be available. The total power is the sum of all the three wattmeters readings.

$$\therefore \qquad W_T = W_1 + W_2 + W_3 \text{ Watts.}$$

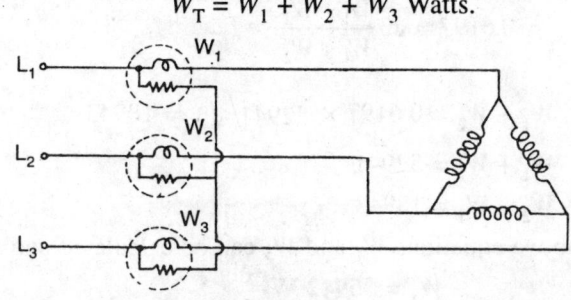

Fig. 16.12. 3 wattmeters method of power measurements.

(b) **By using single wattmeter of three phase.** In this system of power measurement it can be directly read from the meter. It can be used for balanced or unbalanced loads and also for three phase four wire or three phase three wire systems as shown in Fig. 16.13.

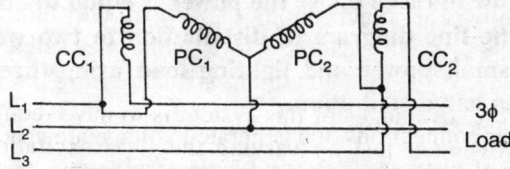

Fig. 16.13. Single wattmeter or 3 ϕ.

Example. *A three phase 400 V 50 c/s induction motor has an output of 45 kW and operating at a power factor of 0.85 lagging with an operating efficiency 0.85. Calculate the readings of both the wattmeters connected to measure the input.*

Solution. The motor has an output

$$= 45 \text{ kW}$$

So the input $\qquad = \dfrac{\text{Output}}{\text{Efficiency}} = \dfrac{45}{0.85} = 52.941 \text{ kW}$

Here the total input = 52.941 kW

In other wards it is the total power

i.e. $W_1 + W_2 = 52.941$ kW

Now the power factor is 0.85 lagging

so $\cos \phi = 0.85$

∴ $\phi = \cos^{-1} 0.85 = 31.79°$

and $\tan \phi = 0.6197$

$\tan \phi$

∴ $0.6197 = \sqrt{3} \dfrac{W_1 - W_2}{W_1 + W_2}$

∴ $W_1 - W_2 = 0.6197 \times 52941 / \sqrt{3} = 18943$

$W_1 + W_2 = 52941$

$W_2 - W_1 = 18943$

From the above equations W_1 and W_2 can be calculated and thus

$$W_1 = 35942 \text{ W}$$
$$W_2 = 16999 \text{ W. Ans.}$$

Q. (*a*) **Prove in star connections the line voltage is equal to $\sqrt{3}$ phase voltage.**

(*b*) **Show how in three phase the power is equal to $\sqrt{3} \cdot V_L I_L \cos \phi$.**

(*c*) **Draw the line diagram of distribution to two workshops, for boiler, small power and lighting load using three phase four wire bus bar distribution.**

Ans. (*a*) In star connections, the generated voltage are spaced 120° elect. apart *i.e.* if phase A, is starting from zero degree, the phase B, will start from 120° elect. and phase C from 240° elect. as shown in Fig. 16.14.

Here all three phases voltages are shown by E_{P_1}, E_{P_2} and E_{P_3} voltages spaced 120° apart. Now to find out the resultant line voltage the phase Voltage E_{P_2} is reversed as $-E_{P_2}$ and the resultant is obtained. Both the voltages are 60° apart so the resultant can be obtained according to the parallelogram of forces method and is

$$E_L = \sqrt{E_{P3}^2 + E_{P2}^2 + 2E_{P_3} E_{P_2} \cos \phi}$$

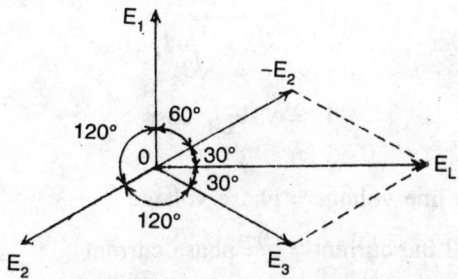

Fig. 16.14. Star connections.

Here all the phase voltages are equal because the phase windings are identical so let,

$$E_{P1} = E_{P2} = E_{P3} = E_P$$

$$\therefore \qquad E_L = \sqrt{E_P{}^2 + E_P{}^2 + 2E_P \cdot E_P \cos\phi}$$

$$= \sqrt{E_P{}^2 + E_P{}^2 + 2E_P E_P \times \frac{1}{2}} \qquad \left(\because \cos 60 = \frac{1}{2} \right)$$

$$= \sqrt{3E_P{}^2}$$

or $$\mathbf{E_L = \sqrt{3}\, E_P.} \text{ Proved.}$$

(b) The power in three phase system is equal to the power consumed by all the three circuits so the total power is

Total power = 3 × phase power

When the voltage and currents are taken to have the same phase values *i.e.* $V_{P_1} = V_{P_2} = V_{P_3} = V_P$ and $I_{P_1} = I_{P_2} = I_{P_3} = I_P$ and the power factor being same on all the phases then,

$$\text{Total power} = V_{P_1} I_{P_1} \cos\phi_1 + V_{P_2} I_{P_2} \cos\phi_2 + V_{P_3} I_{P_3} \cos\phi_3$$

$$= V_P I_P \cos\phi + V_P I_P \cos\phi + V_P I_P \cos\phi$$

$$= 3 V_P I_P \cos\phi.$$

When this phase power is considered in line values then
In star.

$$\text{The line voltage} = \sqrt{3} \text{ phase voltage}$$

$$\text{Line current} = \text{phase current}$$

So power $\qquad = \sqrt{3} \times \dfrac{V_L}{\sqrt{3}} I_L \cos\phi$

$$W = \sqrt{3} V_L I_L \cos\phi$$

In delta.

The line voltage = phase voltage

Line current = $\sqrt{3}$ phase current

So power = $3V_L \cdot \dfrac{I_L}{\sqrt{3}} \cdot \cos\phi = \sqrt{3} V_L I_L \cos\phi$

So in both the cases the power in three phase system,

$$W = \sqrt{3}\ V_L I_L \cos\phi$$

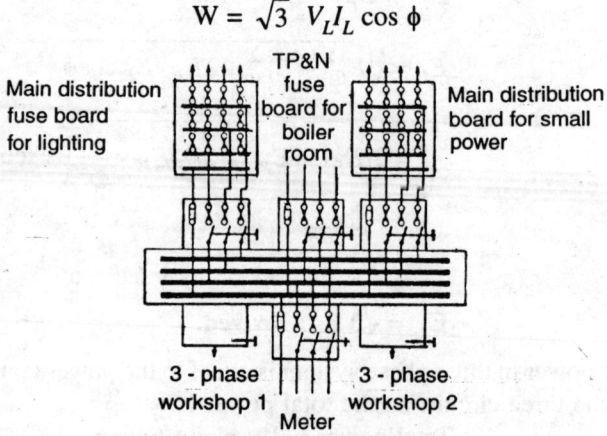

Fig. 16.15. 3 ϕ distribution.

(c) The three phase four wire system is very important and mostly used for general distribution purposes, on the same line the lighting and power load can be connected as shown in Fig. 16.15.

Q. Show how the power in an inductive circuit can be measured by voltmeter and ammeter methods.

Ans. The power which is a product of voltage current and power factor of the circuit, can be measured not only by wattmeter but can be measured by ammeter and voltmeter methods too.

(a) By using three voltmeters method
   The connections are made as shown in Fig. 16.16(a).
   Here $\qquad L$ = The inductive load
   $\qquad\qquad R$ = Known value of resistance
   $\qquad\qquad V_1$ = Voltage measured across inductive load

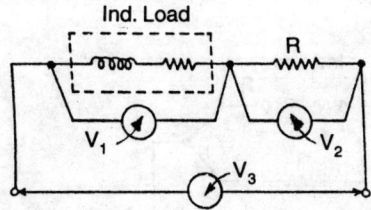

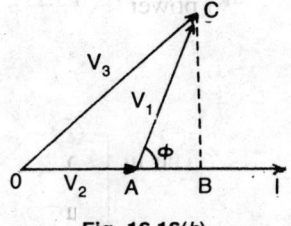

Fig. 16.16(a). Three voltmeters method.　　　**Fig. 16.16(b).**

V_2 = Voltage across the resistance
V_3 = Supply voltage

The vector diagram is shown in Fig. 16.16(b) here in the ΔOBC

$$OB = OA + AB = V_2 + V_2 \cos\phi$$
$$BC = V_1 \sin\phi$$

Now　　　$$OC^2 = OB^2 + CB^2$$

$$V_3^2 = \left(V_2 + V_1\cos\phi\right)^2 + \left(V_1\sin\phi\right)^2$$

$$= V_2^2 + V_1^2\cos\phi^2 + 2V_1V_2\cos\phi + V_1^2\sin^2\phi$$

$$= V_2^2 + V_1^2\left(\sin^2\phi + \cos^2\phi\right) + 2V_1V_2\cos\phi$$

$$= V_2^2 + V_1^2 + 2V_1V_2\cos\phi$$

or　　　$$2V_1V_2\cos\phi = V_3^2 - V_1^2 - V_2^2$$

The series circuit current

$$I = \frac{V_2}{R} \text{ amp}$$

$V_2 = IR$ substituting in $2 V_2 I_1 \cos\phi$, we get

$$\alpha = 2V_1 \cdot IR\cos\phi \text{ or } 2R\left(V_1 I\cos\phi\right)$$

or　　　$$2R\left(V.I.\cos\phi\right) = V_3^2 - V_1^2 - V_2^2$$

$\therefore$　　　$$V_1 I\cos\phi = \frac{V_3^2 - V_1^2 - V_2^2}{2R}$$

$V_1 I \cos\phi$ is the power consumed by the inductive load so this power can be given by the formula

Load power, $$\mathbf{W} = \frac{1}{2\mathbf{R}}\left(\mathbf{V_3^2 - V_1^2 - V_2^2}\right) \textbf{ Watts}$$

(b) **By using three ammeters method.** The connections are made as shown in Fig. 16.17.

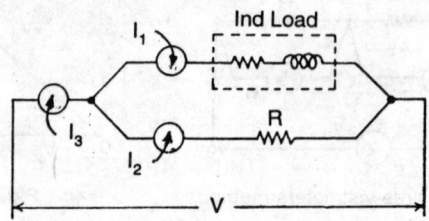

Fig. 16.17. Three ammeters method.

The power consumed by the inductive load can be given as
Load power

$$= \frac{R}{2}\left(I_3^2 - I_1^2 - I_2^2\right) \text{ Watts.}$$

where I_1 = Current through the inductive load

I_2 = Current through a non-inductive resistance

I_3 = Total current taken.

Example. *A non-inductive resistor of 10 Ω is connected in series with an inductive load and the combination is placed across a 100 V supply. A voltmeter connected across the load and then across the resistor indicates 49 V and 70 V respectively. Calculate the*

(a) *Impedance of the load.*

(b) *Impedance of the combination.*

(c) *Power absorbed by the load.*

(d) *Power absorbed by the resistor.*

(e) *Power taken from the mains.*

(f) *P.f. of the load and the circuit as a whole.*

Solution. The current taken

$$= \frac{70}{10} = 7 \text{A}$$

(a) Impedance of the load

$$= \frac{\text{Voltage across load}}{\text{Current taken}}$$

$$= \frac{49}{7} = 7 \Omega \quad \textbf{Ans.}$$

(b) Impedance of the combination

$$= \frac{V}{I} = \frac{100}{7} = 14.3\,\Omega \quad \textbf{Ans.}$$

(c) Power absorbed $= \dfrac{1}{2R}\left(V_3^2 - V_1^2 - V_2^2\right)$

$$= \frac{1}{2 \times 10}\left(100^2 - 49^2 - 70^2\right)$$

$$= 134.95 \approx \textbf{135 W.} \quad \textbf{Ans.}$$

(d) Power absorbed by resistor

$$= I^2 \times R = 7^2 \times 10 = \textbf{490 W.} \quad \textbf{Ans.}$$

(e) Total power taken

$$= 135 + 490 = \textbf{625 W.} \quad \textbf{Ans.}$$

(f) P.f. of the load $= \dfrac{135}{49 \times 7} = \textbf{0.394 Ans.}$

P.f. of the combination

$$= \frac{625}{100 \times 7} = \textbf{0.893} \quad \textbf{Ans.}$$

Q. What do you understand by the phase sequence? Give formula for converting the converting the star to delta and delta to star connections.

Ans. The phase sequence is meant by the order in which the three phases attains their maximum and minimum values. When it is said the phase sequence is RYB, it means phase Y is lagging behind phase R by 120° elect. and phase B is lagging behind R by 240 elect. and to Y phase 120° elect. This rotation is said to be correct and the phase sequence is correct, otherwise if these are so connected to attains their values other than the stated the sequence is not correct.

Let there be the following resistances connected in star and in delta. Let us take three R_1, R_2, and R_3 resistances connected in star are required to be converted into delta connections as shown in Fig. 16.18.

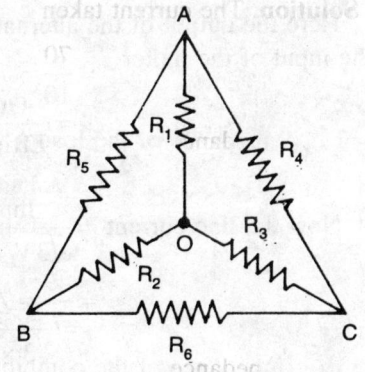

Fig. 16.18.

Then

$$R_{AB} = R_5 = R_1 + R_2 + \frac{R_1 R_2}{R_3} \, \Omega$$

$$R_{BC} = R_6 = R_2 + R_3 + \frac{R_2 R_3}{R_1} \, \Omega$$

$$R_{CA} = R_4 = R_3 + R_1 + \frac{R_3 R_1}{R_2} \, \Omega$$

Similarly if R_4, R_5 and R_6 are in delta and required to be converted into star connections, then

$$R_{0A} = R_1 = \frac{R_5 R_4}{R_4 + R_5 + R_6} \, \Omega$$

$$R_{0B} = R_2 = \frac{R_6 R_5}{R_4 + R_5 + R_6} \, \Omega$$

$$R_{0C} = R_3 = \frac{R_4 R_6}{R_4 + R_5 + R_6} \, \Omega \quad \textbf{Ans.}$$

Example. *A three phase 400 V 50 c/s star connected alternator is supplying a 3 phase 100 h.p. delta connected induction motor. The motor efficiency is 85% and the power factor is 0.8. Calculate the current in*
  (a) *Motor per phase*
  (b) *Alternator per phase.*
Solution. The load on the alternator is the motor of 100 h.p. *i.e.*
$$100 \times 746 = 74600 \text{ W}$$

Here the output of the alternator will be the input to the motor. Now the input of the motor

$$= \frac{\text{Output}}{\text{Efficiency}} = \frac{74600}{0.85} = 87764.7$$

Now the line current $= \dfrac{\text{Input}}{\sqrt{3} \, V_L \cos \phi}$

$$= \frac{100 \times 746}{\sqrt{3} \times 400 \times 0.85 \times 0.8} = \textbf{158.35 A.}$$

Now this is the current flowing in the line, the alternator is connected in star, so the line current will be equal to the phase current. The motor is delta connected so the phase current

$$= \frac{\text{Line current}}{\sqrt{3}} = \frac{158.35}{\sqrt{3}} = 91.42 \text{ A.}$$

So the current per phase in motor = **91.42 A**

Current per phase in alternator = **158.35 A. Ans.**

REVIEW QUESTIONS

1. What is meant by the polyphase supply system?
2. What are the advantages of polyphase over the single phase supply system?
3. How many sliprings are used for three phase star connected alternator and why?
4. What are the different methods of inter-connecting the three phase windings.
5. What are the relationship between the line and phase values of current and voltage in star and delta connections?
6. How the three phase power can be measured? Describe two wattmeters method of power measurement.
7. Compare the star and delta connections.
8. Fill up the blanks:
   (*i*) The .... system is commonly used for distribution purpose.
   (*ii*) The line voltage is .... times the phase voltage in star.
   (*iii*) The line current is .... the phase current in star.
   (*iv*) In delta connection only .... wire are used.
   (*v*) In .... connections the line voltage is equal to phase voltage.
   (*vi*) The power in 3 ϕ circuit is ....
   (*vii*) In balanced load the neutral wires carries .... current.
   (*viii*) Two wattmeters method is suitable for .... load.
   (*xi*) In case of two wattmeters method the load power factor = (W_1-W_2) ....
9. How will you measure the power of an inductive load using two wattmeter method. Draw connection diagrams.
10. Will it be useful to generate six phase supply instead of three phase supply?

NUMERICAL PROBLEMS

1. The full load efficiency of a delta connected three phase induction motor is 90% and power factor 0.92. The output of the motor is 5 H.P. (metric) at 400 V. Calculate the line and phase currents. (Hint. 1 H.P. = 735.5 W)

   [Ans. 6.48 A, 3.75 A]

2. Three coils each having a resistance of 10 ohm and inductance of 0.02 Henry are connected in star to a three phase 500 V 50 c/s mains. Calculate the line and phase currents and the power absorbed.

   [Ans. 24.44 A, 24.44 A, 17.93 kW]

3. A 440 V 50 c/s squirrel case induction motor has a full load output of 10 h.p. the power factor being 0.83, the full load efficiency 82%. The power is measured by two wattmeter method. Find the readings of the wattmeters.

[**Ans. 6.315 kW, 2.785 kW**]

4. Three identical coils, each having a reactance of 20 Ω and resistance 20 Ω are connected in delta across a 3 φ 440 V 50 c/s mains. Calculate the line current and phase currents. Also calculate the readings of two wattmeters if the power consumed is measured by two wattmeter method.

[**Ans. 15.5 A, 26.9 A, 11430 W, 3060 W**]

5. A motor generator set running from 3 phase 440 V mains at 0.85 power factor delivers 17.5 A, D.C. at 230 V. Assuming the efficiency of the set to be 70%. Find the line current on the A.C. side and phase current if the motor windings are connected in delta.   [**Ans. 8.876 A, 5.125 A**]

6. The power taken by 3 φ 400 V 50 c/s induction motor is measured by the two wattmeter method. The readings of the wattmeters are 2.5 kW and 1.25 kW respectively. Find out the input power and power factor of the motor.

[**Ans. 3.75 kW, 0.866**]

7. Two wattmeters are connected to measure the power of a 3φ balanced load, the readings are 20 kW and 30 kW. Find out the total power and the angular displacement between current and voltage.   [**Ans. 50 kW, 19.1°**]

8. Two wattmeters are connected to a 3 φ 500 V motor and indicates the total input as 10 kW the power factor is 0.78. Find the readings of individual wattmeters.   [**Ans. 7.3094 kW, 2.6906 kW**]

9. A 3 φ 400 V 20 H.P. delta connected motor draws a line current of 30 A. Determine the voltage across each phase and current flowing through each phase of the motor.   [**Ans. 400 V, 17.32 A**]

10. Three identical coils each having a reactance of 20 Ω and a resistance of 20 Ω are connected in (a) star (b) delta, across 440 V 3 φ mains. Calculate the line and phase values of current in both the cases. Also calculate the readings of the two wattmeters used for measuring the power consumed, when the load is connected in star.

[**Ans. 8.98 A, 8.98 A, 26.94 A, 15.56 A, 3.816 kW, 1.023 kW**]

11. The power input to a 3φ 2000 V 50 c/s motor running on full load at an efficiency of 90%; is measured by two wattmeter method, which indicate 300 kW and 100 kW respectively. Calculate ((a) input (b) the angle of power factor (c) line current and the (d) the h.p. output.

[**Ans. 400 kW, 40.9° lagging, 152.7 A, 490 H.P. (metric)**]

17

Alternators

The machine which converts mechanical energy into A.C. electrical energy, is known as alternator.

Q. Describe with the help of sketches the working of an alternator.

Or

State the working principle of an alternator, also illustrate how e.m.f. is induced?

Ans. An alternator works on the principle of Faradays Law's of electromagnetic induction. The *Working principle* can be stated as, "Whenever a conductor cuts the lines of force, an e.m.f. is induced in that conductor". This induced e.m.f. is directly proportional to the rate of change of flux and the number of conductors.

Let there are two magnetic poles 'N' and 'S' as shown in Fig. 17.1 and a loop of conductor *ABCD* is placed in between the magnetic poles North and South. Let the loop ends are connected to two sliprings. Now whenever the conductor loop is rotated, it experiences the magnetic flux of different nature *i.e.* North and South. Let the direction of rotation be clockwise, the conductor be at position number 1, where the direction of motion and direction of the flux are parallel so there will not be any linkage between the magnetic field and the conductor, resulting no e.m.f. When the conductor moves to position number 2, it experiences some linkage and producing the e.m.f. as shown according to the Fleming's right hand rule the direction can be found as outwards. At position number 3, the conductor has maximum linkage and resulting the production of e.m.f. maximum in outward direction. Thus there will be some e.m.f. induced in 4th position, and in fifth position, the motion of conductor and flux are parallel so no e.m.f. will be produced as shown in Fig. 17.1.

In the sixth position, the direction of motion is reversed or say changed, the e.m.f. 'induced will be inward as shown. At position seventh the maximum cutting angle will cause maximum production of e.m.f. as shown, at position 8th the e.m.f. will be comparatively less and at position number 1 same cycle will again start.

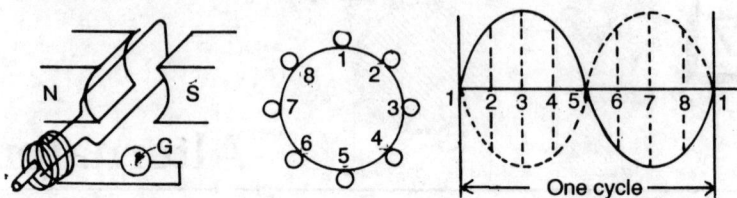

Fig. 17.1 Production of e.m.f.

Now as both conductor are connected to slip rings and if we plot a graph for the values of e.m.f. thus obtained a curve will be obtained generally known as sine wave.

So the e.m.f. thus obtained will have the magnitude periodically changing after a fixed interval of time, such e.m.f. is known as A.C. and the machine producing A.C. is called alternator. In alternator we are having more number of conductors, which are systematically placed over the armature to obtain a smooth curve.

Q. Describe briefly the salient parts of an alternator and their function. *(N.C.V.T. 1967, 73, 74, 78 ; W/m Delhi 64)*

Ans. The alternator has the following parts:
 (*i*) Stator.
 (*ii*) Rotor.
 (*iii*) Exciter.

 (*i*) **Stator.** It is stationary portion consisting of stator frame and stempings etc. The stator frame is used for holding the armature stempings and the stator winding in position.

 Depending upon the capacity some holes are provided or casted in the frame to circulate the air for cooling. In big frames the horizontal and vertical ducts are made for the air to circulate or heat dissipation *i.e.* for cooling purpose.

 The stempings are made of laminations of silicone steel. Each lamination is insulated from each other by means of paper or varnish layer or oxide coating or enamel coating. There are number of slots provided in the laminations to house the winding. The slots are of these types, *open type*, having same opening from top to and bottom, *semi-closed*, having small opening and *closed type* which are totally closed as shown in Fig. 17.2. The slots, which are cut in the inner periphery of the laminations are generally open types or semiopen types of slots.

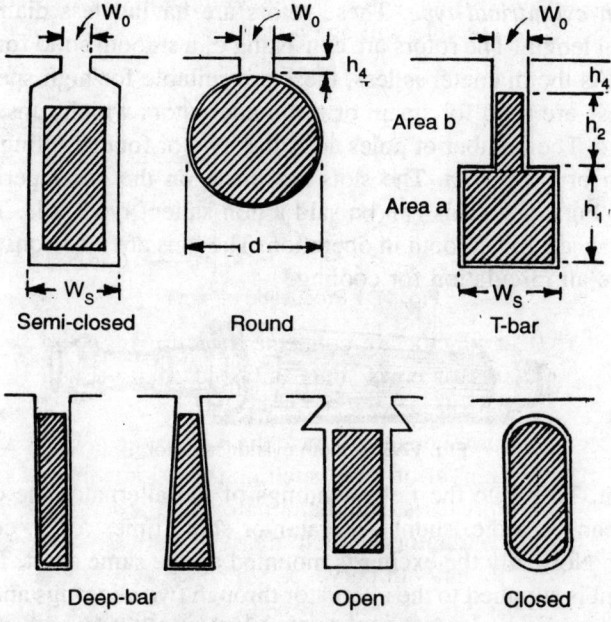

Fig. 17.2. Types of slots

The winding is placed in the slots and the conductors are held in position by means of the fiber or bamboo wedges.

Rotor. It is a rotating portion of the alternator. The rotors are of two types:

    (*i*) Salient pole type.   (*ii*) Smooth cylindrical type.

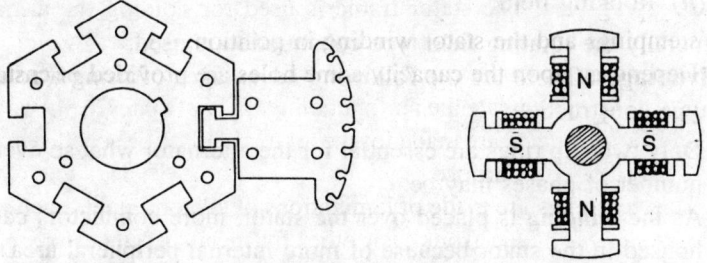

Fig. 17.3. Dove tailed arrangement.   **Fig. 17.4** Salient pole rotor

    (*i*) *Salient pole type.* In this type of the rotor, the poles are projected having large diameter and less axial length. These are suitable for low and medium speed. As the speed is less so the number of poles can be more *i.e.* from 6 to 40. The poles cores assembly is bolted or devetailed joined to the cast iron or steel wheel.

Smooth cylindrical type. These rotors are having less diameter and more axial length. The rotors are consisting of a smooth solid forged steel cylinder. As the diameter is less, these are suitable for high speed. Generally these are used for steam or turbo alternators which runs at a very high speed. The number of poles are either two or four resulting speed as 3000 rpm or 1500 rpm. The slots are made on the outer periphery as shown in Fig. 17.5. This can be said a non-salient type pole. These are much balanced and smooth in operation. The fans are also constructed to enable the air circulation for cooling.

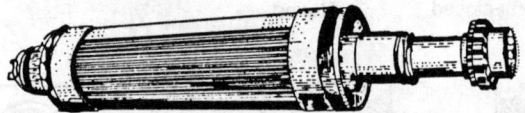

Fig. 17.5 Smooth cylindrical rotor.

Exciter. To excite the field windings of the alternator the exciter in used. It can be a d.c. shunt generator or some times a d.c. compound generator. Normally the exciter is mounted on the same shaft. The exciting current is supplied to the alternator through two slip rings and brushes etc. The exciting voltage varies from 125 V to 250 V with the power rating 0.3 to 1% of power rating of alternator.

Q. Which type of the field construction is preferred? What are the advantages of using rotating field type construction?

Ans. The field of an alternator can be of two types:

    (*i*) Stationary field.

    (*ii*) Rotating field.

Generally the rotating field type construction is used.

Advantages. There are the following advantages of using a rotating field type construction:

    (*i*) Only two slip rings are essential for the alternator what so ever the number of phases may be.

    (*ii*) As the winding is placed over the stator, more conductors can be housed in the stator because of more internal peripheral area.

    (*iii*) As the winding in which e.m.f. is induced, is stationary, so there is no possibility of breaking or loosening the winding and its joints.

    (*iv*) There is no sliding contact between the armature and external (load) circuit, only two slip rings are provided for field excitation, thus less sparking and less possibility of faults.

(v) The winding being stationary can be easily and effectively insulated and the insulation cost will also be less (less dielectric strength insulation can work).

(vi) Armature can have more number of conductors so e.m.f. induced will be more.

(vii) Less maintenance and possibility of faults.

(viii) As the rotor is light so can be driven at higher speed.

(ix) The current of rotor winding is independent of the capacity of the alternator, hence lighter rotor can be used than the rotating armature type alternators.

Q. (a) **Write down the formula used for calculating the frequency of the induced e.m.f.**

(b) **What will be the speed of the alternators having 2, 4 and 6 poles at 25 c/s, 50 c/s and 60 c/s.**

Ans. (a) A conductor has to cover an angular distance equal to twice the pole pitch to produce one cycle. Thus if the number of poles of an alternator are P, the e.m.f. induced by this alternator can have $P/2$ cycles in one revolution. Obviously the number of cycles per second is known as the frequency.

So frequency = Number of pairs of poles · Number of revolution in one second.

$$\therefore \qquad f = \frac{PN}{2 \times 60} = \frac{PN}{120} \; c/s.$$

were P = Number of poles of the alternator

N = Speed in r.p.m.

(b) The speed of the alternator is

$$N = \frac{120 f}{P}$$

So for a machine having 25 c/s and 2 poles

$$= \frac{120 \times 25}{2} = 1500 \; r.p.m$$

Similarly we can calculate for the other alternators and can tabulate them.

Table 17.1: Speed and Poles

Number of poles	2	4	6	8	10	12
R.P.M. for 25 c/s	1500	750	500	375	300	250
R.P.M. for 50 c/s	3000	1500	1000	750	600	500
R.P.M. for 60 c/s	3600	1800	1200	900	720	600

Q. Deduce the e.m.f equation of an alternator.

Ans.   Let there be

ϕ = Useful flux/pole in Wb

Z = Number of conductor per phase

f = Frequency of induced e.m.f. in c/s

P = Number of poles

N = Speed in r.p.m.

As we know that e.m.f. induced is directly proportional to the rate of change of flux, so the flux cut in one revolution

$$= \frac{\text{Total flux}}{\text{Time taken}}.$$

The speed is N r.p.m. so the time taken for one revolution is 60/N sec. So the e.m.f. induced per conductor

$$= \frac{\phi P}{60/N} = \frac{\phi NP}{60} \text{ Volts}$$

Obviously $f = \dfrac{PN}{120}$ or $NP = 120\,f$

Substituting in the equation, we have, the average e..f. induced

$$= \frac{\phi}{60} \times 120 f = 2\phi f \text{ Volts.}$$

If there are Z conductors per phase then the e.m.f. induced per phase

$$= 2\phi Z f \text{ Volts.}$$

The r.m.s. value of e.m.f. induced per phase

$$= k_f 2 \cdot \phi \cdot Z \cdot f = 2.22 \cdot \phi \cdot Z \cdot f \text{ Volts}$$

The winding done, is not always full pitched winding but the fractional pitched winding *i.e.*, long pitch or short pitch windings are also used. The conductors are also concentrated in one slot instead of being distributed in the slots under one pole. Hence two more factors comes into picture the *distribution factor* (k_d) and *coil span* factor k_c.

Now the e.m.f. induced per phase,

$$E_{\text{phase}} = 2 \cdot k_f \cdot k_d \cdot k_c \cdot \phi \cdot Z_{ph} \cdot f \text{ Volts}$$

Where k_f the form factor 1.11, k_d = the distribution factor and k_c = the coil span factor, ϕ = flux in Wb, Z_{ph} = conductors per phase, f = frequency in c/s.

The e.m.f. for a single phase alternator is

$$E = 2 \cdot k_f \cdot k_d \cdot k_c \cdot \phi \cdot Z \cdot f \text{ Volts}$$

for three phase alternator.

$$E_{\text{ph}} = 2 \cdot k_f \cdot k_d \cdot k_c \cdot \phi \cdot Z \cdot f \text{ Volts}$$

and as we know that $E_L = \sqrt{3}\, E_{ph} \cdot (\text{in star})$

$$\therefore \qquad E_L = \sqrt{3}\, E_{ph} = \sqrt{3} \cdot 2 \cdot k_f \cdot k_d \cdot k_c \cdot \phi \cdot Z \cdot f \text{ Volts}$$

Example. *Calculate the e.m.f. induced (line voltage) in a 3ϕ, 8 pole star connected alternator. The stator has 160 slots with six conductor per slot. Coil span factor being unity and k_d = 0.85. The speed is 750 r.p.m. and ϕ is 19 m Wb.*

Solution. The frequency of induced e.m.f.

$$= \frac{PN}{120} = \frac{8 \times 750}{120}$$

$$= 50 \text{ c/s.}$$

Here coil span factor = 1 and k_d = 0.85

E.m.f. being in star = $\sqrt{3}\, E_p$

$$= \sqrt{3} \times 2 \times k_f \times k_d \times k_c \times \phi \times Z \times f \text{ Volts}$$

$$= \sqrt{3} \times 2 \times 1.11 \times 0.85 \times 1 \times 19 \times 10^{-3} \times \frac{160}{3} \times 6 \times 50$$

$$\mathbf{E_L = 993.6 \quad Ans.}$$

Example. *Calculate the speed and open circuited line and phase voltage of a 4 poles 3 phase 50 c/s, star connected alternator with 36 slots and 30 conductors per slot. The flux per pole is 0.05 Wb sinusoidally distributed.*

Solution. Speed $= \dfrac{120 \times \text{frequency}}{\text{poles}}$

$$= \frac{120 \times 50}{4} = \textbf{1500 r.p.m. Ans.}$$

$$E_{ph} = 2.22 \, k_d \cdot k_c \cdot \phi \cdot Z \cdot f \text{ Volts.}$$

and
$$\alpha = \frac{180}{\text{No. of slots/pole}} = \frac{180 \times 4}{36} = 20°$$

$$n = \text{No. of slots/pole/phase} = \frac{36}{4 \times 3} = 3$$

Let the winding is fully distributed $k_c = 1$

$$\therefore \quad k_d = \frac{\sin \dfrac{n\alpha}{2}}{n \sin \dfrac{\alpha}{2}} = \frac{\sin \dfrac{3 \times 20}{2}}{3 \times \sin \dfrac{20}{2}} = \frac{\sin 30}{2 \sin 10} = 0.96$$

Let the winding is fully distributed $\therefore \; k_c = 1$

$$E_{ph} = 2.22 \times 0.96 \times 1 \times 0.05 \times \frac{36}{3} \times 30 \times 50$$

$$= \textbf{1918 V. Ans.}$$

$$E_L = \sqrt{3} \, E_{ph} = 1.732 \times 1918 = \textbf{3322 Volts Ans.}$$

Example. *A 3 phase 440V alternator is running at 1000 r.p.m. There are 6 poles and the flux per pole is 0.069 Wb. Calculate the number of conductor per slot if the stator has 36 slots. The winding is connected in star. The $k_c = 0.96$ and $k_d = 0.96$.*

Solution. The alternator is running at 1000 r.p.m. and the poles are 6.

Now frequency $= \dfrac{PN}{120} = \dfrac{6 \times 1000}{120} = 50$ c/s.

Given
$$k_c = 0.96 \text{ and } k_d = 0.96.$$

$$E_{ph} = 2.22 \, k_d \cdot k_c \cdot \phi \cdot Z \cdot f \text{ volts.}$$

$$\frac{440}{\sqrt{3}} = 2.22 \times 0.96 \times 0.96 \times 0.069 \times Z_{ph} \times 50$$

$$Z_{ph} = 36 \text{ conductor.}$$

Now number of slots per phase $\dfrac{36}{3}$ = 12 slots

Conductor/phase = 36.

Conductor/slot = $\dfrac{36}{12}$ = 3 **Conductors. Ans.**

Q. Explain clearly what is meant by distribution factor and the pitch factor? Deduce the expressions for both of them.

Ans. Both these factors are concerned with the winding type, nature and the induced e.m.f. of the alternator.

Distribution factor. It is also known as the breadth factor or winding factor.

If the coil sides of any one phase under one pole are concentrated in one slot, the total e.m.f. induced will be equal to the arithmetic sum of the e.m.f. induced in all the coils of one phase under one pole. But in order to obtain a sinusoidal wave of alternating e.m.f., the coils are spread into a number of slots under each pole forming a polar group in Fig. 17.6. In what case the resultant e.m.f. will not be the arithmetic sum of the individual coil side e.m.f. but it will be the vector sum of the e.m.f.s induced in different slots under one pole as shown in Fig. 17.7.

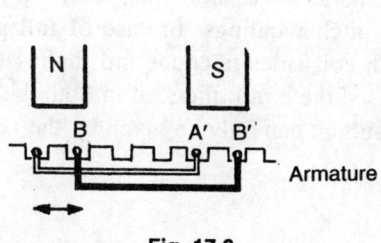

Fig. 17.6

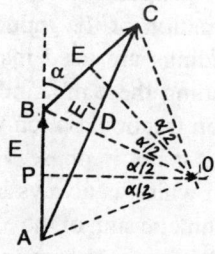

Fig. 17.7 Distribution factor.

Hence the distribution factor is defined as the ratio of the vector sum of the e.m.f. induced in all the coils under one pole to the arithmetic sum of the induced e.m.f. if the winding would have been in one slot.

i.e. $k_d = \dfrac{(\text{Vector sum of the e.m.f.s in all the coils under one pole})}{(\text{Arithmetic sum of induced e.m.f. if the winding would have been in one slot})}$

Let there be

n = Number of slots per pole per phase.

E_c = E.m.f. induced in each coil side.

α = Angular displacement between the adjacent slots

$$= \frac{180}{n}.$$

Obviously $\qquad k_d = \frac{E_r}{n \cdot E_c} \qquad\qquad (\because n = 2)$

Here $\qquad\qquad E_c = 2AP = 2OA \sin \alpha/2$

and $\qquad\qquad E_r = 2AD = 2OA \sin 2\alpha/2$

Substituting we have,

$$k_d = \frac{E_r}{2E_c} = \frac{2OA \sin 2\alpha/2}{2 \times 2OA \sin \alpha/2}$$

$$= \frac{\sin 2\alpha/2}{2 \sin \alpha/2}$$

$$\therefore \qquad \mathbf{Kd} = \frac{\sin n\dfrac{\alpha}{2}}{n \sin \dfrac{\alpha}{2}}$$

Coil span factor. It is also known as pitch factor. To improve the wave formation of the induced e.m.f. generally the short pitch or long pitch windings are used instead of full pitch windings. In case of full pitch winding the e.m.f. induced in both coil sides is equal and inphase, but when fractional pitch winding is used the e.m.f. induced in both the coil sides is not in phase. Hence the resultant can only be given by the vector sum which is always less than the arithmetic sum of the e.m.f. of both coil sides as shown in Fig. 17.8.

Now the ratio of the vector sum to the arithmetic sum of the induced e.m.f. per coil is known as the coil span factor or pitch factor. It is always less than unity so,

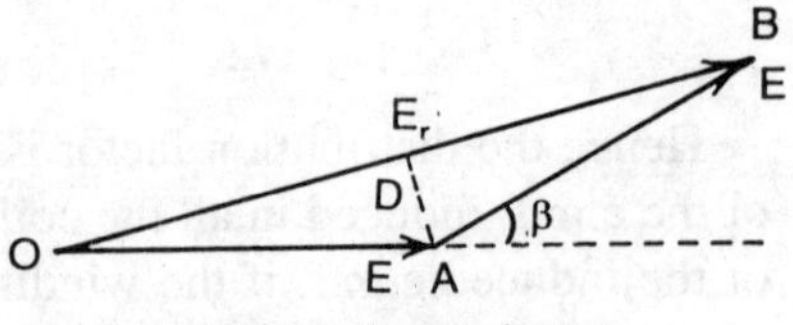

Fig. 17.8. Coil span factor

$$k_c = \frac{\text{Vector sum of the voltage in two sides of a coil}}{\text{Arithmetic sum of the voltages in two sides of a coil}}$$

Let the coil sides are shortened by one slot, the angle of displacement between e.m.f.'s of two coil sides.

$$\beta = \frac{180}{3n} \text{ for } 3\phi \text{ machines}$$

when $\quad\quad n = Number\ of\ slots/pole/phase$

$$k_c = \frac{Resultant\ e.m.f.}{2 \times e.m.f./coil\ side}$$

$$\therefore \quad\quad k_c = \frac{OB}{OA+AE} = \frac{OD+DB}{OA+AE}$$

$$= \frac{E\cos\beta/2 + E\cos\beta/2}{E+E} = \cos\frac{\beta}{2}$$

$$(\because \angle AOD = \angle ABD = \beta/2)$$

$$\therefore \quad\quad \boxed{k_c = \cos\frac{\beta}{2}}.$$

If the pitch is given in fractions then also the k_c can be calculated

$$k_c = \cos\frac{180(1-p)}{2}.$$

Example. *Calculate the distribution factor and coil span factor for a 3 phase winding with 3 slots per pole per phase and coil span 8 slots.*
Solution. Number of slots/pole/phase = 3
Number of slots/pole = No. of slots/pole/phase × No. of phases
$$= 3 \times 3 = 9.$$
Now the angular displacement between the slots is

$$\frac{180}{3n} = \frac{180}{9} = 20°$$

Now $\quad\quad k_d = \dfrac{\sin\dfrac{n\alpha}{2}}{n\sin\dfrac{\alpha}{2}}$

$$= \frac{\sin\dfrac{3 \times 20}{2}}{3\sin\dfrac{20}{2}} = \frac{\sin 30}{3\sin 10} = \mathbf{0.96} \quad \mathbf{Ans.}$$

Now coil span factor $= \cos\dfrac{\beta}{2}$

$$= \cos\dfrac{20}{2} = \cos 10 = 0.985 \quad \textbf{Ans.}$$

or $\qquad \cos\dfrac{180(1-p)}{2} = \cos\dfrac{180\left(1-\dfrac{8}{9}\right)}{2} \qquad \left(\because p = \dfrac{8}{9}\right)$

$$= \cos\dfrac{180(1)}{9 \times 2} = \cos 10° = 0.985 \quad \textbf{Ans.}$$

Q. What are the different types of alternators? Describe briefly with the diagram.

Ans. The alternators are classified as under

(a) **According to their field construction.** As the field excitation of the alternators is concerned, these are classified as under:

    (i) Stationary field type.

    (ii) Rotating field type.

The alternators in which the field winding is kept stationary and armature rotating, are called *stationary field type alternators*. The alternators in which the armature is stationary and field is rotating are known as *rotating field type* alternator, as shown in Fig. 17.9. Generally rotating field type alternators are commonly employed.

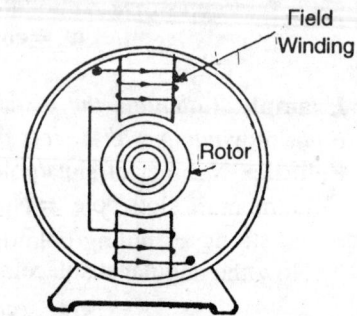

Fig. 17.9. Stationary field type alternator.

(b) **According to their pole constructions.** These are classified as under:

    (i) Salient pole type   (ii) Smooth cylindrical pole type.

The alternators in which poles are projected having more diameter and less axial length are known as *salient pole type* alternators. The other which has got less diameter and more axial length and are smooth, are called *smooth cylindrical pole type* alternators.

(c) **According tot he number of phases.** These are as follows:

    (i) Single phase.

    (ii) Three phase.

Generally poly phase alternators are employed.

(*d*) **According to their field excitation.** The alternators can be classified as follows:

(*i*) One, in which the field winding is excited from a separate source *i.e.* battery etc.

(*ii*) Other, in which the filed winding is excited from a D.C. generator mounted on the same shaft. The generator can be either a D.C. shunt generator or compound generator. A variable resistance is connected in series with the field winding to regulate the generated exciting voltage.

Q. Describe briefly the armature reaction in the alternators.

Ans. Whenever the alternator is supplying the load, the load current will flow through the conductors, producing the magnetic field, proportional to the load current.

The armature reaction is defined as "the effect of the magnetic field produced by the armature conductors on the distribution of magnetic flux under the main poles".

The load on the alternator is not only of one nature but it is of resistive, inductive or capacitive nature.

Purely resistive load. Whenever the load is of purely *resistive nature i.e.* having unity power factor, the armature coil will carry maximum current, in which maximum e.m.f. is induced. Then the magnetic field produced by the conductor will drift the main magnetic field as shown in Fig. 17.10. The net effect will be the lagging of armature flux wave behind the main flux by $\pi/2$. Thus the flux will be weak at leading pole tips and strong at trailing pole tips and the main flux is distorted.

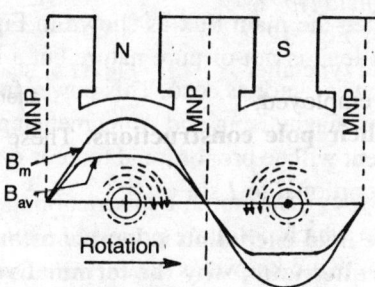

Fig. 17.10. Purely resistive load.

Purely inductive load. Whenever the inductive load is connected, it means the current lags behind the voltage by 90°. The armature coil will carry the maximum current which is 90°, behind the coil inducing maximum e.m.f. Otherwise also we can say that if the coil at magnetic axis inducing maximum e.m.f., the coil at *MNP* will carry maximum current, and the net result will be the demagnetising as shown in Fig. 17.11.

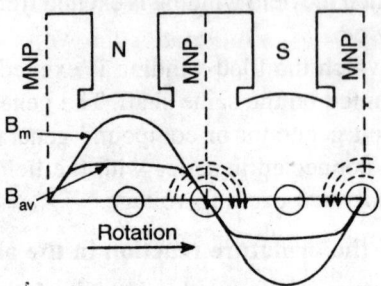

Fig. 17.11 Purely inductive load.

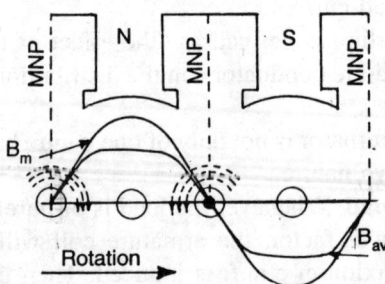

Fig. 17.12 Purely capacitive load.

Purely capacitive load. In this case the current will lead the voltage by 90°, so the coil ahead by 90° will have maximum current then the coil producing maximum voltage. Thus the current will establish the flux which will strengthened the main flux as shown in Fig. 17.12.

But in practice the load is not of pure nature but a mixed one. Let us have a load having power factor as $\cos\phi$. This power factor will cause the armature to have cross magnetising and demagnetising fluxes. The cross magnetising component will be proportional to $I\cos\phi$ and the demagnetising component proportional to $I\sin\phi$.

Q. (*a*) **Whenever the field excitation is kept constant and the load is increased on the alternator, why the terminal voltage decreases?**

(*b*) **Define the voltage regulation of the alternator.**

Ans. (*a*)Whenever the alternator is loaded keeping the field excitation constant, the terminal voltage decreases because of the following reasons:

(*i*) Voltage drop because of the armature resistance and current, *i.e.* IR_a.

(ii) Voltage drop because of the armature leakage reactance and current *i.e.* IX_c.

(iii) Less induced voltage because the armature reaction.

(b) The voltage regulation of the alternator is defined as, "the increase in terminal voltage when full load is thrown off, keeping the filed excitation and speed constant; shown in the percentage of full load voltage."

$$\text{So \% Regulation} = \frac{E_0 - V}{V} \times 100$$

where E_0 = e.m.f. induced in volts
V = Full load terminal voltage.

Whenever the load is thrown off, the difference in voltage is positive, the case will be of lagging power factor load. If the difference is negative the load will be leading power factor and the regulation is positive and negative respectively.

Now the voltage regulation with different load can be stated, as shown in Fig. 17.13 (a) and (b).

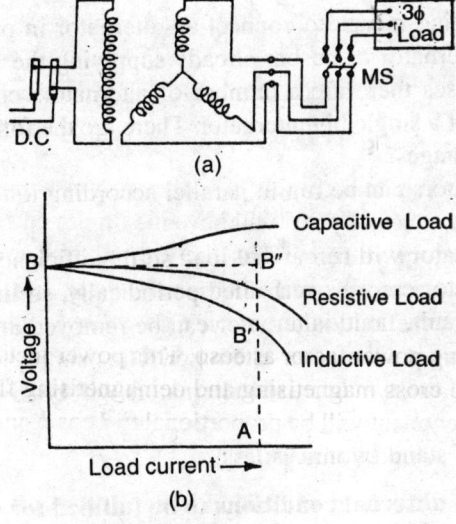

(a)

(b)

Fig. 17.13

% Regulation $= \dfrac{E_0 - V}{V} \times 100$

$= \dfrac{AB - AB'}{AB'} \times 100$ \hspace{1cm} (Resistive load)

$$= \frac{AB - AB''}{AB''} \times 100 \qquad \text{(inductive load)}$$

$$= \frac{AB - AB'''}{AB'''} \times 100 \qquad \text{(capacitive load)}$$

The e.m.f. at any particular power factor can be calculated as

$$E = \sqrt{V(\cos\phi + IR_e)^2 + (V \sin\phi \pm IX_s)^2}$$

where
V = terminal voltage

$\cos\phi$ = The power factor

I = The load current

R_e = The effective resistance of armature winding

X_s = Synchronous reactance

± sign are used positive for inductive load and negative for capacitive load.

Q. What do you mean by synchronizing and why it is required?

Ans. Synchronizing means to connect an alternator in parallel with the another alternator or busbar already supplying the load.

In power houses there are a number of alternators connected to the busbar in place of a single big alternator. There are the following requirements and advantages:

(*i*) The alternators can be run in parallel according to the requirement of the load.

(*ii*) Each alternator will run at full load so the efficiency is good.

(*iii*) The alternator can be overhauled periodically, so life is more.

(*iv*) In case of fault, faulty alternator can be removed and the load can be transferred on the other alternator. Thus supply will continue.

(*v*) The load on the synchronized alternators will be divided automatically.

(*vi*) The cost of stand by unit is less.

Q. What are the different conditions to be fulfilled for synchronizing the alternators?

Ans. There are the following conditions to be fulfilled before synchronizing the alternators:

(*i*) The frequency of the e.m.f. induced by the incoming machine must be equal to the busbar frequency or the frequency of already supplying alternator.

(*ii*) The voltage of the incoming machine should be equal to the busbar/ already supplying alternator voltage.

(*iii*) The phase of the incoming machine voltage must be the same to that of the phase of the busbar/machine already supplying the voltage.

The frequency of the induced voltage can be checked by means of the frequency-meter. The voltage can be checked by means of the voltmeter. The voltage can be adjusted by adjusting the field regulator of the incoming machine. The frequency can be regulated by regulating the speed of the prime mover. The synchroscope is used for checking the phase sequence etc. and this single instrument can be suitably used for frequency and phase sequence test for synchronizing the alternators.

Q. What are the different methods of synchronizing the alternators? Describe with neat sketch the dark and bright lamp method of synchronizing.

OR

What do you understand by parallel operation of two alternators? Describe in detail one method of putting 3 phase alternators in parallel. (*N.C.V.T. 1971, 73, 79, 81; A.I.C. 1971; Foreman Delhi 1950; App.. 1979*)

Ans. To connect the alternators into parallel is called synchronizing. This can be performed by these following methods:

(*i*) Dark lamp method.

(*ii*) Bright lamp method.

(*iii*) Dark and bright lamp method.

(*iv*) Synchroscope method.

Out of these four methods, the synchroscope method is most suitable. The dark and bright lamp method is next to synchroscope method.

Dark and bright lamp method. In this method three sets of the lamps are used for synchronizing. Two sets are connected across different phases at the terminals of main switch of the incoming machine. The third set is connected across the same phase of the busbar or the alternator already supplying and the incoming machine as shown in Fig. 17.14.

Whenever the proper synchronism is done, the lamps which are connected across the opposite phases will be bright and the lamp set which is connected across the same phase will be dark. As soon as this situation is obtained, the main switch of the incoming machine, which was open, is closed.

Alternators will operate satisfactorily in parallel and will share the additional incoming load in proportion to their ratings, if the machines have the same **terminal voltage, frequency and similar characteristics.**

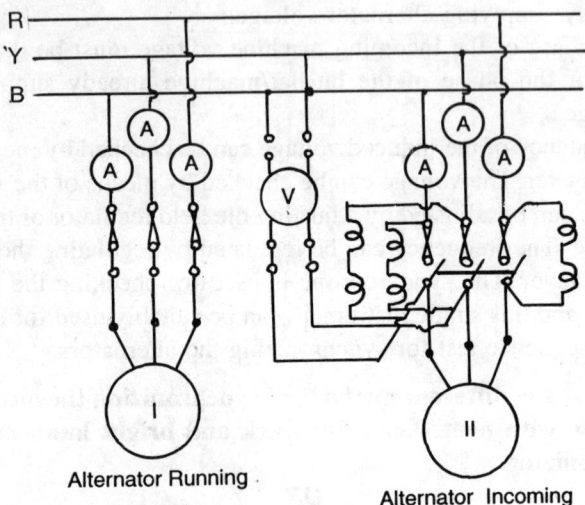

Alternator Running

Alternator Incoming

Fig. 17.14. Synchronizing of alternator dark and bright lamp method.

It should be noted that the change of excitation of an alternator operating in parallel with other machines, has no effect upon the effective load delivered by that machine. It only effects the *power factor* of the machine on the *kVA load* delivered by the machine. The *effective load* delivered by the machine can be regulated by regulating the *power input* to the prime mover, only increased input will make the machine to deliver more load.

Q. How the alternators are synchronized with the help of synchroscope?

Ans. The synchroscope is an instrument used for synchronizing the alternators, with the busbar or the alternator or incoming machine.

There are two parts—the stator and the rotor. The stator and rotor have individual windings. The stator winding can be connected across two phases of the busbar through a potential transformer, if needed otherwise directly. The rotor winding is connected across two phases of the incoming machine through a potential transformer or directly as the case may be *via* the synchronizing busbar.

As both the windings are done, these are connected in such a way, as one winding has a rheostat in series and other an inductance in series as shown in Fig. 17.15.

The resistances and inductances are connected in both the circuit of rotor and stator. Whenever supply is given two magnetic fields are set up,

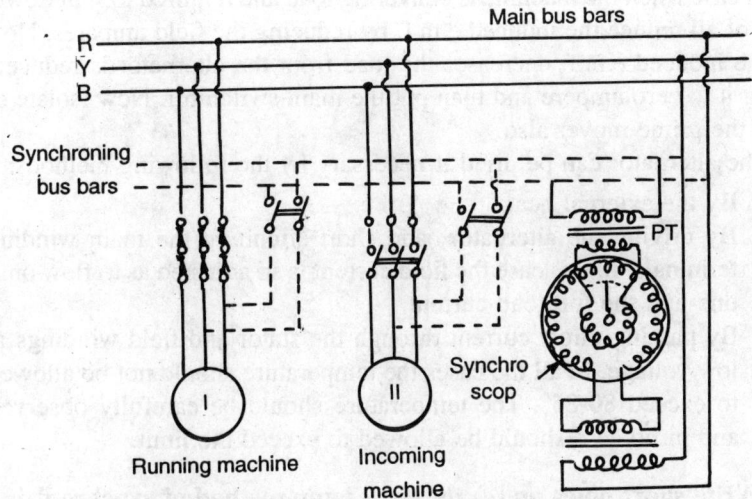

Fig. 17.15. Synchroscope method.

one of rotor circuit other of stator circuit. Now if the speed of rotor field is more than the stator field the rotor will indicate towards fast side. If the speed of rotor magnetic field is slow than stator field then the rotor will indicate towards slow side. Exact synchronism is obtained when the pointer of the rotor disc does not give any deflection. The synchronism is obtained by adjusting the speed of the prime mover of the incoming machine.

This method is very suitable for synchronizing the alternators.

Q. Whenever the alternators are synchronized, how will you control voltage and frequency? How the alternators can be started and shut down? How the alternators are dried out if needed?

Ans. Whenever a number of alternators are running in parallel,the busbar *voltage* can be regulated by regulating the excitation of all the machines. Changing the field excitation of a single machine, will nearly affect the reactive load delivered by that machine.

Similarly the supply *frequency* can be regulated by regulating the power input to all machines. Power input increasing to one machine only will cause it to take more load at the expense of the other machines with possibility of damage to itself.

Whenever the alternator is to be started, the circuit breaker or the main switch of incoming machine is opened. The voltage and frequency both are checked and when it is equal to the busbar voltage, the machine is synchronized.

In case when the machine is delivering load and required to shut down, first of all reduce the induced e.m.f. by reducing the field amperes. Now as the induced e.m.f. decreases the load from the alternator is reduced, bring it to zero ampere and than put the main switch off. Now isolate or stop the prime mover also.

The alternator can be dried if necessary by the following methods:

(*i*) By the external heat.

(*ii*) By driving the alternator and short circuiting the main winding terminals. In this case the field current is so adjusted as to flow only one-half the full load current.

(*iii*) By passing direct current through the stator and field windings at low voltage. In all the cases the temperature should not be allowed to exceed 80-85°. The temperature should be carefully observed and in no case should be allowed to exceed the limit.

Q. Write short notes on (*a*) the dark lamp method of synchronizing (*b*) Bright lamp method of synchronizing (*c*) Rating of alternators.

Ans. (*a*) **Dark lamp method.** In this case the set of lamps depending upon the voltage generated by the alternator are connected in such a way so that each set is across same phases. Whenever the lamps are having same voltage on both sides, the resultant voltage working across will be zero, causing the lamps not to illuminate as shown in Fig. 17.16.

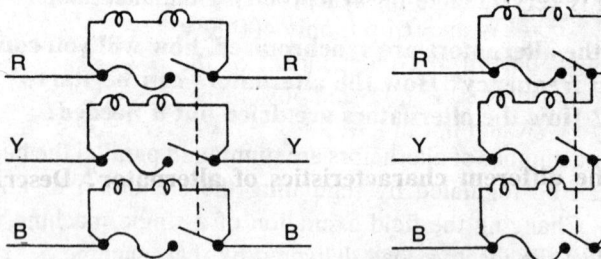

Fig. 17.16. Dark lamp method.   **Fig. 17.17.** Bright lamp method.

Here one thing should be very much clear that both voltage busbar and incoming voltage are inphase relative to external circuit and regarding local circuit these are directly inphase opposition. The voltage can be set equal by adjusting the field regulators. regarding frequency the speed of the prime mover is the main governing factor and can be adjusted by regulating the speed of incoming machine. In that case the lamps will be alternatively bright and dim. The flickering will be the difference of the frequencies of both the supplies. The darkness shows that two voltages are in exact opposition relative to local circuit. Synchronizing is done in the middle of the dark period. That is why it is called the dark lamp method.

There are certain drawbacks, the middle of the dark period for the synchronizing is difficult to judge and the little voltage may not glow the lamp hence that voltage may cause some difficulty for smooth synchronizing.

(b) **Bright lamp method.**Owing to the drawbacks of the dark lamp method, the bright lamp method is used. In this method the lamps or sets of lamps are connected across the opposite phases. The lamps will be bright, as these are connected across the phases of both supplies. The working voltage will also be twice the voltage of each supply, as shown in Fig. 17.17.

The synchronizing is done when the lamps are bright. Here also the correct position is difficult to judge.

(c) **Rating of alternators.** The alternators are rated in kVA, not in kW, because the power factor of the load is not always constant. The power factor depends upon the electrical load connected on the supply. This load will certainly not be having a constant power factor.

The machines are designed to carry the maximum current independent of the power factor of the load. A machine will carry less useful power at low power factor than the useful power at higher power factor.

A machine of 1000 kVA on full load will carry useful power at 0.8 p.f. 800 kW, at 0.6 p.f. 600 kW and 0.4 p.f. only 400 KW. The copper losses and iron losses will be full and independent of the power factor and efficiency will decrease as 80%, 60% and 40% respectively, so the alternators are rated in kVA and not in kW.

Q. What are the different characteristics of alternator? Describe briefly.

Ans. There are three main characteristics:

(a) Magnetisation or open circuit characteristic.

(b) Load characteristic.

(c) Short circuit characteristics.

(a) **Magnetisation or open circuit characteristic.** As we know that the e.m.f. induced in the alternator is very closely related with the field current, *i.e.* $E \propto I_f$ and can be stated as if the field current is increased the voltage induced is increased and *vice versa*. The o.c.c. is a curve obtained between the field current and the voltage induced. The connection diagram and curve are shown in Figs. 17.18 and 17.19.

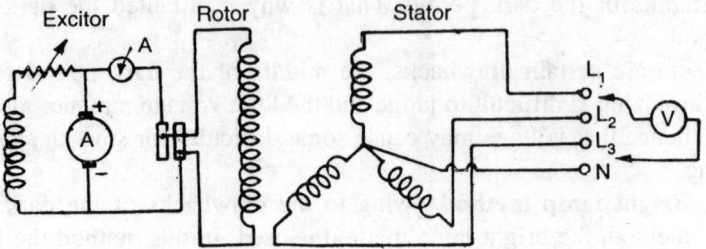

Fig. 17.18. Open circuit characteristics.

It is usual to plot the magnetisation characteristic of an alternator for both no load and full kVA conditions. The kVA rating is because of power factor, as we are concern the full load current not a particular power factor. Also it should be a point to note that the operating power factor of the generating station is decided by the nature of the load, it is not the characteristic of the alternators.

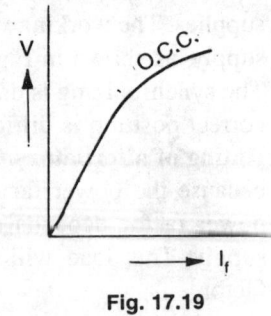

Fig. 17.19

(b) **Load characteristic.** This characteristic shows the relationship between the terminal voltage and the load current of different nature. In case of resistive load the terminal voltage decrease a little, in case of inductive load decrease much but in case of capacitive load, it increases with the increasing of load as shown in Fig. 17.13(b).

(c) **Short circuit characteristic.** it is a graph shown in between the exciting current and the short circuit current. The armature is not suddenly short circuited. It is obtained by short circuiting the stator winding through ammeter as shown in Fig. 17.20 and then plotting the stator current against the field current *i.e.* exciting current. The excitation is gradually increased from zero amperes to such a value so that the short circuited armature current may not exceed more than (120%) of the full load current capacity of the alternator.

The graph obtained is a straight line S.S.C., as shown in Fig. 17.21, the reason is that on short circuit an armature winding is a circuit of almost zero power factor (lagging) because of the very low armature winding resistance.

The above characteristics are very much useful for determining the voltage regulation. From O.C.C and S.C.C. the synchronous reactance can be determined. It is an essential factor for calculating the induced voltage in an alternator at any load.

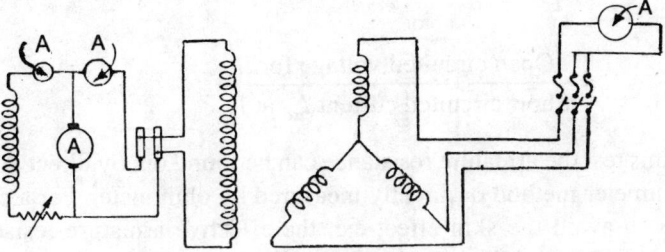

Fig. 17.20. Short circuit characteristic of alternator.

Q. What are the different methods of determining the voltage regulation of the alternators? Describe the synchronous impedance method of determining the regulation.

Ans. The voltage regulation can be determined by these following methods:

(*i*) **Direct load test.** In this test the alternator is directly put under the normal load conditions and all the E_0 and V etc. are measured and the regulation is calculated. It is only suitable for small alternators not for big or very big alternators.

Other methods are based on calculations and different values obtained by O.C.C. and S.C.C. etc.

(*ii*) **Synchronous impedance method** (*e.m.f. method*)

(*iii*) **Ampere turn method** (*m.m.f. method*)

(*iv*) **Zero power factor method** (*Potier method*)

Synchronous impedance method. In this method the O.C.C. and S.C.C. of the alternator are obtained. The O.C.C. is a curve obtained by plotting the terminal voltage against the field current. Similarly the S.C.C. is drawn by the given data. It is also plotted against the same field current. It is a straight line passing through the origin as shown in Fig. 17.21.

Now consider a field current I_f which on O.C.C. shows E voltage and on S.C.C. it shows a current I amp. which is the short circuiting current. In this it is assumed that the whole of the voltage E is being consumed to overcome the impedance and circulating current I in the armature, this impedance is called synchronous impedance Z_s,

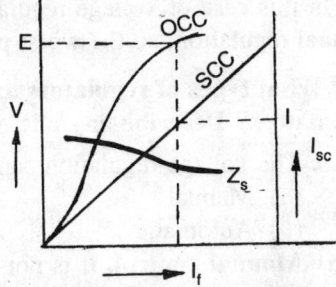

Fig. 17.21. Synchronous impedance

and $\qquad Z_s = \dfrac{E}{I}$

$$= \dfrac{\text{Open circuited voltage for } I_1}{\text{Short circuited current } I_{sc} \text{ at } I_{f_1}}$$

In this test the armature resistance can be found out by direct voltmeter and ammeter method or directly measured by ohm meter. For a.c. calculations to avoid the skin effect etc. the effective armature resistance is increased 60% and thus R_a is calculated or measured.

We get now the Z_S and R_a, the armature synchronous reactance X_s can be calculated as

$$X = \sqrt{Z_s^2 - R_a^2}.$$

The induced voltage can be calculated as shown in Fig. 17.22 here, the resistive, inductive and impedance drops are shown by AB, BC and AC respectively. So the induced voltage OC is equal to

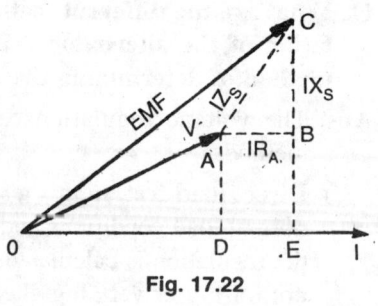

Fig. 17.22

$$OC = \sqrt{(OD + DE)^2} + \sqrt{(BE \pm CB)^2}$$

$$E = \sqrt{(V\cos\phi + I_a R_a)^2} + \sqrt{(V\sin\phi \pm I_a X_s)^2}$$

When V is known, E is calculated at any power factor whether leading (−ve sign) or lagging (+ sign) the regulation can be now calculated as

$$\% \text{ Regulation} = \dfrac{E - V}{V} \times 100.$$

In this case of voltage regulation, the value is always more than the actual regulation; so the worst possible case is achieved.

Q. What types of regulators are used for voltage regulation in alternators? Describe any one of them, with neat sketch.

Ans. The voltage regulation can be obtained by these two ways:

    (*a*) Manual

    (*b*) Automatic.

  (*a*) **Manual control.** It is not a convenient method of controlling the voltage, because for manual operation it is difficult. In this method

Alternators **417**

we control the field current flowing through the field winding of the alternator, it is difficult to control manually because the load on the alternator is not known at a particular time it depends upon the consumers and their needs.

(b) **Automatic control.** Whenever a sudden change of alternator load occurs, demanding a change in the field excitation to arrange the constant voltage under the new load conditions, the regulator must operate for the exciter field to give sufficient current to develop same voltage under all load conditions. When this arrangement is done automatically the system is called automatic control.

These regulators are of two types:

(i) Vibrating contact type.

(ii) Sliding contact type.

Brown Boveri Automatic regulator.

It is a sliding contact type voltage regulator.

Construction. The Brown Boveri regulator is mainly consists of a split phase induction motor, having two windings placed at 90° to each other. Due to the angular displacement between the current of both the windings a rotating magnetic field is produced and resulting a rotating torque which causes a rotation in the segments. The movement is controlled by providing the combination of two springs. There is a regulating resistance, permanently connected with the studs in pair as shown in Fig. 17.23. The contact between the segment and the resistance sector is made by making use of the spring which presses the segment outside. The

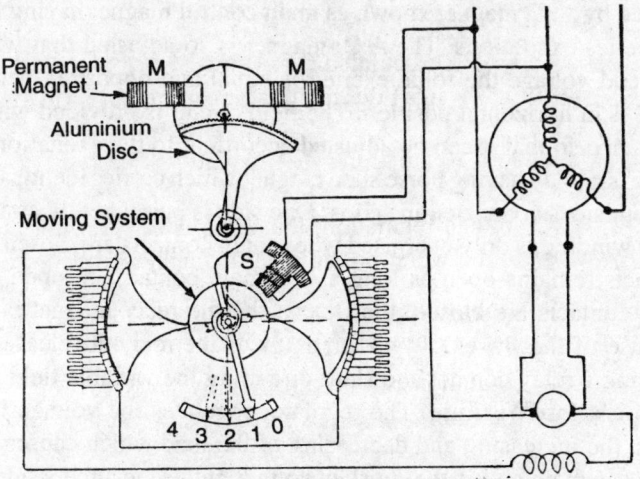

Fig. 17.23. Brown Bovery regulation for alternator.

sector moves in anticlockwise or clockwise direction depending upon the change in voltage from the set value. There is a pointer to indicate the position. A damping arrangement is also provided so that the oscillations in field circuit are damped and voltage is regulated.

Working. In case of normal working of the regulator, the whole system is balanced. In that case the sectors are so connected that some variable resistance is in the exciter field circuit and alternator producing the sufficiently required voltage. Now when the load increases, the terminal voltage will fall. The induction motor will now come into function and will develop a torque so that the segments move in such a way so that the resistance in the field circuit is reduced. This will result more field current and hence increasing the voltage induced by the alternator. The movement is balanced by the damping mechanism, provided on the top of the regulator as shown and the voltage will be maintained automatically. Secondly if the load is less and voltage increases in that case the torque developed will move the whole segment in such a way so that the connected field resistance is increased. It will result less field current and less voltage generation. Thus the voltage is controlled automatically.

Q. Describe briefly the construction and working of Tirril regulator.

Ans. It is also an automatic regulator, generally known as the vibrating type regulator.

Construction. Mainly there are two levers at the top, carrying main contacts. The lever No.1 is controlled by the exciter magnet which carries the exciting current proportional to the exciter voltage. The lever No. 2 is excited by A.C. magnet known as main control magnet having both shunt and series excitations. The A.C. magnet is so adjusted that with normal load and voltage the force exerted is equal and opposite, then the lever No. 2 is in horizontal position. The field circuit is provided with variable resistance so that it can be adjusted according to the excitation required. There is a relay having horse shoe magnet which carries identical windings on both sides connected in series. One side is permanently connected the other winding is only energised when main contacts are closed. The relay contacts remains open as long as the main contact are open. When the main contacts are closed, the flux in the the relay magnet is destroyed because of the fluxes of two windings of the relay. it releases the lever carrying a relay contact and short circuiting the variable field resistance.

Operation/Working. The regulator regulates the voltage in both the cases, the increasing and decreasing of the load which causes the disturbances in maintaining the constant voltage at the consumers side, or within the specified limit. Generally the increasing load, causes a voltage drop and difficulty also.

Whenever the load increases then A.C. control system is pulled because of the series excitation. Then the contacts of the two levers are closed and finally the variable field resistance is short circuited, increasing the exciting voltage hence the excitation of the alternator. The voltage of the alternator depends upon the field excitation and thus increases.

Now the excitation of the exciter control magnet increase due to increase in the exciter voltage pulling the lever No. 1 and opening the main contacts which results against he inserting of the resistance in the exciter circuit and reducing the voltage before it could be too much.

Uses. This method is not suitable for long lines.

Q. What will happen, when:

(*a*) **alternators are running with slightly unequal voltages?**

(*b*) **changing the field excitation of the alternator in synchronised condition?**

(*c*) **changing the fuel of alternators in synchronised condition?**

Ans. (*a*) **Unequal voltage.** Whenever both alternators are running with slightly unequal voltage, there will be a voltage working for local circuit. Let E_1 be the voltage of alternator no 1 and E_2 of alternator no. 2 and $E_1 > E_2$. Then E_r will be circulating a current 90° lagging E_1 and leading E_2 as shown in Fig 17.24 which will result the demagnetising effect for E_1 and magnetising effect for E_2 thus reducing the voltage of machine no. 1 and increasing of machine no. 2 both the effects are simultaneous and hence resulting a stable condition.

(*b*) **Change in field excitation.** Change in field excitation will result only the change in reactive component of the power, not the actual or active power. When both alternators are working in parallel, with the increasing of field excitation, the voltage will increase. Let us take m/c-*I* inducing E_1; whose field excitation is increased, and m/c-II with E_2 voltage. Then $E_1 \sim E_2 = E_r$

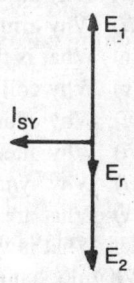

Fig. 17.23

will cause a current $= E_r/2Z_S$. This current will be added up vertically in m/c-I and subtracts from m/c-II resulting the different current I_1 and I_2 and naturally the power factors ϕ_1, will decrease and ϕ_2 will increase but the kW supplied will be same.

So only by changing the excitation the kVARs are changed or effected not the kW.

(*c*) **Change in steam** (to the prime mover). It will effect the kW supplied not he kVAR. Thus we can say that the load shared by the alternators is directly proportional to the driving power by the alternator.

REVIEW QUESTIONS

1. What do you understand by an alternator?
2. Describe the main constructional parts of an alternator with their functions.
3. What is meant by an exciter? Why it is necessary for an alternator?
4. What are the types of the alternators according to the construction of rotor?
5. Why stationary field type construction is not so popular in comparison of rotating field type of alternators?
6. Which type of rotor construction is used for Hydro Electric power station.
7. Explain the distribution factor and coil span factor.
8. What is the e.m.f. equation of the 3ϕ alternator?
9. Describe an expression for the e.m.f. equation of an alternator.
10. What do you understand by the term "armature reaction". Describe the armature reaction in case of alternator.
11. What is meant by voltage regulation? Describe any one method of determining the voltage regulation of an alternator.
12. What are the methods of synchronising the alternators? Why it is necessary to synchronise the alternators.
13. Describe any one most efficient method of synchronizing the alternators.
14. Describe any one type of automatic voltage regulator used for alternator.
15. Explain the followings:
    (*i*) Why stator cores of an alternator are laminated?
    (*ii*) Why an exciter is needed for an alternator?
    (*iii*) Why armature windings are chorded?
    (*iv*) What is the distribution factor?
    (*v*) Why coil span factor is used in the e.m.f. equation of the alternator?
    (*vi*) Why cylindrical rotors are used in turbo alternators?
    (*vii*) Why alternators are rated in kVA?
    (*viii*) Why synchronizing of alternators is needed?
    (*ix*) What are the causes of voltage drop in alternator?
    (*x*) What is voltage regulation, why it is necessary to study?
16. What do you understand by the parallel operation of two alternators? Describe in details one method of putting 3 phase alternators in parallel.

(N.C.V.T. 1971, 73, 79, 81; A/C. 1971;
Foreman Delhi 1980, April 1979)

NUMERICAL PROBLEMS

1. Calculate the number of poles of an alternator running at 1500 r.p.m. on 50 c/s main. **[Ans. 4 poles]**
2. A three phase 50 c/s alternator is running at 66.6 r.p.m. Calculate the number of poles. **[Ans. 90 poles]**

3. Determine the synchronous speed of the generating set having
   (a) 4 poles 25 c/s,   (b) 8 poles 60 c/s,
   (c) 25 c/s 8 poles and   (d) 6 poles 40 c/s.
                    [Ans. 750 r.p., 900 r.p.m., 375 r.p.m. and 800 r.p.m.]

4. Calculate the e.m.f. induced in a 3ϕ star connected alternator when it has 8 poles, driven at 750 r.p.m. There are 3 slots per pole per phase having 15 conductors per slot, the flux being 0.0536 Wb. (Assume $k_c = 1$)
                    [Ans. $V_L = 3565$ V]

5. An eight pole A.C. single phase alternator running at 750 r.p.m. has 144 slots 2/3 of which are wound with 8 conductors per slot. The flux per pole is 42 mWb sinusoidally distributed $k_d = 0.829$ and $k_c = 1$. Find the frequency and the e.m.f. generated.
                    [Ans. 50 c/s 2969V]

6. A 3ϕ 1000kVA 11 kV 50 c/s star connected alternator has armature resistance of 1.5 Ω per phase and synchronous reactance 25 Ω per phase, determine the percentage regulation for a load of 600 kW at 0.8 p.f. leading.
                    [Ans. −7.72%]

7. A two pole three phase star connected alternator has 24 stator slots and wound with 40 conductors per slot. The flux per pole is 0.063 Wb. If the speed is 3000 r.p.m. Calculate the frequency and generated e.m.f. ($k_d = 0.966$, $k_c = 1$).
                    [Ans. 50 c/s 3743 V]

8. Find the generated e.m.f. when an alternator is delivering 50A at 500 V at (a) unity power factor (b) 0.87 lagging power factor. The effective resistance and synchronous reactance being 0.2Ω and 2.2 Ω respectively.
                    [Ans. 522 V, 571 V]

9. The generated e.m.f. of 3ϕ alternator is 11480 V at the rated speed. If it supplies the 10 MW load at 11 kV, calculate the power factor of the load if the resistance per phase is 0.1 Ω and synchronous reactance is 0.66 Ω.
                    [Ans. 0.85]

10. A lighting load of 1500 kW at unity power factor and 2200 kW at 0.707 lagging power factor; is supplied by two alternators running in parallel. If one alternator is supplying 1700 kW at 0.924 lagging power factor, calculate the load shared and power factor of the second alternator.
                    [Ans. 2000 kW, 0.8 lagging]

18

Transformer

Q. What is a transformer? Write down the different causes of popularity of a transformer.

Ans. The transformer is an electromagnetic static device, which is used to transfer the electrical energy from one level to another level without changing the frequency. It can increase or decrease the voltage with the corresponding decrease and increase in current keeping the power of transformation as same. A transformer can change high voltage to low voltage and low voltage to high voltage but in both the cases the frequency remains unchanged.

Step up transformer. If a transformer changes low voltage to high voltage, it is known as the step up transformer .

Step down transformer. The transformer which changes the high voltage into low voltage is known as the step down transformer.

The energy transformation in the transformer is magnetically. Truely speaking there is no electrical and mechanical connections (in a two winding transformer) between the circuit energised by the supply and the circuit to which load is connected, but magnetically these are connected. The winding where supply is connected is known as primary winding. This winding may be H.V. winding, having more number of turns or L.V. winding having less number of turns. The winding to which load is connected is known as secondary winding and that too can be H.V. or L.V. depending upon the number of turns.

These are the following reasons for the popularity of the transformer:

(i) The transformer is a static device, there is no moving part in it, so it requires less maintenance.

(ii) It can be designed and can be made of any size and power and can be installed anywhere, even at the poles in open area.

(*iii*) Being stationary unit the losses are considerably less than the rotating machines, hence the efficiency is very good, say 97-98%.

(*iv*) It is a self regulating device. The primary winding draws current automatically if the load is increased on secondary winding.

(*v*) Being stationary and self regulating, no attendant is required.

(*vi*) The voltage can be stepped up or stepped down whenever needed according to the requirement and nature of load.

(*vii*) The construction is easily.

(*viii*) Less possibility of faults.

Q. State and illustrate the working principle of the transformer.

Ans. A transformer works on the principle of Faradays Laws of Electromagnetic Induction. As the transformer is a static device and no rotating part, so the type of induction is also the static induction that too the *mutual induction*. "Whenever a conductor links the changing flux an e.m.f. is induced in that conductor". This induced e.m.f. is proportional to the rate of change of flux and the number of turns.

Consider there are two coils '*A*' and '*B*' wound on the iron cores as shown in Fig. 18.1 (*a*). Coil *A*, is connected across the supply *i.e.* it is the primary winding and coil *B*, where load is connected is known as secondary winding. If the coil '*A*' is energised by a.c. mains the magnetic flux of alternating nature will be produced. This changing flux is also realised by the iron cores. The coil '*B*' which is wound on the cores will also link this changing flux, through iron cores which is a path of low reluctance. A result the e.m.f. is induced in coil '*B*'.

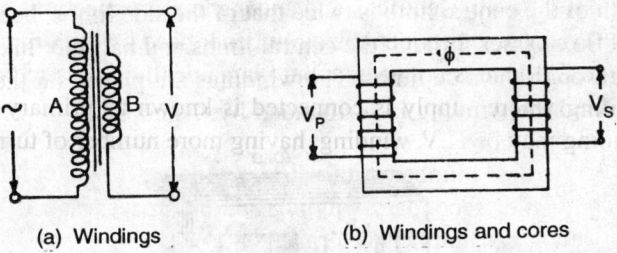

(a) Windings   (b) Windings and cores

Fig. 18.1

This phenomenon is known as mutual induction, the basic principle of transformer. Both windings are shown in Fig. 18.1(*b*), the winding which is fed and the winding in which e.m.f. is induced.

Q. Describe briefly the construction of the transformer.

Ans. The transformer has the following parts:

(*a*) Cores, (*b*) windings, (*c*) terminals and bushings, (*d*) tank (*e*) transformer oil, (*f*) conservator, (*g*) breather and (*h*) cooling system.

(*a*) **Cores.** The aim of the cores is to offer an easy path for the flux. The cores are made of laminations of silicone steel to minimise the eddy current and hysteresis losses. Each lamination is insulated from each other by means of varnish or impregnated paper or enamel to minimise the eddy currents and silicone is added to minimise the hysteresis losses. The thickness of the lamination varies from 0.35 to 0.5 mm. It must be kept in mind that the assembly of the lamination should be such that there is no air gap left.

According to the design of the cores these are:

(*i*) *Core type.* In this type of cores the magnetic flux has only one magnetic path. The two windings are placed on the two limbs. Sometimes the windings are placed on the single limb to avoid the magnetic leakage. In this case the cores are surrounded by the winding as shown in Fig. 18.2. In case of three phase core type construction there are three limbs of equal width.

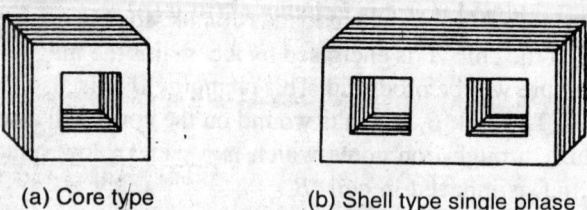

(a) Core type   (b) Shell type single phase

(*ii*) *Shell type.* In this type of cores there are two magnetic paths. The width of the central limb is twice that of the side limbs, because the total flux passes through the central limbs and half the flux is passing through the side limbs. The windings are placed on the central limb and the winding is surrounded by the cores.

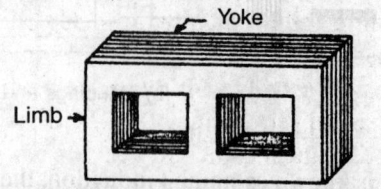

(c) Core type three phase care assembly

Fig. 18.2.

In case of shell type construction of three phase transformer, the windings are placed on the three limbs, but the outer two limbs are there to shell it as shown in Fig. 18.9. The size of each limb is same.

In five limb construction the cross-section in yoke can be reduces to 30% than the limbs but the cross-section of limbs is increased by a little percentage.

(*iii*) *Berry type.* In this case the distributed paths of the magnetic field are used. Generally the core type construction is used. One limb of all the cores passes through the center of the windings. The length of the core inside the coil is less than the length of limb outside the winding. These are used to obtain the variable voltage.

In case of the small transformers, simple rectangular limbs and coils are used. The coils can be either rectangular or circular. The rectangular has more number of weak points (bends) than circular one. To avoid the air gap in case of circular coil, the cruciform of cores setup is used. It demands the use of two or more size stempings. The different sizes depends upon the number of core stepings.

(*b*) **Winding.** The windings are placed on the cores. The winding is done with the insulated copper conductors. The winding which is connected to supply is the *primary winding* and to which load is connected is known as *secondary winding.*

According to the construction the windings are classified as under:

(*i*) *Cylindrical type winding.* The winding in which the length of the coil is equal to the length of the limb is known as cylindrical windings as shown in Fig. 18.3 (*a*). The coils of high tension or low tension or say primary and secondary are wound keeping L.T. near the cores.

(*ii*) *Sandwitch type winding.* The term is almost self explanatory in this winding the primary and secondary are placed one over the other alternately as shown in Fig. 18.3 (*b*).

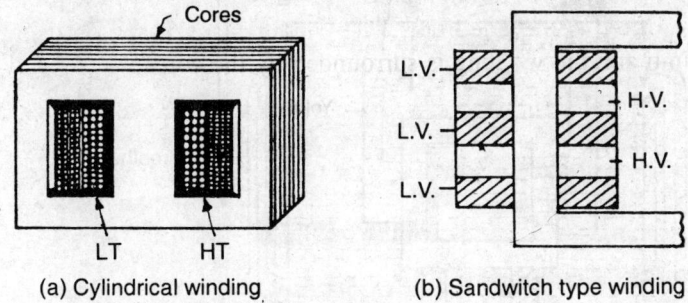

(a) Cylindrical winding   (b) Sandwitch type winding

Fig. 18.3

These are suitably employed for the shell type transformers. The reactance is minimised in this type of winding to an appreciable limit. These are commonly used with large H.T. transformers.

(c) **Terminals and bushings.** The leads of both windings are connected to the terminals so that the supply can be taken and connected to. In a small transformer of low voltage the terminals are fixed on the bakelite plate. For the transformers of high voltage ordinary porcelain insulators and bushings are used. Sometimes the oil filled bushings and condenser type bushings are also used for high voltage transformers.

(d) **Tank.** The tank is used to accumulate the windings, cores etc. in it. The tank of small transformer is made with iron sheets having the provision of ventilation and provision for connections to the load and supply.

The tank of large transformers are made with boiler plate sheet properly welded so that the oil may not come out of the tank. If the surface area of the tank is not sufficient to dissipate the heat produced, the surface area is increased by providing extra means. Elliptical tubes, round tubes are attached with the tank to increase the dissipating area. Sometimes these are also provided with the radiator facility.

(e) **Conservator.** It is a small tank mounted over the top of the main tank as shown in Fig. 18.4. It is sometimes called as the expansion tank. A level indicator is fitted to check the level and colour of the transformer oil. The main tank is completely filled with the transformer oil and the conservator partially. With the increase and decrease of current (load) the heat produced is also increased and decreased; as a result the expansion and contraction of the oil takes place; so the conservator is not filled completely to facilitate the expansion etc. An instrument is also attached to indicate the temperature of the oil.

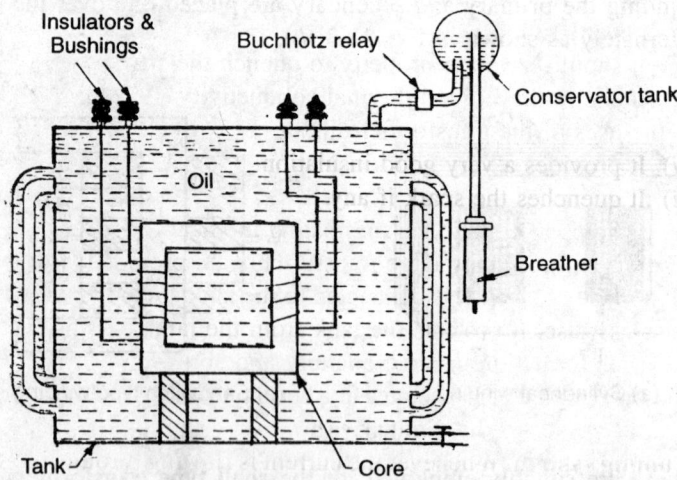

Fig. 18.4. Conservator and breather.

(*f*) **Breather.** It is a pipe attached with the conservator as shown in Fig. 18.4. With the expansion of the oil, when it is heated, the hot air comes out from the conservator through this pipe. The movement it is cooled, the air is being sucked in via the same pipe, this action is known as the breathing action of the transformer. Now the moisture is the main factor which effects the dielectric strength of the transformer oil. So a provision to absorb the moisture from the air to be breathed in is provided just on the mouth of the breathing pipe as shown in Fig. 18.4. The $CaCl_2$ or silicazal is kept there to absorb the moisture and the clean and moistureless air is allowed to come in, thus this air will not effect adversely on the dielectric strength of the transformer oil.

The color of the silicazal or $CaCl_2$ becomes blue while it absorbs moisture in sufficient quantity, now it should be replaced. A filter is also provided with the breather which prevents the dust etc. entering the transformer tank.

(*g*) **Transformer oil.** The mineral oil which is used in the tank of a transformer is called transformer oil. The main aim of this oil is to provide a better insulation and a better medium for heat conduction and finally dissipation. The transformer oil should have the following properties:

  (*i*) It should have good dielectric strength at least 35 kV/mm, or as per the maker's instructions specified for the particular transformer.
  (*ii*) It should be cleaned and free from the dust.
  (*iii*) It should not have any moisture.
  (*iv*) The flash point should not be less than 160°C and freezing point 10°C.
  (*v*) It should have the property to quench the spark if any.
  (*vi*) It should have good thermal conductivity.

In brief the uses of the transformer oil, are

  (*i*) It provides a very good insulation.
  (*ii*) It quenches the spark if any.
  (*iii*) It helps in conducting dissipating the heat.

(*h*) **Exhaust pipe.** Many large transformers are provided with the exhaust pipe made of steel. It is connected to the tank and covered from the top by a glass disc. It protects the tank from the large expansion of the accidental gas formation in large quantity and abrupt rise in pressure in case of short circuit etc. It operates at a certain pressure when the formation of gases comes to that limit.

(*i*) **Cooling system.** Whenever the current is flowing through the winding inherently some heat is produced; which should be dissipated; otherwise the heat may spoil the windings and insulation etc. There are a number of methods used for this purpose. Natural air cooling, oil immersed forced air, water and oil cooling and the air blast cooling. (These are discussed in the next question.)

Q. Name the different gauges and relay used in a common transformer.

Ans. Almost every transformer is provided with some measuring, safety, protective and informative devices. Some of them are:

(*a*) **Temperature gauge.** Whenever current is flowing in the windings the heat is produced and that heat will ultimately rise the temperature of the oil, cores and winding etc. So a temperature gauge is provided to indicate the temperature of the hot oil in the tank. Nowadays special devices one designed and fitted inside the tank so that sitting in control room we can measure the temperature of oil, stempings and winding separately.

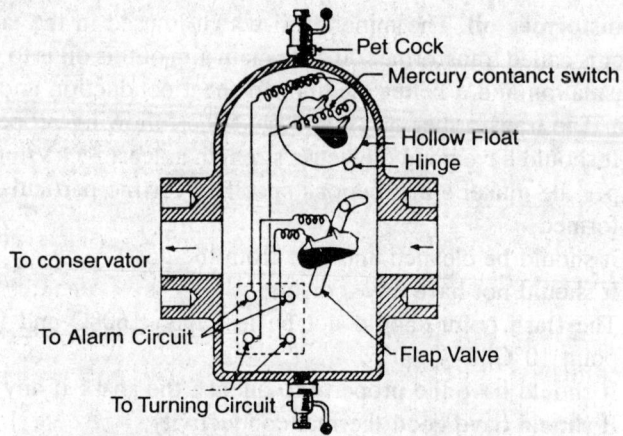

Fig. 18.5. Sectional view of a Buchholz relay.

(*b*) **Oil gauge.** The level of the oil should be maintained in the conservator so that the shortage of the oil in the tank may be sufficiently equipped. So a oil gauge is used. It is provided with
a alarm to inform the operator or control room. If the level of the oil goes down there the alarm will caution the operator and the operating circuit to equip the conservator.

(*c*) **Buchholz relay.** It is a safety relay provided with the oil immersed transformer. The main function of this relay is to indicate the presence of gas in case of some inner fault. or the level of transformer oil. It also isolate the transformer in case of serious fault or continuous fault. It is connected between the main and conservator tank.

Generally it is observed that the formation of gas in the transformer oil starts if the local temperature exceeds 150°C that is the main cause that the movementary faults does not operate the Buchholz relay. In case of serious fault the gas is generated and lower float is pushed towards the trip circuit contacts and finally cut off the input. In case of minor faults the gas is generated slowly which is not able to operate the lower float but the gas accumulates at the top and thus pressing the upper float which operates the alarm circuit to alarm the operator and control room as shown in Fig. 18.5.

Q. Write a short note on the cooling systems of the transformers.

Ans. Cooling systems. Whenever the transformer is loaded, the current in the windings inherently produces the heat ($\alpha I^2 Rt$). This heat, if not dissipated, will cause damage to the windings cores and other accessories in the tank. Sometimes nearly the tank surface is enough to dissipate the heat without extra provisions but it is only for small transformers. In large capacity transformer extra means are also employed.

These are the following methods of cooling:

(*i*) *Natural air cooling.* It is also known as dry type cooling. In this method the outer container dissipates the heat. This type of cooling is only suitable for the transformers up to 5-10 kVA, extra means of cooling are not provided, only the natural air is sufficient.

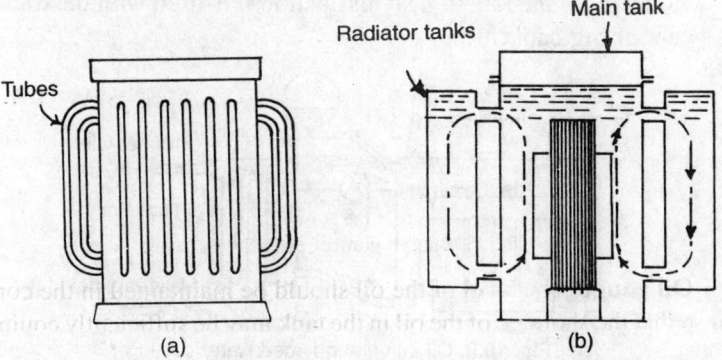

Fig. 18.6. Natural air cooling.

(*ii*) *Oil immersed natural air cooled.* Transformers up to 500 kVA ratings are designed for oil immersed natural air cooled. In this case the cores and winding are immersed in the tank of oil. The heat developed in the windings and cores, is conducted by the circulation of oil to the surface of the tank. This heat is then dissipated to the atmosphere. The area of heat dissipation is also increased by providing the corrugated sheets and other tubular constructions. The oil after getting hot, circulates and comes up, naturally the cold oil from the walls and tubes of the tank will occupy the empty space. Thus the circulation of oil will help in cooling the transformer.

The oil also provides the additional insulation for the windings. For small transformer the smooth surface is sufficient, for medium transformer the ribs are provided to increase the dissipating area and for large transformers the dissipating area is increased by providing the tubes and radiators.

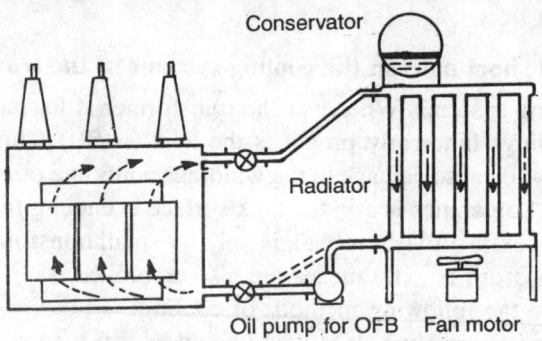

Fig. 18.7. Oil immersed forced air cooling.

(iii) *Oil immersed forced air cooling.* In this system the windings and cores are immersed in oil and cool air is forced on the tank surface to increase the rate of heat dissipation It is used with the transformers of big capacities.

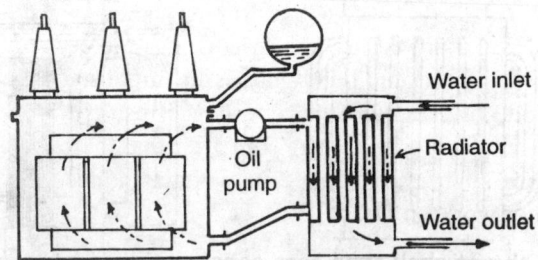

Fig. 18.8. Oil immersed forced water cooling.

(iv) *Oil immersed forced water cooling.* In this case also the cores, windings are immersed in oil. The cooled water is then circulated in the tank through the circular tubes provided. Thus the cold water brings out the heat from the oil and oil is cooled which in other words will control the temperature of oil, core and windings.

In this case the main drawback is that the leakage may cause the serious effect over the dielectric strength of the transformer oil. Obviously the dielectric strength is inversely proportional to the moisture.

(v) *Oil immersed forced oil cooled.* In this case the cores and windings are immersed in the oil. In this type instead of the cold water the cooled oil is forced through the circular tubes. Thus this oil brings out the heat and cooling is obtained. In this case the danger of water leakage is also eliminated because the oil is also having the same dielectric strength, which will not effect the transformer oil in case of any leakage if it is caused.

(vi) *Air blast cooling.* In this case air is passed through the cores and windings of the transformer. The air brings out the heat because the air velocity is high enough. Thus heat is dissipated. It is a dry transformer. In this case air should be dry and free from moisture and dust particles.

The transformers up to 10 MVA are oil immersed radiator type natural air cooled, and above 10 MVA generally employed with air blast cooling radiator type.

Q. Why a transformer should not be connected on D.C.?

Ans. The transformer is a static device which works on A.C. The windings of the transformer are having the resistance and inductance which actually contributes the impedance and controls the current drawn on A.C. In D.C. only the contribution of current is because of resistance and not impedance. The resistance is less this will cause the heavy current in the windings and will cause the burning out of the winding. Therefore the transformer should never be connected on D.C., secondly the current in D.C. has the constant magnitude resulting the constant flux. Only in the initial stage the current, causes the rate of change of current, hence results the rate of change of flux, and inducing voltage. But after that there is no rate of change of flux, so there will not be any voltage induced. So the transformer cannot work on D.C.

Q. Draw the 3ϕ shell type core construction.

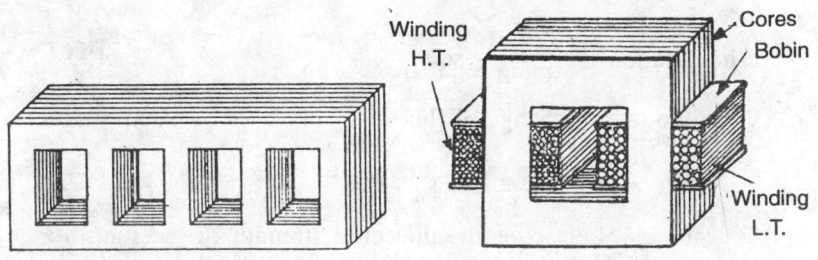

Fig. 18.9. 3ϕ-Shell type core construction.

Fig. 18.10. Winding and core 1ϕ transformer.

Q. Show how the windings are placed on the core type construction of single phase transformer.

Ans. Figure 18.10 shows the windings placed on the cores.

Q. Derive the e.m.f. equation of the transformer.

Ans. Consider a transformer having the following data:

ϕ_m = The maximum flux in Wb.

B_m = The maximum flux density in Wb/m^2.

A_i = Area of cross section of cores in m^2.

f = The frequency of reversal of flux in c/s.

E_1 = The voltage in primary winding in volts.

E_2 = The voltage in secondary winding in volts.

N_1 = The number of turns in primary winding.

N_2 = The number of turns in secondary winding.

The transformer obviously works on the principle of Faraday's Laws of Electromagnetic induction so the e.m.f. induced is given as

$$e = \frac{d\phi}{dt} \times N \text{ Volts}$$

The flux is directly proportional to the current which is sinusoidally distributed. So the wave of the flux as shown in Fig. 18.11 has the maximum flux in positive and negative directions both.

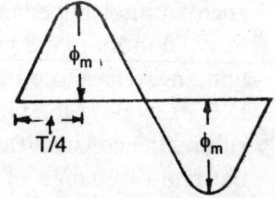

Fig. 18.11. Flux wave.

The actual change in flux (irrespective of the direction) is four times during one complete cycle *i.e.* from zero to $+\phi_m$, from $+\phi_m$ to zero, from zero to $-\phi_m$ and from $-\phi_m$, to zero. The time is $T/4$ for each interval.

The duration $= \dfrac{T}{4} = \dfrac{1}{4f}$ sec $\qquad \left(\because T = \dfrac{1}{f} \right)$

Now the rate of change of flux

$$= \frac{\phi_m}{\dfrac{1}{4f}} = 4\,\phi_m \cdot f, \qquad \left(\because e = \frac{d\phi}{dt} \times N \right)$$

So e.m.f. = Rate of change of flux × No. of conductors

$$= 4 \, \phi_m \times f \times N \text{ Volts}$$

because N = Number of conductors

$\therefore \qquad E = 4 \, \phi_m \times f \times N \text{ Volts.}$

It is the average induced voltage but the A.C. values are always mentioned in r.m.s. terms so it is to be multiplied by the form factor. So the r.m.s. value of the induced voltage

$$E = 4 \times 1.11 \times \phi_m \times f \times N \text{ Volts}$$

or

$$\boxed{\mathbf{E = 4.44 \, \phi_m \cdot f \cdot N \text{ Volts}}}$$

It is known as the fundamental e.m.f. equation of the transformer.

The voltage in primary winding is

$$E_1 = 4.44 \, \phi_m \cdot f \cdot N_1 \text{ Volts}$$

Similarly

$$E_2 = 4.44 \, \phi_m \cdot f \cdot N_2 \text{ Volts}$$

is the secondary induced voltage.

In case the flux density and area of cross section is to be taken into account.

$$\phi_m = B_m \cdot A$$

$\therefore$

$$E = 4.44 \, B_m \cdot A \cdot f \cdot N \text{ Volts.}$$

IInd method. The e.m.f. induced in primary winding

$$e_1 = -\frac{d\phi}{dt} \times N_1 \text{ Volts}$$

The value of flux $\phi = \phi_m \sin \omega t$ when the sinusoidal wave has the angular velocity as ω.

$\therefore$

$$e_1 = -\frac{d}{dt}(\phi_m \cdot \sin \omega t) \times N_1$$

$$= -\omega \cdot \phi_m \cdot N_1 \cos \omega t \text{ Volts.}$$

The primary induced voltage is V_1 is equal and opposite to this induced voltage *i.e.*

$$V_1 = -e_1 = \omega \cdot \phi_m \cdot N_1 \cos \omega t$$

$$= \omega \cdot \phi_m \cdot N_1 \sin(\omega t - \pi/2)$$

If

$$\sin(\omega t - \pi/2) = 1$$

then it will be the maximum induced voltage *i.e.*

$$V_{1\max} = \omega \cdot \phi_m \cdot N_1 \text{ Volts}$$

It is the maximum value but the A.C. is represented in r.m.s. value so this is to be divided by the peak factor *i.e.* 1.414

$$\therefore \qquad V_{1\,r.m.s} = \frac{\omega}{1.414}\phi_m \cdot N_1 = \frac{2\pi f}{1.414}\phi_m \cdot N_1$$

$$V_{1\,r.m.s} = 4.44\,\phi_m \cdot f \cdot N_1 \text{ Volts}$$

or

$$\boxed{V_1 = 4.44\,\phi_m \cdot f \cdot N_1 \text{ Volts}}$$

Q. Define the transformation ratios.

Ans. There are the following types of transformation ratios of a transformer:

    (*i*) Voltage transformation ratio,

    (*ii*) Turn transformation ratio.

    (*iii*) Current transformation ratio and

    (*iv*) Impedance transformation ratio.

Let the primary winding has got N_1 turns and secondary winding N_2 turns then the voltage induced can be given as

$$E_1 = 4.44\,\phi_m \cdot f \cdot N_1 \text{ Volts}$$

and $\qquad\qquad E_2 = 4.44\,\phi_m \cdot f \cdot N_2$ Volts

Now the ratio of secondary voltage to primary voltage is defined as the *voltage transformer ratio i.e.*

$$= \frac{V_2}{V_1}$$

Now if both the voltages are represented in the form of transformer e.m.f. then,

$$\frac{V_2}{V_1} = \frac{4.44\,\phi_m \cdot f \cdot N_2}{4.44\,\phi_m \cdot f \cdot N_1}$$

$$\therefore \qquad\qquad \frac{V_2}{V_1} = \frac{N_2}{N_1}.$$

The ratio of secondary number of turns to primary number of turns is defined as the *turn transformation ratio.*

As in the case of an ideal transformer the power input is equal to the power output so,

$$V_1 \cdot I_1 = V_2 \cdot I_2$$

or
$$\frac{V_2}{V_1} = \frac{I_1}{I_2}.$$

hence the ratio of primary winding current to secondary winding current is known as the *current transformation ratio*. Similarly if the ratio of secondary winding impedance and primary winding impedances are taken into account the ratio is said *the impedance transformation ratio*.

Let Z_1 = the primary winding impedance

$$= \frac{V_1}{I_1}$$

and Z_2 = sec. winding impedance $= \dfrac{V_2}{I_2}$

so
$$\frac{Z_2}{Z_1} = \frac{V_2}{I_2} \times \frac{I_1}{V_1}$$

$$= \frac{V_2}{V_1} \times \frac{I_1}{I_2}$$

$$= \left(\frac{N_2}{N_1} \times \frac{N_2}{N_1} \right) \qquad \left(\because \rho = \frac{V_2}{V_1} = \frac{I_1}{I_2} = \frac{N_2}{N_1} \right)$$

$$\therefore \qquad \frac{Z_2}{Z_1} = \left(\frac{N_2}{N_1} \right)^2 = \rho^2 .$$

So the *impedance transformation ratio* is the ratio of secondary impedance to primary winding impedance. It is equal to the square of voltage or turn or current transformation ratio.

Example. *Calculate the voltage induced per turn in a single phase transformer operating at 50 c/s supply having 0.08 Wb flux.*

Solution. $E_t = 4.44 \cdot \phi_{max} \cdot f$ Volts

$$= 4.44 \cdot 0.08 \cdot 50$$

$$= 17.8 \text{ V. Ans.}$$

Example. *Calculate the voltage induced in a transformer secondary, if it is operating at 50 c/s mains having flux density 0.01 Wb/m^2, area of cross section 0.67 m^2 and the number of turn in secondary are 220.*

Solution. The e.m.f. induced in a transformer is

$$E = 4.44 \, \phi_{max} \cdot f \cdot N \text{ Volts}$$

and
$$E = 4.44 \cdot B_{max} \cdot A \cdot f \cdot N \text{ Volts}$$

$$(\because = B \cdot A)$$

Substituting in this equation,
$$E = 4.44 \times 0.01 \times 0.67 \times 50 \times 220$$
$$= 327.3 \text{ V. Ans.}$$

Example. *A single phase 6600/250 V 50 c/s transformer has 100 turns in low voltage side, calculate the value of flux in the cores.*

Solution. $E = 4.44 \cdot \phi_{max} \cdot f \cdot N$ Volts

and
$$\phi_{max} = \frac{E}{4.44 \times f \times N} \text{ Wb}$$

$$= \frac{250}{4.44 \times 50 \times 100}$$

$$= 0.01126 \text{ Wb. Ans.}$$

Example. *A 11000/250 V 50 c/s single phase transformer has 25 A as the primary winding current. Determine (a) kVA rating (b) secondary winding current.*

Solution. The primary voltage = 11000 V
Secondary voltage = 250 V
Primary current = 25 A.

Now the voltage and current transformation ratios are

$$\frac{V_2}{V_1} = \frac{I_1}{I_2}$$

$\therefore$
$$I_2 = I_1 \times \frac{V_1}{V_2}$$

$$= 25 \times \frac{11000}{250} = 1100 \text{ A. Ans.}$$

The *kVA* rating
$$= \frac{\text{voltage} \times \text{current}}{1000}$$

$$= \frac{11000 \times 25}{1000} = 275 \text{ kVA Ans.}$$

Example. *A single phase core type transformer has 300 as primary turns the net area of cross section is 80 sq cm. The length of magnetic path is 0.9 metre. The primary winding is connected across 500 V 50 c/s. Calcu-*

late the maximum flux density in the cores. Assuming the relative per-meability 2500 at this flux density, also find out the magnetising current.

Solution. The e.m.f. equation of the transformer

$$E = 4.44 \, B_{max} \cdot A \cdot f \cdot N \text{ Volts}$$

and

$$B_{max} = \frac{E}{4.44 \times A_i \times f \times N} \text{ Wb/m}^2$$

$$= \frac{500}{4.44 \times 80 \times 10^{-4} \times 50 \times 300}$$

$$= \textbf{0.938 Wb/m}^2. \textbf{ Ans.}$$

The magnetising current

$$I_m \cdot N = \frac{B_m \cdot I}{\mu_0 \, \mu_r}$$

Here as given, $N = 300$, $l = 0.9$ m, $B_m = 0.938$ Wb/m^2, $\mu_0 = 4\pi \times 10^{-7}$ and $\mu_r = 2500$.

So

$$I_{mag} = \frac{B_m \cdot l}{\sqrt{2} \cdot N \cdot \mu_0 \, \mu_r}$$

$$= \frac{0.938 \times 0.9}{\sqrt{2} \times 300 \times 4\pi \times 10^{-7} \times 2500}$$

$$= \textbf{6.633 A. Ans.}$$

Example. *A 6600/500 V 50 c/s single phase transformer has a maximum flux density of 1.35 Wb/m^2 in its core. It has net cross sectional area of the core as 250 cm^2. Calculate the number of turns in primary and secondary windings of the transformer.*

Solution. The e.m.f. equation

$$= 4.44 \, B_{max} \cdot A \cdot f \cdot N \text{ Volts}$$

Here $B_{max} = 1.35$ Wb/m^2,

Primary voltage = 6600 V

Secondary voltage = 500 V

Frequency 50 c/s and area of cores

$$= 250 \text{ cm}^2.$$

The number of turns in primary

$$= \frac{E_P}{4.44 \times B_{max} \times A \times f}$$

$$= \frac{6600}{4.44 \times 1.35 \times 250 \times 10^{-4} \times 50}$$

$$= 880.9 \simeq 881 \text{ turns}$$

Similarly secondary turns

$$= \frac{V_2}{V_1} \times N_1$$

$$= \frac{500}{6600} \times 881$$

$$= 66.74 \simeq 67 \text{ turns}$$

The current is 5 A in primary side so the kVA rating

$$= \frac{6600 \times 5}{1000} = 33 \text{ kVA.} \quad \textbf{Ans.}$$

Q. Explain with the help of vector diagram when a transformer is said to be on No load. What are the different components of no load current of the transformer?

Ans. Whenever load is not connected on the secondary winding of the transformer and the primary winding is energised, the condition is said to be *no load condition* of the transformer. The primary winding having high inductance and low resistance draws a small current lagging behind the voltage by nearly 90° electrical. The current is said no load current, I_0, and the power factor resulted is said the no load power factor, $\cos \phi_0$. The no load current can be resolved into two components:

(a) Magnetising current (I_μ).

(b) Wattful current (I_ω).

(a) **Magnetising current (I_μ).** The magnetising component is lagging behind the voltage V_p by 90°. It is responsible for the magnetic flux and both I_μ and the flux are in-phase. It is also called the reactive component of the load current. It is equal to $I_0 \sin \phi_0$ and is the wattless component of no load current.

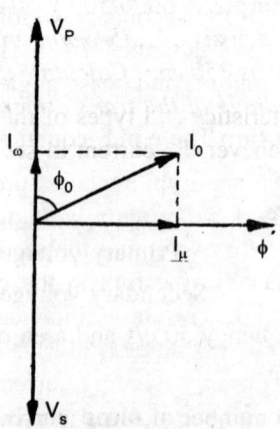

Fig. 18.12. No load condition of the transformer.

(*b*) **Wattful component (I_ω).** It is known as the working component or active component of the no load current. It is in phase with the voltage applied and is $I_0 \cos \phi_0$. The component is used to feed the necessary real power to supply the hysteresis and eddy current losses in the iron cores. Here as shown in Fig. 18.12 $\cos \phi_0$ is the power factor at no load and ϕ_0 sometimes is called the hysteresis angle of advances.

The magnetising component produces a changing flux, in the iron cores of the transformer, which is actually responsible for producing the voltage in primary winding, the self induced primary voltage and in secondary winding, the mutually and self induced voltages.

Q. What are the different information obtained, from the "no load" condition of the transformer?

Ans. These following informations can be obtained from the 'no load' conditions of the transformer:

   (*i*) the voltage transformation ratio,

  (*ii*) the no load current I_0,

 (*iii*) the hysteresis angle of advance ϕ_0.

 (*iv*) X_0 and R_0 can be calculated for the equivalent circuit of the transformer.

Q. What do you understand from the "on load" condition of the transformer, explain clearly, with the different type of load, the vector representation of the transformer on load. Also prove that the transformer is a self regulating device.

Ans. The primary winding is connected across the main and load is connected on the secondary winding, the condition is called the 'on load' condition of the transformer. Thus the I_s current in secondary is flowing. The magnitude and phase angle of the I_s, is different depending upon the characteristics and types of the load.

Whenever the current in secondary winding is flowing, the magnetic flux ϕ_s is set up in the cores proportional to the secondary current. Thus ϕ_s opposes the primary winding flux as a result the net flux in the cores reduces. To overcomes the secondary winding flux, the primary winding will maintain the flux in the cores by drawing more current from the supply. This current will induce the flux which will balance the ϕ_s. Thus the flux in the cores is always constant in every condition of load.

The primary winding current is the vector sum of primary balancing current and the no load current I_0, of the transformer. As the load increases on the secondary winding consequently the current increases in the primary winding. Thus with the increase of current on the secondary winding, the primary winding current will automatically increase, hence it is said that the transformer is a *self regulating device.*

The angle between the voltage and current of the primary winding depends upon the secondary current and power factor. The load on the transformer can be of the following types:

(*a*) **Purely resistive load.** If there is a purely resistive load on the transformer, the current and voltage on secondary will be in phase as shown in Fig.1 18.13 (*a*). The antiphase or reversed secondary current I_s' is drawn on the primary and now there are two currents the I_0 and I_s' the vector sum will be the primary winding current, I_P as shown in Fig. 18.13.

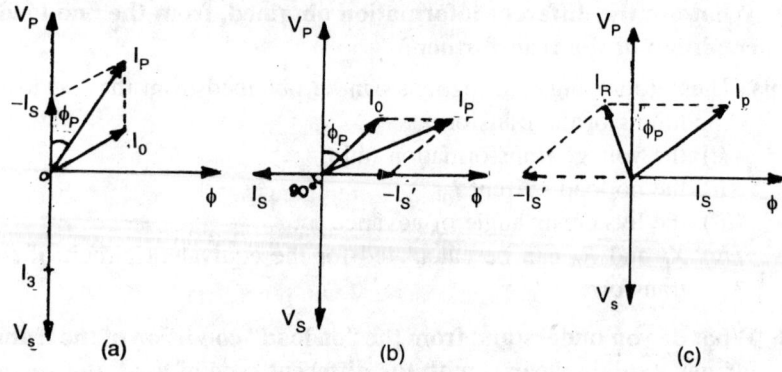

Fig. 18.13. (*a*) Purely resistive load, (*b*) Purely inductive load, (*c*) Purely capacitive load.

(*b*) **Purely inductive load.** In case of purely inductive load the current lags behind the voltage by 90° elect. Now the resultant current will be the vector sum of I_0 and I_2' as shown in Fig. 18.13(*b*).

(*c*) **Purely capacitive load.** In that case when the current is leading the voltage by 90° elect. the primary winding current I_P is drawn as the vector sum of I_0 and I_2', as in Fig. 18.13 (*c*).

In actual condition the load on the secondary winding is of the mixed nature. Generally the power factor is of lagging nature, in that condition the primary winding current I_P will be as shown in Fig. 18.14. It should be clearly under-

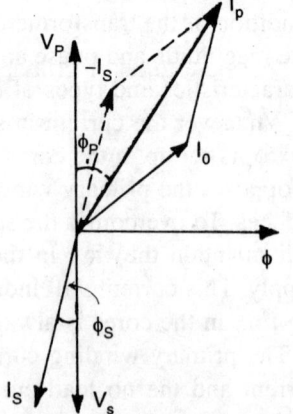

Fig. 18.14. Mixed load on transformer.

stood that the power factor on primary side of the transformer and secondary side is not identical, but for the simplicity of calculations it is taken as same.

Q. What do you understand by the primary and secondary leakage fluxes? Also state the equivalent resistance and equivalent impedance of the transformer.

Ans. In case of an ideal transformer it is assumed that the transformer is not having the resistance and reactance, but in an actual transformer both are present. The primary winding is having some resistance and reactance, similarly the secondary winding is also having both. Whenever current is flowing in any one of the winding say primary winding, the flux is produced. This flux is linking with secondary winding and producing some voltage, but a part of this flux is not linking with the secondary at all and only links the primary winding alone. So the flux which links the primary alone and not with secondary is known as primary leakage flux. Similarly, the secondary leakage flux is also defined as that flux which links the secondary winding but not the primary winding. The effect of these leakage fluxes is to induce an e.m.f. which opposes the cause responsible for the production of this emf that is the current in primary winding and in secondary windings.

Equivalent resistance and impedance. Let the primary winding and secondary winding of a transformer be having R_1 and R_2 resistances respectively.

The copper losses W_c are equal to the primary winding copper losses and secondary winding copper losses.

So $$W_c = I_1^2 R_1 + I_2^2 R_2 \qquad \qquad ...(i)$$

where I_1 and I_2 are the currents in primary and secondary windings. The ampere turns of primary and secondary windings are equal.

i.e. $$I_1 \cdot N_1 = I_2 \cdot N_2$$

or $$I_1 = \left(\frac{N_2}{N_1}\right) I_2$$

Substituting this value in (*i*)

$$W_C = \left[\left(\frac{N_2}{N_1}\right) I_2\right]^2 \times R_1 + I_2^2 R_2$$

$$= I_2^2 \left[\left(\frac{N_2}{N_1}\right)^2 \times R_1 + R_2\right]$$

Let the total resistance referred to secondary winding be R_{02} Ω. which in other words can be said that there is no resistance in primary winding but to compensate that the secondary winding resistance has been increased from R_2 to R_{02} ohm. So the copper losses, then

$$I_2^2 \times R_{02} = I_2^2 \left[\left(\frac{N_2}{N_1} \right)^2 \times R_1 + R_2 \right]$$

or

$$R_{02} = R_1 \times \left(\frac{N_2}{N_1} \right)^2 + R_2 \ \Omega.$$

This value R_{02} is called the equivalent resistance of the transformer referred to secondary winding.

Similarly the equivalent resistance of the transformer referred to primary winding is,

$$R_{01} = R_1 + R_2 \left(\frac{N_1}{N_2} \right)^2 \Omega.$$

Both the windings of the transformer also possess the reactances, because of the primary and secondary leakage fluxes. In the same way the whole of the reactance is confined to either primary or secondary winding.

Let X_1 = reactance of primary winding,

X_2 = reactance of secondary winding,

X_{01} = equivalent reactance referred to primary winding,

X_{02} = equivalent reactance referred to secondary winding.

then

$$X_{01} = X_1 + X_2 \left(\frac{N_1}{N_2} \right)^2 \Omega.$$

$$X_{02} = X_2 + X_1 \left(\frac{N_2}{N_1} \right)^2 \Omega.$$

Now when the equivalent values of resistance and reactances are known, the equivalent impedance can be calculated as,

$$Z_{01} = \sqrt{R_{01}^2 + X_{01}^2} \ \Omega.$$

and

$$Z_{02} = \sqrt{R_{02}^2 + X_{02}^2} \ \Omega.$$

Example. *A 55 kVA 4400/220 V 50 c/s transformer has a primary resistance and reactance of 3.85 ohms and 5.4 ohms respectively. The secondary resistance and reactance are 0.0085 ohm and 0.018 ohms respectively. Determine:*

(a) *Equivalent resistance referred to primary,*
(b) *Equivalent resistance referred to secondary,*
(c) *Equivalent resistance referred to both primary and secondary,*
(d) *Equivalent impedance referred to both primary and secondary.*

Solution. The transformer has 55 kVA capacity and voltages are 4400/220 V.

Now current $\qquad I_1$ = the primary winding current

$$I_1 = \frac{55 \times 1000}{4400}$$

$$I_1 = 12.5 \text{ A}$$

Secondary full load current $= \dfrac{55 \times 1000}{220}$

$$= 250 \text{ A}.$$

The transformation ratio $\dfrac{V_2}{V_1} = \dfrac{220}{4400} = \dfrac{1}{20} = \dfrac{N_2}{N_1}$

(a) The equivalent resistance referred to primary winding

$$R_{01} = R_1 + R_2 \left(\frac{N_1}{N_2} \right)^2$$

$$= 3.85 + 0.0085 \, (20)^2$$

$$= 3.85 + 3.4 = \textbf{7.52 } \boldsymbol{\Omega}.$$

(b) The equivalent resistance referred to secondary winding

$$R_{02} = R_2 + R_1 \left(\frac{N_2}{N_1} \right)^2$$

$$= 0.0085 + 3.85 \times \left(\frac{1}{20} \right)^2$$

$$= \textbf{0.01813 } \boldsymbol{\Omega}.$$

(c) Equivalent reactance referred to primary winding

$$X_{01} = X_1 + X_2\left(\frac{N_1}{N_2}\right)^2$$

$$= 5.4 + 0.018\ (20)^2 = \textbf{12.6 } \Omega.$$

It is the equivalent reactance referred to secondary winding, so

$$X_{02} = X_2 + X_1\left(\frac{N_2}{N_1}\right)^2$$

$$X_{02} = 0.018 + 5.4\left(\frac{1}{20}\right)^2$$

$$= \textbf{0.0315 } \Omega.$$

(*d*) The equivalent impedance of the transformer referred to primary winding.

$$Z_{01} = \sqrt{R_{01}^2 + X_{01}^2}$$

$$= \sqrt{7.25^2 + 12.6^2} = \textbf{14.54 } \Omega$$

$$Z_{02} = \sqrt{R_{02}^2 + X_{02}^2}$$

$$= \sqrt{0.01813^2 + 0.0315^2}$$

$$= \textbf{0.03634}\Omega$$

7.25, 0.01813, 12.6, 0.0315, 14.54 and 0.03634 Ω. Ans.

Q. Draw the equivalent circuit of the transformer.

Ans. In an actual transformer both resistance and reactances are there. When it is connected across the line, it draws the no load current I_0, which has two components the $I\omega$, the wattfull component $I_0 \cos \phi_0$ because of the resistive component say R_0 and is inphase with the voltage, the other I_μ, the magnetising component $I_0 \sin \phi_0$ because

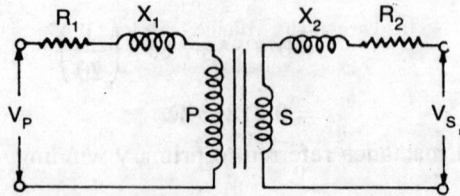

Fig. 18.15. Equivalent circuit of transformer.

of X_0, and lagging behind the voltage by 90°. Now the equivalent circuit can be illustrated in the following ways:

(*i*) With the actual values of R_1, X_1 and R_2, X_2 in both the windings Fig. 18.15.

(*ii*) With the total values in secondary winding the not in primary winding Fig. 18.16.

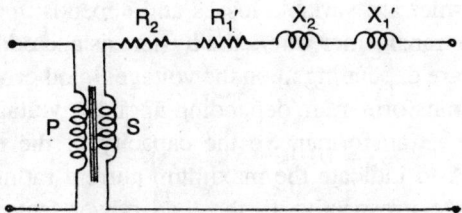

Fig. 18.16. Simplified equivalent circuit (secondary side).

(*iii*) With the total values in primary winding the not in secondary winding Fig. 18.17.

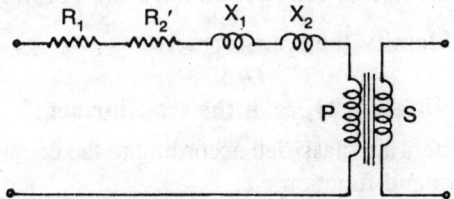

Fig. 18.17. Simplified equivalent circuit (primary side).

Q. Explain the voltage regulation of the transformer.

Ans. Whenever the secondary of the transformer is loaded, the secondary voltage falls down with the increasing of load. The drop in voltage is because of the impedance drop in the secondary winding.

Now the voltage at the consumers end is to be maintained constant; for that the regulation is needed.

The voltage regulation of a transformed is defined as the change in secondary voltage from no load to full load to the secondary no load voltage. Generally it is expressed in percentage.

So % regulation = $\dfrac{\text{(No load secondary voltage - full load secondary voltage)}}{\text{No load secondary voltage}} \times 100$

i.e. % regulation = $\dfrac{V_0 - V_2}{V_0} \times 100$

The voltage regulation can also be calculated at any desired load. The regulation varies with the power factor of the load. In normal case the full load voltage regulation of a normal transformer is about 3 to 5%.

Q. State the unit in which the capacity of a transformer is expressed; and why?

Ans. The rating of the transformer is expressed in kVA. The copper losses of a transformer are variable losses and depends upon the current rating of the transformer where the hysteresis and eddy current (the iron) losses are depending upon the voltage. In other words the total losses of a transformer are depending upon the voltage and current ratings of the transformer. So the capacity of the transformer is rated in kVA to indicate the maximum current rating on that particular voltage irrespective to the load power factor. In A.C. the power is the product of voltage current and power factor. The power factor depends upon the load. So the $V \cdot I \cos\phi$ will not give any constant value and will depend upon the power factor. That is the reason the transformer are rated in kVA and not in kW.

Q. How will you classify the transformers?

Or

What are the different types of the transformers?

Ans. The transformers are classified according to the construction, cores, magnetic path and function etc.

(a) **According to the magnetic paths.** These are as follows:
   (i) *Core type.* Having only one magnetic path.
   (ii) *Shell type.* Having two magnetic paths.
   (iii) *Berry type.* Having distributed magnetic paths.

(b) **According to the transformation ratio.** These are classified as follows:
   (i) *Step up.* The transformers which changes the low voltage into high voltage are known as step up transformers.
   (ii) *Step down.* The transformers which changes the high voltage into low voltage are known as step down transformers.

(c) **According to the number of phases.** These can be classified according to the number of phases as under:
   (i) Single phase transformer.
   (ii) Two phase transformer.
   (iii) Three phase transformer.
   (iv) Six phase transformer.

(d) According to the principle:
   (i) *Two winding transformer.* In this transformer the one winding which is known as primary is energised and voltage due to

mutual induction induces in the secondary winding to which load is connected.

(*ii*) *Single winding transformer.* In this case the transformer has only one winding, a portion of which is common to primary and secondary both, and emf is induced due to self induction.

(*e*) According to the function:

(*i*) *Transmission transformer.* These are used in the power generating stations to step up the voltage for transmission.

(*ii*) *Distribution transformer.* These are used in the substation for distributing the supply to the different lighting and power loads.

Q. What is meant by the three phase transformer? Write the advantages and disadvantages of star and delta connections.

Ans. A three phase transformer is that which operates on three phase supply. The voltage may be 440 V, 6. kV, 11 kV, 66 kV, 132 kW, 200 kV, 400 kV and even more. Generally the core type construction is used. The winding is done only on ;the internal three limbs. Each limb carries the primary and secondary both. So there are three sets of primary winding and three sets of secondary windings. The windings can have either cylindrical or sandwitch type of coils.

They can either be connected in star or in delta. These are the possible connections as shown in Fig. 18.18.

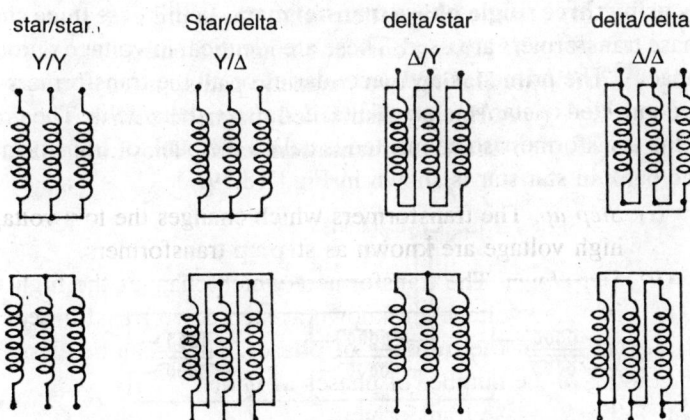

Fig. 18.18. Different connections of 3φ transformers.

Advantages of star connections:

(*i*) In star connections the line voltage is equal to the $\sqrt{3}$ · phase voltage. Therefore the insulation required for phase winding will be of less dielectric strength.

(*ii*) Less phase voltage means less number of turns *i.e.* less copper is required *i.e.* cost is less.

(*iii*) In star the neutral point is obtained, so three phase four wire supply can be taken to distribute the lighting and power load simultaneously. The secondary of the distribution transformer is connected in star.

Advantages of delta connections:

(*i*) In delta connections if one phase fails the three phase supply will be continued to the load but with a reduced efficiency.

(*ii*) In open delta connections the efficiency decreases to 57.7% of the full load output of the transformer.

(*iii*) These connections are efficiently used for transmission transformer, only three wires are required to transmit the power thus reducing the cost of the transmission lines.

Q. What do you understand by the three phase transformation? How it is obtained?

Ans. To transform the three phase voltage from one level to the another level, is known as three phase transformation. It can be obtained by these following methods:

(*a*) By using three single phase transformers.

(*b*) By using one transformer of three phase.

(*a*) **By using three single phase transformers.** In this case three single phase transformers are used. These are identical in voltage ratio and capacity. The primaries and secondaries of all the transformers can be connected in star/star, star/delta, delta/star, delta/delta. The group of the transformers so connected is called the bank of transformers. The bank in star/star is shown in Fig.1 18.19.

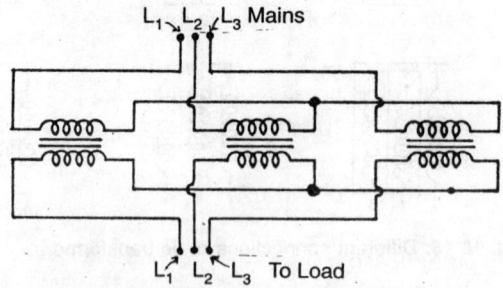

Fig. 18.19. 3φ transformation three 1φ transformer.

Advantages

In case of any fault in one phase only one transformer is required to change. Thus one transformer is required in spare.

Disadvantages

(*i*) More cost of the three transformers.

(*ii*) Requires more space and more maintenance.

(*iii*) Has more weight and the installation cost of the bank is more. Nowadays these methods are not so popular.

(*a*) **Using single unit of three phase transformer.** This system of transformation is commonly used at present. In this system only one transformer is used. The windings can be connected either in star or in delta depending upon the requirement.

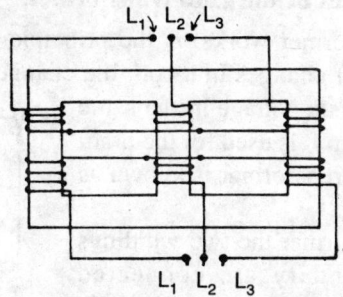

Fig. 18.20. 3ϕ transformer single 3ϕ transformer.

Advantages:

(*i*) It requires less maintenance.

(*ii*) Less cost than three transformers of single phase.

(*iii*) Less space is required and have less weight.

Disadvantages. In case of any fault and break down, even in one phase the complete set in required in spare to replace.

Q. Describe Scott connections of 3 phase transformation into two phase.

Ans. In case of welding and furnaces etc. two phase supply is needed. So the three phase supply can be transferred into two phase by means of the Scott connection, as shown below in Fig. 18.21.

In this case two single phase transformers having identical ratings and provided with different tappings are used. A middle tapping is taken from one transformer's primary winding. It

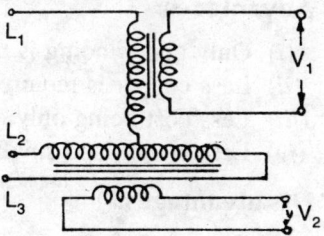

Fig. 18.21. Scott connection.

is known as the main transformer. The other single phase transformer has tapping at 86.6% and is known as the teaser. These are connected as in star making the connection as T. Three phase supply is given to the L_1, L_2 and L_1 terminals.

As the primaries are connected in tee form so these are also known as Tee connections. The secondaries are as usual. The two phase supply can be obtained from these two windings. The less or more voltage can be obtained from these two windings by connecting them in parallel or in series.

Q. Describe the autotransformer. Also state the advantages, disadvantages and uses of the auto transformer.

Ans. An autotransformer works on the principle of self induction *i.e.,* whenever the current changes in a coil, the change of flux linking with

the same coil induces a voltage in the same coil. This induced e.m.f. is used for the practical purpose. Such transformer is known as the auto transformer.

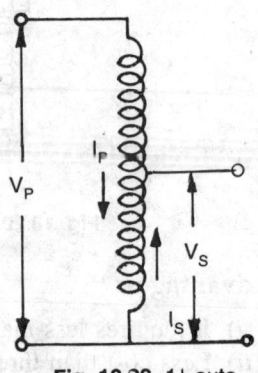

In the auto transformer the two windings primary and secondary are connected electrically as well as mechanically. Truely speaking a part of the single continuous winding is common for both the primary and secondary purposes. An autotransformer has one continuous winding with tappings for different voltage as shown in Fig. 18.22. The secondary voltage depends upon the number of turns *i.e.*

Fig. 18.22. 1ϕ auto transformer.

$$N_S = N_P \times \frac{V_S}{V_P}$$

This transformer can be used for step up and step down transformer.

Advantages:

(*i*) Only one winding is used for primary and secondary.

(*ii*) Less copper is required.

(*iii*) Less cost being only one winding.

(*iv*) Less weight and small in size.

Disadvantages:

(*i*) Two windings are not electrically separate so the danger of shock is apparent.

(*ii*) Secondary may give shock even if on load and without load.

Uses. The transformers are very commonly and efficiently used where the transformation ratio is nearly equal to one. These are used:

(*i*) In starter for starting the three phase induction motors. When a low voltage is required at the time of starting and full voltage in running condition.

(*ii*) As a booster in the feeders to raise the line voltage.

(*iii*) As a stabilizer for getting constant voltage for the load like T.V. sets, refrigerators etc.

(*iv*) As a continuously variable voltage device, it finds useful application in electrical testing laboratories.

(*v*) As a balancer to obtain a neutral in three wire A.C. distribution as in case of D.C. three wire distribution circuits.

Q. (*a*)**Prove that for the same capacity and voltage ratio an auto transformer requires less copper than an ordinary two winding transformer.**

(*b*) **What is a 3φ autotransformer?**

Ans. The copper required for an auto transformer is obviously less because of one winding and that too of different gauges of wires.

The cross-section of the conductor is proportional to the load current or current carrying capacity, simultaneously the length of the conductor is proportional to the number of turns in the winding.

The weight of the copper required α Current $\times$ Number of turns

$$\alpha \, A \times T$$

Now the weight of the copper for an autotransformer and for an ordinary transformer has got the ratio as

$$= \frac{\text{Weight of copper for autotranformer}}{\text{Weight of copper for two winding transformer}}$$

$$= \frac{\text{Copper volume required for autotransformer}}{\text{Copper volume required for two winding transformer}}$$

$$= \frac{\text{Total ampere turns in an autotransformer}}{\text{Total ampere turns in an ordinary transformer}}$$

$$= \frac{(N_P - N_S)I_P + N_S(I_S - I_P)}{N_P I_P + N_S I_S}$$

$$= \frac{N_P I_P - N_S I_P + N_S I_S - N_S I_P}{N_P I_P + N_S I_S}$$

$$= \frac{N_P I_P + N_S I_S - 2N_S I_P}{N_P I_P + N_S I_S}$$

$$= 1 - \frac{2N_S I_P}{N_P I_P + N_S I_S}$$

$$= 1 - \frac{2\dfrac{N_S I_P}{N_P I_P}}{\dfrac{N_P I_P}{N_P I_P} + \dfrac{N_S I_S}{N_P I_P}}$$

$$= 1 - \frac{2\dfrac{N_S}{N_P}}{1 + 1} = 1 - \frac{2\rho}{2}$$

$$= 1 - \rho.$$

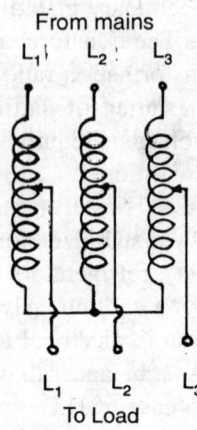

From mains

L₁ L₂ L₃

L₁ L₂ L₃
To Load

Fig. 18.23. 3ϕ auto-transformer.

Thus the saving of copper depends upon ρ, the transformation ratio. The practical aspects shows that the copper saving is maximum say 75% when $\rho = 0.75$, 50% when $\rho = 0.5$.

 (b) **Three phase autotransformer.** In case of three phase autotransformer the three sets of windings are placed over each limb. The tappings are taken from the suitable points as shown in Fig. 18.23. The secondary phase voltage can be calculated as

$$E_S = \frac{E_P}{N_P} \times N_S.$$

Then the line voltage will be

$$E_L = E_2 \times \sqrt{3}$$

The windings are connected with a common star and finally the voltage is obtained. These are generally used with the three phase induction motor for starting purpose. The tappings are so adjusted so that the suitable voltage for starting of the motor is obtained.

Automatic Voltage Stabilizer

The automatic stabilizer is used to keep the secondary load voltage as constant say 220 ± 9%, or so, whatsoever the input voltage may be (Generally ranging from 180 V to 250 V). It employs a voltage sensing circuit comprising of zenor diode and BC-148. This circuit is used for SK-100 and the relay. The zener diode does not function during normal voltage.

This keeps BC-148B off while second BC-148, SK-100 and relay is functioning in ON state.

While the supply voltage is high the zener diode operates and off the relay. Now *A.C.* can be obtained from the upper relay point which is actually 200V point. It can be operated to function with the help of preset 100Ω. If the supply voltage is always on high voltage side the middle and top connections of the transformer may be interchanged to obtain the constant voltage on secondary side.

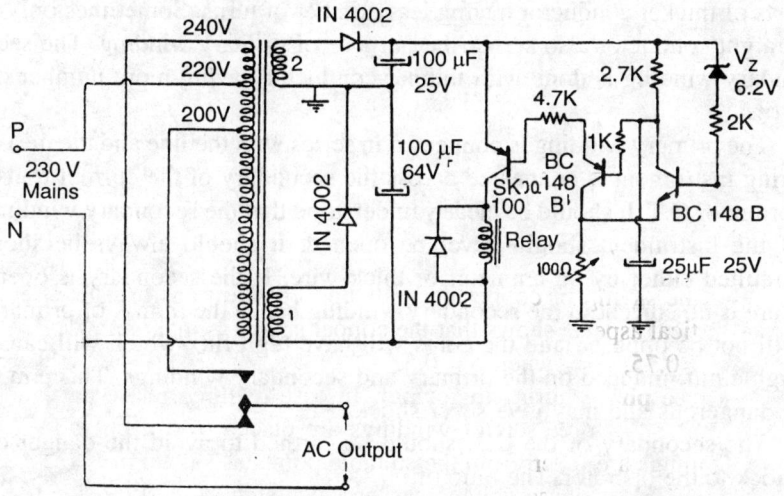

Fig. 18.24. Automatic Voltage Stabilizer. Input—180 V—250 V, Output—220 ± 9% V

Q. Is it advisable to connect the measuring instruments directly on H.T. line? What are the instrument transformers? Explain the working of C.T.

Ans. It is always against safety to connect instruments or control apparatus directly to high voltage line. The normal range of instrument may also not be sufficient to measure high values of voltage and currents. In case of d.c., shunts and multipliers are used to increase the range of the instruments. In a.c., the instrument transformers are used to reduce high voltage and currents of safe and lower values.

The main function of instrument transformer are as follows:

(*i*) These act as *ratio devices,* so that with their uses the use of low range instrument is possible.

(*ii*) These also act as the instrument devices to protect the instrument and apparatus from high voltage.

The transformer in conjunction with the measuring instrument for measurement purposes, is known as the *instrument transformer.* These are:

(i) *Current transformer.* The transformer used for measuring heavy current is called current transformer or C.T.

(ii) *Potential transformer.* The transformer used for measuring high voltage is known as the potential transformer or P.T.

Current transformer. The current transformer is used for measuring the heavy currents. It is a step up transformer. The primary winding consists of thicker conductor having less number of turns. Sometimes only a straight conductor also serves the purpose of primary winding. The secondary winding is done with thinner conductor having more number of turns.

The primary winding is connected in series with the line and the measuring instrument is connected across the secondary of the current transformer or C.T. It should be clearly understood that the secondary winding of the instrument should never be opened. It should always be short circuited either by an ammeter or thick wire. If the secondary is open, there is no current in the secondary winding hence the m.m.f. of primary will not be opposed and the cores will have high flux which will cause high e.m.f. induced on the primary and secondary windings. This e.m.f. is dangerous and may give sever shock.

The secondary of the C.T. should be earthed to avoid the danger of shock to the operator. The ratio of turns of both the windings is given. Sometimes the dial of the instrument is given with a multiplier to obtain a correct reading, but nowadays the dials of the instruments are calibrated and marked to have the direct reading.

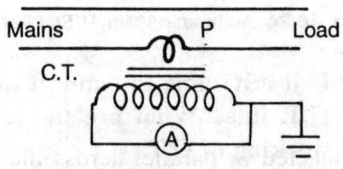

Fig. 18.25. Current transformer.

Sometimes nearly a coil is placed around the current carrying conductor. In this case the conductor itself behaves like the primary and the coil as secondary. The ammeter is connected across the coil. The current flowing in the conductor sets up the magnetic flux which links the secondary and thus inducing voltage in the coil, this e.m.f. causes the current to flow through the meter to deflect. Thus the reading is obtained as in Tong Tester, or Clip-on-meter etc.

In case of three phase line the readings are taken separately and individually for all the three phases. The reading can be taken without disturbing the conductor by means of a split core type current transformer.

In case of three phase line if all three phases are placed inside the split core, the reading will be zero; because in 3φ the vector sum of all the three line fluxes will be zero. Thus the induction will not be possible to induce the voltage and meter will not give any deflection.

Q. Explain the working of potential transformer. What are the advantages of instrument transformers?

Draw a connection diagram for measuring power and energy consumption in H.T. line using C.T., P.T. for single phase supply.

Ans. The instrument transformer which is used to measure the high voltage, is known as potential transformer (P.T.) as shown in Fig. 18.25.

The primary of the P.T. is having more number of turns of fine wire and secondary is having less number of turns. So the P.T. in principle is a step down transformer. The primary winding is connected across the line and secondary across the meter to measure the line voltage. The primary winding when connected to line, carry some current, which produces the magnetic flux. The secondary winding is linked with flux causing the induction of some voltage, (generally 110 V in case of P.T.). This voltage defect the voltmeter on the secondary of the P.T.

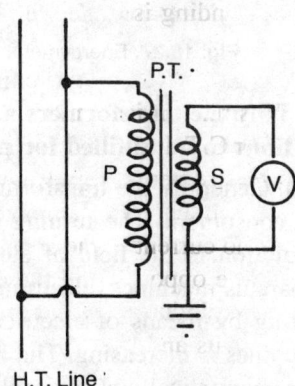

Fig. 18.26. Potential transformer.

The scale is directly calibrated to obtain the actual voltage.

The secondary of the P.T. is always connected to earth.

If two or more instruments are to be connected then the voltmeters are connected in parallel across the secondary of the P.T. and ammeter in series with the secondary of C.T. as shown in Fig. 18.26.

Advantages of instrument transformer. There are the following advantages:

(*i*) The size of the instrument is reduced or say moderate because the secondary of C.T. is designed for 5 A and of P.T. for 110 V.

(*ii*) The replacement of damaged instrument is easy.

(*iii*) Several instruments can be operated from a single instrument transformer.

(*iv*) Low consumption of metering circuit.

(*v*) Accessibility on H.T. is easy.

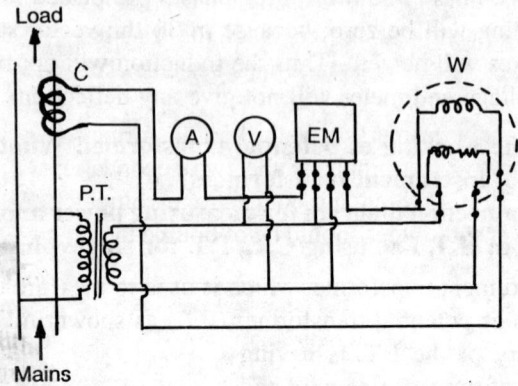

Fig. 18.27. Energymeter and Wattmeter on H.T. with C.T. and P.T.

Q. Why the transformers are connected in parallel. State the conditions to be fulfilled for parallel operation. *(N.C.V.T. 1981, 84)*

Ans. Generally the transformers are installed to distribute the energy to the consumers. The turning of this century has brought a tremendous revolution in the field of electrical engineering specially the electrical apparatus machines, electronic apparatus and devices etc. which are operating by means of electrical energy. So day by day the load on the machines is increasing. The replacement of the new transformer is neither economical nor advisable, so these are connected in parallel.

Necessity of parallel operation:

(*i*) The transformers can be connected according to the load. If load is more, more number of transformers can meet the demand and if load is less, then one transformer can serve, resulting better efficiency of operation, so the transformers are connected in parallel to share the load.

(*ii*) In case of fault in any transformer, the faulty transformer can be taken out from the bus bar and the load can be transferred to other transformer, thus the supply will continue.

(*iii*) For periodic overhauling the transformers can be taken off, thus the life is increased.

Conditions. These are the following conditions to be fulfilled before parallel operation.

(*i*) The transformer should have the same voltage ratings.

(*ii*) The percentage impedance drop should be same.

(*iii*) The leads which are to be connected together must be of the same polarity.

(*iv*) The phase sequence must be same.

The voltage ratings can be easily checked and the percentage imped-ance from the characteristic curve of the transformer can be obtained. The polarity and phase sequence must be same. The polarity can be tested as under:

(a) Connect the primary of both the transformer across the line.

(b) Connect two terminals of the two secondaries in such a way as shown in Fig. 18.28 and connect one voltmeter across the second-aries. The voltmeter should be of double the range of one secondary voltage.

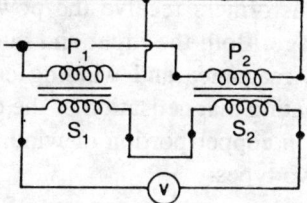

Fig. 18.28. Polarity test.

If the voltmeter shows double voltage than that of one secondary then the connections are suitable for series connections and if the voltmeter reading is zero then the two wires of same polarity are joined. Now these two wires can also be connected together to connect the transformer in parallel and the transformers are suitably ready for parallel operation.

In case of three phase transformer the secondaries are tested sepa-rately in the same manner.

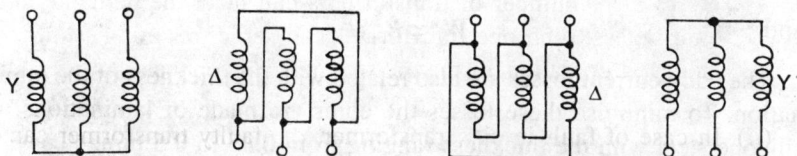

Fig. 18.29. Transmission transformer.   **Fig. 18.30.** Distribution transformer.

Q. What are the transmission and distribution transformers?

Transmission transformer. It is step up transformer which changes the induced voltage of the alternator to the high and sometime extra high voltage for transmission. These transformers are designed for high volt-age of the order of 220 kV, 400 kV or even more. These are having high capacity. These voltages are 6.6 kV, 11 kV, 22 kV, 33 kV, 66 kV, 132 kV, 200 kV, 400 kV and even 500 kV. Their primary is connected in star and secondary is delta as shown in Fig. 18.29 to save the number of conduc-tors for transmission line.

Distribution transformer. It is the step down transformer. It transforms the high voltage of the transmission line into low voltage for distribution purpose. The secondary is rated for 11 kV, 440 V for distribution purpose. Generally the primary is in delta and secondary in star as shown in Fig. 18.30. Thus three phase four wire system is obtained so that the lighting and power load can be connected simultaneously.

Q. What are the losses in a transformer, define the constant and variable losses?

Ans. Basically the transformers receive the power at one voltage and deliver at another voltage. Both the input and output are electrical. The presence of iron core, reactance and winding contributes some losses which are due to alternating magnetisation of the core, leakage reactance and the power wasted in copper portion of windings.

The losses are of two types:

(*i*) **Constant losses.** These are independent of load current.

(*ii*) **Variable losses.** These are proportional to the load current.

Constant losses. These includes the hysteresis and eddy current losses. As in case of a transformer, the core flux remains constant from no load to full load, so these losses remains constant from no load to full load. These losses are the hysteresis and eddy current losses.

The hysteresis losses are related to the area of B.H. loop of the material used. The losses in a given time is proportional to the number of times the loop travels. According to the steinmetz, these are

$$W_h \propto B_m^{1 \cdot 6} f$$

and

$$W_s \propto B_m^2 f^2$$

The eddy current losses are also related with the thickness of the lamination. To minimise these losses the cores are made of laminations of silicone steel with the thickness ranging from 0.35 to 0.5 mm.

Variable losses. These are the copper losses. These are due to the resistance of the transformer windings. These losses are proportional to the square of the load current.

The total copper losses are $= I_1^2 R_1 + I_2^2 R_2$ Watts.

Additional load losses. Apart from these losses, there are some additional losses. These are due to the induction of current in the parts of the transformer other than the cores such as tank, structural steel and even in copper winding. This happens only on load hence the name is given the additional load losses.

Q. 8. Describe the different methods for determining the iron and copper losses of a transformer. *(April, 1979)*

Ans. These are two methods for determining these losses:

(a) Open circuit test.   *(N.C.V.T. 1979)*

(b) Short circuit test.

(a) **Open circuit test.** This test is performed to find out the iron losses of the transformer. In this test one winding is connected to the supply of normal voltage and frequency. The circuit of the other winding is kept open. A wattmeter, ammeter, and voltmeter is connected to the supply side. In this test the H.T. winding is kept open circuited and L.T. is connected to the mains. It is because the metering is easy on low tension side.

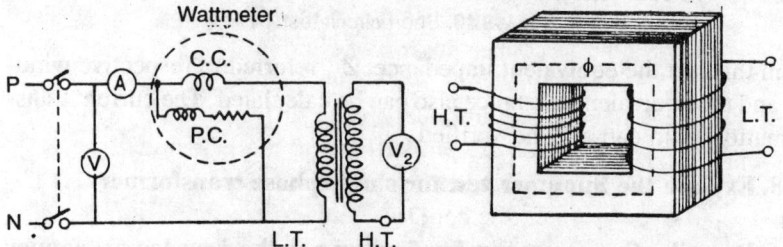

Fig. 18.31. Open circuit   **Fig. 18.32.** Single phase transformer.

Under these circumstances the flux will be produced in the core and it is obvious that the magnitude of the flux under the full load condition or on no load condition remains the same. So all the parts and cores of the transformer are under going to the same magnetic stresses.

As shown in Fig. 18.31 the connected wattmeter indicates the no load power losses or the iron losses. The ammeter indicates the no load current of the transformer. These losses remains constant from no load to full load, from the datas obtained by this test, the R_0, X_0, cos ϕ_0 and I_0, I_ω, I_μ and the transformation ratio can be determined.

(b) **Short circuit test.** This test is performed to determine the copper losses at any desired load. In this test the low voltage side is short circuited through an ammeter or by simply a thick copper conductor. The ammeter on primary and secondary side are suitably selected according to the capacity of the transformer. A low voltage is applied usually 5 to 10% of the normal primary voltage with the help of a variable transformer as shown in Fig. 18.33. The voltage is so adjusted as to have desired load current through the windings. Since the applied voltage is small enough so the flux is also reduced that of the normal value. Hence the core losses are negligible and the total reading obtained from the wattmeter is the copper losses of the transformer at that particular load.

In this test generally the low voltage side is short circuited, because metering on low side will be unusual because of much current in L.T. side than H.T. The copper losses are proportional to the square of current.

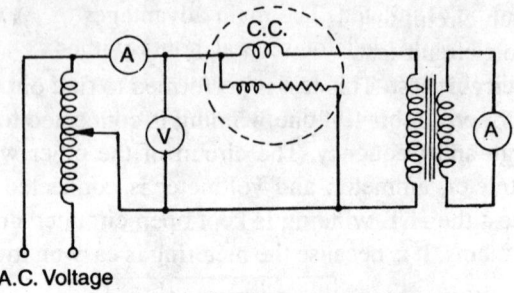

A.C. Voltage

Fig. 18.33. Short circuit test.

From this test the equivalent impedance, Z_{01} referred to respective winding and the equivalent reactance also can be calculated. The current transformation ratio can also be verified.

Q. 8. Explain the Sumpner test for single phase transformers.
Or

Describe the single test for finding out the iron losses, copper losses and the rise in temperature for single phase transformers.

Ans. This test is also known as the Regenerative test, back to back test or sumpner test. In this test two transformers of identical ratings are taken. The primary of both the transformers are connected in parallel across the rated voltage and frequency. The wattmeter W_1, gives the total core losses of both the transformers.

The secondaries are connected in such a way that their terminals of same polarity are short circuited. Both secondaries are connected as their polarities are in opposition. Connect B and C terminals and a voltmeter across A and D. The voltmeter should be of double the range of one secondary winding. The double reading indicates that the connected terminals are having opposite polarity. In that case connect B to D and the voltmeter will now give zero reading. The zero reading indicates that the secondaries are in opposition and the terminals of same polarity can be joined.

In that case there is no secondary circulating current, because the voltage are in opposition having same potential difference. In order to flow some current in secondary winding the injecting transformer is used and the required voltage is injected to circulate the full load or any other desired load in the secondary windings as shown in Fig. 18.34. The circulating current will have the direction as marked by dotted lines. The current will not effect the wattmeter W_1 reading. These are the core losses of both the transformers. Wattmeter W_2 will give the copper losses of the both transformers. Thus when copper losses and core losses are known, the efficiency can be calculated. The main advantages of this test is that the transformer can be loaded for several hours without consuming much power, as only the losses are fed. The temperature rise and behaviour of the transformer at that particular temperature, losses and efficiency can be observed and calculated.

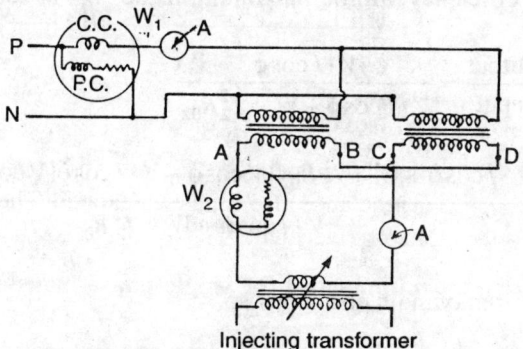

Injecting transformer

Fig. 18.34. Sumpner test

Q. Define efficiency of the transformer? State the conditions for maximum efficiency.

Ans. The efficiency is the ratio of output to input generally shown in percentage.

i.e.
$$\% \text{Efficiency} = \frac{\text{Output}}{\text{Input}} \times 100$$

$$= \frac{\text{Input} - \text{losses}}{\text{Input}} = \frac{\text{Output}}{\text{Output} + \text{Losses}} \times 100$$

$$= 1 - \frac{\text{Losses}}{\text{Input}} \text{ or } 1 - \frac{\text{Losses}}{\text{Output} + \text{Losses}} \times 100$$

The efficiency depends upon the load and load power factor, the efficiency is maximum when variable losses are equal to the constant losses. It decreases with the decreasing of the load and power factor.

Condition for maximum efficiency. the power output of the transformer is

$$= V.I. \cos\phi.$$

The input if the W_i = the iron losses and $I_2^2 R_{02}$ = the to copper losses referred to secondary side.

$$\therefore \qquad \text{Input} = V \cdot I \cos\phi + W_i + I_2^2 R_{02}$$

where R_{02} is the equivalent resistance referred to secondary side of the transformer.

Now the efficiency will be maximum, if the $\dfrac{d\eta}{dI}$ is zero.

$$\therefore \quad \eta = \frac{\text{Output}}{\text{Input}} = \frac{V \cdot I \cos\phi}{V \cdot I \cos\phi + W_i + I_2^2 R_{02}}$$

$$\therefore \quad \frac{d\eta}{dI} = \frac{\left(V \cdot I \cos\phi + W_i + I_2^2 R_{02}\right) V \cos\phi - V \cdot I \cos\phi \left(V \cos\phi + 2I_2 R_{02}\right)}{V \cdot I \cos\phi + W_i + I_2^2 R_{02}}$$

$\dfrac{d\eta}{dI} = 0$ for maximum efficiency so,

$$0 = \frac{\left(V \cdot I \cos\phi + W_i + I_2^2 R_{02}\right) V \cos\phi - V \cdot I \cos\phi \left(V \cos\phi + 2I_2 R_{02}\right)}{\left(V \cdot I \cos\phi + W_i + I_2^2 R_{02}\right)^2}$$

or $\quad \left(VI \cos\phi + W_i + I_2^2 R_{02}\right) V \cos\phi - VI \cos\phi \left(V \cos\phi + 2I_2 R_{02}\right) = 0$

or $\quad VI \cos\phi + W_i + I_2^2 R_{02} = I\left(V \cos\phi + 2I_2 R_{02}\right)$

or $\quad VI \cos\phi + W_i + I_2^2 R_{02} = VI \cos\phi + 2I_2^2 R_{02}$

or $\qquad W_i = I_2^2 R_{02}$

i.e., the iron losses equal to copper losses so the efficiency will be maximum if the iron losses one equal to copper losses of the transformer.

Q. Define all day efficiency of the transformer. What is the necessity of this efficiency

Ans. The transformers are connected to line to give service around the clock irrespective of the load, whether these are fully loaded or even without load. The primary winding is always connected across the line and draws some current from the mains. Under these circumstances the ordinary efficiency, the ratio of output to input, does not give any correct picture of the performance of the transformer. So it leads to find out some other efficiency taking time into consideration, generally the time is 24 hrs.

The copper and iron losses are inherent when the transformer is supplying. The iron losses remains constant where the copper losses depends upon the load current. The copper losses vary as the square of current.

The all day efficiency is defined as the ratio of output in kWh over a state a period to the intake power in kWh over the same period. This duration is usually 24 hrs.

$$\text{So all day efficiency} = \frac{\text{kWh output in stated period}}{\text{kWh input in same period}}$$

$$\therefore \quad \text{\% All day efficiency} = \frac{\text{kWh output in 24 hrs.}}{\text{kWh intake in 24 hrs.}} \times 100.$$

Example. *Find out the all day efficiency of a 50 kVA transformer when it is loaded as under:*

 (i) 100 W lamps 40 in number working for 6 hours a day.

 (ii) 2000 W heater five in number working for 4 hrs a day.

(iii) 45000 VA load at 0.8 P.F. working for 10 hrs a day.

(iv) 50 kVA load at 0.9 P.F. working for 4 hrs a day.

The transformer has maximum efficiency 96% at full load at unity power factor.

Solution.

$$\text{The all day efficiency} = \frac{\text{kWh output in 24 hrs.}}{\text{kWh intake in 24 hrs.}} \times 100$$

So the kWh output:

(*i*) 40 lamps 100 W each working for 6 hrs.	= 40 × 100 × 6
	= 24000 Wh
(*ii*) 5 heaters 2000 W each working for 4 hrs	= 2000 × 5 × 4
	= 40000 Wh
(*iii*) 45 kVA load at 0.8 p.f. for 10 hrs	= 45000 × 0.8 × 10
	= 360000 Wh.
(*iv*) 50 kVA load at 0.9 p.f. for 4 hrs	= 50000 × 0.9 × 4
	= 180000 Wh
The total Wh load output	= 604000 Wh
	= 604 kWh

The capacity is 50 kVA and maximum efficiency is 96%, so the losses are 4%.

$$\text{The total losses} = \frac{50 \times 4}{96 \times 100} = 2.08 \text{ kW}$$

The maximum efficiency is only possible when the copper losses are equal to the iron losses.

$$W_c = W_i = \frac{2.08}{2} = 1.04 \text{ kW}.$$

The core losses are constant from no load to full load so the power consumption in 24 hrs. = 1.04 × 24 = 24.96 kWh.

The copper losses will be proportional to the load currents.

The power factor of lamps and heater is unity, so

$$\text{In case of lamp load} = 1.04 \times \left(\frac{4000}{50000}\right)^2 \times 6 = 0.040 \text{ kWh}$$

$$\text{heater load} = 1.04 \times \left(\frac{10000}{50000}\right)^2 \times 4 = 0.1664 \text{ kWh}$$

$$\text{45 kVA load} = 1.04 \times \left(\frac{45}{50}\right)^2 \times 10 = 8.424 \text{ kWh}$$

$$\text{50 kVA load} = 1.04 \times \left(\frac{50}{50}\right)^2 \times 4 = 4.16 \text{ kWh}$$

Total copper losses during 24 hrs = 12.79 kWh

Total losses during 24 hrs = $W_c + W_i$ = 12.79 + 24.96
= 37.75 kWh.

Now % all day efficiency

$$= \frac{604}{604 + 37.75} \times 100$$

$$= \frac{604}{641.75} \times 100 = \textbf{94.11\%} \quad \textbf{Ans.}$$

Q. What is the necessity of testing the transformer? Name the different tests to be conducted for the transformer.

Ans. Before putting any engineering component on the job, it has to go for several tests for its satisfactory performance. The transformers are in no way exempted. In fact a transformer after designing, manufacturing and finally assembling, has to go for the tests to assess its performance and predetermination of the losses and efficiency.

These are the following tests:

(*a*) **Insulation resistance test.** Megger is used to perform this test. The megger should be taken of suitable range. These tests are:

    (*i*) *Insulation resistance between the windings and body of the transformer.* Connect '*E*' terminal of the megger with the body and '*L*' to the windings separately. The result, after rotating the handle up to the required speed, should be either infinity or at least one mega ohm or as per specifications of the manufacturer.

    (*ii*) *Insulation resistance between the windings.* Check the continuity of the windings. Short the terminals of primary winding and secondary winding, separately. Connect '*L*' terminal of megger to one winding and '*E*' to another winding. On rotating the handle the megger should show infinity, or at least one mega ohm or as per the specifications of the manufacturer.

(*b*) **Core loss test.** It is also known as the open circuit test. In this test the primary is connected to the rated voltage and rated frequency through ammeter and wattmeter as shown in Fig. 18.30. The secondary winding (generally h.t. is taken in this test) is left open. As the core losses are independent of the load, so the power shown by wattmeter with secondary as open circuited are the core losses. The copper losses being very less are neglected.

In case of three phase transformer the core test is performed in the same manner with secondary opened. In this test i.e., the open circuit test the X_0, R_0, I_μ, I_ω, I_0 and cos ϕ_0 can be determined.

(*c*) **Copper loss test.** This test is also known as the short circuit test. It is used to find out the impedance, the copper losses, equivalent impedance and current transformation ratio also. In this test generally the low tension side is short circuited; with a thick conductor or a suitable ammeter. The reduced voltage is applied on the primary side to enable the full load or any desired load current in the transformer windings as shown in Fig. 18.33. In this test being low voltage the core losses are neglected and the wattmeter reading gives away the copper losses at that load.

In case of three phase transformer the test is carried out and losses are measured by means of two wattmeters method.

(*d*) **Load test.** To check the performance of the transformer on load, the load test is conducted. By this test, the efficiency, temperature conditions etc. are determined. In transformers of low capacity the arrangement of load is easy but for big transformer it is difficult and the power consumed is nearly a wastage.

(e) **Sumpner test.** It is also known as the Regeneration test or back to back test. In this test two similar transformers are taken. The primary windings are connected in parallel across the supply. The secondaries are connected in opposition to enable the transformer for parallel operation. In ideal conditions when the secondaries are connected there will not be any circulating current in secondary windings of the transformer. Now a voltage from outside is injected by means of an injecting transformer as shown and the current starts flowing in the transformer. The regulation is stopped as the required current is obtained. The wattmeter as connected in Fig. 18.34, W_1 shows the core losses and W_2 gives away the copper losses of both the transformers. The transformer can be loaded for several hours without consuming much power only the losses are feed from the supply.

In this single test the core losses, copper losses and temperature rise for the performance and efficiency of the transformer are known.

(f) **High voltage test.** It is a test by direct application of the voltage from a suitable source or an induced voltage test in which the transformer is operated at a voltage and frequency sufficiently in excess of normal values.

(g) **Impulse test.** This test is carried out in order to determine the ability of transformer to withstand the effects of high unidirectional voltage resembling the surge lightening.

(h) **Phasing out test.** In this test all phases are short circuited except a primary and the corresponding secondary. A voltmeter is connected across the secondary and a small direct current is circulated in primary. A momentary deflection of voltmeter with the making and braking of primary current, confirms that these two windings are belonging to the same phase.

(i) **Polarity test.** Normally terminals are distinguished by suffix numbers in such a way that the same sequence of number represents the same direction of induced e.m.f. in both primary and secondary windings.

(j) **D.C. resistance test.** Any suitable method for testing the d.c. resistance may be employed.

(k) **Voltage ratio test.** It is directly found out from the readings of the voltmeters connected to the primary and secondary windings.

Q. What are the common safety devices used in a transformer?

Ans. There are the following safety devices:

(*a*) **Circuit breaker.** There are two circuit breakers, one on primary side and other on secondary side. These C.B. will disconnect the supply in case of overloading or short circuit or any other fault or abnormal conditions. In big transformers automatic circuit breakers are used.

(*b*) **Earthing.** All the transformers are connected to earth. For maintaining good earthing connections the double earthing is done.

(*c*) **Lightning arrester.** A lightning arrester is attached with the body of the big transformer to save it from the sky lightning.

(*d*) **Relays.** The buchholz Relay is fitted between the tank and the conservator. The relay operates an alarm or trips the circuit breaker when the gases increases. It's main function is to indicate the pressure of the gas in case of some minor faults and isolate the transformer in case of serious fault.

Some other safety devices are also employed viz. temperature gauge and oil gauge etc.

REVIEW QUESTIONS

1. What do you understand by the transformer? What are the causes of popularity of a transformer?
2. What is the working principle of the transformer?
3. Describe the different parts of a transformer.
4. What is the transformer? How does it transform electrical energy form one circuit to another circuit?
5. Describe an expression for the e.m.f. of an ideal transformer.
6. Explain the active and reactive components of the no load current of the transformer.
7. Draw the vector diagrams for the different loadings of the transformer.
8. What do you understand by the on load transformer, explain with diagram?
9. Develop an equivalent circuit of a single phase transformer.
10. Explain how iron and copper losses are determined in a single phase transformer.
11. What is the voltage regulation of the transformer, how it calculated in a single phase transformer?
12. What is Sumpner Test? Draw the circuit diagram and also mention the unique characteristics of this test.
13. What are the method of 3φ transformation?
14. What are the application and uses of the auto transformer?
15. What is scott connection and where these are used?
16. Give connection diagram of two single phase transformer running is parallel.
17. What are the different tests to be carried out before commencing the new transformer?
18. What are the conditions for satisfactory parallel operation of the two 3φ transformer?

19. Write short note on the followings:
    (a) Magnetising current of the transformer.
    (b) Leakage reactance.
    (c) Voltage regulation.
    (d) Hysteresis and eddy current losses of a transformer.
20. Fill up the blanks:
    (a) A transformer is a ....
    (b) The primary no load current comprises of ....
    (c) The magnetic flux in primary and secondary windings is ....
    (d) Open circuit test of the transformer gives away ....
    (e) Sumpner test is also known as .... test.

NUMERICAL PROBLEMS

1. A single phase 50 c/s transformer has 25 turns in primary winding and 300 turns in secondary winding. The net area of cross-section of cores is 400 cm^2. If the primary is connected to 250 volts, find the peak value of flux density in the cores and secondary induced voltage.
   [Ans. 1.126 Wb m^2, 3000 V]

2. A 5 kVA 110/220 V 50 c/s transformer has 150 turns in its low voltage side. Neglecting losses, calculate the number of turns in high tension side and currents in primary and secondary windings.
   [Ans. 300, 22.7 A and 45.4 A]

3. A single phase 10 kVA 50 c/s transformer has turns ratio 300/23. The primary is connected to 1500 V mains. Neglecting losses, calculate the voltage on open circuit and the currents in primary and secondary windings.
   [Ans. 115, 6.67 A and 86.96 A]

4. A single phase transformer has 525 turns in primary winding and 35 turns in secondary winding. If the primary is connected to 6600 V, find the secondary p.d. Neglecting losses, find the primary winding current if the secondary winding current is 125 A. [Ans. 440 V 8.33 A]

5. A 200 kVA 6600/400 V 50 c/s single phase transformer has 80 turns in its secondary winding, neglecting losses, calculate the primary winding current, secondary winding current and number of turns in primary winding.
   [Ans. 30.3 A, 500 A, 1320]

6. A 2200/200 V 50 c/s single phase transformer takes 0.6 A at 0.3 power factor on open circuit. Calculate the magnetising and wattful component of the current. [Ans. 0.57 A, 0.18 A]

7. A transformer takes 0.8 A when the primary is connected to its normal supply 200 V 50 c/s and secondary is open circuited. The wattmeter reading is 64 watts. Calculate the losses, magnetising current and wattful current.
   [Ans. 64 W, 0.733 A, 0.32 A]

8. A transformer takes 1.1 A when the primary is connected to normal voltage and frequency, 250 V 50 c/s secondary being open. If the wattmeter reading is 90 watts, Calculate the I_μ and cos ϕ.
   [Ans. 1.04 A, 0.33]

9. A single phase 1000/250 V 50 c/s transformer takes 4 A at 0.2 power factor lagging. When the secondary is open circuited calculate the R_0.

[**Ans. R_0 = 12.5 Ω**]

10. A 2200/200 V 50 c/s transformer has the primary resistance 3.2 Ω and secondary resistance 0.02 Ω. Calculate the equivalent resistance referred to primary winding and equivalent resistance referred to secondary winding.

[**Ans. 5.62 Ω, 0.0464 Ω**]

11. A 5 kVA 2000/250 V transformer has a primary winding resistance 1.8 Ω and secondary winding resistance 0.013 Ω, calculate total resistance referred to secondary winding. [**Ans. 0.0411 Ω**]

12. A 200 kVA step down transformer has a voltage ratio 11000/250 V. The open circuit iron losses are 800 W. When a voltage of 250 V at normal frequency is applied to high voltage side, the secondary being short circuited, a full load current is produced in the secondary and the wattmeter indicates 2000 W. Calculate:

(*i*) The efficiency.

(*ii*) The secondary potential difference when taking 150 kW at 0.8, lagging power factor.

[**Ans. 98.28%, 216 V**]

13. A 75 kVA transformer has iron losses 500 W and full load copper losses 1200 W find the efficiency at full load and half load at:

(*i*) Unity power factor.

(*ii*) 0.8 power factor lagging.

[**Ans. 97.8%; 97.9%, 97.24%, 97.4%**]

14. A 50 kVA, 6600/220 V step down transformer has the resistance 10 Ω and secondary resistance 0.01 Ω, find:

(*a*) total resistance referred to secondary winding.

(*b*) full load copper losses.

(*c*) efficiency if the core losses are equal to copper losses at unity power factor at that load.

[**Ans. 0.021 Ω, 1090 W, 95.8%**]

19

A.C. Three-Phase
Induction Motors

INDUCTION MOTOR

In case of an induction motor the e.m.f. is not given to the rotor from outside but it is induced by induction. So it is called as induction motor.

Q. Describe the construction of an induction motor.

Ans. The induction motor essentially consists of two parts (*a*) stator, (*b*) rotor.

(*a*) **Stator.** The stator is a stationary part of the induction motor. It has hollow cylindrical shape. It is made of laminations of silicone steel highly compressed and riveted together to reduce the eddy current and hysteresis losses. There are semi-enclosed types of slots in which three phase winding is done for two poles, four poles etc. as required. This winding can be permanently connected either in star or in delta or sometimes in star and delta externally.

The stator is fixed in the body of the motor. The body is made of cast steel. On giving three phase supply the three phase rotating magnetic field is produced in the stator.

In case of big motors, the slots are of open types and ready made coils are inserted with proper insulation; but in case of medium and small motors the ready made coils are inserted turn by turn as the opening of the slot is narrow on the top.

(*b*) **Rotor.** It is a rotating part of the induction motor. These are of the following types:

(*i*) Squirrel cage (*ii*) Phase wound rotor.

Squirrel cage rotor. These are called as cage rotors. These are very common because of their easy and robust construction. These are of two types:

Fig. 19.1. Squirrel cage induction motor.

1. *Single Cage rotor.* These are made of the laminations of silicone steel compressed and riveted. There are number of closed type slots in which copper or aluminium bars are either placed or casted. These bars are short circuited at both the ends by means of the end rings as shown in Fig. 19.2 (*a*). In single cage rotor only one line of bars is used. The rotor bars and end rings forms a closed circuit in itself resembling to the cage of squirrel, that is why these are called the squirrel cage rotors.

   In small and medium size of motors these bars are casted but for high power these are inserted or wedged and are welded to form the end rings.

2. *Double squirrel cage rotor.* The double squirrel cage rotor has two cages. The outer cage near the outer periphery, having low resistance made of brass or bronze and less cross-sectional area. The inner cage below the outer cage is in the deep of the rotor. It is having more cross-sectional area than the outer cage and copper

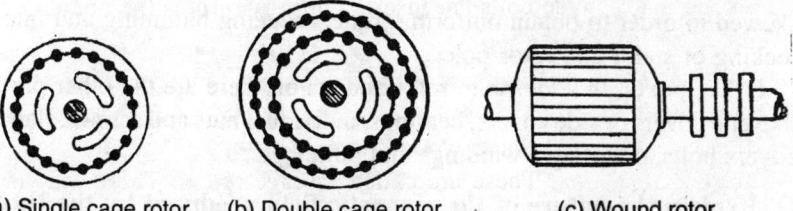

(a) Single cage rotor   (b) Double cage rotor   (c) Wound rotor

Fig. 19.2. Types of rotor.

bars are used as shown in Fig. 19.2(*b*). The flux will be more near these bars hence having high inductance. Both the bars are short circuited by their individual end rings.

At the time of starting the flux is more near the inner cage and the frequency of flux linking is the supply frequency because the rotor is in stand still condition, so the reactance of the inner cage is higher than that of the outer cage. The current depends upon the reactance so a major portion of the current flows through the cage having low reactance *i.e.* the outer cage. The outer cage has low reactance and more resistance so the power factor is good as a result of this high starting torque is produced. In running condition the rotor frequency is less so the reactance decreases considerably and the major portion of the current flows through the cage having less resistance *i.e.* the inner cage and low current in the outer cage. Thus the starting torque and power factor is improved.

(*ii*) *Phase wound rotor.* In this case also the rotor is made of the laminations of silicone steel to reduce the eddy current and hysteresis losses. Generally open type slots are used in the rotor. In rotor the three phase winding is done having same number of poles as that of the stator winding. The winding is connected in star, so that external resistance can be added in the rotor circuit. The three ends are brought to the phospher bronze sliprings as shown in Fig. 19.2(*c*) mounted on the same shaft. The brushes which carry the current from and to the rotor winding are held in position in a box type brush holder mounted with a loading spring for tension, on the insulated rod which is attached with the side cover. These brushes are connected to a star connected external rheostat. At the time of starting the complete resistance is in the rotor circuit and gradually taken out as the motor attains speed and finally the three sliprings are short circuited. This external resistance is used for starting and speed control. Approximately 5% speed controlling is obtained by this method.

The starting torque in this type of motor is high because of more rotor resistance; so the slipring induction motors are used where high starting torque and low starting current is required.

The slots in all types of rotors, are not parallel to the shaft, but are skewed in order to obtain uniform torque, reducing humming and interlocking of stator and rotor poles.

Other parts. In addition to rotor and stator, there are the other parts, like end covers or side covers, bearings, individual nuts and screws, flanch covers bolts, couplings, windings and pulley etc.

Q. Explain the nature of the magnetic field produced by the three phase supply system.

Ans. Whenever three phase supply is given to the stator of a three phase motor the phase currents vary in time and phase. Every phase current will produce the magnetic field of pulsating or alternating in nature. However due to the orientation of the windings and time phase of the currents, the flux produced by each phase winding combines to form a resultant flux that moves around the stator surface at a constant speed having constant magnitude. This revolving or rotating flux is called the rotating magnetic field.

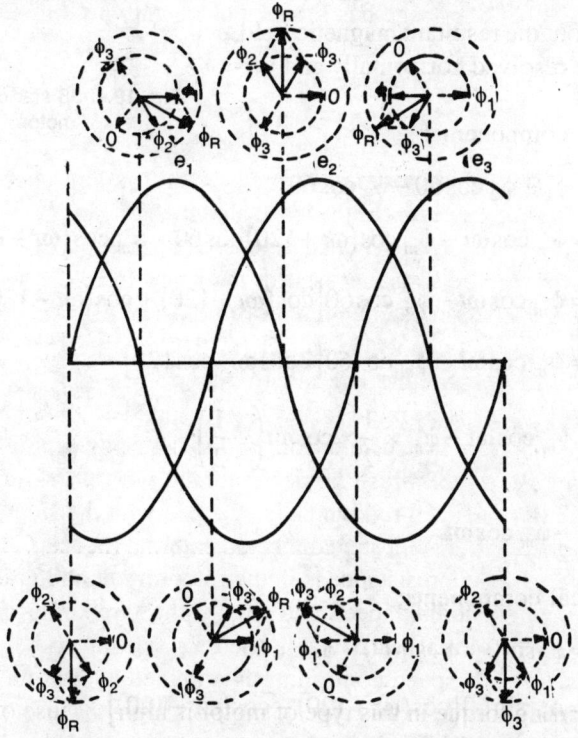

Fig. 19.3. Rotating magnetic field in 3ϕ induction motor.

The illustration shows the production of the rotating magnetic field by a three phase motor supplied by three phase sinusoidal waveform. The Fig. 19.3 shows the different instants at 60° interval moving in the same direction as that of the wave. The direction of the resultant magnetic field is also shown at the same intervals. It is observed that the resultant magnetic field is also moving with the same angular movement. It indicates that the resultant magnetic field completes its one cycle during one cycle of waveform and having the constant magnitude.

So the magnetic field produced by a 3ϕ sine wave is a rotating magnetic field having constant magnitude and speed.

IInd method. Let there are three coil *A, B* and *C* equally spaced 120° electrical apart. When three phase supply is given ϕ_1, ϕ_2 and ϕ_3 fluxes are produced as shown in Fig. 19.4.

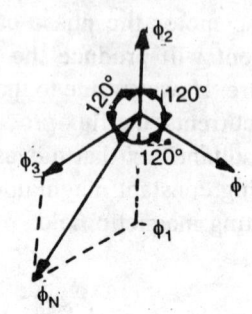

So
$$\phi_1 = \phi_m \cos \omega t$$
$$\phi_2 = \phi_m (\omega t + 120)$$

and
$$\phi_2 = \phi_m (\omega t + 240)$$
$$= \phi_m \cos (\omega t + 120)$$

To find out the resultant magnetic field, let these be resolved horizontally and vertically.

Horizontal components:

Fig. 19.4. 33 star connected motor.

$$\phi_H = \phi_1 - \phi_2 \cos 60 - \phi_3 \cos 60$$

$$= \phi_m \cos \omega t - \phi_m \cos(\omega t + 120)\cos 60 - \phi_m \cos(\omega t - 120)\cos 60$$

$$= \phi_m \cos \omega t - \phi_m \cos 60\left[\cos(\omega t + 120) + \cos(\omega t - 120)\right]$$

$$= \phi_m \cos \omega t - \phi_m \cos 60\left[2\cos \omega t \cdot \cos 120\right]$$

$$= \phi_m \cos \omega t - \phi_m \times \frac{1}{2} \times \cos \omega t \left(-\frac{1}{2}\right)$$

$$= \frac{3}{2}\phi_m \cos \omega t$$

Now vertical components:

$$\phi_v = \phi_1 \sin 0 + \phi_2 \sin 60 - \phi_3 \sin 60$$

$$= \phi_m \sin 60\left[\cos(\omega t + 120) - \cos(\omega t - 120)\right]$$

$$= \phi_m \sin 60\left[2\sin \omega t \sin 120\right]$$

$$= \phi_m \times \frac{\sqrt{3}}{2} \times 2 \times \sin \omega t \times \frac{\sqrt{3}}{2}$$

$$= \frac{3}{2}\phi_m \sin \omega t$$

For resultant:

$$\phi_R = \sqrt{\phi_v^2 + \phi_H^2}$$

$$= \sqrt{\left(\frac{3}{2}\phi_m \sin\omega t\right)^2 + \left(\frac{3}{2}\phi_m \cos\omega t\right)^2}$$

$$= \frac{3}{2}\phi_m \sqrt{\sin^2\omega t + \cos^2\omega t}$$

$$= \frac{3}{2}\phi_m = 1.5\,\phi_m$$

So the resultant magnetic field has $1.5\,\phi_m$ magnitude

and $\tan\theta = \dfrac{\phi_v}{\phi_H} = \dfrac{3 \times \phi_m \times \sin\omega t \times 2}{2 \times 3 \times \phi_m \times \cos\omega t}$

$$= \tan\omega t$$

hence the magnetic field at any other instant also, has the same $1.5\,\phi_m$ magnitude revolving at a particular speed *i.e.* ωt radian/sec. So it is clear that the magnetic produced by three phase has a constant magnitude rotating at a constant speed.

Q. What is the working principle of an induction motor? Explain how the torque is developed.

Ans. The action of the induction motor depends upon the production of the rotating magnetic field. In the rotor the e.m.f. is not given from outside but it is induced by induction, so it is called the induction motor.

Working principle. Whenever a current carrying conductor is placed in the magnetic field, a force is exerted on it, which tends to rotate it.

The rotor of an induction motor has a number of copper bars which are short circuited at both the ends by means of end rings. When the rotor is placed in the rotating magnetic field, an em.f. is induced in the rotor bars. These bars are short circuited so a current will flow depending upon the impedance of the rotor winding, hence producing a torque to rotate the rotor. Figure 19.4 shows a rotor placed in the stator magnetic field at any particular instant. Let the stator field is rotating in clockwise direction now the relative speed of the magnetic field with respect to the rotor bars is in anticlockwise direction. Due to induction, an e.m.f. is induced in the bars. Now the bars are short circuited hence a current will flow through the bars, the direction of current can be determined by the Fleming's right hand rule. It is understood, that the current carrying conductor is placed in the magnetic field produced by the stator, so a torque as shown in Fig. 19.5 is excreted on the bars and the rotor rotates in the clockwise direction. The direction can be determined by the Fleming's left hand rule.

Thus the rotor will follow the same direction in which the rotating magnetic field is rotating.

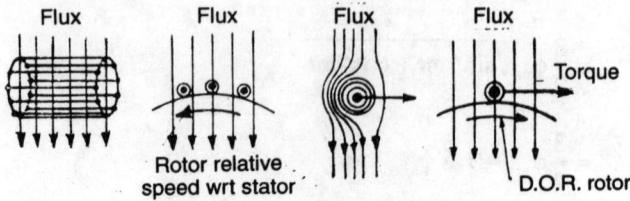

Fig. 19.5. Prodution of torque.

Q. What do you understand by the terms synchronous speed, rotor speed, slip speed and slip?

Ans. Synchronous speed. The speed of the rotating magnetic field is known as the synchronous speed. It depends upon the frequency and the number of poles of the motor. It is denoted by N_s and its unit is r.p.m.

$$N_s = \frac{120f}{P}$$

Where N_s = the synchronous speed in r.p.m.

f = the supply frequency in c/s.

P = the number of poles

The synchronous speed decreases with the increasing of the poles and increases with the increasing of supply frequency or *vice versa*. For a motor running at 50 c/s mains having two poles the synchronous speed will be 3000 r.p.m. for four pole – 1500 r.p.m. and so on.

Rotor speed. It is also known as the actual speed. The speed at which the rotor rotates is known as the rotor speed or actual speed. It is represented by N_r and measured in r.p.m. It is measured by means of tachometer or the speedometer. It is always less then the synchronous speed for the induction motors. It is maximum at no load and decreases with the increasing of load.

Slip speed. Obviously the rotor of an induction motor follow the direction of the rotating magnetic field produced by the stator. The speed of the rotating magnetic field is the synchronous speed. So the difference between the synchronous speed and rotor speed is known as the slip speed.

∴ Slip speed = $(N_s - N_r)$ r.p.m.

Slip. The ratio of the slip speed and the synchronous speed is known as the fractional slip but if shown in percentage, it is called the slip.

$$\therefore \quad \text{Fractional slip} = \frac{\text{Synchronous speed} - \text{Rotor speed}}{\text{Synchronous speed}}$$

$$= \frac{N_s - N_r}{N_s}$$

and % slip

$$= \frac{N_s - N_r}{N_s} \times 100.$$

At no load the slip is very less but increases with the increasing of load. The greater will be the load, more will be the slip to enable the motor for necessary torque to over come the load. The full load slip of normal induction motor varies from 1 to 6%.

Example. *A four pole 440 V 50 c/s induction motor is running at 1440 r.p.m., calculate the synchronous speed and the percentage slip.*

Solution. The synchronous speed

$$N_s = \frac{120f}{P}$$

$$= \frac{120 \times 50}{4} = 1500 \text{ r.p.m.}$$

Now the %slip

$$= \frac{N_s - N_r}{N_s} \times 100$$

$$= \frac{1500 - 1440}{1500} \times 100$$

$$= 4\%$$

Speed = 1500 r.p.m. and slip is 4%. Ans.

Example. *Find the full load slip and number of poles of a motor running at 2880 r.p.m. at 50 c/s mains.*

Solution. In case of the induction motor the rotor speed is nearly equal to the synchronous speed, so let the poles be calculated on 2880 r.p.m.

$$\therefore \quad P = \frac{120f}{N}$$

$$= \frac{120 \times 50}{2880} = 2.083 \text{ poles}$$

The poles cannot exists in fraction, so the poles are two. Now the synchronous speed

$$N_s = \frac{120f}{P} = \frac{120 \times 50}{2} = 3000 \text{ r.p.m.}$$

and %slip

$$= \frac{N_s - N_r}{N_s} \times 100$$

$$= \frac{3000 - 2880}{3000} \times 100$$

$$= 4\%$$

Slip is 4% and Poles are two. Ans.

Example. *An eight pole induction motor is supplied by a two pole alternator running at 3000 r.p.m., calculate the percentage slip if the motor is running at 725 r.p.m.*

Solution. The number of poles of the alternator is 2 and speed is 3000 r.p.m. so the frequency.

$$f = \frac{PN}{120} = \frac{2 \times 3000}{120} = 50 \text{ c/s}$$

Now the motor is running at 725 r.p.m. *i.e.* $N_r = 725$ r.p.m.

So %slip

$$= \frac{N_s - N_r}{N_s} \times 100$$

Now the

$$N_s = \frac{120f}{P}$$

$$= \frac{120 \times 50}{8} = 750 \text{ r.p.m.}$$

$\therefore$ $\%\text{slip} = \dfrac{750 - 725}{750} \times 100 = 3.33\%$ **Ans.**

Q. Derive the relationship between the rotor current frequency and supply frequency, also state the different effects of rotor frequency.

Ans. The frequency of the rotor current depends upon the rate of cutting of the flux by the rotor bars. In other words the relative speed of the rotor and rotating magnetic field.

Hence rotor e.m.f. frequency $= \dfrac{\left(\begin{array}{l}\text{Relative speed in r.p.m. of the rotor}\\ \text{and rotating magnetic field} \times P\end{array}\right)}{120}$

$$= \frac{(N_s - N_r)P}{120}$$

Now multiplying and dividing the numerator by N_s.

$$= \frac{\left(\dfrac{N_s - N_r}{N_s}\right) \times PN_s}{120}$$

$$= \left(\frac{N_s - N_r}{N_s}\right) \times \frac{PN_s}{120}$$

Here $\dfrac{PN_s}{120}$ is known as the supply frequency and $\dfrac{N_s - N_r}{N_s}$ the slip.

So rotor frequency = supply frequency × slip.

$$f_r = f_s \times s$$

or $$f_r = sf_s$$

There are the following effects of the rotor frequency:

(*i*) With the increase of this frequency the rotor reactance increases. Hence the rotor impedance is proportional to the rotor frequency.

(*ii*) With the increase in impedance the power factor of rotor circuit decreases so the rotor power factor is inversely proportional to the rotor frequency.

Q. Derive an expression for:

(*a*) **Starting torque of an induction motor.**

(*b*) **Running torque of an induction motor.**

(*c*) **Condition for maximum torque.**

Ans. At the time of starting, when the rotor is stationary, the e.m.f. induced in the rotor is because of the transformer action. The frequency in the rotor will also be the supply frequency.

Let R_2 = Rotor resistance per phase in Ohms.

X_2 = Rotor reactance per phase in Ohm.

Z_2 = Rotor impedance per phase in Ohm.

I_2 = Current in rotor in Amp.

$$\cos \phi_2 = \text{Rotor power factor.}$$

$$E_2 = \text{Induced e.m.f. in rotor in Volts.}$$

The movement of the rotor depends upon the production of torque which is the turning and twisting moment of force about its axis. So the torque obtained at the time of starting is known as the *starting torque* and torque in running condition is known as the *running torque*.

The torque $\propto$ flux $\times$ rotor current $\times$ rotor power factor.

$$\propto \phi \cdot I_2 \cdot \cos \phi_2$$

But $\qquad E_2 \propto \phi$

So $\qquad T \propto E_2 I_2 \cos \phi_2$

In stand-still condition the rotor impedance

$$Z_2 = \sqrt{R_2^2 + X_2^2}$$

and $\qquad \cos\phi_2 = \dfrac{R_2}{\sqrt{R_2^2 + X_2^2}}$

as E_2 is the voltage in rotor, so the current

$$I_2 = \dfrac{E_2}{\sqrt{R_2^2 + X_2^2}}$$

Now substituting in the torque equation,

$$T_s = K \cdot E_2 I_2 \cos\phi_2$$

$$= K \cdot E_2 \cdot \dfrac{E_2}{\sqrt{R_2^2 + X_2^2}} \cdot \dfrac{R_2}{\sqrt{R_2^2 + X_2^2}}$$

$$= K \cdot \dfrac{E_2^2 \cdot R_2}{R_2^2 + X_2^2}$$

$$= K \cdot E_2^2 \dfrac{R_2}{Z_2^2}$$

The supply voltage being constant

$$= K_1 \dfrac{R_2}{Z_2^2} \text{ where } K_1 \text{ is another constant}$$

or $\qquad T_s \propto \dfrac{R_2}{Z_2^2}$

The starting torque considerably depends upon the rotor resistance.

(*b*) **In running condition.** In running condition the speed of the flux linking will change corresponding to the slip of the motor.

hence

$$E_r = sE_2$$

and

$$I_2 = \frac{sE^2}{\sqrt{R_s + (sX_2)^2}}$$

$$\cos\phi_2 = \frac{R_2}{\sqrt{R_2^2 + (sX_2)^2}}$$

Now torque,

$$T = K \times E_2 \times \frac{sE_2}{\sqrt{R_2^2 + (sX_2)^2}} \times \frac{R_2}{\sqrt{R_2^2 + (sX_2)^2}}$$

$$= \frac{K \cdot sE_2^2 \cdot R_2}{R_2^2 + (sX_2)^2}$$

$$= \frac{K \cdot sE_2^2 R_2}{R_2^2 + s^2 X_2^2}$$

Thus the running torque depends upon the slip.

(*c*) **Condition for maximum torque.** The starting torque is

$$T_s = \frac{K_1 R_2}{R_2^2 + X_2^2}$$

Now differentiating T_s w.r.t. rotor resistance

$$\frac{dT_s}{dR_2} = K_1 \left[\frac{1}{R_2^2 + X_2^2} - \frac{R_2 \cdot 2R_2}{\left(R_2^2 + X_2^2\right)^2} \right] = 0$$

or

$$\frac{1}{R_2^2 + X_2^2} = \frac{2R_2^2}{\left(R_2^2 + X_2^2\right)^2}$$

or

$$2R_2^2 = R_2^2 + X_2^2$$

or $$R_2^2 = X_2^2$$

or $$R_2 = X_2$$

So the starting torque will be maximum when the rotor reactance per phase will be equal to the rotor resistance/phase.

Relation between the torque and slip, torque and power factor. At no load the speed is nearly equal to the synchronous speed and decreases with the increasing of load. So at no load the slip is less and torque is also less. When load is put on the motor the speed decreases and slip increases. With the increase of slip more current is drawn by the motor and thus more torque is produced.

So $\qquad$ torque $\propto$ slip.

Torque and power factor. At no load the power factor of the motor is low. So the torque is also less, with the increasing of load the power factor and torque also increases, so the torque

Torque α Rotor power factor.

Q. State the characteristics, advantages, disadvantages and uses of the squirrel cage and slipring induction motors.

Ans. Followings are the characteristics:

(a) **Squirrel cage induction motor:**

  (i) *Starting current.* The starting current is about 5 to 6 times the full load current, at no load the current is less and increases with the increasing of load.

  (ii) *Starting torque.* The starting torque of single cage rotor is less than that of the double squirrel cage induction motor. The torque increases with the increasing of load. The maximum torque is obtained at about 87% of the load.

  (iii) *Speed.* The speed at no load is nearly equal to the synchronous speed and slightly falls down with the increasing of load. The fall in speed is about 1 to 5% of the synchronous speed.

  (iv) *Power factor.* The power factor is low and increases with the increasing of the load.

  (v) *Application.* The single cage induction motors are commonly used and are the general purpose motors. The single cage motor should not be started with load. These motors are used in lathe machine, drill machine, and other machine tools, fans etc. The double cage motor is used where high starting torque and low starting current is required, as in case of conveyors, compressors etc.

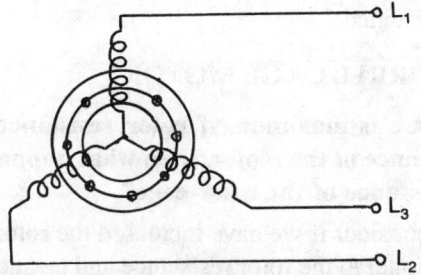

Fig. 19.6. Squirrel cage motor.

Advantages. Following are the advantages:
  (*i*) Easy and robust construction.
 (*ii*) Low cost.
(*iii*) No rubbing part on the rotor, so maintenance cost is less.
(*iv*) Starting is easy.

Disadvantages. Followings are the disadvantages:
  (*i*) High starting current.
 (*ii*) Speed cannot be adjusted.
(*iii*) Cannot be started with full load except double cage induction motor.

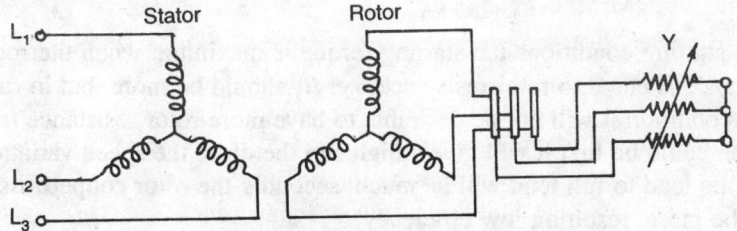

Fig. 19.7. Slipring Induction motor.

(*b*) **Slipring induction motor.** This motor has high starting torque and low starting current. Its speed is maximum at no load and drops with the increasing of load. The speed can be controlled up to 5% by connecting the external resistance in the rotor circuit.

Application. It is widely used for high starting torque and low starting current as in centrifugal pumps, elevators, etc.

Advantages. Followings are the advantages:
  (*i*) High starting torque.
 (*ii*) Low starting current.
(*iii*) Speed control is obtained up to 5%.

Disadvantages. Followings are the disadvantages:
  (*i*) It has more cost.
 (*ii*) More maintenance, because the rotor is to be connected to the external circuit.

(*iii*) Starting is not so easy as that of squirrel cage motor because of external resistance.

DOUBLE S QUIRREL CAGE MOTOR

Q. Starting torque is maximum if rotor resistance is equal to the starting reactance of the motor, than what happens if we increase the rotor resistance of the rotor cage?

Ans. Now let us consider if we have increased the rotor resistance, since the slip is proportional to the rotor resistance and torque too, than the slip will increase and the full load speed will fall considerably causing rotor current more power loss in rotor and considerably less efficiency of the motor.

Q. What is the actual conflict in this solution, explain?

Ans. From the torque equation,

$$T_{ST} \propto \frac{R_2}{R_2^2 + X_2^2} \quad \left(\text{it leads to } R_2 = X_2\right)$$

and

$$T_{run} \propto \frac{S \cdot E_1^2 \, R_2}{R_2^2 + s^2 \cdot X_2^2}$$

In starting conditions the starting torque is maximum when the rotor reactance is equal to rotor resistance, i.e. R_2 should be more, but in running condition it will not be desirable to have more rotor resistance (the reason could be first it will give a high slip therefore the speed variation from no load to full load will be much; secondly the rotor copper losses will be more, resulting low efficiency).

If we keep R_2 small it will give small slip thereby small variation from no load to full load, secondly low rotor copper loss, high efficiency so the conflict how to achieve this in a single motor is obvious.

Q. How this problem is over come?

Ans. It could be possible by having an arrangement of introducing some external resistance in the rotor circuit at the time of starting and by taking out from the circuit in running conditions of the motor. In case of wound rotor it is possible by introducing external resistance in rotor circuit at the time of starting and taking out in running conditions, with the help of slipring induction motor starter. But in case of squirrel cage induction motor it is not possible to introduce any external resistance into the squirrel cage, rotor winding which is permanently short circuited at both the ends.

Thus if we have low resistance, it is not good for starting and if more it is good for starting but poor for running condition. This type of intermittent starting and running is tolerable but for industrial operation it is not desirable.

So when a high starting torque and a small slip on load is required with the squirrel cage rotor a double cage rotor can be used. Secondly with a wound rotor, a low rotor resistance is designed to suit maximum running efficiency and other desirable properties; with an external resistance so that it could be introduced in rotor circuit at the time of starting and taken out step by step as the motor attains speed.

Q. Explain the construction and working principle of double squirrel cage motor.

Ans. It is a type of squirrel cage induction motor with a speciality of different type of rotor having not only one cage but two different independent cages of different resistances and cross sectional areas.

Construction

The squirrel cage motor has two main parts the stationary and rotory parts. The stationary parts are those which remain stationary during the operation of motor viz, body or yoke, side covers, bed sheet, flench covers, terminal box etc. The rotory parts are those which rotates during the working of motor viz. rotor, fan, shaft, bearings, etc.

The stator of the motor has slots and in them a three phase winding whether in star or in delta is done according to the designing of the motor winding.

The rotor of this type of motor is specially designed. There are two distinct rotor windings say two different cages. The stempings are respectively designed having provision of two separate windings, one having rotor bars close to the surface and other having some distance i.e., deep into the rotor stempings. Both the cages are short circuited independently at both the ends, i.e., in the same fashion as that of single squirrel cage rotor winding.

The resistances of both inner cage and outer cage are different. The outer cage have high resistance compared to inner cage. It is generally made of iron or brass bars. Since this cage is near the periphery, the reactance of the windings due to magnetic leakage will be less. The inner cage is made of low resistance material wire such as copper deeply embedded, as a result more flux as compared to upper cage will be linking and resulting more reactance. The resistance of both cages (apart from the selection of material) is made different by choosing different cross sectional area of the conductors. The outer cage is made of less cross sectional area resulting more resistance (A is less, ρ is more so R-will be more). The inner cage is made of larger cross sectional area resulting less resistance (A - is more ρ is less so R is also less).

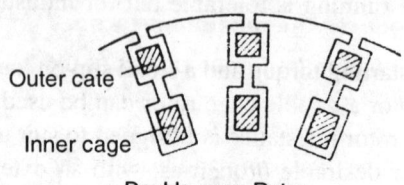

Double-cage Rotor

Fig. 19.8

Working

When the three phase supply is given to the stator a rotating magnetic field is produced. The flux links the rotor conductors and according to the Faradays Laws of electromagnetic induction an e.m.f. is induced in the rotor bars. Initially when the rotor is stand-still, the magnetic flux linking the bars, will have the same frequency as that of supply voltage. The flux links both the windings *i.e.*, cages inducing an emf. As both the squirrel cages are short circuited, by the independent end rings, the circulating current starts flowing. The quantity of current will be different in both the cages. The inner cage have maximum reactance resulting more impedance and so less current in this cage bars. The upper cage being on top results less reactance and hence the impedance, as compared to inner cage will be considerability less as a result more current will flow through the upper cage. Thus the starting torque produced by the outer cage will be more (the outer cage has more resistance). There will be less current flowing through innercage (which has less resistance) thus the torque produced will also be less.

As the motor speeds up, the slip speed comes into existence and this the flux linking the rotor conductors reduces its speed and hence frequency. The reactance of the inner cage becomes less resulting less impedance and more current. The outercage has more resistance as compared to inner cage causes more impedance and as a result less current and less torque.

Thus in starting the motor is started mainly because of outer cage (having more resistance) and in running condition the torque is produced by the inner cage (having less resistance). Thus we get high starting torque and less starting current with high running efficiency.

Q. Can we have different characteristics of performance by the motor, if yes how?

Ans. Yes, we can have up to some extent by changing or having variation into the types of slots their depth, their size, types of conductors, their materials and their placement too.

Q. Which cage produces the maximum torque?

Ans. Obviously the maximum torque occurs at the fractional slip which is equal to the ratio of R_2/X_2.

The outer cage will have $R_2/X_2 \approx 1$ so that maximum torque will be developed by outer cage *in starting* and least by the inner cage, and when the slip is minimum the inner cage will have R_2/X_2 for the inner cage nearly equal to one, so the maximum torque will be produced by the inner cage.

Q. Do we have a option to have a total characteristic of almost any desired performance?

Ans. Yes by varying the ratios, the resistance of the cages; the reactance, shape and position of the slots and conductors we can have almost any desired performance.

Q. Can you illustrate how the power factor and the quantity of the load is co-related?

Ans. The power factor of an induction motor depends upon the condition of the load. The power factor without load is low and it increases with the increasing of load on the motor as shown in Fig. 19.9.

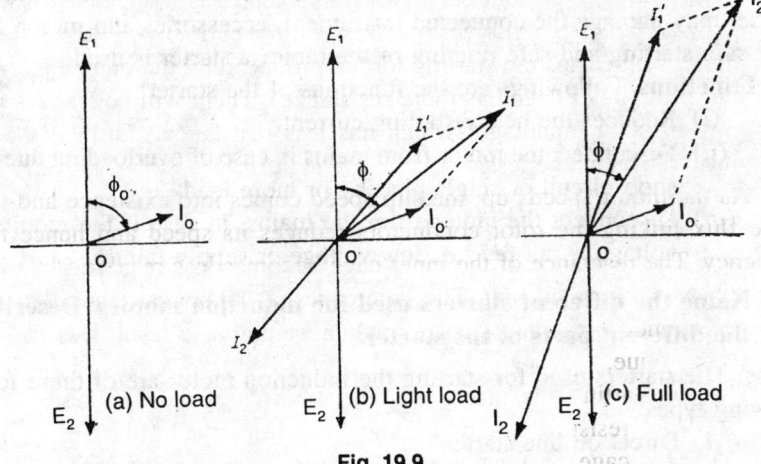

Fig. 19.9

Here as shown in Fig. (*a*) the motor is running without load so mainly the magnetising current which is a no load current is flowing. This current has more lagging power *i.e.*, the power factor is very poor. When the motor is loaded half the current I_1 will be as shown in Fig. (*b*) and the power factor will be better i.e. the angular displacement will be less than the no load condition; resulting better power factor. In figures when full load current is flowing, than the sum of the full load current and no load current will lead the angle i.e. the angular displacement will be lesser than the previous value hence we can say that only by putting full load on the motor the power factor is improved.

Q. What is the starter, explain the function and necessity of starter for induction motors?

Ans. The starter is a device which is used for the safety of A.C. motors. A starter is used to reduced and heavy starting current. An induction motor behaves like a transformer whose secondary winding, is short circuited. The stator winding, which is fed by the supply behaves like a primary winding of the transformer and rotor winding as the secondary of the transformer. Obviously if the secondary winding of a transformer is short circuited it draws heavy current from the mains. Similarly in case of induction motor the stator winding draws heavy current at the time of starting because the rotor is stand still. The induction motor of small horse power can be started by connecting across the line without damage to the motor. However because of the voltage disturbance created on the line by their heavy starting current, the motors larger than 5 H.P. are usually started with reduced voltage. The heavy starting current, otherwise, may damage the connected instrument, accessories and motor. So for safe starting and safe running of the motor a starter is used.

Functions. Followings are the functions of the starter:
   (*i*) Reduces the heavy starting current.
   (*ii*) Disconnect the motor from mains in case of overloading due to short circuit or single phasing or more load.
   (*iii*) Disconnect the motor from the mains, in case if the required voltage is not fed *i.e.*, low voltage or supply failure.

Q. Name the different starters used for induction motors. Describe the different parts of the starter.

Ans. The starters used for starting the induction motor are of these following types:
   1. Direct on line starter.
   2. Star-delta starter (manual and automatic).
   3. Auto-transformer starter.
   4. Slipring induction motor starter.
   The different parts of the starter and as follows:
  1. **Contacts.** There are the following types of the contacts:

(*i*) Fixed contact.

(*ii*) Moving contact.

These contacts are made of a good conductor, as copper. The size of these contacts depends upon the current carrying capacity of the starter. The surface of these contacts is silver plated. The surface is coated to minimise the sparking. A special shape generally oval is given to the contacts to avoid any possibility of sparking etc.

These contacts are fixed over an insulated material as bakelite or fiber. According to the position these contacts are of two types:

(*i*) The contacts which are close in normal condition and opens while operated, are known as normally *closed type contacts.*

(*ii*) The contacts which are open in normal condition and closes while operated, are known as normally *opened type contacts.*

(*a*) *Fixed contacts.* The contacts which are fixed on the frame and do not move are known as the fixed contacts. The supply and motor connections are made to them.

(*b*) *Moving contacts.* These contacts are not fixed but move during the operation of the starter, so these are called and moving contacts. These are mounted over the insulated construction which is attached with the cores of no volt coil. Some springs are provided with the plunger due to which the moving system *i.e.* cores of the no volt coil remains in normal position. The whole assembly is known as plunger. When the no volt coil is energised the moving core is attracted the circuit of the starter is completed.

2. **No volt release coil.** This coil is also known as the holding magnet. It is wound with the thin conductors having more number of turns over a bobin and generally operates on two phases *i.e.* 440 V. One line to this coil comes directly and second through the stop button, start button, over-load tripping contact and other safety and controlling accessories and devices. There are two cores of *E* shape, one is fixed and other is moving which is connected with the assembly of moving contacts. When the coil is energised it attracts the plunger as a result of this, the line goes to the motor and in case of de-energised condition the plunger remains in open position due to springs and sometimes due to weight of the contacts.

3. **Over load release coil.** These are used in the starter for the purpose of avoiding the motor from taking an over load current. These coils are connected in series with each line. These over load coils are of the following types in different types of the starters:

(*i*) Thermal relay.

(*ii*) Magnetic relay.

(*iii*) Desh pot oil type relay.

(i) *Thermal relay.* In this type of the relay a heating element is placed over a bimetalling strip. The element is connected in series with the line. When the normal current flows through the coil, heat is developed but does not exceeds the limit to bend the strip. In case of over loading the heat ($\alpha I^2 Rt$) developed bends the strip in a definite direction. It opens the over load tripping contact and finally opens the not volt release coil. The angle of bend depends upon the value of current in the heating element. The is proportional to the current carrying capacity of the motor *i.e.* h.p. of the motor. In starters of more current capacity the strip is itself connected in series with the line.

These are also known as the bimetallic thermal relay as shown in Fig. 19.10.

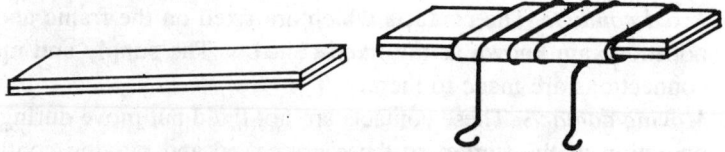

Fig. 19.10. Bimetallic relay.

(ii) *Magnetic relay.* In this relay a coil of wire having less number of turns is wound over a bobin. An iron rod is placed under the coil so that this rod may pass through the centre of the coil. The distance between the coil and iron rod can be adjusted. This coil is connected in series with the line and motor circuit, the motor current flows through the coil. Whenever current flows through the coil, it is energised and produces the flux and the rod is attracted. The force of attraction depends upon the amount of current, more current more force less current less force. In case of more current through the coil, the predetermined value of the force exceeds and rod is attracted as a result the rod strikes the tripping contact and the circuit of no volt release coil is disconnected demagnetising the coil as a result the plunger is released to off position. This relay is used in starters of more current capacity such as star delta starter, slip ring induction motor starter etc.

(iii) *Desh pot oil type relay.* It is also a magnetic relay with time leg arrangement. A number of motors are over loaded for a little time as in the case of the rolling mills etc. The over loading is for a small period. The starter may not trip for this over loading an arrangement of time leg is provided. In this case a pot containing oil is attached under the magnetic relay. The iron rod containing a disc and sometimes disc with different holes is provided in the oil. When the over load comes and the force exceeds, the rod is attracted but because of the oil viscosity and friction it takes sometimes to trip the starter but during that period the over loading is cleared. If the over loading continues for some more time, the force pulls out the rod and finally the tripping contacts are opened, demagnetising the holding magnet.

The time leg can be increased or decreased by the adjustment of round plates having different diameter holes or sometimes by adjusting the distance of rod. In that adjustment the pot containing the oil is threaded up or down depending upon the requirement. These relays are used mostly in big starters of more current capacity e.g. in rolling mills, drawing mills etc.

(4) **Overload tripping contact and current adjusting system.** The tripping contact are connected in the circuit of no volt release coil and are normally closed type contact. One contact is fixed and the other is connected with the tripping contact rod. In case of over loading the contact is opened and no volt is demagnetised. The distance between the tripping contact and the thermal strip or rod in magnetic relay can be adjusted, more current more distance less current less distance.

(5) **Push button.** These are of two types the start button and stop buttons. The start button is normally open type contact and is closed only at the time of starting. The stop button is the normally closed type contact and is opened only in case of stopping the motor.

In case of remote control, all the start buttons are connected in parallel and the stop buttons in series of the main controls.

(6) **Box.** Generally a metal box is provided to accommodate all the parts of the starter with the provision of making connections of the line and motor. A screw is also provided to earth the box.

Q. How the phase sequence of an induction motor is determined?

Ans. The phase sequence test is performed to determine the starting and ending terminals of the windings. Phase sequence means the correct orientation of phases *i.e.* let phase *A* be started from zero degree than phase *B* should be started from $(0 + 120)°$ ahead and similarly phase *C*, from $(120 + 120)°$ ahead.

There are three sets of windings and six terminals are taken out; from these, three sets of windings can be easily determined by means of either series test lamp or megger, as shown in Fig. 19.9.

Then two sets of windings are connected in series as shown in Fig. 19.8. Let *A* and *B* sets are connected. A low A.C. voltage is supplied to one winding. Then the voltage is measured across one winding *i.e.* V_1 and across both windings *i.e.,* V_2. Now if V_2 is greater than V_1, then the connecting ends are of similar polarity *i.e.* A_2 and B_2. If V_2 is less then V_1 it means the connected terminals are of opposite polarity *i.e.* A_2B_1. Thus the polarity of two sets are known, similar test is conducted for third winding.

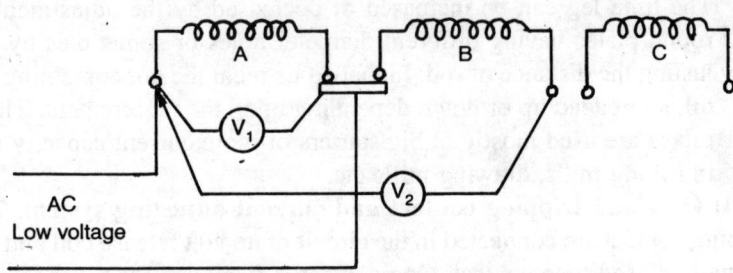

Fig. 19.11. Phase sequence test.

Once, if the polarities are known the motor can easily be connected in star or delta or star-delta.

Q. Draw the internal connection diagram of a D.O.L. starter and explain its working. **(N.C.V.T., 1983)**

Ans. This type of starters are used with the three phase induction motor up to 5 H.P. In this case the three windings of the motor are connected permanently either in star or in delta. These three ends are connected to the starter at M_1, M_2 and M_3.

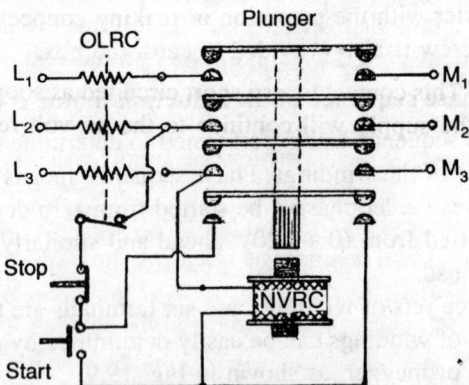

Fig. 19.12. (a) D.O.L. Starter

This starter connects the motor across full line voltage at the time of starting, so it is known as direct on to the line starter as shown in Fig.

19.12. The line to the motor goes through safety devices like over load release coil etc. The D.O.L. starters are available of different company, size and capacity.

The no volt coil is energised by two phases *i.e.* 440 V, here as shown L_3 line is directly connected to the no volt coil where L_2 line through the tripping contact, stop button and start button to the no volt release coil. The auxiliary contacts are connected across the start button.

Control circuit line diagram:

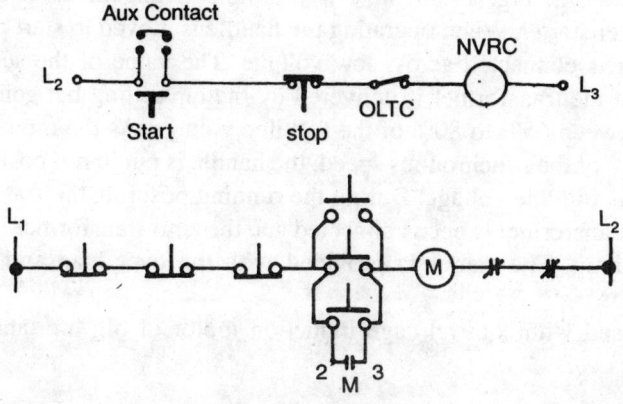

Fig. 19.12 (b).

Working. Press the start button to start the motor, L_3 is directly connected to no volt release coil, L_2 is going through tripping contact, stop and now through the start button to no volt release coil. The N.V.R.C. is energised and pulls the plunger so that the L_1, L_2 and L_3 are connected to M_1, M_2 and M_3 terminals respectively. The full line voltage is impressed to the motor which starts. To maintain the L_2 to no volt coil, an auxiliary contact is used. This contact is also short circuited as soon as the plunger is pulled thus the supply will continue to the no volt release coil even when the start button is released. Now the supply will continue and motor will be working.

To stop the motor press the stop button L_2 will be interrupted and the no volt release coil will de-energise releasing the plunger to off position.

In case of faults say over loading, the bimetallic relay will operate and operate the tripping contact as a result L_2 will be disconnected releasing the plunger to off position. In case if the voltage supply is not coming the no volt release coil will not magnetise and ultimately plunger will remain in off position.

Q. Draw the internal connection diagram of the three phase auto-transformer starter and explain its working. (*N.C.V.T., 1981*)

Ans. This starter is used to start the motor which are permanently connected in star or in delta. It has three phase auto-transformer. The low voltage tappings are taken for the required voltage, so that in starting the motor can be connected across low voltage. The auto-transformer is employed just for starting purpose. When the motor attains the speed full line voltage is impressed to the motor.

As shown in Fig. 19.13 there is a simple arrangement of an auto-transformer starter. When operating the handle is moved to start position, the motor is connected across low voltage. The value of the secondary voltage of the transformer is delivered by the tap setting but generally it varies between 65% to 80% of the full line voltage. As the motor attains about 75% of the synchronous speed, the handle is put in run position and motor gets full line voltage. During the running position, the star point of the auto-transformer is open connected and the auto-transformer becomes out of circuit. The starter is provided with the over load and no volt protections.

It is used with squirrel cage induction motor of big h.p. and rotory converters etc.

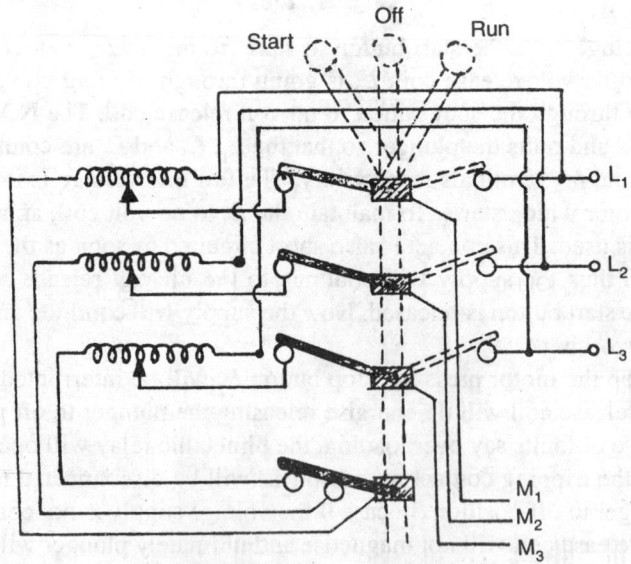

Fig. 19.13. Auto-transformer starter.

Q. Describe the starter used for three phase slipring induction motor with proper controllings.

Ans. The rotor of three phase slipring induction motor is wound. The rotor winding is connected in star and the three ends are brought out to the phospher bronze sliprings. The external resistance which is also a star connected resistance with a number of steps, is introduced in the rotor circuit through these sliprings as shown in Fig. 19.14. At the time of starting full line voltage is impressed to the stator and full external resistance in the rotor circuit. As the motor attains speed, the external resistance is cut off gradually and finally the three sliprings are short circuited.

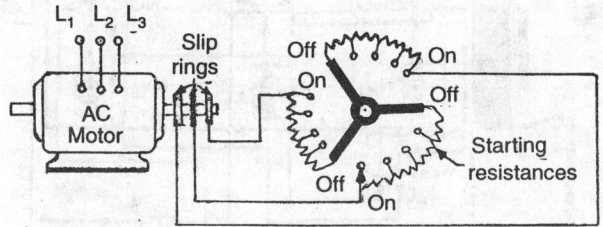

Fig. 19.14. Slipring Induction motor starter.

Mostly face plate type starters are used. The main safety is the interlocking that will not allow the contactor to 'ON' unless full external resistance is in the rotor circuit. The over load and no volt release coils are also provided with this starter. The big motors are sometimes provided with drum type oil immersed starters.

Q. Draw the connection diagram of star delta starter. What are the advantages of star delta starter?

(*N.C.V.T. 1963, 75, 76, 80, 81, 84*)

Ans. The Fig. 19.15 shows the connection diagram of a star delta starter.

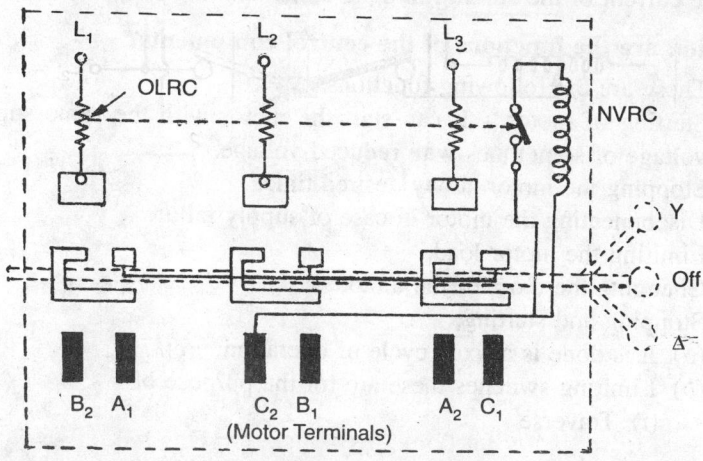

Fig. 19.15. Star delta starter.

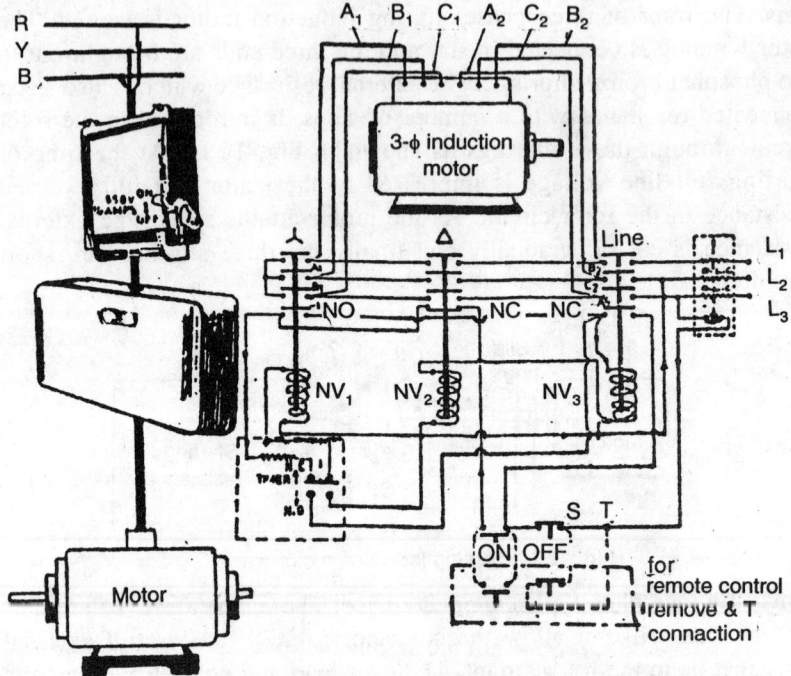

Fig. 19.16. 3φ plunger Automatic star Delta starter.

Advantages. In this starter, full line voltage is not fed to the motor but the voltage fed is $V_L/\sqrt{3}$ times. By connecting the motor in star the starting current is reduced to 1/3 of the starting current if it would be started in delta.

Disadvantages. The starting current is reduced to 1/3 times the full starting current of the starting in delta so the starting torque is less.

Q. What are the functions of the control equipments?

Ans. These are the following functions:

(*i*) Starting of motor – It can start the motor with the same supply voltage or sometimes with reduced voltage.

(*ii*) Stopping the motor at any desired time.

(*iii*) Disconnecting the motor in case of supply failure.

(*iv*) Limiting the motor load.

(*v*) Changing the d.o.r. of the motor.

(*vi*) Stopping and starting.

  (*a*) It is done is a fixed cycle of operation.

  (*b*) Limiting switches these are for the purpose of

    (*i*) Traverse

  (*ii*) Temperature

  (*iii*) Pressure

  (*iv*) Fluid level.

  This application is used for multispeed motors only.

(*vii*) Changing the speed.

(*viii*) Starting the motor with a definite sequence of speed.

Q. Write a short note on the speed control of A.C. induction motors.

(*N.C.V.T.*, *1970, 79*)

Ans. The induction motors are practically a constant speed motors but wound motors even then have some variations. Following are the methods of speed control:

(*a*) Rheostatic control method.

(*b*) Pole changing method.

(*c*) Frequency changing method.

(*d*) Carcade control – by controlling the value of current in the rotor circuit of the motor.

(*a*) **Rheostatic control.** The speed of the induction motor can be controlled below normal by this method, at constant load. It can be performed by these two ways:

  (*i*) *Rheostat in stator circuit.* In this method the voltage applied to the stator is controlled by connected the rheostat in the stator circuit as shown in Fig. 19.17. The speed depends upon the voltage applied to the stator. The speed will less *i.e.*, below normal.

  (*ii*) *Rheostat in rotor circuit.* It is only applicable in case of slipring induction motor. The variable resistance is connected in rotor circuit as shown in Fig. 19.7. At a constant load when the rotor resistance is increased the current will decrease but to maintain the rotor current, the rotor will slow down to increase the rotor voltage and maintaining the rotor current and thus the torque.

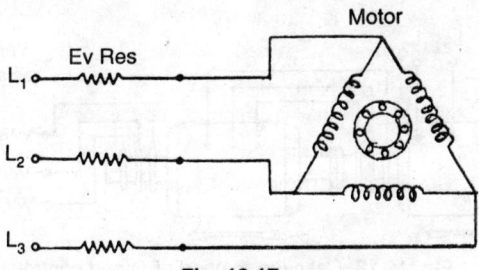

Fig. 19.17.

Disadvantages. This method has got some disadvantages:

(*i*) Speed cannot be raised above normal.

(*ii*) More power loss in the rheostat.

(*iii*) Less efficiency of the machine.

(*iv*) Wide range of speed controlling is not possible.

(*b*) **Pole changing method.** The speed is inversely proportional to the number of poles. More pole less speed, less pole more speed. The stator winding for this particular method is done in such a way that the number of poles can be varied. When supply is connected to more number of pole connections, the speed will automatically be less and *vice versa*. There is a change over switch which actually controls the line to the different pole connections. Generally variation of speed is in the ratio of 1:2.

(*c*) **Frequency changing method.** Obviously the speed of the motor is directly proportional to the frequency. If the frequency is less, the speed is also less and if frequency is more the speed is also more. A special arrangement is done to change the frequency of the supply, generally for this the frequency changer is used. This method is costly and naturally the efficiency of this operation is also less.

(*d*) **Cascade control method.** In this method two induction motors are mechanically coupled. The rotor of at least one motor is wound. The stator of first motor is connected to the supply and the rotor of the second motor is connected to the external rotor circuit. The speed in this case will be

$$N = \frac{120 \times f}{(P_1 + P_2)}$$

where N is the speed of the combination

f = Supply frequency

P_1 = Number of poles of first motor

P_2 = Number of poles of second motor.

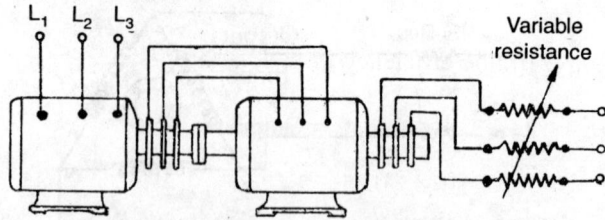

Fig. 19.18. Cascade method of speed control.

Q. What are the synchronous watts? Also state the different power stages of the induction motor.

Ans. It is a term generally used in case of an induction motor. It is defined as the torque which develops a power of one watt at the synchronous speed of the motor. It is the new unit of torque. It can be stated as

$$\text{Rotor input} = T_{\text{synch}} \times 2\pi N_s$$

or

$$T_{\text{synch}} = \frac{\text{Rotor input}}{2\pi N_s}$$

The different power stages can be stated as, the input given to the motor overcomes the stator copper and iron losses and the remaining power to the rotor. It is called as power input to rotor. This power overcomes the rotor copper losses and remaining the rotor output. It is the mechanical power developed, this power overcomes the windage and friction losses and finally the motor torque *i.e.* B.H.P. available on the shaft or pulley.

Different power stages of motor

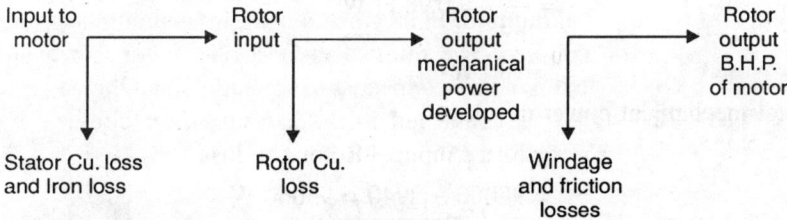

Example. *Calculate the line current of a 5 H.P. motor running at 440 V 50 c/s mains. The motor has 85% efficiency and 0.8 power factor.*

Solution. The output of the motor = 5 H.P.

$$= 5 \times 746 = 3730 \text{ W}$$

The power in three phase $$W = \sqrt{3} \, V_L I_L \cos \phi$$

$$\eta, \text{ the efficiency } = \frac{\text{Output}}{\text{Input}} \times 100$$

or

$$\text{Input} \times \eta = \text{Output}$$

$$\sqrt{3} \times 440 \times I_L \times 0.8 \times 0.85 = 5 \times 746$$

$$I_L = \frac{5 \times 746}{\sqrt{3} \times 440 \times 0.8 \times 0.85} = 7.2 \text{ A } \textbf{ Ans.}$$

Example. *The input to the particular induction motor is 100 kW. The stator copper losses amounts to 2000 W. Find (a) Rotor copper loss/phase. (b) The total mechanical power developed, if the slip is 3%.*

Solution. Stator input as given

$$= 100 \text{ kW}$$

The stator copper losses

$$= 2 \text{ kW}$$

Now stator output $= 100 - 2 = 98$ kW

The stator output is the rotor input

$$= 98 \text{ kW}$$

∴ Rotor copper loss

$$= \text{Slip} \times \text{Rotor input}$$

$$= \frac{3}{100} \times 98 = 2.94 \text{ kW}$$

Now rotor copper loss/phase

$$= \frac{2.94}{3} = 0.980 \text{ kW}$$

$$= 980 \text{ W}$$

Total mechanical power developed

$$= \text{Rotor input} - \text{Rotor Cu. loss}$$

$$= 98000 - 2940 = 95060 \text{ W}$$

$$= \frac{95060}{746} = \textbf{127.43 B.H.P. Ans.}$$

Example. *The rotor input to the rotor of a 400 V 50 c/s six poles three phase induction motor is 80 kW. The rotor e.m.f. is observed to make 100 complete revolution per minute. Calculate:*

   (i) *Slip*

  (ii) *Rotor speed*

 (iii) *The mechanical power developed.*

 (iv) *The rotor copper loss.*

  (v) *The rotor resistance/phase, if the rotor current is 65 A.*

Solution. The motor is having six poles running at 400 V 50 c/s mains.

Now $$N_s = \frac{120 f}{P} = \frac{120 \times 50}{6} = 1000 \text{ r.p.m.}$$

   (i) The rotor frequency

$$= \text{Slip} \times \text{supply frequency.}$$

and $$s = \frac{f_r}{f_s}$$

Here rotor makes 100 c/s is one minute so

$$f_r = \frac{100}{60} \text{ c/s}$$

$$\therefore \qquad s = \frac{100/60}{50} = \frac{100}{60 \times 50}$$

$$= 0.033$$

or in % $s = 0.033 \times 100 = \textbf{3.3\%. Ans.}$

(*ii*) Slip $$= \frac{N_s - N_r}{N_s} \times 100$$

$$3.3 = \frac{1000 - N_r}{1000} \times 100$$

$$1000 - N_r = 33$$

or $$N_r = 1000 - 33 = \textbf{967 r.p.m. Ans.}$$

(*iii*) Total mechanical power developed

$$= (1 - s) \text{ Rotor input}$$

$$= (1 - 0.033) \, 80$$

$$= \textbf{77.333 kW. Ans.}$$

(*iv*) Rotor copper loss

$$= s \times \text{rotor input}$$

$$= 0.033 \times 80$$

$$= 2.667 \text{ kW.}$$

The rotor total copper losses are 2.667 kW so the copper loss per phase

$$= \frac{2667}{3} = 0.889 \text{ kW}$$

or $$= \textbf{889 W. Ans.}$$

(*v*) The current is given 65 A and the rotor copper losses per phase.

$$= 889 \text{ W}$$

So $$I_2^2 R_2 = 889$$

$$R_2 = \frac{889}{65 \times 65} = \textbf{0.21}\,\Omega \quad \textbf{Ans.}$$

Q. (*a*) **What do you understand by** (*i*) **no load and** (*ii*) **short circuit tests of A.C. induction motor?**

(*b*) **What are the different methods of determining the slip of an induction motor?**

Ans. (*a*) (*i*) **No load test.** This test is performed to determine the no load current, no load power factor, windage and friction losses, no load core losses, no load resistance R_0, reactance X_0, and finally the no load input to the motor.

This test is performed by applying different voltages above and below the normal value. In this case the motor is run without load. Different instruments are connected to note down the values like current, voltage, power and power factor also.

One should clearly understand that the no load power factor will be below 0.5, so for obtaining positive readings the two wattmeter method of power measurement is applied that too the total power will be $(W_1 - W_2)$ watts.

(*ii*) **Short circuit test.** This test is performed to determine the short circuit current with normal voltage applied to stator winding; power factor on short circuit, total and equivalent resistance leakage reactance and impedance referred to stator.

In this case the rotor (slip ring Induction motor) is short circuit and variable low voltage is applied to the stator winding as shown in Fig. 19.19. This test is just similar to the short circuit test of the transformer. The voltage for the stator winding is so adjusted as to circulate full load current in the rotor which is held stationary. The voltage is gradually increased from zero value.

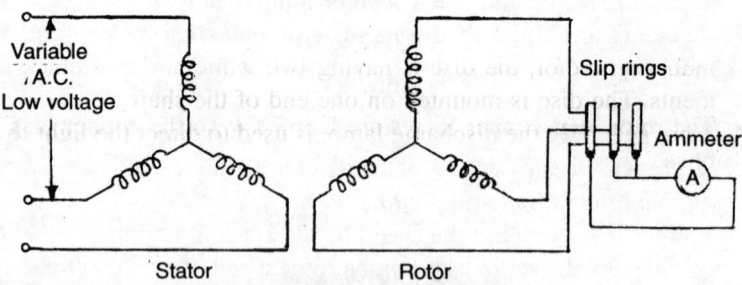

Fig. 19.19. Short circuit test.

(*b*) **Slip Measurement:** The slip of an induction motor can be found out by the following methods:

(*i*) By actual measuring the rotor speed.

(*ii*) By measuring the rotor frequency.

(*iii*) By means of stroboscopic method.

(*i*) **By actual measuring the rotor speed.** In this method the rotor speed is measured by means of the tachometer or speedometer. The synchronous speed can be known from the name plate data of the motor, then using the formula, the slip can be calculated at any load *i.e.*

$$s\% = \frac{N_s - N_r}{N_s} \times 100$$

(*ii*) **By measuring the rotor frequency.** In this test the rotor frequency which is very less, is determined by means of a centre zero galvanometer. This test is performed with wound rotor motor. In case of squirrel cage motor, it is not accessible. The deflection of the galvanometer is seen and the time is recorded. The number of oscillations during a particular time span is noticed to find the frequency of rotor current.

As the time of starting the frequency of rotor current is same as that of the supply frequency. So this test is only carried out when all the three sliprings are short circuited. Then the galvanometer is placed across two sliprings to note down the frequency of induced e.m.f.

$$f_r = s \times f_s$$

So $s = \dfrac{f_r}{f_s} = \dfrac{\text{rotor frequency}}{\text{supply frequency}}$

(*iii*) **Stroboscopic method.** This method is based on the stroboscopic effect of the light. In this method a circular disc having the same number of white and black segments alternately, is used. The total segments are equal to the number of poles e.g. for a four pole induction motor, the disc is having two white and two black segments. The disc is mounted on one end of the shaft.

A lamp, generally the discharge lamp, is used to direct the light to the disc as shown in Fig. 19.19. The discharge lamp will project the stroboscopic effect efficiently.

The motor is started. Let the speed be the synchronous speed then the disc will appear to be stationary, but the rotor speed of an induction motor is always less than the synchronous speed. Now because of the slip, the disc will appear to be rotating backward with some revolutions. If the motor is rotating in clockwise direction, the slip will appear to be rotating in the anticlockwise direction.

Thus by counting the number of revolutions taken during one minute will give away the slip speed of the motor. The synchronous speed can be determined from the name plate data. So the slip can be calculated.

$$\% \text{ slip} = \frac{\text{Slip speed}}{\text{Synchronous speed}} \times 100.$$

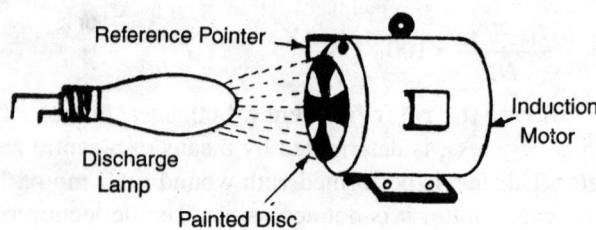

Fig. 19.20. Stroboscopic effect.

Q. Define: (*a*) **cogging,** (*b*) **crawling,** (*c*) **noise, in A.C. induction motors.**

Ans. (*a*) **Cogging.** When the induction motors are started, the rotor is stand still and stator winding is fed. The action at this time is similar to that of the transformer. There is a definite relationship between the stator and rotor slots; but if the number of stator slots and number of rotor slots are equal the machine may refuse to start because of internal magnetic locking between stator and rotor. This phenomenon is termed as *Cogging.* The cogging in induction motors is overcomed by making the rotor slots prime to the stator slots.

(*b*) **Crawling.** Sometimes the squirrel cage induction motor runs at a very slow but stable speed about one seventh of the synchronous speed. This phenomenon of the motor is termed as *crawling.* It is because of the different harmonics, but specially the seventh harmonic plays as important role. The seventh harmonic torque reaches to maximum positive torque just before the one seventh of the synchronous speed and the motor does not accelerate to the synchronous speed and crawl only at 1/7th of the synchronous speed.

(*c*) **Noise.** Sometime a motor exhibits a tendency to exhibit the excessive vibrations at a particular speeds below the normal speed of the motor. This sound resemble with a roaring, and is termed as noise.

REVIEW QUESTIONS

1. What do you understand by an induction motor? Why it is called so?

2. What are the constructional features of an induction motor?
3. What is meant by torque and how the torque is developed in case of 3ϕ squirrel cage induction motors?
4. What are the different types of the 3ϕ induction motors?
5. What is the working principle of an induction motor? Describe with neat sketch.
6. Describe the general principle of action of an induction motor. State how it got the name induction motor?   (*N.C.V.T. 1978, 79, 81*)
7. What are the conditions for getting good starting and running torques in induction motors?   (*N.C.V.T. 1972*)
8. Describe the constructional details of a slipring induction motor. How it is started?
9. What are the differences between the squirrel cage and slipring induction motors? Which type is more efficient and why?
10. What do you understand by the synchronous speed of an induction motor? How it relates with the rotor speed and why?
11. What do you understand by the slip in induction motor? What are the different methods of determining the slip, describe any one of them?
12. Write short notes on the following:
     Synchronous speed, rotor speed, slip, fractional slip, rotor frequency and rotor power factor.
13. What do you understand by the phase sequence of an induction motor and how to identify?
14. What is the necessity of a starter for starting the electric motor?
15. What are the different types of starter used for starting the 3ϕ induction motor? Describe in details D.O.L. starter with neat sketch.
16. Differentiate between D.O.L. starter and star delta starters. How star delta starter reduces the voltage at the time of starting the motor?
17. What are the different types of over load relays used in starters for the production of induction motor from over loading?
18. Define, cogging, crawling, noise, and rotor frequency in an induction motor.

NUMERICAL PROBLEMS

1. Calculate the synchronous speed of a 4 pole 440 V 50 c/s star connected A.C. motor.   [**Ans. 1500 r.p.m.**]
2. Calculate the % slip of an induction motor running at 1440 r.p.m. If the machine is having 4 poles and operating at 50 c/s.   [**Ans. 4%**]
3. Calculate the number of complete revolutions in two minutes by the rotor current of an induction motor if the percentage slip is 5% and the rotor speed is 482 r.p.m.   [**Ans. 105**]
4. A 12 pole 3-phase alternator driven at a speed of 500 r.p.m. supplies power to an 8 poles 3 phase induction motor. If the slip of the motor at full load is 3%. Calculate the full load speed of the motor.   [**Ans. 727.5 r.p.m.**]

5. A six pole 3φ induction motor runs at a speed of 960 r.p.m. and the shaft torque is 135.7 Nm. Calculate the rotor Cu losses if the friction and windage losses amount to 150 W. The frequency of supply is 50 c/s.

[Ans. 574 W]

6. A star connected 3φ induction motor draws 4 A when connected across a 400 V 50 Hz supply. If the power factor is 0.8 and efficiency of motor 90% how much B.H.P. will the motor develop. (*N.C.V.T. 1980*)

[Ans. 2.675 H.P.]

20

Synchronous Motor

The motor which always works on the synchronous speed, is known as the synchronous motor.

Q. Explain the working and construction of the three phase synchronous motor. *(N.C.V.T, 1960, 69, 70, 74, 76)*

Ans. The synchronous machine can work as an alternator and a motor both in similar way as that of the D.C machine can work as D.C. motor and D.C. generator both. The synchronous motor is that which runs at a constant speed i.e., synchronous speed. The speed depends upon the supply frequency and number of poles.

$$\text{Synchronous speed} = \frac{120 \times \text{frequency}}{\text{Number of poles}} \text{ r.p.m.}$$

i.e.
$$N_s = \frac{120 \times f}{P} \text{ r.p.m.}$$

Working. When the armature is connected to A.C. three phase supply, the magnetic field of the rotating nature is produced. The magnetic field produced by D.C. is the stationary field. Now when the current carrying conductors are placed in the magnetic field, so a torque is developed on the conductors.

The armature is wound for A.C. supply producing rotating magnetic flux. The rotor let us consider is wound for D.C. producing a steady flux. Let the first half of the a.c. sine wave is considered, the direction of torque be in anticlockwise, but during the next half cycle the torque will change to clockwise direction because of the change in direction of current in the second half cycle; thus a pulsating torque is developed and the rotor does not accelerate in any of the directions. So the synchronous motor is *not a self starting* one.

If by any of the external means the rotor is rotated and brought to the synchronous speed and the poles are excited the magnetic interlocking results. The north pole of rotor and south pole of stator will get magnetically locked and so on. The rotor which is excited will now rotate with the speed of rotating magnetic field *i.e.* the synchronous speed. Thus the motor works on the principle of magnetic interlocking and the speed will be the constant *i.e.* the synchronous speed.

Construction. The construction of the synchronous motor is the same as that of an alternator. It is also of two types the stationary field type and rotating field type as shown in Fig. 20.1. An excitor is also mounted on the same shaft to excite the field winding (for more details please see the construction of alternator on page 402).

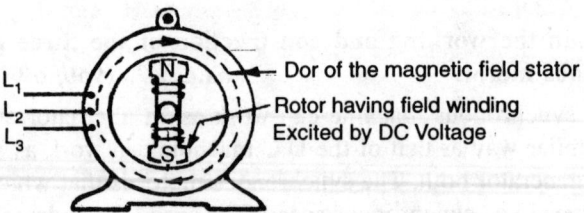

Fig. 20.1 Synchronous motor.

Q. How the stationary field type synchronous motor runs at synchronous speed?

Ans. In case of stationary field type A.C. synchronous motor, the d.c. supply is given to the stator and being stationary field construction, the field produced will be stationary in space. Now the three phase supply is given to the rotor through the three sliprings mounted on the same rotor as shown in Fig. 20.2. The winding is connected in star. Now the magnetic field produced by the winding is of rotating in nature at the synchronous speed.

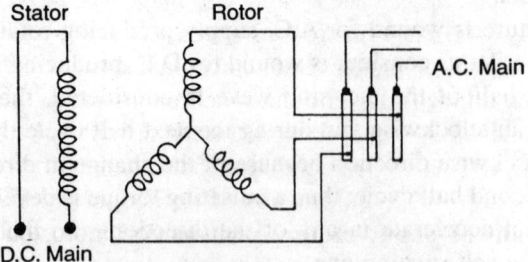

Fig. 20.2. Stationary field type synchronous motor.

There are two magnetic fields, the one of stationary in space and other rotating at synchronous speed now let the rotor is brought to the synchronous speed and then the rotor is excited, it means the relative speed of the magnetic field with respect to rotor is zero, hence the magnetic poles will appear to be stationary in space. At this time the field winding is also excited and these two magnetic fields are interlocked. The poles because of relative speed of rotor and field being zero, are stationary in space, thus the magnetic attraction will interlock both the fields. The rotor poles are stationary only as long as the rotor speed is synchronous speed. So to keep the inter-locking the motor will work on the synchronous speed. Then the external source of accelerating the rotor to synchronous speed is taken off and the motor will automatically run to the synchronous speed.

Q. Describe the different methods of starting the synchronous motor.

Ans. The synchronous motor is not a self-starting motor, so there are the various methods of starting the synchronous motor.

(a) **Starting as an alternator.** In this case the synchronous motor is started as an alternator by the external source. The e.m.f. is induced in the armature and is synchronised with the bus bar. After synchronising both the voltages the prime mover is taken off. The machine now will run as a synchronous motor. The motor can be started by means of diesel engine or d.c. motor etc.

(b) **Starting by means of d.c. source.** Generally for this purpose the d.c. compound motor is used, when the d.c. voltage is available easily. The d.c. motor is coupled with the synchronous motor. The speed of d.c. motor is controlled and brought nearly equal to the synchronous speed.

Now the stator winding is energised by the three phase supply, and the field switch is also on. There are now two magnetic fields, one of the stator winding rotating at synchronous speed, other because of the rotor field winding which rotating at synchronous speed by the external source. These fields are interlocked and the rotor will continue to run at synchronous speed.

The motor prime mover either can be taken off or if the d.c. motor it can be utilized to excite the synchronous motor.

Sometimes the synchronous motor can also be started by the excitor mounted on the same shaft. Then after synchronising it will again perform its main function of excitation.

(c) **By means of ponny motor.** An induction motor called as ponny motor, is coupled with the shaft of the synchronous motor. This motor has less number of poles than the synchronous motor; so that the running speed of the combination can be raised above than the synchronous speed of the motor. The field winding and armature both are energised as a result of which the magnetic interlocking results and the motor continues to run at the synchronous speed. Then the ponny motor is taken off.

(*d*) **Self-starting method.** So far it is discussed that extra means are adopted to start the synchronous motor, but this can be made self-starting by providing the following means:

(1) *By means of damper windings.* The damper winding consists of short circuited copper bars embedded in the face of field poles. Whenever three phase supply is given to the stator, a rotating magnetic field is produced in space. This field will link the dampers and will induce some e.m.f. in them. These dampers being short circuited causes a current to circulate through the damper winding. Now the motor will start as a squirrel cage induction motor. The no load speed of the motor is nearly the synchronous speed. The excitor is also rotating at the same speed, producing some voltage. Now when the speed reaches to the synchronous speed the field is excited. Now because both the rotor and stator are energised, so the magnetic interlocking will result and the motor will be running at the synchronous speed. In this condition there is no linkage, so there will not be any e.m.f. in the damper winding.

Following are the advantages and disadvantages of this method.

Advantages:

(*i*) No need of any starting means.

(*ii*) The motor will continue to run even after losing the synchronism. If the speed falls down the damper winding will come into operation and the torque will develop to run the motor.

(*iii*) It hunts the oscillations which are due to the loading conditions i.e. the damper are used to prevent the *hunting effect*.

Disadvantages:

(*i*) The motor starts as the squirrel cage induction motor so the starting current is more.

(*ii*) To avoid the danger of shock the field winding should be carefully handled. There are more number of turns in the field winding, so the voltage induced due to induction is much larger. It requires special arrangement for insulation and the field winding should be short circuited through a suitable resistance to limit the current and distribute the induced voltage.

(2) *By means of slipring induction motor action or by external resist-ance method.* The motor if started by using the damper winding procedure, it takes much current, to avoid this large current, the motor is started by using the slipring induction motor starting method as shown Fig. 20.3 and Fig. 20.4. The rotor resistance is introduced at the time of starting and thus current is reduced and limited with a good starting torque. The external resistance is then taken off step by step.

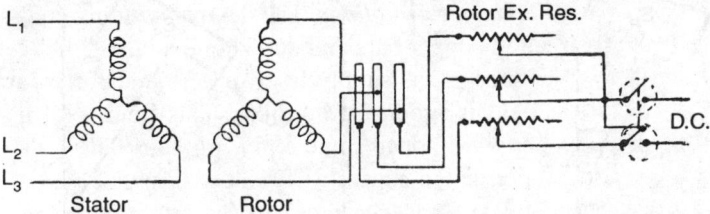

Fig. 20.3. Slipring induction motor action-I.

The no load speed is nearly equal to the synchronous speed. Now both operations, taking out the rotor external resistance and switch-ing on to d.c. source take place simultaneously. Thus the magnetic interlocking results and machine continues to run at the synchro-nous speed.

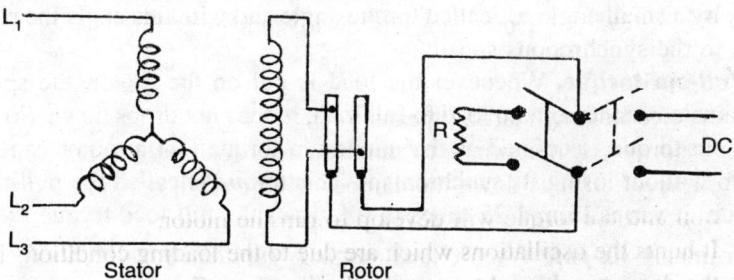

Fig. 20.4. Slipring induction motor action II.

Q. Define the pull-in-torque and pull-out-torques in the synchro-nous motor.

Ans. Whenever the synchronous motor is started using any of the starting method, the field winding is excited as soon as it comes nearly-to the synchronous speed. Now it is not essential that both the poles of stator and rotor are enlined. The rotor pole might be trailing or leading the stator opposite pole. Thus an angular displacement is observed. In that case the force of attraction between the opposite poles comes into existence and the rotor pole is pulled into enline as shown in Fig. 20.5.

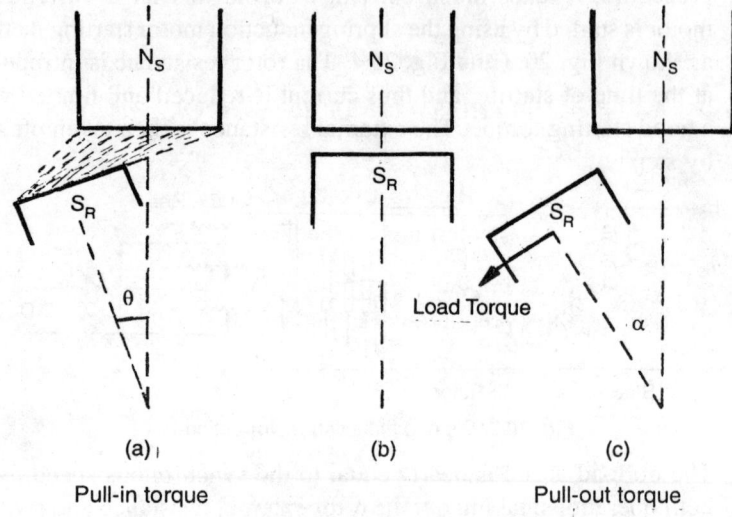

(a)

Pull-in torque

(b)

(c)

Pull-out torque

Fig. 20.5. Magnetic locking.

So the amount of the torque that a motor exerts when pulling into synchronism is called the pull-in-torque. Its magnitude differs motor to motor. As the driving source is taken off the rotor is retarded or pulled back by a small angle α_0, called torque angle and with this angle the rotor runs to the synchronous speed.

Pull-out-torque. Whenever the load is put on the motor, the speed remains constant from no load to full load, it does not drops down. So the pull-out-torque is defined as the maximum torque that a motor can develop without losing its synchronism. This torque is called the pull-out-torque. It varies from 1.25 times to 3.5 times the full load torque.

Example. *A six pole synchronous motor is supplied 1100V 50 c/s. The field excitation is adjusted so that the e.m.f. induced is equal and opposite to the voltage supplied. The synchronous reactance is 0.4Ω. If the angle of retard is 2° mech. at a certain load, find the armature current.*

Solution. The angle of retard is 2° mech. so the equivalent electrical angle can be determined.

Total electrical angle
$$= 360 \times \text{Number of pair of poles}$$

Here mechanical degrees are 2° so
$$= 2 \times 3 = 6° \text{ elect.}$$

The resultant e.m.f. at 2° mech. or 6° elect

$$E_r = \sqrt{V^2 + E^2 - 2E \cdot V \cos\theta}$$

$$= \sqrt{1100^2 + 1100^2 - 2 \times 1100 \times \cos 6°}$$

$$= 115.139 \simeq 115.14 \text{ V}$$

The armature current

$$= \frac{\text{Resultant voltage}}{\text{Synchronous impedance}}$$

Here the synchronous impedance is not given, the synchronous reactance be assumed as the synchronous impedance. So

$$I_a = \frac{115.14}{4} = 28.785$$

$$= \textbf{28.8 A. Ans.}$$

Q. What are the merits, demerits and uses of the synchronous motor?

Ans. Following are the merits and demerits of the synchronous motor:

(a) **Merits.** These are the following merits:

   (i) It gives the constant speed from no load to full load.

   (ii) The motor can be run as an alternator also.

   (iii) The power factor of the factory or organisation can be improved.

   (iv) It has higher efficiency specially low speed and unity power factor range.

Demerits. There are the following demerits:

   (i) Speed cannot be adjusted.

   (ii) More cost and maintenance.

   (iii) These are not self-starting unless special arrangements are made.

   (iv) It cannot be started with load.

   (v) External source of d.c. is required.

   (vi) There is a tendency to produce the hunting effect.

   (vii) If it is out of synchronism, this tends to stop unless some other arrangement is provided to continue.

Uses.   (i) It is used where a constant speed is required.

   (ii) It is used for power factor correction in the factories, subtractions and in power houses. For this purpose these motors are run even without load in over excited condition.

   (iii) It is used in rubber mills, cement factories and other big industries for driving the air compressors, driving fans, driving common shaft, blowers and pulverizers etc.

Q. Explain the behaviour of the synchronous motor on no load and on load conditions. Draw the vector diagram in support of the statements.

Ans. Whenever the motor is accelerated and brought to the synchronous speed, the field is excited as a result the magnetic interlocking results to drive the motor to the synchronous speed. The movement the driving source is taken off the rotor retard back by small angle, generally known as the torque angle say α_0 as shown in Fig. 20.6. It differ with different motors.

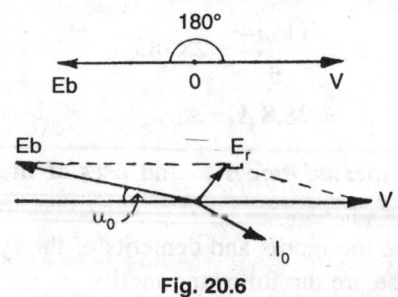

Fig. 20.6

The no load current of the motor is given by the formula

$$I_0 = \frac{E_r}{Z_s}$$

where E_r = The resultant voltage in Volts.

Z_s = The synchronous impedance in Ohms.

and the power will be, $W_p = V_L I_0 \cos \phi_0$ per phase.

It is used to meet the losses and keeps the motor in running condition. The behaviour of the motor can be studied on load also.

Motor on load. When the load is put on the motor, the speed remains the same it does not drop down, but the angle of coupling falls as shown in Fig. 20.7 and ultimately the relative shift between the rotor and stator poles increases, but continues to run even on the same synchronous speed.

Now because the field excitation is constant, so the magnitude of the back e.m.f. is same and will not change. The angle of coupling will certainly be effected and will change to new value *i.e.* α_1. Thus the result-ant voltage will change to E_{r_1} and the current to I_1 but the speed will be same as the synchronous speed.

The power required will be $V_p I_p$, $\cos\phi_1$ watts/phase and this power will adjust the motor to meet the load. If the load is further increased, the speed will be constant but the angle of coupling will change to new value α_2, the current I_2 and power required to drive the load per phase

$$= V_p \cdot I_2 \cos\phi_2$$

There is a limitation beyond which the load cannot be pulled. The synchronism because the force of attraction between rotor and stator is not sufficient will not be maintained. The maximum torque is known as the pull out torque beyond which the speed will not be the synchronous speed. The angle of coupling is different with different motors. The maximum torque varies from 150% to 350% of the full load torque.

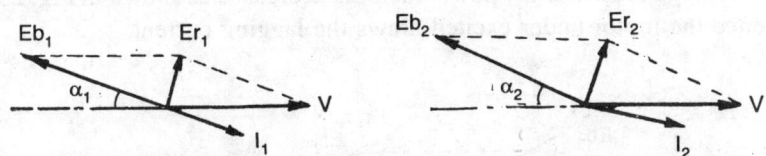

Fig. 20.7 Vector diagram at different loads.

Q. Explain the effects of the varying field excitation of a synchronous motor.

Ans. The unique and outstanding characteristic of a synchronous motor is in fact the vide range of operation for power factor. It is possible by the adjustment of the field excitation. The field excitation can be changed in these following two ways:

    (a) Under excitation.

    (b) Over excitation.

    (a) **Under excitation condition.** Whenever the exciting current is less than the normal exciting current the motor is said to be running under excited. In that condition the generated e.m.f. is less than the E_b.

Let V = The voltage supplied in Volts

          E_b = back e.m.f. with normal field excitation in Volts.

          E_{b1} = New back e.m.f. with new excitation in Volt.

          I_0 = No load current in Amp.

          I_1 = Current at excitation in Amp.

          ϕ_0 = No load power factor

          ϕ_1 = Power factor at new load current

          E_{r0} = Resultant voltage at no load in Volts.

E_r = Resultant voltage at new excitation in Volts.

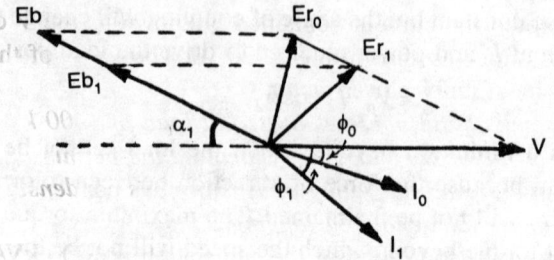

Fig. 20.8 Under excited synchronous motor.

Thus it is seen that if the excitation is further decreased the current further legs behind or the power factor is decreased as shown in Fig. 20.8. Hence the motor under excited draws the lagging current.

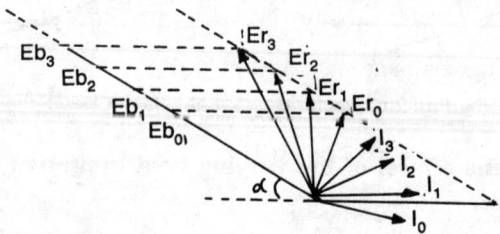

Fig. 20.9. Over excited synchronous motor.

(b) **Over excitation condition.** If the load on the motor is constant and the excitation is changed, the e.m.f. induced will change as shown in Fig. 20.9. Let us take the example of induced excitation E_b and the voltage fed V volts, the resultant voltage is Er_0 which results in current I_0, which is on lagging side. Now increasing the field excitation, the induced voltage changes from E_{b0} to E_{b1} and the resultant current from I_0 to I_1 Amp. It results the change of power factor from lagging to leading side. Here different positions are shown and it should be clearly understood that with increasing of the excitation current start shifting from lagging side to unity and then towards the leading side.

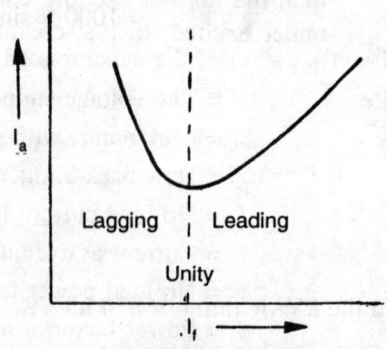

Fig. 20.10. V-curve.

Thus the over excited synchronous motor takes leading current.

If the value of exciting current and the load current both are noted and a curve is ploted the curve so obtained is known as the *V* curve of the synchronous motor as shown in Fig. 20.10. The value of the minimum current will be at unity power factor.

Example. *A substation operates at its full load of 1000 kVA supplies a load of power factor 0.71 lagging. Calculate the permissible additional load at this factor and the rating of synchronous condenser to raise the substation power factor to 0.87 lagging.*

Solution. The full load rating of the sub station is 1000 kVA

The full load output at 0.71 power factor
$$= 1000 \times 0.71 = 710 \text{ kW}$$

The full load output as required at 0.87 power factor
$$= 1000 \times 0.87 = 870 \text{ kW}$$

The increase in kW from 0.71 p.f. to 0.87 p.f.
$$= 870 - 710 = 160 \text{ kW}$$

Now the addition load supplied at 0.71 p.f.

$$= \frac{\text{kW}}{\cos\phi} = \frac{160}{0.71} = \textbf{225.35 kVA Ans.}$$

The reactive power drawn by the load
$$= \text{New load in kW} \times \tan\phi_1$$
$$= 870 \times \tan(\cos^{-1}0.71)$$
$$= 870 \times \tan 44.77 = 862.9 \text{ kVAR}$$

The reactive power supplied by the power station
$$= \text{Total power supplied} \times \sin\phi_2.$$
$$= 1000 \times \sin(\cos^{-1}0.87)$$
$$= 1000 \times \sin 29.54° = 493 \text{ kVAR}$$

Now the rating of the synchronous condenser
$$= \text{Reactive power drawn by the load}$$
$$- \text{The reactive power supplied by the}$$
$$\text{generating station}$$
$$= 862.9 - 493 = 369.9 \text{ kVAR}$$
$$\simeq 370 \text{ kVAR}$$

So the kVAR rating = **370 kVAR. Ans.**

Example. *A 3 phase 11000 V 50 c/s synchronous motor has effective armature resistance 0.95 Ω and reactance 29 Ω per phase. Calculate the* *power supplied to the motor and e.m.f. induced if it draws a load current of 50 A at 0.8 p.f. lagging.*

Solution. The power factor is 0.8 lagging

The power input

$$= \sqrt{3} \cdot V_L I_L \cos\phi$$

$$= \sqrt{3} \times 11000 \times 50 \times 0.8$$

$$= 762102 \text{ W} = 762.102 \text{ kW}$$

The voltage is 11000 V, line voltage so the phase voltage

$$= \frac{11000}{\sqrt{3}} = 6250.85 \simeq 6351 \text{ V}$$

Now

$$\phi = \cos^{-1}(0.8) = 36.87°$$

and

$$\theta = \tan^{-1}\left(\frac{X_s}{R_s}\right) = \tan^{-1}\frac{29}{0.95}$$

$$= 88.12°$$

Now the impedance drop,

$$= IZ_s = 50 \times \sqrt{0.95^2 + 29^2}$$

$$= 50 \times 29.015 = 1450.8 \text{ V} \simeq 1451 \text{ V}$$

Now the induced e.m.f. can be calculated as

$$E_{b_p} = \sqrt{V^2 + E_r^2 - 2V \times E_r \times \cos(\theta - \phi)}$$

$$= \sqrt{\left[6251^2 + 1415^2 - 2 \times 6351 \times 1451 \times \cos(88.2 - 36.87)\right]}$$

$$= 5561 \text{ V}$$

This voltage is the pahse voltage so the line voltage induced

$$= V_P \times \sqrt{3}$$

$$= 5561 \times \sqrt{3} = \mathbf{9631.9 \text{ V}}$$

$$\simeq \mathbf{9632 \text{ V}. \text{ Ans.}}$$

Example. *An 11000 V three phase 50 c/s star connected synchronous motor takes 60 A. The effective armature resistance and reactance per phase is 1 Ω and 30 Ω respectively. Find* (*i*) *The power supplied to the motor.*

(*ii*) *The induced e.m.f. for a power factor 0.8 lagging.*

(*Allahabad Univ. leading and 1997 Suppl.*)

Solution. The line voltage is 11000 V so the phase voltage, when the motor is connected in star

$$= \frac{V_L}{\sqrt{3}} = \frac{11000}{\sqrt{3}} = 6351 \text{ V}.$$

The effective armature resistance and reactance are 1 Ω and 30 Ω so the synchronous impedance is

$$Z_s = \sqrt{R_e^2 + X_s^2}$$

$$= \sqrt{1^2 + 30^2} = 30.017 \approx 30\,\Omega$$

and
$$\tan\theta = \frac{X_S}{R_S} = \frac{30}{1} = 30$$

$$\theta = \left(\tan^{-1} 30\right) = 88.1°$$

and
$$\phi = \left(\cos^{-1} 0.8\right) = 36.9°$$

Now the impedane drop

$$= I \times Z = 60 \times 30$$

$$= 1800 \text{ V}$$

(*i*) The power supplied to the motor

$$= \sqrt{3}\, V_L I_L \cos\phi$$

$$= \sqrt{3} \times 11000 \times 60 \times 0.8$$

$$= \mathbf{914.52\ kW.\ Ans.}$$

(*ii*) The resultant voltage when the power factor is 0.8

$$E_r = \sqrt{V^2 + E^2 - 2VE \cos(\theta \pm \phi)}$$

The + sign is used for leading power factor and (–) sign for lagging power factor. So for leading power factor

$$E_r = \sqrt{6351^2 + 1800^2 - 2 \times 6351 \times 1800 \times \cos(88.1 + 36.9)}$$

$$= 7529 \text{ V}$$

It is the phase voltage so the line voltage will be

$$= \sqrt{3} \times \text{phase voltage}$$

$$= \sqrt{3} \times 7529 = \textbf{13040.6 V} \simeq \textbf{13041 V. Ans.}$$

(*ii*) For lagging power factor

$$E_r = \sqrt{6351^2 + 1800^2 - 2 \times 6351 \times 1800 \times \cos(88.1 - 36.9)}$$

$$= 5408 \text{ V}$$

So the line voltage is

$$= V_P \times \sqrt{3} = 5408 \times \sqrt{3} = \textbf{9357 V.} \quad \textbf{Ans.}$$

Q. What do you understand by the three phase variable speed motors? Describe the Scharge motor.

Ans. This motor is known as variable speed motor. It was introduced by Sh. H.K. Scharge, so it is named as Scharge motor. It is a 3 ϕ commutator motor of which the speed can be increased or decreased beyond normal effectively. The basic principle of the speed control is that the speed can be controlled by injecting the voltage into the secondary circuit of the motor. When more current flows through the secondary circuit the speed will increase and if the current is less then the speed will be low. These motors are generally classified as

    (*a*) Scharge motor.

    (*b*) Susceptor type.

    (*a*) **Scharge motor.** It is a variable speed commutator motor. The stator is would as three phase induction motor. The stator has two windings: (*i*) the primary winding and (*ii*) the secondary winding or adjustable or regulating or compensating winding. The primary winding is connected across the three phase supply through three sliprings. A separate D.C. armature winding known as adjustable or regulating winding is housed in the rotor slots. The leads of the secondary coils are connected to the commutator. There are three sets of brushes which are placed on the commutator as shown in Fig. 20.11 and each stator winding is connected across two brushes. The brushes can be moved in any of the direction around the commutator by the lever attached or by means of a small motor. The magnitude of the e.m.f. injected and its slip frequency injected in the secondary circuit, depends upon the spacing between the two sets of brushes and the angular displacement depends upon the angular position with respect to the centre of the rotor winding.

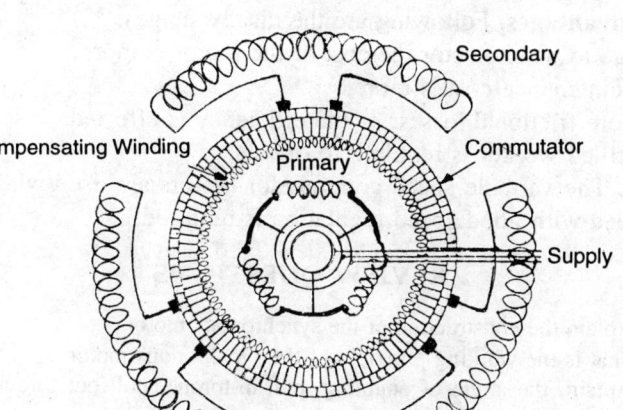

Fig. 20.11 Scharge motor

Action. Let three phase suplpy is given to the sliprings. A rotating magnetic field is produced by the rotor winding. Initially let the stator winding brushes are placed on the same segment, it means the stator winding is short circuited. Due to electromagnetic induction the voltage is induced in the stator winding which is short circuited and the motor is started as an induction motor.

In the compensation or regulating winding the e.m.f. is induced due to transformer action.

Now as the sets of the brushes are placed on the commutator and on the same segment, it means the stator winding carries only the stator winding current and not the injected current. Now if the brushes are separated then the e.m.f. induced in that particular portion of the compensating winding is injected in the stator circuit. The speed depends upon the e.m.f. injected, if the injected voltage is in the phase opposition to the secondary voltage then the speed will fall. In other position if the votlage injected is obtained by moving the brushes in the opposite direction to that of first, the voltage will be in phase with the secondary voltage and the speed will increase. Thus by changing the direction of the current in the stator circuit the speed of the motor can be controlled.

The power factor can also be controlled, if the injected e.m.f. is in quadrature leadng the induced rotor e.m.f. the power factor of the motor will be improved.

Advantages: Following are the advantages of this motor:
   (*i*) Good control and regulation of the speed is obtained. The speed can be varied from zero r.p.m. to above normal speed.
   (*ii*) Power factor can be improved.

Disadvantages. Following are the disadvantages:

(*i*) Due to wound rotor the cost of the motor is more.

(*ii*) Maintenance cost is more.

(*iii*) More frictional losses, so the efficiency is effected.

(*iv*) Skilled worker is required to operate.

Uses. The variable speed commutator motors are used where a variable speed with good speed regulation is required.

REVIEW QUESTIONS

1. Explain the construction of the synchronous motor.
2. What is the working principle of the synchronous motor?
3. Explain, the angle of coupling, pull-in-torque, pull out torque and load angle.
4. What are the effects of load on the performance and behaviour of a synchrobnous motor?
5. Describe with vector diagrams, the behaviour of synchronous motor is under and over excited conditions.
6. Why a synchronous motor in over excited condition is known as the synchronous capacitor, explain clearly with diagrams?
7. What do you understand by the V curve of the synchronous motor?
8. What are the characteristics of the synchronous motor?
9. Explain the hunting in synchronous motor. How these are overcome?
10. How the synchronous motors are started? Describe any one of the method.
11. What do you understand by the 3ϕ variable speed commutator motor? Describe any one of them.

NUMERICAL PROBLEMS

1. A 440 V 7.46 kW 3ϕ synchronous motor has negligible armature resistance and a synchronous reactance of 10 Ω/phase. Determine the minimum current and the corresponding induced e.m.f. for full load conditions assuming an efficiency of 85%. **[Ans. 263.4 V]**
2. A 400 V synchronous motor has an effective armature resistance and synchronous impedance 0.5 Ω and 4 Ω respectively. To what voltage must the motor be excited to give its full load output 50 H.P. at unity power factor assuming armature efficiency 90%. **[Ans. 463.86 V]**
3. A 3ϕ 11000 V 50 c/s synchronous motor has effecting armature resistance 0.95 Ω and reactance 29 Ω per phase, calculate the power supplied to the motor and e.m.f. induced per phase, if it draws a load current of 50 A at 0.8 P.F. lagging. **[Ans. 762.1 kW, 9633 V]**
4. A 400 V 3ϕ star connected synchronous motor delivers an output of 40 kW at 0.8 leading power factor with an efficiency of 86.5%. The motor has a synchronous reactance per phase of 2 Ω. Find the magnitude of the induced e.m.f. per phase. **[U.P.S.C., I.E.S, Electrical Engg., 1972]**
[Ans. 356 V]

21

A.C. Single-Phase Motors

Q. What do you understand by the A.C. single phase motors and name the different types?

Ans. The motors which are working on single phase supply are known as A.C. single phase motors. This motor converts A.C. single phase energy into mechanical energy. These motors are also having fractional horse power, because of low power or say fractional horse power these are also sometimes called fractional horse power motors.

These motors are of the following types:

- (A) Induction motors.
- (B) Commutator motors.
- (C) Synchronous motors.

These motors are further classified as;

(A) Induction motors:
- (a) Split phase motors:
    - (i) Resistance start motor.
    - (ii) Reactance start motor.
    - (iii) Capacitor start motor.
    - (iv) Capacitor start capacitor run motor.
    - (v) Permanent capacitor motor.
- (b) Shaded pole motor.

(B) Commutator motors:
These are further divided into
- (a) Series or universal motor.
- (b) Repulsion motor.
    - (i) Repulsion motor.
    - (ii) Repulsion induction motor.
    - (iii) Repulsion start induction run motor.

Q. Why the A.C. single phase motors are not self-starting motors.

Ans. The production of the torque in A.C. induction motor depends upon the production of rotating magnetic field, where in A.C. single phase motors the magnetic field produced is not of rotating in nature but is of alternating in nature.

In A.C. single phase induction motors, the rotor is of squirrel cage type. The e.m.f. in the rotor induces because of the induction. The single phase winding is done in the stator and connected to the mains. The magnetic field produced will change its polarity and value after a fixed interval of time. Because of the changing flux, the e.m.f. is induced in the rotor bars, the bars being short circuited causes a current to flow. During half cycle the torque developed will be in one direction and during the next half cycle the direction of torque will be reversed. Thus resulting a pulsating torque; and the pulsating torque will conclude zero torque during one complete cycle. The rotor will not move in either of the direction. So the single phase induction motors are not self-starting motors.

The explanation of the single phase induction motor can be easily and clearly understood by the revolving field theory. According to the *revolving field theory* the pulsating field is resolved into two opposite rotating fields each of half the magnitude as shown in Fig. 21.1. Each field induces a torque but being equal and opposite the net torque during the complete cycle being equal and opposite, the net torque during the complete cycle is the sum of both the torques hence no torque is achieved as shown in Fig. 21.2. and rotor is standstill. So the motors are not self-starting one. According to the *cross field theory* the m.m.f. produced by stator fed by A.C. single phase, is stationary in space, but pulsates in magnitude and vary sinusoidally with time. Now the e.m.f. induced is because of the transformer action resulting no torque. During running condition, in addition to the transformer e.m.f. there is one more e.m.f. known as speed e.m.f. The generated rotor e.m.f. vary in space with the stator current and flux. Now both the fields are at right angle to each other resulting the rotating field revolving at synchronous speed.

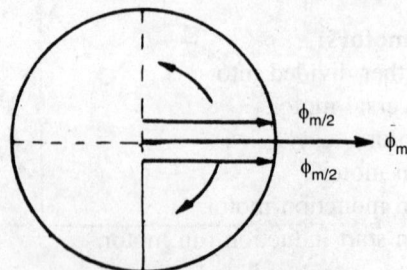

Fig. 21.1. Field revolving theory.

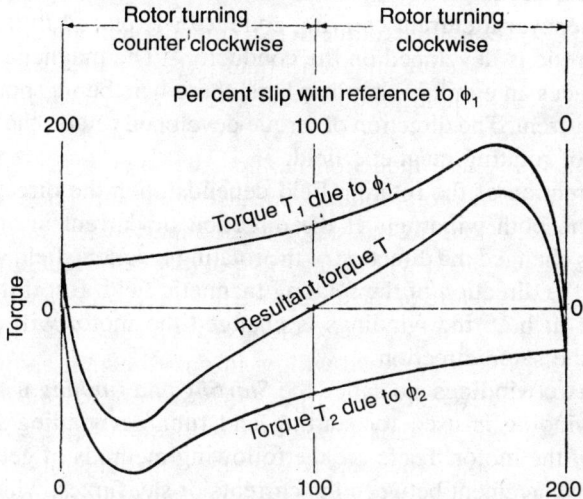

Fig. 21.2. Torque – slip curve for a single phase motor.

So if an induction motor is switched on and given some starting torque it will continue to run in the same direction in which the torque is given.

Q. Explain the working principle of the single phase induction motor. How these motors are made self-starting?

Ans. The A.C. single phase induction motors are not self-starting. To make these motor a self starting one a rotating magnetic field is essential. So two windings are done in the stator at 90° electrical to each other to obtain a rotating magnetic field as shown in Fig. 21.3. These windings are so designed as to have two different currents with some angular displacement. More angular displacement between the currents more torque and *vice versa*.

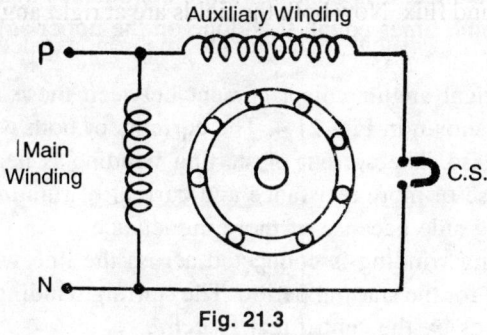

Fig. 21.3

Working principle. The motor works on the principle of magnetic drag, "whenever a current carrying conductor is placed in the magnetic field, a torque is developed on the conductor". The magnetic flux of the stator induces an e.m.f. in the rotor bars, these bars being short circuited, results a current. The direction of torque developed rotates the rotor in the direction of rotating magnetic field.

The direction of the rotating field depends upon the direction of currents in the both windings. If the direction of current in either of the winding is changed the direction of the rotating magnetic field will change, changing the direction of the rotating magnetic field. But if the direction of current in both the windings is changed the motor will continue to rotate in the same direction.

These two windings are called the *Starting* and *running windings.* The starting winding is used for starting and running winding for running purpose of the motor. There are the following methods of getting a good angular displacement between the currents or say fluxes, which are classified as under and are termed as the starting methods.

(*a*) Split phase methods.

    (*i*) Resistance start motors.

    (*ii*) Reactance start motors.

    (*iii*) Capacitor motors.

(*b*) Shaded pole induction motors.

Q. Explain the working and characteristics of the plane split phase induction motor.

Ans. It is a type of A.C. single phase motor. It has two windings placed at 90° to each other. These are starting winding and running windings. The running winding has low resistance and high inductance. It is wound with thick conductor having less number of turns then the starting winding. It is embedded in the slots. The starting winding has more resistance and low inductance. It is done with thin conductor having more number of turns or some times equal. It is done on the upper side of the running winding.

The electrical angular displacement between these two windings is 90° elect. as shown in Fig. 21.4. The currents of both windings differ in value and phase. The current of starting winding is near to the voltage vector because of more resistance and current of running winding is towards lagging side because of more inductance.

The running winding is connected across the line where the starting winding only for the starting period. The starting winding is disconnected from the mains by the centrifugal switch.

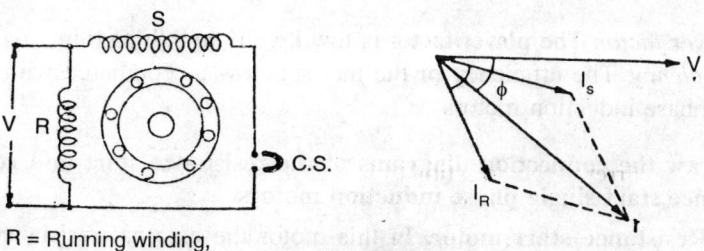

R = Running winding,
S = Starting winding, CS = Centrifugal switch.

Fig. 21.4. Split phase motor.   **Fig. 21.5.** Vector diagram.

Operation. When the motor is connected across the supply the starting winding because of more resistance draws current at a phase difference near to the voltage vector. The running winding because of more inductance draws current on the lagging side, away from the voltage vector as shown in Fig. 21.5. There are two currents, resulting two magnetic fields which produces the magnetic field of rotating in nature. Hence the motor starts. As the motor attains its 75% of the synchronous speed, the centrifugal switch disconnect the starting winding and now motor is running on running winding only.

Characteristics:

Torque. The starting torque is low. It is about 1.3 to 1.7 the full load torque. The torque increases with the increasing of load as shown in Fig. 21.6.

Current. The starting current is approximately five times the full load current. Without load, current is minimum and increases with the increasing of load.

Speed. The speed at no load is approximately equal to the Synchronous speed and slightly falls down with the increasing of road.

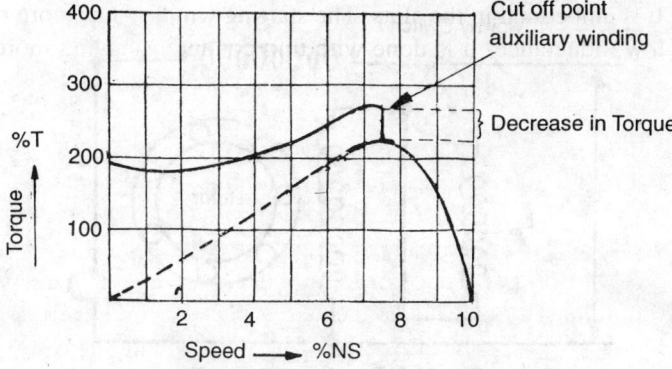

Fig. 21.6. Characteristics curve.

Power factor. The power factor is low *i.e.* 0.6 to 0.8 lagging.

Efficiency. The efficiency of the motor is low as compared with the three phase induction motors.

Q. Draw the connection diagrams of the resistance start and reactance start single phase induction motors.

Ans. Resistance start motor. In this motor the external resistance is introduced in the starting winding because it has more resistance as shown in Fig. 21.7. With the addition of extra resistance the vector of current will come more closer to the voltage vector hence increasing the angular displacement between the currents. More angle good starting torque.

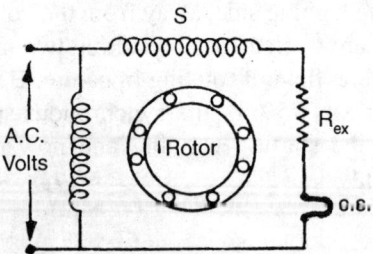

Fig. 21.7. Resistance start motor.

Reactance start motor. In this case the external reactance is added in main or running winding as shown in the Fig. 21.8. It has more reactance, with the addition of more inductances in running winding circuit the current vector of the running winding will be lagging more, increasing the angular displacement between the currents of starting and running windings hence improved starting torque.

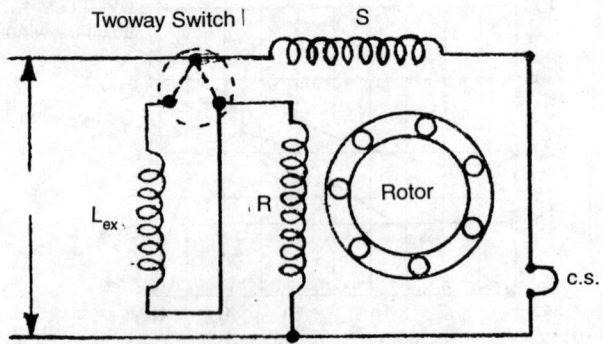

Fig. 21.8. Reactance start motor.

Q. Explain the working and characteristics of the A.C. capacitor motors.

Ans. These motors are the type of split phase motors. The capacitor is used to make the splitting of the phase. The angular displacement between the currents is increased by introducing the capacitor in starting winding and finally to obtain a good starting torque.

The capacitor motors are of the following types:
(*a*) Capacitor start induction run motor.
(*b*) Permanent capacitor motor.
(*c*) Capacitor start capacitor run motor.

(*a*) **Capacitor start induction run motor.** In this type of motor a capacitor is connected in series with the starting winding at the time of starting as shown in Fig. 21.9. As usual there are two windings the main or running winding and auxiliary or starting winding. The starting winding resistance is more than the running winding resistance. The inductance of running winding is more than the starting winding reactance. The current of strating winding leads the voltage because of the capacitor in the starting winding circuit. The current of running winding lags behind the voltage. The angular displacement between the current of starting and running winding is much larger say about 70-80° elect. as shown in Fig. 21.10. It causes the production of good starting torque. So this motor has good starting torque. When the motor attains the 75% of the synchronous speed, the centrifugal switch disconnect the starting winding circuit and the motor remains working on running winding only. The capacitor rating depends upon the h.p. of the motor.

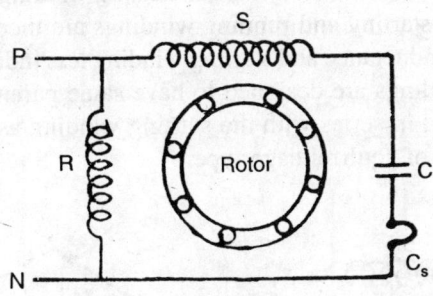

Fig. 21.9

Characteristics. It has the following characteristics:
(*i*) *Starting torque.* The starting torque is better than single phase plain split phase induction motor. It is about three times the full load torque of the motor as shown in Fig. 21.11.

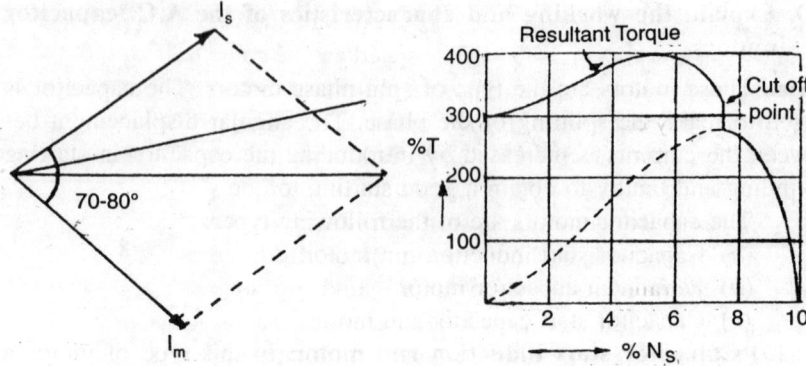

Fig. 21.10. Vector diagram.   **Fig. 21.11.** Characteristic curves.

(*ii*) *Current.* The starting current is less than the starting current of plain split phase motor. The current is less at no load and increases with the increasing of load.

(*iii*) *Speed.* The speed at no load is nearly equal to the synchronous speed and slightly falls down with the increasing of load.

(*iv*) *Power factor.* The power factor at the time of starting is good and in running condition equal to the split phase induction motor.

Uses. It is used where the good starting torque is required as in refrigerator, air compressors, drill machines, lathe machine and compressors. These are available from 1/6 H.P. to 1 H.P. motor.

The direction of rotation can be changed if the current in either of the windings is changed.

(*b*) **Permanent capacitor motor.** In this type of capacitor motor, there is no centrifugal switch to disconnect the starting winding circuit. In this motor as usual the starting and running windings are there. The running winding has more inductance and starting winding less inductance. Sometimes both the windings are designed to have same parameters. The capacitor is connected in series with the starting winding as shown in Fig. 21.12. The rotor is of squirrel cage type.

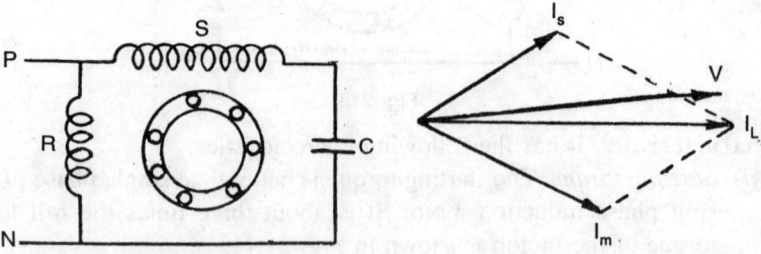

Fig. 21.12. Permanent capacitor motor.   **Fig. 12.13.** Vector diagram.

The *starting torque* is low, it is about half the full load torque of the motor as shown in Fig. 21.14. The speed can be controlled by connecting an external resistance in series with the motor circuit. The speed without load is nearly equal to the synchronous speed. The power factor is good.

These motors are used in ceiling fan, table fans and oil burner, where high starting torque is not so important.

(*b*) **Capacitor start capacitor run motor.** In this type of motor two capacitors are used as shown in Fig. 21.15. Both the capacitors are connected in parallel at the time of starting and one capacitor is disconnected through the centrifugal switch after starting the motor. The motor will now be running as a permanent capacitor motor. As usual the running and starting windings are there. The capacitors are connected in series with the starting winding circuit. The two capacitors are of different ratings and types. The starting capacitor is of short rating. It has large capacity and is of electrolyte type. The other capacitor is of continuous rating, small capacity and of impregnated paper type.

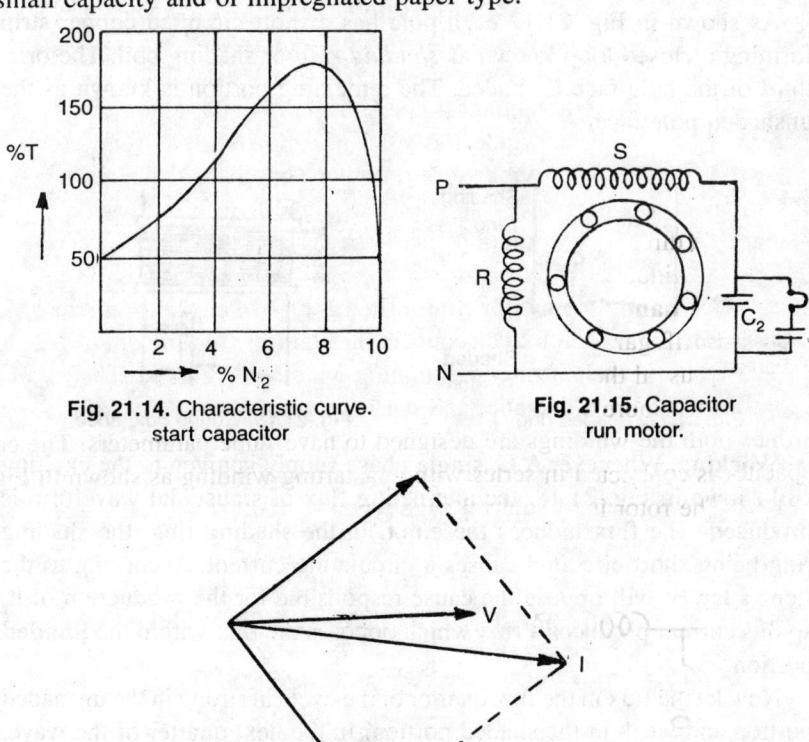

Fig. 21.14. Characteristic curve.
start capacitor

Fig. 21.15. Capacitor
run motor.

Fig. 21.16. Vector diagram.

Characteristics. The *starting torque* is better than the other types of capacitor motors. It is about three to four times the full load torque of the motor as shown in Fig. 21.16. The starting current is not so much. It increases with the increasing of load. The *power factor* is good in starting and in running conditions both. It has higher efficiency and can be over loaded up to 125%.

Uses. It is used where a high starting torque is required. It has good and smooth operation. Generally these motors are used for air compressors, refrigerator and air-conditioner etc.

Q. Describe the construction and working of a shaded pole motor.

Ans. It is also a form of an induction motor. The rotor is of squirrel cage type and stator has salient pole type construction. The number of salient poles depends upon the speed required. The poles are excited by their respective exciting coils. All exciting coils are connected in series to make the required number of poles.

As shown in Fig. 21.17 each pole has a short circuited copper strip forming a closed loop known as *shading ring* or shading coil. The one-third of the pole face is shaded. The remaining portion is known as the unshaded pole face.

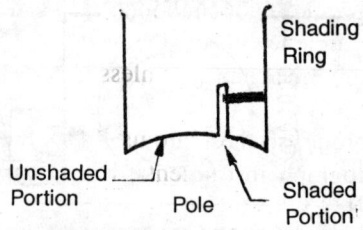

Fig. 21.17. Shaded ring.

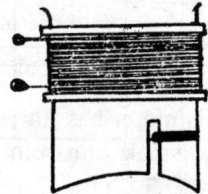

Fig. 21.18. Wound pole shoe.

Working. Whenever A.C. single phase supply is given to the exciting coil shown in Fig. 21.18, the alternating flux of sinusoidal waveform is produced. The flux induces the e.m.f. in the shading ring, the shading ring being short circuited causes a circulating current. According to the Lenz's law, it will oppose the cause responsible for the production of it, so this current produces a flux which opposes the flux within the **Shaded** portion.

Now let the flux in the first quarter of the cycle in strong in the unshaded portion and weak in the shaded portion. In the next quarter of the wave, when current from maximum value starts to zero, the flux will be weak in the unshaded portion and naturally strong in the shaded portion. Thus it is said that the flux has swept away from the unshaded portion to the shaded portion. This action of the flux in shaded pole motor is known as the *sweeping action* of the magnetic flux as shown in Fig. 21.19.

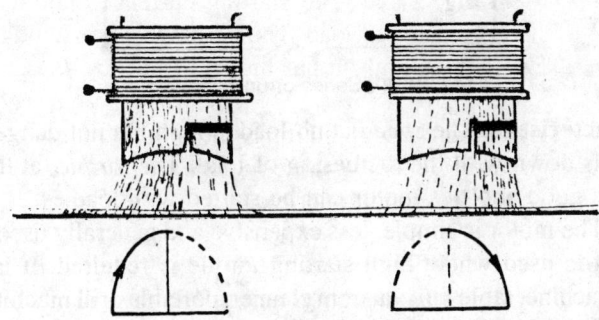

Fig. 21.19. Sweeping action of motor.

In the next half cycle also the flux is swept from unshaded portion to the shaded portion in opposite polarity. In this way a magnetic field of rotating in nature is produced. For the production of torque the rotating magnetic field is essential. So the torque is achieved and motor accelerates.

The direction of rotation of the rotor will be from unshaded portion to the shaded portion.

Characteristics. The motor has a low starting torque. The speed falls down with the increasing of load. The efficiency is low. the power factor is also low. The direction of rotation cannot be changed unless the direction of the shaded portion is changed.

Uses. This motor has low starting torque, so these are used for small outputs; like small fans, clocks, phonograph instruments, hair dryers, electric shaves etc.

Q. Explain A.C. series or universal motor.

Ans. The A.C. series motor or universal motor is designed to work on A.C. and D.C. both. The stator is made of laminations of silicone steel to reduce the hysteresis and eddy current losses.

The armature of the universal motor is the same as that of the ordinary series motor. In some large motors the compensating winding is used to improve the commutation. It is always connected in series with the field and the armature at the same time as shown in Fig. 21.20. High resistance brushes are employed to improve the commutation. Whenever supply is given, the current carrying conductors are placed in the magnetic field and a torque will develop to rotate in a particular direction. The d.o.r. can be changed by changing the direction of current in series field or in armature winding.

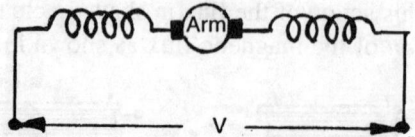

Fig. 21.20. Series or universal motor.

Characteristics. The *speed* at no load is high but not dangerous. The speed falls down with the increasing of load. The *torque,* at the time of starting is good and this motor can be started with load.

Uses. The motor is simple, less expensive and generally used for lower ratings. It is used where high starting torque is required. It is used for sewing machine, table fan, vacuum cleaner, portable drill machines, blowers and kitchen appliances etc.

Q. (*a*) **Describe construction and working principle of the repulsion motor.**

(*b*) **What are the different types of the repulsion motors?**

Ans. Construction. The repulsion motor works on the principle of magnetic repulsion. The construction is same as that of usual motor except the rotor and stator are inductively coupled. The stator winding has distributed A.C. single phase winding for two or four or any other number of poles. There is only one winding in the stator. The rotor is wound as that of D.C. motor, but it is not fed by the supply. The rotor current is obtained by transformer action. The commutator in this motor is short circuited by brushes. A number of brushes equal to the number of poles are placed on the commutator, which are short circuited. The brushes can be moved around the commutator to change their position.

Working principle. The direction of the current in the armature conductors depends upon the position of the brushes. It the brushes are placed to right angle to the main magnetic field there will not be any current in the short circuited armature. Hence no torque is developed. Now when the brushes are placed on the magnetic axis, the current will flow in the armature producing torque in such a way so that the sum of the entire torques developed will be zero and rotor will not rotate in either direction as shown in Fig. 21.21 (a, b). But when the brushes are placed in between the magnetic and neutral axis, the torque developed will have a definite direction as shown in Fig. 21.21(*c*). The sum of clockwise torque is greater than the sum of the anticlockwise torques, as a result of which the armature rotates in a particular say clockwise direction in this case. The d.o.r. depends upon the position of brushes. The d.o.r. and speed can be changed by changing the position of the brushes. There is no resultant torque on the magnetic axis and magnetic neutral axis and in between both these axis the torque is achieved.

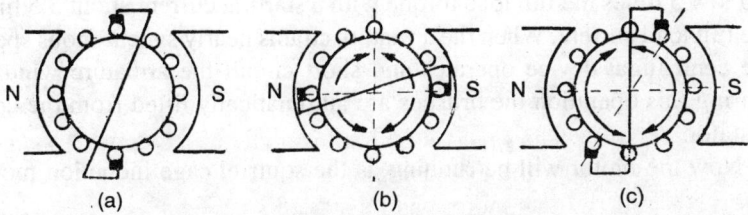

(a)   (b)   (c)

Fig. 21.21. Repulsion motor principle.

Characteristics. The motor has high starting torques with normal starting current.

The power factor and efficiency is low.

Uses. The motor is used where high starting torque is required with adjustable speed arrangement, for example in the coil winding where the operator can adjust the speed. A lever mechanism is provided to shift the position of the brushes on the commutator.

Repulsion start Induction run motor. This is also a form of the repulsion motor as the name implies the motor starts as a repulsion motor and continues to run as an induction motor. The armature winding is brought to the commutator as shown in Fig. 21.22. The commutator is short circuited in the running condition to enable the change of type of motor. This is accomplished by a centrifugal switch or device, that operates as the motor attains nearly synchronous speed. That device short circuit the commutator segments and motor continues to run as a squirrel cage induction motor. The brushes are automatically lifted to reduce the wear of the brushes and commutator bars to make the operation more smooth and noiseless.

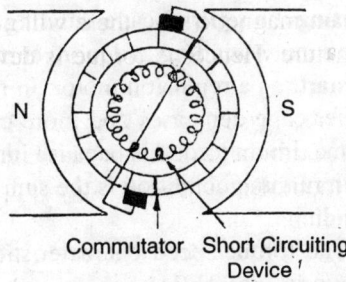

Commutator Short Circuiting
Device

Fig. 21.22. Repulsion start Induction run motor.

When the motor is connected across the supply it has all the characteristics that of the repulsion motor. The current is induced in the rotor; the rotor being short-circuited by the brushes causes a current in the rotor winding and resulting high starting torque. The starting torque is about 2.5 to 4.5 times the full load torque with a starting current about 3.5 times the full load current. When the armature attains nearly synchronous speed, the centrifugal device operates and short circuit the armature winding. During this operation the brushes are automatically lifted from the commutator.

Now the motor will be running as the squirrel cage induction motor.

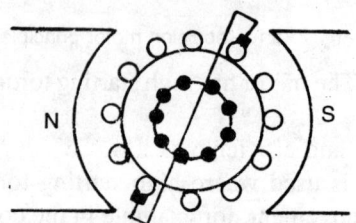

Fig. 21.23. Repulsion Induction motor.

Uses. It is used for compressors, pumps, refrigerator, ventilating fans etc., where high starting torque and approximately constant speed is required. The direction of rotation can be reversed by changing the position of brushes on the commutator.

Repulsion Induction motor. This is also a type of induction motor. It has a single phase stator winding but it has two separate windings on the rotor. One winding is main winding which is done in the upper slots just as d.c. armature winding. A squirrel cage winding is done below the armature winding which is permanently short circuited as shown in Fig. 21.23. The motor exhibits the characteristics of repulsion and induction motor in a common motor. So sometimes it is called as squirrel cage repulsion motor.

At the time of starting the squirrel cage winding has high inductance. So less current through inner winding and more current through the outer winding. The motor starts as a repulsion motor. In running condition the inductance of the inner cage decreases and more current flows through the inner cage. Now maximum torque is because inner cage winding and thus the total torque in running condition is the sum of the torques developed by both the windings.

Characteristics. The torque/speed characteristic for a repulsion induction motor is shown in Fig. 21.24.

The *starting current* of the motor is about 3 to 4 times the full load current. The starting torque of the motor is about 1.5 to 3 times the full load torque.

The main disadvantages of this motor are the high cost, careful maintenance and tendency of sparking.

The efficiency is good and torque is uniform from starting to running positions.

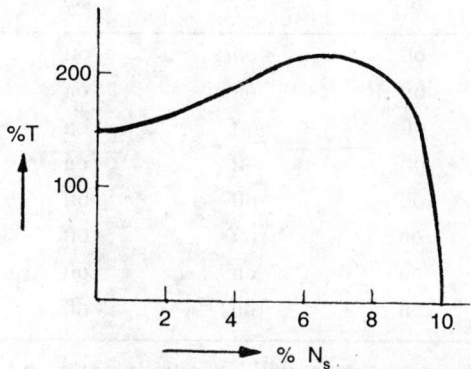

Fig. 21.24. Characteristic curve.

Uses. It is used where high starting torque is required. The speed is constant at less current. The common size is 1/6 to 4 kW for household appliances like air compressors etc.

Q. What do you understand by a stepper motor? can you give a brief idea of its working?

Ans. A stepper motor is unique in characteristics, construction and operations. A stepper motor essentially a brushless d.c. motor whose rotor rotates in discrete angular increments when its stator windings are energised in a programmed manner. In these motor, the rotation is obtained because of the magnetic interaction between rotor poles and poles of the sequentially energised motor stator winding. In these type of motor the rotor has no electrical winding but it has (Salient and/or magnetised) poles.

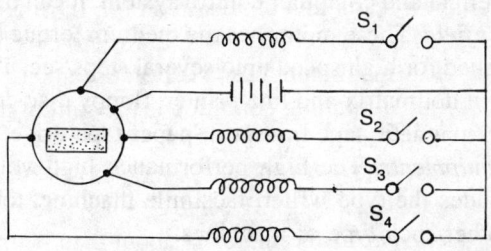

Fig. 21.25

In a simple manner, this motor operation can be understood by following diagram.

Sequential steps

Steps	S_1	S_2	S_3	S_4
1	off	on	off	off
2	off	on	on	off
3	off	off	on	off
4	off	off	on	on
5	off	off	off	on
6	on	off	off	on
7	on	off	off	off
8	on	on	off	off

Thus the lamination (rotor) will be initially energised and a complete rotation in obtained. This sequential steps are arranged in order adopting a control circuit.

Q. What are the types of stepper motors? Write a the application of stepper motor.

Ans. Thus we can say that the stepper motor is a brushless and almost always operates without shaft position sensing. The stepper motors are classified into the followings.

(*i*) Without permanent magnet
(*ii*) With permanent magnet

(*i*) **Without permanent magnet.** These are clow pole (CPM), hybrid (PMH), enhanced Hybrid (EHYB) Disc magnet (D.M.)

With permanent magnet. These are variable reluctance (VR) motors.

It is a motor in which the input is in the form of digital pulses which actually actuate the motor which increases the wide suitability in the field of digital, numerical and computer control system. It can be elaborated as

(*a*) *Computer field.* These motor has the medium torque (1 to 2 kg. cm) and designed for high speed upto several steps/sec. The wide application is in dot matrix and line printer, floppy disc drive, digital X-Y platter, magnetic tape transport, paper tape drive.

(*b*) *Office Equipments.* The high performance high volume category, they includes the type writer, facsimile machine, telex, xerox and several other such type of machines.

(*c*) *Machine tool application.* Numerical control for example CNC (Computer NC) and DNC (Direct NC) are the important areas of its application. Here three axis machine operation three different step-per motor for X, Y and Z axis are employed. Then motors have high torque 10-100 kgm which are required for controlling machines tools.

(d) *Robotics.* It is unique and fast developing area where the application is encouraged and appreciated stepper motor is successfully operational in large number of robotic application. It is used as an actuator for activating the joints of robots. Even in India a six degree of freedom robot was developed. It is vitally used in space for research works.

(e) *Electromedical.* In this field also for X-ray machine CAT scanner, ultrasound scanner, radiations therapy etc. the stepper motors find their suitable application.

(f) *Instrumentation.* In this type of application these includes the used in quartz watches, synchronised clocks, camera shutter operation etc.

Q. Explain in detail with diagram the working of stepper motor.

Ans. Working. In case of variable reluctance stepper motor there is no permanent magnet either on stator or rotor. The rotor is made of laminations of soft iron strips and does not have any winding. The stator is also made of laminations of soft iron. The stator has salient poles carrying stator winding. The stator number of poles are as an even multiple of number of phases of stator windings. The number of phases on the stator be at least three for bidirectional control of stepper motor. As shown in fig. the stepper motor has 12 poles on stator and 8 poles rotor.

Here when let phase A is energised which is wound on a salient construction results the production of maximum flux and the reluctance offered will be minimum on the rotor teeth then the rotor teeth will align themself with this teeth of phase A. Subsequently when phase B is energised the teeth having phase B winding will produced maximum flux and the rotor teeth will align the teeth of stator offering minimum reluctance and rotor will shift to new position in phase B. Similarly when phase C is energised the rotor will shift to the teeth having phase 'C' winding.

The operation of P.M.H. stepper motor can be easily understood.

Let phase 'A' is energised with positive voltage so that pole 1 be North and pole 2 be south. The stator flux will increases and shown by Φ_A. Now the rotor being soft iron lamination will be automatically magnetised to align with Φ_A as shown in fig. Now both A and B are energised causing Φ_B at 90° to Φ_A. Both will result

$$\dot{\phi}_R = \dot{\phi}_A + \dot{\phi}_B \text{ i.e. } 45°, \text{ ahead from the } F_A.$$

Now let B is only energised and A is de-energised the rotor will now accelerate to Φ_A resulting further 45° in the same direction. Now the phase A is energised reverse and causes a N pole at 2 and south poles at 1 i.e. (reversed polarity). Φ_A is reversed and now both Φ_A and Φ_B will result for the acceleration of 45°. Now phase B is de-energised and only Φ_A will align the rotor 45° ahead. Further the phase A energised and phase B is reversed carrying $-\Phi_A$ and $-\Phi_B$ resulting acceleration of resultant field for the 45° as shown. Than $-\Phi_A$ is de-energised and the $-\Phi_B$ will accelerates the 45° alignment than under phase A, is positive and phase B is negative than again the resultant will step for the 45° and if $--\Phi_B$ is de-energised Φ_A is positive the rotor resultant field will align 45° ahead to it. Then a complete cycle of rotation is obtained.

Φ_A = Flux produced by phase A winding

Φ_B = flux produced by phase B winding

Φ_R = resultant flux (alignment of rotor)

$(-\Phi_A$ and $-\Phi_B$ are reversed flux)

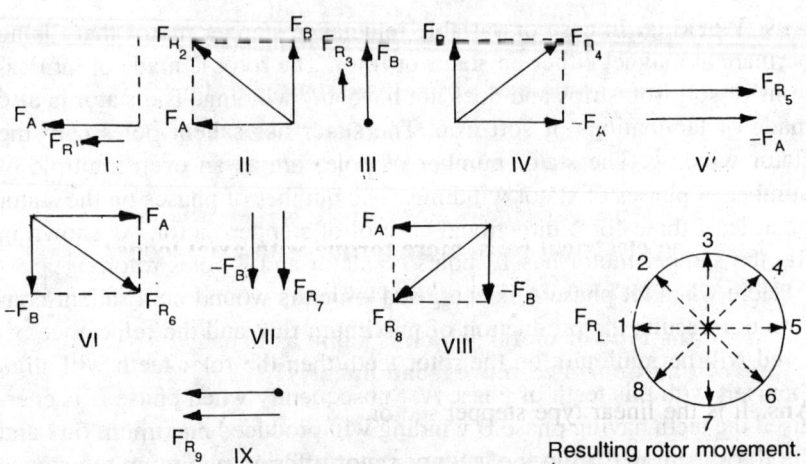

Resulting rotor movement.

Resultant direction of magnetic fields produced

Here this constitutes eight step sequence in which rotor moves 45° per step and stator moves four step sequence i.e. 90° per step the difference between them is in the equilibrium position of the rotor. Here both phases are alternatively energised this mixed is known as hybrid sequence.

The d.o.r. can be reversed by adopting the reverse magnetisation cycle. The stepper motors can even be designed for high torque to weight and high torque to inertia ratios but are limited in top speed and power to weight ratio. These can even be designed upto for example 100 kg cm torque at 2000 steps/Sec and 28.8 kg cm torque at 5000 steps/sec.

Stepper Motors

It is an electro magnetic actuator. It is a digital or incremental drive and is actuated in a fixed angular steps. It actually receive a digital pulse to rotate to a specific number of degrees in rotations each step of rotation is the actual response of the input digital commands or pulses. As the steps are synchronized with the pulses fed, so a systematic synchronised rotation in accordance with the pulses supplied is obtained. No step is missed. These motors have emerged as the cost effective alternatives for DC servomotors in high speed, motion controlled applications except the high torque-speed characteristics.

These are widely used in computer peripherals, machine tools, medical equipments, automotive devices and small business machines etc. These stepper motor are usually operated in open loop mode. These motors are available in different types as brushed, brushless, step motors, piezo (ultrasonic) and linear types.

Q. Can you give some examples of brushed stepper motor.

Ans. Yes, these are in toys, battery powered tools, electric machines and the places where apply power and go sequence is desired.

Q. What is the significance of brushless motors?

Ans. These stepper motors are less common; have higher efficiency, less friction, less electrical noise and requires electronic driver.

Q. Name the types of stepper motor which is relatively new, requires driver, no electrical coils, more torque with axial loads?

Ans. It is stepper motor Piezo (ultrasonic).

Q. Name the type of motor which is same as brushed or step but is opened and unrolled moves load linearly?

Ans. It is the linear type stepper motor.

Q. What are advantages of stepper motors?

Ans. There so many advantages for example,

(*i*) A high accuracy of motion is obtained. The position error is eliminated even under open loop control.

(*ii*) Large savings in sensor and controller cost is achieved.

(*iii*) Because of digital nature of commands these are widely used in digital control applications.

(*iv*) The torque capacity and power requirements can be optimised and response is controlled by electronic switching or pulses.

(*v*) It has excellent response to starting, stopping reversing operations.

Q. What are the disadvantages?

Ans. The motor instead of having numerous utilities but have certain disadvantages too

   (*i*) These motors are very low capacity motor.

   (*ii*) These motors have limited speed i.e. limited by torque capacity.

   (*iii*) There can be pulse missing problem due to faulty switching system.

   (*iv*) Resonances can be possible if not properly controlled.

   (*v*) It is not carry to operator this type of motor at extremely high speeds.

Q. The rotation is understood by steps of movement, can you give some idea about the step angle?

Ans. The angular position of rotor can be controlled in units of step angle by switching process. If the switching is carried out in sequences, the rotor will rotate with a stepped motor and thus controls the average speed.

The step angel the angle per step and is equal

Step angle $\theta_s = \dfrac{360}{m.Nr.}$

where θ_s – angle i.e. step angle.

m = Number of phases.

N_r = Number of rotor teeth.

REVIEW QUESTIONS

1. Is a single phase induction motor self-starting one, if not why?
2. State and discuss the double revolving field theory of a single phase induction motor.
3. How the single phase motors are made self-starting? Describe any one method.
4. What are the types of a.c. single phase motors? Describe the capacitor start induction motor with sketch.
5. Discuss with neat sketches the working of shaded pole motor.
6. Explain the construction and working of a.c. series motor or universal motor.
7. Describe the construction and working of the repulsion motor.
8. What are the classifications of the repulsion motor? Describe repulsion start induction run motor.
9. Write short note on the followings:
   (*a*) Starting of a.c. single phase motor.
   (*b*) Shaded pole motor.   (*d*) Reactance start motor.
   (*c*) Capacitor motors.   (*e*) Universal motor.

10. Fill up the blanks:
    (*a*) The a.c. single phase motors are ………
    (*b*) Starting winding has ………
    (*c*) The reactance of running winding is ……… than the starting winding.
    (*d*) The a.c. single phase magnetic field is ……… in nature.
    (*e*) The brushes in case of repulsion motor are placed ……….the m.n.p. and magnetic axis.

22

Armature Winding

The arrangement of coils on the armature in a systematic way is known as the armature winding.

22.1. WINDING MATERIAL

Q. Describe the different winding materials used for the winding purpose.

Ans. The materials used for the winding of the armature are of the following types:

(a) Wire or current Carrying conductor.

(b) Insulation and insulating materials.

(a) **Wire or current carrying conductors.** The winding is done by means of copper or aluminium insulated wires. There are as follows:

(i) **Enamelled copper wire.** These wires are made of copper conductors with an insulating covering. The insulating is known as enamel. These are available in weight and unit is kg or lb. The sizes are available from 12 S.W.G. to 48 S.W.G.

(1) *Enamelled wire.* These winding wires are having only one layer of enamel insulation coating and are used for low voltage machines.

(2) *Super enamelled copper wire.* There are two layers of enamel coating over the bare conductor. The super enamelled copper wire are used for making field coils of small and medium size machines and for other winding purposes in general.

The other type of insulation is the cotton or silk coverings, these are as follows:

(1) *Single silk covered (S.S.C.) wire.* It has a single covering of silk on the conductors.

(2) *Double silk covered (D.S.C.) wire.* These are having double silk

covering. The silk covered wires are available from 16 S.W.G. to 47 S.W.G. These are also available in weight and are used to make the field coils of medium and big machines. These wires are also used for winding the armatures of medium size machines.

For big machines the conductors are the solid or hollow without any insulation coating but are insulated with the proper insulation coverings and finally given a shape of coil for example the armature winding in turbo alternators, hydro-generators etc.

(*b*) **Insulation and insulation materials.** The insulation is used for the following purposes:

(*i*) to avoid the short circuit in wires and short circuit between the coils, etc.,

(*ii*) to avoid earth fault,

(*iii*) to avoid leakage,

(*iv*) to have safety against moisture,

(*v*) to withstand against temperature rise,

(*vi*) to have good mechanical and dielectric strength.

There are the following materials used as the insulating materials:

(*i*) **Empire cloth.** It is highly flexible and non-absorbent of moisture. It has good mechanical and dielectric strengths. It is available in thickness of 7, mil 10 mil, etc. in the shapes of roll of length in metres. Generally it is one metre wide in different colours yellow and black. It is used as the slot insulation.

(*ii*) **Leatheroid or elephantide.** It has good dielectric and mechanical strengths. It is black and sometimes grey is colours. These are available in sheets of 60 × 80 cm and rolls of metres. These are available in different sizes as 5, 7, 10 and 15 mil thickness. It is available in weight. It has a wide range of its applications as slot insulation, field coil insulation, etc.

(*iii*) **Pressphan paper.** It has good dielectric strength which is about 300 V/mil. It is available in sheets and in weight. It is available in thickness 3, 5, 7, 10 and 15 mils. It is also used as the slot insulation and field coil insulation, etc.

(*iv*) **Flexible micanite sheets.** It is flexible at all the temperatures. It has good dielectric strength and can be easily cut into pieces. Thickness available is 2 mm to 6 mm. It is available in sheets of one metre × one metre. It consists of soft mica with an adhesive mixed with either paper or cloth base. It is generally used as the insulation for the big turbo alternators as a slot insulation prepared with a definite treatment and shape.

(*v*) **Fibre.** It is used for making wedges and for making winding tools etc. It is available in square bars and in sheets of thickness 1.5 mm to 6 mm in weight. It is hard and water proof.

The above stated insulation materials are used as slot insulation in winding works.

(*vi*) **Cotton tape.** It is used to tape the coils. The coils are taped to give them a particular shape. These are available in rolls of 50 m and 100 m. These are available width wise as 10 mm, 15 mm, 20 mm, 25 mm etc. width.

(*vii*) **Silk tape.** These are also used to tape the coils. These are available in rolls of 25 m, 50 m and 100 m and 10, 15, 20 and 25 mm width.

(*viii*) **Rubber tape.** These are available in rolls of 50 m and according to width of 10 mm, 15 mm, 20 mm and 25 mm.

(*ix*) **Empire tape.** These are used for binding the coils and are available in length and width sizes as 50 m, 100 m and 10, 15, 20 and 25 mm. The empire tape is also available according to the thickness as 7 mil, 10 mil etc. These are black and yellow in colour. It is used for medium and high voltage machine.

(*x*) **Sleeves.** The sleeves are used for covering the leads of the coils. These are also used to cover the soldered joints. These are as follows:

(1) *Cotton sleeve.* It is available in different colour and lengths of 100 metres. These are available according to the diameter ranging from 1 mm to 4 mm.

(2) *Empire sleeve.* These are available in length of one metre in different colours and diameters from 1 mm to 6 mm. It is used to guard the leads coming out from the slots.

(3) *P.V.C. Sleeve.* This sleeve is specially used for the machines which are to be used in automobile or oil mills. The sleeve is highly oil resisting. It is available in different colour and different sizes from 1 mm to 6 mm. It is available in rolls and in length of one metre also.

(*xi*) **Solder.** The solder is used to solder the joints or to solder the leads to the commutator segments. Mostly resin cored solder is used for this purpose. It is available in 10, 12, 14, 16, 18 S.W.G. etc. in the rolls of 300 gm, 400 gm, 500 gm or even more.

(*xii*) **Soldering flux.** It is used to solder the joints properly. It spreads and clean the joints, so that the oxidation is checked and uniformality of the solder on the joint is maintained. The melting temperature of the flux should be low than the melting temperature of the solder. It is also available in the form of soldering paste. Generally it is available in weight of 50 gm, 100 gm and so on.

(*xiii*) **Varnish.** After completing the winding, it is impregnated in the varnish. The shellac with spirit and the black insulating varnish is available in the container of one litre, five litre etc. The main purposes and aims of the varnish are as follows:

(1) To increase the insulation resistance of the winding.

(2) The moisture does not enter the winding.

(3) The mechanical strength of the overhangs and winding is increased.

The air drying insulating varnish is directly applied to the armature winding and dried up, in the air. The stoving varnish is heated up first and then applied and dried in the air.

(*xiv*) **Binding thread.** Sometimes it is also known as twin thread. It is used for binding the coils of the winding. It is available in balls.

(*xv*) **Bamboo.** It is used for closing the slots as a wedge. The wedge are prepared from the bamboo and then these are inserted above the coil in the slot so that the conductors may not come out.

22.2. D.C. ARMATURE WINDING

Q. Describe the different terms used in the D.C. armature winding.

Ans. There are the following terms used for the armature winding:

(*a*) **Active side.** The conductor which lies in the sloted portion of the machine, is known as active side. These are also known as the coil sides or inductors. The effective induction takes place only in this portion.

(*b*) **Inactive side.** The part or the portion of the coil which does not lie into the slot is known as the inactive side. It is also known as the overhang. It only connects two active sides of a coil.

(*c*) **Turn.** It is the combination of active and inactive side of the coil. It allows the complete closed path to the current.

(*d*) **Coil.** The combination of so many turns in series is known as the coil. It consists of so many active and inactive sides.

(*e*) **Pole pitch.** The distance between the centre of two adjacent unlike poles, is known as the pole pitch. It is measured in terms of conductors or slots.

$$\therefore \text{ Pole pitch, } Y_P = \frac{\text{Number of slots}}{\text{Number of poles}} \text{ in terms of slots.}$$

and $$Y_P = \frac{\text{Number of conductors}}{\text{Number of poles}} \text{ in terms of conductors.}$$

(*f*) **Coil pitch.** It is also known as winding pitch. It is defined as the distance between the two sides of a coil. It may also be in terms of slots or conductors.

(g) **Back pitch.** It is generally denoted by Y_B. The distance between two coil sides of a coil at the back of the armature, is defined as the back pitch. It may also be in terms of slots or conductors as shown in Fig. 22.1.

(h) **Front pitch.** It is denoted as Y_F. The distance between two sides of a coil at the commutator side is known as front pitch. It is measured in terms of slots or conductors.

The back pitch and front pitch must differ by two conductors or one slot. Back pitch (Y_B) can be greater or lesser than front pitch (Y_F) by two conductors and should be nearly equal to the pole pitch.

(i) **Pitch factor.** It is defined as the ratio between the winding pitch and pole pitch so

$$\text{Pitch factor} = \frac{\text{Winding pitch}}{\text{Pole pitch}}$$

It is nearly equal to unity.

(j) **Resultant pitch.** The distance between two adjacent coils is known as resultant pitch.

$$Y_R = Y_B - Y_F \ (\text{For lap winding})$$
$$= Y_B + Y_F \ (\text{For wave winding})$$

(k) **Commutator pitch.** The distance between the two segments where the two ends of a coil are connected is known as commutator pitch.

$$Y_c = 1 (\text{in lap winding})$$
$$Y_c = \frac{\text{Number of segments} \pm 1}{\text{Number of pair of poles}} \quad (\text{in wave winding})$$

(l) **Full pitch winding or whole chorded winding.** In this type of winding the pitches are so chosen that the coil pitch is equal to the pole pitch. The pitch factor for this type of windings is unity, so here is this type of winding.

(m) **Long chorded winding.** The winding in which the coil span is greater than the pole pitch, is known as the long chorded winding, so in this type of winding.

∴ Coil span > Pole pitch.

(n) **Short chorded winding.** The winding in which the coil span is less than the poled pitch, is known as the short chorded winding.

Coil span > Pole pitch.

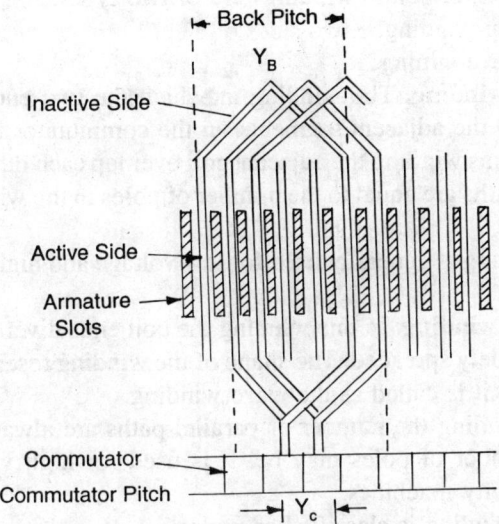

Fig. 22.1. Lap winding and pitches.

(*o*) **Progressive winding.** The winding which progress in the clockwise direction when viewed from the commutator side, is known as the progressive winding. Sometimes it may be stated that the winding is moving in the direction of coil. In this type of winding the back pitch is greater than the front pitch.

$$\therefore \qquad Y_B > Y_F.$$

(*p*) **Retrogressive winding.** The winding in which the winding progress in the anticlockwise direction when viewed from the commutator side, is known as the retrogressive winding or the winding is progressing in the opposite direction of the coil. In this type of winding the front pitch is greater than the back pitch *i.e.*

$$Y_F > Y_B.$$

(*q*) **Single layer winding.** In a single layer winding each slot has only one coil side. The total number of coil is half the total number of slots.

(*r*) **Double layer winding.** In this type of winding each slot has two coil sides and the number of coils are equal to the number of slots.

22.3. TYPES OF D.C. ARMATURE WINDING

Q. What are the different types of windings? Describe the simplex lap winding and its uses.

Ans. The D.C. armature windings are of two types:

    (*a*) Lap winding.

    (*b*) Wave wining.

(*a*) **Lap winding.** The winding in which the two ends of a coil are connected to the adjacent segments on the commutator is known as lap winding. In this winding the adjacent coil over lap each other. The number of parallel paths are equal to the number of poles in the winding as shown in Fig. 22.1.

This winding is generally used for low voltage and high current capacity machines.

(*b*) **Wave winding.** In this winding the coil ends diverse and go to the segments widely spreaded. The shape of the winding resembles the shape of a wave so it is called as the wave winding.

In this winding the number of parallel paths are always two whatsoever the number of poles may be. It is used for high voltage and low current capacity machines.

The lap winding is classified as under:

    (*i*) Simple lap winding.

    (*ii*) Multiple lap winding.

    (*iii*) Multiplex winding.

Simple lap winding. This winding is done where the number of slots are equal to the number of segments. In this winding the total number of coils will be equal to the number of segments and number of conductors will be equal to the twice the number of coils.

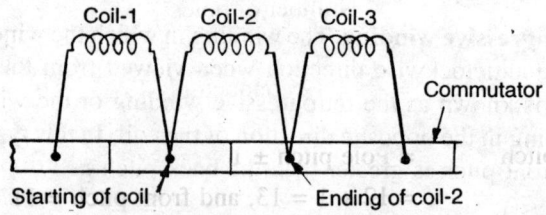

Fig. 22.2. Simplex winding.

In this winding the beginning and the ends of one coil are connected to the adjacent segment of the commutator as shown in Fig. 22.2. Thus the beginning of the second coil and end of first coil are connected to the one segment and the beginning of third coil and end of second coil are connected to the next segment and so on. This winding is used for medium size machine.

Conditions for simple lap winding. These are the following conditions:

    (*i*) The number of slots are generally in even number.

(*ii*) The back pitch and front pitches are nearly equal to the pole pitch.

(*iii*) The back pitch and front pitch must be in odd number in terms of conductors.

(*iv*) The back pitch and front pitch must differ by two conductors.

(*v*) The winding must be reentrant, *i.e.* it must finish from where it was started.

(*vi*) The commutator pitch is equal to one segment.

(*vii*) Each slot must have equal number of conductor.

The simplex lap winding is mostly used for small and medium size machine. If there is only one winding element in one slot it is said single layer simplex lap winding and if two, four or six etc. winding elements in one slots then it is called multiplex simple lap winding.

Example. *Draw a development diagram of a D.C. armature lap winding having, number of slots 12, number of coils 12, number of segments 12 and number of poles are two.*

Solution. The total number of conductors

$$= 2 \times \text{Number of coils}$$

$$= 2 \times 12 = 24 \text{ conductors}$$

Number of conductor/slot

$$= \frac{24}{12} = 2 \text{ Nos.}$$

Now pole pitch $= \dfrac{\text{Number of conductor}}{\text{Number of poles}}$

$$= \frac{24}{2} = 12 \text{ conductors}$$

The back pitch   = Pole pitch ± 1

Back pitch   = 12 + 1 = 13, and front pitch = 11

$$Y_B = 13, \quad Y_F = 11 \text{ (for progressive winding)}$$

$$Y_B = 11, \quad Y_F = 13 \text{ (for retrogressive winding)}$$

But in normal practice the windings are progressive type. So the back pitch is 13 and front pitch is 11. The commutator pitch is 13 − 11 = 2 conductors as shown in Fig. 22.3.

Example. *Draw the winding diagram of a simple lap winding having the following datas:*

Number of slots = 8 No.   *Number of segments = 8 No.*

Poles   = 2 No.   *Number of coils   = 8 No.*

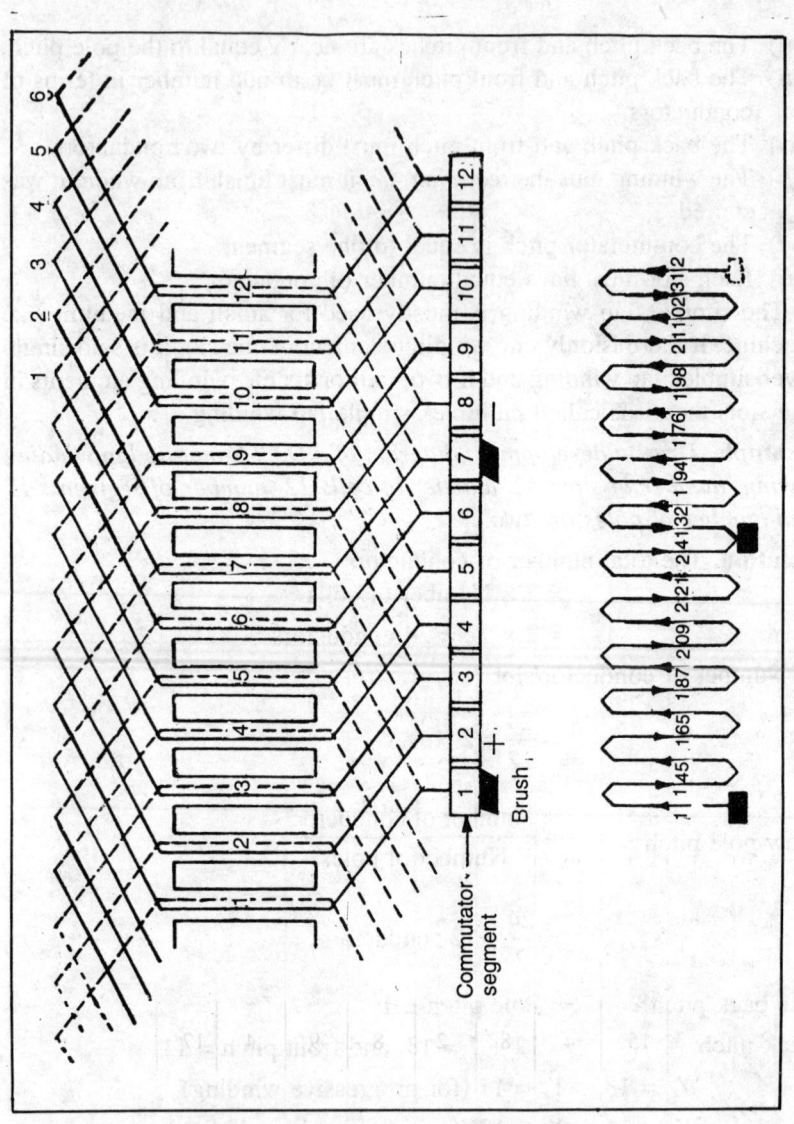

Fig. 22.3. Development diagram.

Solution. Number of conductors

$$= 8 \times 2 = 16 \text{ conductors}$$

The Number of poles = 2 No.

$$\therefore \text{ Pole pitch} \quad = \frac{\text{Number of conductor}}{\text{Number of poles}} = \frac{16}{2} = 8$$

The coil pitch = 8,

Pole pitch ± 1 or 8 ± 1 = 9 or 7

so the back pitch is 9 and front pitch is 7 for progressive winding.

The commutator pitch

$$= 9 - 7 = 2 \text{ conductors}$$

It can be tabulated as under and development diagram as shown in Fig. 22.4.

Table 22.1

S. No.	Coil No.	Conductor Y_B		Slot No.		Segment No.		Conductor Y_F		Remark
		T	B	T	B	T	B	B	T	
1.	1	1	14	1	7	1	2	14	3	
2.	2	3	16	2	8	2	3	16	5	
3.	3	5	18	3	9	3	4	18	7	
4.	4	7	20	4	10	4	5	20	9	
5.	5	9	22	5	11	5	6	22	11	
6.	6	11	24	6	12	6	7	24	13	
7.	7	13	2	7	1	7	8	2	15	
8.	8	15	4	8	2	8	9	4	17	
9.	9	17	6	9	3	9	10	6	19	
10.	10	19	8	10	4	10	11	8	21	
11.	11	21	10	11	5	11	12	10	23	
12.	12	23	12	12	6	12	1	12	1	

Table 22.2

S No.	Coil No.	Conductor Y_B		Slot No.		Segment No.		Conductor Y_F		Remark
		T	B	T	B	T	B	B	T	
1	1.	1	10	1	5	1	2	10	3	
2	2.	3	12	2	6	2	3	12	5	
3	3.	5	14	3	7	3	4	14	7	
4	4.	7	16	4	8	4	5	16	9	
5	5.	9	2	5	9	5	6	2	11	
6	6.	11	4	6	10	6	7	4	13	
7	7.	13	6	7	11	7	8	6	15	
8	8.	15	8	8	12	8	1	8	1	

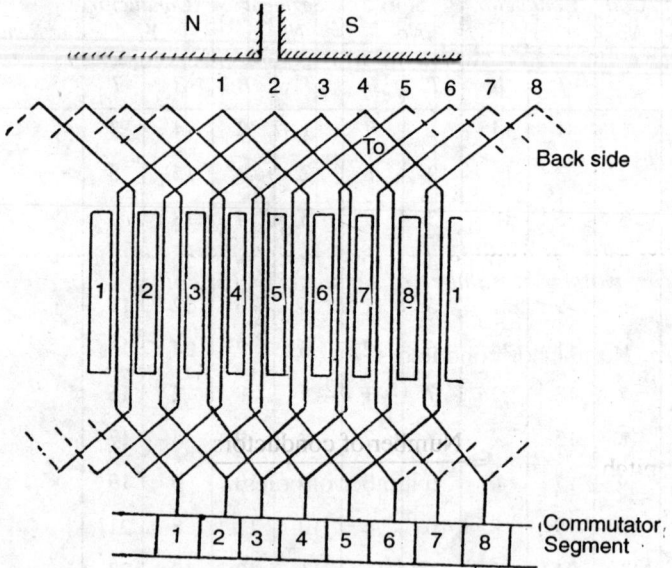

Fig. 22.4 Development diagram

Table 22.3

S No.	Coil No.	Conductor Y_B		Slot No.		Segment No.		Conductor Y_F		Remark
		T	B	T	B	T	B	B	T	
1	1.	1	18	1	5	1	2	18	3	
2	2.	3	20	1	5	2	3	20	5	
3	3.	5	22	2	6	3	4	22	7	
4	4.	7	24	2	6	4	5	24	9	
5	5.	9	26	3	7	5	6	26	11	
6	6.	11	28	3	7	6	7	28	13	
7	7.	13	30	4	8	7	8	30	15	
8	8.	15	32	4	8	8	9	32	17	
9	9.	17	2	5	1	9	10	2	19	
10	10.	19	4	5	1	10	11	4	21	
11	11.	21	6	6	2	11	12	6	23	
12	12.	23	8	6	2	12	13	8	25	
13	13.	25	10	7	3	13	14	10	27	
14	14.	27	12	7	3	14	15	12	29	
15	15.	29	14	8	4	15	16	14	31	
16	16.	31	16	8	4	16	1	16	1	

Example. *Draw a winding table of a D.C. armature lap winding having number of slots = 8, number of segments = 16, coil = 16 and the poles are two.*

Solution. Number of conductors

$$= 2 \times 16 = 32$$

∴ Pole pitch $= \dfrac{\text{Number of conductors}}{\text{Number of poles}}$

$$= \frac{32}{2} = 16$$

∴ Coil pitch $\qquad = $ Pole pitch ± 1

∴ $\qquad\qquad\qquad Y = 16 \pm 1$

∴ $\qquad\qquad\qquad Y_B = 17$

and $\qquad\qquad\qquad Y_F = 15$ for progressive type of winding.

Number of conductor per slot

$$= \frac{32}{8} = 4.$$

Q. Define duplex, triplex, quadruplex and multiplex lap windings.

Ans. Duplex winding. It is defined as the winding in which the end leads of coil are connected two bars away from beginning leads as shown in Fig. 22.5. In this particular type, the end leads of first coil is placed on the same commutator segment where the starting of the third coil is placed and so on.

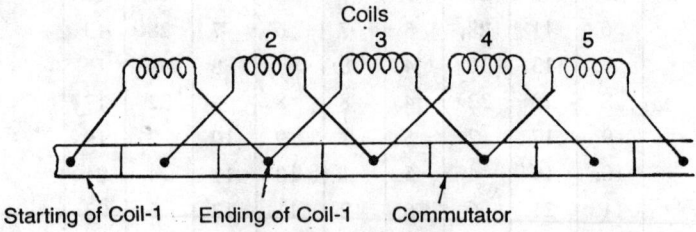

Fig. 22.5. Duplex winding.

Triplex lap winding. It is defined as the lap winding in which the end leads of a coil is connected to the three segments away from the beginning lead. Thus the end of first coil is connected with the starting of the fourth coil as shown in the Fig. 22.6.

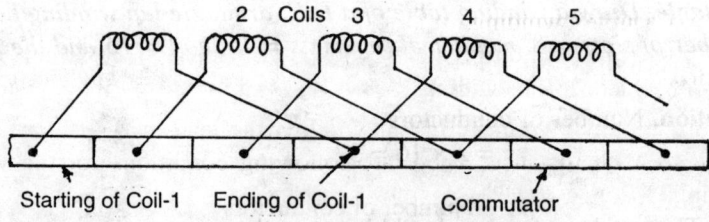

Fig. 22.6. Triplex winding.

Quadruplex lap winding. The winding in which the ends of a coil are connected to four bars or segments away from the beginning lead, is known as the quadruplex winding. In this winding the end leads of the first coil is connected to the starting lead of the fifth coil as shown in Fig. 22.7 and the end of fifth coil with starting of the ninth coil and so on.

Multiplex winding. The multiplex winding is that in which two or more than two simplex independent windings are done on the same armature.

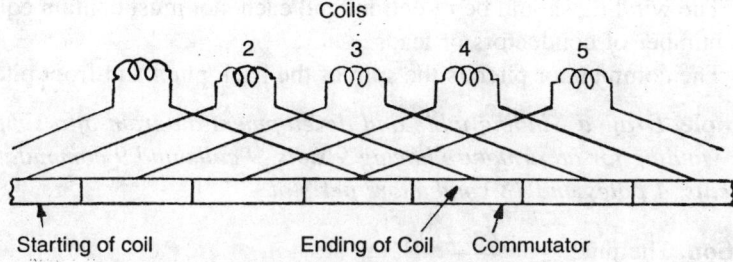

Fig. 22.7. Quadruplex winding.

Q. What is meant by wave winding? Explain the simplex wave winding.

Ans. In this type of winding the leads of a coil are not connected to the adjacent segments but the leads diverge and are connected to the segment widely spread *i.e.* the segments at a far distance.

In this winding the conductor starts from the slots under one pole and goes to the slot under the other adjacent opposite pole and after completing one slot under each pole returns to the slot under the same pole. The front end connections do not overlap. It is done in the machines of high voltage and less currents capacity.

The wave windings are classified as:

(*a*) Simplex winding.
(*b*) Multiplex winding.
   (*i*) Duplex winding.
   (*ii*) Triplex winding.
   (*iii*) Quadruplex winding.

Simplex wave winding. In this winding the leads of a coil are connected to a far distance. There are following conditions for the wave winding:

(*i*) The number of poles should be more than two.
(*ii*) The number of brushes should always be two, what so ever the number of poles may be.
(*iii*) The number of coil should be in odd number.
(*iv*) The back pitch and front pitch should be nearly equal to the pole pitch and should be in odd number.
(*v*) The back pitch and front pitch may be equal or may differ by two conductors.
(*vi*) The commutator pitch $Y_c = \dfrac{\text{Number of segment} \pm 1}{\text{Number of pair of poles}}$.

(*vii*) The winding should be re-entrant and each slot must contain equal number of conductors or leads.

(*viii*) The commutator pitch is the sum of the back pitch and front pitch.

Example. *Draw a winding table and development diagram of a simple wave winding for an armature having 9 slots, 9 coils and 9 commutator segments, 4 poles and 18 conductors per slot.*

Solution. The pitch $Y = \dfrac{Z \pm 2}{2P}$

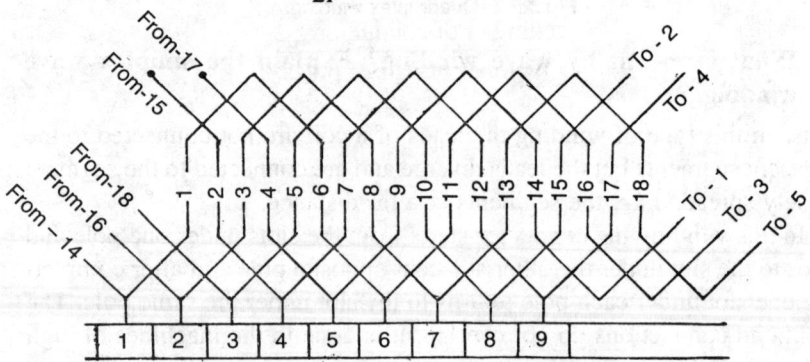

Fig. 22.8 Development Diagram.

$$= \frac{18 \pm 2}{2 \times 2} = 5 \text{ or } 4$$

Now taking coil as 5, the coil throw will be 1–6.

Table 22.4. Winding Table.

S.No.	Y_B	Y_F	Remark
1.	1–6	6–11	
2.	11–16	16–3	
3.	3–8	8–13	
4.	13–18	18–5	
5.	5–10	10–15	
6.	15–2	2–7	
7.	7–12	12–17	
8.	17–4	4–9	
9.	9–14	14–1	

Example. *Draw the winding diagram and table for a wave wound armature, having, number of slot = 17, number of segments = 17, number of coils = 17, number of poles = 4.*

Solution. Total number of conductors = 2 × 17 = 34
Now conductor per slot

$$= \frac{34}{17} = 2 \text{ conductor/slot}$$

Pole pitch $\qquad = \dfrac{17 \times 2}{4} = 8.5 \approx 9 \text{ conductor}$

∴ $\qquad\qquad Y_B = 9 \text{ and } Y_F = 7$

$$Y = \frac{\text{Number of conductor} \pm 2}{\text{Number of pair of poles}}$$

$$Y = \frac{34 \pm 2}{2} = 9 \quad \text{or} \quad 8$$

Table 22.5. Winning Table.

Coil No.	Conductor Y_B		Slot No.		Segment No.		Conductor Y_F		Remark
	T	B	T	B	T	B	B	T	
1.	1	10	1	5	1	9	10	17	
2.	17	26	9	13	9	17	26	33	
3.	33	8	17	4	17	8	8	15	
4.	15	24	8	12	8	16	24	31	
5.	31	6	16	3	16	7	4	13	
6.	13	22	7	11	7	15	22	29	
7.	29	4	15	2	15	6	2	11	
8.	11	20	6	10	6	14	20	27	
9.	27	2	14	1	14	5	1	9	
10.	9	18	5	9	5	13	18	25	
11.	25	34	13	17	13	4	34	7	
12.	7	16	4	8	4	12	16	23	
13.	23	32	12	16	12	3	32	5	
14.	5	14	3	7	3	11	14	21	
15.	21	30	11	15	11	2	30	3	
16.	3	12	2	6	2	10	12	19	
17.	19	28	10	14	10	1	28	1	

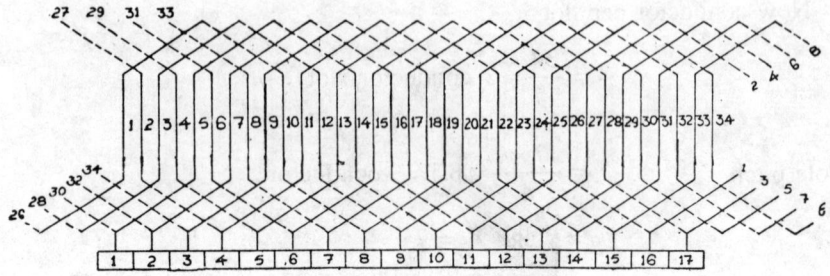

Fig. 22.9. Development diagram.

Example. *Draw the winding table for a simplex progressive winding having the following datas:*

Number of slots = 13,   Number of coils = 13,

Number of segments = 13,   Number of poles = 2.

Solution. Number of conductor = 2 × 13 = 26

Number of poles   = 2

∴   Pole pitch $= \dfrac{\text{No. of conductor}}{\text{Number of poles}}$

Table 22.6.

S. No.	Coil No.	Back pitch		Front pitch		Remark
		Top	Bottom	Top	Bottom	
1	1.	1	16	16	3	
2	2.	3	18	18	5	
3	3.	5	20	20	7	
4	4.	7	22	22	9	
5	5.	9	24	24	11	
6	6.	11	26	26	13	
7	7.	13	2	2	15	
8	8.	15	4	4	17	
9	9.	17	6	6	19	
10	10.	19	8	8	21	
11	11.	21	10	10	23	
12	12.	23	12	12	25	
13	13.	25	14	14	1	

$$= \frac{26}{2} = 13$$

Coil pitch   = Pole pitch ±1

$$= 13 \pm 1 = 14, 12$$

$$Y = \frac{Z \pm 2}{2P} = \frac{26 \pm 2}{2 \times 1} = 14, 12.$$

Coil throw for progressive winding.

$$Y_B = \text{back pitch} = 15$$
$$Y_F = \text{front pitch} = 13.$$

Q. What do you understand by the lead connections and brush position? How the brush positions are determined?

Ans. The lead connections means to connect the leads of the coil to the commutator segment. The lead connections basically depend upon the position of the brushes on the commutator.

Mainly there are two positions of the brushes:

(*a*) **The brushes are at right angle to the main magnetic field.** Whenever the brushes are so placed as shown in Fig. 22.10(*a*) the lead coming out from a slot is connected to the segment in front of that slot. In this case, take a thread and connect the two ends to two centres of the armature by passing through any slot. Mark the segment lying under the thread and the slot also. The lead coming out from the slot will be soldered to this in front slot and so on.

(*b*) **Brushes are parallel to the main magnetic field.** In this case the connections areas shown in Fig. 22.10(*b*). Take a thread, observe any coil, pass the thread through the middle of that coil joining the two centre of the shaft on opposite sides. Mark the segment under the thread and connect one lead of the coil on that segment the second lead to the adjacent segment.

Fig. 2.10. Brush Positions.

Method of finding out the brush position. There are two methods of finding the brushes positions on the commutator.

(*i*) Ring method. (*ii*) Degree method.

(*i*) *Ring method.* In this case the direction of the current lying under different pole is imagined. As the conductors lying under north pole are marked upward and conductors under south pole are marked downwards. The direction of current is marked on the diagram then the brushes are placed at those segments where the direction of current in the conductors are same.

Let us say there are 16 conductors draw 16 lines as shown in Fig. 22.11 so that the $Y_B = 9$ and $Y_F = 7$. Then connect the conductors joining according to the winding diagram. Mark upward direction from 1 to 8 and downward from 9 to 16. Here the direction in the commutator segments having 1 & 8 and 16 & 9 conductors are same so the brushes are placed at both the places.

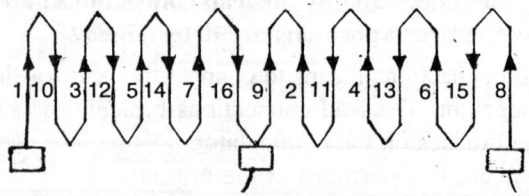

Fig. 22.11. Ring method

(*ii*) *Degree method.* In this method the electrical degrees are the main consideration:

Elect. degree = 360 × Number of pair of poles

For a two pole machine the commutator has 360° so the brushes are placed at 180° Elect. apart. In case of four pole machine, total electrical degrees are 720° Elect. so the brushes are placed 90° mech. apart which comes to 180° Elect.

Q. What may be the possible faults in a newly rewound armature and how will you test them?

Ans. The armature winding may have the following possible faults:

(*a*) **Short circuit.** If the two leads of a coil are joined directly, it is known as direct dead short circuit. If the few turns are short circuited then partial short circuit.

(*b*) **Open circuit.** If the wire of any coil is broken either at the segment or inside the coil side it is known as open circuit.

(*c*) **Reverse connections.** In case of reverse connections the leads of a coil are interchanged and are connected to wrong segments.

(*d*) **Earth fault and leakage.** In this case the wire comes in contact with the body or armature core. In this case the insulation resistance between the conductor and earth becomes defective or reduced.

There are the following methods of locating the faults in d.c. armature winding.

(*d*) **Ammeter method.** In this case the current is passed through coils
individually. The value of the current is observed and results are
obtained. The diagram as shown in Fig. 22.12 shows the connection
for the test. The supply is controlled by means of an external resistance to the safe value; so that the current is restricted to safe limit.
The results are concluded as under:

(*i*) Normal reading across the coils—winding is alright.

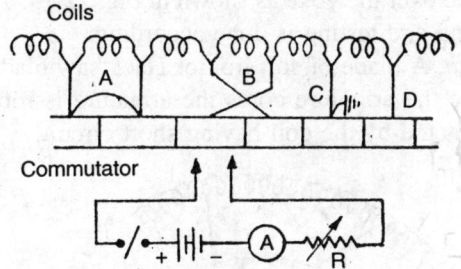

Fig. 22.12. Ammeter method testing.

(*ii*) Reading maximum as adjusted—short circuit in the coil.

(*iii*) Reading between maximum value and normal value. There is a
possible short circuit.

(*iv*) Reading is zero or approximately zero—Open circuit.

(*v*) The reading is half of the normal reading—Reverse connection.

(*vi*) In case of earth fault testing, connect one lead to the shaft and other
to the different segments. Maximum reading with any segment there
is earth fault and if the reading is between the maximum and normal readings then a partial short circuit and no reading means O.K.

(*b*) **Drop method or Voltmeter method.** In this case d.c. low voltage
is applied to the armature. A voltmeter of low voltage is taken and its
leads are connected to the individual coils. In case of lap winding the
leads connected to the adjacent segments, but in case of wave winding
these are connected to the ends of a coil (Y_B and Y_C are taken into account). The voltmeter will read the voltage drop in that particular coil as
shown in Fig. 22.13 where it is connected. The readings are observed as
follows:

(*i*) Same reading across each coil—winding is alright.

(*ii*) Zero reading—short circuit in the coil.

(*iii*) Reading between zero and normal—partial short circuit in the coil.

(*iv*) Reading double the normal—Reverse connections.

(*v*) Reading high enough as set by the external resistance—open circuit.

(*vi*) While connecting one end to the shaft other with the individual coil—zero reading means no earth fault.

In this method the voltage applied is very low so that half the normal current is passed through the armature winding.

(*c*) **Growler test.** For testing d.c. armature it is the effective and very sensitive apparatus. The growlers are of two types, the inside growler and outside growler. The d.c. armature (rotor) winding is tested on the outside growler. For stator say A.C. machines, the inside growler is used. The armature is placed over the yoke as shown in Fig. 22.14. The instructions given are followed and testing is done according.

(*i*) *Short circuit.* A blade of soft iron or Hack saw blade is kept about 1 cm above the armature cores the armature is rotated, the blade will be attracted by the coil having short circuit.

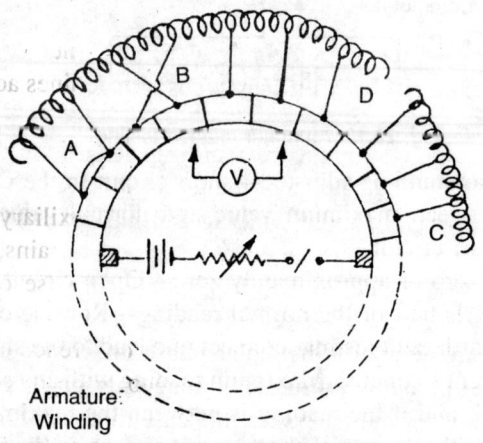

Armature Winding

Fig. 22.13. Voltmeter testing.

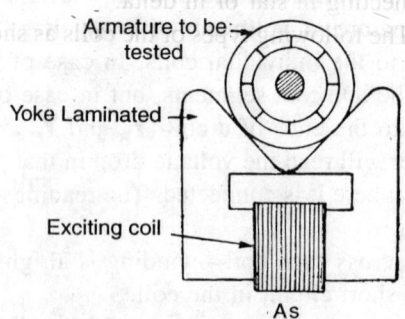

Armature to be tested

Yoke Laminated

Exciting coil

As

Fig 22.14. Growler test.

(*ii*) For other test, a voltmeter is provided with the growler, and two leads are brought out. The results and faults are described above as in voltage drop method or as per given instructions.

(*d*) **By means of the capacitor.** In this method a lamp and a condenser are connected in series. The two leads are brought out from the condenser. These two leads are connected across the individual coils. The observations are as follows:

(*i*) Sparking at commutator—short circuit.

(*ii*) Low sparking and normal light of the lamp—open circuit.

(*iii*) The light is dim and no sparking on the commutator—winding is alright.

22.4. A.C. Armature winding

Q. What do you understand by a.c. armature winding? What types of coils are used for the winding?

Ans. The winding which is done on the A.C. machines is known as A.C. armature winding. There are two main types of machines according to the number of phases.

(*a*) **A.C. single phase armature winding.** It is done on A.C. single phase machines like fan, a.c. induction motors etc. The windings are of two types the main winding and the auxiliary winding. The main winding is always connected across the mains, where auxiliary winding for auxiliary purposes only. In case of shaded pole motor only one winding is done on the stator but for other types two windings are done at a particular phase difference of 90° elect.

(*b*) **A.C. three phase winding.** It is done only on the stator of three phase motors or alternators. In this case three windings are spaced 120° elect. apart and then either six terminals are brought out or there by connecting in star or in delta.

Types of coils. The following types of the coils as shown in Fig. 22.15, are mainly used for the windings of A.C. machines.

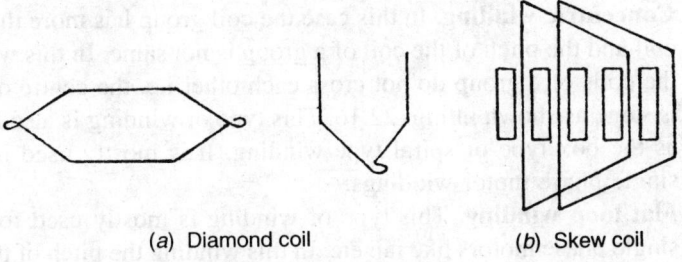

(*a*) Diamond coil   (*b*) Skew coil

Fig. 22.15. Types of the coils

(*i*) *Mush coils.* In this type of coils the active sides and level of the sides both are same. The coils are wound in a single former or a stepped former in case of concentric winding. These coils are according to their shapes, as U-shape, V-shape and rectangular shape.

(*ii*) *Diamond coil.* In this type the two coil sides are not in the same level, one side is down. These coils are made in two operations. First coil is prepared and then given the shape. These coils are mostly used in H.T. and high capacity motors and alternators. The appearance of the coil and winding is good. These are easy to house.

(*iii*) *Skew type coil.* In this case the length of the inactive side is not same. It is also used for H.T. machine winding. These coils provides better cooling effect.

Q. What are the main types of A.C. windings?

Ans. The A.C. winding is classified according to the shapes, these are:

(*a*) Basket winding.

(*b*) Concentric winding.

(*c*) Flat loop winding.

(*d*) Skeen winding.

(*a*) **Basket winding.** After the completion of the winding, the shape of the winding just resemble the weave of a basket so it is named as the Basket winding. In this case the inactive sides of the coils cross each other. The pitch of the coil is same. The winding is done mostly in three phase motors.

The basket winding can also be classified according to the coil used:

(*i*) *Diamond coil winding.* In this type of windings the coils are of the diamond coils, so the name is given as diamond coil winding.

(*ii*) *Skew coil winding.* In this winding the skew coils are used so the winding is known as the skew coil winding.

(*iii*) *Involute coil winding.* In this case the shape of the coil is as the diamond coil type but the end turns are pressed as the involute coil hence this type is known as involute coil winding.

In this type of coils the checking of the connections is easy.

(*b*) **Concentric winding.** In this case the coil group has more than one coil and the pitch of the coil of a group is not same. In this winding the coils of a group do not cross each other i.e. the centre of coils is same as shown in Fig. 22.16. This type of winding is also known as the box type or spiral type winding. It is mostly used in A.C. single phase motor windings.

(*c*) **Flat loop winding.** This type of winding is mostly used for A.C. single phase motors like fan etc. In this winding the pitch of the coil is same. If the winding has only one winding the coil do not over lap each other, as shown in Fig. 22.17 and the winding is known as flat loop non over lap winding. But if the windings are more than one, then the coils of the group over lap each other then the winding is flat loop over lapping winding. This winding forms a chain shape and winding is known as *chain winding* and if concentric group over lap then if is the *concentric chain winding*.

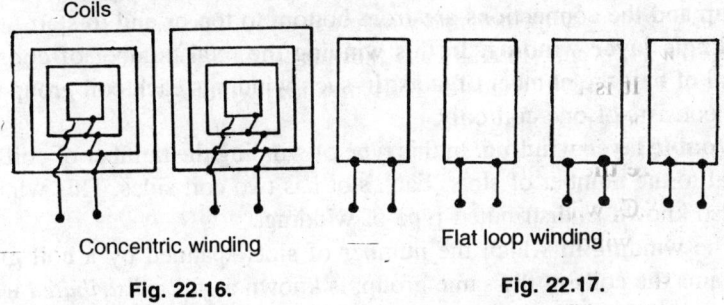

Fig. 22.16.   **Fig. 22.17.**

Q. Explain the important terms used in A.C. winding.

Ans. There are the following important terms:

Coil group. A winding has a number of coil group. In an ordinary winding the coil groups are equal to the number of phases multiplied by the number of poles. In each group there may be one or more than one coil connected in series and forming a pole.

No. of coil in a group

$$= \frac{\text{Total number of coils}}{\text{No. of phases} \times \text{No. of poles}}$$

$$= \frac{\text{Total number of coil per phase}}{\text{Number of poles}}$$

Coil connections. The connections which joins two coils of a group in series. It is also known as stub or end connection. In a coil group the connections are end to start or bottom to top.

Group connections. In these connections, the coil groups are joined phase wise. The coil group of one phase is connected with the another coil group of the same phase.

In a whole coiled winding the group connections are bottom to bottom and top to top or end to end and start to start.

In a half coiled winding the group connections are from bottom to top or end to start.

Phase connections. These are the connections which joins the three phase winding either is star or in delta.

Whole coil winding. A whole coil winding is that which has one or more than one coil in a group. In this winding the number of coil in a group is always in whole number.

In this winding the group connections are from bottom to bottom and top to top.

Half coil winding. In this winding each group has only half coil in a group and the connections are from bottom to top or end to start.

Single layer winding. In this winding the total number of coils are equal of half the number of slots. In such windings each coil group usually consists of one coil only.

Double layer winding. In this type of winding the number of coils are equal to the number of slots. Each slot has two coil sides. This winding is also known as distributed type of winding.

The winding in which the number of slots spanned by a coil group contains the coils of the same group, is known as *fully distributed winding*. Secondly if the slots spanned by the coil group do not contain the coils of the same group then the winding is known as *partially distributed* winding.

Balanced winding. A balanced winding is that which has same number of coils in every group as in case of 36 slots 3 phase 4 pole will have

$$\frac{36}{3 \times 4} = 3 \text{ coils in each group, so it is known as balanced winding.}$$

Unbalanced winding. If the number of coils in each group are not same then the winding is known as unbalanced winding. For example the

three phase four pole and 30 coils winding has only $\frac{30}{3 \times 4} = 2.5$ coils,

here the coils can not be divided, so these are arranged as 3 + 2 + 3 + 2 = 10 coils per phase. Now the coils are same for every phase; this type of winding is known as unbalanced winding. In this case the coils per group are different by coil per phase are same.

Pole pitch. It is generally taken in terms of slots. It is defined as the number of slots per pole, so

$$\text{Pole pitch} = \frac{\text{Total number of slots}}{\text{Number of poles}}.$$

Coil pitch. It is the distance between two sides of the coil. It is also taken in terms of slots.

Pitch factor. It is the ratio of coil pitch to the pole pitch.

$$\therefore \qquad \text{Pitch factor} = \frac{\text{Coil pitch}}{\text{Pole pitch}}.$$

Q. Explain the followings:

    (*i*) **How to find out the number of coils in a group?**

    (*ii*) **How to find out the degree per slot?**

    (*iii*) **Method of making winding table and diagram.**

Ans. (*a*) **Coil per group.** In this case the number of coil per group is determined by the following method:

Here total number of groups in a winding

           = Number of pair of poles × Number of phases.

$$\therefore \qquad \text{Coils per group} = \frac{\text{Total number of coils}}{\text{Number of poles} \times \text{Number of phases}}$$

If it is a full number the winding is a balanced winding and if not in whole number then the winding is unbalanced winding. In that case by division, there will be one part as the quotient and other as remainder. Thus the groups equal to remainder will contain quotient number of coil per group equal to quotient.

Example. *A motor has 24 slots 12 coils, the wound for three phase four poles. Calculate the coil per group.*

Solution. The total coil groups

    = No. of phases × No. of pair of poles

    = 3 × 2 = 6

Here coil group per phase

$$= \frac{6}{3} = 2 \text{ coils}$$

Here two complete coils are there so it is a balanced winding.

Example. *Calculate and arrange the coils per group in a motor having 36 slots, 18 coils 4 pole wound for three phase.*

Solution. Coils per phase

$$= \frac{18}{3} = 6 \text{ coils}$$

Coil groups   = No. of phases × No. of pair of poles

                    = 3 × 2 = 6

$$\text{The coil per group} = \frac{\text{Total number of coils}}{\text{No. of poles} \times \text{No. of phases}}$$

$$= \frac{18}{4 \times 3} = \frac{6}{4} = 1\frac{2}{4}$$

Here one is the quotient and two as the remainder. So two coil groups will have two coils and the remainings two groups will contains only one coil and each phase will have (1 + 2 + 1 + 2) coils.

The coils will be arranged 1 + 2 + 1 + 2, fashion.

(b) **Degree per slot.** In three phase winding each phase is displaced 120° elect. apart. In single phase and two phases the windings are displaced at 90° electrical.

So to calculate degree per slot take the following steps:

(i) Find out the total electrical degrees in the stator.

Total elect. degree = 360 × Number of pair of poles.

(ii) Now divide the total degrees by the total number of slots

$$\frac{\text{Degree}}{\text{Slot}} = \frac{360 \times \text{Number of pair of poles}}{\text{Total number of slots}}$$

Here if the number of slots are required for a specific degree than divide the electrical degrees by the degree per slot, the required number of slots will be obtained.

(c) **Making the winding table and diagram.** Take the following steps:

(i) Calculate the number of poles of the motor from the speed and frequency given (in case if the poles are not given).

(ii) Find the total number of coils.

In single layer winding the coils are equal to half the total number of slots and in double layer winding the number of coils are equal to the number of slots.

(iii) Calculate the number of coil per phase

$$= \frac{\text{Total Number of coils}}{\text{Number of phases}}$$

(iv) Calculate the total number of coils per phase per pole, so that the type of winding whether balanced/unbalanced, whole coil/half coil will be known.

(v) Find the pole pitch and if it is given in the old winding take that pitch only.

(vi) Be acquainted with the type of winding.

(vii) Find the number of slots for the given phase difference *i.e.* 120° elect. for three phase.

(*viii*) Draw a winding table showing coil number and coil throw also. In single layer the coils will be started from alternate slots and for double layer winding from every adjacent slot.

(*ix*) Arrange the coils in groups according to the number of coils per group.

(*x*) Make the group and connect phase wise independently showing coil and group connections.

(*xi*) Reverse the centre phase leads and connect the winding in star or in delta or bring all the six leads as required.

Example. *Make the winding table and draw the development diagram of a 3 phase 4 pole induction motor having 24 slots, 12 coils operating on 440 V 50 c/s mains.*

Solution. Obviously the number of coils are 12 and number or slots are 24, so the winding is a single layer winding.

Now number of coils/phase

$$= \frac{12}{3} = 4 \text{ coils}$$

Number of coil per phase per pole

$$= \frac{4}{4} = 1 \text{ coil}$$

So the coil per group is one it is a balanced winding.

Here total electrical degrees

$$= 360 \times \frac{P}{2}$$

$$= 360 \times \frac{4}{2} = 720° \text{ elect.}$$

$$\therefore \quad \text{Degree per slot} = \frac{720}{24} = 30° \text{ elect.}$$

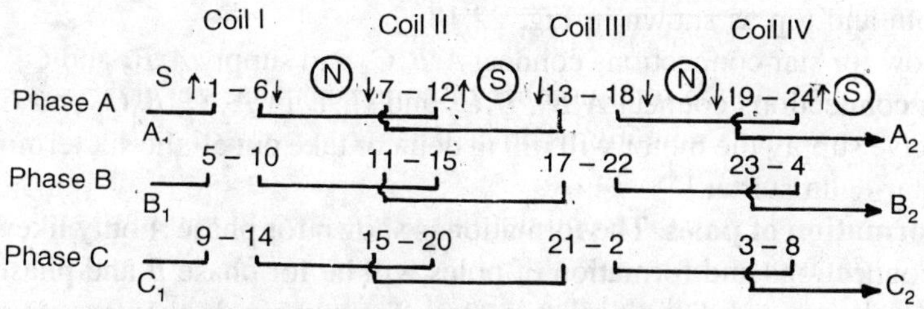

Fig. 22.18. Coil and group connections.

In three phase winding the phase displacement is 120° elect. So the number of slots for 120° elect.

$$= \frac{120}{30} = 4 \text{ slots}$$

So the first phase will start from slot number 1, the second phase will start from slot number $1 + 4 = $ 5th and the third phase will start from $5 + 4 = $ 9th slot.

$$\text{The pole pitch} \quad = \frac{\text{Number of slots}}{\text{Number of poles}} = \frac{24}{4} = 6$$

So in this winding the coil throw is 1 to 6.

Table 22.7. Winding Table

S.No.	No. of coil	Coil throw	Phases	Remark
1.	1	1 — 6	A	The three phases are
2.	2	3 — 8	C	taken as
3.	3	5 — 10	B	Phase-A
4.	4	7 — 12	A	Phase-B
5.	5	9 — 14	C	Phase-C
6.	6	11 — 16	B	
7.	7	13 — 18	A	
8.	8	15 — 20	C	
9.	9	17 — 22	B	
10.	10	19 — 24	A	
11.	11	21 — 2	C	
12.	12	23 — 4	B	

Phase and group connections. There are four coils per phase and winding has got four poles, so the group connections are from bottom to bottom and top as shown in Fig. 22.18.

Now for star connections connect $A_1 B_1 C_1$ and supply $A_2 B_2$ and C_2. For delta connections connect $A_1 B_2$, $B_1 C_2$ and $C_1 A_2$ or $A_2 B_1$, $B_2 C_1$ and $C_2 A_1$ and give supply the motor will run in delta or take out all the six terminals for star-delta starter.

Formation of poles. The formation is shown for phase A only likewise the connections and formation of poles will be for phase B and phase C.

The downward direction let us mark for north pole then the upward direction will be for south pole, thus the winding has four poles as shown in phase A of Fig. 22.18.

Example. *Draw the development diagram of a three phase six pole 400 V 50 c/s motor having 36 slots, 18 coils connected in star.*

Solution. The coils are 18 and 36 slots so it is a single layer winding.

Now the pole pitch $= \dfrac{\text{Number of slots}}{\text{Number of poles}}$

$$= \frac{36}{6} = 6$$

Now the coil throw is 1–6

Table 22.8. Winding Table

S.No.	Coil No.	Coil throw	Phases	Remark
1.	1	1 — 6	A	
2.	2	3 — 8	C	
3.	3	5 — 10	B	
4.	4	7 — 12	A	
5.	5	9 — 14	C	
6.	6	11 — 16	B	
7.	7	13 — 18	A	
8.	8	15 — 20	C	
9.	9	17 — 22	B	
10.	10	19 — 24	A	
11.	11	21 — 26	C	
12.	12	23 — 28	B	
13.	13	25 — 30	A	
14.	14	27 — 32	C	
15.	15	29 — 34	B	
16.	16	31 — 36	A	
17.	17	33 — 2	C	
18.	18	35 — 4	B	

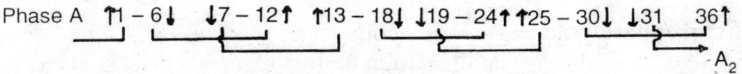

Phase A ↑1 – 6↓ ↓7 – 12↑ ↑13 – 18↓ ↓19 – 24↑ ↑25 – 30↓ ↓31 36↑ →A₂

Fig. 22.19 Coil and group connections.

The number of coils per phase $= \dfrac{18}{3} = 6$ coil

The coil/phase/pole $= \dfrac{6}{6} = 1$

So it is a whole coil single layer winding.

Now the total electrical degrees

$$= 360 \times \dfrac{6}{2} = 1080° \text{ Electrical}$$

So the degree per slot

$$= \dfrac{1080}{36} = 30°$$

Now slots required for 120° electrical displacement

$$= \dfrac{120}{30} = 4 \text{ slots}$$

So the phase A will start from slot No. 1 and phase B from slot 1 + 4 = 5th and phase C from 5 + 4 = 9th slot.

Similarly both B and C phase will also be connected. A_1, B_1, and C_1 are connected together to form a star point and A_2, B_2 and C_2 are connected across the line.

Example. *Draw the connection and development diagram of a three phase 400 V 50 c/s 4 pole A.C. squirrel cage induction motor having 24 slots, 24 coils running at 1440 r.p.m.*

Solution. The number of coils and number of slots are equal, so this is a double layer winding.

Pole pitch $\qquad = \dfrac{24}{4} = 6$

Here coil throw is 1–7

No. of coils per phase $= \dfrac{24}{3} = 8$

No. of coils/phase/pole $= \dfrac{24}{3 \times 4} = 2$ coils

So the group connections are bottom to bottom and top to top.

Degree per slot $\qquad = \dfrac{360 \times 2}{24} = 30°$ Electrical

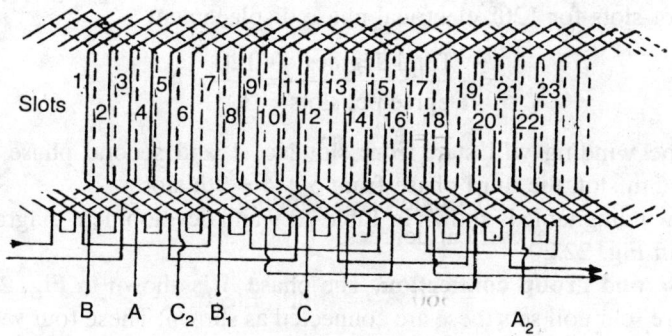

Fig. 22.20 Developed diagram.

Table 22.9. Winding Table.

S.No.	Coil No.	Coil throw	Phase	Remark
1.	1	1 — 7 ⎤	A	
2.	2	2 — 8 ⎦		
3.	3	3 — 9 ⎤	C	
4.	4	4 — 10 ⎦		
5.	5	5 — 11 ⎤	B	
6.	6	6 — 12 ⎦		
7.	7	7 — 13 ⎤	A	
8.	8	8 — 14 ⎦		
9.	9	9 — 15 ⎤	C	
10.	10	10 — 16 ⎦		
11.	11	11 — 17 ⎤	B	
12.	12	12 — 18 ⎦		
13.	13	13 — 19 ⎤	A	
14.	14	14 — 20 ⎦		
15.	15	15 — 21 ⎤	C	
16.	16	16 — 22 ⎦		
17.	17	17 — 23 ⎤	B	
18.	18	18 — 24 ⎦		
19.	19	19 — 1 ⎤	A	
20.	20	20 — 2 ⎦		
21.	21	21 — 3 ⎤	C	
22.	22	22 — 4 ⎦		
23.	23	23 — 5 ⎤	B	
24.	24	24 — 6 ⎦		

No. of slots for 120° electrical phase displacement

$$= \frac{\text{Total degrees}}{\text{Degree per slots}} = \frac{120}{30°} = 4$$

So the winding will start from slot No. 1 and second phase from $1 + 4 = 5$th slot, the third phase from $5 + 4 = 9$th slot.

The winding table is shown in Table 22.9 and development diagram as shown in Fig. 22.20.

Phase and group connection. The phase A is shown in Fig. 22.21. There are four coil sets these are connected as shown. These four sets are connected top to top and bottom to bottom to form four poles. This way phase 'B' will also have $B_1 B_2$ terminals and phase 'C' will have C_1 and C_2. So to have star connection $A_1 B_1 C_1$ are connected to form a star point and line is given to $A_2 B_2$ and C_2. It can also be connected in delta or all six leads may be taken out for star-delta starter.

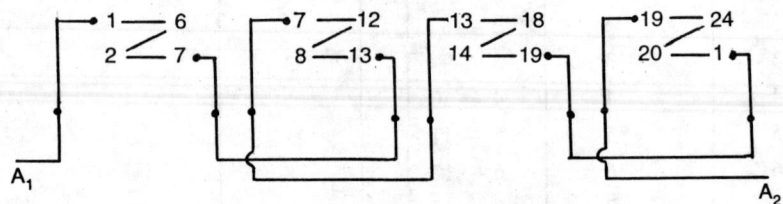

Fig. 22.21. Coil and group connections.

Example. *Draw the development diagram of a three phase 440 V 50 c/s four poles motor having 36 slots, 18 coils, six groups for concentric single layer half coil balanced winding.*

Solution. There are 36 slots and 18 coils so it is a single layer winding.

The coil per phase $\qquad = \dfrac{18}{3} = 6$ coils

No. of coil per phase/group $\qquad = \dfrac{6}{2} = 3$ coil

Now the group connections are from bottom to top.

Total electrical degrees $\qquad = 360 \times 2 = 720°$ Elect.

$\therefore$ Degree per slot $\qquad = \dfrac{720}{36} = 20°$ Elect.

Slots required for 120° Elect. phase displacement

$$\frac{\text{Total degree required}}{\text{Degree per slot}} = \frac{120}{20} = 6 \text{ slots}$$

So phase *A* will start from slot No. 1, phase *B* will start from slot No. 1 + 6 = 7th and phase *C* will start from 7 + 6 = 13th slot.

The pole pitch $= \dfrac{36}{4} = 9$

Number of coils per group are three. It is a **concentric winding**, so the pitch will be 9 ± 2.

∴   Pitch for outer coil   9 + 2 = 11

      Pitch for centre coil   9 + 0 = 9

      Pitch for inner coil   9 − 2 = 7

Now connect $A_1 B_1 C_1$ to form a star point and give supply to A_2, B_2, and C_2, the motor will be running in star.

Table 22.10. Winding Table

Phase	Coil group-I	Coil group-II	Remark
A	1 — 12	19 — 30	
	2 — 11	20 — 29	
	3 — 10	21 — 28	
B	7 — 18	25 — 36	
	8 — 17	26 — 35	
	9 — 16	27 — 34	
C	13 — 24	31 — 6	
	14 — 23	32 — 5	
	15 — 22	33 — 4	

Coil, group and phase connections:

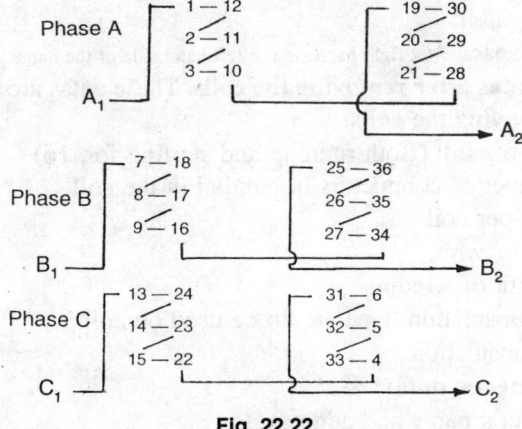

Fig. 22.22.

Q. What is the procedure for rewinding a motor?

Ans. Some datas should be collected from the motor itself for the rewinding purpose. These datas are very helpful while rewinding the motor. No change should be made in design of the motor winding. It should be wound as it was. The procedure can be classified as under:

1. Collecting the datas

(A) **Name plate data.** The following datas should be taken from the name plate of the burnt out motor.

   (i) Manufacturer's name.

   (ii) Type of motor

   (iii) Motor number

   (iv) Volts

   (v) Current

   (vi) HP/kVA/kW

   (vii) Phases

  (viii) Frequency

   (ix) Speed in r.p.m.

   (ix) Phase connections.

(B) **Stator datas before removing the coils:**

   (i) Number of slots.

   (ii) Number of coils.

   (iii) Type of winding. (whether concentric, basket, single layer or double layer)

   (iv) Number of poles.

   (v) Number of coil groups.

   (vi) Whole coil winding or half coil winding.

   (vii) Coil connections.   (Group connection)

  (viii) Coil per group.

   (ix) Coil pitch.

   (x) End space. (at right hand side or left hand side of the name plate)

(C) **Stator datas after removing the coils.** These datas are noted down after removing the coils:

   (i) Size of coil. (Both running and starting for, 1ϕ)

   (ii) Number of conductors in parallel in the coil.

   (iii) Turn per coil.

   (iv) Size of wire.

   (v) Weight of winding.

   (vi) Coil insulation.(tape etc. to be used on coils)

   (vii) Slot insulation.

(D) **Miscellaneous datas:**

   (i) Owner's name and address.

    (*ii*) Date of taking the motor.

   (*iii*) Date of delivering.

   (*iv*) Job card number.

    (*v*) Remark if any......

After taking these informations the motor is wound, then the following operational sequences are adopted.

2. Dismantling operation

    (*i*) Loose the pulley screws. Remove the key and then remove the pulley or gear with the help of pulley puller. Avoid hammering on pulled or gear etc.

   (*ii*) Mark the end covers of the machine for left and right covers and position of the covers with the help of centre punch etc.

  (*iii*) Loose the screws of the bearing covers and take out.

  (*iv*) Remove the end covers.

   (*v*) Take out the rotor and examine its surface condition, if wound then examine the winding condition and if squirrel cage then the condition of the bar connections.

  (*vi*) Check the teeth of the stator.

 (*vii*) Clean the stator with air gun etc.

3. Winding diagram

Take down all the datas and make connections and winding diagrams and remove its old winding.

4. Slot insulation and coils

    (*i*) Arrange slot insulation.

   (*ii*) Arrange wire for winding purpose.

  (*iii*) Make required number of coils.

  (*iv*) Clean the slot and insulate them.

5. Rewinding and connections

    (*i*) Arrange the coils in the stator.

   (*ii*) Connect the coils.

  (*iii*) Check coil connections, group connections, phase connections and phase sequence of the winding.

6. Varnishing, backing and reassembling

    (*i*) Reassemble the motor.

   (*ii*) Test the motor winding for earth fault, phase connections and star or delta connections etc. and arrange for trial.

  (*iii*) Again dismantle the motor and varnish the motor winding.

(*iv*) Bake the winding the after drying, measure the insulation resistance and reassemble the motor.

(*v*) Check the motor on the supply and for a certain period to see the working condition.

22.5. SINGLE PHASE WINDING

Example. *Draw a winding table to rewind a single phase motor having 24 slots, 12 coils (8 coils for running winding, 4 coils for starting winding). Number of poles –4, coil pitch for running winding 5, 3 coil pitch for starting winding –5, running at 1440 r.p.m.*

Solution. There are four poles; because it is an induction motor. The supply frequency is 50 c/s speed.

So the number of poles

$$= \frac{120f}{N} = \frac{120 \times 50}{1440} = 4.16 \text{ poles.}$$

The poles cannot be in fractions so there are four poles.

The pole pitch = 5, 3 for running winding

5 for starting winding

Degree per slot $= \dfrac{360 \times 2}{24} = 30°$ Elect.

Number of slots for 90° phase displacement

$$= \frac{90°}{\text{degree per slot}} = \frac{90}{30} = 3 \text{ slots,}$$

Running winding. Coil pitch 5, 3.

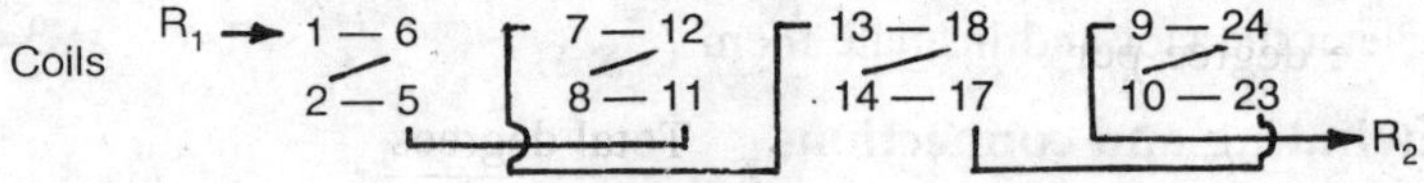

Fig. 22.23.(*a*)

Starting winding. Coil pitch 5.

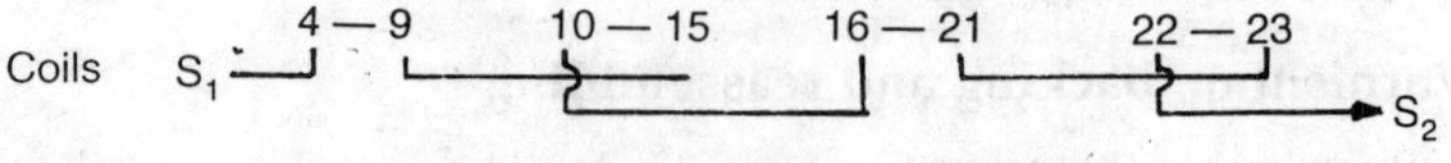

Fig. 22.23.(*b*)

Connect S_1A_1 and S_2A_2 and give supply to start the motor. Winding diagram.

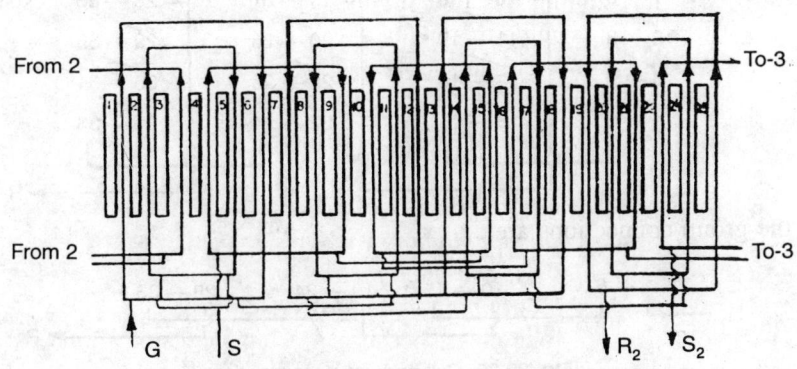

Fig. 22.24. Development diagram.

Example 2. *Draw the winding table for a single phase motor having 36 slots, 26 coils (16 coils for running winding, 10 coils for starting winding) coil pitch for running winding 8, 6, 4, 2 and for starting winding 9, 7, 5 and 7, 5.*

Solution. The pole pitch $\qquad = \dfrac{36}{4} = 9$

Coil groups are 4×1 $\qquad = 4$ groups

Now number of coil groups in running winding

$$= \frac{16}{4} = 4$$

and Number of coil groups in starting winding

$$= \frac{10}{4} = 2.5$$

so is starting winding the coils will be arranged in 3 and 2 coils per group. Now the degree per slot

$$= \frac{\text{Total degrees}}{\text{Slots}}$$

$$= \frac{360 \times 2}{360} = 20° \text{ Elect.}$$

So the number of slots required for 90° phase displacement

$$= \frac{90}{20} = 4.5$$

Running winding. Coil pitch 8, 6, 4, 2.

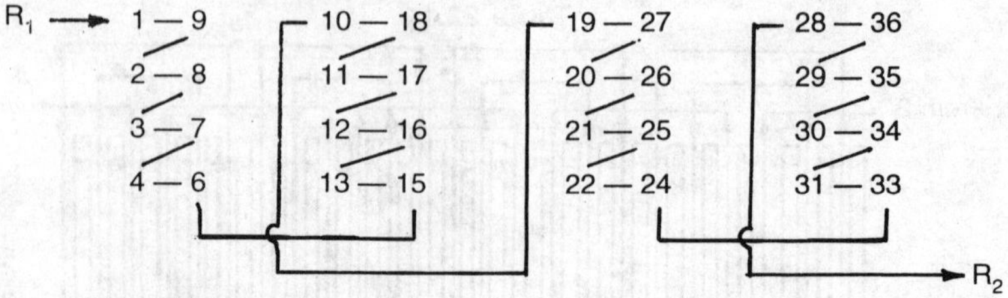

R₁ ⟶ 1 — 9   10 — 18   19 — 27   28 — 36
         2 — 8   11 — 17   20 — 26   29 — 35
         3 — 7   12 — 16   21 — 25   30 — 34
         4 — 6   13 — 15   22 — 24   31 — 33
                                                        ⟶ R₂

the group connections are

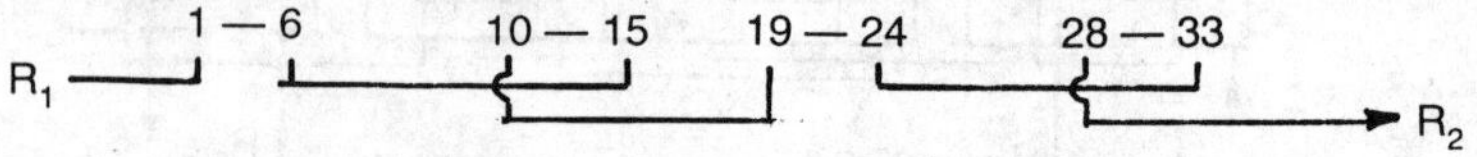

     1 — 6   10 — 15   19 — 24   28 — 33
R₁ ⟶   ⟶ R₂

Fig. 22.25. Coil and group connections.

Starting winding. Coil pitch 9, 7, 5 and 7, 5.

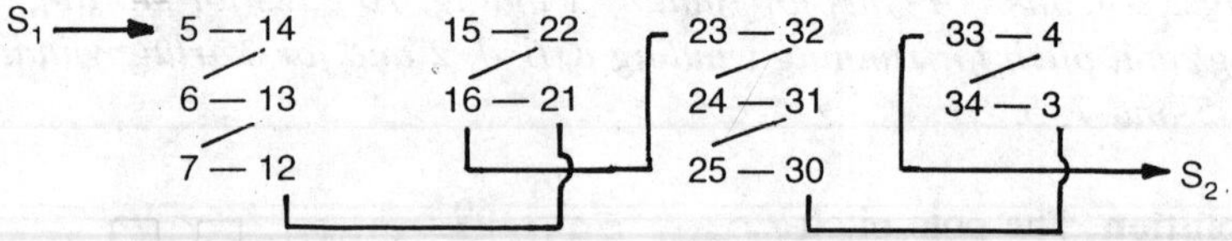

S₁ ⟶ 5 — 14   15 — 22   23 — 32   33 — 4
         6 — 13   16 — 21   24 — 31   34 — 3
         7 — 12   15 — 21   25 — 30   ⟶ S₂

or it can be stated as per the group connections.

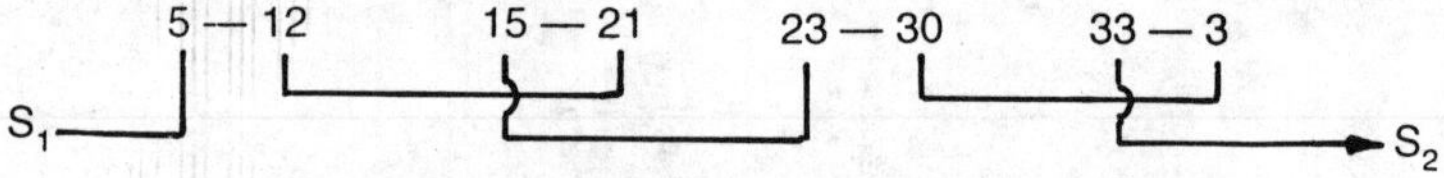

     5 — 12   15 — 21   23 — 30   33 — 3
S₁ ⟶   ⟶ S₂

Fig. 22.26. Coil and group connections.

Connect S_1A_1 and S_2A_2 and give supply to these terminals.

Single phase fan motors

The single phase fan motor have two windings and low speed. These are of two types the shaded pole and capacitor type motors.

Examples. *Draw a winding table for a shaded pole motor having 24 slots 12 coils, 24 poles, turn/coil 250 and wire gauge 24 S.W.G.*

Solution. The pole pitch $= \dfrac{\text{No. of slots}}{\text{No. of poles}} = \dfrac{24}{24} = 1$

The coil pitch is one so the coil through is $(1 - 2)$

Degree per slot $= \dfrac{360}{24} \times \dfrac{24}{2} = 180°$ Elect.

Table 22.11

S.No.	Coil No.	Coil throw
1.	1	1 — 2
2.	2	3 — 4
3.	3	5 — 6
4.	4	7 — 8
5.	5	9 — 10
6.	6	11 — 12
7.	7	13 — 14
8.	8	15 — 16
9.	9	17 — 18
10.	10	19 — 20
11.	11	21 — 22
12.	12	23 — 24

Winding Diagram

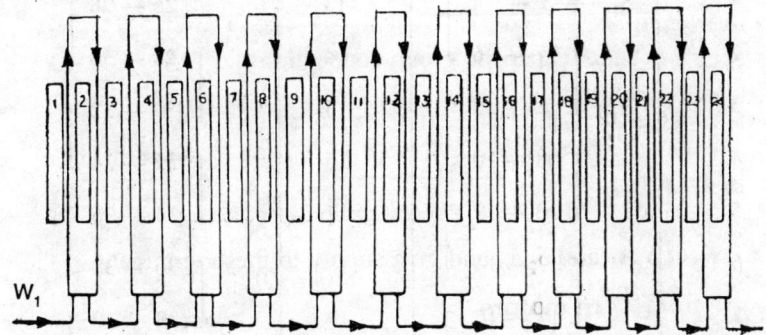

Fig. 22.27. Flat loop winding.

The coil connections will be from bottom to top and finally two leads are taken out for supply. (Winding diagram as in copy).

Example. *Draw a winding table for rewinding a fan having the following datas:*

Number of slots—25   *Coils for running winding—12*
Number of coils—36   *Coils for starting winding—24*
Number of poles—24   *T/coil starting—200—30 S.W.G.*
Capacitor—2.5 μF   *Running—460—30 S.W.G.*

Solution. The pole pitch $= \dfrac{48}{24} = 2$

The coil pitch is 2 and coil throw is 1 – 3.

Table 22.12. Running winding

Coil No.	Coil throw	Coil No.	Coil throw
1.	R_1 —1 — 3	7.	25 — 27
2.	5 — 7	8.	29 — 31
3.	9 — 11	9.	33 — 35
4.	13 — 15	10.	37 — 39
5.	17 — 19	11.	41 — 43
6.	21 — 23	12.	45 — 47 —— R_2

Table 22.13. Starting winding.

Coil No.	Coil throw	Coil No.	Coil throw
1	S_1 —— 2 — 4 ⌐	13	⌐ 26 — 28 ⌐
2	⌐ 4 — 6 ⌐	14	⌐ 28 — 30 ⌐
3	⌐ 6 — 8 ⌐	15	⌐ 30 — 32 ⌐
4	⌐ 8 — 10 ⌐	16	⌐ 32 — 34 ⌐
5	⌐ 10 — 12 ⌐	17	⌐ 34 — 36 ⌐
6	⌐ 12 — 14 ⌐	18	⌐ 36 — 38 ⌐
7	⌐ 14 — 16 ⌐	19	⌐ 38 — 40 ⌐
8	⌐ 16 — 18 ⌐	20	⌐ 40 — 42 ⌐
9	⌐ 18 — 20 ⌐	21	⌐ 42 — 44 ⌐
10	⌐ 20 — 22 ⌐	22	⌐ 44 — 46 ⌐
11	⌐ 22 — 24 ⌐	23	⌐ 46 — 48 ⌐
12	⌐ 24 — 26 ⌐	24	⌐ 48 — 2 ⌐ —— S_2

Degree per slot $= \dfrac{360 \times 12}{48} = 90°$ Elect

The slots for 90° displacement

$$= \frac{90}{90} = 1 \text{ slot}$$

Number of coil group in running winding

$$= \frac{12}{24} = \frac{1}{2}$$

Number of coil groups in starting winding.

$$= \frac{24}{24} = 1$$

Now there are four terminals $S_1 S_2$ and $R_1 R_2$, as shown in Tables 22.12 and 22.13. $S_1 R_1$ and $S_2 R_2$ are connected together. The starting winding has a capacitor in series then the supply is given to these terminals for starting the motor.

22.6. TESTING

Q. Describe in brief the testing of stator winding?

Ans. There are the following tests to be preformed for testing the stator winding.

- (*a*) Checking by winding diagram.
- (*b*) Continuity test or open circuit test.
- (*c*) Ground test or earth test.
- (*d*) Short circuit test.
- (*e*) Balance test.

(*a*) **Checking by winding diagram.** In this test the coil connections, group connections and phase connections are checked by the third person, who did not wind that motor. He should be an electrician and he should check it by the winding diagram.

(*b*) **Open circuit test.** In this test the continuity of each phase winding is checked by means of a series testing lamp or megger. Two winding leads are connected with the two terminals of each winding. If lamp gives light it means that there is no break in the winding and if it remains dark it means there is an open circuit.

(*c*) **Ground test.** This test is performed whether any winding is coming in contact with the body of the motor. If there will be earth fault in the winding the motor may give shock in case of improper earthing and fuse will blow in case of proper earthing. This test can be performed by series testing lamp or megger.

(*d*) **Short circuit test.** It is performed to check whether there is any short circuit in the coils of the winding. The short circuit can be in coil group connections, coil connections, phase connection. In case of short circuit the motor will be heated up or the fuse will blow; this can be done by means of AVO meter or by means of growler.

(*e*) **Balance test.** In this test the impedance per phase is checked. In this method a low voltage is applied to each winding through an ammeter and the current in each winding is measured. If the ammeter shows equal reading with every coil, then the impedance of each coil is same and windings are balanced. If the ammeter reads more then there is some short circuit in that winding. Secondly in case if the ammeter gives no reading it means there is an open circuit in that winding.

Q. Describe the procedure of phase sequence test of an a.c. squirrel cage induction motor.

Ans. The phase sequence test is performed to determine the beginning and ending terminals of the phase windings.

First of all, each phase winding terminals are checked by means of series testing lamp or megger. The two windings are connected in series as shown in Fig. 19.8. Winding A and B are in series. Now low voltage a.c. is applied to one winding. The voltage are measured across one winding say V_1 and to the two extreme ends say V_2.

Now if V_2 is more than V_1, the terminals connected are of same polarity *i.e.* A_2B_2 and if V_2 is less than V_1 then the terminals connected are of the opposite polarity say A_2B_1.

Similarly the test is carried over for C-phase winding and the terminals are known.

Q. Draw a free hand sketch of placing concentric coils in stator.

Ans.

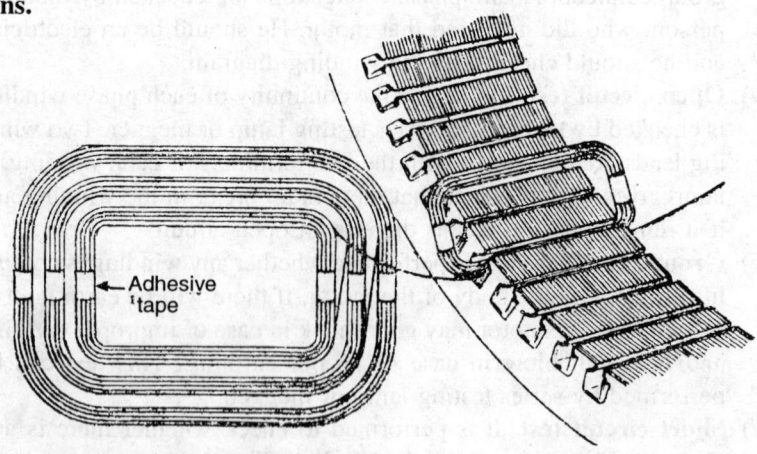

Adhesive
tape

REVIEW QUESTIONS

1. What do you understand by the armature winding?
2. What are the different types of coils and which material is used for rewinding an armature?
3. What are the classification of D.C. armature windings?
4. What is the lap winding? Describe the further classification of the lap winding.
5. Describe the wave winding. Explain its classifications.
6. (a) Describe the followings:
   Back pitch, coil pitch, pole pitch, commutator pitch, resultant pitch.
   (b) Compare the progressive and reprogressive types of windings.
7. Draw a development diagram of a simple lap winding having the following datas:

   Number of slots—8,   Number of segments—8.

   Number of poles—2,   Number of coils—8.
8. How will you determine the brush position on the commutator of an armature?
9. What are the classification of A.C. windings?
10. Give one example of a 3ϕ Basket winding and draw the development diagram.
11. What are the different methods of testing the newly wound 3ϕ stator.

23

Electrical Measuring Instruments

23.1 SYMBOLS

The following symbols are used for measuring instruments:

Symbol

1. Moving coil Instruments for D.C. Voltmeters, Milli Voltmeters, Ammeters, Milli Ammeters, Pointer Galvanometers Centre Zero Instruments.

2. Moving Coil Instruments with rectifiers for A.C. measurements accurately.

3. Moving Coil Instruments with self-contained. Thermocouple, for measuring quantities at high frequencies.

4. Moving Iron Instruments for A.C. Voltmeters and Ammeters.

5. Electrodynamic Instruments for reactive current measurement in A.C. balanced and unbalanced load – Also for Power Factor meters and Wattmeters.

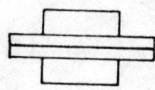

6. Vibration Reed type for Reed type for Reed type Frequency meters.

7. Electrostatic Instruments for voltmeters in A.C. and D.C.

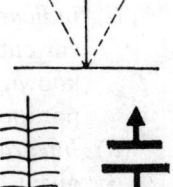

23.2. CLASSIFICATION AND TYPES

Q. How the instruments are classified?

Ans. The instruments are classified according to the standard of accuracy and exhibiting properties.

According to the standard of accuracy these are of the following types:

(i) *Standard instruments.* These instruments are having 100% accuracy. These are used as the standard and are kept in the laboratories.

(ii) *Sub-standard instrument.* These instruments are calibrated from the standard instruments. These are used to calibrate the commercial instruments.

(iii) *Commercial instrument.* These instruments are used for commercial work. These are calibrated from the sub-standard instruments. The accuracy may be ±3%.

Secondly, according to the exhibiting property to measure the quantity these are:

(i) *Absolute instrument.* The instrument which gives away the value of quantity to be measured in terms of deflection and instrument constant; are known as the absolute instruments. Such instruments do not require to be calibrated from the standard instruments. For example the tangent galvanometer it gives away the deflection and quantity to be measured in terms of tangent of the angle of deflection radius, turns used and horizontal component of earth magnetic field. These instruments are very precise and are kept in laboratories.

(ii) *Secondary instruments.* Those instruments which exhibit the quantity to be measured in terms of direct deflection, are known as the secondary instruments. Such instrument has their calibrated scale and an attached pointer. The scale is calibrated from the standard instrument. These instruments are used for commercial purpose.

Q. What are the different types of electrical measuring instruments?

Ans. The electrical measuring instruments are of the following types:

(a) **According to the utility.** According to the utility these instruments are classified as:

(*i*) *Indicating type.* The instruments which indicate the value of current, voltage, power etc., directly on the calibrated scale are known as the indicating type instrument. The deflection, proportional to the quantity is only seen as long as these are in use.

(*ii*) *Integrating type.* These instruments measure the quantity in a given time; such as energy-meter and ampere-hour meter. In this type of instruments the rate of use during that period is not mentioned; only these can sum up the quantity used.

(*iii*) *Recording type.* The instrument which actually register the value of quantity in a given time, are called the recording type instruments. There is a pointer attached which moves over a graph paper and record the value. In this way the quantity can be checked for any particular time and date. For example, maximum demand indicator, load graph and power factor recorder etc.

(*b*) **According to the working principle.** The measuring instruments are classified according to the working principle as under:

(*i*) Electromagnetic moving coil type.

(*ii*) Electromagnetic moving iron type.

(*iii*) Electromagnetic dynamometer type.

(*iv*) Shaded pole type.

(*v*) Hot wire type.

(*vi*) Electrostatic type.

(*vii*) Rotating magnetic field type.

(*c*) According to the standard:

(*i*) Standard type.

(*ii*) Sub-standard type.

(*iii*) Commercial type.

(*d*) According to the exhibiting properties:

(*i*) Absolute instrument.

(*ii*) Secondary instrument.

(*e*) **According to the shape.** These instruments are also named according to the shape it acquires:

(*i*) Portable instruments.

(*ii*) Pannel board.

(1) Surface mounting. (2) Pannel mounting.

Q. What are the properties of good instruments? Also states the factors to be considered for specifying the instrument.

Ans. Following are the properties of a good instrument:

(*a*) **Dead beatness.** The pointer or the needle should be dead beat and its pointer should not oscillate. Its pointer should take its final position very quickly according to the value of the quantity to be measured.

(*b*) **Applicability.** The instrument should be able to read on A.C. and D.C. both. It should work properly and with maximum accuracy.

(*c*) The consumption of the instrument should be very low.

(*d*) It should be accurate according to the standard and the error in no case should exceed more than ±3%.

(*e*) The scale should be calibrated in such a way as to have quick and accurate reading.

(*f*) The range of the scale should be accurate.

(*g*) It should be free from the external atmospheric effects.

(*h*) Should be portable and easy to handle.

(*i*) The pointer should be light enough.

There are the following factors to be considered for specifying the instrument:

(*i*) The scale should be mentioned as 50 mm, 100 mm or 150 mm etc.

(*ii*) The degree of deflection should be specified as 90°, 120° or any other degree.

(*iii*) The range of the instrument for example single range, double range or multi-range etc. should be given.

(*iv*) Whether centre zero or not.

(*v*) In case of multi-range, regarding the resistance with internal or external fixing.

(*vi*) Regarding the anti-parallex mirror, whether it is required or desirable.

23.3. FORCES EMPLOYMENT

Q. What are the different forces employed in an indicating type instruments? Describe with neat sketch.

Ans. In case of an indicating type instrument the pointer moves over a calibrated scale. So there are the following forces employed in the instrument:

(*a*) **Deflecting force.** The force which compel the pointer to move over a calibrated scale is known as the deflecting force. In all the electrical measuring instruments, the force of deflection is obtained from the different effects of electric current. These can be electro-magnetically, electrodynamically, heating effect, electrostatically.

(*b*) **Controlling force.** The force which controls the deflecting force is known as the controlling force. This force controls the movement of the pointer on the scale in a proper position according to the value of the quantity which may be zero or any other value.

The controlling torque performs two purposes:

(*i*) The controlling torque opposes the deflecting torque and increases with the deflection of the moving system.

(*ii*) It brings back the pointer to zero position when there is no deflecting force.

There are the following two ways of producing the controlling torque or force.

(*i*) Gravity control.

(*ii*) Spring control.

(*c*) **Gravity control.** In this system the small adjustable weights are attached with the spindle. The weights are attracted by the earth gravitational pull, hence producing the controlling torque. To obtain a uniform torque the instrument is used vertically. The controlling torque can be varied by changing the position of the weight on the arm. Let the pointer be moved in clockwise direction, the weight will then move to offer a controlling and mechanical balancing force to the system. Due to gravitational pull the weight will try to come to its original position [*i.e.* vertical position as shown in Fig. 23.1(*a*)]. Generally these types of the instruments are centre zero instruments.

There are some merits and demerits of this system.

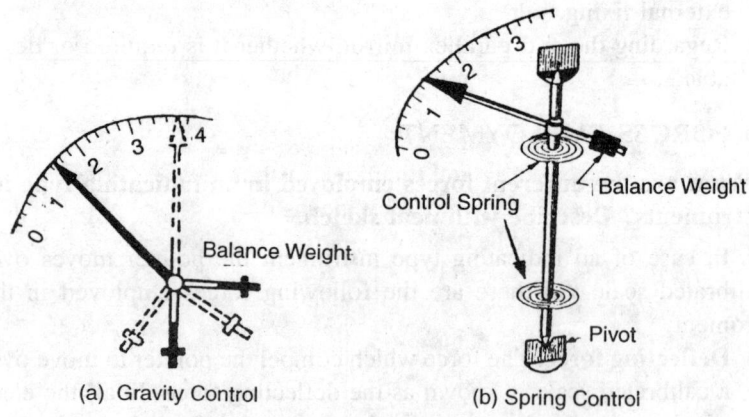

(a) Gravity Control   (b) Spring Control

Fig. 23.1. Controlling force.

Merits

(*i*) It is very cheap.

(*ii*) It is easy to control.

(*iii*) It is not effected by the temperature variation.

(*iv*) It is not subjected to fatigue.

Demerits

(*i*) To obtain a uniform torque the instrument has to be vertical.

(*ii*) The pull depends upon the angle of deflection hence the uniform torque cannot be obtained.

Spring control. It is a very common method of providing the controlling torque in electrical meaning instruments. The controlling torque in observed due to the twisting force of the spring and this force is proportional to the angle of twist. The springs are made of phospher bronze and are attached with the spindle as shown in Fig. 23.1(*b*). These springs are made of such material which are non-magnetic, low specific resistance, low temperature coefficient and not subjected too much fatigue. There are two springs attached to the spindle. The two springs, wound in opposite directions avoids the possibility of error.

If only one spring is used it may cause error because of the increase/decrease in length due to temperature, the aging factor and the stiffness constant. The spring serves two purposes:

(*i*) The spring produces the controlling torque uniformly as the spring offer uniform torque.

(*ii*) The spring helps in leading in and out the current to the moving coil in M.C. instruments.

(*iii*) It brings back the pointer to zero position when the deflecting force is removed.

Damping force. The force which damp the oscillations is known as the damping force or the damping torque. When the deflecting system is acted upon by the deflecting and controlling force, the pointer due to inertia will oscillate about the final position. These oscillations causes undesirable delay in reading and the quick vision. Therefore these must be prevented and these are damped by means of damping force.

An instrument is said to be dead beat when the pointer takes its final position quickly without any oscillations. The degree of damping decides the behaviour of moving system. The different degree of damping is shown in Fig. 23.2.

There are the following methods of obtaining the damping force:

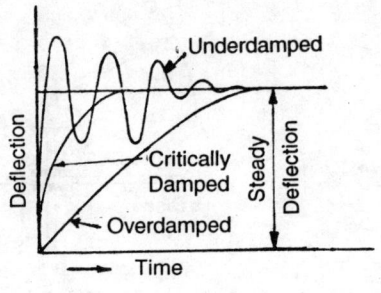

Fig. 23.2. Damping

(*a*) Air damping,

(*b*) Fluid damping, and

(*c*) Eddy current damping.

(*a*) **Air damping.** In this system a light vane is attached with the spindle. The vane moves in the air chamber having only a little clearance. The chamber may be a circular or rectangular and similar the vane as shown in Fig. 23.3, so that it can move freely. The compressed air offer a resistive force in the opposite direction to the movement of the pointer. If the pointer is moving in clockwise direction the direction of the resistive force offered by compressed air will be in anti-clockwise direction and *vice versa,* thus the oscillations are damped.

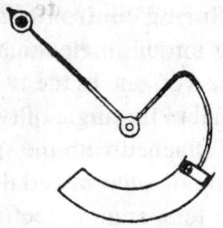

Fig. 23.3. Air damping.

(*b*) **Fluid damping.** In this system the damping disc or vane or piston mores through the oil, whenever the moving system moves in any of the direction the vane also moves thus the friction of the oil causes a damping torque on the movement and the oscillations are damped. This system is not suitable for the portable instrument because of the fluid.

(*c*) **Eddy current damping.** It is the most effective and efficient method. In this system the oscillations are damped by utilizing the principle of electromagnetic induction. Whenever the copper or aluminium disc is rotated between the two magnetic poles as shown in Fig. 23.4, the eddy currents are induced in the disc. According to Lenz's law these eddy currents will oppose the cause which is responsible for the production. Hence the force caused by the eddy currents will oppose the movement and thus oscillations.

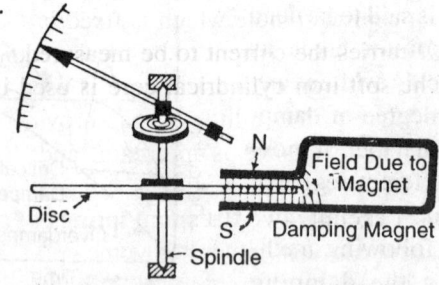

Fig. 23.4. Eddy current damping.

This principle of damping is adopted into a number of instruments. The moving system of the instrument has an aluminium or copper disc which rotates between the two permanent magnetic poles. Due to movement, the eddy currents are induced, which offer a damping torque over the disc to damp the oscillations of the moving system.

Q. Describe the construction and working of a moving coil type instrument. State the merits, demerits and uses of the instrument.

Ans. In this instrument, a coil moves between the two poles of a permanent magnet, so it is known as moving coil type instrument.

Working principle. The working principle of M.C. type instrument is that whenever a current carrying conductor is placed in the magnetic field a torque is developed on the conductor which tends to move at right angle to the magnetic field, with the production of the torque the pointer moves over a calibrated scale and the reading is obtained.

Construction. It has the following parts as shown in Fig. 23.5:

(*i*) *Permanent magnet.* It consists of a permanent magnet. The magnet is made of special forged and hardened tungsten, hard steel, cobalt chrome steel. It is generally of horse shoe or *U*-shape magnet, with soft iron pole pieces, in order to achieve a good range of deflection say about 270°-300° special shapes of magnets and pole pieces are used. The function of the permanent magnet is to provide a uniform radial field in the air gap.

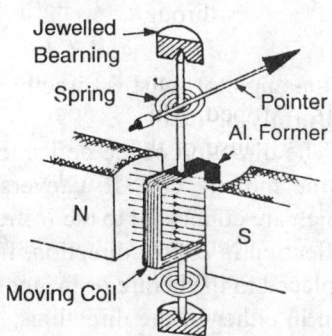

Fig. 23.5. Moving coil instrument.

(*ii*) *Moving coil.* The coil which rotates between the poles has a number of turns of fine wire. The coil is wound over an aluminium former. It is fixed on a central spindle which is fixed in the jewelled bearings. The coil carries the current to be measured.

(*iii*) *Iron cores.* The soft iron cylindrical core is used in case of horse shoe magnet. There is only a little air gap provided in between the poles. So that the reluctance is minimum and the main field be strong and uniform.

(*iv*) *Spiral springs.* Generally two flat spiral springs of phospher bronze are used one on the top and other below the coil. The spring serves two purposes, it leads the current to the coil and offers the controlling force to the deflecting system.

(*v*) *Pointer and calibrated scale.* A light pointer is attached with the spindle. The pointer moves over a calibrated scale. Sometimes an antiparallex mirror is also provided for accurate reading. The balance weight is also provided along with the pointer.

Working. When the instrument is in use the current flows through the coil and a torque is produced. The torque developed is proportional to the value of current in the coil. Due to this torque the coil moves and the pointer deflects over the calibrated scale. When the coil moves, the springs are twisted or rewound, thus the spring offer a controlling torque to the moving system, when the controlling and deflecting forces are balanced, the pointer stops as the torque developed

$$T_d \propto \phi \, I_a.$$

Here ϕ is constant because of permanent magnet.

So $$T_d \propto I$$

and $$T_c \propto \theta \quad \text{(angle of twist)}$$

The meter will read and gives a stable deflection when the controlling and deflecting torques are equal

$$T_c = T_d$$

$$\therefore \qquad \theta \propto I$$

as the angle of twist is directly proportional to the current, so the sale is uniformly divided.

The direction of the deflection depends upon the direction of current in the moving coil. If a reverse deflection is obtained, the two wires which are connected to the instrument must be interchanged to obtain the deflection in proper direction. In case of centre zero instruments the coil is placed in the centre or magnetic neutral axis, so it may give the deflection in either of the directions.

The aluminium former offers the eddy current damping.

Merits. Following are the merits of the type instruments:

(*i*) The scale is uniformly divided.

(*ii*) It is free from the hysteresis effect.

(*iii*) It is almost free from the errors due to stray fields.

(*iv*) Efficient damping is obtained.

(*v*) Very reliable and sensitive.

Demerits. Following are the demerits:

(*i*) These are costly than M.I. type instruments

(*ii*) These can only be used on D.C.

(*iii*) The aging of permanent magnet may effect the reading.

(*iv*) The control springs may loose their resilience which will effect the reading.

Uses. The M.C. type instruments are used for D.C. as an ammeter, voltmeter and ampere hour meter etc.

As an ammeters. The ammeters are connected in series with the load to measure the current. If the load current is more, then the suitable value of shunt is used. The excess current is passed through the shunt and only a definite part of it through the ammeter.

As a voltmeter. The coil of the instrument has low resistance so if it is to be used as a voltmeter the series resistance is used in series with the instrument to limit the current through the instrument.

Q. Write a short note on the extension of range of the M.C. type instrument.

Ans. The M.C. type instruments are used to measure the quantity of current and voltage to a wide range. It is possible with the conjunction of devices known as shunt and multipliers.

Shunt. The shunt is essentially a low value resistance connected in parallel to the ammeter. It is used to extend the range of an ammeter.

Value. Let there be an instrument as shown in Fig. 23.6 connected in series with a load and supply.

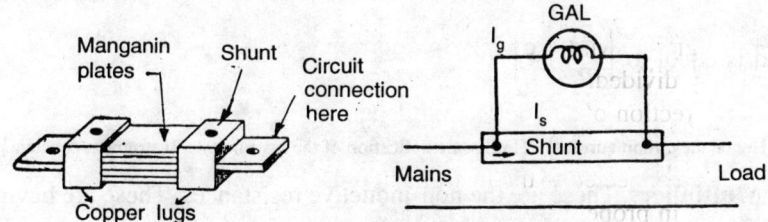

Fig. 23.6.

R_g = Resistance of the galvanometer in Ohm,

R_s = Resistance of the shunt in Ohm,

I_s = Current in shunt in Amp,

I_g = Current in the galvanometer in Amp.

Let the total current is I Amp, which is

$$I = I_g + I_s \text{ Amp}$$

or $\qquad I_s = (I - I_g) \text{ Amp}$

The voltage drop in the shunt and galvanometer is same so

$$I_g \cdot R_g = I_s \cdot R_s$$

Now substituting for I_s

$$I_g \cdot R_g = (I - I_g) R_s$$

and $\qquad R_s = \dfrac{I_g \cdot R_g}{(I - I_g)} \text{ Ohm}$

$$= \frac{R_g}{\left(\dfrac{I}{I_g} - 1\right)} \text{ Ohm}$$

$\dfrac{I}{I_g}$ is known as the *multiplying factor*

Similarly $I_g \cdot R_g = IR_s - I_g \cdot R_s$

or $I \cdot R_s = I_g \cdot (R_g + R_s)$

or $$\frac{I}{I_g} = \left(\frac{R_g + R_s}{R_s}\right) = \left(\frac{R_g}{R_s} + 1\right)$$

This is also known as the multiplying factor or the instrument constant

and is equal to $\left(\dfrac{R_g}{R_s} + 1\right)$

(Hence the circuit current) = (Full scale deflection of the instrument × Instrument constant)

Multipliers. These are the non-inductive resistances. These are having more resistance and are connected in series with the instru-
ment as shown in Fig. 23.7. The
material of the multipliers should
be of low temperature coefficient
and high specific resistance so
that the meter may be used to
measure the voltage of high
range.

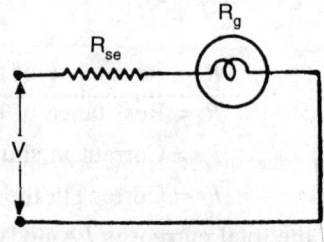

Fig. 23.7. Multipliers.

Value. The multipliers are always connected in series with the instrument. Let

R_{se} = Resistance of the multiplier in Ohm,

R_g = Resistance of the galvanometer in Ohm,

I_g = Current through the galvanometer in Amp,

v = Voltage drop across galvanometer in Volts,

and V = The voltage to be measured in Volts.

It is a series circuit, so the supply voltage will be the sum of the
voltage drop across the galvanometer plus the voltage drop across the
multiplier.

Now
$$v = V_g = I_g \cdot R_g \text{ Volts}$$

and
$$I_g = \frac{V_g}{R_g} \text{ Amp}$$

$$V = V_g + V_{ex}$$
$$= I_g \cdot R_g + I_g \cdot R_{ex}$$
$$= I_g [R_g + R_{ex})$$

or
$$\frac{V}{I_g} = R_g + R_{ex}$$

or
$$R_{ex} = \left(\frac{V}{I_g} - R_g\right) \text{ Ohm}$$

Similarly it can be stated as

$$\frac{V}{I_g \cdot R_g} = 1 + \frac{R_{ex}}{R_g}$$

$$\frac{V}{v} = \left(1 + \frac{R_{ex}}{R_g}\right)$$

The $\dfrac{V}{v}$ is the multiplication factor and can be given as $\left(1 + \dfrac{R_{ex}}{R_g}\right)$

Example. *A.M.C. type instrument gives full scale deflection 15 mA and has a resistance of 3 Ω. Calculate the resistance to be connected to use this meter (i) up to 500 mA ammeter and (ii) up to 100 V voltmeter.*
Solution. The galvanometer has the resistance of 3 Ω and the full scale deflection is 15 mA.

(*i*) When it is used as an ammeter 0 – 500 mA. The total current is 500 mA so the current to bypass through shunt is
$$500 - 15 = 485 \text{ mA}$$

Here
$$I_g \cdot R_g = I_s \cdot R_s$$

The voltage drop across both will be same, so $R_s = \dfrac{I_g \cdot R_g}{I_s}$

$$= \frac{15 \times 10^{-3} \times 3}{485 \times 10^{-3}} = 0.093 \, \Omega \text{ Ans.}$$

(*ii*) When using as a wattmeter up to 100 Volts

$$\text{The resistance} = \frac{\text{Voltage applied}}{\text{Current through galvanometer}}$$

$$R = \frac{V}{I_g} - R_g$$

$$= \frac{100}{0.015} - 3 = 6663.67 \ \Omega \quad \textbf{Ans.}$$

Example. *A M.C. type instrument gives full scale deflection with one volt and has a resistance of 110 Ω. It is desired to use it up to 300 Volts. Find the value of the multiplier.*

Solution.
$$R_{ex} = \frac{V}{I_g} - R_g$$

$$= R_g \left(\frac{V}{I_g \cdot R_g} - 1 \right)$$

$$= R_g \left(\frac{V}{v} - 1 \right)$$

$$= 110 \left(\frac{300}{1} - 1 \right) = 328.90 \ \Omega \quad \textbf{Ans.}$$

Example. *A M.C. instrument has a resistance of 10 Ohm and gives full scale deflection of 1 mA. Show how it can be adopted to measure voltage up to 250 V and current up to one amp.*

Solution. The external resistance
$$R_{ex} = \frac{V}{I_g} - R_g$$

$$= \frac{250}{0.001} - 10 = 249990 \ \Omega \ \textbf{Ans.}$$

While it will be used for one amp as an ammeter,

$$R_s = \frac{R_g}{\left[\dfrac{I}{I_g} - 1 \right]} = \frac{10}{\left[\dfrac{1}{0.001} - 1 \right]}$$

$$= 0.01001 \ \Omega. \quad \textbf{Ans.}$$

Q. What are the different types of the moving iron type instruments? Describe with the help of neat sketch the construction, working, merits, demerits and use of any one of them.

Ans. In this type of instruments the current carrying coil is stationary, but the core of soft iron rotates, so it is called as moving iron instrument. These are of the following type:

    (*a*) Attraction type.   (*b*) Repulsion type.

Repulsion type instrument. These instruments are based on the principle of magnetic repulsion. If two soft iron pieces are magnetised in the same magnetic field and field intensity, these will be magnetised with similar polarities and pole strengths. Thus because of same poles, these will experience a force of repulsion. If any of these two soft iron strips is free to move, it will move.

Construction. The diagram as shown in Fig. 23.8 shows the construction of the repulsion type instrument. It consists of a coil wound over an insulated bobin. The coil is wound with a number of turns of super enamelled copper wire. There are few number of turns if the instrument is used as an ammeter and more number of turns of fine wire when the instrument is used as a voltmeter. Two strips of soft iron and fixed inside the bobin. One is fixed with the bobin and is known as the fixed iron, the other is fixed with the spindle and is known as moving iron. It has a tongue shape or some other peculiar shape fixed strip to improve the scale to be more uniform and linear. The spindle carries a pointer and an arm with an aluminium vane to move in the air chamber for air damping. To obtain the controlling torque the flat spiral shape springs are used, these are made of phospher bronze and exert the controlling torque on the moving system. Generally a damping chamber consisting of covered quadrant shape is used.

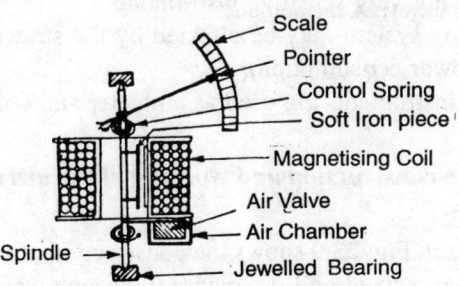

Fig. 23.8. Repulsion type M.I. instrument.

Working. Whenever a current is passed through the coil of the instrument (connecting in series for an ammeter and in parallel for a voltmeter) the magnetic field is produced. The both soft iron pieces are in the same magnetic field, so these are magnetised having the similar pole strength and polarity. The repulsion will take place exerting a repulsive force; which causes the moving iron to move. The pointer also deflects as it is fixed on the spindle where the soft iron piece is attached. The required controlling torque is produced by the control springs. The damping is obtained by air damping.

The force of attraction or repulsion is directly proportional to the product of the pole strengths. The both pieces are having the same pole strength so,

$$T \propto m_1 m_2 \propto m^2 \propto I^2$$

So the deflecting torque, $\qquad T_d \propto I^2$

Now controlling torque, $\qquad T_c \propto \theta$

and $\qquad T_c = T_d$

so $\qquad \theta \propto I^2$.

Since the deflection is proportional to the square of current through the coil, therefore the scale of the instrument is not *linear*. It is crowded in the beginning and spreaded in the middle. The scale is made somewhat linear by using the tongue shape or peculiar type of iron piece.

Merits. Following are the merits:

  (*i*) These instruments can be used for A.C. and D.C. both.

  (*ii*) These are cheap, robust and simple in construction and use.

  (*iii*) Can be used in any position.

  (*iv*) These are reasonably accurate.

  (*v*) The coil is fixed so the less possibility of faults.

Demerits. Following are the demerits:

  (*i*) The scale is not uniform.

  (*ii*) The reading up to 1/10th of the scale is not reliable.

  (*iii*) These are not very sensitive instruments.

  (*vi*) The moving system may be effected by the stray magnetic field.

  (*v*) Higher power consumption.

Uses. These instruments are used as ammeter and voltmeters for A.C. and D.C. both.

Q. Describe the construction and working of the attraction type M.I. instruments.

Ans. The diagram, Fig. 23.9 shows the construction of an attraction type instrument. It has a fixed coil or solenoid; through which the current is passed. It has only one iron rod or disc generally of oval in shape. Oval shape disc is used to make the scale a linear one. The disc is attached with the spindle. The spindle carries a pointer and an arm having an aluminium vane which moves in the air chamber; thus the air damping is obtained. The spindle is pivoted in jewelled bearings at both the ends. Two flat spiral springs are attached with the spindle to offer the controlling torque.

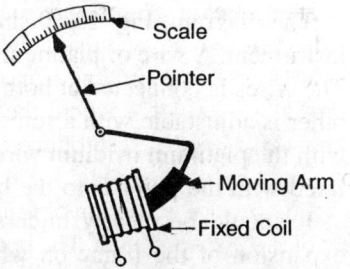

Fig. 23.9. Attraction type M.I. instrument.

Working. Whenever current flows through the coil, the electromagnetic field is set up by the coil. Now because of the induction an opposite pole is formed on the soft iron oval disc and is attracted towards the coil; thus producing the deflecting torque over the deflecting system. The deflecting force depends upon the value of magnetic flux or in other words the current flowing through the coil. The pointer moves over the calibrated scale. The springs offers the controlling torque which depends upon the angle of twist. When the controlling and deflecting forces are equal the pointer will stop. The damping is obtained by means of the air damping. The value of magnetic flux does not increase after the saturation, so the scale is not *uniform*. It is uneven, crowded in the beginning and end; because the deflection is directly proportional to the square of current.

If the current is reversed, the direction of the flux developed will also change, causing a change in polarity of the soft iron piece, but because of the induction and attraction, the iron piece will be attracted towards the centre of the coil, thus the meter can be used on A.C. and D.C. both.

The moving iron type instruments are having the following characteristics:

(*i*) It can be used on A.C. and D.C. both.

(*ii*) It has the spring controlling system.

(*iii*) Air damping is used.

(*iv*) The construction is robust and easy.

(*v*) It has high operating torque.

(*vi*) These are cheap.

(*vii*) The scale is not uniform.

(*viii*) It can be overloaded momentarily.

So because of these reasons these are used as an ammeter and voltmeter for A.C. and D.C. both.

Q. Explain the construction and working of a hot wire instrument.

Ans. The principle of working of these instruments is "when-ever the current is passed through the metallic wire, it increases in its length (principle of expansion of metals) due to heating effect".

The diagram, Fig. 23.10 shows the simple construction of a hot wire instrument. A wire of platinum-irridium of about 16 cm in length is used. The wires is connected at both *A* and *B* contacts one contact is fixed and other is adjustable with a screw. A fine phospher bronze wire is attached with the platinum irridium wire at a point *C*. The other end of the wire is fixed with the point *D* to the base of the instrument.

It should be clearly understood that the temperature coefficient of expansion of the frame on which the hot wire is mounted and the hot wire, be made of an exact match to avoid the error in exhibiting the reading and uneven expansion between wire and its fixture. A fine silk thread is attached at a point, this thread is passed over a pulley and through a spring. A permanent magnet and aluminium disc is also provided for eddy current damping.

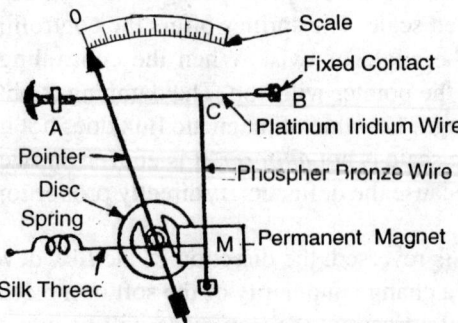

Fig. 23.10. Hot wire instrument.

Whenever the current is passed through the platinum irridium wire it expands in length. The increase in length is proportional to the square of current ($H \propto I^2$). The sag of the hot wire will result the slack in the phospher bronze wire. The silk thread which is provided with a spring, will be pulled by the spring and the pulley through which it passes, will move. Now due to this tension the pointer will move over the calibrated scale.

As the aluminium disc moves in between the permanent magnet, the eddy currents are induced in the aluminium disc and the eddy current damping is obtained. Generally gravity control is used in this instrument.

Following are the advantages and disadvantages of the instrument.

Advantages

    (*i*) The expansion is proportional to the square of current, so the instrument can be used on A.C. and D.C. both.

    (*ii*) There is no induction effect so it can be effectively used for higher frequencies.

Disadvantages

   (*i*)   The readings are not obtained quickly.

  (*ii*)   On higher current the wire may burn out.

 (*iii*)   Zero position is to be adjusted frequently.

 (*iv*)   Power consumption is relatively high.

  (*v*)   The scale is not uniform.

    Uses. It is used as an ammeter, voltmeter. These are appreciably used for higher frequencies.

Q. Describe the essential features of construction of an electro-dynamometer type instrument also show how these instruments can be used for ammeter, voltmeter and wattmeter.

Ans. These are also known as the dynamometer type instruments. These instruments are very much similar in construction to that of permanent magnet type instruments accept that the permanent magnet is replaced by the electromagnet as shown in Fig. 23.11.

    Working principle. Whenever the current carrying conductor is placed in the magnetic field a torque is produced on the conductor which tends to rotate it.

    Construction. These instruments essentially consist of two set of coils. It has one fixed coil which is divided into two halves, both halves are connected in series. These are wound over a thick former having less number of turns of thick conductor. These halves are so connected that the opposite poles are formed in the air gas.

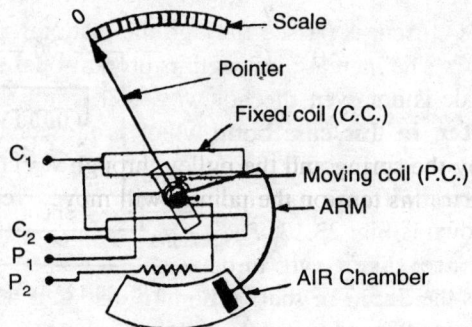

Fig. 23.11. Dynamometer type instrument.

    The other is moving coil, which is placed in between these coils is free to rotate about its axis. The coil is wound with thin conductor having more number of turns. Two phospher bronze springs are mounted on the spindle to offer a controlling force and also make the current flow in the moving coil. An external resistance is connected in series with the moving coil to reduce the current to the safe Value. For damping purpose, air damping or eddy current damping is used. In eddy current damping an aluminium disc is used on the spindle which rotates in between the two poles of a permanent magnet.

When current flows through the fixed and moving coil a torque is produced on the moving coil proportional to the value of current in both the coils.

This instrument can be used as an ammeter, voltmeter and wattmeter.

As an ammeter. The fixed coil and moving coil are connected in series and the external resistance is not connected in series as shown in Fig. 23.12. The value of external resistance (shunt) is designed to limit the current in the moving coil for the safety of the instrument. As an ammeter the scale is not uniformly divided, the external resistance always remains out of circuit as shown in Fig. 23.12.

Let T_d = deflection torque

T_c = controlling torque

I_f = moving coil current

q = angle of twist.

$\therefore$ $T_d \propto I_f \times I_m$

and $T_c \propto \theta$

Since both are in series, so

$I_f = I_m$

$\therefore$ $T_d \propto I^2$

or $\theta \propto I^2$

Hence the scale is not even.

As a voltmeter. In this case both the fixed and moving coils are connected in series with the external resistance, as shown is Fig. 23.13. The two terminals are taken out and connected across the line. The scale is not uniformly divided because $\theta \propto I^2$.

As a wattmeter. Whenever this instrument is used as a wattmeter, the moving coil is connected across the supply and the fixed coil in series with the load as shown in Fig. 23.14. The torque developed is proportional to the value of current in fixed coil and

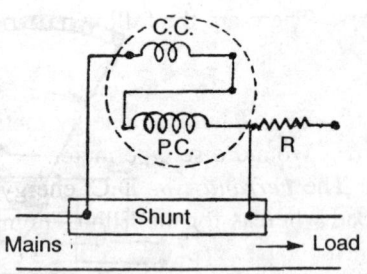

Fig. 23.12. As an ammeter.

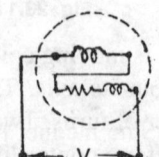

Fig. 23.12. As a voltmeter.

moving coil. As the voltage is constant (approximately) so the current in the moving coil, now as a pressure coil, is constant. In this way the torque developed is proportional to the current in the fixed coil, here known as current coil.

$$T_d \propto I_m \cdot I_f$$

Here current in moving coil is proportional to the voltage which is constant.

So the flux of the moving coil is constant hence it is stated that

$$T_d \propto I_f,\ T_c \propto \theta$$

or

$$\theta \propto I_f$$

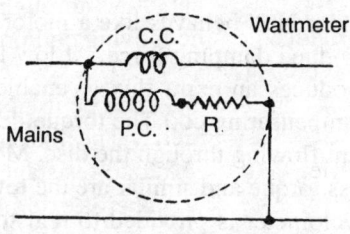

Fig. 23.14. As a Wattmeter.

Hence the scale of the wattmeter is uniformly divided.

In a single phase wattmeter there is only one pressure coil and one current coil. In three phase three element wattmeter, three current coils and three pressure coils. In three phase two element wattmeter, there are two pressure coils and two current coils. The direct reading is obtained due to the resultant torque produced by two pressure coils.

Uses. The dynamometer type instruments are used as an ammeter, voltmeter and as a wattmeter for A.C. and D.C. both.

Q. What are the different types of instruments used for D.C. energy measurement? Describe the construction and working of the Ferranti type ampere hour meter.

Ans. There are the following types of instruments used for D.C. energy meters:

(*a*) Ferranti type meter.

(*b*) Elihu Thomson energy meter.

(*c*) Wound disc type meter.

The *Ferranti type* D.C. energy meter in connected in series with the load whereas for the Elihu Thomson and Wound Disc types the current and voltage considerations are essential. Both current coil and pressure coils are used in later type of instruments.

Ferranti type meter. These are known as the Ferranti type mercury motor ampere hour meters. These are commonly used for D.C. energy measurements.

Construction. The construction of the meter is shown in Fig. 23.15. The meter basically consists of two permanent magnets with poles as *N*, *S* and *N′, S′*. The magnet has mild steel pole pieces. The pole pieces are so set around the circular brass plate that these are separated from each other with the help of fibre rings. Between the pole pieces of the magnet, a copper disc '*D*' rotates in the mercury bath *M*. The mercury bath is in between the two insulated rings. The load current flows through the copper disc. The copper disc is insulated from both sides. The current comes to the centre and leaves outwards as shown, resulting the development of torque on the disc which tends to rotate. The half portion of the disc *i.e.* under N'S' behaves like a motor and half under the poles NS is used to produce damping force. At low load the disc may not rotate so a coil *G* produces an extra flux to enable the movement. This coil is known as compensating coil. The torque developed is proportional to the load current flowing through the disc. More current more torque and less current less torque and similar are the revolutions of the disc. A train of wheel or cyclometer is provided to register the reading.

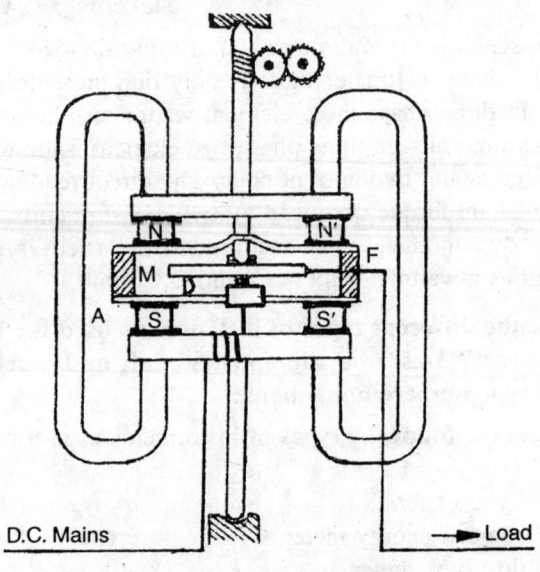

Fig. 23.15. Ferranti type meter.

Damping torque. It is developed, whenever the copper disc rotates between the *N.S.* magnet. The eddy currents are developed in the copper disc, resulting the eddy current damping. This force developed is proportional to the speed.

The *controlling torque* is produced by the friction of mercury, gears and jewels.

Here the deflecting torque is proportional to the flux and current flowing through the copper disc.

∴
$$T_d \propto \phi \cdot I$$

As it is a permanent magnet so ϕ is constant

$\therefore$ $$T_d \propto I$$

The disc rotates between the permanent magnets, the eddy currents are induced which produces the braking torque. So the braking torque is proportional to production of eddy currents or in other words the speed of the disc.

$\therefore$ $$T_b \propto N$$

At steady speed the braking torque equals the driving speed

So $$T_b = T_d, \quad N \propto I, \quad \overline{N} = KI.$$

$\therefore$ So it is clear that the total number of revolutions in a given time are proportional to the quantity of electricity passed in a specified time. Mathematically, the total number of revolution in a given time,

$$\int N dt, = \int KI \cdot dt, = K \int I \cdot dt.$$

The $I \cdot dt$ is the quantity of electricity in a given time.

Advantages. Following are the advantages of this instrument:

(*i*) Simple construction.

(*ii*) The voltage drop across the mercury is small.

(*iii*) More current capacity without shunt.

Disadvantages

(*i*) Cannot be used on A.C.

(*ii*) If a part of the mercury is less or dirty the instrument will not work properly.

Errors. The errors can be because of friction, change in magnetic flux, the change in temperature and change in supply voltage.

Uses. It is used for D.C. ampere hour measurement.

Q. Write short notes on:

(*i*) **Wound disc type meter.**

(*ii*) **Elihu Thomson energy meter.**

Ans. (*a*) **Wound disc type meter.** It is also known as the ampere hour meter. It is connected in series with the load.

Construction. As shown in Fig. 23.16 it consists of an aluminium disc which contains the winding. The winding ends are brought to the commutator. Two silver plated brushes are provided to send the current through the winding of the disc. A shunt is connected across the circuit of the disc to limit the current to the safe value. Two permanent magnets are provided to produce the magnetic field in which the disc rotates and on the spindle, the revolution counter or train of wheels is attached to register the reading *i.e.* the energy consumption.

This instrument works on the principle that whenever the current carrying conductor is placed in the magnetic field a torque is developed which tends to rotate it. The load current flows through the shunt and the disc, so a torque is developed on the disc and it rotates. The torque developed is proportional to the current flowing through the disc and in other words to the load current. The eddy current damping is used in this instrument.

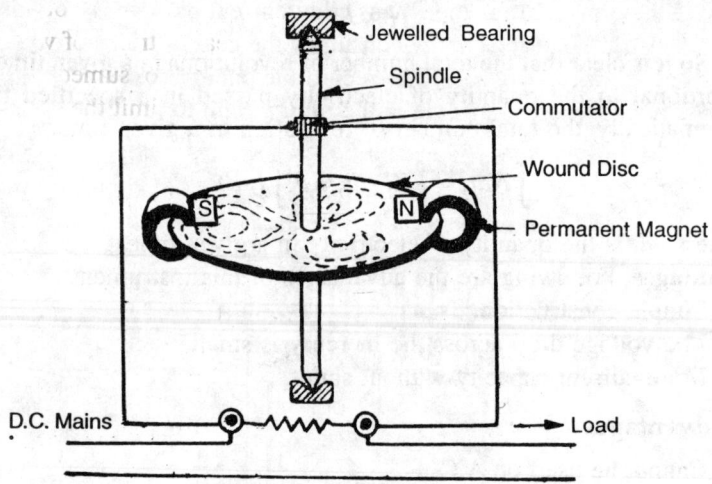

Fig. 23.16. Wound disc type meter.

Advantage. This instrument is cheap.

Disadvantages:

(*i*) It cannot be used on A.C.

(*ii*) The winding may burn out if heavy current flows.

(*iii*) The contacts between the commutator and strips may become loose or dirty due to time and load.

Uses. It is used for light load in D.C. circuits.

(*b*) **Elihu Thomson energy meter.** It is type of motor meter; in which the armature circuit is connected across the supply and field coils carry the main load currents, so it works on the principle of motor.

Construction. The construction of this meter is shown in Fig. 23.17. It has current coil divided into two halves as shown. The magnetic flux is produced by these halves, depending upon the current of the load circuit. A light armature is wound and placed in between these two coils. The winding of the armature is connected to the commutator.

The armature circuit is fed through the brushes and commutator. The armature is connected across the supply. A compensating winding is also provided in this meter to compensate the low flux because of small load current. The compensating winding is done with thin wire and is placed on the axis of current coils. This coil produces an extra flux so the torque on the armature. The extra torque also balances the controlling torque which is due to friction of gears and commutator. So the armature can rotate even at light load.

A disc of aluminium is attached on the spindle and rotates between the poles of the permanent magnet. Thus eddy current damping is obtained. For the registration of the energy consumed the gear or trains of wheel is attached as shown in Fig. 23.17 to record the energy consumed. A suitable resistance is also connected in the rotor circuit to limit the armature current.

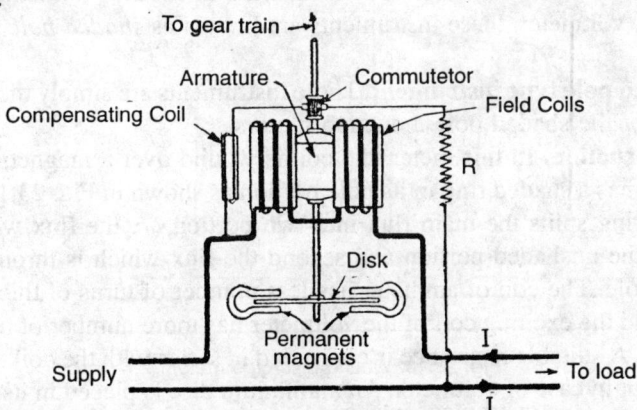

Fig. 23.17. Elihu Thomson Energy Meter.

When current flows through the armature winding current coils the torque is produced and the armature rotates. This torque is proportional to the armature current and field current. Here as the armature is connected across the voltage which is approximately constant, so the flux developed will also be constant and the torque constant, so the flux developed will also be constant and the torque developed will be proportional to the current in the field coils or current coils is other words the load current. So the torque increases with the increasing of the load current and *vice versa.*

The revolution counter will register the reading of energy consumption.

Advantages. The meter can be used on A.C. and D.C. both for energy measurement, only the commutator and brushes may cause some problem.

Q. Explain the working principle of induction type instruments.

Ans. The induction type instruments are used for A.C. measurements. In these instruments the deflecting force is developed by the interaction of the alternating fluxes developed and the current induced in the moving element of the instrument. The secondary winding induces some e.m.f. and the current is not given from outside to the secondary circuit. The working principle is same, the current carrying conductor placed in the magnetic field experiences a torque which tends to rotate it.

The required field can be obtained by the following ways:

(*i*) By placing two windings of different inductances in the magnetic field.

(*ii*) By placing a shaded ring in the magnetic field.

The first method is used in wattmeter, energy meter, frequency meter, power factor meter etc. The second method is used in induction type ammeter, voltmeter, these instruments are known as *shaded pole instruments.*

Shaded pole type instrument. These instruments are simply the modification of the shaded pole induction motors.

Construction. In this meter, the coil is wound over a magnetic core. The core has a shaded ring in its pole portion as shown in Fig. 23.18. The shading ring splits the main flux into two portion *i.e.* the flux which is through the unshaded portion and second the flux which is through the shaded pole. The coil of ammeter has less number of turns of thick conductor and the exciting coil of the voltmeter has more number of turns of fine wire. A suitable resistance is connected in series with the coil to limit the current in case of voltmeter. An aluminium disc is placed in its centre pivoted at both the ends in jewelled bearings. A pointer and phospher

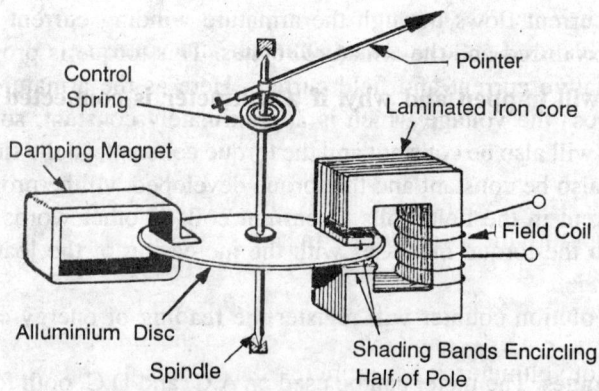

Fig. 23.18. Shaded pole instrument.

bronze spring is also attached on the spindle. To obtain a regular or constant torque, the disc is not round but oval in shape. The permanent magnet is also placed for producing the damping force. The scale is not uniformly divided.

Working. Whenever current flows through the exciting coil, a rotating magnetic field is developed. Due to this field the disc rotates. The deflecting force is controlled by means of the spring control method. The place where the deflecting and controlling forces are equal the pointer will stop on the calibrated scale. The aluminium disc rotates between the permanent magnets which causes the eddy current damping. The torque developed in proportional to the square of the current.

$$\therefore \qquad T_d \propto I^2$$

The controlling force

$$T_c \propto \theta$$

So $$\theta \propto I^2.$$

Hence the scale is uneven, because the deflection is directly proportional to the square of current.

Advantages. There are the following advantages:

(*i*) These are simple and robust.

(*ii*) The deflection up to 250° can easily be obtained.

(*iii*) Efficient damping is obtained.

(*iv*) These are not much effected by the stray magnetic effect.

Disadvantages. There are the following disadvantages:

(*i*) These can only be used on A.C.

(*ii*) These are costly.

(*iii*) Scale is not uniformly divided.

(*iv*) More power consumption.

Uses. This meter is used as an ammeter and voltmeter.

Q. What will happen and why, if a voltmeter is connected is series with the line and an ammeter in parallel with the line?

Ans. The ammeter has low resistance. It is wound with less number of turns of thick wire, so the ammeter is always connected in series because it has to measure the load current. In series the current remains the same in all components. If it is connected in parallel, in that case it has to withstand the full line voltage. The current drawn because of low resistance will be very high and sufficient to burn the ammeter.

In case of voltmeter, it has more resistance. It is wound with fine wire having more number of turns, thus increasing the resistance. It is able to withstand the voltage with normal current. Now if it is connected in series, depending upon the value of resistance in series it will give reading.

It the resistance is low the reading will be approximately the supply voltage and if the value of external resistance is in kΩ then the reading will be less depending upon the resistance because in series the voltage will divide depending upon the resistances.

Q. Describe the construction, working principle and the working of an Induction type wattmeter.

Ans. The induction type wattmeter is used to measure the a.c. electrical power.

Construction. The construction as shown in Fig. 23.19 has got the different parts like, series and shunt electromagnets, permanent magnet, spindle, the aluminium disc, springs and pointer. The series electromagnet which has low resistance wound with few number of turns of thick conductor. It is always connected in series with the load. It is also known as current coil. The shunt coil has more number of turns, wound with fine wire. It is also known as the pressure coil.

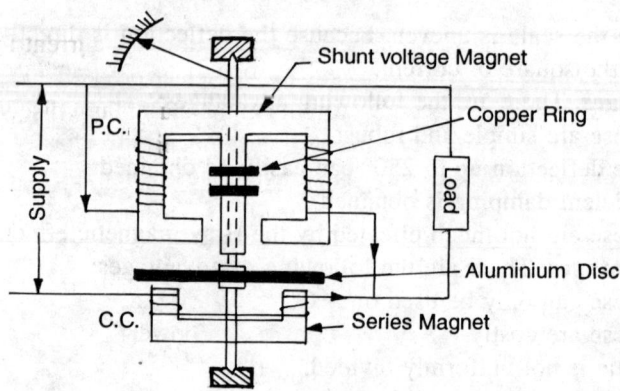

Fig. 23.19.

An aluminium disc is placed in between both the electromagnets. The disc is mounted over the spindle. The spindle is fixed in the jewelled bearings. One bearing is spring loaded. The permanent magnet is also used to offer the damping torque. The shading ring is provided on the central limb to maintain the angular displacement between voltage and current as 90° elect. The spindle carries the controlling springs for controlling torque.

Working principle. The working principle of the wattmeter is same as that of induction type ammeter and voltmeter except that instead of one coil two (current and pressure) coils are used. The eddy currents are induced in the aluminium disc and the disc rotates.

Whenever the wattmeter is used to measure the wattage of the connected load. The connections are made as shown in Fig. 23.19. The current coil is in series with the load, the pressure coil is connected across the mains. The magnetic field produced by the current coil has the same angular displacement and relation as that of the load current. The current and voltage of the pressure coil are at 90°, to each other, current lags behind the voltage. if the angle is not 90°, it is adjusted by the adjustment of the shading ring.

The deflecting torque is obtained due t the fluxes of current coil and pressure coils, the permanent magnets provides the necessary damping torque. The controlling torque is obtained by the spiral spring attached to the spindle.

The deflecting torque can be seen as,

Let V = the applied voltage in Volts

I_p = current in pressure coil in Amps

I_c = current in current coil in Amp

$\cos \phi$ = lagging power factor of the Load.

Whenever the meter is connected across line, the current in pressure coil lags behind the voltage vector by 90° elect producing the flux ϕ_p. This flux will induce the eddy currents in the aluminium disc which lags behind ϕ_p by 90°, similarly current of current coil also produces the flux ϕ_c and induces eddy currents which again lags behind by 90°.

Now mean deflecting torque

$$T_d \propto \phi_p \times \phi_c \sin(90 - \phi)$$

Here $\phi_p \propto V$ and $\phi_c \propto I$

So $T_d \propto V \cdot I \cos \phi$ (power)

or $T_d \propto$ A.C. power

The controlling torque obtained by the springs

$$T_c \propto \theta$$

and for steady deflection

$$T_c = T_d$$

or θ = A.C. power

So angle of twist is directly proportional to the A.C. power hence the scale in uniformly divided.

Advantages:

(*i*) It has long scale.

(*ii*) It is free from stray magnetic effect.

(*iii*) It has good damping.

Disadvantages:

(*i*) Change in temperature will effect the operating torque hence reading.

(*ii*) Change in frequency will affect the reactance hence the reading.

(*iii*) It can used to measure the A.C. power consumed by A.C. single phase circuit.

Uses. It is used to measure the A.C. power consumed by A.C. single phase circuit.

Q. What are the differences between the dynamometer and induction type wattmeter?

Ans. There are the following similarities and differences between the induction type and dynamometer type wattmeters.

Induction type	Dynamometer type
1. It can only be used A.C.	It can be used on A.C. and D.C. both.
2. It has uniform and long scale.	It has uniform scale.
3. These are simple and robust in construction.	These are not so robust.
4. These are having efficient and effective eddy current damping.	The air damping is used
5. Power consumptions relatively high.	Power consumption is comparatively low.
6. It has weaker torque.	It has strong torque.
7. These are not so accurate the temperature and frequency effect the reading.	These are very accurate if designed carefully.
8. These are not much effected by the stray magnetic field.	The reading may be effected by the stray magnetic effect if not properly shielded.

Q. Explain the construction and working of an induction type single phase energy meter.

Ans. The construction of the A.C. single phase energy meter is as shown in Fig. 23.20. These are used to measure the A.C. single phase energy.

Construction. The construction of the instrument is sub-divided into certain parts or systems.

(*a*) *Moving system.* The moving system, consists of the spindle, aluminium disc, warm and train of wheel. The spindle is fixed is cup shaped jewelled bearings, one is fixed nearly to support and the other with a spring journal provision at the top.

The pointer is replaced by worm to have a provision of registering the value of energy consumed.

(b) *Operating system.* The operating system mainly consists of the electromagnet and permanent magnets. There are two electromagnets, one which is known as series magnet, has few number of turns of thick conductor. The other which is the shunt magnet has winding of fine wire of more number of turns. The series magnet is wound over the U-shape laminations. It is also known as current coil. It is connected in series with the load. The shunt magnet has M shape laminations. It is always connected across the mains. The current because of voltage is approximately constant. On the central limb there is a shading ring, which is used to make the current of P.C. 90° lagging behind the voltage.

There is a permanent magnet provided on one side of the aluminium disc. It is used for eddy current damping. As the disc rotates between the magnetic field the eddy currents are induced which according to Lenz Law, opposes the cause responsible for its production, i.e., the rotations. These eddy currents are induced depending upon the rotation. More speed more current and less speed less current. The magnet provides a braking torque which controls the speed and also provide damping. The torque developed is proportional to the speed, so

The braking torque μ speed.

(c) *Recording system.* Over the spindle there is worm where the recording mechanism is fixed. The train of wheel or sometimes the cyclometer is provided to register the reading of energy consumed.

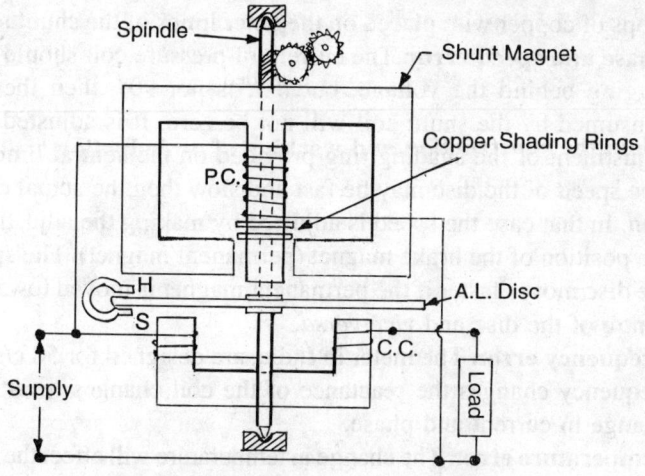

Fig. 23.20. Single phase Energy meter.

Working. Whenever the meter is connected for energy measurement, the current coils are connected, so that the load current may flow through the current coils. The pressure coil is connected across the line. Now there are two fluxes, one of the current coil which is proportional to the load current in value and displacement, the other of pressure coil, which is lagging behind the voltage by 90° elect. Now in the aluminium disc the eddy currents the produced, as a result the torque is developed which tends to rotate the aluminium disc. So the spindle moves and the reading is recorded in the recording mechanism. As the disc rotates between the permanent magnet, the controlling and damping torques are developed over the moving system.

The speed of the disc is proportional to the load current.

Q. Name the different efforts in the induction type energy meter. Also state any I.S.I. rule for the installation of the energy meter.

Ans. There may be the following errors to which the energy meter is subjected.

(*a*) **Creeping error.** The term creeping is used for the error when the disc of energy meter moves slowly even when there is no load. These rotations are continuous and slow. This error may be because of the incorrect friction compensation, excessive voltage supply and stray magnetic field etc. This error is overcomed by providing two opposite holes in the disc. Thus the disc tends to remain stationary when one of the holes comes under one of the poles.

(*b*) **Frictional error.** It may be because of the friction in bearing of the rotating system. The amount of the error depends upon the speed (*i.e.* load). This error is compensated by providing the short circuited loops of copper wire placed on the outer limbs of the shunt magnet.

(*c*) **Phase and speed error.** The current of pressure coil should be 90° lagging behind the voltage, but if it is not 90°, then the power consumed by the shunt coil will not be zero. It is adjusted by the adjustment of the shading ring provided on the central limb.

The speed of the disc may be fast and slow then the actual calculation. In that case the speed is adjusted by making the adjustment in the position of the brake magnet (permanent magnet). The speed of the disc moves faster if the permanent magnet is moved towards the centre of the disc and *vice versa*.

(*d*) **Frequency error.** The meter in India, are designed for 50 c/s. If the frequency changes the reactance of the coil changes resulting the change in current and phase.

(*e*) **Temperature error.** The change in temperature will effect the change in resistance etc. and thus resulting some error in the reading of the instrument.

I.S.I. has laid down some precautionary rules regarding the installation of the energy meter. According to *I.S.: 4648-1968, 5 : 2*, energy meter shall be installed at such a place which is readily accessible to both the owner of the building and authorised representatives of the supply authority. These should be installed at a height where it is convenient to note the meter reading. It should not preferably be installed below one metre from the ground level. The energy meters should either be provided with protective covering, enclosing it completely except the glass window through which the readings are noted or should be mounted inside a completely enclosed panel provided with hinged or sliding doors with arrangement for locking it.

Q. Describe in brief the three phase energy meter.

Ans. The three phase energy meter are used to measure the energy consumed by three phase loads. The three phase load may be balanced or unbalanced. The three phase meter have different pressure coils and current coils as under:

(*i*) One pressure coil and two current coils for balanced 3ϕ load measurements.

(*ii*) Two pressure coils and two current coils for meaning the balanced 3ϕ load.

(*iii*) Three current coils and three pressure coils for unbalanced and balanced 3ϕ loads. The neutral is available so one pressure coil can be connected across one phase and neutral.

Construction. The construction of a three phase energy meter with two pressure coils and two current coils is shown in Fig. 23.21. One pressure coil and one current coil is mounted over one side and the other set of current coil and pressure coil is connected on the other side of the meter as shown. The pressure coil has high inductance, and is connected across two phases. The current coil is connected in series with the load. The value and angular displacement of the current in current coil is same as that of the load connected. An aluminium disc is placed in the magnetic field produced by current coils and pressure coils. A revolution counter is attached with the worm of the spindle of the disc to record the energy consumed.

A permanent magnet is also provided for the eddy current damping. Sometimes it has one permanent magnet and sometimes two magnets on either side for effective and efficient damping.

Working. As shown in Fig. 23.21, when the meter is connected across the line and to the load, a rotating magnetic field is produced by these current and pressure coils. The aluminium disc is placed in the magnetic produced by current coils and pressure coils. As a result the currents are induced in the Al. disc, and a torque is developed which tends to rotate the disc. The torque developed is proportional to the load current. The power factor compensator is placed on the cores of C.C. and P.C. to obtain zero torque at zero power factor.

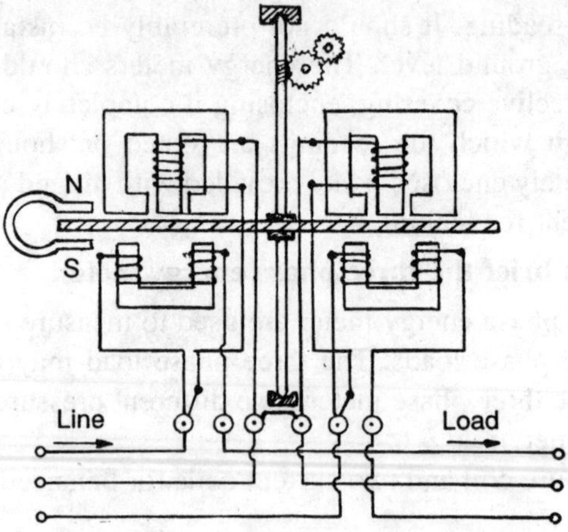

Fig. 23.21. 3ϕ two element energy meter.

Precautions

(*i*) Before giving supply or connections, the meter should be tested.

(*ii*) It should be used in vertical position.

(*iii*) Carefully the connections should be made for current coil and pressure coils.

Q. Describe the construction, working and use of megger.

Ans. Megger is the commercial name given to an instrument which is used to measure the insulation resistance of the installation of machines or appliances in mega ohms. It is a dynamometer type instrument in which the current carrying coil rotates in a magnetic field produced by the two permanent magnets.

Construction. The diagram shows the construction of the megger. In the generating portion of the megger an armature rotates in between two permanent poles. The D.C. supply is induced in D.C. generator by rotating the handle.

The moving system of the megger has two coils attached on the same spindle at a particular angle. One coil is known as current coil and the other as pressure coil. The pressure coil is connected across the generated voltage through a high resistance to limit the current in the pressure coil. The direction of current in this coil is such that when current flows through it, the torque developed is in anticlockwise direction. The current coil is also connected across the generated voltage as shown in Fig. 23.22, through the external resistance between line and earth and limiting resistance. When current flows through is coil the direction of torque is clockwise. The voltage and the current coils are so connected that the torques produced by them are in opposition. An iron ring is placed in between the two coils, so that both coils can rotate over this ring. The ring produces an easy path for the flux. A pointer is attached with the spindle which moves the scale. The scale is divided from zero to infinity.

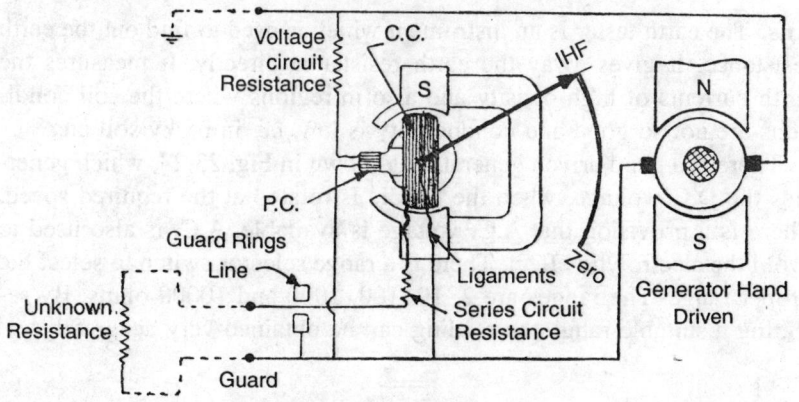

Fig. 23.22. Megger.

Working. Whenever the handle is rotated at a given speed, generally 160 r.p.m., the designed D.C. voltage is induced. The pressure coil is connected across the generated voltage. Whenever there is no current in the current coil or there is no resistance in between the L and E the current flows only through pressure coil. The torque is produced in anticlockwise direction and the pointer indicates towards the infinity as indicated in Fig. 23.22. When L and E are connected together through the external resistance, the current flows through the current coil also. Now two torques in opposite directions and produced. If both torques are similar the torque developed will cause the pointer to be in between zero and infinity depending upon the value of external resistance. In case if there is no external resistance in between L and E and the short circuited, the pointer will indicate zero on the scale.

The pressure coil is divided into two parts one is threaded over the cores while the other moves outside the cores, known as compensating coil. There are certain advantages of using the compensating coil.

(*i*) It cancel out the effect of external field on the moving system.

(*ii*) To improve the division of the scale.

Cares. (*i*) The megger of suitable range should be used.

(*ii*) The megger should not be used on live line.

(*iii*) It should be checked carefully before use.

(*iv*) Handle should be rotated at 160 r.p.m. or as per the given instructions.

Ranges. These are available from 100 V to 2000 V. The scale is divided from zero to infinity.

Q. Explain the principle of working and the method of earth resistance with earth tester.

Ans. The earth tester is an instrument which is used to find out the earth resistance. It gives away the earth resistance directly. It measures the earth currents of high density and also in regions where the soil conditions are not so good and conductivity is low, *i.e.* in rocky soil etc.

There is a hand driven generator as shown in Fig. 23.23, which generates the D.C. voltage, when the handle is rotated at the required speed. There is a provision that A.C. voltage is available. A.C. is also used to avoid the electrolytic effect. There is a range selector switch to select the proper range. The ranges are 2, 10, 100, 1000 and 10000 ohms. By selecting a suitable range the reading can be obtained very accurately.

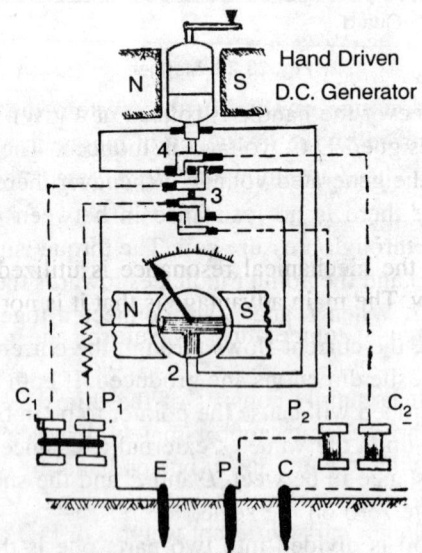

Fig. 23.23. Earth tester.

There is a ohmmeter provided in the instrument. It has current coil and pressure coil, mounted at a particular angle. The current coil carry current which flows through the earth circuit. The current in the pressure coil is proportional to the voltage.

The reading obtained is the direct value of the earth resistance, moreover this instrument can be used to measure the resistance of this liquid conductor.

For the use and application of the meter there are four terminals $C_1 C_2$ and P_1, P_2 $C_1 C_2$ are of current coil circuit and $P_1 P_2$ of the pressure coil, out of which two are short circuited. There are three spokes, one common of P_1 and C_2 terminals other are current and pressure coil as shown in Fig. 23.23. The spokes are suitably installed in the earth at a particular distance generally 25 metre from earth electrode under test. The P_1 and C_1 are short circuited and connected the earth electrode under test. The P_1 and C_1 are short circuited and connected the earth electrode under test. The handle is rotated and direct reading of the earth resistance is obtained.

Q. What are the types of frequency meter? Describe with neat sketch the construction and working of one of the frequency meter.

Ans. The number of cycle per second of the alternating quantity is known as the frequency. There are the following methods of determining the frequency.

(*i*) By mechanical resonance method.

(*ii*) By electrical resonance method.

(*iii*) By weston frequency meter.

(*iv*) By ratio type frequency meter.

Mechanical resonance method. In this system the mechanical frequency is synchronized by the frequency of magnetic field produced by alternating current. These are known as *Reed Type* frequency meter.

Reed type frequency meter

In this instrument the mechanical resonance is utilized to find out the electrical frequency. The main advantage is that it is not subjected to the errors due to change in temperature, waveform and magnitude of the vibrations.

Construction. It is mainly consists of the thin steel strips generally known as reeds. The reeds are approximately 4 mm wide and 0.5 mm thick. These are having a white flag at one end and the other end is fixed with an armature. The reeds are not exactly same. These are slightly different in dimension or carry different weights or flags at their free end. The natural frequency of vibration depends upon the dimension and weight of the strip. As these strips are different in either of their parameters, so their natural frequency is different for 50 c/s instruments these strips are

calibrated and arranged for 0.5 cycle interval ranging from 47 to 53 c/s.
To increase the amplitude of vibra-
tion the white flag is mounted and it
exhibits the reading properly.
Working. Whenever the exciting
coil is energised by the supply, the
magnetic flux produced has got a
certain frequency. The frequency depends upon the frequency of the
main supply. The core will experience the same magnetic field and
will start vibrating. The force of attraction between the reeds and electromagnet is proportional to the

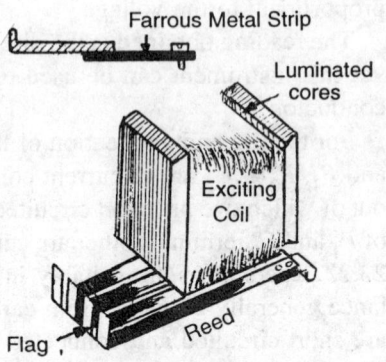

Fig. 23.24. Reed type frequency meter.

square of current, therefore the force varies twice the supply frequency
or double the supply frequency. Now the reeds which are under the influence will vibrate. The reed whose natural frequency is twice that of the
supply will be in resonance and will vibrate maximum which can be seen
from the face of the meter.

Range. For reed type frequency meter the reeds are ranging from 47
to 53 c/s for 50 c/s mains.

Advantages. Following are the advantages of using this type of meter:
   (*i*) The vibration is independent of the waveform and magnitude of the
       supply frequency, provided that the voltage should not be too low
       to vibrate the strip.
  (*ii*) These reeds are not effected by the temperature variation.

Disadvantages. Following are the disadvantages:
   (*i*) The variation is of 0.5 cycle per second so less that 0.5 c/s cannot
       be observed by this meter. So it cannot be so precise.
  (*ii*) To design the reeds for 0.5 c/s interval is not so easy.

**Q. What do you understand by the electrical resonance frequency
meter (Indicating type frequency meter)? Draw the neat and clean
diagram and explain.**

Ans. The instrument in which for the measurement of electrical frequency
the phenomenon of electrical resonance is adopted, is known as the electrical resonance frequency meter.

Construction. It consists of the fixed coil which is connected across
the mains; the frequency of which is to be measured, the magnetising coil
is placed over the laminations of silicone steel. The laminations are designed with a varying cross-section. It has maximum area near the coil

and minimum on the free end. A moving coil which carries pointer is pivoted in jewelled bearings. The moving coil is connected across a capacitor as shown in Fig. 23.25.

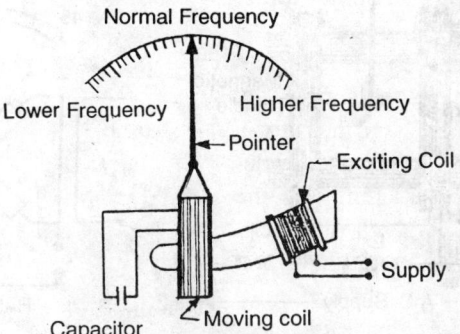

Fig. 23.25. Frequency meter.

The scale is calibrated as to have normal frequency in the centre and lower at one side and higher on the other side.

Working. Whenever the magnetising coil excited, the magnetic field is produced. The flux being alternating in nature will produce some e.m.f. lags behind the voltage by 90°. The current in the coil depends upon the inductance of the coil and capacitance connected. The deflecting torque will be zero, if the inductive and capacitive reactances are equal. The capacity of the capacitor is so designed that the moving coil occupy a convenient position on the iron cores where the frequency is normal. The inductance of the coil is not constant and it depends upon the value of the flux linkage, so the cores are designed having varying cross-section. The inductance depends upon the position of the coil occupied on the laminated strip.

When the supply frequency is normal it occupy the normal position and the normal reading is obtained. If the frequency is less the capacitive reactance becomes more than the inductive reactance, the torque developed moves the coil where the inductive reactance is equal to the capacitive reactance and the coil deflects towards the exciting coil or more area. In other words when frequency is more the capacitive reactance is less and the coil takes a position on the strip where both reactances are equal. Thus the coil moves away towards the less cross sectional area of the strip. If the scale is carefully calibrated the value of the frequency can be very accurate and precise. The sensitivity of the instrument is very good and effective.

Q. Draw the connection diagram of the Weston and Ratio-meter type frequency meter.

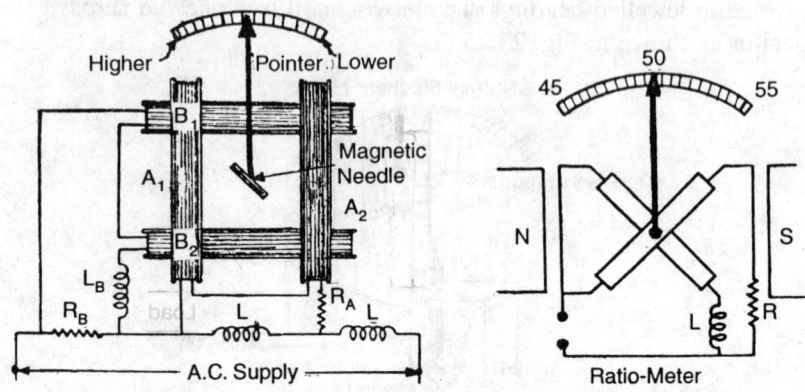

Fig. 23.26. Weston type frequency meter.　　　**Fig. 23.27.** Ratiometer

Q. What do you understand by the power factor of a.c. circuit? Explain the construction and working of a single phase power factor meter.

Ans. The power factor of a.c. circuit plays an important role in calculating the actual work done by the supply. It is defined as the ratio of real power to apparent power or the cosine of the angular displacement between current and voltage. The instrument which is used to measure the power factor of the circuit is known as the *power factor meter.* The instrument which is used for a.c. single phase circuit, is known as single phase power factor meter.

Single phase power factor meter

Construction. The working principle is same as that of dynamometer type instruments. It consists of the fixed coil which is splited into two halves as shown in Fig. 23.28. These sets are connected in series and in series with the load. Therefore the magnetic flux produced by these coils will be proportional to the load current having the same angular displacement. There are two moving coils which are fixed on the spindle. These are placed at 90° to each other. The spindle carries the pointer. There is a calibrated scale and a pointer to indicate the reading. These moving coils are connected across the mains through inductance or capacitance and resistance as shown in Fig. 23.28. The value of resistance R and inductance L is so chosen that for the normal frequency in pressure coils current is same. Thus the magnetic flux produced by both the coil is maximum and spaced by 90°. Both these coils move together as these are fixed on the same spindle.

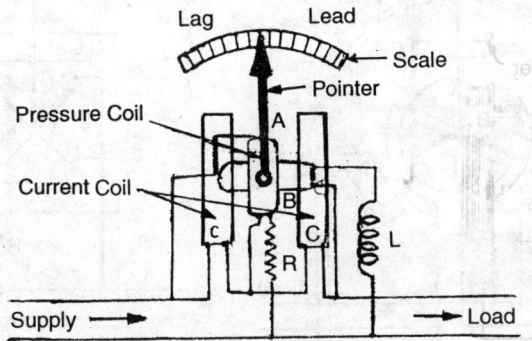

Fig. 23.28. Power factor meter.

For the measurement of power factor on h.t. line, the meter is connected through the instrument transformers.

Working. The current in the current coil is same as that of the load having the same angular displacement. But current in moving coils, will depend upon the external resistance and inductances. The current in the coil having resistance will be in phase with the voltage and in inductance lagging by 90° and in case if capacitance, it will lead by 90°.

Let the load is purely resistive in nature, the current and voltage in current coil will be in phase. The current in moving coil having resistance will be in phase and in other coil having capacitance will be leading by 90°. Thus the torque will be experienced on coil having resistance, and will enline the flux produced by the current coil. Thus the pointer will indicate unity, secondly if the load in inductive the torque is mainly contributed by the coil having reactive component and the pointer will indicate to lagging side. In case of capacitive load the pointer will indicate towards the opposite direction *i.e.* leading side.

The power factor meter have no springs for controlling torque. the controlling torque is automatically obtained because of the moving coils. Generally the scale is centre zero. The left side indicates the lagging power factor and right side the leading power factor.

Uses. It is used to find out the power factor of single phase circuits.

Q. Draw the connection diagram of a three phase dynamometer type power factor meters, also explain in brief.

The three phase power factor meter is used to find out the power factor of three phase circuits. In case of three phase circuits the load may be balanced or the unbalanced. There can be then two possibilities that the meter for balanced load cannot measure the power factor effectively on unbalanced load. Where the power factor meter suitable for unbalanced load can measure even on balanced load.

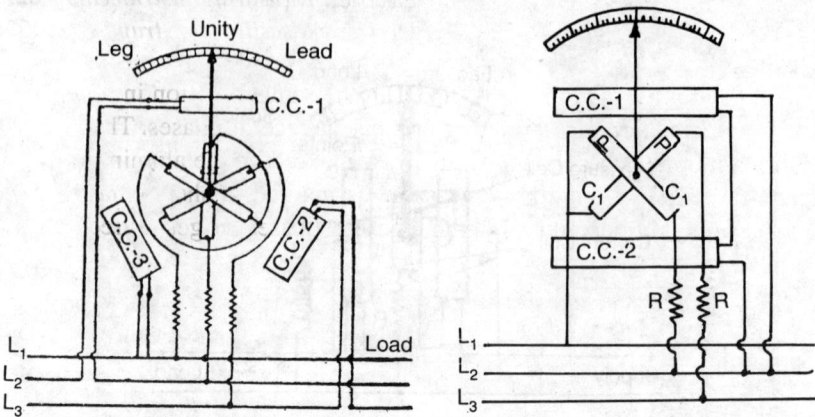

Fig. 23.29. 3φ power factor meter.   **Fig. 23.30.** 3φ power factor meter.

In case of meter suitable for unbalanced load, there are three current coils and three pressure coils. The current coils are connected in series with the load. The pressure coils are connected across line, generally in star as shown in Fig. 23.29. Now there are two rotating magnetic fields one due to current coils and other due to pressure coils. Both will be rotating at synchronous speed. If both fields are moving in the same phase the pointer will indicate unity. Due to the angular displacement of the magnetic fields of current coil and pressure coil, the pointer will take its respective positions depending upon the capacitive and inductive components on leading or lagging side. Figs. 23.29 and 23.30 show the connection diagrams of the meter suitable for 3φ loads; whether it is unbalanced or balanced.

Q. What do you understand by the three phase sequence indicator? Describe how it works.

Ans. The instrument used to determine the phase sequence of three phase supply, is known as the phase sequence indicator as shown in Fig. 23.31. The working principle of this instrument is same as that of the induction motor.

It consists of three coils mounted 120° electrical apart in space. The three ends are brought out and are connected to the three phase mains (RYB). Generally the coils are connected in star. The three ends are connected to the mains, the polarity of which is to be tested. An aluminium disc is mounted on the top of the coil. The phase supply produces the three phase rotating magnetic field and then the eddy currents are induced in the aluminium disc. The torque is developed over the current carrying conductor *i.e.* the aluminium disc and the disc rotates. The net torque developed is produced by the interaction of the magnetic fluxes.

The direction of rotation of the disc is in the same direction in which the rotating magnetic field is rotating or the sequence of phases. The indication of the rotation is shown by an arrow marked on the aluminium disc. If the direction of rotation is not same as marked, it shows that the sequence is not proper then interchange the phases to get correct phase sequence.

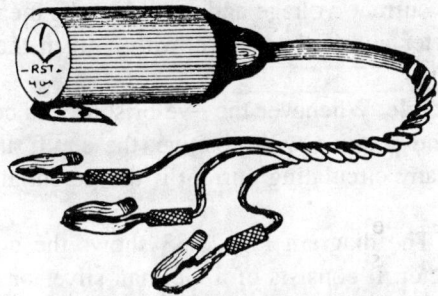

Fig. 23.31. Phase sequence indicator.

Q. Write short notes on the following:

    (*i*) **Ohm meter**

    (*ii*) **Potentiometer**

Ans. (*i*) **Ohm meter.** The meter is used to measure the value of unknown resistance in ohms. Fig. 23.32 shows an arrangement of a simple Ohm meter. It contains a moving coil type meter. The scale is calibrated in ohms. The meter is energised by a dry cell of 1.5 V. A variable resistance is connected in series with the moving coil instrument for zero adjustment. The deflecting torque of the instrument depends upon the value of the current in the moving coil, the current is maximum when both the terminals are short circuited. When an external resistance is connected in

M.C. Type Inst.

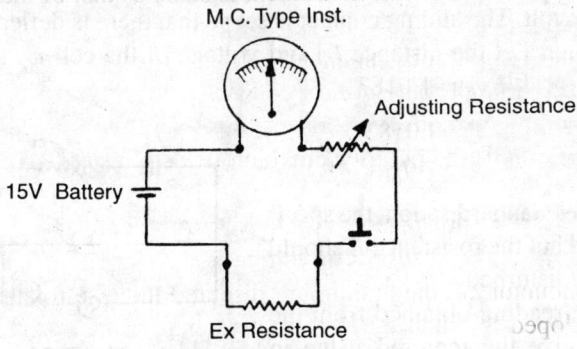

Adjusting Resistance

15V Battery

Ex Resistance

Fig. 23.32. Ohm meter.

series the value of current depends upon the resistance and current automatically decreases and the reading is obtained on the calibrated scale in ohms.

(*ii*) **Potentiometer.** It is an instrument used for measuring the unknown e.m.f., or potential difference by balancing it wholly or partially by a source of current in the network of known characteristics. The e.m.f.s are measured directly with the potentiometer in terms of the e.m.f. of a standard cell. The current, voltage and power can be measured by means of the potentiometer and if time is known or given then the electrical energy also.

Working principle. Whenever the −ve of source is connected to −ve of other source and positive with positive, the e.m.f. are opposed, and there will not be any circulating current if the potential differences are same.

Construction. The diagram Fig 23.33 shows the construction of a simple potentiometer. It consists of a German silver or manganin wire, which has uniform cross-section and stretched between two terminals *A* and *B*. A mm scale is also attached near the wire usually of one metre in length. The ends *A* and *B* are connected to a battery through an external resistance *R*. The cell which is to be tested is also connected. The galvanometer indicates the balancing stage. The key is also used to make and break the contact.

Working. Let a cell *S* be connected with its positive connected to *A* and negative to *B* through a galvanometer and a key. Since the wire *AB* is of uniform cross-section so the voltage drop per mm will be constant thus by using and moving the sliding contact on the wire a point can be located where there will not be any current in galvanometer. This point will have the same potential difference as that of the cell under measurement.

For giving the correct reading first of all the potentiometer is standardised. A standard cell is used for that purpose whose e.m.f. is a constant one 1.0183 volt. The sliding contact is set so that there is deflection of the galvanometer. Let the distance l_0, and voltage of the cell v_0.

Then $\qquad l_0 v_0 = 1.0183$

$$v_0 = \frac{1.0183}{l_0} \text{ v/cm.}$$

Thus after standardisation, the specimen is used but the resistance *R* should no case be disturbed.

Now the reading obtained from the scale will give the required value in

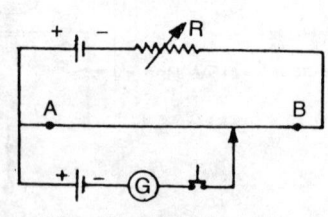

Fig. 23.33.

terms of length. For more precise readings the length of wire may be increased.

Applications and Use:

(*i*) It is used for the measurement of e.m.f.s up to 2 V and can be used up to 250 V.

(*ii*) It is used for comparing the e.m.f.s.

(*iii*) It is used for measuring unknown resistance.

(*iv*) It can be used for measuring current and for calibration of ammeter also.

(*v*) It is used for calibrating voltmeter.

(*vi*) It is used for wattmeter calibrations.

Q. What is the post office box and how it is used to find out the value of a unknown resistance?

Ans. The post office box is the another example of wheatstone bridge principle. In this instrument all the three unknown resistances *P*, *Q* and *R* are suitably arranged in a resistance box of special design as shown in Fig. 23.24. The unknown resistance forms the fourth arm of the bridge. In the top row of the instrument, there are six resistances in two sets of three resistances of 10, 100 and 1000 ohms. These two sets are arranged symmetrically round the centre of row as shown. These sets are called the ratio arms. It represents *P* and *Q* of the wheatstone bridge.

The remaining resistance of 1 to 5000 Ω are arranged as an ordinary resistance box. These resistances corresponds to the third unknown resistance *R*. The unknown resistance *X* is connected between the *A* and *C*. The galvanometer is connected between point *B* and *D*. The battery is connected between *A* and *C*. There are two keys provided for galvanometer and cell, known as Galvanometer key and cell key respectively.

Working. To make the determination of the unknown resistance, make the connection diagrams as shown in Fig. 23.24. Take out two plugs from

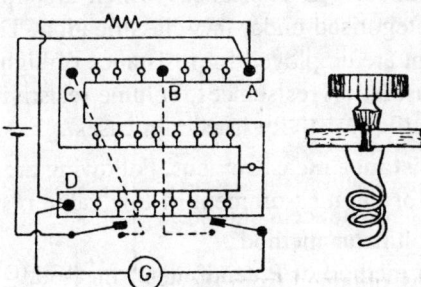

Fig. 23.34. Post office box.

the resistance arm, 10 and 10. Press cell key and then galvanometer key, see the deflection, now remove the infinity plug and press the keys and note the deflection. It should now be in reverse direction. It indicate the correctness of the connections. The resistance in the portion R of the post office is adjusted till the null point in the galvanometer. If no null point is achieved then, find out two resistances differing by one ohm which will give opposite deflection in the galvanometer. Now change the ratio from 10 : 10 to 10 : 100, and the reading will be now between the multiple of 10. First say the deflection was changing its direction between 1 and 2. Now find the position between 10 onward, let it be coming at 15 and 16. Again change the ratio arm from 10 : 1000 the reading can be obtained between 150 and 160. Find the null point let it is 153 and 154. Now the average of the readings is 153.5 and the correct value is

$$= \frac{10}{1000} \times 153.5 = \mathbf{1.535\ \Omega.}$$

Thus we can find out the unknown resistance.

Q. Classify the resistances according to the measurement and state the different methods of measuring low resistances. Describe any one method with neat sketch.

Ans. The different resistances according to the measurements are classified as under:

(*i*) **Low resistance.** The low resistance are those which are of about one ohm or even less. For example the measurement of armature resistance of an armature, series field winding, shunts, cable length and contact resistance etc.

(*ii*) **Medium resistance.** The resistances which are ranging from 1 ohm to 100 kilo ohms are classified under this categories. Most of the electrical apparatus are under this classification.

(*iii*) **High resistance.** The resistances which are more than 100 kilo ohms are categorised under this classification. The high resistance measurement are employed for resistance of high resistance circuit elements, insulation resistances, volume resistivity of the material and for specific resistivity in certain cases.

Low value resistance measurement. Following are the methods generally employed for the measurement of low value resistances:

(*i*) Ammeter-voltmeter method.

(*ii*) Comparison method or Potentiometer method.

(*iii*) Kelvin double wire bridge method.

(*iv*) Ohmmeter method.

Ammeter-voltmeter method. It is a very common, simple and quick method of measuring the low value resistances.

As shown in Fig. 23.55 the connections are made and both the current and voltage drop across the resistance are simultaneously measured by the ammeter and voltmeters. The accuracy mainly depends upon the accuracy and range of the instruments employed for the measurement of current and voltage.

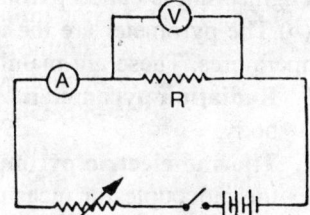

Here are current through the ammeter (I) is the sum of the current in voltmeter (I_v) and resistance (I_R), so

$$I = I_v + I_R$$

or $$I_R = I - I_v$$

The true value of the resistance

Fig. 23.35. Ammeter-voltmeter method.

$$= \frac{\text{Voltage across the resistance}}{\text{Current in the resistance}} = \frac{V}{(I - I_v)} \text{ Ohm.}$$

Thermocouple

Q. (a) What is thermo-electricity? Explain the construction and use of thermocouple.

(b) What are the different types of pyrometers?

Ans. In 1824 scientist Seebeck of Bertain, observed that when the junction of two dissimilar metals is heated up an electric current flows in the circuit. Thus current can be easily measured by means of a galvanometer or millivoltmeter. This e.m.f. is known as thermo-electric e.m.f., the combination is known as the thermocouple and the phenomenon of producing the e.m.f. is known as the Seebeck effect.

Consider two dissimilar metals copper and iron joined at A and B as shown in Fig. 23.36. The junction is heated up. The e.m.f. is produced. The magnitude depends upon the temperatures between the cold and hot junctions. At hot junction the current flows from copper to iron while at cold junction current flows from iron to copper.

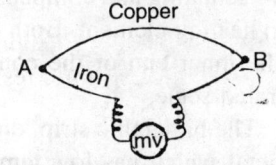

Combinations of thermocouple. The materials used in practice depends upon the temperature range to be measured. The platinum with 10% rhodium is suitable for 1400°C chromel to alumel is suitable for 1100°C. Iron to constantan is suitable for 900°C and copper to constantan is suitable for 300°C.

Fig. 23.36. Thermocouple.

Uses. The thermocouples are used for measuring temperature. Such devices are employed with the instrument, the thermocoupling will supply the energy and the instrument will work as the load. The *pyrometer* are the examples, the pyrometer is cheap, easy and can be used to measure the temperature up to 1400°C and even higher. The pyrometer is a thermometer which is used to measure the high temperature and the branch of engineering is called pyrometery.

(*b*) The pyrometer are the instruments used for measuring the higher temperatures. These are mainly of four types:

1. **Radiation pyrometer.** Which detect the heat radiation from a hot body.
2. **Thermo-electric pyrometer.** Which makes use of the principle of thermocouple for measuring temperature.
3. **Platinum resistance thermometer.** Which are based on the principle that resistance of the platinum increases with rise of temperature.
4. **Optical pyrometer.** Which measures temperature by estimating the quantum of light emitted by a body in narrow wavelength range.

Q. What do you understand by the maximum demand indicator, describe any one of them?

Ans. The maximum demand indicator actually measure the maximum demand of consumer. The main characteristic is that these should record to maximum power taken by the consumer during the particular period, say 15 or 30 minutes. These instruments are designed, as not to take into consideration the movementary overloading because of short circuit or motors starting etc.

These maximum demand indicators are of the following types:

(*i*) Recording demand indicator.

(*ii*) Average demand indicator.

(*iii*) Thermal type maximum demand indicator.

Thermal type maximum demand indicator. In its simple construction, it consists of two similar flat coils of bimetallic strips. These coils are actuating and compensating coils. The actuating coil is surrounded by an heating element. Both the coils are connected as shown in Fig. 23.37. The inner end of the compensating coil drives the pointer over a calibrated scale.

The bimetallic strip, obviously has the property of curling towards the metal which has low temperature coefficient. The bimetal is heated by means of the current in heating element, thus the heat developed is proportional to the square of current flowing through the heating element.

Thus the curling or bending of the bimetallic element is transferred to the compensating coil and finally it actuates the pointer. The pointer hits the needle of the demand indicator and move along. The set reading of the maximum demand, once moved cannot come along the needle but will rest there only. The indicator of maximum demand will retain there only until the larger current then the earlier one causes it to advance or it is brought back by the authorised person of supply company.

The compensating coil is used to prevent the effect on demand because of change in air temperature etc.

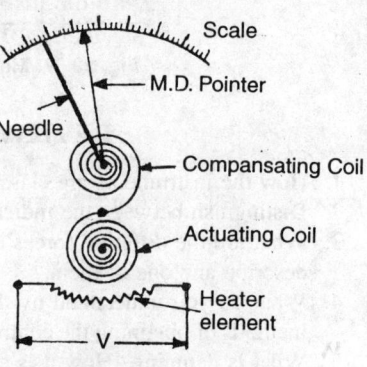

Fig. 23.37. Thermal type maximum demand indicator.

Advantages. Followings are the advantages:

(*i*) It is suitable for A.C. and D.C. both.

(*ii*) It is robust and cheap.

(*iii*) It can withstand the momentary overloading.

Disadvantages. Followings are the disadvantages:

(*i*) The readings are based on the production of heat, so the voltage should be constant for that current.

(*ii*) The scale is not uniform because the heat is proportional to the square of the current.

(*iii*) It is not so sensitive.

Q. What is the necessity of Kvarh metering? Draw a circuit diagram for Varh measurement in polyphase system.

Ans. Followings are the main reasons for the measurement of reactive kilo volt ampere hours.

(*i*) To keep a limit on the power factor and to check bulk consumers for failing to maintain it.

(*ii*) For the measurement of VAR.

(*iii*) It is an essential factor when there is an exchange of energy from one station to the another station.

VARH measurement in polyphase system. It can be measurement by means of an induction type energy meter with their scheme as shown in Fig. 23.36.

In case of three phase balanced load the meter will register the total reactive volt ampere hours.

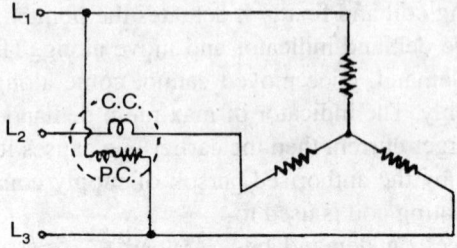

Fig. 23.37. Measurement of KVARH.

REVIEW QUESTIONS

1. How the instruments are classified?
2. Distinguish between the indicating and integrating type instruments.
3. What are the different forces employed in an indicating type instruments, describe any one of them.
4. What do you understand by the controlling force? Describe the different methods of obtaining the controlling force in an indicating type instrument.
5. What is damping? How it is achieved? Describe different methods of obtaining damping force.
6. Name the deflecting force used in indicating type electrical instrument. Describe eddy current damping.
7. What happens if
    (a) The control spring is removed from the indicating type instrument,
    (b) The instrument is over damped,
    (c) The gravity control instruments are used in horizontal position,
    (d) Commercial instruments are used as a standard?
8. How the instruments are classified according to the working principle.
9. Describe with the help of neat diagram the working of a moving coil type instrument.
10. Sketch and describe the M.C. type instrument. How it differs from the M.I. type instruments?
11. Describe the construction and working principle of attraction type M.I. instruments.
12. What are the classification of the M.I. type instruments? Describe the repulsion type M.I. instrument.
13. Compare the M.I. and M.C. type instruments.
14. Write a short notes on shunt and multipliers.
15. Describe the essential features of a electrodynamometer type instrument. How it is used as an ammeter, voltmeter and wattmeter?
16. What is the working principle of the dynamometer type instrument? Describe with neat sketch.
17. Give reasons of
    (a) Why a M.C. type meter is not suitable for A.C.?
    (b) Why the scale of M.C. instrument is evenly divided?

(c) Why the scale of M.I. instruments is uneven?

(d) Why the ammeter is connected in series?

18. Describe with neat diagram the working and construction of a hot wire instrument.

19. (a) What is the wattmeter? Describe the induction type single phase watt-meter.

(b) Explain the working of a single phase energy meter.

(N.C.V.T. 1976)

20. Draw a neat sketch of a 3 ϕ 4 wire, A.C. energy meter and explain its working. *(N.C.V.T. 1968)*

21. Name the different types of D.C. energy meter. Describe with neat sketch any one of them.

22. What is the thermocouple, how it is used?

23. What is the maximum demand indicator? Describe with neat sketch the working principle and construction of thermal type maximum demand indicator.

24. Write short notes on the followings: (a) Megger, (b) Earth tester, (c) Potentiometer, (d) Reed type frequency meter, (e) Frequency meter, (f) 1 ϕ power factor meter, (g) Three phase power factor meter, (h) kVAR metering, (i) Phase sequence indicator.

SOUND

1. The best propagator of sound is
   (a) Metals   (b) gas   (c) Air   (d) Mercury

2. Photo Electric cell is a device which converts
   (a) Chemical energy into light energy.
   (b) Light energy into electrical energy.
   (c) Electrical energy into light energy.
   (d) Electrical energy into magnetic energy.

3. The Wheatstone bridge is most sensitive when
   (a) the ratio arm is equal to 100.
   (b) the ratio arm is equal to one.
   (c) they are having any value.
   (d) all above are correct.

4. If a hole is bored into the earth along its diameter and a stone is dropped into the hole.
   (a) the stone reaches on the other side of the earth and escape into space.
   (b) the stone reaches the centre of the earth and stops there.
   (c) the stone reaches other side of earth and stops there.
   (d) the stone executes simple harmonic motion about the centre of earth.

5. The time period of a freely suspended magnet does not depend upon:
   (a) The length of the magnet.
   (b) The length of the suspended thread.
   (c) The pole strength of the magnet.
   (d) None of the above

6. A metal getting magnetised by the orientation of atomic magnetic moments in external magnetic field are called
   (a) diamagnetics
   (b) ferro magnetics
   (c) antimagnetics
   (d) paramagnetics

7. To send 10% of the main current through a moving coil instrument of resistance 99Ω, the shunt should have resistance of
   (a) 10 Ω
   (b) 11 Ω
   (c) 9 Ω
   (d) 99 Ω

8. Whenever a current carrying conductor is placed in a uniform magnetic field than
   (a) a torque is produced.
   (b) an e.m.f. will be induced.
   (c) Both are (a) and (b) correct.
   (d) Both (a) and (b) are incorrect.

9. A steady current is flowing into a wire, it is turned to form a coil, now it is again turned to form a coil of smaller diameter forming double loop, can you tell the magnetic field at the centre by the same current will be .... of its first value.
   (a) Half
   (b) Quarter
   (c) Unaltered
   (d) 4 times.

10. An electric heater operated on 220 V heats certain amount of water in 10 minutes, can you tell that heater on 110 V will heat the same water in ....
    (a) 15 min
    (b) 10 min
    (c) 20 min
    (d) 5 min.

11. Two heaters of 2000 W are connected in series their total wattage will be ....
    (a) 500 W
    (b) 2000 W
    (c) 4000 W
    (d) 1000 W

12. A electric bulb and heater are marked as 200 W 200 V and 200 W 200 V respectively can you tell the resistances of bulb will be .... as that of the heater.
    (a) same
    (b) double
    (c) one fourth
    (d) four times.

13. A person has 5 resistances of 10W each. What is the maximum resistances, he can make with them.
    (a) 5 Ω
    (b) 10 Ω
    (c) 50 Ω
    (d) 2 Ω

14. A battery has e.m.f. 15V and its internal resistance is 0.07 W if it is delivering 10A current. What will be its terminal voltage?
    (a) 22 V
    (b) 14.05 V
    (c) 14.3 V
    (d) 12.1 V.

15. Which of the following is volt.
    (a) Joules per coulombs
    (b) Ergs per ampere.
    (c) Newton/(coulomb × m²)
    (d) None of the above.

16. Ampere hour is the unit of ....
    (a) power
    (b) Energy
    (c) current
    (d) quantity of charge.

17. If two heater elements of 1000 W each are connected in series and than in parallel across same voltage than the ratio of heat produced in two cases will be
    (a) 1:2   (b) 1:4
    (c) 2:1   (d) 4.1.
18. Fill in the blanks:
    (i) A magnet, no matter how long is it, will have .... poles.
    (ii) In a magnet all molecules points towards the .... direction.
    (iii) If a permanent magnet is brought near an iron piece as .... pole is induced on that piece, this property is called as magnetic ....
    (iv) An electromagnet is a .... magnet.
    (v) In an iron bar all the molecules are in .... manner.
    (two, same, opposite, induction, temporary, random).
19. .... is not attracted by magnet.
    (i) Cobalt   (ii) Iron
    (iii) Al   (iv) Nickle
20. If a magnet is rolled in brass filings, the filings will get attracted to
    (i) north pole   (ii) south pole
    (iii) nowhere   (iv) at the centre.

NUMERICAL PROBLEMS

1. A.M.C. type instrument has 2Ω resistance and reads 100 mA. What resistance must be placed in series to enable it to use as
   (i) an ammeter up to 10 A and
   (ii) a voltmeter up to 10 V?   [Ans. **0.0202 Ω and 98 Ω**]
2. A.M.C. type instrument having a coil resistance as 2Ω gives a full scale deflection when a current of 50 mA is passed through the coil. Calculate the value of shunt and multiplier to use it as
   (i) an ammeter up to 5 A.
   (ii) as a voltmeter up to 500 V.   [Ans. **0.0202 Ω, 9998 Ω**]
3. A moving coil instrument gives a full scale deflection with 15 mA and a resistance of 5Ω. Calculate the resistance to be connected in
   (i) in parallel to enable the instrument to read up 1 A and
   (ii) in series to enable it to read up to 100 V.
              (*U.P.S.C., I.E.S. Elect. Engg. 1972*)

$$\left[\text{Ans. } \frac{10}{1999}\Omega, 19{,}999 \,\Omega\right]$$

4. A moving coil instrument requires 5 mA for full scale deflection. Show how it can be used as an ammeter up to 15 A and as a voltmeter up to 300 V. The coil resistance of the instrument is 5 Ω

$$\left[\text{Ans. } \frac{5}{2999}\Omega, 59995 \,\Omega\right]$$

24

Illumination

Q. What do you understand by the illumination? Describe different factors which effects the correct illumination.

Ans. The illumination means the light of proper colour, quantity and so directed as to allow quick and accurate vision without discomfort to the eyes. There are certain factors which effects the correct illumination.

(a) **Degree of illumination.** It decides the power of the illuminating source. For that determination it is to be observed:

  (i) The size of the object to be seen.

  (ii) The distance from the observer.

  (iii) Contrast between the object and ground.

  (iv) The duration and state, moving or static.

(b) **Colour of light.** The colour of the body also depends upon the colour of the incident rays.

(c) **Nature of lighting.** Whether it is coming directly from the source or indirectly or any other type of lighting system is used.

(d) **Mounting height.** The height of the illuminating source decide the proper illumination.

(e) **Space between illuminaries.** Correct spacing between the illuminating source will effect the illumination and uniform light.

(f) **Colour of the surroundings.** The correct and accurate illumination in any room depends upon the light reflected from the walls and ceiling, white colour of the walls reflects more light as compared to coloured one.

There are certain points which effects the *characteristics* of the good illumination:

  (i) The light should not strike the eyes of the worker *i.e.* the light should fall on the job not on the workers eyes.

(*ii*) Correct size and type of the lamp.

(*iii*) Proper location of the lamp.

(*iv*) Proper reflective equipment.

Advantages of good illumination. There are the following advantages:

(*i*) Production is increased.

(*ii*) Less number of accidents.

(*iii*) Less wastage of raw material.

(*vi*) Efficiency of production and the benefit to the organisation will increase.

Q. Define the following terms:

(*a*) **Light,** (*b*) **Luminous flux,** (*c*) **Luminous intensity,** (*d*) **Lumen,** (*e*) **Candle power,** (*f*) **Foot candle,** (*g*) **Meter candle,** (*h*) **M.H.C.P.** (*i*) **M.S.C.P.:** (*j*) **M.H.S.C.P.,** (*k*) **Brightness,** (*l*) **Coefficient of utilization** (*m*) **Depreciation factor,** (*n*) **Reflection factor,** (*o*) **Solid angle,** (*p*) **Space height ratio and** (*q*) **Light efficiency.**

Ans. (*a*) **Light.** The light is defined as the energy radiated in the form of waves which produces the sensation of vision to the eyes. It may be the natural light from the sun or the artificial light from the means created by human beings.

(*b*) **Luminous flux.** It is the total quantity of light emitted by the source of light per second. It is represented by symbol ϕ or F. It is measured in Lumen.

The total luminous flux emitted by a source of I candela, is

$$= 4\pi \times I \text{ lumen}$$

where I is the luminous intensity of the source.

(*c*) **Luminous intensity.** The luminous intensity in any given direction is the luminous flux emitted by a source per unit solid angle. It is represented by I and measured in candle power.

If $d\phi$ is the flux emitted by a source within a solid angle $d\omega$ in any particular direction, then

$$I = \frac{d\phi}{d\omega} \text{ lum./cd.}$$

(*d*) **Lumen.** It is the unit of luminous flux and is defined as the luminous flux emitted per unit solid angle by a point source of one candle power.

(*e*) **Candle power.** The candle power is the light radiating capacity of a source in a given direction. It is defined as the number of lumens given out by a source in a unit solid angle in the given direction. It is denoted by symbol C.P.

(*f*) **Foot candle.** It is the unit of illumination. It is defined as the luminous flux falling per square foot on a surface which is perpendicular to the rays of light from a source of one C.P. and is one ft. away from it.

It can otherwise be defined as, "if one C.P. source is suspended in a hollow sphere of one ft. radius then the luminous flux falling on a surface of one square ft" is known as one foot C.P.

(g) **Flux or meter candle.** It is the unit of illumination and is defined as the luminous flux falling per square metre on a surface which is every where perpendicular to the rays of light from a source of one C.P. and one metre away from it.

It can otherwise be defined as that if a source of one C.P. is placed in the centre of a sphere of one metre radius, then the flux falling on the inner surface of one square metre in area is known as one meter C.P.

(h) **M.H.C.P.** It is known as the mean of candle powers in all directions in the horizontal plane containing the source of light.

(i) **M.S.C.P.** It is defined as the mean of the candle powers in all directions and in all planes from the source of light.

$$\text{M.S.C.P.} = \frac{\text{Total flux in lumen}}{4\pi}$$

(j) **M.H.S.C.P.** It is the mean hemispherical candle power in all directions above or below the horizontal plane passing through the source of light.

(k) **Brightness.** The brightness is defined as the luminous intensity per unit projected area of either a surface source of light or reflecting surface. It is denoted by B and the unit is 'stib' which is nothing but candle/cm^2 other unit is 'nit' which is candle/m^2.

(l) **Coefficient of ulitization.** It may be defined as the ratio of the lumens actually received by a particular surface to the total lumens emitted by the luminous source.

$$= \frac{\text{Total lumens actually received by the working plane}}{\text{Total lumen emitted by the light source}}$$

(m) **Depreciation factor.** It is a factor which is actually related to the cleanness of the lamp, shade, reflector etc. similarly the walls and ceiling. It is obvious, that the luminous intensity of the lamp deteriorates due to blackening, dirt and dust etc. so the depreciation factor is defined as the ratio of illumination when everything is perfectly cleaned in order to the illumination under actual conditions *i.e.* Depreciation factor = $\dfrac{\text{Illumination when everything is clean}}{\text{Illumination under an actual condition}}$

The D.P. is more than one *i.e.* 1.3, 1.4.

It should be clearly understood that the depreciation factor and utilization factors can also be defined in reverse order so the D.F. can be less than one in that case.

(*n*) **Reflection factor.** Whenever the light rays strikes the surface and reflects back at the same angle to the normal (incident ray and reflected rays with normal). A certain amount of flux is absorbed by the surface so the ratio is defined as the reflection factor *i.e.,*

$$\text{The reflection factor} = \frac{\text{Reflected light}}{\text{Incident light}}$$

It is also less then one and depends upon the characteristics and colour of the surroundings *i.e.* walls and ceiling etc.

(*o*) **Solid angle.** It is defined as the angle subtended at a point in space by an area. In plane angle it is the area which subtained but here it is the volume which is enclosed by a number of light rays in a surface as shown in Fig. 24.1 and meeting at a point. The solid angle is measured in steradians and represented by ω.

∴ ω (solid angle)

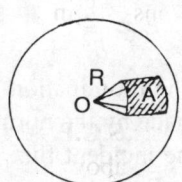

Fig. 24.1. Solid angle.

$$= \frac{\text{Area}}{(\text{radius})^2} = \frac{A}{R^2}$$

Since the area of the surface is $4\pi R^2$ and the total solid angle subtended by the point source is

$$\omega = \frac{4\pi R^2}{R^2} = 4\pi.$$

(*d*) **Space height ratio.** It is defined as the ratio of horizontal distance between lamps and the mounting height of illuminating source. So this is represented as the space height ratio

$$= \frac{\text{The horizontal distance between lamps}}{\text{Mounting heing of the lamps}}.$$

There are certain permissible values of space/height ratio, for ordinary industrial fitting 1.5, concentrating light 1.0, indirect lighting 1.5, standard fluorescent lamps 1.5.

(*q*) **Light efficiency.** It is the ratio of luminous output to input power of the lamp in watts.

$$\therefore \quad \text{Light efficiency} = \frac{\text{Lumen output}}{\text{Watts taken}}$$

The *life* of the lamp is the number of hours it operates to burn out or there is a change in efficiency or light output.

Q. What are the laws of illumination?

Ans. The illumination on a definite surface depends upon certain laws, which are known as the laws of illumination. These can be stated as

(*i*) The illumination (*E*) on a surface is directly proportional to the luminous intensity (*I*) of the illuminating source *i.e.*,

$$E \propto I$$

(*ii*) The illumination on a surface is inversely proportional to the square of distance between the illuminating source and the surface *i.e.*,

$$E \propto \frac{1}{d^2}.$$

(*iii*) The illumination is directly proportional to the cosine of the angle made by the normal to the illuminated surface with the direction of the incident flux.

$$E \propto \cos\theta$$

Thus when all the factors are considered, the illumination,

$$E = \frac{I}{d^2} \cdot \cos\theta$$

where
E = the illumination
I = the intensity of illumination
d = the distance between the surface and source, and
$\cos\phi$ is the angle as shown in Fig. 24.2.

Here point B is directly under the source A and C, a point at x from B say in a corner.

Here illumination directly below the lamp at 'B'.

$$E_B = \frac{I}{d^2}$$

and illumination at point C.

$$E_C = \frac{I}{(\text{distance})^2}(\cos\theta)^3$$

Here $$\cos\theta = \frac{AB}{AC} = \frac{d}{\sqrt{d^2 + x^2}}$$

because the distance $$AC = \sqrt{d^2 + x^2}$$

so $$E_C = \frac{I}{AC^2} \times \frac{d}{AC}$$

$$= \frac{I \times d}{AC^3} = \frac{I}{d^2} \times \frac{d^2 \times d}{AC^3}$$

$$= \frac{I}{d^2} \times \frac{d^3}{AC^3}$$

$$= \frac{I}{d^2} \cdot (\cos\theta)^3$$

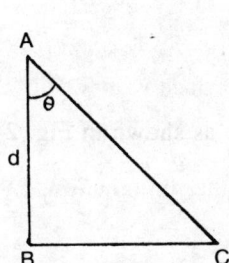

Fig. 24.2. Laws of illumination.

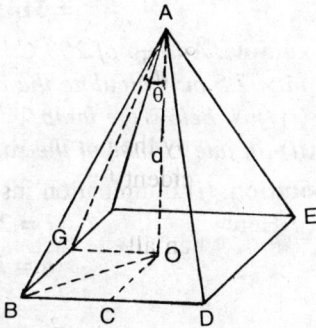

Fig. 24.3.

or $$E_C = E_B \cos^3\theta \qquad \left(\because E_B = \frac{I}{d^2}\right)$$

It is also known as $\cos^3\theta$ *law i.e.,* the illumination of a surface at any point is dependent upon the cube of the cosine of the angle between the line of incident flux and the normal on that point.

Example. *A lamp is rated 250V and takes 0.4A current. It produce a total flux 2000 lumens. Calculate the*

(a) M.S.C.P. of the lamp   (b) Efficiency of the lamp.
Solution. The mean spherical candle power

$$= \frac{\text{Total lumens}}{4\pi}$$

$$= \frac{2000}{4\pi} = \mathbf{159.15 \ Ans.}$$

The efficiency of the lamp

$$= \frac{\text{Lumens produced}}{\text{Wattage of the lamp}}$$

$$= \frac{2000}{250 \times 0.4} = \mathbf{20 \ Ans.}$$

Example. *A lamp has mean spherical candle power of 25, calculate the total flux of the light from the lamp.*

Solution. The mean spherical candle power is the candle power in all the directions and in all the planes from the source of light so,

Total flux $= 4\pi \times \text{M.S.C.P.}$

       $= 4\pi \times 25$

       $= \mathbf{314 \ lumens \ Ans.}$

Example. *A lamp of 200 C.P. is placed at the centre of the room 5 m × 5 m × 2.5 m. Calculate the illumination*

 (*i*) *just below the lamp,*

 (*ii*) *in one corner of the room.*

Solution. (*i*) illumination just below the lamp as shown in Fig. 24.3 is

 Here $I = 200$ C.P.

        $d = 2.5$ m

$$\therefore \qquad E_0 = \frac{200}{(2.5)^2}$$

        $= \mathbf{32 \ lux. \ Ans.}$

 (*ii*) Here the distance of the point from the source is *AB*

$$\therefore \qquad AO = 2.5, \ OC = 2.5$$

$$BO = \sqrt{BC^2 + CO^2} = \sqrt{2.5^2 + 2.5^2} = 3.536 \text{ m}$$

$$AB = \sqrt{AO^2 + BO^2} = \sqrt{2.5 + 3.536^2} = 4.33 \text{ m}$$

$$\cos\theta = \frac{AO}{AB} = \frac{2.5}{4.33}$$

Now $E_B = \dfrac{I}{(\text{distance})^2} \times (\cos\theta)^3$

$$= \frac{200}{4.33^2} \times \left(\frac{2.5}{4.33} \right)^3$$

$$= 6.159 \simeq \textbf{6.16 lux. Ans.}$$

Q. State the various types of lighting schemes.

Ans. The interior lighting scheme may be classified as,

(*i*) Direct lighting.

(*ii*) Indirect lighting.

(*iii*) Semi-direct lighting.

(*iv*) Semi-indirect lighting.

(*v*) General lighting.

Direct lighting. It is the most commonly used type of lighting scheme. In this lighting system 90% of the light falls directly on the job. It is mostly used for industrial and general outdoor lightings. This type of lighting is liable to cause glare and shadows.

Indirect lighting. In this system the light more than 90% is reflected upwards to the ceiling from there it reaches to the object. In this system ceiling acts as the source and the glares are reduced to minimum. The light is more softer and diffused. Generally there is a opaque reflector below the lamp. This type of lighting is used for decoration purpose in cinema halls, dancing halls, hotels and in workshops where large machines and other obstructions would cause troublesome shadows if direct lighting would have been used.

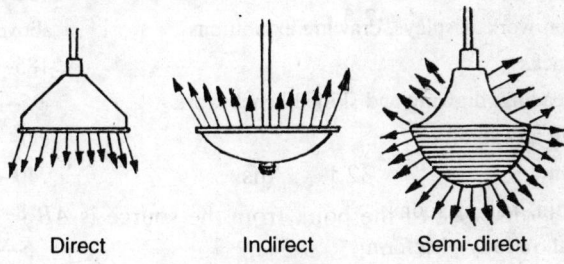

Direct   Indirect   Semi-direct

Fig. 24.4. Lighting scheme.

Semi-direct lighting In this system 60-90% of the light is made to fall downwards directly with the help of semi-direct reflectors, remaining light is used to illuminate the ceiling and walls, such lighting system is mostly used with the room of high ceiling. Glares are avoided by using the diffusing globes.

Semi-indirect lighting. In this lighting scheme 60-90% light is thrown upwards to the ceiling for diffuse reflection and remaining reaches to the object directly. It is mainly used for indoor decoration purpose.

General lighting. In this scheme lamps made of diffusing glass are used and are illuminated with a property of uniform illumination in all directions.

Q. State the level of illumination.

Ans. It is an important factor which actually involves the correct sight, colour, contrast, state of object and moreover the duration of the visuability. So the level or degree of illumination is defined as the luminous flux which gives away necessary brightness resulting minimum fatigue and maximum output in term of quality and quantity.

There is a illumination or range of brightness for every type of work, which can give the correct illumination without eyes discomfort.

Recommended illumination level

Purpose and place	lm/m²
Industrial purpose	
1. Precision machine room etc.	240—500
2. Lathe machine, sewing machine, bench work	140—189
3. Other types of industrial work	60—150
4. General lighting factory	18—40
General purpose	
5. Boxing rings	1750—2700
6. Precision work, displays, drawing exhibitions	above 500
7. Race tracks	185—280
8. Proof reading, drawing and skilled bench work	95—185
9. Drawing an exhibition	60—100
10. Museum	40—60
11. Bed room, waiting room	18—32
12. Hospital, railway plateform	5—10
13. Railway shunting yard	2—5
14. Aerodromes, landing ground	1—2

Example. *A room of (30 × 15) metres is to be illuminated by the 15 lamps to give an average illumination 40 lum/m². The utilization factor is 4.2 and depreciation factor 1.4. Find the M.S.C.P. of each lamp.*

Solution. The room is to be illuminated with 40 lumen/m² with 15 bulbs.

∴   Area of the room = 15 × 30 = 450 m²

The utilization factor $\dfrac{1}{4.2}$ and depreciation factor is 1.4.

So the total lumens required = Area × illumination

$$= 450 \times 40 = 18000 \text{ lumens}$$

Now the total flux required

$$= \frac{\text{Total flux required} \times \text{Depreciation factor}}{\text{Utilisation factor}}$$

$$= \frac{18000 \times 1.4}{1/4.2} = 105840 \text{ lumens}$$

There are 15 lamps so lumen per lamp

$$= \frac{105840}{15} = 7056 \text{ lumen}$$

Now $\qquad$ M.S.C.P.$= \dfrac{\text{Lumen per lamp}}{4\pi}$

$$= \frac{7056}{4\pi} = \textbf{561 candela. Ans.}$$

Q. State the different types of lamps used for illumination purpose.

Ans. The electric lamp is a source which converts electric energy into heat energy and then lighting energy. Whenever a body is heated up beyond a certain temperature it radiates the light energy. The visible radiation is directly proportional to the temperature of the body, more temperature more radiation and less temperature less radiation.

Types of lamps. Following are the types of the lamps:

$\quad$ (*a*) Incandescent lamp. $\qquad$ (*b*) Gas discharge lamp.

$\quad$ (*a*) *Incandescent lamp.* The lamps in which the light is produced directly by the heating of the filament are known as incandescent lamps.

These are further classified as

$\quad\quad$ (*i*) Carbon filament lamps.

$\quad\quad$ (*ii*) Metal filament lamps. These are further classified as *gas filled* and *vacuum type* lamps.

$\quad\quad$ (*iii*) Carbon arc lamp.

$\quad$ (*b*) *Gas discharge lamp.* The lamps in which the light is produced by discharging through the luminous material. In this lamp the light is produced by the excited atoms of the gas. These discharge lamps are, fluorescent tube, high pressure mercury vapour lamp, sodium vapour lamp and neon signs. The gas lamps are considered superior than the metal filament lamps, but even then these have some drawbacks, for example:

(*i*) high initial cost and low power factor.

(*ii*) starting is not so quick and easy.

(*iii*) can be used only in particular direction.

(*iv*) the light flickers and the stroboscopic effect is produced.

Q. Write short notes on the followings:

(*i*) **Carbon filament lamp.**

(*ii*) **Types of filaments used in incandescent lamps.**

(*iii*) **Effect of atmosphere on the working filament.**

(*iv*) **Advantages and disadvantages of the vacuum and gas filled lamps.**

Ans. (*i*) Carbon filament lamp.

In carbon filament lamps the filament is made of carbon. It was the first material to be used for making the filaments of the lamps. The melting point of carbon is about 3500°C, and the state working temperature is 1800°C. At higher temperature the carbon starts evaporating. The glass container of the lamp becomes black and thus the light decreases. Moreover at higher temperature, the carbon being a negative temperature coefficient material, the resistance decreases and the lamp starts taking more current and it will again boost the temperature.

Advantage. The carbon filament lamp has the advantage:

(*i*) These are mainly advantageous in battery charging circuit as the resistance for voltage drops.

Disadvantages. Following are the disadvantages:

(*i*) The carbon evaporates, so the inside of the glass container becomes black.

(*ii*) The life is less, it is about 800-860 hrs.

(*iii*) Light efficiency is less it is about 3.5 W/C.P.

Use. These are used for battery charging purpose as a resistance. These are available according to the wattage and voltage, as 15 W, 25 W, 40 W, 60 W, 250 V lamps.

(*ii*) Types of filaments (*N.C.V.T. 1982*)

The carbon filament because of the drawbacks was replaced by the tantalum. The working temperature was 1850°C but the efficiency of the filament was less than the tungsten filament. Nowadays the filaments are made of tungsten, because of the following properties:

(*i*) Its melting point is about 3400°C.

(*ii*) It can satisfactorily work up to 2000°C.

(*iii*) It has good mechanical strength.

(*iv*) The wires of tungsten can be drawn up to a very thin diameter as 0.0005″ with a efficiency of 10 lumens per watt.

There are two main types of the filaments:

 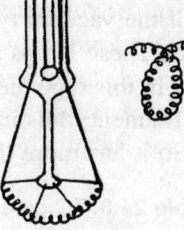 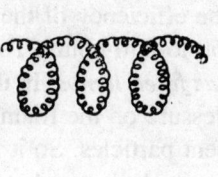

Straight filament   Coiled coil filament

Fig. 24.5. Types of filaments.

(*a*) **Straight filament.** In this case the filament is made of a resistance coil. The filament is supported by the number of supports as shown in Fig. 24.5. Such types of filaments are used for low and high wattage bulbs.

(*b*) **Coiled coil filament.** In this case the filament is in the shape of a coiled coil as shown in Fig. 24.5.

These types of the filaments are used for the lamps of medium wattage as 60 W, 100 W and 200 W etc.

Advantages

(*i*) In a small portion more length is used.

(*ii*) As the turns are very nearer to each other so a higher working temperature is obtained.

(*c*) Effects of atmosphere on the working filament

If the filament is heated up in the open atmosphere *i.e.* air, it will burn out with white fumes. The atmospheric oxygen will react and oxidise the filament and it will born out.

(*d*) Advantages and disadvantages of gas filled and vacuum type lamps

The incandescent lamps are of two types the gas filled and vacuum type:

(*i*) *Vacuum type.* In this type of lamps there is no air or inert gas but a vacuum. It has the following advantages:

Advantages

1. Loss of heat is reduced.

Disadvantages

(*i*) As there is no pressure on the filament so it evaporates and the filament becomes weak and life is reduced.
The efficiency of the vacuum type lamps is 1.5 W/C.P. and life is 750 to 850 hours. These lamps are generally of low wattage.

(*ii*) *Gas filled lamps.* In this type the gas is filled thus the gas has got a pressure on the filament; thus it avoids the evaporation of the filament particles. So it has more life.

Table 24.6. Tungsten filament lamps

S.No.	Power in watts	Luminous flux in lumens	Efficiency lumen/watts	Remark
1.	10 W	80	8	
2.	40 W	460	11.5	
3.	60 W	840	14.0	
4.	100 W	1630	16.3	
5.	200 W	3660	18.3	
6.	500 W	9250	19.9	

Advantages

(*i*) The heat is dissipated therefore the temperature of the glass is not very high.

(*ii*) The gas, which is nitrogen or argon, neither helps in burning nor burns,so it creates a pressure on the filament. Thus the evaporation is restricted; hence the life of the lamp is increased.

In this lamp, generally the filament is of coiled coil type. The efficiency of the gas filled lamp is about 0.5 to 0.6 W/C.P. and average life is 1000 hours.

Q. Describe with the help of neat sketches the working of the carbon arc lamp.

Ans. The carbon arc lamp is the example of an incandescent lamp. The light is because of the arc between two carbon electrodes. The lamps are of two types:

(*i*) D.C. carbon arc lamp.

(*ii*) A.C. carbon arc lamp.

D.C. carbon arc lamp

There are two electrodes which are connected across D.C. supply of 42-50 volts. For producing an arc, the two electrodes are touched together and then separated, when the rods are touched current starts flowing and as soon as these are separated the current starts flowing through the air, *i.e.* the arc is produced. The space between the electrodes behaves like a conductor because there are certain vapours of carbon. The carbon particles move from negative to positive electrode and results a creter on the positive electrode. Now because of the bombardment of these particles, the temperature of the crater is high 3500-4000°C and contributes maximum light −85%. The temperature of the negative electrode is about 2500°C and only 5% light is produced. As the temperature of positive electrode is high so the consumption is very fast, it is twice that of the negative electrode. So to keep its consumption rate equal the positive electrode is made of approximately double the diameter then negative electrode as shown in Fig. 24.6.

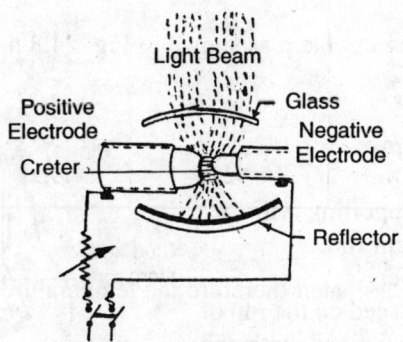

Fig. 24.6. D.C. carbon arc lamp.

An automatic or sometimes manual arrangement is provided to keep the arc length constant. If the arc length is increased, the arc may go off. The length of the arc is about 3 mm to 6.5 mm. There is a reflector to reflect the rays and for the protection a glass is provided as shown.

A.C. carbon arc lamp

In this case the two electrodes of same dimensions are used. These are connected across the voltage of 50-55 volts A.C. single phase. As A.C. is alternating in nature, so the electrodes rate of consumption is same for both phase and neutral electrodes as shown in Fig. 24.7. The arc length is 3 mm to 6.5 mm, other parts are same as that of D.C. carbon arc lamp.

Uses. The carbon arc lamps are used in cinema projects, search lights etc.

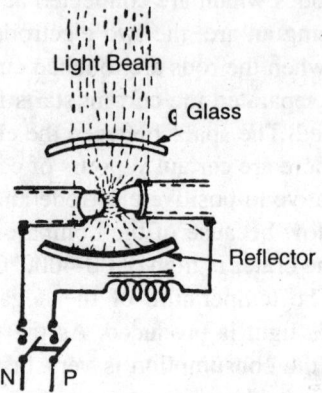

Fig. 24.7. A.C. carbon arc lamp.

Q. Describe wit the neat sketch the different parts of an incandescent lamp. *(N.C.V.T. 1981)*

Ans. The incandescent lamp as shown in Fig. 24.8 has got the following different parts:

(*a*) Cap.
(*b*) Glass container.
(*c*) Filament.
(*d*) Filament supporting wire.
(*e*) Stud and stem rod.
(*f*) Exhaust tube.

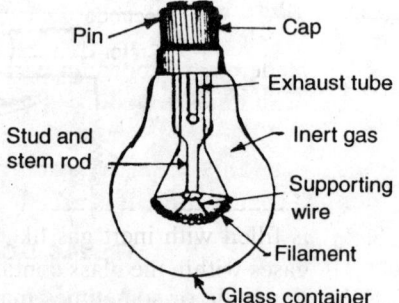

Fig. 24.8. Parts of incandescent lamp.

(*a*) **Cap.** It is placed on the top of the lamp and fixed with the neck of the container by means of a compound. The cap is made of brass or aluminium. Two main leads of the filament are brought out at the top, either the two leads are at the top or one at the top and other at the cap.

The caps can be of two types"

(*i*) **Bayonet cap.** In this particular cap there are two pins and two contact plates at the top of the cap. The contact plates and the pins are at right angle to each other. For this type of lamp the holder is also of the same type. Such caps are used for lamps of 15 W, 25 W, 40 W, 60 W, 100 W and 200 W etc.

(*ii*) **Screw cap.** The screw cap is used for the lamps of larger power. This cap has got threadings for this type of cap, the holder should be of screw type. Out of the filament leads one is connected with the top and other with the cap.

Edison screw cap is used for the lamps 150 W and 200 W. The Goliath cap is used for more wattage lamps as 300 W, 500 W and 1000 W etc.

(*b*) **Glass container.** The container is made of glass, so that the light rays may come out easily. The size of the container depends upon the power of the lamp. It is so designed that the heat dissipation is proper and the temperature of the envelop does not rise much. The glass of the lamp are of different shape, size and colour. The glass material is either coloured glass or painted glass. These are clear, frosted, milky, blue, green, red, white, yellow umber and shapes are pearls. pygmy, candle, twisted candle, flame and tumbler. Generally different shapes and colour are used for decoration purposes.

(*c*) **Filament.** The filament is made of tungsten. The light energy is radiated from it, when the current is allowed to flow. The filaments are of two types, the straight and coiled coil.

The filament is connected across the two mains, the wires are known as the lead in wires.

(*d*) **Filament supporting wire.** These wires are used to support the filament and are made of a metal which has high melting temperature than the tungsten. For tungsten filament supporting wires are made of molybdenum wire.

(*e*) **Stud and stem rod.** The supporting wires are attached with the stud and stem rod. These are made of solid glass and filament is supported by means of stem rod.

(*f*) **Exhaust tube.** It is used to exhaust the air and the evacuated glass is filled with inert gas like nitrogen or argon. The pressure of the gases within the glass container will decrease the evaporation of the filament. Thus the life is increased. It is sealed after filling the gas.

Discharge through gas or vapour

Let a tube containing two electrodes: the anode and cathode is evacuated and filled with some gas or vapour. The gas particles under normal temperature and pressure say normally are not conductive and will not contribute to the flow of electrons. Now if the electrodes are charged the electrons will start emitting from cathode and will collide with the gas particles, if the collision is violent enough the outer most electron will be detached and the particle say atom will behave like an ion having positive charge. Here we can easily understand that such a collision will result three charged particles, two electrons and one cation (positively charged atom). In this case the reaction will be cumulative and if the charges are continued the electrons while attracted towards cathode will further collide with atoms and again detain the electron and make more cations, the heavy charged particles. These charged particles which coming to cathode will attract the electron and thus help in conducting in the gas column and helping electrons to flow that is the conduction of current. Thus the ions (cations) will strike the cathode, their kinetic energy will be converted into heat energy, raising the temperature which again give rise to additional electron emission. The ions getting one electron from cathode will again be neutral (having equal number of electron and protons) and will take part in the process again and again. With the bombardment of heavy particles the cathode can be hot or even sometimes very hot as in the case of mercury arc rectifier, because of heavy. Hg^+ ions, the temperature rises and a hot spot is maintained on the mercury surface. It again helps in vapourising the mercury.

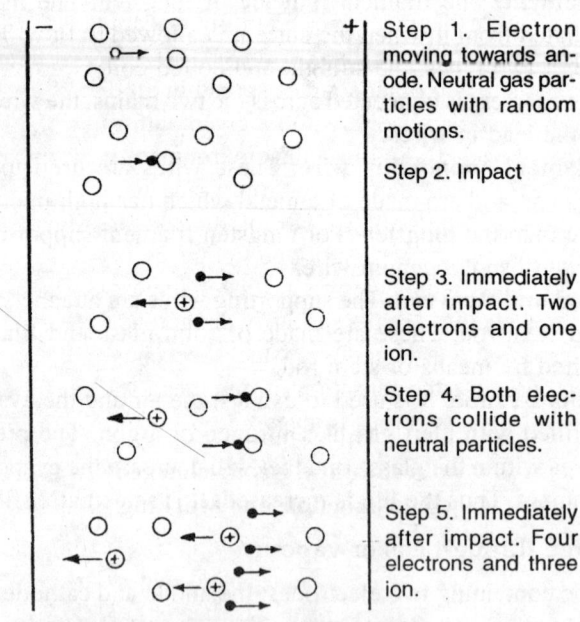

Step 1. Electron moving towards anode. Neutral gas particles with random motions.

Step 2. Impact

Step 3. Immediately after impact. Two electrons and one ion.

Step 4. Both electrons collide with neutral particles.

Step 5. Immediately after impact. Four electrons and three ion.

Fig. 24.9. Stages in the Process of Ionization by Collision

Laser

There are some suitable materials whose energy labeled diagram has the valence band, conduction band and also possesses an another band generally known as trapping band. This trapping band is in between the conduction and valence band and an intermediate band. These materials are suitably used for 'Laser'. The Laser is a short term used for "Light amplification by the stimulated emission of radiation". There is another analogous device generally known as 'maser' *i.e.*, microwave amplification by the stimulated emission of radiation." The laser is actuated by the visible light of very much shorter wavelength. The maser is actuated by very short electromagnetic waves and are generated by means of electrical oscillatory circuit.

An important fact of laser is a suitably shaped piece of material whose energy level diagram contains, as above stated an intermediate band known as trapping band. The atoms of the valence band are excited or raised to conduction band and than fall back with specific radiation. In first-stage from conduction band to intermediate band and in second succession from intermediate band to valence band. In both successions the radiation has its unique characteristics. But they are certain materials in which the intermediate band width is very thin like a line only and there the radiation is of monochromatic. This effect has almost a revolutionary effect on a wide range of fields like medicine, scientific researchers, communications industries and in defence also.

The light from sun and other sources spread in all directions but laser light is very directional and confined to a given path a narrow beam and a parallel beam. Its penetration into the environment is much faster and long, for example a laser beams of 13 mm. wide with travelling of 1.5 km, the spreading will be only 7.5 cm. or so. So it is very useful. The monochromatic emission has light of one frequency only where the general light has so many frequencies on it. It is of single colour very bright. Thus looking directly into laser is seriously dangerous to eyes because of their enormous power. A laser cutter for steel sheets has a power of 10,000 watts but it can produce the sensation of one mega watt per square centimetre which is truly dangerous to eyes.

Q. How the light is produced in the gas discharged lamps? Describe with neat sketch the construction and working of the high pressure mercury vapour lamp.

Ans. The lamps in which the light is produced due to the flow of electrons through the gas are known as the gas discharge lamps.

Ionization of gases. In a gas discharge lamp the gas must be ionized to produce the light. In normal condition the gas is an insulator and current cannot flow through it. The atoms are neutral. A gas can be ionized by decreasing the pressure and applying the voltage. The electron will flow, the movement can even be accelerated by the potential gradient and thus high velocity and kinetic energy is possessed. The electrons collide with neutral atom of gas or vapours. This collision will excite the atom. The excited state is unstable and after 10^{-8} second it comes to original state. Thus the acquired energy is released in the form of electro-magnetic radiation which are in other words the light waves.

It is the basic principle of the discharge lamps. The colour of the glow depends upon the gas or vapours. The glow is red with neon, bluish with mercury, yellow with sodium and whitish with helium and white with carbon dioxide.

High pressure mercury vapour lamp (H.P.M.V. Lamp.)

It is a gas discharge lamp. It has high operating pressure with mercury vapours, so it is called as the high pressure mercury vapour lamp.

Construction. The Fig. 24.10 shows the construction of a H.P.M.V. lamp. It mainly consists of a discharge tube enclosed in an outer bulb of ordinary glass. The inner tube is made of hard glass or quartz. It contains two main electrodes at both sides of the tube, the main electrodes are made of tungsten. These are oxide coated to emit the elec-trons. There is an auxiliary electrode near the main electrode. This electrode is connected with the high resistance (50 kΩ) as shown. The inner tube con-tains the argon and a certain amount of mercury. It is placed inside the outer tube. The space between both tubes is partially or completely evacuated to prevent heat loss from the inner tube by convection.

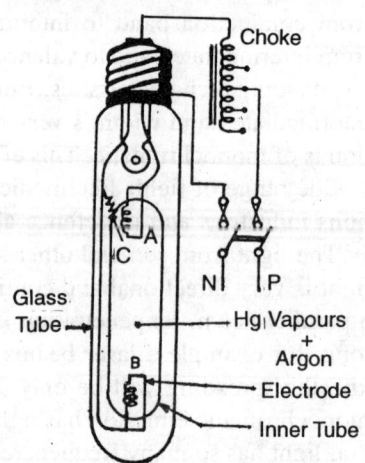

Fig. 24.10. Mercury vapour lamp.

The two lead from the electrodes are connected to a bayonet or screw type cap. A choke is connected in series with the lamp to limit the current. The choke being inductive lowers the power factor, so a capacitor is connected in parallel to improve the power factor.

Working. In the normal condition the gas is cool and mercury is in liquid form, so when the switch is on, the discharge between the main and auxiliary electrode takes place through the organ gas. The gas is heated up due to the discharge and becomes conductive. Now the mer-cury is also vaporised due to heating of argon gas. Moreover the mercury vapours behaves like a conductor and discharge takes place between the main electrodes. The current will not flow through the auxiliary electrode because of the high resistance (50 kΩ). Now the complete column will be heated up. The mercury vapours and gas increases the internal pressure of the tube. This lamp now works at a high pressure (2 to 3 atmospheric pressure) because of mercury vapours the light is whitish in colour.

Function of choke. The main function of the choke is to control the flow of current in the lamp. When the gas is cool its resistance is high and a low current flows through the gas. When it is heated up the resistance decreases and more current will flow. This more current may further increase but as choke is connected in series so a voltage is droped in the choke and the lamp actually works at low voltage and current is regulated to a safe value.

Power of the lamp. The lamps are available in different wattage, 80 W, 125 W, 250 W and 400 W etc. The shape is either pearl or tumbler.

Uses. The lamps are used for decoration purpose in big halls and for street lighting also.

General instructions:

(*i*) The working temperature of the lamp is 600°C.

(*ii*) It can be used only in vertical position with cap up.

(*iii*) It can be used only on A.C.

(*iv*) It requires sometimes to come to full light.

(*v*) It must be used with a choke of suitable wattage.

(*vi*) It has whitest and smooth light.

(*vii*) The correct identification of colour is not possible in this light.

Q. Draw the diagrams and give brief idea about the other types of H.P.M.V. lamps.

Ans. The high pressure mercury vapour lamps are of the following types which are shown in Fig. 24.11.

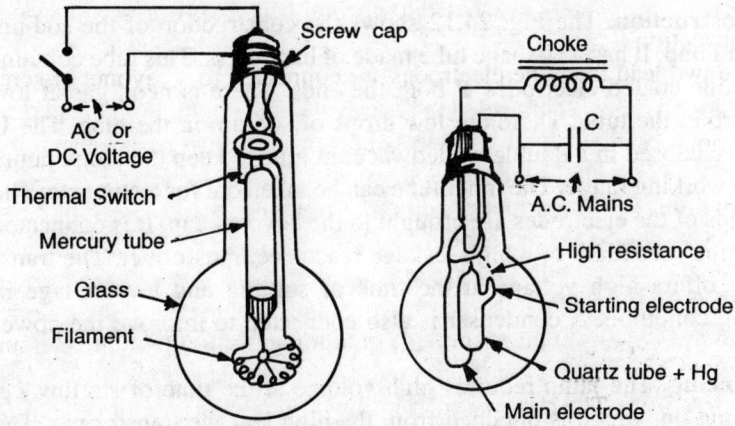

Fig. 24.11. MAT and M.B. type mercury vapour lamps.

(*i*) **M.A. type.** Mercury vapour lamp with auxiliary electrode these lamps are used in vertical position with cap up (Fig. 24.10).

(*ii*) **M.A.T. type.** Mercury vapour lamp with auxiliary electrode and tungsten filament. It can be used on A.C. and D.C. both. There is no choke for this type of lamp. The tungsten filament is in series with the electrode and works as the blast. The light is the combination of both M.A. type and a filament bulb type. The power factor is 0.95 hence capacitor is not used.

M.B. Type. Mercury vapour lamp with bayonet cap these are having very high pressure up to ten atmospheric pressure. These are operated as the M.A. type lamps except that the starting resistance in series with starting electrode is large and outer bulb is made of quartz, in order to withstand the high temperature. It can be used in any position.

Q. Describe the construction and working of the sodium vapour lamp.
(N.C.V.T. 1973, 76)

Ans. It is also the gas discharge lamp in which sodium is used, so this is called as the sodium vapour lamp.

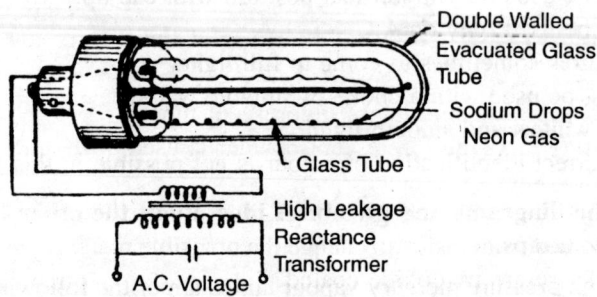

Fig. 24.12. Sodium vapour lamp.

Construction. The Fig. 24.12 shows the construction of the sodium vapour lamp. It has a U-shape tube made of hard glass. This tube contains two oxide coated electrodes at both the ends. There is neon gas at low pressure in the tube. There are few drops of sodium in the tube. The U tube is enclosed in a double walled vacuum tube to keep the temperature within working range. The inner tube can be taken out for testing etc. The two ends of the electrodes are brought to the bayonet cap. It is connected across the secondary of a high leakage reactance transformer. The transformer offers high voltage at the time of starting and low voltage in running condition. A condenser is also connected to improve the power factor.

Working. The lamp requires high voltage at the time of starting *i.e.* switching on, which is obtained from the high leakage transformer. The discharge takes place through the neon gas and is of pink colour. The gas is heated up and sodium is vaporised. A yellowish light is produced when sodium vapours also takes part. Now the resistance of the lamp decreases and the current increases but the voltage is dropped and controlled by the high leakage transformer. The lamp works at low voltage and the working temperature is about 300°C.

General informations:

(i) The lamp has yellowish light.

(ii) It takes about ten minutes to give full light.

(iii) It must be used in horizontal position otherwise the sodium drops may come to the bottom of the tube.

(iv) The transformer must be high leakage transformer and its power should be according to the lamp.

(v) In its light the colour cannot be judged properly.

(vi) It can work on A.C. only.

Uses. The lamps are available in different wattage 40 W, 60 W etc. and are used for decoration purpose and street lighting purpose etc.

Q. Explain the construction and working of the fluorescent tube.
(N.C.V.T. 1965, 72, 76, 80)

Ans. It is a gas discharge lamp. Fluorescent tubes are widely used for lighting purpose. As the fluorescent powder is used in the tube, so it is known as the fluorescent tube.

Construction. As shown in the Fig. 24.13 it has a long tube which is phospher coated from inside to change the unvisible radiation into visible radiation. There are two oxide coated tungsten filament to emit the electrons. The two ends of the filaments are connected to the pins. Each side has got two pins to be connected electrically. The tube contains organ gas at 2.5 mm pressure of mercury and a small amount of mercury.

Fig. 24.13. Fluorescent tube.

Different accessories. There are the following other accessories.

(a) **Base plate.** Generally it is made of thin mild-steel sheet. It has two holders in which the tube is held in position and electrical connections are made. Sometimes it has a transparent cover.

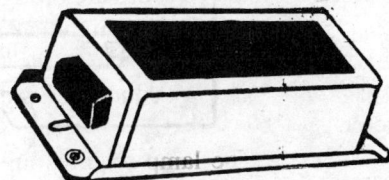
Fig. 24.14. Choke.

(*b*) **Choke.** As shown in Fig. 24.14, it is an electromagnetic device. It is connected in series. It is wound with super enamelled copper or aluminium wire having a large number of turns on the cores. It is used for the following purposes.

(*i*) It gives a surge voltage at the time of starting *i.e.* about 1000 V.

(*ii*) In running condition, it drops the voltage.

(*iii*) It limits the current.

The power of the choke and the tube must be same.

(*c*) **Starter.** These are automatic starters and are of the following types.

(*i*) Glow type starter (*ii*) Thermal type starter.

(*i*) *Glow type starter.* These are widely used at present. It has two bimetallic strips in the glass tube. The tube contains helium at low pressure. A condenser is used to minimise the sparking. At the time of starting full line voltage puncture the gas and a discharge is produced in the helium. As the gas is heated up, the bimetallic strips are also heated up and bends in a particular direction short circuiting the contacts. Now the current is flowing through the strips and not through gas, so gas is cooled and the strips open. Thus the make and break is achieved. In normal working of the tube, the voltage across the starter is less, so it does not operate.

(*ii*) *Thermal type starter.* In its simple construction, the bimetallic strips and a small heater is provided to heat the gas. In normal condition the contacts are closed and opens when the supply is on the heating element is in the circuit so it heat the gas and contacts remain opened. These are not used at present.

Condenser. Sometimes a condenser is also provided with the circuit to improve the power factor.

Working. In normal condition the terminals of the starter are opened as shown in connection diagram 24.15. When the switch is on, the phase through choke and element reaches to the starter and the neutral through element to the other terminal of the starter. The electrical stresses developed because of this voltage, produces the discharge through the helium. The gas is heated up and the bimetallic strips are short circuited. The

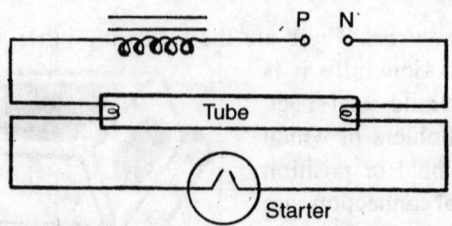

Fig. 24.14. Connections of fluorescent tube.

circuit of the current is completed through choke, element, starter and other element. The current heats the element and these are ready to emit the electrons. In the mean time the gas in starter is cooled, and bimetallic strips are opened. Now the choke because of current interruption produces self induced voltage about 1000 V, which acts across the elements and produces the discharge in the tube. The fluorescent power is used to change the ultra violet light into the visible radiation *i.e.,* light and the powder also helps in spreading the light uniform.

In running condition the choke drops the voltage and tube gets low voltage (110 V), generally half the line voltage.

Uses. These are commonly used in domestic lighting, street lighting, industrial lighting in medical apparatus and decoration etc.

Q. What are the advantages of the fluorescent tube, and also mention the sizes of the tube?

Ans. Following are the advantages:

(*i*) The light efficiency is very good 40 lumen/watt.

(*ii*) The life of the lamp is about 4000 hrs.

(*iii*) It can be used on A.C. and D.C. both.

(*iv*) Its light is uniform.

(*v*) It has pleasant light.

(*vi*) It can be used in any position.

(*vii*) It consumes 50% less energy than the incandescent lamp of the similar wattage.

(*viii*) It comes to full light earlier than the other type of discharge lamps.

Sizes. These are available in different wattage diameter and lengths.

Table 24.1. Tube wattage, length and diameter

S.No.	Wattage	Length	Diameter	Resistance in ohm
1.	15 W	12", 18"	$\frac{3}{4}''$, $1\frac{1}{2}''$	660 Ω
2.	20 W	24"	1", $1\frac{1}{2}''$	528 Ω
3.	30 W	36"	1"	380 Ω
4.	40 W	48"	$1\frac{1}{2}''$	297 Ω
5.	80 W	60"	$1\frac{1}{2}''$	147 Ω
6.	125 W	96"	$1\frac{1}{2}''$	–

Table 24.2. Fluorescent coating and colour

S.No.	Coating	Colour
1.	Calcium tungstate	Blue
2.	Magnesium tungstate	Blue white
3.	Cadmium borate	Pink
4.	Cadmium silicate	Yellow pink
5.	Zinc silicate	Green
6.	Zinc beryllium silicate	Yellow white

Table 24.3. Lumens output and colour of fluorescent lamp

S.No.	Gas or vapour	Colour	Theoretical lumens/watt	Actual lumen/watt
1.	Sodium vapour	Yellow	475	40-50
2.	Mercury vapour low pressure	Bluish-green	248	15-20
3.	Mercury vapour high pressure	Bluish white	298	20-30
4.	Neon	Red	198	15-40

Q. Draw the connection diagrams of:

(a) **Fluorescent tube with thermal type starter,**

(b) **Fluorescent tube on D.C. voltage,**

(c) **Two fluorescent tube of 20 W on one choke (40 W) and**

(d) **Two fluorescent tube (40 W) operating in parallel.**

Ans.

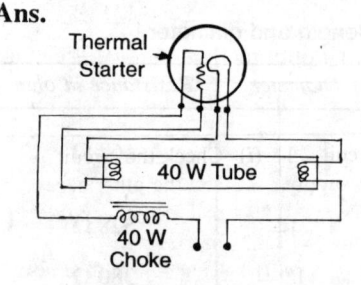

(a) Tube with thermal type starter

(b) Tube on D.C.

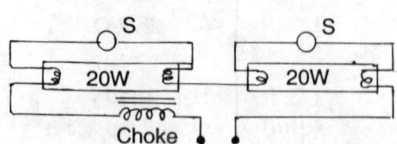

(c) 20 W tube and 40 W choke

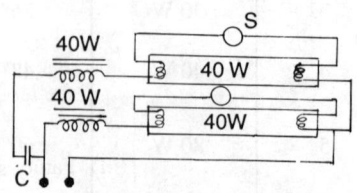

(d) 40 W tube parallel.

Fig. 24.15.

Q. Compare the fluorescent tube and the filament lamps.

(*N.C.V.T. 1966*)

Ans. The comparisons between the filament lamp and the fluorescent tube are as follows:

S.No.	Tungsten filament lamp	Fluorescent tube
1.	Its light is close to the natural light.	Its light is not close to the natural light.
2.	Actual colour can be judged.	Actual colour cannot be judged.
3.	Initial cost is low.	Initial cost is high.
4.	Life is about 1000 hrs.	Life is about 4000 hrs.
5.	Maintenance cost is more.	Maintenance cost is low.
6.	Brightness is more.	Its light is cool and pleasant.
7.	Light output is reduced with time.	It is also but very less.
8.	Heat radiation loss is there.	As the temperature is less so less radiation losses.
9.	Less lumen output watt.	More lumen output/watt it is 40 lumen/watt.
10.	Lumen efficiency is poor because of coloured glass etc. in case of coloured light.	It is not effected because the colour of light depends upon the gas.
11.	No stroboscopic effect.	It has the stroboscopic effect

Q. List out the possible faults, causes and their remedies of the fluorescent tube.

Ans. These may be following different faults in fluorescent tubes:

	Fault	Causes	Remedies
1.	Tube appear quite dead when switch is on.	(*i*) Break in the circuit or failure of the supply. (*ii*) Broken filament. (*iii*) Starter fail to start.	(*i*) Check the circuit and the other accessories and also check the supply mains. (*ii*) Change the accessory if necessary.
2.	Tube flickers on and off.	(*i*) The useful life of the tube is over. (*ii*) Low supply voltage. (*iii*) Faulty starter.	(*i*) Check and replace if necessary. (*ii*) Check the voltage. (*iii*) Check and replace if necessary.

Contd...

	Fault	*Causes*	*Remedies*
3.	Light appear moving in spiral shape.	It usually occurs with a new tube and it is automatically removed and sometimes.	If not rectified after sometimes then replace the tube.
4.	Slow starting.	Low line voltage, choke defective.	Check voltage, check tube and replace if necessary.
5.	Slow flickering	Useful life is over.	Change the tube.
6.	Filament glows but tube does not strike up.	Starter contacts short circuited.	Check and replace the starter.
7.	Tube filament burns out when switch is on.	The choke is defective.	Check and replace if necessary.
8.	Short life of the tube.	Excess voltage.	Check voltage and the choke, replace if necessary.

Q. Describe with the neat sketch the neon lamps.

Ans. The neon lamps are used for indication purpose.

Construction. The neon lamp has got two electrodes, sometimes these are flat or sometimes spiral. The electrodes are placed closely together in a glass tube or bulb. The glass contains neon gas with small amount of Helium at low pressure. A limiting resistance is placed in series with the electrode as shown in Figs. 24.17(*a*) and (*b*). The resistance may be inside or outside the cap. It can work on low voltage also say 110-150 V.

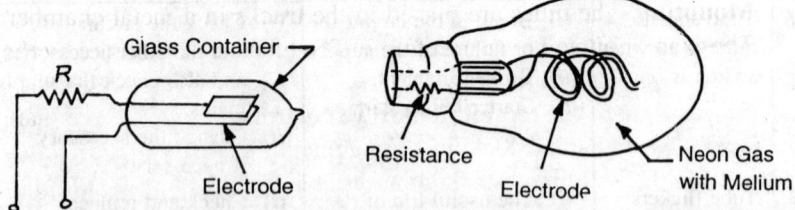

Fig. 24.17. Neon lamps.

Working. When the supply is given the gas between the electrodes is ionized and produces a discharge through neon gas. In A.C. both the electrodes glow equally but on D.C. only the negative electrode glows, hence the negative electrode is made somewhat larger. The series resistance of 2 or 3 kΩ limits the current. The light is of pink colour. The efficiency is low and power consumption is also low 0.5 W to 5 W only.

Uses. These are commonly used as the indicating, lamps and night lamps also.

Q. Describe briefly the construction and working of the neon sign tubes.

Ans. The neon sign tubes are used for advertising and decoration purpose also. Figure 24.17 shows the various parts of the neon signs.

(a) **Tubes.** These tubes are made of glass. The diameter is 10 to 20 mm. These can be given any shape according to the letter or object. The length of the tubes varies from 2 m to 4 m.

(b) **Gas and colour.** The neon tubes are filled with some conducting gas or the mixture of gases for different colours, for example, for red or orange colour neon gas, for blue colour-mercury in neon argon gas, for green colour-yellow glass and mixture of neon and mercury, for yellow colour-yellow glass and helium.

(c) **Electrodes.** The electrodes are provided on both ends of each letter. The electrodes may be of iron, nickle or copper in cylindrical shape. The electrodes are joined by the fine nickel wire.

(d) **Voltage and current.** High voltage is required to start the tubes about 15% more than the operating voltage. But the voltage depends upon the length of the tube required. The current depends upon the diameter of the tube, so far 12 mm, 15 mm and 20 mm diameter tubes, the current is 25 mA, 35 mA and 60 mA respectively.

The high voltage is obtained from the transformer. The ends of the tube are connected with the transformer by means of H.T. cable and from letter to letter with Ni-wires.

A condenser is used to improve the power factor.

(e) **Mounting.** The tubes are placed in the trucks in a metal chamber. These are mounted over iron frame. The tubes are supported by the clips of phospher bronze.

(f) **Controlling of the circuit.** It must be controlled properly. The main switch is with inter-locking arrangement. There is a fire switch outside the building to make it accessible to operate in case of fire or other danger.

(g) **Tubes on D.C.** For that purpose the inverters are used to convert D.C. into A.C.

(h) **Flashing sign.** Sometimes a flasher is also used for flashing purpose. It is connected in primary circuit.

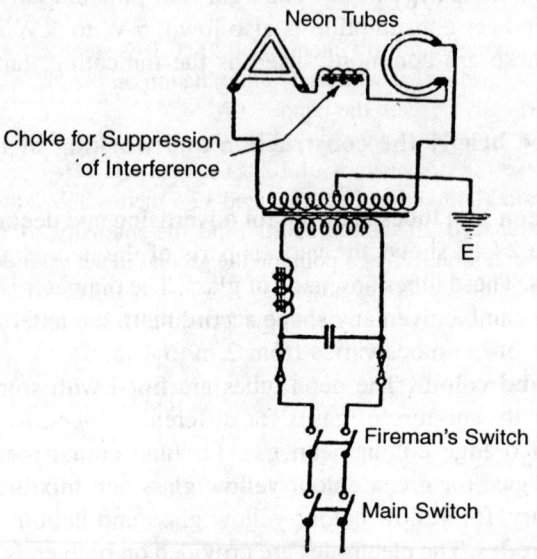

Fig. 24.17. Neon sign tube.

(*i*) **General precaution.** In case of fault the tube must not be tested on live line. First cut off the main and then check.

REVIEW QUESTIONS

1. What do you understand by the illumination? How does it effect the workman, production and quality of the product?
2. State the laws of illumination.
3. Define, candle power, lux, luminous flux, luminous intensity, cosine law, M.H.C.P. and brightness.
4. (*a*) Define, depreciation factor, reflection factor, and coefficient of utilization.
   (*b*) What are the kinds of lighting scheme?
5. What are the types of lamps? Describe carbon arc lamp?
6. Describe the different part of the incandescent lamp.
7. Describe the working and construction of the fluorescent tube.
8. What is the function of choke? How it is useful for the discharge lamps?
9. Describe the construction and working of high pressure mercury vapour (HPMV) lamp.
10. How the ionization in sodium vapour lamp is achieved? Describe the construction and working of the sodium vapour lamp.
11. What are the different types of the discharge lamp? Describe their construction and working of the neon signs.
12. Compare the incandescent lamp and discharge lamps.

NUMERICAL PROBLEMS

1. A lamp having a uniform intensity of 200 C.P. is suspended 7 metre above the street level. What will be the illumination on the ground
   (a) Vertically beneath the lamp.
   (b) 7 metre away from it?   **[Ans. 4.08 lux, 1.42 lux)]**

2. Two powerful street lamps of 1000 Candela and 800 Candela (assumed uniform in all directions) are mounted 12.5 metres above the ground level and are spaced 25 metres apart. Find the intensity of the horizontal illumination produced at a point on the ground in between the lamps posts and just below the lamps posts.   (*AMIE, Sec., B, U.E. 1975*)
   [Ans. 4.07 lux, 6.857 lux, 5.69 lux]

25

Converting Machines

Q. What are the different purposes and methods of converting of A.C. into D.C.?

Ans. At present A.C. energy is generated in large quantity. In spite of the fact that the A.C. system has proved to be superior in generation, transmission, distribution and utilization, even then there are certain applications where D.C. is essential, superior and effective. These are as follows:

(a) **Traction purpose.** The D.C. series motor is more suitable for traction purpose then any other type of motors. For example in trains, trams and trollies the D.C. motors are widely used.

(b) **Speed control.** The speed of D.C. motors can be controlled easily and precisely.

(c) **Electrochemical works.** D.C. is only used for electro-chemical works such as electroplating, electrolysis and battery charging etc.

(d) **Telephone and telegraphic purposes.** D.C. is used for telephone and telegraphic purposes in telephone industries etc.

(e) **Relays operation.** D.C. is more precisely used for the operations of relays in control panels, switch gears in sub-station, power stations and so many other places.

(f) **Arc lamps.** D.C. is more suitable for the arc lamps, search lights, cinema projectors and arc welding etc.

Because D.C. is not generated in large quantity and is essential for certain applications, so the A.C. is converted into D.C. by the following methods.:

(i) Motor generator set.

(ii) Motor converters.

(iii) Rotary converters.

(*iv*) Mercury arc rectifier.

(*v*) Metal and other rectifiers.

Q. Describe the Motor Generator set method of conversion. Also state the merits and demerits of this system.

Ans. In this system of conversion, the A.C. motor is coupled with D.C. generator. A.C. supply is given to the motor which rotates the armature of the D.C. generator, where D.C. is generated. The D.C. generated voltage can be controlled by means of shunt field regulator.

The motor generator set is the most flexible type of converting system. This system has got the following merits and demerits.

Merits:

(*i*) Easy construction,

(*ii*) Voltage can be controlled easily, and

(*iii*) Easy starting and controlling.

Demerits. Following are the demerits:

(*i*) Two machines are required,

(*ii*) More maintenance cost,

(*iii*) More space and more cost,

(*iv*) Less efficiency,

(*v*) It cannot be used to covert D.C. into A.C., and

(*vi*) High D.C. output cannot be controlled.

Uses. This system is not used for more output. It is used to obtain D.C. voltage of less output such as for electroplating, electrolysis, battery charging etc.

Q. What do you understand by the motor converter? What will be the speed of a converter if the number of poles of induction motor and converter are given?

Ans. It mainly consists of an induction motor which is coupled both electrically and mechanically with a rotary converter. The wound rotor of the motor through the shaft to the tappings on A.C. side of the rotory converter the tappings are not brought out to the sliprings. Thus it runs as the cascade for speed and a part of the energy input to the induction motor is converted through the shaft (mechanically) and the other portion as an electrical energy. It is not widely used at present.

Let there are P_1 = Number of poles of induction motor.

P_2 = Number of poles of the converter.

f_1 = Supply frequency to motor.

$$f_2 = \text{Converter frequency.}$$

$$N_2 = \text{Speed of the set in r.p.m.}$$

Obviously the synchronous speed

$$= \frac{120 \times \text{Frequency}}{\text{Number of poles}}$$

So

$$f_1 = \frac{N_1 P_1}{120}$$

$$f_2 = \frac{N \cdot P_2}{120}$$

Here the N, the converter speed. The converter frequency is the frequency of the rotor i.e.,

$$f_2 = s \times f_1$$

$$= \left(\frac{N_1 - N}{N_1} \right) \times f_1 \qquad \left(\because \text{slip} = \frac{N_1 - N}{N_1} \right)$$

Now substituting f_2 in the above expression

$$\frac{N \cdot P_2}{120} = \frac{N_1 \cdot P_1}{120} \times \left(\frac{N_1 - N}{N_1} \right)$$

$$N \cdot P_2 = P_1 \cdot (N_1 - N) = P_1 \cdot N_1 - P_1 \cdot N$$

or

$$N \cdot (P_1 + P_2) = P_1 \cdot N_1$$

$$N = \frac{P_1 \cdot N_1}{P_1 + P_2}$$

or

$$= \frac{120 \cdot f_1}{(P_1 + P_2)}$$

So the net speed of the combined set is

$$= \left(\frac{120 \cdot f_1}{P_1 + P_2} \right) \text{ r.p.m.}$$

Q. Describe with neat sketch, the construction and working of a rotory converter.

Ans. The rotory converter is a single machine which converts the A.C. into D.C. and D.C. into A.C.

Construction. It is very well known that inherently A.C. is produced in the armature of a generator. This A.C. is converted into D.C. by means of commutator, whereas alternating current is obtained from the sliprings, so if A.C. is supplied to the armature having the commutator which is rotating, D.C. will be obtained at the commutator.

The rotory converter has got the following parts:

(*i*) *Armature.* The armature of rotory converter is the same as that of D.C. machine except the fact that thewound armature has oneside the commutator and on the other side of **its slipring** for A.C. for D.C. The leads of the winding are brought to the commutator and on the other side are connected to the sliprings as shown in Fig. 25.1. The cores are made of laminations of silicone steel.

(*ii*) *Stator.* The stator has a number of salient poles having wound for a certain number of poles. The field winding can be of shunt or compound type damper winding is also placed in the pole shoes of the converter. The interpoles are also housed to improve the commutation and armature reaction.

Tappings to sliprings. The number of tappings depends upon the type of winding, number of poles and number of phases. If the armature is wave wound each slipring will have one tapping. The number of conductors between any two sliprings will be equal. In case of lap winding, each slipring has the number of tappings equal to the number of pair of poles. So for a three phase four-pole lap wound armature each slipring will have two tappings and six tappings are brought out from the winding for three slipring.

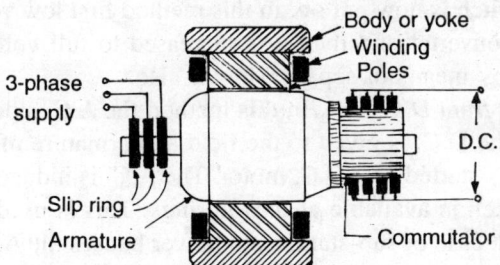

Fig. 25.1. Rotory converter.

Voltage and current ratio. There is a fixed ratio between the A.C. and D.C. values. The table 25.1 shows the various voltages and currents of A.C. and D.C.

Table 25.1. Voltage and Current Values

S.No.	D.C.	Single ϕ	Two phase	Three phase	Six phase	Twelve phase
1.	1 V	0.707 V	0.707 V	0.612 V	0.354 V	0.183 V
2.	1 A	1.414 A	0.707 A	0.543 A	0.472 A	0.236 A

Generally for *m* phase converter

$$E_p = \frac{1}{\sqrt{2}}, E_{d.c.} \sin\frac{\pi}{m}$$

where   *m* = 2 for single phase, 4 for two phases.

                = 3 for three phases and 6 for six phases.

Starting of rotory converter. The rotory converter convert either A.C. into D.C. or D.C. into A.C., so it can be started from either sides.

(*a*) *Starting from A.C. side.* The very common method of starting from A.C. side is by the *taps on transformer.* In this method at the time of starting the A.C. side switch is left open. The shunt field current is disconnected and a low voltage is applied to the sliprings of the converter. Due to this low voltage, a rotating magnetic field of low strength will be produced in the .armature. The e.m.f. will be induced in the damper winding and the converter is started as the inverted squirrel cage induction motor. The speed will be approximately equal to the synchronous speed. The D.C. voltage is given to the field winding. The stator and rotor poles will be inter-locked and remains in the synchronism. Thus the armature will run at synchronous speed and D.C. will be available at the commutator. The D.C. switch is now put on. In this method first low voltage is given to the converter and then it is increased to full voltage. It can be started by means of separate motor also.

(*b*) *Starting from D.C. side.* In this method the A.C. side switch is left open. The D.C. is given to the field and armature of the converter. Now it is started as a D.C. motor. The A.C. is induced in the armature which is available at the sliprings. This is used for load purposes. In case of sub-stations on power houses the A.C. is synchronized with the busbar voltage and then D.C. switch is put on.

Advantages of rotory converter. Following are the advantages:

(*i*) Only one machine is required to convert A.C. into D.C. and D.C. into A.C.

(*ii*) It can convert A.C. into D.C. or D.C. into A.C.

(*iii*) Good efficiency.

(*iv*) Reliable machine.

(*v*) Less space is required.

(*vi*) Less heating effect.

(*vii*) Can be build in large capacity.

(*viii*) Maintenance cost is less.

(*ix*) If the rotor is rotated by some external means A.C. and D.C. can be obtained at the same time.

Q. What do you understand by the rectifier and name their types? Why these are so popular?

Ans. The rectifier is a static device which converts A.C. into D.C. The rectifier only permits the current to flow in one direction only and offer a high resistance in reverse direction. These are very popular because of the following reasons:

(*i*) These are compact in size.

(*ii*) These are cheap.

(*iii*) These are static hence no rotating parts. So it requires less maintenance.

(*iv*) No noise during operation.

(*v*) Higher efficiency.

Type C: The rectifiers are of the following types:

(*i*) Copper oxide rectifier.

(*ii*) Selenium rectifier.

(*iii*) Electrolytic rectifier.

(*iv*) Tunger bulb rectifier.

(*v*) Mercury arc rectifier.

Q. Explain the different classification of the single phase rectifiers.

Ans. The single phase rectifiers are classified as under:

(*a*) Half wave rectifier.

(*b*) Full wave rectifier.

The full wave rectifier may have either

(*i*) Centre tap circuit.　　　　(*ii*) Bridge circuit.

(*a*) **Half wave rectifier.** A rectifier which rectifies only one half of each A.C. cycle is called half wave rectifier.

In this rectifier the current flows during positive half cycle: During negative cycle no current is conducted hence no current is rectified during the negative cycle. Thus only positive cycle is rectified, so it is called half wave rectifier as shown in Fig. 25.2.

(*b*) **Full wave rectifier.** A rectifier which rectifies both the half cycles of A.C. input is known as full wave rectifier. In this case the con

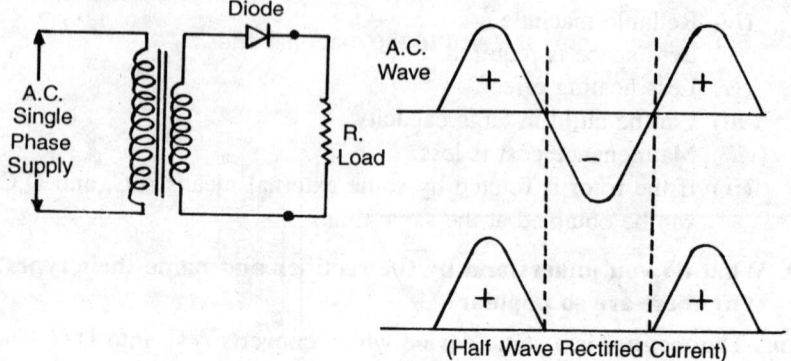

Fig. 25.2. Half wave rectifier.

duction appears for positive and negative halves of the A.C. input. So *pulsating D.C.* is obtained and this rectification is for the entire full wave. It can be by the following two types:

(*i*) Centre tap circuit   (*ii*) Bridge circuit

(*i*) *Centre tape circuit.* Figure 25.3. (*a*) shows a circuit with centre tap provision. Two diodes are used in this connection and a centre tapping is brought out for the load. During half cycle when the upper end is positive, the current start flowing from A and terminal 'D' behaves like positive for the external load 'R', but the 'B' is negative, so there will not be any conduction from this end. During next half cycle the end 'B'is positive, so the current will flow from 'B' to 'D' and during this period also 'D' is positive. But there will not be any conduction from 'A' to 'D'. Thus during the positive and negative half cycles the terminal 'D' behaves like a positive terminal and 'C' as negative terminal.

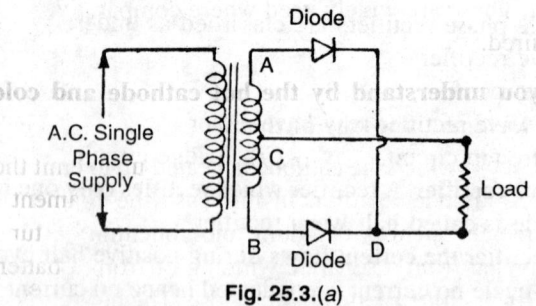

Fig. 25.3.(*a*)

(*ii*) *Bridge type full wave rectifier.* It is an another type of full wave rectification. There are four semi-conductor diodes connected in the shape of a bridge. The connection diagram is shown in Fig.

25.4. A.C. supply is given to the diagonal ends. The D.C. is taken from the remaining two ends.

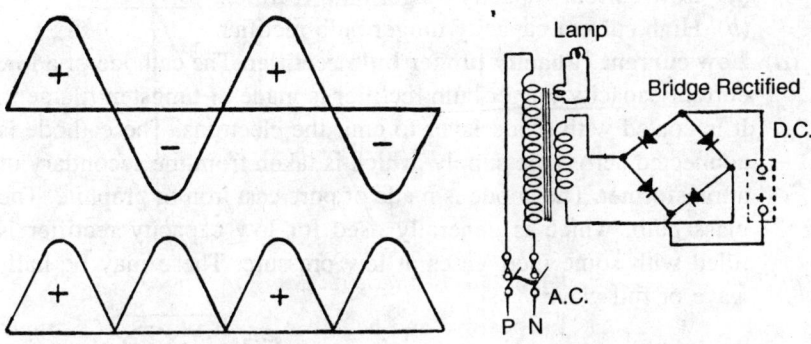

Fig. 25.3.(*b*) Rectified A.C.   **Fig. 25.4.** Bridge rectifier.

Here let terminal 'A' is positive, during the half cycle the current will flow from 'A' to diode '2' and then to the terminal 'C' which behaves like a positive terminal and current returns from 'D' to 'A' terminal *via* diode '1'. During next half cycle the terminal 'B' is positive and the current is rectified from 'B' to diode '3' and then to 'C'. Thus, even in positive or negative cycles the terminal 'C' is always positive and 'D' as negative. Thus the complete cycle is rectified into pulsating D.C. The rectified wave diagram is as shown in Fig. 25.3(*b*).

Advantages. Following are the advantages of bridge rectifier:

(*i*) The output voltage is double than that of the centre tapped secondary winding.

(*ii*) It does not require tap on the secondary winding.

Disadvantages. It is requires four diodes for the circuit.

Application. These are widely used where comparatively high output voltage is required.

Q. What do you understand by the hot cathode and cold cathode rectifiers? Describe any one of them.

Ans. The rectifiers in which the cathode is heated up to emit the electrons is known as hot cathode rectifier. In this case the filament is of high melting point metals such as tantalum, molybdenum or tungsten. The cathode is energised from a separate winding or from a battery or other D.C. source of 4 to 6 volts for heating purpose. The heated filament emits electrons due to thermionic emission. On the other side, if the cathode is not heated up for example the mercury arc rectifier is the example of a cold cathode rectifier. In fact the heating of the cathode is required for the emission of the electrons is localised at the cathode spot and is derived from the bombardment of the cathode by the positive ions.

The tunger bulb rectifier is a hot cathode rectifier. These are of two types:

   (*a*) Low current capacity tunger bulb rectifier.

   (*b*) High current capacity tunger bulb rectifier.

(*a*) **Low current capacity tunger bulb rectifier.** The cathode of a low current capacity tunger bulb rectifier is made of tungsten filament. It is coated with oxide layer to emit the electrons. The cathode is connected across the supply, which is taken from the secondary of a transformer. The anode is made of pure cast iron or graphite. The glass bulb, which is generally used for low capacity rectifier is filled with some inert gases at low pressure. These may be half-wave or full-wave.

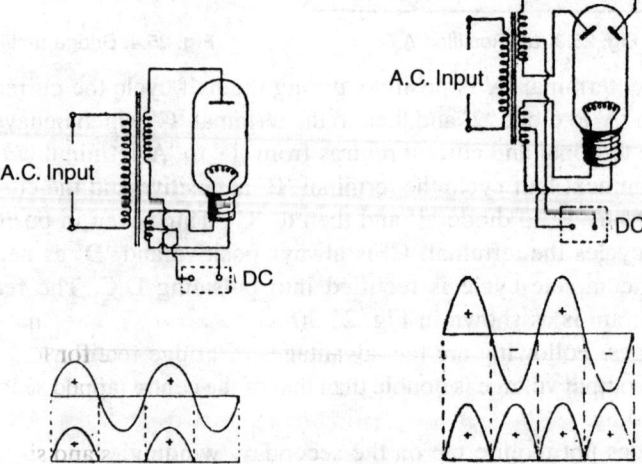

Fig. 25.5. Half-ware rectification.   Fig. 25.6. Full-wave rectification.

(*b*) **High current capacity tunger bulb rectifier.** It is known as tunger bulb mercury vapour type rectifier. The current capacity is up to 100 A. The anode is in the shape of metal plate. The cathode is oxide coated to emit the electrons. This is connected across the supply from the secondary of a transformer. The glass container contains no air but some mercury is placed. When mercury is heated up it is vaporised and the conducting path is obtained from cathode to anode. There vapours decreases the more resistance path.

When cathode is connected across the supply, it is heated up to emit the electrons. These free electrons are attracted towards a plate which is at higher potential, from where the electrons flows through the secondary winding of the transformer and through D.C. circuit back to the cathode. The voltage drop inside the rectifier is about 10-15 volts. During the half cycle one anode collects the electrons and passes through the circuit, during the next cycle the circuit is completed from the second anode. Thus the full-wave is rectified.

Uses. It is used for battery charging and cinema projectors etc. For obtaining smooth D.C. a *filter circuit* is used.

Q. Describe the construction, working and uses of the single phase mercury arc rectifier.

Ans. In mercury arc rectifier the cathode is a pool of mercury. The mercury will emit the electrons, if it is heated up. The electrons will move towards a plate which is at higher potential. These electrons will collide with the ions of mercury vapour. The free electrons of these ions are attracted towards the anode. Now the heavy positive ions bombard the cathode surface and produces a white spot from where the electrons are emitted. The hot spot will be maintained as long as the bombardment is continued.

These are of two types:

(*i*) Single phase mercury arc rectifier.

(*ii*) Three phase mercury arc rectifier.

Construction. The container is made of glass up to 500 kW capacity and for more than 500 kW, it is made of steel tank. The bulb is highly evacuated of other gases so that when it is finally sealed it contains low pressure mercury vapour obtained from the mercury pool cathode. There are a number of main electrodes made of either graphite for low capacity and for high current capacity this is made of cylindrical cast iron for rectification. In single phase rectifier there are two anodes, for three phase there are three anodes, for six phase there are six anodes and similarly for twelve phase these are twelve in numbers. In the bottom of the container there is a pool of mercury. The mercury is vaporised by some means, then the electrons can pass from the cathode to anode. Thus the *uni-directional current* is obtained. For steady and smooth D.C. rectification the multi-anode and filter circuit is used. A transformer with different secondary windings for indication, anodes, exciting circuit etc. is used as shown in Fig. 25.7.

Excitation circuit. The purpose of this circuit is to start the rectifier. The anodes maintain the hot spot even at no load. This circuit is energised by the separate secondary winding on the transformer. It contains an ignition switch which makes and brakes the contacts under soft iron strip generally named as ignition electrode as shown in Fig. 25.7 by '*d*'.

Operation. To start the rectifier the switch in put on. The solenoid switch helps in making and braking the contact with the upper layer of

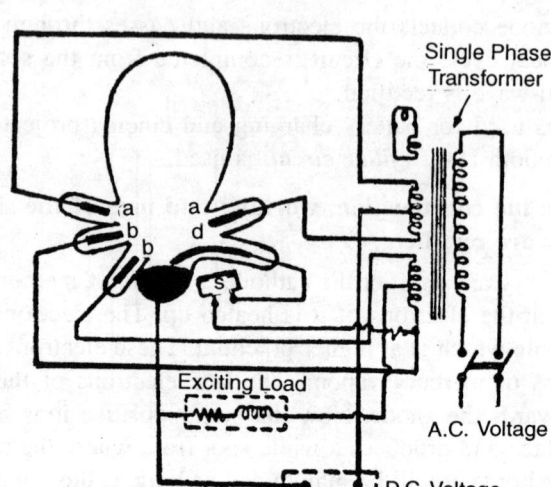

Fig. 25.7. Single phase mercury arc rectifier.

mercury pool. So a sparking results and the mercury is heated up. It is a good thermal conductor so it is vaporised. The free electrons are attracted towards the main electrodes. Once the arc is started and electron stream is emitted from the hot sport, the hot spot is maintained at the higher temperature necessary for electron emission by the bombardment of the positive ions.

The ions are attracted towards the auxiliary electrodes; and the hot spot is maintained even at without load. After starting the ignition switch is put off but the excitation anode and excitation load remains in the circuit for all the time. The diagram shows the connections.

For cooling the rectifier a fan is mounted near the bottom of the container. The D.C. voltage output depends upon the number of main anode and the D.C. voltage. There is a voltage drop of approximately 20-volts inside the rectifier.

Uses. The mercury arc rectifiers are used for the following purposes:

(*i*) D.C. power and lighting purpose.

(*ii*) Electrochemical works.

(*iii*) Battery charging.

(*iv*) Control circuits of substation and station equipments.

(*v*) Power supply to the projectors arc lamps.

Q. Draw the connection diagram of a 3ϕ mercury arc rectifier.

Ans. The diagram is as shown in Fig. 25.8.

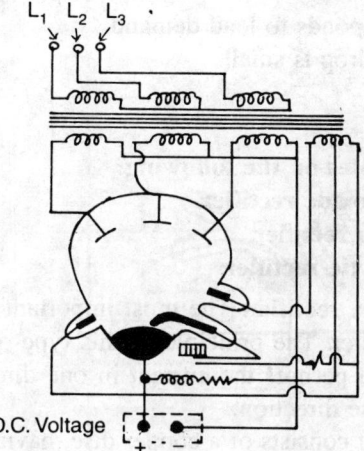

Fig. 25.8. 3φ Mercury Arc Rectifier.

Q. Why mercury pool is used as the cathode of the mercury arc rectifier and what are the advantages of this rectifier?

Ans. The mercury is used as the cathode because of the following reasons:

(*i*) It is a material having low specific heat, so the thermal sensitivity is very good. It has small latent heat of vaporisation and high atomic weight.

(*ii*) There is no damage to mercury cathode on account of violent bombardment by positive ions. But if some solid cathode was used, the damage would have occurred.

(*iii*) The mercury vapour after condensed drains back into pool and so it is self restoring.

(*iv*) It remains liquid at ordinary temperature.

(*v*) It requires a very low voltage to ionise about 10.4 V.

(*vi*) Once the arc is started and electron stream is emitted from the hot spot, it is maintained automatically by the bombardment of heavy positive ions of Hg.

There are the following advantages of the mercury arc rectifier:

(*i*) It is simple in operation as compared other type of conversions.

(*ii*) It has noiseless operation.

(*iii*) It is lighter in weight.

(*iv*) Less space is required.

(*v*) As there is no rotating part, so less maintenance and care is required.

(*vi*) It has higher efficiency.

(*vii*) It has high over loading capacity.

(*viii*) It quickly responds to load demand.

(*ix*) The voltage drop is small.

Rectifiers:

Q. Write short notes on the following:
   (*a*) **Copper oxide rectifier.**
   (*b*) **Selenium rectifier.**
   (*c*) **Electrolytic rectifier.**

Ans. Copper oxide rectifier. The most important metal rectifier is the copper oxide rectifier. The principle of this type of rectifier is same as that of a valve and permits the current in one direction and offer high resistance in reverse direction.

Construction. It consists of a copper disc, having one side a layer of copper oxide, obtained by the oxidation of pure copper as shown in Fig. 25.9. The layer of cuprous oxide has greater resistance in one direction then in reverse direction for the current to flow. These rectifiers for small output consists of copper disc usually 1 mm thick and 25 mm in diameter having one side a cuprous oxide layer. The layer has a coating of graphite. To make the contact with copper and the oxide layer glass washers are used. The current can easily flow from copper oxide to copper and not in reserve direction.

Rating. The average current setting for 3/4″ diameter disc is about 0.33 A and safe reverse voltage is about 8 V. For obtaining a certain voltage more number of rectifiers are connected in series. Similarly for more current the discs are connected in parallel.

All the plates are mounted over the insulated rod. For big rectifiers cooling arrangements are provided, generally cooling wings are provided with every rectifier. Full-wave and half-wave rectification can be easily obtained by using these rectifiers.

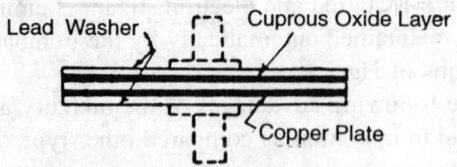

Fig. 25.9. Copper oxide rectifier.

(*b*) **Selenium rectifier.** In this type of rectifier a disc of steel or aluminium is deposited by a film of 0.05 mm. thick selenium. Contacts are made by the ring of special alloy. The rectifier is gray in colour. These rectifiers are more reliable. These are efficient and for the same output are more compact than the copper oxide rectifier. The working voltage is about 4 volts. The current can flow from selenium to steel but a high resistance is offered in reverse direction. It can operate upto 75°C. The efficiency is about 80%.

A number of rectifiers are connected in series for more voltage and for more current in parallel. These rectifiers are available in the capacities upto 50 kW or 100 kW etc. The range can even be extended. Their application is limited to a voltage of 100 volt; for more voltage these units are assembled in series. The age is also about 12-15 years.

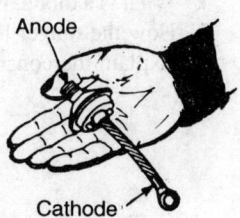

Fig. 25.10. Selenium rectifier.

(*c*) **Electrolytic rectifier.** These are the liquid rectifiers. The electrolyte generally used is sodium phosphate or sodium bicarbonate or amonium phosphate. The solution is filled in the container with one plate of aluminium and the other a lead plate. The current is allowed to flow towards aluminium, a thin layer of aluminium oxide is formed which offers a easy path for the current in one direction and high resistance in other direction. If only one plate of aluminium is used it works as half-wave rectifier and by using two aluminium plates it will work as the full-wave rectifier. These are used for low capacity rectifiers. The connections are shown in Fig. 25.11.

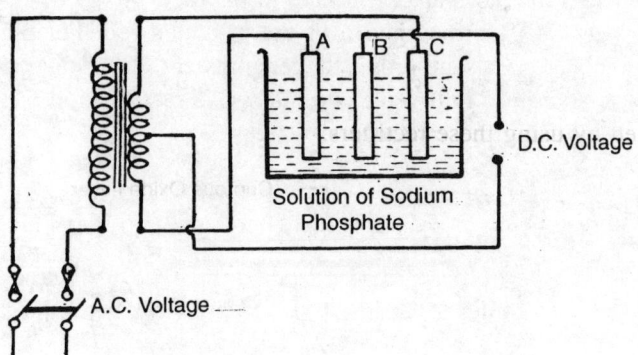

Fig. 25.11. Electrolytic rectifier.

REVIEW QUESTIONS

1. What is the necessity of converting machines?

2. What are the different types of converting machines? Explain any one of them.
3. What do you understand by the rotory converter? Explain its function.
4. What is a diode, how it is useful as a half wave and full wave rectifications?
5. How the tunger bulb rectifier is used as a full wave rectifier?
6. Explain the construction and working of the mercury arc rectifier.

26

Electronics

The branch of engineering which deals with the flow of electrons through vacuum or gas or semi-conductor is called electronics. The flow of electrons is the base of electronics. The electron has the following characteristics:

(i) It has a charge $e = 1.602 \times 10^{-19}$ coulombs

(ii) It has a mass $m = 9.1 \times 10^{-31}$ kg

(iii) It has a radius $r = 1.9 \times 10^{-15}$ metre

(iv) The ratio of charge/mass $= 1.77 \times 10^{11}$ coulomb/kg which indicates the necessity of electric and magnetic fields.

(v) The final one, the flow of electrons constitutes the electric current.

There are some important signs and symbols used in electronics:

Particulars	Symbols
1. Bridge Rectifier	
2. Crystal detector	
3. Moving coil	
4. Tube filament	

Particulars	Symbols
5. Triode	
6. Pentode	
7. Double head phone	
8. *PN* diode.	Actual. Device
9. *PNP* transistor	
10. *NPN* transistor with transverse biased base	
11. *PNIP* transistor	
12. *PNIP* transistor with ohmic connection to the intrinsic region	
13. *PNIN* transistor with ohmic connection to the intrinsic region	
14. Unijunction transistor with *N* type base	

Particulars	*Symbols*
15. Unijunction transistor with P-type base	
16. Field-effect transistor with *N*-type base	
17. Field-effect transistor with *P*-type base.	

26.1. ELECTRONIC EMISSION

Q. What do you understand by the electronic emission? Describe in brief the different methods of electronic emission.

Ans. The electronic emission can be defined as the liberation of electrons from the surface of a substance. In normal state the free electrons are in random motion and cannot leave the surface *i.e.* these electrons may transfer from one atom to the another atom within the metal but cannot come out from the surface to enable the electronic emission. In other words the metal surface offer a barrier and is called the *surface barrier.* The surface barrier is defined as the force exerted at the surface of a metal to prevent the free electron from escaping.

In normal stage none of the electron has the energy equal to or greater than the amount necessary to overcome the potential energy barrier. However if sufficient external energy is imparted to the free electron to overcome the potential barrier and leave the surface or metal, this additional energy required by the electron to cross over the surface barrier is called the **work function** of the metal.

The work function of the metal depends upon the nature of the metal and surface conditions. Therefore different metal has different work functions. The material used for electronic emission should have low work function. The different methods of electronic emission are as under:

(*i*) **Thermionic emission.** It is based upon the thermal energy given to the metal. It can be achieved by heating the metal sufficiently high, so the process of electron emission from a metal surface by supplying thermal energy to it, is known as *thermionic emission.* It is employed in electronic tubes etc.

(*ii*) **Field emission.** The field emission is defined as the escaping of electrons by the application of strong electrostatic field on the metal surface.

(*iii*) **Secondary emission.** It is defined as the process of emission by the application of bombardment of high velocity electrons on the metal surface. It is used in electron multipliers, cathode rays tube etc.

(*iv*) **Photo emission.** It is defined as the emission of electrons by the application of light on a certain surface. The photo emission is used in photo electric cell and lamps etc.

(*v*) **Radioactive disintegration.** The radioactive materials disintegrate with time and during the process of disintegration the electrons in the form of β-rays (beta rays) are liberated and is known as radioactive disintegration.

Q. What is a cathode explain some commonly used filamentary cathodes.

Ans. The cathode or emitter is an essential part of the electron tube. The substance which is used for electron emission is called a *cathode* or *emitter.* The cathode should have the following properties.

(*i*) The melting point should be high since the electron emission starts at a very high temperature, therefore the substance used as a cathode should have high melting temperature.

(*ii*) **Low work function.** The material should have low work function so that the energy required for escaping the electron is less.

(*iii*) **Sufficient mechanical strength.** The cathode should have enough mechanical strength to withstand the bombardment of the electrons. Whenever the cathode is heated up electrons are released. The different forms of the cathodes can be distinguished from each other by manner in which these are heated. These are of the following types:

(*a*) Directly heated cathode or filamentary cathode.

(*b*) Indirectly heated cathode or heater cathode.

(*a*) *Directly heated or filamentary cathode.* In this type the filament is directly heated as shown in Fig. 26.1 by the flow of current through it. The cathode is generally oxide coated nickel ribbon. In these type the heat energy is liberated in the form of thermionic emission. Nowadays these are widely used. These are made of:

(*i*) *Tungsten.* It is a pure metal which is used for the emitter or cathode, because it is well suited for the thermionic emission. It produces the maximum emission between 2000-2700°C temperature.

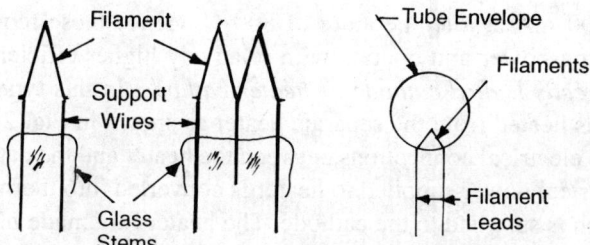

Fig. 26.1. Directly heated filament with symbol.

The melting point is also sufficiently high *i.e.* 3300°C. Tungsten has high mechanical strength to withstand the bombardment of fast moving positive ions. The work function (4.54 eV) of tungsten is practically suited for high power transmitting tubes. The tungsten emitter produces a current of 2 mA/Watt of heating power.

(*ii*) *Thoriated tungsten.* Thoriated tungsten is obtained by adding 1% to 1.5% thorium oxide to pure tungsten. The heating power required by the pure tungsten is high than the thoriated tungsten. The presence of monoatomic layer of thorium on tungsten surface lowers its work function. Hence the thermionic emission is obtained at a moderate temperature (1700°C). The filament produces electrons about 10,000 times that of ordinary tungsten. At normal operating temperature it produces 30 mA/ Watt of power consumed. These are used in the valves operating ranging from 500 to 5000 V.

(*iii*) *Oxide coated filaments.* Oxide coated filaments are more efficient than the thoriated, but its surface is less stable as compared to thoriated tungsten. These are prepared by coating a wire or ribbon like cobalt, nickel, iron or titanium with a 50 : 50 mixture of carbonates of barium and strontium. The work function is low of the order of 1 eV. It produces the current

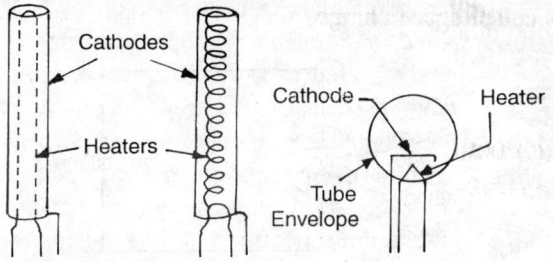

Fig. 26.2. Indirectly heated filament and symbol.

of 100 mA at a temperature of 800°C. Hence these require small heating power and operate with relatively higher efficiency.

(b) *Indirectly heated cathode or heater cathode.* In this type the cathode is heated from the separate heater as shown in Fig. 26.2. There is no electrical connections between the heater and the cathode. The electrical power supplied to heater is converted into thermal energy which is supplied to the cathode. The heaters are made of tungsten, molybdenum and its alloy wire.

26.2 DIODE

Q. Describe the construction and working of vacuum diode.

Ans. It is the most common type of electron valve.

Construction. It is the most elementary type of valve. It consists of a glass envelope from which air is taken out. It has two electrodes, the *Cathode* from where the electrons are emitted. It is generally in the form of a nickel cylinder coated with metal oxide. The cathode or emitter is surrounded by the collector for electrons. The cathode may be either of directly or indirectly heating type. The *plate* or *anode* is generally a hollow metallic cylinder made of nickle, molybdenum, graphite, tantalum, monel or iron etc.

Sometimes two diodes are located in one bulb such diodes are known as the double diode or dual diode, *viz* 5Y3-GT, and 5Y4-G etc.

Operation. The element is heated up and a number of electrons are emitted because of thermionic emission. The behaviour and characteristics of the action depends upon the potential with respect to cathode. The different stages can be illustrated as under.

(i) *When anode is at zero potential with respect to the cathode.* In this case the electrons emitted cannot go to the anode or plate and hence *no current.* But the electrons start accumulating near the cathode, thus a cloud is formed as shown in Fig. 26.3 which is known as space charge. So the dense cloud of electrons surrounding the cathode is called space charge.

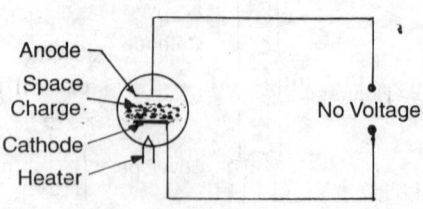

Fig. 26.3.

(*ii*) *When the plate is positive with respect to the cathode.* In this case the electrons are attracted towards the plate hence plate current is set up. The electrons on reaching to plate travel through connecting wires, milliammeter and battery. The terminals are marked as shown in Fig. 26.4 if the positive potential of the plate is increased the plate current increases. Since more electrons are pulled from the space charge to anode.

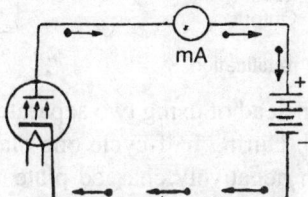

Fig. 26.4. Flow of current.

(*iii*) *When the anode is negatively charged with respect to cathode.* Since the anode is at negative potential therefore it repels the electrons back and hence no current flows through the circuit as shown in Fig. 26.5. It is to be noted that no current flows in opposite direction.

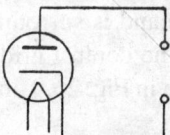

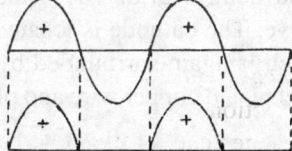

Fig. 26.5. **Fig. 26.6.** Half wave rectification of A.C.

Diode as half wave rectifier. The diode can serve as a half wave rectifier. During half cycle when the plate is at positive potential the electrons are attracted towards the plate which constitutes the flow of current. But during the next half cycle when the plate is at negative potential, the electrons being negatively charged will be repelled back and there will not be any current. Thus during complete cycle the current will only flow during half the cycle. Thus only the half cycle will be rectified as shown in Fig. 26.6 and the rectifier will be known as half wave rectifier.

Diode as full wave rectifier. Obviously the negative plate repel the electrons if there are two diodes connected as shown in Fig. 26.7 across the winding. The anodes will be positive one by one during every alternative half-cycle and the complete cycle will be rectified.

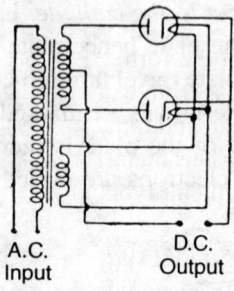

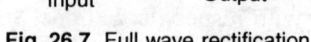

A.C. Input   D.C. Output

Fig. 26.7. Full wave rectification.

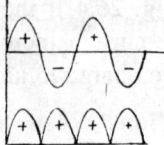

Fig. 26.8.

In this modern age instead of using two separate diodes, a single diode with two anode is used. During half-cycle one plate works and the electrons are repelled from negatively charged plate. Thus during next half the polarity will be reversed and current will flow through the second one. Thus the complete cycle is rectified. For further smoothing the filter circuit is used and a steady D.C. is obtained.

26.3 TRIODE

Q. Explain the construction and working of vacuum triode.

Ans. The triode as the name implies contains three electrodes, the cathode, anode and control grid. The cathode and anode are similar as that of diode valve. The cathode is located at the centre and is surrounded by a grid, which is again surrounded by the anode. The control grid is a fine thin nickel plated copper wire and placed as shown in Fig. 26.9, in between

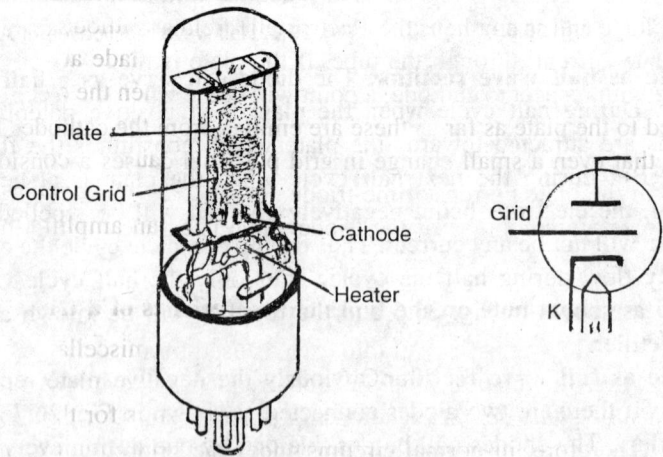

Plate

Control Grid

Cathode

Heater

Grid

K

Fig. 26.9. Triode-valve.

the cathode and anode. The grid may be in the form of helix or perforations through which electrons can easily pass from cathode to anode. The grid is used to serve as an electrostatic screen and thus partially shield the cathode from electrostatic field of anode. The metals used for grid are nichrome, iron, nickel, tungsten, tantalum, molybdenum and various alloys.

Operations. Whenever the cathode is heated it starts thermionic emission, since anode is at higher potential with respect to cathode, so electrons are attracted by it, in other words it constitutes the plate current. The flow of electron is entirely determined by the conditions residing on the grid. When the control grid is at zero potential with respect to cathode, the triode valve just behaves as a diode because the presence of the grid do not effect the electric field set up by the cathode and anode. Secondly, if the grid is at negative potential with respect to cathode, the passing of electrons (being negatively charged) is opposed resulting the reduction in the plate current. It is because the control grid is very near to the cathode and the negative potential has the characteristic to decrease the plate current.

Thus it is concluded that the triode valve behaves like a diode if the grid is at zero potential w.r.t. cathode and even a less voltage supplied to grid affects the plate current considerably. When the grid is negative the current start decreasing and the more negative voltage decreases the plate current considerably. If the negative potential is further increased it may lead the cut off point.

Now if the grid is given positive potential w.r.t. cathode it aids the plate voltage and strengthens the electrostatic field at cathode resulting in large plate current, through the tube. If the plate is made at sufficiently positive with respect to cathode, a point will come when the electrons are attracted to the plate as far as these are emitted from the cathode. Thus it is seen that even a small change in grid potential causes a considerable change in the plate current of the triode.

The most common use of the triode valve is as an amplifier and the oscillator, etc.

Q. Write a short note on the types characteristics of a triode valve.

Ans. Valve characteristics. A radio wave is used for miscellaneous work *e.g.*, detection, modulation, amplification, frequency converter, etc., hence there must be some definite relationship and constants for the input and outputs. Therefore, in normal circumstances the grid circuit of a valve forms the input circuit and anode circuit as output circuit. It is observed that even a small change of grid voltage causes a considerable change in anode current.

It has the following characteristics:
(a) **Static characteristic.** It shows the sub-characteristic:
    (i) Mutual characteristics.   (iii) Constant current characteristics.
    (ii) Anode characteristics.
(b) **Dynamic characteristic.** The behaviour of valve under the practical circuit conditions is termed as the dynamic characteristic.
(i) *Mutual characteristics.* This characteristic gives away all the information regarding the working of the tube. The most important constant is that which describes the rate of change of small anode curves (∂I_a) with the small change in grid voltage (E_g) keeping the (E_a) anode voltage constant. It is termed as mutual or transfer characteristic.

The plate voltage is kept constant at a certain fixed value and the change in anode current or plate current (I_a) due to change in E_g is observed in milliamperes. When grid is positive, the electrons are attracted hence I_a is increased. If the grid is negative the electrons are repelled, hence I_a is reduced and if E_g is reduced too much then the plate current will lead to zero ampere as shown in Fig. 26.10. From Fig. 26.8, it is clear that the mutual or transfer characteristic is
(a) practically straight over an appreciable portion of the curve length,

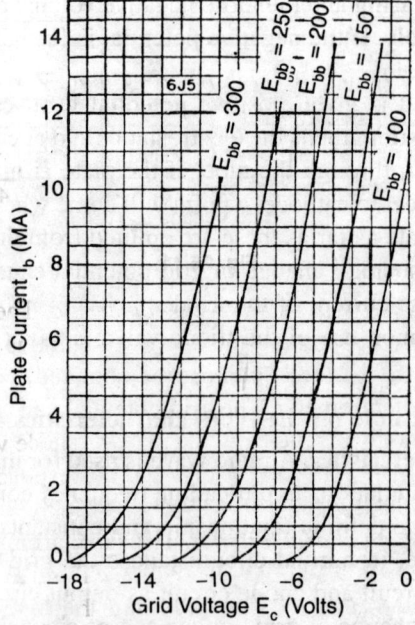

Fig. 26.10. Static characteristics.

(*b*) equally spaced for equal interval of the plate voltage,

(*c*) that the cut off potentials for different curves are different.

This characteristic also leads the information about the behaviour when A.C. signal is fed to the valve. When there is an external impedance in the anode circuit the voltage drop in the impedance alter with the change in anode current producing corresponding change in the anode and filament. It is represented by G_m, g_m or s and is defined as change of plate current with respect to change of grid potential.

i.e.

$$g = \frac{\partial I_a}{\partial E_g}$$

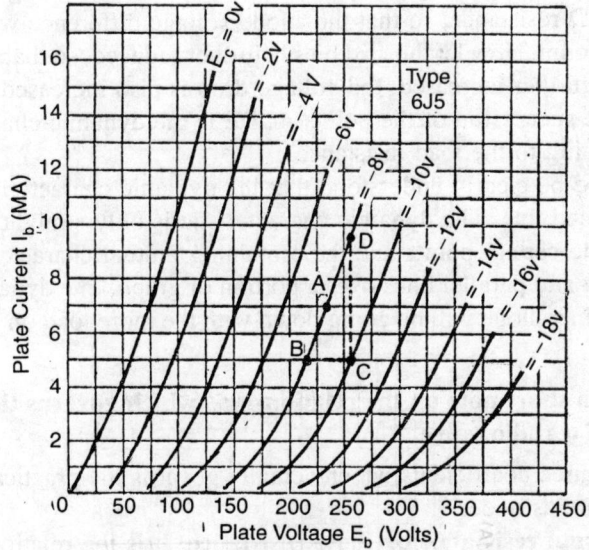

Fig. 26.11.

(*ii*) *Anode characteristic or plate characteristic.* The arrangement for changing and maintaining the anode potential is made so that any desired polarity of the anode can be obtained, generally it is done by using a potentiometer. The grid potential can also be changed by using a reversing switch. Now the plate or anode voltage is changed keeping the grid voltage constant and corresponding value of I_a is noticed as shown in Fig. 26.11.

(*iii*) *Constant current characteristics.* The same arrangement is used for this characteristic also. The plate voltage is set at a particular value and then grid voltage is changed until the highest current is obtained for which the characteristic is desired. The plate and grid voltages are decreased and readjusted respectively to obtain the same plate or anode current. The datas so obtained are plotted.

(*b*) **Dynamic characteristic.** The characteristic drawn between the anode or plate current and grid voltage when the plate resistance is not zero is known as dynamic characteristic, *i.e.* the behaviour of the tube is studied under actual practical conditions. The dynamic characteristics are usually drawn over the mutual characteristic curves.

Let the load in the anode circuit is RΩ. When grid bias is sufficiently negative to prevent the flow of anode current, the voltage of anode is same as that of h.t. battery. Now the I_a, current is allowed to pass in the anode circuit by reducing the grid bias, there will be a voltage drop I_aR Volts in RΩ resistance, so that the anode voltage difference will be less by that amount. Now let the grid bias is further reduced, so that the anode current is further increased, the voltage drop is also increased. The line joining the succession of the points *ACDE* is the dynamic characteristic corresponding to the load resistance.

It should be clearly understood that the dynamic characteristic differ from the load line. The dynamic characteristic is only straight line only between the certain points and the remaining mutual characteristic line are straight and parallel lines over a portion of graph. The dynamic characteristic lines slope will be going down with the more load on the anode circuit.

Q. Write a short note on the relationships which governs the behaviour of a radio valve.

Ans. The three coefficients which actually governs the practical aspects of a tube are as under:

(*a*) **Internal resistance or Plate resistance.** It is the relationship between the change in small anode voltage and small change in anode current keeping the E_g as constant. It is expressed in ohms and represented by R_a

$$\therefore \qquad R_a = \frac{\partial E_a}{\partial I_a} \Omega.$$

It is important that the A.C. resistance is concerned only with the ratio of change of anode voltage to the change in anode current and not with the ratio of total anode voltage to total anode current.

(*b*) **Mutual conductance.** It is the relation between the change in anode current and the change in grid voltage keeping the E_a as constant. It is expressed in mA/volt and is denoted by *S* or g_m or *P*

$$g = \frac{\partial I_a}{\partial E_g}$$

(c) **Amplification factor.** It is given as the relation between the change in anode voltage and the change in grid voltage. It is expressed by μ. In fact it is the ratio of relative effectiveness of the grid voltage to that of the voltage in controlling the anode current.

$$\mu = \frac{\dfrac{\partial I_a}{\partial E_g}}{\dfrac{\partial I_a}{\partial E_a}}$$

$$= \frac{\partial E_a}{\partial E_g}$$

So the amplification factor $= \dfrac{\text{Change in anode voltage}}{\text{Change of grid voltage}}$

keeping the anode current as constant. It is a ratio hence do not have any unit.

Q. State the Barkhansen's law.

Ans. Barkhansen's law. It is the law which relates all the valve constants. It is the product of the initial resistance and mutual conductance *i.e.,* amplification factor, which is called Barkhansen's law. According to Barkhansen's law

$$\mu = R_g \times g_m$$

or $$\mu = \frac{\partial E_a}{\partial I_a} \times \frac{\partial I_a}{\partial E_g} = \frac{\partial E_a}{\partial E_g}$$

It is equal to the amplification factor *i.e.* μ. So Barkhansen's law is the product of R_i and g_m is nothing but the amplification factor.

Example. *If the mutual conductance of a certain valve is 2.8 mA/V and the amplification factor is 22. Find the internal resistance.*

Solution. The mutual conductance $g_m = \dfrac{\partial I_a}{\partial E_g}$

$$g_m = 2.8 \text{ mA/V}$$

and the amplification factor, $\mu = \dfrac{\partial E_a}{\partial E_g}$

$$\mu = 22$$

Now the relationship $\qquad \mu = R_i \times g_m$

or $\qquad\qquad R_i = \dfrac{\mu}{g_m} = \dfrac{22 \times 10^3}{2.8} = 8\,\mathrm{k\Omega}$ **Ans.**

Example. *In a certain valve of a set, the 0.64 V grid voltage causes a production of change in anode current of 1.6 mA. Calculate the mutual conductance.*

Solution. The mutual conductance, $g_m = \dfrac{\partial I_a}{\partial E_g}$

$$= \dfrac{1.6 \times 10^{-3}}{0.64} = 2.5 \ \mathrm{mA/V}. \ \mathbf{Ans.}$$

26.14 TETRODE AND PENTODE

Q. Write short notes on the following vacuum tubes (a) Tetrode, (b) Pentode.

Ans. (a) Tetrode.
Tetrode means the valve having four electrodes as shown in Fig. 26.12 to reduce the inter-electrode capacitance and to overcome undesirable feed back at higher frequency a second grid is inserted in triode between grid and plate. Thus a tube having two grids, control and screen grid is called

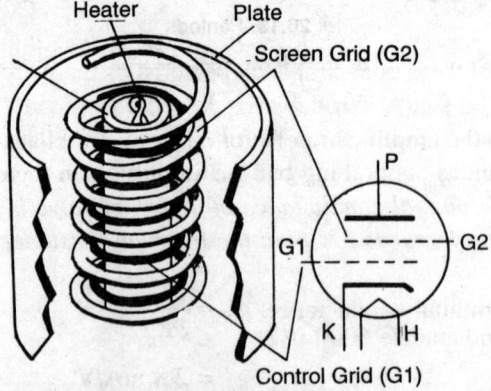

Fig. 26.12. Tetrode.

tetrode. The only construction difference between a tetrode and triode is the presence of additional grid known as screen grid, having same construction but loosely wound. It reduces the effect of plate voltage upon space charge, thus increasing the gain considerably to hundred times than that of triode. It also helps in minimising the capacitance to considerably limit between the grid and plate.

When a proper voltage is applied to the tetrode, electrons are attracted, the screen being positively charged with respect to cathode attracts the electrons screen grid produces strong electro-static force and the flow accelerates. Thus it can give much amplification factor then triode. The transconductance being the ratio of amplification factor and plate resistance is expected to be about the same tetrode as for triode. The screen grid causes a drawback *i.e.* the secondary emission, where it does not affect the operation in diode and triode. This effect lowers the plate current and limit the working range of the tube.

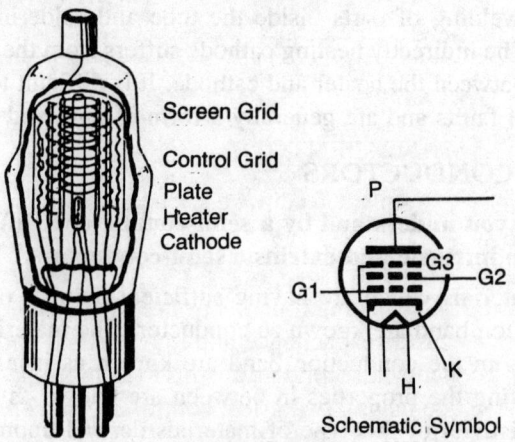

Screen Grid

Control Grid
Plate
Heater
Cathode

Fig. 26.13. Pentode.

But because of these undesirable characteristics these are seldom used.

Pentode. It is a five electrode valve. In order to retain the advantages of screen grid and eliminate the harmful effect of secondary emission, the third grid, known as suppressor grid is inserted between the screen grid and the plate. The five electrodes are a cathode, a control grid, a suppressor grid, the screen grid and a plate as shown in Fig. 26.13. The screen grid voltage is fixed, so control grid voltage is a major factor to control the plate current, amplification may be 400, the plate resistance 1 MΩ and the transconductance 5000 μΩ.

Q. What are the reasons of valve failure in electronics?

Ans. The valve failure may be due to various reasons, but practically, those are due to failure of vacuum, failure of cathode emission and other miscellaneous valve defects.

(a) **Failure due to vacuum.** The hard glass may becomes soft due to crack in the envelope, a leak seal or by excessive heating of electrodes. The first evidence of failure of vacuum is a blue glow due to the bombardment of ions seen near the filament.

(b) **Failure due to cathode emission.** The filament gradually wear thin due to volatilisation of the surface as well as to the effects of ionic bombardment. The decrease in diameter causes an increase in the resistance with a corresponding decrease in heating current and in cathode emission.

In case of oxide coated filaments, loss of emission is usually the result of the excessive voltage.

(c) **Miscellaneous.** There can be some manufacturing defects. The most common fault in the intermittent contact or open circuit due to faulty welding of parts inside the tube and soldering of external leads. The indirectly heating cathode suffers from the internal insulation between the heater and cathode. It is difficult to locate these types of faults and are generally a form of hum and rattle.

26.5 SEMI-CONDUCTORS

Q. What do you understand by a semi-conductor? Differentiate between the intrinsic and extrinsic semi-conductors.

Ans. The materials which are having sufficient number of electrons in their conduction band are known as conductors, the materials having no free electrons in the conduction band are known as insulators and the materials having the properties in between are known as semi-conductors. The resistivity of this type of materials depend upon the temperature, illumination and magnitude of the dielectric field. These are the non-linear resistors.

The resistivity can be very well govern and control by the introduction of some other substances known as impurities. These are used as rectifier, detector, transistor, thermistor etc. There are so many semi-conductors such as Germanium (Ge) Silicon (Si), Sulpher (Si), Selinium (Se), Teliurium, (Te), Boron (B), Carbon (C), Arsenic (As), Sulpher (S) and Iodine (I) etc. But Germanium and Silicon are widely used at present.

The *intrinsic semi-conductors* are those which are in pure form and exhibits roughly equal number of free electrons and holes. Their conductivity is very less and little flexibility in its conductivity.

The *extrinsic semi-conductors* are those which are prepared by adding some impurities. The introduction of impurity overcomes the drawbacks of the intrinsic semi-conductors. So the improved semi-conductors thus obtained are called extrinsic semi-conductors and are widely used.

Q. Write a short note on the p-type and n-type semi-conductors.

Ans. The atom of silicon has a positive nucleus and four valence electrons outside its nucleus. It has covalent bonds and there is no free electron to carry the current. It is thus a high semi-conductor material.

Now if an impurity is added, the drifting of electron can be obtained. The impurity is of the order of one in 10^9 to 10^{10} having five valence electrons. The fifth atom does not fit into the lattice and readily available to carry the current. This type of doped crystal is known as *n type semi-conductors.*

Now if an impurity having three electrons is added, it causes a deficiency of electron and the deficit bond is known as *hole*. It may be filled by the electron of the neighbouring atom if a suitable potential is applied. Thus the hole is transferred from one atom

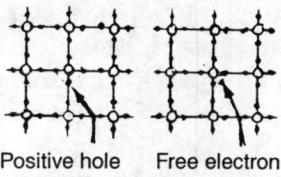

Positive hole   Free electron

Fig. 26.14.

to the another and the semi-conductor is known as *p-type semi-conductor* as shown in Fig. 26.14.

Energy Level

There are certain terms used in semiconductors for example:

(*a*) *Doping:* Adding of impurity atoms to pure semiconductor material is called doping. The addition helps in increasing the conductivity of the semiconductor material. The impurity material used is called dopant.

(*b*) *Donar:* The impurity which result in a surplus of free electrons, which conduct current for example phosphorous or any other pentavalent material used as a dopant in any crystal, such crystals are called or N-type crystal.

(*c*) *Acceptor:* The dopant which causes the deficiency of electrons creating a hole which being mobile causes conduction of electric current for example Boron or any other trivalent impurity in a crystal such crystals are called acceptor or P-type crystals.

The important semiconductor silicon can behave both ways; as a donar if the impurity may be any of these phosphorous, arsenic, or antimony, and as a acceptor if the impurity is any of these boron, gallium and Indium. A semiconductor is described by these three parameters, the mobility of free electrons, the mobility of holes and the energy gap.

Energy Level

As shown in Fig. there are two bands, the valence band that is a filled band, the conduction band that is an empty band. In empty band or conduction band the movement or drifting of electron constitute the current. When the electron after gaining some energy jumps out from the valence band and comes into the conduction band it may take part in constituting a flow of electrons i.e., electric current.

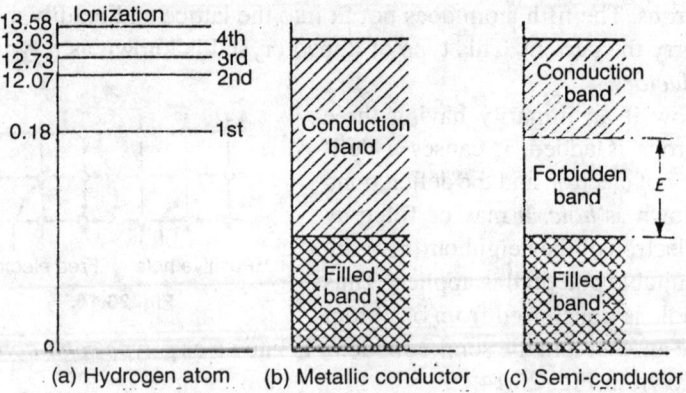

(a) Hydrogen atom   (b) Metallic conductor   (c) Semi-conductor

Fig. 26.15. Energy Level Diagrams

In the second figure there is a space left between the conduction band and valence band known as forbidden space. It shows that the electrons from the valence band cannot jumps out with that little amount of the energy but a substantial amount of energy is required to jump out the forbidden space. Truly speaking these materials are insulator at room temperature and do not conduct. The energy given may be much higher than the conductors i.e. upto 10 eV (instead of 0.7 eV for germanium and 0.06 eV for tin). The thermal energy supplied may not be sufficient to enable atoms to be raised from the filled band to that of conduction band unless it is produced by high temperature or with other rich means.

Conductors metal   – 10^{-3} Ω cm

Insulator   – 10^6 to 10^{15} Ω cm.   Resistivity of the materials

Semiconductors   – $10^3 – 10^{-3}$ Ω cm.

The materials which are inherently semicouductors and no impurity is added to make them donar or acceptor, has very narrow forbidden space (band). The energy difference between the valence band and new conduction band resulted by an impurity activator is comparatively less, so less thermal energy is required to excite electrons into the conduction band. So a semiconductor is such a material in which such activation can be obtained either thermally or optically (by reception of radiant energy).

Table 26.1. Substances with electron mobility

S. No.	Substance	Energy gap	Mobility $cm^2/v/sec.$	
		eV	Electrons $cm^2/v/cm$	Holes
1.	Carbon	5.6	400	200
2.	Silicon	1.12	1200	250
3.	Germanium	0.75	3600	1700
4.	Tin	0.07	3000	–
5.	Indium Antimonide, antimonide		50,000	

Q. Explain the working of PN-junction semi-conductor, and show that it may be used as a diode like vacuum diode.

Ans. The *P-type* semi-conductor, because the added impurity having three electrons, has deficit bond known as hole. It has same number of fixed negative ions carrying exactly the same number of total charge as that of total positive charge represented by holes, similarly the *n*-type semi-conductors has one extra free electron and have same number of fixed positive ions carrying the same total charge, hence the region is neutral.

Thus when an element is constructed by forming a junction of *p* and *n*-type materials in a single thin slice of silicon crystal, it is known as *pn-junction* or *pn-diode*. There some of the holes diffused across boundary into *n*-type semi-conductor and some free electrons diffuse into the *p*-type semi-conductor and thus forming a potential barrier between the two regions.

The barrier can be reduced or increased by applying the external potential difference. Let a battery is connected in such a way as shown in Fig. 26.15, it weakens the barrier and electrons from *n*-type and holes from *p*-type crosses the barrier easily and results a conductive stage thus if the source permits current to flow it is known as *Forward biasing*. If the source terminals are reversed and electrons and holes cannot cross the boundary as shown in Fig. 26.16. Here there is no flow of current through the barrier, so it is known as *reverse biasing*.

It is now clear that the junction now offer a conducting path when forward biased and high resistance when reverse biased. Thus it is concluded that the *pn*-junction permits the current to flow only in one direction as that of the diode so it acts like a rectifier too.

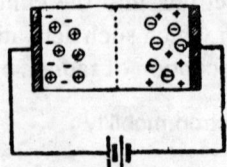
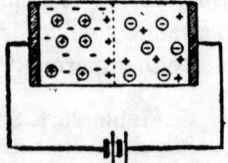

Fig. 26.16. Reverse.   **Fig. 26.17.** Forward.

Q. Compare the silicon and germanium junctions diodes.

Ans. Both silicon and germanium are used as the diode rectifiers but differ in the following ways:

Table 26.2. Silicon and Germanium Diodes

Silicon	Germanium
1. The forward voltage drop is approximately double that of the corresponding germanium diode.	It has half the forward voltage drop than silicon.
2. The reverse current is hundredth of the germanium diode.	It is hundred times the silicon diode.
3. It can stand much higher reverse voltage.	It cannot.
4. It can operate upto 50-200° C.	It can operate only upto 70 to 90°C.
5. Reverse current for a given voltage, double for every 8°C rise in temperature.	It is double for every 10°C rise in temperature.

Q. What is a transistor, how it differs from the valves? Give brief idea of the theory of transistor action.

Ans. The transistor is a combination of semi-conductors in a systematic manner. The main difference between a triode and transistor is that the valves are the voltage control devices and transistors are the current controlling devices.

It consists of a suitably prepared thin section of silicon or germanium with a deficiency of electrons materials sand-witched between two layers or excess of electrons materials. If the centre section is as *p*-type and alternates are *n*-type then it is known as *npn-type transistor*. The centre portion correspond to the grid of a valve and the remaining *n*-sections as emitter and collector respectively, corresponding to cathode and anode of the valve.

Secondly, if centre section is *n*-type and alternately *p*-type then it is known as *pnp-type transistor* as shown in Fig. 26.18.

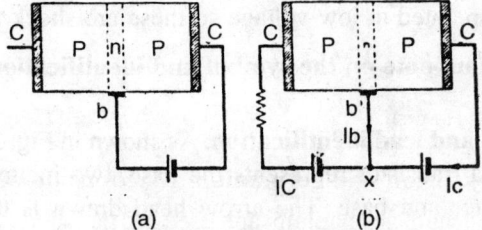

(a)   (b)

Fig. 26.18. *pnp*-transistor.

In the junction transistor structure, here *e, b* and *c* indicates emitter, base and collector respectively. It comprises of two *pn*-junctions are biased in reverse direction. The length of the arrow indicates the electric field at that junction as a result collector current is very small. In Fig. 26.18 the field across the barrier is in the forward or conducting direction is reduced, so as to have a movement of electron or holes. The hole moves across the emitter base junction into the *n*-region, which has a width about 0.005 mm. Thus the application of the voltage to the emitter causes an increase in collector current. The amount of the emitter current may be several milliampere resulting much larger collector current.

Q. Write short note on the following:

    (*a*) **Majority and minority carrier.**

    (*b*) **Hall effect.**

    (*c*) **Merits of semi-conductors in electrical engineering.**

Ans. (*a*) **Majority and minority carrier.** In *n*-type semi-conductors, the conduction takes place through the free electrons, created mostly by doping. The small number of holes created because of the thermal generation, moves in opposite direction. Since the number of free electrons are more than the holes, so these are called majority carrier and holes being less are known minority carrier.

    In *p*-type semi-conductors holes are the majority carrier and the electrons are the minority carrier.

    (*b*) **Hall effect.** Whenever current flows through a semi-conductor bar, placed in the magnetic field, a voltage is developed at right angle to both the current and magnetic field. This voltage is proportional to the current and the magnetic flux. It is called as the *hall effect.*

    (*c*) **Merits of semi-conductors in electrical engineering.** Following are the merits:

        (*i*) These are much smaller and light.

        (*ii*) Heater is not used so the operation is easy.

    (*iii*) Low power consumption and high efficiency.

    (*iv*) Long life.

    (*v*) As operated at low voltage so these are shock proof also.

Q. Write a short note on the symbol and identification of transistor leads.

Ans. Symbols and lead identification. As shown in Fig. 26.19, the horizontal line of a transistor represents the base, two inclined lines represents the emitter and base. The arrow head drawn is the emitter and indicates the direction of current. In *PNP* type the arrow head is pointed towards the plate and in *NPN* type the arrow is drawn pointing away from the base.

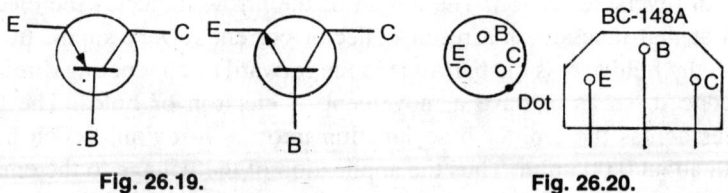

Fig. 26.19.   **Fig. 26.20.**

The identification of the leads is done by the following ways:

  (*i*) When the leads of a transistor are in the same plane but unevenly placed there are identified by the spacing. The centre lead is the base lead, the emitter and collectors are on either sides. The collector lead is identified by the larger space between it and the base and the remaining is the emitter.

 (*ii*) When the leads are in the same plane and evenly spaced, the centre is the base, the lead identified by a dot is the collector and remaining the emitter.

Q. Describe basic circuit configuration of common base, common emitter and common collector transistor circuits.

Ans. The transistor operated as an amplifier may be connected into the following three ways:

  (*i*) Common base emitter input connections.

 (*ii*) Common emitter base input connections.

(*iii*) Common collector base input connections.

    Each of the above connections are having their own characteristics.

  (*a*) **Common base emitter connections.** The common base circuit correspond to the grounded grid valve circuit. The base circuit terminal is common to both input and output circuit (Fig. 26.21). This circuit has a low input impedance and a high output impedance. In this case the phase of input signal remains the same in output. The base common circuit is not very much in use because it has very low power gain.

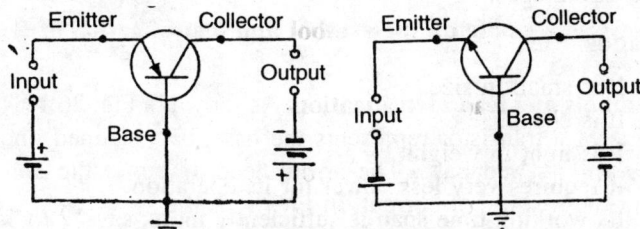

Fig. 26.21. Common base emitter input connections.

(*b*) **Common emitter base connections.** The common emitter circuit is mostly used. The emitter terminal is common to both input and output circuits as shown in Fig. 26.22. It has moderate input and output impedances similar to the vacuum tube it gives a phase reversal to the input signal.

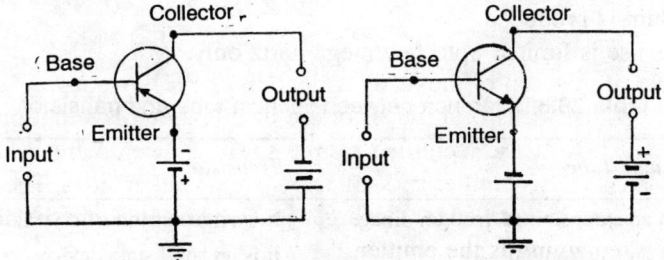

Fig. 26.22. Common emitter base input connections.

(*c*) **Common collector connections.** The common collector circuit correspond to grounded plate or cathode in vacuum tube connections. The collector terminal is common to both input and output circuit. The base common collector circuit is shown in Fig. 26.23. The phase shift angle remains the same in both output and input circuits.

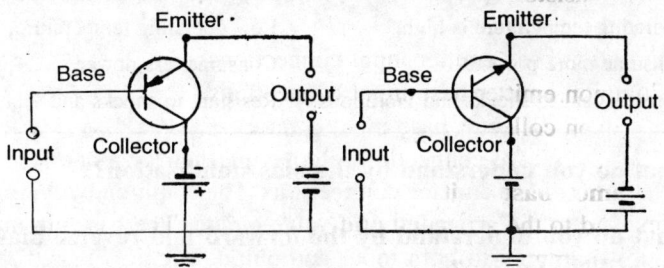

Fig. 26.23. Common collector base input connection.

Q. What are the advantages of a transistor? Differentiate between the transistor and the vacuum valve.

Ans. The transistor is the extension of crystal diode. It is having the following advantages:

Advantages

(*i*) It is small in size.

(*ii*) It is cheap.

(*iii*) It is light in weight.

(*iv*) It requires very less power for its operation.

(*v*) Its working time span is sufficiently more, say 12 to 15 years and even more.

(*vi*) It requires very less power for its operation.

(*vii*) It is rigid being a solid state device and can be taken to any-where.

(*viii*) Power dissipation is very less.

Disadvantages. It has the following disadvantages:

(*i*) Beyond 80°C the operation is not possible.

(*ii*) Hum is produced.

(*iii*) Its use is limited upto few mega hertz only.

Table 26.3. Difference between vacuum tube and transistor

Vacuum Tube	Transistor
1. It is an extension of vacuum diode.	It is an extension of crystal diode.
2. It is an electronic devices.	It is an solid state device.
3. It has high biasing voltage.	Low biasing voltage.
4. It has low efficiency.	It has high efficiency.
5. It is costly.	It is cheap.
6. It has comparatively short life.	It has long life.
7. It occupies more space.	It requires less space.
8. Heater element is required.	It does not require any heater.
9. Operating temperature is high.	Low operating temperature.
10. Consume more power.	Consume less power.
11. Non-resistant to shocks and vibrations.	Resistant to shocks and vibrations.

Q. What do you understand by the bias stabilization?
Or
What do you understand by the forward and reverse biasing in case of PN-junction?

Ans. The term biasing relates with the connections of the battery to a diode in a particular fashion. If can be biased into two ways:

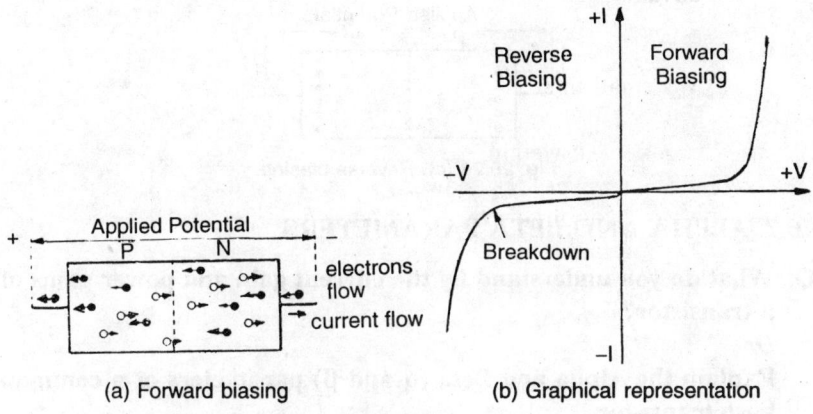

(a) Forward biasing   (b) Graphical representation

Fig. 26.24.

(*a*) Forward biasing.

(*b*) Reverse biasing.

(*a*) **Forward biasing.** The diode is forward biased, when the positive terminal of battery is connected to *P*-type and negative to *N* type across a *PN*-junction (see Fig. 26.24). In this condition, the electrons in *N*-type, the holes in *P*-type are repelled by battery potential to the junction between *N* and *P* type semi-conductor, say germanium etc. There is a potential barrier at *PN*-junction which prevent electrons and holes to move and to combine but under electric potential the electrons and holes combine and *PN*-junction and neutralises each other causing a flow of electric current, depending upon the potential of the battery applied to the function.

A very few volts are sufficient to break the barrier as soon as this barrier is crossed, there is a rapid increase in current for even a small increase in voltage as shown, hence connections of the battery to the *PN*-junction are known as *forward biasing*.

(*b*) **Reverse biasing.** The diode is reversed biased when the positive terminal of the battery is connected to *N*-type and negative terminal with *P*-type across a *PN*-junction diode. Under this influence, the electrons are attracted by the positive charge and holes towards the negative terminal of the battery, as shown in Fig. 26.24(*c*), this type of connections are known as *reverse biasing*.

Now the applied potential, if increased, causes a production of heat which distort the covalent bonds of the semi-conductor because of the production of heat. The amount of current is very small. As the potential applied is increased, the junction breakdown and current increases very rapidly as shown in Fig. 26.24(*b*).

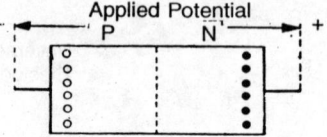

Fig. 26.24 (c). Reverse biasing.

26.7. ALPHA AND BETA PARAMETERS

Q. What do you understand by the current gain and power gains of a transistor?
Or
Explain the Alpha and Beta (α and β) parameters of a common base transistor.

Ans. Current gain. The ratio of change in collecter current to the change in emitter current for a constant collector voltage is known as current gain. It is denoted by letter alpha '*a*' for a common base circuit

i.e.,
$$\alpha = \frac{\Delta I_C}{\Delta I_E} \; V_{CB} \text{ constant}$$

In case of common emitter circuit, the current gain is the ratio of the change in collector current to the change in base current for a constant collector voltage. It is denoted by the letter β.

i.e.,
$$\beta = \frac{\Delta I_C}{\Delta I_B} \; V_{CE} \text{ constant}$$

Now $$I_E = I_B + I_C$$
or $$I_B = I_E - I_C$$
Obviously $$I_C = a \, I_E \qquad\qquad ...(i)$$
So $$I_B = I_E - \alpha \, I_E = I_E \, (1 - \alpha) \qquad ...(ii)$$
Dividing (*i*) by (*ii*)

$$\frac{I_C}{I_B} = \frac{\alpha I_E}{I_E (1 - \alpha)}$$

$$= \frac{\alpha}{1 - \alpha}.$$

The ratio of $\dfrac{I_C}{I_B}$ is the 'β' parameter so

$$\beta = \frac{\alpha}{1-\alpha}$$

Secondly $\qquad I_E = I_B + I_C$

$$= I_B + \beta I_B \qquad\qquad \left(\because \beta = \frac{I_C}{I_B} \right)$$

$\therefore \qquad\qquad I_E = I_B (1 + \beta) \qquad\qquad$ ...(iii)

Now dividing (i) by (iii)

$$\frac{I_C}{I_E} = \frac{\beta I_B}{I_B(1+\beta)} \qquad\qquad (\because I_C = \beta I_B)$$

The ratio $\dfrac{I_C}{I_E}$ is the α parameter so

$$\alpha = \frac{\beta}{1+\beta}$$

where $\qquad\qquad I_C$ = Collector current

I_B = Base current

I_E = Emitter current

V_{CB} and V_{CE} Collector Voltages.

Power gain: The power gain in the circuit is the ratio of the power output delivered to the load to the power available from the generator.

The power delivered to the load is $\dfrac{V^2}{R_L}$ and input is $\dfrac{Vg^2}{R_L}$.

For common emitter circuit

$$G = \frac{\text{Power output}}{\text{Power input}} = \frac{I_C^2 \cdot R_L}{I_B^2 \cdot R_i} = \beta^2 \frac{R_L}{R_i}.$$

For common base circuit

$$G = \frac{P_{\text{out}}}{P_{\text{in}}} = \frac{I_C^2 \cdot R_L}{I_E^2 \cdot R_i} = \alpha^2 \frac{R_L}{R_i}$$

For common collector circuit

$$G = \frac{P_{out}}{P_{in}} = \frac{I_E^2 \cdot R_L}{I_B^2 \cdot R_i} = \left(\frac{I_E^2}{I_C^2} \times \frac{I_C^2}{I_E^2} \right) \cdot \frac{R_L}{R_i}$$

$$= \left(\frac{\beta}{\alpha} \right)^2 \cdot \frac{R_L}{R_i}.$$

Q. Compare the common base, common emitter and common collector transistor amplifiers.

Ans. The comparison and characteristics can be tabulated as under:

Table 26.4. Comparisons

S. No.	Items	Common base amplifier amplifier	Common emitter amplifier	Common collector	Remarks
1.	Input R	30-150 Ω	500-1500 Ω	20-500 k Ω	
2.	Output R	300-500 k Ω	30-50 k Ω	50-1000 Ω	
3.	Voltage gain	500-1500	300-1000	less than one	
4.	Current gain	less than one	25-50	25-50	
5.	Power gain	20-30 db.	25-40 db.	10-20 db.	

Table 26.5. Characteristics of amplifiers

S. No.	Item	Common base amplifier	Common emitter amplifier	Common collector amplifier	Remarks
1.	Current gain	Nearly one (α)	Large (β)	Largest (β + 1)	
2.	Voltage gain	Large	Very large	Approx. one	
3.	Power gain	Moderate	Largest	Lowest	
4.	Input R	Lowest	Moderate	Highest	
5.	Output R	Highest	Moderate	Lowest	
6.	Signal phase shift between output and input	No	Yes, 180°	No	

26.8 H-PARAMETERS

Q. What are the hybrid parameters, give the expression of h-parameters of common base amplifier?

Ans. The hybrid means the mixture of distinctly different items (for example a hybrid set, *i.e.,* a combination of valves and transistors). A different approach to the derivation of a transistor equivalent circuit comes from the choice of hybrid or *h*-parameters, to describe the transistor characteristics. The a.c. signals limited to quasi-linear region leads to these constants, as

$$V_1 = h_i \cdot I_1 + h_r \cdot V_2$$
$$I_2 = h_f \cdot I_1 + h_o \cdot V_2$$

The *h* co-efficients are known as the hybrid parameters, since both open and short circuit terminations are used.

When the open circuit, $I_1 = 0$, and short-circuit ($V_2 = 0$) conditions are applied then the parameters are as under:

$$h_i = \frac{V_1}{I_1},$$ short circuit input impedance, $V_2 = 0$

$$h_r = \frac{V_1}{V_2},$$ open circuit reverse voltage ratio, $I_1 = 0$

$$h_f = \frac{I_2}{I_1},$$ short circuit forward current ratio, $V_2 = 0$

$$h_0 = \frac{I_2}{V_2},$$ open circuit output admittance, $I_1 = 0.$

Similarly for a common base amplifier these parameters are as under:

$$h_{ib} = \frac{V_{cb}}{I_e},$$ short circuit input $V_{eb} = 0$

$$h_{rb} = \frac{V_{eb}}{V_{cb}},$$ open circuit reverse voltage $I_e = 0$

$$h_{fb} = \frac{I_c}{I_e},$$ short circuit forward current ratio, $V_{eb} = 0$

$$h_{ab} = \frac{I_c}{V_{eb}},$$ open circuit output admittance, $I_c = 0.$

26.9 POWER SUPPLY

Q. What are the necessities of using power supply? Explain a full wave rectifier circuit used in a power supply.

Ans. Most of the electrical energy generated is the A.C. supply but there are certain applications where the D.C. is only essential, *viz.*, battery charging electroplating, electron tubes, transistors, etc. The efficient and most convenient method of getting D.C. is by using the rectifiers. The apparatus or source by which D.C. can be obtained by using the electronic components, is known as power supply.

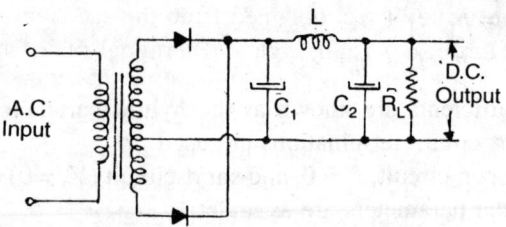

Fig. 26.25. Full wave rectifier power supply circuit.

With the help of two diode the full wave rectification is obtained as shown in Fig. 26.25. When terminal A is positive the current of positive half will be rectified by D_1 and in the second half cycle when B will be positive current will flow from D_2 to external circuit. The condenser filter circuit is used to smooth the rectified voltage. In this full wave rectification a smooth D.C. is obtained on P and Q terminals.

Q. Write a short note on the voltage doubler.

Ans. A d.c. voltage doubler circuit is used whenever high d.c. voltage and small current is required. It is capable of delivering a d.c. output voltage twice the peak value of the applied a.c. input voltage.

(*a*) **Half wave voltage doubler.** As shown in Fig. 26.26, there are two diodes D_1 and D_2. Whenever the a.c. is fed let terminal A is positive, there is flow of current through the capacitor C_1 charging it; to the maximum voltage, *i.e.*, peak value of the applied voltage. In the

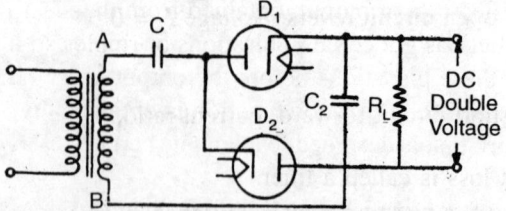

Fig. 26.26. Half wave voltage doubler.

next half cycle when the polarity is reverse, the voltage across C_1 is added to the line so that double the voltage is applied to the plate of diode D_2. Now current flow will charge C_2 capacitor, to twice the value of C_1 so the working voltage of capacitor should be twice as that of capacitor C_1. Thus twice the peak value voltage is obtained at *CD* terminals.

(*b*) **Full wave voltage doubler.** The circuit diagram of the full wave doubler is shown in Fig. 26.27. Whenever terminal '*A*' of the transformer winding is positive with respect to *B* terminal, diode D_1 comes into function, thus and charges the capacitor C_1 upto the maximum value of a.c. obtained from the secondary of the transformer. During next half cycle '*B*' terminal of the secondary becomes positive with respect to terminal '*A*'. Thus the D_1 remains in charged condition except for a small amount that leaks off through R_1. Now the valve D_2 comes into function and charge the lower plate of the capacitor C_2 up to the peak value of the secondary of the transformer, except a small leak through C_2. Now the voltage at *P* and *Q* is equal to the voltage across C_1 and C_2 which are connected in series. Thus resulting twice the peak secondary voltage of the secondary of the transformer so it works as the device to double the voltage *i.e.*, full wave diode doubler.

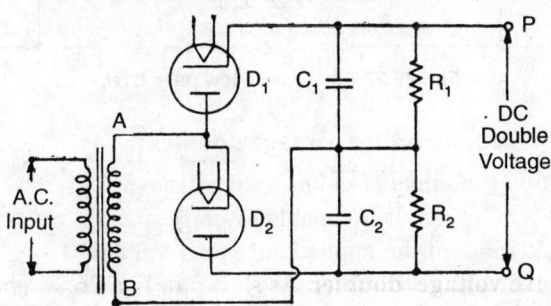

Fig. 26.27. Full wave voltage doubler.

Q. What do you understand by the filter circuit, describe in brief?

Ans. The rectified d.c. or output, obtained from the electron tubes is not a smooth d.c. but has got certain pulsations or ripples so it is very essential to smooth these pulsations before the output is applied to any plate, screen grid circuit etc. The smooth action is obtained from the filter circuit so a network that is designed to attenuate certain frequencies but pass others without loss is called a filter.

A filter therefore possesses at least one 'band pass' and at least one 'attenuation band'. The circuits which are generally used for filtration are the choke input filter and capacitor input filter circuit. A capacitor stores the energy as electrostatic field, when the voltage is applied to the capacitor, the fundamental action is to oppose the change, so when rising the voltage it acquires charge and it releases in case when voltage falls. Similarly in the inductor coils, a magnetic field is produced and stores the energy in the shape of charge and releases in case of decrease in voltage. Thus minimising the pulses and smoothing the rectified output voltage of a rectifier.

There are the two main types of the filters:

    (*a*) Choke input filter   (*b*) Capacitor input filter.

In accordance with the signal bypassings these filters are again classified as *low pass* where low frequency signals are to be by-passed, *high pass* where high frequency signals are to be bypassed, *band pass* where a particular band of frequencies are to be allowed to bypass and remainings to be attenuated *band suppression,* where the particular band of frequencies are to be attenuated and remaining are bypassed.

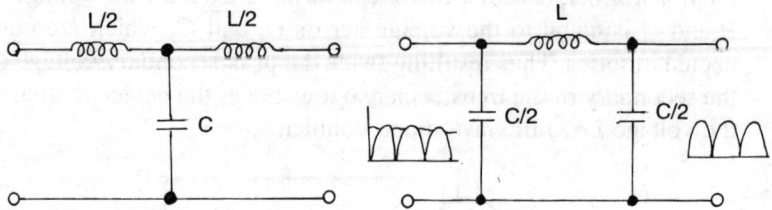

Fig. 26.28.(*a*). *T* and π low pass filter.

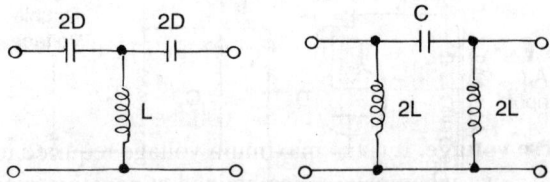

Fig. 26.28.(*b*). *T* and π high pass filter.

Q. Define ripple factor, ripple frequency, voltage regulation applied to a rectifier, peak inverse voltage.

Ans. Ripple factor.   *(N.C.V.T., R/TV 1974)*

The ripple factor is the ratio of the fundamental r.m.s. A.C. component of the rectified D.C. output of rectifier to the d.c. component.

i.e., Ripple factor = $\dfrac{\text{Effective value of the A.C. component in the output}}{\text{d.c. component in the output}}$

The effective value of the load current is

$$I_{r.m.s.} = \sqrt{I_{d.c.}^2 + I_{a.c.}^2}$$

and
$$I_{a.c.} = \sqrt{I_{r.m.s.}^2 - I_{d.c.}^2}$$

Now as in the ripple factor

$$I_{a.c.} = I_{d.c.} \sqrt{\frac{I_{r.m.s.}^2}{I_{d.c.}^2} - 1}$$

or
$$\frac{I_{a.c.}}{I_{d.c.}} = \sqrt{\left(\frac{I_{r.m.s.}}{I_{d.c.}}\right)^2 - 1} = \sqrt{F^2 - 1}$$

where F is the ratio of effective value to average value.

Let in a rectified voltage d.c. component is of 200 V and a.c. r.m.s. component as 100 V, then the ripple factor is $100/200 = 0.5$ and in percentage as $0.5 \times 100 = 50\%$. For better performance and effectiveness the ripple factor should be as low as possible even of the order of 0.00003.

Ripple frequency. The ripple frequency of a rectifier is the fundamental frequency of the a.c. component appearing in the output of the rectifier circuit, higher the ripple frequency better the output.

Voltage regulation. It is the rise in voltage when the full load on the rectifier is thrown off expressed in terms of the full load voltage. It is expressed in percentage.

i.e.,
$$\% \text{ Regulation} = \frac{E - V}{V} \times 100$$

where E = the voltage of the rectifier without load.

V = the voltage of the rectifier at load.
smaller rise in voltage better will be the voltage regulation.

Peak inverse voltage. It is the maximum voltage required to make the diode not to conduct when plate is maintained at negative potential. The peak inverse voltage would be twice as the amplitude of the plate voltage assuming a large reservoir capacitor. It can be determined by the A.C. r.m.s. voltage multiplied by 1.414.

Example. *Deduce an expression of the ripple factor for the output voltage of a half wave rectifier.*

Solution. The mean square value of a half wave output will be

$$\frac{\left(\dfrac{V_m}{\sqrt{2}}\right)^2}{2} = \frac{V_m^2}{4} \text{ V.}$$

Because if V_m is the peak inverse voltage than r.m.s. value

$$= \left(\frac{V_m}{\sqrt{2}}\right)^2 = \frac{V_m^2}{2}.$$

The d.c. average value of the output voltage for half wave rectification will be V_m/π.

The ripple factor

$$= \frac{\text{Effective value of the a.c. component in the output}}{\text{d.c. component in the output}}$$

The r.m.s. value of the ripple voltage

$$= \sqrt{\left(\frac{V_m}{4}\right)^2 - \left(\frac{V_m}{\pi}\right)^2} = \frac{V_m\sqrt{\pi^2 - 4}}{2\pi}$$

so ripple factor

$$= \frac{\dfrac{V_m\sqrt{\pi^2 - 4}}{2\pi}}{\dfrac{V_m}{\pi}} = \frac{\sqrt{\pi^2 - 4}}{2}$$

$$= 1.21. \quad \textbf{Ans.}$$

Example. *A full wave rectifier uses a double diode with each element having a constant internal resistance of 400 Ω. The transformer r.m.s. secondary voltage from the centre tape to each plate is 250 V. The load has a total resistance of 3000 Ω. calculate:*

(a) *D.C. output power.*
(b) *A.C. input power.*
(c) *Rectification efficiency.*
(d) *Voltage regulation.*
(e) *Ripple factor.*

Solution. The r.m.s. value of transformer voltage = 250 V.

So the $\quad E_{\max} = \sqrt{2} \times E_{r.m.s.} = 250\sqrt{2}$ V.

Now forward resistance $R_0 = 400$ Ω
and Load resistance $\quad\quad R_L = 3000$ Ω

(a) $\quad\quad\quad\quad$ D.C. output $= \dfrac{4E_m^2 R_L}{\left(R_0 + R_L\right)^2 \times \pi^2}$

So Power d.c., $P_{d.c.} = \dfrac{4\left(250\sqrt{2}\right)^2 \times 3000}{\left(400 + 3000\right)^2 \times \pi^2}$

$= 13.15$ **W. Ans.**

(b) A.C. input power, $P_{a.c.} = \dfrac{E_m^2}{2\left(R_0 + R_L\right)} = \dfrac{\left(250\sqrt{2}\right)^2}{2\left(400 + 3000\right)}$

$= 18.4$ **V. Ans.**

(c) Rectification factor $= \dfrac{P_{d.c}}{P_{a.c.}} \times 100 = \dfrac{13.15}{18.4} \times 100$

$= 71.5\%$ **Ans.**

(d) Output voltage at no load $= E_{d.c.} = \dfrac{2E_m}{\pi}$

$= \dfrac{2 \times 250\sqrt{2}}{\pi} = 225$ V

Output voltage at full load, $V_{d.c.} = \dfrac{2E_{max}}{\pi\left(1 + \dfrac{R_0}{R_L}\right)}$

$= \dfrac{2 \times 250 \times \sqrt{2}}{\pi\left(1 + \dfrac{400}{3000}\right)}$

$= 198.599 \simeq 199$ **V**

Now % voltage regulation $= \dfrac{E - V}{V} \times 100$

$= \dfrac{225 - 199}{199} \times 100 = 13.3\%$ **Ans.**

(e) Ripple factor $= \sqrt{\left(\dfrac{I_{r.m.s}}{I_{d.c.}}\right)^2 - 1}$

$$= \sqrt{\left(\frac{\pi}{2\sqrt{2}}\right)^2 - 1} \qquad \left(\because \frac{I_{r.m.s}}{I_{d.c.}} = \frac{\pi}{2\sqrt{2}}\right)$$

$$= \sqrt{(1.11)^2 - 1}$$

$$= \mathbf{0.48.} \quad \mathbf{Ans.}$$

Example. *Find the (i) ripple factor of full wave rectifier,*
(ii) the output voltage at no load,
(iii) the voltage regulation of the given circuit. If the no load current is
100 mA at 18 V d.c.

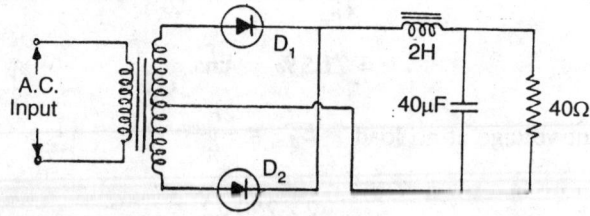

Fig. 26.29.

Solution. Ripple factor $= \dfrac{0.83}{LC} = \dfrac{0.83}{2 \times 40} = 0.0104 \times 100$

$$= 1.04\%.$$

$$V_{d.c.} = 0.637 \, V_m - I_{dc} \cdot R_L$$
$$18 = 0.637 \, V_m - 40 \times 100 \times 10^{-3}$$

$$V_m = \frac{18 + 40 \times 1000 \times 10^{-3}}{0.637} = 34.5 \text{ V}$$

At no load $\qquad V_{d.c.} = 0.637 \, V_m$

$$= 0.637 \times 34.5 = 21.97 \simeq 22 \text{ V}$$

Now percentage regulation $= \dfrac{E - V}{V} \times 100$

$$= \frac{22 - 18}{18} \times 100 = 22.2\%.$$

$$\mathbf{1.04\%, \ 22 \ V, \ 22.2\%.} \quad \mathbf{Ans.}$$

Example. *Calculate the input voltage, ripple voltage and current of a*
power supply operating from 500 Hz power line delivering a voltage 30
$V_{d.c.}$ across 1000 Ω load shunted by a capacitor of 5000 μF.

Solution. $E_{d.c.} = E_{in} = 30$ V

$$= \frac{1}{4\,f.R.C.} = \frac{1}{4 \times 500 \times 1000 \times 5000 \times 10^{-6}}$$

$$= 0.0001 = 1 \times 10^{-4}.$$

$$E_{r(\text{peak})} = 30 \times 1 \times 10^{-4} = 3 \text{ mV}$$

$$E_{r(r.m.s)} = \frac{3}{\sqrt{3}} = \textbf{1.732 mV.}$$

$$I_{r(r.m.s.)} = \frac{1.732}{1000} = 1.732\,\mu A$$

$$I_{r(\text{peak})} = \frac{3 \times 10^{-3}}{1000} = 3\,\mu A$$

$$I_{dc} = \frac{E_{d.c.}}{R} = \frac{30}{1000} = \textbf{3 mA Ans.}$$

$$\text{Ripple factor} = \frac{1.732 \times 10^{-6}}{30 \times 10^{-3}} = 0.5773 = \textbf{5.8} \times \textbf{10}^{-5}\textbf{ Ans.}$$

26.10 OSCILLATORS

Q. What is an oscillator and what are the essential requirements of an oscillator?

Ans. An electrical oscillator is a device which acts as a energy converter. It converts d.c. electrical energy into alternating current energy. It can change a.c. of a considerable frequency ranging from few cycles to thousands of cycle per second. The oscillators are comprehensively used in testing equipments radar and telecommunication.

The essential requirements are as follows:

(a) **Tank circuit.** A resonant circuit usually known as oscillatory, which consists of inductance and capacitance to determine the frequency of oscillation is used, and is known as tank circuit.

(b) **D.C. source.** To compensate the loss of energy in the tank circuit a *d.c. source* is required.

(c) **Feed back network.** It is also known as regenerative feed back. It supplies the energy in the right phase to maintain the oscillations.

Q. How the frequency of an oscillator is determined?

or

How the frequency of oscillation is determined in a tank circuit?

Ans. The tank circuit consists of the inductance and capacitance, hence the frequency is related to both. The natural frequency is inversely proportional to both as given by the formula

$$f = \frac{1}{2\pi \sqrt{LC}}$$

where f = frequency in cycle per second.

L = the inductance in Henry.

C = the capacitance in farads.

π = a constant 3.1416.

thus any frequency can be obtained just by adjusting the values of the inductances and capacitances. Similarly the oscillating frequency of a given circuit can be or calculated.

Q. What do you understand by the Audio Frequency (AF) Oscillators? Describe any one of them.

Ans. The audio frequency oscillators may be constructed on the same principles as that of the radio frequency oscillators, except that the larger inductors and capacitors are used. To obtain the larger capacity the iron core inductors are used, but to avoid the iron core inductors, which produces distortions, the following oscillators are used.

(a) Wien's Bridge *i.e.* R.C. tuned oscillator.

(b) Beat frequency oscillator

(a) **Wien Bridge or R.C. tuned oscillator.** It is an audio sine wave oscillator. It is widely used. The output of it is free from fluctuations, distortions and a pure sinusoidal output can be obtained.

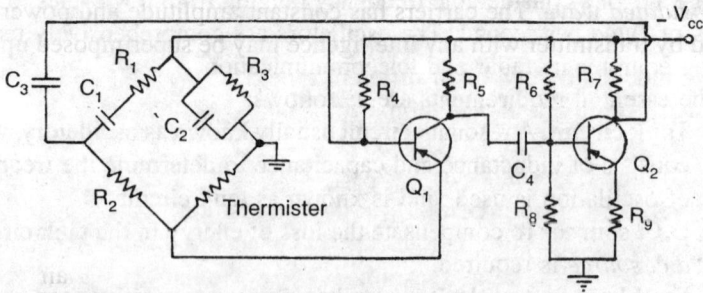

Fig. 26.30. Wien Bridge Transistor Oscillator.

It is also known as inverse feed back oscillator. There are two stages of amplification with bypassed emitter resistors as shown in Fig. 26.29. There is a thermistor provided in the bridge which is used as a sensitive element for temperature increase. As more current flows, greater power is dissipated by the thermistor, and temperature increases which increases the degenerative feed back reducing output. On the other hand if less current flows, the less degenerative voltage is fed resulting an increase in the output. The output appears without disturbances.

The connection diagrams are shown in Fig. 26.30. The oscillator frequency in ideal conditions can be given as

$$f = \frac{1}{2\pi CR}$$

here in this particular case it is

$$f = \frac{1}{2\pi \sqrt{C_1 \cdot C_2 \cdot R_1 \cdot R_3}} \text{ Hz.}$$

26.11 MODULATION

Q. What is the necessity of the modulation? State in brief the amplitude modulation, frequency modulation and phase modulation.

Ans. The clear and audible sound on the receiver side, is the main objective of the broadcasting. The intelligence is transmitted, but it must be propagated through space by some special means. It is well known that the signals about the audio range can be propagated through space very easily in the form of electromagnetic waves, these signals are the radio frequency (R.F.) or the carrier waves. It is possible to use these R.F. waves as a means of transporting audio signals to the receiver, it is done by superimposing the intelligence on R.F. wave. The superimposing of signal on R.F. wave is known as *modulation* and the wave thus obtained as *modulated wave*. The carriers has constant amplitude and power generated by transmitter with any intelligence may be superimposed upon it.

The modulation can be of different types:
- (*i*) Amplitude modulation.
- (*ii*) Frequency modulation.
- (*iii*) Phase modulation:

(*i*) **Amplitude modulation.** It is the modulation in which the amplitude of the R.F. oscillations is varied in accordance with the strength of intelligence to be transmitted. The wave has three main characteristics, the amplitude, the strength and the frequency or phase. When the audio signal or pure sine wave is super-imposed upon the R.F. wave, it is called the *amplitude modulation (A.M.)* as shown in

Fig. 26.31. For example, the modulation of a speech, music, and the information to be conveyed with the radio signal, is accomplished in such a way that the modulated signal alternately increases and decreases its amplitude of the radio signal. Therefore the stronger the audio signal combined with R.F. carrier, greater is the change in the amplitude of the carrier.

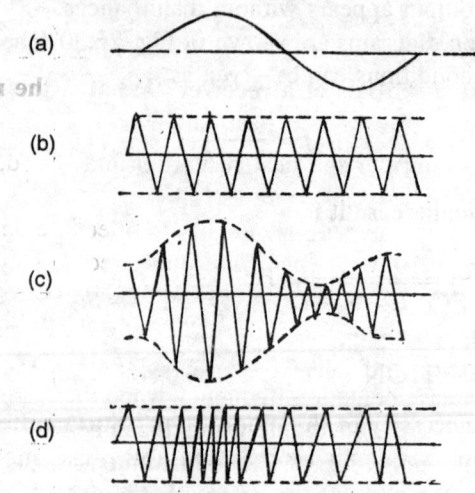

Fig. 26.31. Amplitude modulation.

(b) **Frequency modulation.** It is the type of modulation in which the instantaneous frequency of the R.F. signal is varied in accordance with audio signal to be transmitted keeping the amplitude as shown in Fig. 26.32.

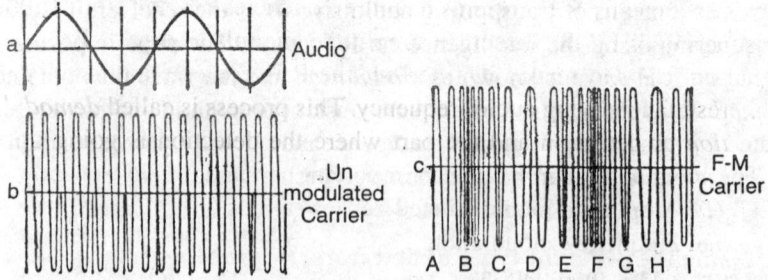

Fig. 26.32. Frequency modulation.

In this system of modulation the frequency of R.F. wave increases and decreases as the A.F. voltage rises or falls. This increase and decrease of the waves called frequency deviation and the process as *frequency modulations (F.M.)*.

(*c*) **Phase modulation.** It is the type of modulation in which the value of reference phase of carrier is varied in such a way that its magnitude is proportional to the instantaneous amplitude of the R.F. signal to be transmitted, so it is the modulation in which the frequency deviation is proportional to the modulating frequency.

26.12 DETECTOR

Q. What are the essentials of a receiver? What is the necessity of a detector?

Ans. In order to consider the best reception and sound, the essential requirements of a receiver are as follows:

(*a*) **Selectivity.** It relates with the quality to select the desired signal to the exclusion of interference from undesired signals. It is nothing but the capacity to distinguish between the desired signal and signals of other frequencies.

(*b*) **Fidelity.** The fidelity of a receiver responds the variation of the output without modulation frequency when the output load impedance is a resistance. In simple words it can be expressed as it must produce different nodes, of all the instruments along with the song to give more charm to the audience and sound be comfortably and clearly audible.

(*c*) **Sensitivity.** It is expressed in terms of the voltage (power) that must be applied to the receiver input to give standard output, in other words the gain of the intermediate frequency amplifier.

(*d*) **Necessity of a detector.** The electro-magnetic propagation may be produced by R.F. currents having frequencies ranging from 16 kHz to 10^9 kHz. To detect an amplitude modulated signal it is only necessary to rectify it, filter out the R.F. contents and supply the resultant varying audio frequency. This process is called *demodulation* or detection and the part where the detection is going on, is known as '*detector*'. It is an important part and used to separate the A.F. from the R.F. modulated carrier signals.

Q. Name the different types of detectors. Describe with the help of a circuit, diagram the diode detector.

Ans. In order to detect the signals many devices have been developed and used. These are coherer, the electrolyte detector, the magnetic detector, the crystal detector and the valve detectors. The diode detector is a valve type detector.

Diode detector. The diode is used as a detector. It is the simplest of the valve detectors. In this case, electrons only flow appreciably from fila-

ment to anode of the diode valve, when anode is positively charged with
respect to cathode; thus it can allow signals only in one way.

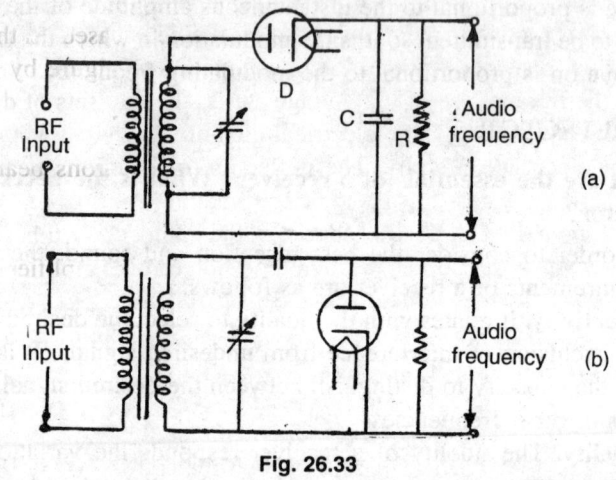

(a)

(b)

Fig. 26.33

A simple circuit of a diode detector is shown in Fig. 26.31. When 'C'
is the small capacitor, R is the relatively high resistance, and the combination R.C. is the load impedance across which rectified voltage E_L of the
diode is developed. At every peak of the positive of the R.F. cycle of the
exciting voltage the capacitance 'C' charge up to a potential that is almost, but not quite equal to the peak of the applied voltage. It fails by a
small amount to reach the peak voltage because of the voltage drop produced in the diode tube by the charging current that flows through the
tube. The presence of R causes C to discharge slightly during the half
cycle that makes the anode negative with respect to cathode. Thus the
D.C. voltage is slightly less than the peak voltage of the carrier and the
carrier frequency will depends upon the time constant of R and C.

If the time constant is large the voltage across C will drop very little
during half cycle and ripple will be small. But on the other hand if low
value of the time constant of C and R as compared to the periodic time
constant of the modulated frequency this voltage across R will be able to
follow the audio frequency variation in the peak voltage. The a.c. component of the voltage across of R is then fed via a d.c. blocking condenser
to the grid of the next stage which will be an audio amplifier. Thus a
diode can be very well used as a detector.

26.12 OSCILLOGRAPH

Q. What do you understand by the oscilloscope? What is the working

principle? Describe in brief the construction, controls and uses of the oscilloscope.

Ans. Oscilloscope. It is an instrument for indicating the wave form of a variable voltage and current. Its working principle is based on the movement of a fluorescent spot which is made stationary figure by properly adjusting the frequencies of the voltage across the two sets of deflection plates. A visible curve of one electrical quantity is plotted as a function of another electrical quantity. The deflection of electrons beam is obtained either by electrostatically or electromagnetically.

The main parts of an oscilloscope are, the cathode ray tube, the time base generator, power supply, vertical and horizontal amplifiers.

The *cathode ray tube* contains an electron gun structure, consisting of cathode, control grid, and the accelerating anodes. The electrons beam is formed by the gun and can be deflected horizontally and vertically by the controls provided. There is a potentiometer to have various voltages tapped from it. The cathode of the tube is indirectly heated nickel cylinder, having oxide coated flat surface. A grid to control electrons is also provided. The accelerating anodes generally three in numbers, are provided to pass high speed electrons through them. These anodes are kept at different potentials. The deflecting arrangement shift controls and other controls are provided to change and excite the waveforms. A screen which has fluorescent powder coating is used to convert the kinetic energy of electrons into the visible radiation. The tube converts the electrical signal (voltage) into a visual one.

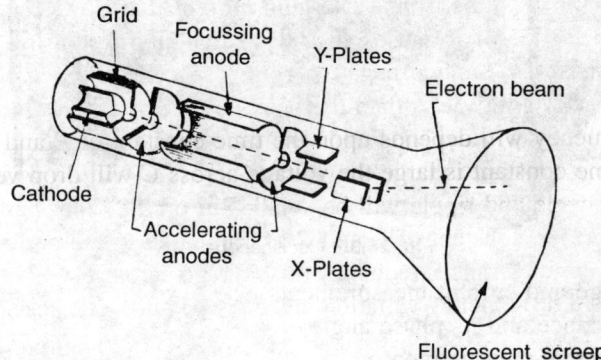

Fig. 26.34. Cathode ray tube.

The voltage wave which is to be observed is applied to the vertical deflection plates and voltage that acts as a *tuning wave* is applied to the horizontal deflection plate circuit incorporated with the oscilloscope generated a saw tooth voltage wave which is used as a tuning wave.

A *sweep oscillator* generates the saw tooth wave. A voltage of constant rate of change to horizontal deflection will cause the beam to move straight across the phase of oscilloscope. Applying a variable voltage to the vertical deflection plate, the waveform will be traced on the fluorescent screen. The voltage supplied, is arranged so that as soon as the tuning voltage moves the beam, all the way across the screen, the beam returned to the starting point and the process is repeated. The movement is called *sweep* and the repeation per second as the *sweep frequency.*

A stationary pattern is obtained on the fluorescent screen when the frequency of sweep voltage is synchronized with the frequency of the voltage to be observed.

The complete operation of the oscilloscope differs with the different manufacturers and models.

Uses. (*i*) It is used for the study and observations of the waveforms.

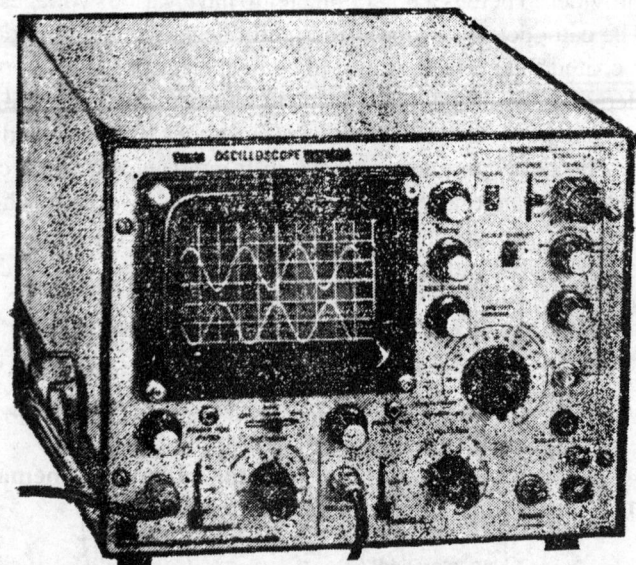

Fig. 26.35. Oscilloscope.

  (*ii*) Voltage and current measurements.

 (*iii*) Impedance and its phase angle.

 (*iv*) Frequency measurement.

  (*v*) It is used to align the I.F. stages of a radio receiver when used in conjunction with a signal generator.

 (*vi*) To align the R.F. stages of a radio receivers when used in conjunction with a signal generator.

(*vii*) To determine the quantity of the A.F. signal by examination of its waveform and various points in audio sections of the receiver.

(*viii*) As a monitor the amplitude modulation envelope of radio transmitter and display frequency modulation output waveforms.

(*ix*) To study the B.H. curves and hysteresis loops.

(*x*) Fault testing of windings of electrical machines.

(*xi*) And so many engineering analysis and study *viz.*, engine analyser, transmission line standing waves, dynamic balancing, radiation of patterns of antenna, microwave spectroscopy lens testing, ultrasonic phenomenon, and so many other places.

Q. What are the precautions to be observed while in handling the C.R.O.?

Ans. Following are the precautions to be observed while handling the C.R.O.:

(*i*) Never temper with the C.R.O. and handle carefully.

(*ii*) Study first the face plate controls and check it before use.

(*iii*) It should never be overloaded, hence the proper controllings of the gain control is essential.

(*iv*) The synchronization applied should be sufficient to lock the display on the screen.

(*v*) Handle carefully that too after knowing the controls and their functions connection, should be tight and right.

(*vi*) When the range of voltage or current under measurement is not known, measurement should always be started from the highest range.

(*vii*) The instrument must not be exposed long in high temperature and moisture.

26.13 SIGNAL INJECTOR

Q. What is the function of signal injector? Draw a schematic diagram of multivibrator signal generating source.

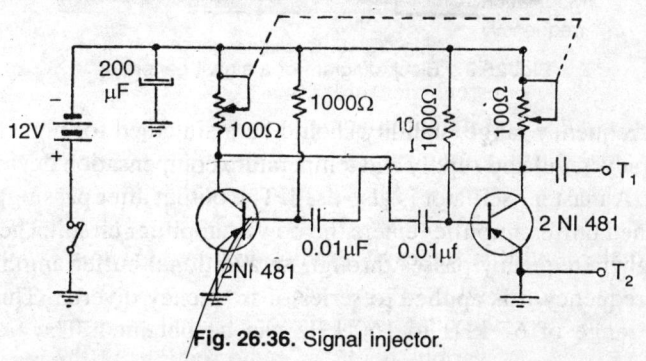

Fig. 26.36. Signal injector.

Ans. It is a device used for the checking of the radio receiver sets. The multivibrator circuit is generally used. The multivibrator circuit produces a square wave signal at an audio-frequency but contains harmonics that extend high into the R.F. region. One head or probe is only required to touch the various points in the circuit to inject the signal. A humming tone will give the operative condition of the circuit. A simple schematic connection diagram is shown in Fig. 26.36; her terminal T_2 is connected to chassis of the radio set and another testing terminal T_1, is touched to various points for testing.

Q. Describe briefly with block diagram the signal generator used in electronics.

Ans. It is a source of generating signals of wide range frequency and amplitude of high degree of accuracy. There is a provision of modulating the carrier frequency and the modulation is indicated by a meter moreover for modulation the sign wave, square wave, triangular wave or the pulses can be used.

The carrier frequency is produced by R.F. oscillator using L.C. tank circuit and are indicated by means of a frequency range control and vernier dial setting. The modulation is done in the output circuit using internal or external fixed frequency oscillators. An amplifier of signal generator is shown in the block diagram 26.37 delivering the output to an attention. A meter and attenuator setting is used to indicate the output.

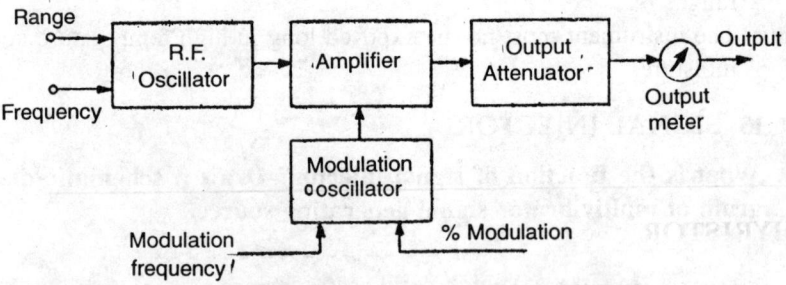

Fig. 26.37. Block diagram of a signal generator.

The frequency range, stability should be maintained for higher degree of accuracy, regulated supply and temperature compensation devices must be used. A master oscillator is also used. The output after passing through an untuned buffer amplifier enters the power amplifier circuit. The output, if of higher frequency passes through an additional buffer amplifier and if low frequency it is applied to series of frequency diverter. Thus a frequency range of 67 kHz to 160 kHz can be obtained. The frequency stability of highest frequency is imparted to low frequency range. The master oscillator may be manual or automatic.

Q. Give the colour coding chart for the resistances.

Ans. Generally the values of the resistances are identified by means of the colour codings. There is a particular arrangement of different colours marked on the resistances to read out the correct value. It can be either by marking a series of colour bands on the resistors or the entire body is coloured and one or two colours are marked on the ends of the resistance which are called tip colours.

The colour coding scheme is given as under table 26.6.

Table 26.6. Colour Code Chart for Resistances

Colour	Band A	Band B		Band C or Spot	Band D Tolerance %
Black	0	0	0	1	–
Brown	1	1	1	10	±1%
Red	2	2	2	100	±2%
Orange	3	3	3	1000	±3%
Yellow	4	4	4	10000	±4%
Green	5	5	5	100000	±5%
Blue	6	6	6	1000000	±6%
Violet	7	7	7	10000000	±7%
Grey	8	8	8	100000000	±8%
White	9	9	9	1000000000	±9%
Gold	–	–	–	–	±5%
Silver	–	–	–	–	±10%
No colour	–	–	–	–	±20%

THYRISTOR

Q. What is a thyristor? Which solid state devices are included?

Ans. The thyristor is a term used to include such devices as SCR and triacs etc. The thyristors can be thought of a solid state switches with three or more PN junctions. These are also called PNPN devices etc. These devices are utilized for fast rate (extremely fast) switching ON and OFF. They includes SCR, Diac, triac, silicon unilateral switch, SBS (silicon bilateral switch etc.

SCR. In its simple construction it has four layers with three junctions transistors as shown in fig. These are made of four layers of alternate layers off P type and N type silicon layers.

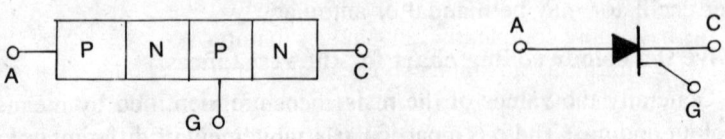

Fig. 26.38

These are extensively used for high speed switching (on & off) in power application in mercury are rectifier etc. It is used for eliminating arcing because of moving contacts and simultaneously it reduces the power losses to a considerable limit.

These are available in very large capacity of voltage and current ratings i.e. 1300 V, 500 A etc. They reduces the power losses so can be effectively used in power gain circuits having power gain more than million.

Since it controls large amount of current with small pulse voltage so it is very useful in relay, switching and control applications. These are widely used in power supply (D.C. to A.C.) invertors, radar modulator, servomedianisim and latching relays etc. The unique characteristic is the extremely rapid switching on and off without any moving contacts.

Diac. It is a three layer and two junction device and resemble to bipolar (PNP) junction. It can also be used for bidirectional switching characteristics, universal motor speed control and heat control etc.

Triac It is vital and five layer bidirectional semiconductor devices and can conduct in either direction under specific conditions. There are vitally used for AC or main power control.

It can otherwise be stated that two SCRs are connected in antiparallel and because of that they are able to conduct in either position and negative halves of sinusoidal wave form.

In addition to this, these are widely used for photodiodes solar cells, photo transistors, photo fet, light emitting diodes (LED) etc.

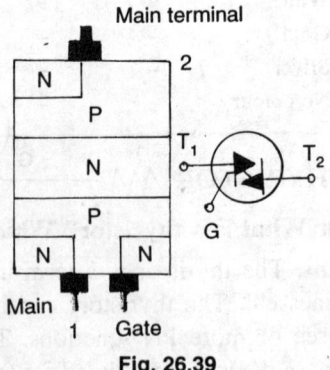

Fig. 26.39

Q. Draw a schematic diagram showing 3 φ six pulse thyristor bridge d.c. motor with a feedback loop.

Ans. The development of thyristor with high current carrying capacity and reliability has enabled thyristor regulators to be designed a d.c. variable drive system. The machines have laminated poles, the smaller machines have laminated cores even. It is to improve the signal response by thyristor regulator. Practical designed have shown a square frame designed for d.c. machine have more comparative advantages in improved power and weight ratio, with enormous advantages. The techogenerator provides a speed related signal for comparison with reference signals. It is the error or difference between the reference signals which advances or retards the firing angle of the thyristor in the regulator to correct the d.c. voltage and hence the speed of the motor.

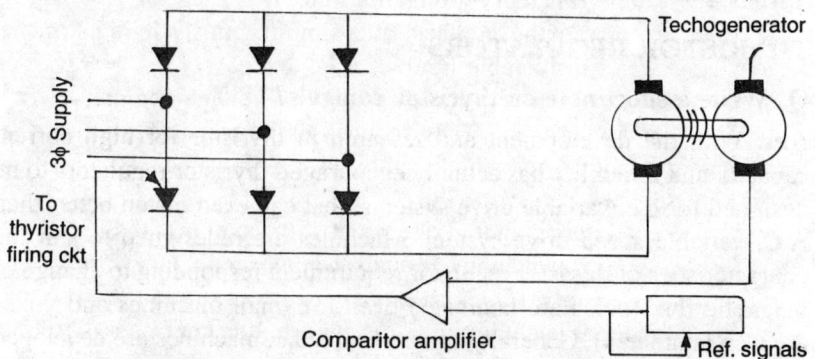

Fig. 26.40. φ six pulse thyristor Bridge d.c. motor and feedback loop.

Q. Draw a practical utility of a SCR used as a switch in control circuit.

Ans.

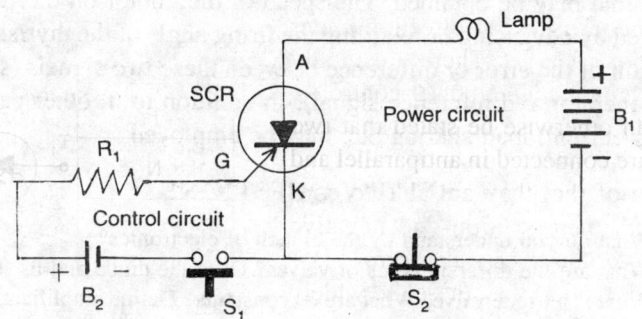

Fig. 26.41. SCR as a switch.

As shown in fig the lamp is not glowing that means SCR is not conducting i.e. SCR is working as an open circuit for the power circuit even the switch S_2 is closed. Now press S_1 (control circuit switch) it will allow a small amount of current to flow in control circuit (grid) and SCR will function as a closed switch for power circuit and lamp will glow. As soon as the switch S_1 is released even the control circuit is off but the power circuit will continue (SCR is working as a latching relay). Now if switch S_2 is pressed and released the lamp will stop glowing but will stay off even the switch S_2 is on again, again SCR is acting as open switch.

Only a holding current in the control circuit keeps the SCR on and even up to a heavy current may be allowed in the power circuit. When the current is interrupted for even an instant the SCR will turns off, with SCR off, the gate signal can turn it on again.

THYRISTOR REGULATORS

Q. Write a short note on thyristor controls?

Ans. With the development and research in thyristor for high current capacity and reliability has actually encouraged thyristor regulators to be designed for d.c. variable drive systems, that can even match better than A.C. variable speed drive system. Machines are redesigned to suit the characteristics of thyristor regulator requirement responding to change of magnetic flux (poles are laminated even for small machines and yokes are also laminated). Generally square frame d.c. machines are developed with much improved designed (power/weight ratio). Generally three phase six pulse thyrister bridge d.c. motor and a feed back loop assembly is used as shown in fig. 26.40.

It is usual for a thyristor controlled d.c. motor to be installed with a tachogenerator so that a speed related signal for comparison with reference signal may be obtained. The speed of d.c. motor on d.c. voltage is corrected by advancing or retarding the firing angle of the thyristor which is a result of the error or difference between these two signals (signals of tachogenerator and reference signal). In addition to it, other parameters such as current, load sharing etc. may be employed.

REVIEW QUESTIONS

1. What do you understand by the branch of electronics?
2. What are the different types of valves? Describe diode and its functions.
3. What is a triode valve? What are its constants? Define amplification factor.
4. Compare the transistor and a valve, their uses and application in modern electronics.
5. What are semi-conductors? Explain the *pn*-junction.
6. What is a voltage doubler? Draw its circuit diagram.
7. What are the functions of the signal inductor, draw its circuit diagram?
8. Write short notes on the followings:

(a) Diode as rectifier.

(b) Characteristics of triode.

(c) P-type semi-conductor.

(d) Amplitude modulation.

NUMERICAL PROBLEMS

1. In an L-type filter circuit compute the various components if it provides 60 $V_{d.c.}$, at 75 mA with a ripple factor of less than 1%.

[Ans. 0.8 H, C = 83 μF)

2. A half wave rectifier has internal resistance of 200 Ω and is to supply d.c. power to 1000 Ω load from 330 $V_{r.m.s.}$ source of supply. Calculate (i) $I_{d.c.}$, (ii) I_{max}, (iii) $I_{a.c.}$ load current, (iv) D.C. tube voltage drop.

(*B.T.E., Raj. May, 1973*)

[Ans. (i) 0.112 A, (ii) 0.3535 A, (iii) 0.1767 A, (iv) 22.5 V]

3. If in a capacitor filter circuit the peak rectified voltage is 190 V, C is 2 μF and load current is 20 mA. Calculate, $V_{r.m.s.}$, $V_{d.c.}$ and % ripple.

[Ans. 24 V, 148.4 V, 16.2%]

27

Electric Power

SOURCES OF ENERGY

A number of energy sources are available some of them are main sources like sun, wind, water, radioactive materials and fuels.

The sun is the prime source of energy. The energy from the sun is achieved in the form of heat, light and some radiations like ultraviolet etc. The heat energy and light energy are utilised for generating electric power. The heat energy is used to change water into steam which in Thermal turbines changes into motion and rotates the generator. Thus electricity is produced. The light energy is also changed into electrical energy which is known as Solar power.

In India the conditions are favourable because about six months the heat and light energies are available in abundance, in raining season the water in plenty helps in generating electricity. The water in hydroturbine rotates the generator producing electricity. The wind (the root cause being sun) have got tremendous amount of energy which compels the wind mills to rotate and produce electricity. The other source is the fuel, which can be in all three states solid (coal) liquid (oil) and gas (coal gas). The coal is used in boiler for producing heat which changes water into steam. This steam rotates the thermal turbine and electricity is produced by the generator. In case of liquid, the diesel engines produce the motion to rotate the generator for producing electricity. In gaseous state the gas engines are used to rotate the gas-turbine and electricity is produced. The atomic energy have got a unique field of producing tremendous amount of heat by fission of nucleous disintegration of uranium. The heat thus librated is used through properly controlling to produce steam from water and thereby rotating the generator and producing the electricity.

Thus different sources are used to convert the source energy into mechanical energy which is converted into electrical energy.

Q. What are the primary considerations for comparing the various sources for power generations?

Ans. The important point to be considered are as follows:

(*i*) Initial cost

(*ii*) Running cost

(*iii*) Availability of prime sources like water coal and gaseous etc.

(*iv*) Pollution

(*v*) Reliability application and accessibility.

Q. What are the advantages of electrical energy over the other forms of energies?

Ans. There are some of these advantages

(*i*) Electrical energy is cheap than the other forms of energies.

(*ii*) It does not pollute.

(*iii*) It is flexible, can be taken to any place.

(*iv*) From its one form, light, power work can be obtained and operated from the same supply.

(*v*) Electric drives and controls are more effective than other types of energies.

(*vi*) It is easy to handle and control than the other forms of energies.

Generation of Electricity

The primary sources like sun, wind, tide etc. are not much in use in India, because of their non-availability for a considerable duration in a year span. So secondary sources like coal, fuel oil, fuel gases, water and atomic fission etc. are used for power generation. The power stations are of these types:

(*i*) Thermal or steam power station.

(*ii*) Hydroelectric power station.

(*iii*) Nuclear power station or atomic power station.

(*iv*) Diesel power stations.

STEAM OR THERMAL POWER STATION

The power stations in which the chemical energy of fuel is converted into electrical energy are known as thermal power stations. The fuel in these station may be solid liquid or gaseous. The solid fuel is like coal may be bituminous coal, brown coal or peat and liquid fuel which is used may be fuel oil, crude oil, petrol, diesel and paraffin oil, the gaseous form may be natural gas, obtained from natural gas well producer gas – manufactured gas and by product gas, these may be from Blast furnace coke ovens, etc. are used for producing steam. The steam of suitable pressure and temperature is used to rotate the steam turbine. The steam is produced in the boiler, send to the turbine for producing the motion which in turn is subjected to the alternator coupled with the turbine to change this so obtained mechanical energy into electrical energy.

The power stations in which steam is used for steam engine or steam turbine as a prime source is called steam power station. In these types of plant the coal is used as a fuel. The burning of coal produce heat which is used to convert water into steam which actually runs the steam engine or steam turbine. The turbine rotates the armature of the alternator and a voltage is generated by the alternator. Generally 11 KV, 3.3 KV, 6.6 KV voltage is generated, which is steped up for further transmission.

The heat energy in these generating stations is obtained by the available fuels which could be coal, (of different types) rice straw, etc.

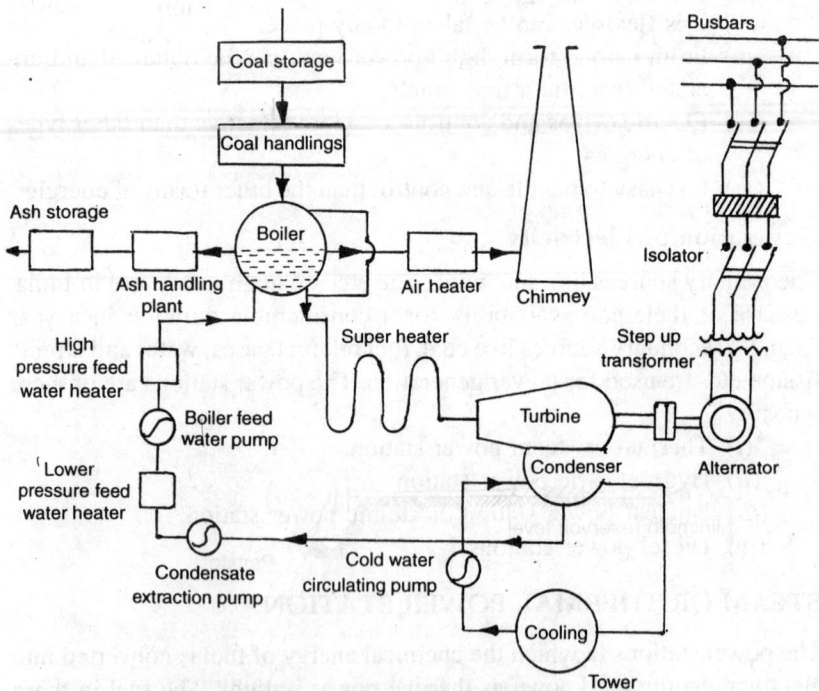

Fig. 27.1. Line diagram of thermal power station.

In these plants the coal is brought to the power station through rails or road and stored in the storage plant. From there it is brought to the coal handling plant where it is pulverised or crushed to small pieces for efficient burning. Now it is fed to boiler where the heat energy of coal is transformed into steam. The coal ash is than removed to the coal handling plant and then delivered to ash storage plant. The removal of ash from boiler is essential for getting proper calorific valve of coal burning.

The steam from the boiler is obtained and is super-heated for proper temperature and pressure it passes through superheater, economiser and air heater. Thus the superheated dry steam is fed to steam turbine through the main value. The impinge of steam rotates the steam turbine thereby causing the rotation of alternator and producing electricity.

These plants are very common because of cheap fuel, less initial cost, less space compared to hydro power plants, but pollution, running cost are the disadvantages. The efficiency of these plants is quite low i.e., approximately 25%.

Hydro Electric power stations:

The power stations which converts the energy stored in water into electrical energy are known as hydraulic electric power station. In this water is stored at a suitable place by constructing dam. Thus a water head is obtained. In some of the cases this difference of level exists naturally as in case of water fall in hilly areas but in other places it may be obtained by constructing the dams.

Thus the water from there is guided to the power house through penstocks which discharge water to the inlet of the water turbine. Now this turbine converts the water energy into electrical energy through the alternator. The mechanical energy obtained from the water is converted into electrical energy with the help of alternator.

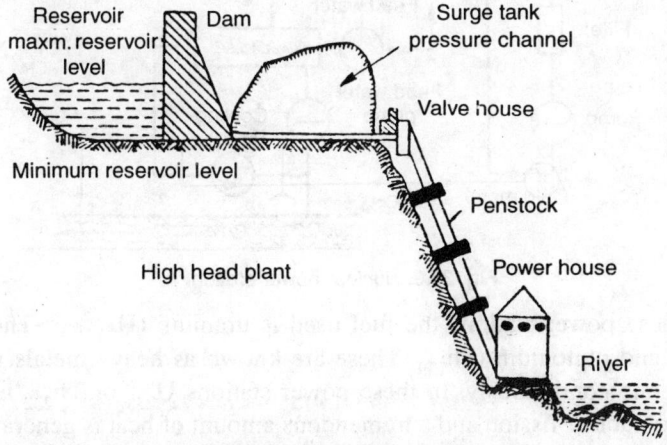

Fig. 27.2. Hydropower station.

The discharge of the water is not wasted but is used through tail race for irrigation purpose in agriculture.

Different types of hydro-turbines are used in these power station. The type of turbine depends upon the water head. In the places of low head i.e., upto 50 M head vertical shaft Francis turbines are used and high head above 50 M to 75 M horizontal turbines with some modifications the Francis turbines are used and for more than 175m the Pelton wheel type, prime movers may be used.

These hydro electric power stations are simple in construction, quite neat and clean, have low maintenance cost, low running charge and generates electricity at very low cost. The disadvantages may be, as high installation charges, appreciable water head, lots of water is required and high cost of transmission lines, so these sites are selected at such a place where adequate water at reasonable water head is available.

Nuclear power stations

The power stations in which the nuclear energy is converted into electrical energy are called nuclear power stations.

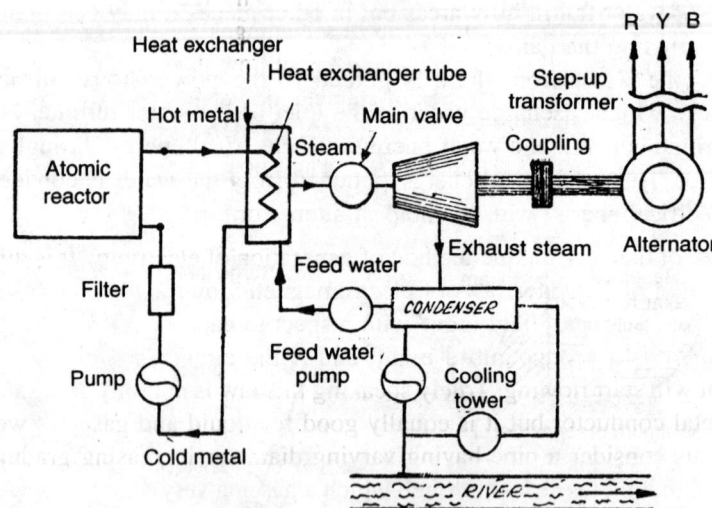

Fig. 27.3. Nuclear power station.

In these power stations the fuel used is uranium (U_{235}) or Thorium (Th_{232}) and plutonium (Ph_{239}. These are known as heavy metals which disintegrate continuously. In these power stations U_{235} or Th_{232} is subjected to nuclear fission and a tremendous amount of heat is generated. It is experienced that if one Kg of U_{235} is splitted by fission, the amount of heat librated by burning a 2500 Tonnes of high grade coal. Thus this heat energy is used for producing steam and this heat after certain processes is utilized for driving the steam turbine which in turn moves the alternator which is coupled with the turbine and electricity is produced. Thus the nuclear energy is converted into electrical energy. It is an example of use of nuclear energy for peaceful purposes.

These power stations can be classified or divided into the nuclear reactor, heat exchanger, steam turbine and alternator. In nuclear reactor the fission process is carried out. It is specially designed and the chain reaction is controlled here. Once the fission starts the splitting will result tremendous amount of heat energy and neutrons which will establish the chain reaction and an explosive stage due to fast release of heat energy will be dangerously difficult if not properly controlled. The fast moving excess neutrons are absorbed into heavy water and controls the chain reaction to establish a controlled heat energy. This heat energy with the help of heat exchanger is extracted and is used for producing steam. This steam is fed to the steam turbine. The turbine rotates along with the coupled alternator and produce electricity.

In the nuclear power stations the amount of fuel required is less, low running cost, economical for large capacity and the cost of electricity generation is considerably low. The disadvantages are the fuel is costly, high capital cost and greater technical know how is required.

In India there are a number of nuclear power stations for example RAPP (Rajasthan Atomic power project) NARORA power plant, Tarapur Atomic Power Plant, Kalpakam power plant near Madras.

MHD (Magneto Hydrodynamics) system of power generation

It is one of the very unique method of generation of electricity. It is direct application of Faradays laws of electromagnetic induction where a conductor and magnetic field move with respect to each other (relative motion) there is a production of e.m.f. and if the circuit is completed the current will start flowing. Truely speaking this law is not only confined to the metal conductor but it is equally good for liquid and gases as well.

Let us consider a pipe having varying diameter increasing gradually from beginning to the end through which a gas at a very high temperature and pressure passes from one end to the other say from left to right. Now let us have a magnetic field perpendicular to the direction of motion of the gas. In this condition, the emf is induced in a third mutually perpendicular direction. Here one should be clear that at high temperature the gas becomes conductive. The conductivity of the gas could be increased just by adding a small amount of vaporised metal such that potassium etc.

It is not a very new idea of generating emf but a lot of experiments and work has to be done for making it a success. There are many advantages like these are simpler, no highly stressed moving part so losses are less, the duct could be of any size so high capacity generations could be possible, the walls cooling effectively increases the capacity of generations. But still work is in progress to make it a great success. The high temperature is a limiting problem, the thermal insulations and conductions is also a problem and some other practical problems are yet to be overcome. A lots of experiments are carried over and scientists are going on to make it convenient and useful.

The liquid MHD generators are also on its way.

Q. Give a live diagram of a typical modern A.C. power station.

Ans. The wide use of a.c. has necessitated to have more power stations. These are located at a suitable places, away from the population. The generated power is then transmitted and distributed to different area as shown by the line diagram in Fig. 27.1.

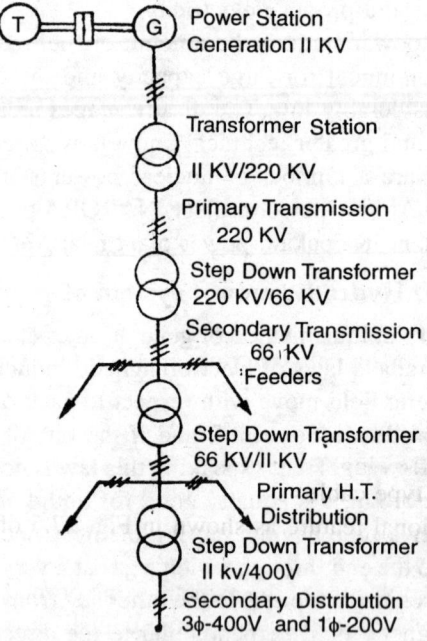

Fig. 27.4. A.C. power system.

27.1 UNDERGROUND CABLES

Q. What do you understand by the underground cables? How these are classified? Give the construction feature of the underground cable.

Ans. The underground cables are used for transmission and distribution of power where it becomes difficult to erect the over head lines. Generally in case of thickly populated areas where the maintenance, conditions, general appearance etc. matters too much; underground cables are used. The type of the cable depends upon the voltage and service requirements.

These cables are laid down in trenches with proper care, so these are named as *underground cables.*

These cables are classified according to the voltage, cores and insulation, these are as follows:

(*a*) **According to the voltage:**
    (*i*) Low tension (L.T.) cables upto 1000 V.
    (*ii*) High voltage cables (H.T.) upto 11 kV.
    (*iii*) Supertension (S.T.) cables upto 33 kV.
    (*iv*) Extra high tension cables upto 66 kV.
    (*v*) Extra super voltage power cables upto 132 kV.

(*b*) **According to the number of cores:**
    (*i*) Single core.
    (*ii*) Two cores.
    (*iii*) Three cores.
    (*iv*) Four core *i.e.,* three and half cores.

(*c*) **According to the insulation shielding:**
    (*i*) P.V.C. cable.
    (*ii*) V.I.R. cables.
    (*iii*) C.T.S. cables.
    (*iv*) Lead sheathed cables.
    (*v*) Oil field cables.

(*d*) **According to the construction:**
    (*i*) Belted cable.
    (*ii*) H-type cable.
    (*iii*) S.L. type cable.
    (*iv*) H.S.L. type cable.

The constructional feature as shown in Fig. 27.5 of the underground cable are as follows:

Cores. The central core of the cable has one or more cores of conductors. The conductors are made of tinned copper or aluminium and are usually stranded, sometimes for low and smaller size cable the solid conductors are used. These are single cores, two cores, three cores and four cores or three and half cores.

Insulation. Each conductor is provided with a suitable thickness of insulation. The thickness and quality of the insulation depends upon the voltage for which it is designed. Generally the insulation used are impregnated paper, varnished cambric tape, vulcanised or rubber minerals compounds.

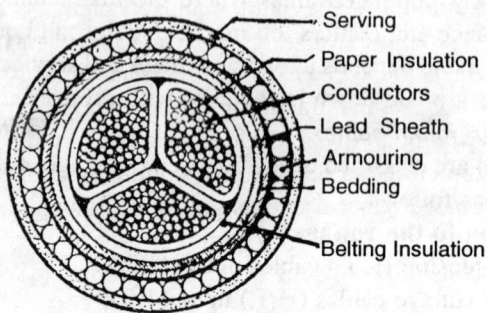

Fig. 27.5. Underground cable.

Metallic sheath. The cable is protected against the moisture, gases, or other damaging liquids (acid and alkalines) under the ground and in the atmosphere by providing a metallic sheathed of lead or lead alloy or aluminium over the insulation.

Bedding. The metallic sheathed is protected against corrosion and mechanical Injury due to armouring by bedding over the sheathed. The bedding consists of a fibrous material like jute or cession tape.

Armouring. It consists of one or more layers of the galvanised steel wires or steel tape. It is provided over the bedding to protect the cable from mechanical damages or injury which may result during laying and handling it.

Serving. To protect the armouring from atmospheric conditions a layer of fibrous material similar to bedding is provided over the armouring. It is known as serving.

Q. What should be the properties of the insulating materials of U.G. cables? Name the different insulation used in manufacturing these cables.

Ans. The insulation provided for the underground cables should have the following properties:

    (*i*) It should have high resistivity.

    (*ii*) It should have good dielectric strength.

    (*iii*) It should have good tensile strength.

    (*iv*) It should be mechanically strong enough.

    (*v*) Its thermal coefficient should be low.

    (*vi*) Its viscosity should be low at the working temperature.

    (*vii*) Non-absorbent of moisture and if, should be low.

    (*viii*) Chemically strong and stable.

(*ix*) Non-inflammable.

(*x*) Not too costly but should be cheap.

No one material can satisfy all the above stated qualities or the properties. So the type of the insulation material to be used depends upon the service for which the cable is required. The various cable insulating materials used for the manufacturing of the underground cables are rubber, V.I.R. paper, polyvinyl chloride, varnished cambric, polyethylene vulcanised bitumen, gutta percha, silk, cotton, enamel etc.

The impregnated paper used has got the following properties as:

Dielectric strength – 20-30 kV/mm.

Insulation resistance – of the order of 10^9 ohm-m.

Permittivity – 3.6.

Maximum safe temperature – 95°C.

Q. Explain with diagram the supertension cable and name the extra high tension cables.

Ans. The super tension cables are of the following types.

H-type cable. It is named after the inventor Hochstadler. There is no belt insulation but each conductor is insulated with paper to the desired thickness and over it there is a layer of metallized paper perforated to facilitate the process of impregnation and avoiding any possibility of air duct or voids. There is an additional layer of the cotton tape with fine wire of copper and wrapped around all the three cores. Over this tape there is no belt insulation. The lead sheath and armouring is also provided. All the shieldings are at zero *i.e.* earth potential, with the result the electrical stresses are entirely radial and dielectric losses are reduced. The biggest advantages of providing a metal foil is to increase the rate of heat dissipation and thus there are no sheath losses. Thus the current carrying capacity is increased. These are used up to 66 kV.

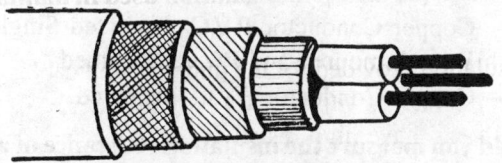

Fig. 27.6. H-type cable.

S.L. type cable. In this type of the cable each core is first insulated separately with an impregnated paper and covered with a metallic sheath. Now all the three cores are just equivalent to three separate cable, each having its own metal sheath. Now these three cables are laid up with fillers, armoured and served small with impregnated hessian tape. No lead covering is provided over all the three cores. This cable has its own advantages:

(i) The bending of the cable becomes somewhat easy owing to the absence of over all lead sheath,

(ii) There is no tendency of oil drainage on hilly routes.

Extra high tension cables. These cables are designed with special dielectrics having sufficient dielectric strength to withstand the higher electrical stresses. Therefore the dielectric used should be homogeneous and no void or space is left that are checked by special means.

The cables are of the following types:

(i) Oil filled cables.

(ii) Gas pressure cables.

(a) Internal pressure cable.

(b) External pressure cable.

Q. What for these designations are used?

PILC, PILCSTA, PILCDTA, AVV, ATRV, AVFV, CVV, CTRV, CVFV.

Ans. These are used for underground cables as under:

PILC – Paper Insulated Lead Covered Cable.

PILCSTA – Paper Insulated Lead Covered Single Taped Armoured Cable.

PILCDTA – Paper Insulated Covered Double Taped Armoured Cable.

AVV – Aluminium Conductor P.V.C. Insulated P.V.C. Sheathed.

ATRV – Aluminium Conductor P.V.C. Insulated Single Round Steel Armoured and P.V.C. Sheathed.

AVFV – Aluminium Conductor P.V.C. Insulated Flat Steel Strip Armoured and P.V.C. Sheathed.

CVV – Copper Conductor P.V.C. Insulated and P.V.C. Sheathed.

CTRV – Copper Conductor P.V.C. Insulated Single Round Steel Wire Armoured and P.V.C. Sheathed.

CVFV – Copper Conductor P.V.C. Insulated.

Q. How would you measure the insulation resistance of a underground cable?

Ans. The most effective and convenient method of measuring insulation resistance of a U.G. cable is by the megger. This instrument is also known as insulation resistance tester. In one instrument there are two departments, one the motoring (the deflecting system) and other the generating system (source of supply). As the supply is not given from outside to this instrument so the generating portion produces the electricity for the deflecting system of the instrument.

It has two coils, the current coil and pressure coil. The current coil is connected in series with the specimen and pressure coil across the voltage induced by the megger through suitable resistances. There are two terminals outside the L and E. The L-terminal is connected to the main conductor of the cable and E-terminal to the insulation of the cable.

The handle is rotated at a uniform specified speed and e.m.f. is induced which circulates the current in the windings depending upon the value under observation. The deflecting system indicates the insulation resistance directly on the calibrated scale.

Distribution system

The distribution of power to the consumers is called the distribution systems. This power is received from the generating station or major substations.

The distribution of power to the consumers with suitable circuit arrangements, controlling etc. will enhance the accessibility. The feeders, distributor and service mains are the parts of distribution system.

(a) Feeder

The feeders are the conductors which connects the substation or generating station to the area where the power is to be distributed. Generally no tappings are taken from the feeders.

(b) Distributor

A distributor is a conductor from which tappings are taken to supply to the consumers. The current distribution in the distributor is different for different distributor depending upon the load.

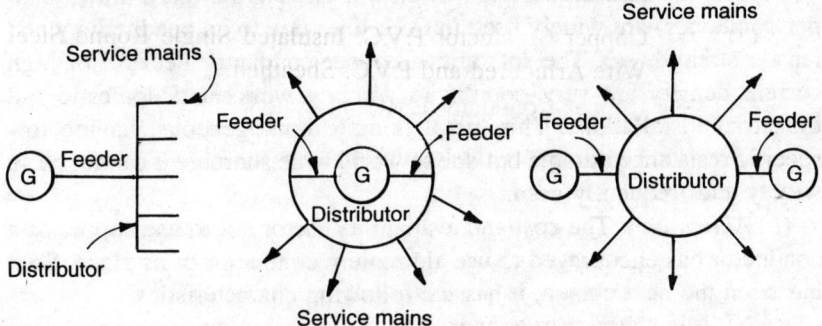

Fig. 27.7. Feeder, distributor, service mains

(c) Service Mains

These are the actual lines connecting the consumers to the distributors. These are for individual consumers and the load in these service mains depend upon the load requirement of the consumers.

Q. Which materials are used for OH lines?

Ans. The materials generally used for OH lines comprises of the followings:

(a) Conductors

(b) Supports

(c) Insulators

(d) Pole fittings and other miscellaneous items like guards, lightning arrestors, anticlimbing devices, phase and danger plates, etc.

Q. What should be the characteristics of line conductors?

Ans. The materials used for line conductors should have the following characteristics:

(i) It should have high conductivity.

(ii) It should have high tensile strength.

(iii) It should have low density.

(iv) It should be free from atmospheric changes and effects.

(v) It should be cheap and easily available.

(vi) It should have longer life and good appearances.

Q. Which types of conductors are used for OH lines?

Ans. In common practice the conductors used for overhead lines are as under:

(a) The most common desire of conductor to be used on OH lines is *Hard drawn copper conductor*. The copper conductors are actually of two types, the hard drawn and soft drawn. The HDCC (hard drawn copper conductor) are widely used in O.H. lines due to its conductivity and tensile strength etc. The soft drawn copper conductor because of high current density are very popular in winding works and domestic and industrial installations. This metal is quite homogeneous, having low specific resistance durable but due to world wide shortage it cannot be so widely and frequently used.

(b) *Aluminium:* The cost and availability factor not to use copper as a conductor has encouraged to use aluminium conductor in its place. So it has been the next choice, it has the following characteristics:

(i) It is cheap than copper.

(ii) It has light weight than copper.

(*iii*) Its conductivity is 60% than that of copper.

(*iv*) For obtaining same resistance (Ohmic value) its diameter has to be increased by 1.26 times that of copper.

(*v*) As far as the weight ratio is concerned it has low density and it is experienced that for same resistance as that of copper the weight of aluminium is half.

(*vi*) The tensile strength is approximately 45% to that of copper. So for same tensile strength as that of copper 1.66 times area will be required. The size of the conductors increases causes the increasing of line supports too.

(*vii*) Jointing is also difficult than that of copper.

(*viii*) As the melting point of Aluminium is less than copper so short circuit could be more dangerous.

To overcame the drawbacks of the aluminium the stranded ACSR (Aluminium Conductor steel reinforced) conductors are used. As the steel wire has more tensile strength.

The different types of poles used for transmission and distribution purposes are of the following types.

1. Wooden poles

The poles are made of wood. Their uses are limited for low voltage unless special care is taken of. These are used in the places where wood is available in plenty like hill areas. They are having good insulating qualities and are cheap. The pole preferably must be straight, strong and gradually taper.

Generally wooden poles are made, I-shaped, A-shaped or H-shaped depending upon the design and requirements. The shape depends upon the load and tension on the conductors which are placed on them. Generally Devdar, chead, sal and teak wood is used for poles.

The main disadvantage of this type of poles is that these are elastic, ant and vermins and climatic conditions effects and reduces their life. The length of the poles depends upon the voltage, ground clearance and setting depth of the pole (approximately 1/5th part of the pole is buried under the ground for erection purpose). The cross arm are generally 50 cm to1 m below the top point. The length of the pole is 40′ to 60′ for transmission purposes and 30′ to 45′ for distribution purposes.

2. Iron Poles

These poles are made of iron and have longer life span. These poles are made of the following types.

(*a*) Tubler poles, (*b*) Rail poles, (*c*) Fabricated poles.

(*i*) *Tubler pole:* These are hollow from the center. Generally these are manufactured with three different diameters for example a 30′ long pole has 8″ diameter for 15 ft., 5″ diameter for 8′ and 3″ diameter for remaining 7′. Each pole is connected with earth and is painted periodically to keep away from rust and for good appearance.

(*ii*) *Rail poles:* The rail poles are made of rails. The shape can be I-type, A-type or H-type. They are also painted periodically to have longer life and good appearance. The length also depends upon the voltage used, ground clearance and span used.

(*iii*) *Fabricated pole:* These are steel fabricated poles; generally made from angle iron and strips riveted together. They have generally two legs. The fabrication can be for I types, A-types or H-type or having three or four legs. These are also properly earthed and painted for good appearance.

3. Concrete pole

The concrete poles are used where the earth is not suitable for wooden poles.

These are either hollow or solid. The hollow poles made in factories the wall thickness carries a web of steel wires to increase their mechanical strength and rigidness. The solid poles are also manufactured in factories. The poles also carries steel wires and the mixture of cement and concrete to increase the mechanical power, having long life and good appearance. Their life is better than wooden poles and the risk of shock is also avoided, but even than the proper earthing is required.

4. Tower

Generally these are used for high voltage transmission purposes. The base of tower is broad so as to have rigidity. These are prepared by angle irons and strips riveted together. The towers are classified as rigid tower and flexible tower.

Generally broad base lattice structured tower are called rigid towers. These are mechanically strong enough and are not effected by the wind, or tension of the conductors, if properly designed these are good indefinitely. These are robust, long life and can serve indefinitely.

Sometimes two legs towers are also used and these are known as flexible tower. It may bend in two directions. These towers are painted and earthed properly for long life. The tower are designed to have either single circuit or double power circuit. The height of the tower, the span, the conductors's ground clearance, the mechanical stresses due to tension of conductors also taken into account when these are erected.

Guarding Type

While crossing the line over road, river, railways crossing, canals, telephone and telegraph lines, over the different voltage line and both sides of the poles, some guarding is applied to safe guard the living things like human beings, animate birds, etc. H.T. and LT lines and telephone lines the guarding is used. These are cradle guarding, cage guarding, bird guarding, bead guarding and trolley guarding.

Q. Can you name the different tests to be conducted for insulator testings?

Ans. Generally for smooth operation of the insulators the following tests are carried over,

1. Flashover test
2. Performance test
3. Routine test.

Q. What is the necessity of guarding on OH lines?

Ans. Once the overhead line is erected, the climatic conditions in India are such that storm and wind, rain, earthquake, very hot and cold session all have their effects on lines too. In case of any eventuality the live wire may break and inherently will come to ground where it may harm any property, living creature directly. So for the safe guard of these the guarding is used, so as if the wire or live conductor breaks first it will touch the guard wire and there after tripping the circuit breaker it may touch the ground and will not be harmful to anybody.

Installation and laying of underground cables

The installation of U.G. cable is not so easy but specially trained technicians are required for proper installation. There are the following methods of layering of U.G. cables:

(*i*) Direct laying
(*ii*) Draw in system and
(*iii*) Solid system.

(*i*) Direct laying method

The most common economical and simple method of laying underground cables is the direct laying system. A trench of about 1 m deep and 50 cm. wide is dug and a layer of about 10 cm. thickness of fine sand is spread so that a bed of cable is prepared. Then the cable is laid over this bed. The sand does not allow the moisture and more over the heat dissipation from the cable is efficient through sand. The sides of the trench are made of bricks and then covered by planks bricks, tiles or concrete from the top to avoid any mechanical damage. The cables are generally laid down in straight line, however loop are formed at the ends or in between to avoid the shortcomings due to any fault. In that case if some cable becomes short, the loop can be opened and length can be increased. This system being simple, cheap and clean, in spite of suffers from the disadvantages

(*i*) The risk of cable damage due to subsequent excavation work.

(*ii*) The fault location is difficult and maintenance cost is more.

(*iii*) The earth condition and nature (chemical) may damage the cable sheathing.

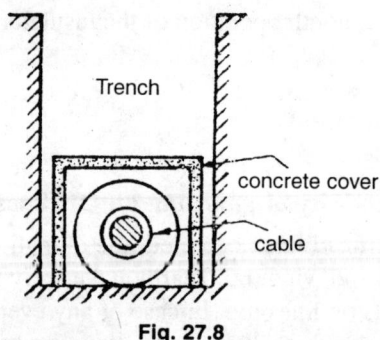

Fig. 27.8

The long trenches are dug out by mean of special rotor type digger or excavator where the length is not so long these are dug manually. In case where the route is curved or zig zag, special care has to observed so as to make it possible the laying of cable with the desired bending radius and not to harm the cable.

This method of laying of cable gives away the best conditions for heat dissipation thus increasing the life and performance too.

Sometimes for laying of cable in industrial plants, generating stations, substations, tunnels, the cables are installed on the racks either fixed in walls or on wheels or supported on ceilings too. The different cables are kept over the shelves and there are systematically arranged. The heat dissipation in this type is easy fault location and repair is also easy. Care has to be taken while dragging the cable on the floor. The kinking of the cable should be properly prevented.

(*ii*) Draw in system

The places in congested city area where excavation is expensive and inconvenient, the drawing system is usually used. In this method the ducts or conduit are laid down in the ground with manholes at suitable places, positions and distances. The conduits, ducts or tubes made up of either iron, glazed stone, water, clay or cement concrete are laid down in the ground. Then the cable are drawn in. The cast iron pipe or concrete ducts or pipes are used because of low price. The biggest advantages secured is that once the conduit is laid, repair alteration or additions to this system can be made without reopening of the ground.

The initial cost of errection is rather high and another disadvantage is that the current carrying capacity of the cable is reduced as the heat dissipation is not so appreciable. The method is suitable for short length cable route such as in workshop, railway bridge crossing, road crossing, where frequent digging is costly and inconvenient.

(*iii*) Solid System

In this system the U.G. cable is laid down in throughing in an open trench or channel dug out in earth along the route and throughing is filled with bituminous or asphaltic compound and covered over. The throughing is of cast iron stone asphalt or treated wood. Cables laid down in this method are usually plain lead covered as the throughing affords good mechanical protection.

This method of laying down is very advantageous as the cable is protected from breakdown due to electrolysis and corrosion but simultaneously it is very expensive and the skilled worker and supervision is essential. Moreover the heat dissipation is also very poor as to have only a very little overloading possibility, this method because of these disadvantages is rarely used.

The underground cables are supplied mostly on reels. The diameter of the cable depends upon the current carrying capacity and voltage rating of the cables. The cable reel are taken from one place to another or to the site by means of cable laying vehicles by manually or otherwise. The cable is reeled off from top to bottom and not from bottom onwards as to have easy accessibility.

27.2 LIGHTNING ARRESTERS

Q. What do you understand by the lightning arresters? Where they are used, what are the different types? Describe them.

Ans. These are used for the safety of the conductors and the equipment connected with the line, from the sky lightning. They sky lightning is of million of volts with very high frequency. If this lightning falls or flows through the live conductor, the equipment or apparatus connected on or to it, will be damaged. The lightning arresters prevent the lightning to flow through the line but sends the lightning to the earth without any damage.

Thus it can be said that the lightning arresters are used to perform the followings:

(*i*) It must not permit the passage of current to the general mass of earth, as long as the voltage is normal.

(*ii*) In case if the voltage rises to the predetermined value, it must allow a path of the general mass of earth, so as to discharge the surge without further rise of voltage.

(*iii*) As soon as the voltage is reduced to normal it should normalize the operation.

(*iv*) It should be accessible and must repeat the function in case of requirement automatically without further delays.

Types of lightning arresters. These are of the following types:

(*i*) Lightning wire or ground wire type.

(*ii*) Horn gap arrester.

(*iii*) Pellet type or oxide film arrester.

(*iv*) Thyrite arrester.

(*v*) Auto valve arrester.

(*i*) **Lightning wire or ground wire.** The overhead lines can be protected from sky lightning by placing a earth wire on the top of the pole or tower. The earth wire is placed above all the wires, so whenever lightning takes place it is earthed through earth wire and line and other equipments are safe.

(*ii*) **Horn gap lightning arrester.** In its simple construction an inductive reactance is connected in series with the equipment to be protected just near the horn gap arrester as shown in Fig. 27.3. The reactance, in most of the cases, consists of a coil of bare copper

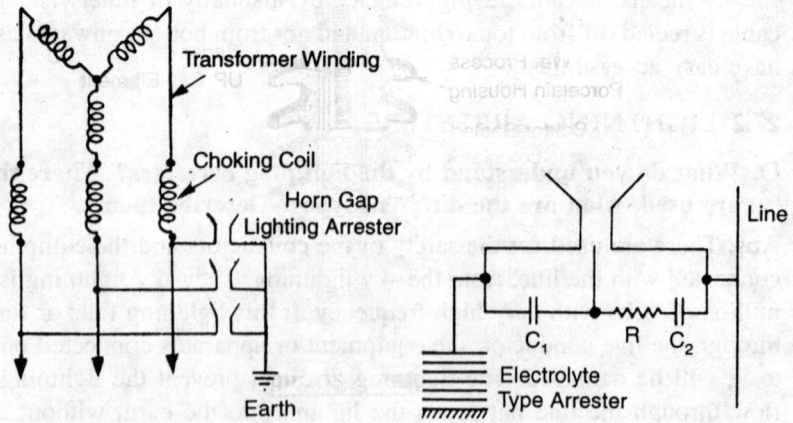

Fig. 27.10. Horn Gap type Arrester   **Fig. 27.11.** Electrolytic arrester.

wire which produces a ray low reactance for the current to pass to the equipment at normal frequency. But it offers a very high reactance for the surge of lightning and that surge is discharged to the earth through horn gap. The sparking goes up and finally stops. The length of the gap may be 1.5 mm for medium pressure line, 10 mm for voltage upto 5000 V, 15 mm for voltage of 10 kV, 2.5 cm for 15 kV, 4.00 cm for 25 kV, 7.75 cm for 40 kV and 10 cm for 50 kV.

.(*iii*) **Pellet type arrester.** In this type of the lightning arrester the balls of lead peroxide are covered with the insulating varnish. A number of these balls are connected in series depending upon the line voltage. In general the lead peroxide pellets are in a column of 56 mm in diameter having a length of 50 mm per kV of rating is used. One wire is connected with the conductor to be protected and on the other side the earth wire with lead peroxide ball. For higher voltage several units are connected in series. The pellets are of 3/32″ in diameter.

At normal voltage the current cannot flow through the lightning arrester to earth, but with the sky lightning it goes to earth through arrester.

(*iv*) **Thyrite arrester.** In this case, the non-porous material which acts as a very good insulator up to certain voltage, but when the voltage rises to predetermined value it becomes a conductor. The thyrite is

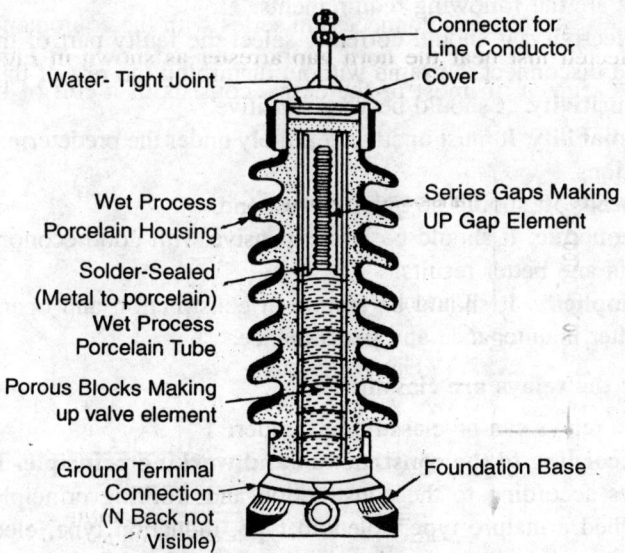

Fig. 27.12. Auto valve lightning arrester.

a dense inorganic compound of a ceramic nature which has high resistance that decreases with the increasing of the electrical stresses, from a high value at low current to a low value at high currents. Generally it is seen that the current increases 13 times when the voltage is doubled. In this way it acts as the insulator for the normal voltage and as conductor for surge voltage. A disc of 15 cm diameter and 2.5 cm thick can withstand a working voltage of 3000 V, so far 11 kV voltage four discs are connected in series. It is also connected with the general mass of earth separately.

Auto valve arrester. It is a very effective robust and cheap lightning arrester. It consists of a number of flat disc of a porous material. The discs are arranged one over the other and separated by a thin layer of mica rings. The material is specially prepared. The capillaries can withstand the voltage of 350 V. The narrow gaps between the blocks are of sufficient total width to prevent flash over due to normal voltage, so that no current flows in the arrester under normal voltage. As soon as the source comes the flash or glow discharge occurs and is discharged to earth.

Q. What are the basic requirements of the different protective relays?

Ans. The protective relay is defined on the electrical device, which protects the circuit, apparatus or machine from any of the abnormality by isolating the faulty element.

These are the following requirements:

(a) **Selectivity.** It should correctly select the faulty part of the circuit and disconnect the same without disturbing the rest of the system.

(b) **Sensitivity.** It should be very sensitive.

(c) **Reliability.** It must operate definitely under the predetermined conditions.

(d) **Speedy.** It should be quick to respond.

(e) **Economic.** It should be less expensive with good economical design and better results.

(f) **Simplicity.** It should be simple in construction and operation for better maintenance and performance.

Q. How the relays are classified?

Ans. The relays can be classified as under:

(a) **According to the construction and working principle.** These relays according to the construction and working principle are attached armature type, solenoid type, induction type, electromagnetic type, moving coil type, thermal type. All these relays are working with their respective principles.

(*b*) **According to the application.** These are classified as over voltage, over current and over power relay, under voltage, under current, and under power relay, directional or the reverse current relay, reverse power relay, differential relay, distance relay etc.

(*c*) **According to the timing characteristics.** There can be easily divided as instantaneous relay, definite time lag relay, inverse time relay, and inverse definite minimum time lag relay.

All these relays operate and protect the circuit under faulty circuit conditions and isolate it.

Q. What do you understand by the circuit breakers. Also mention the basic requirements of a circuit breaker.

Ans. The electrical circuits are protected by means of fuses etc. for low voltage and high voltage when frequent operations are not expected. The protection by fuses is advantageous because of the cost, but it is not appreciable as the fuses are to be replaced. In the circuits for higher voltages, *i.e.* 3.3 kV upwards the isolation is done by using the circuit breakers. The main difference between the fuse and circuit breaker is that under faulty conditions the fuse will melt and requires only the replacement for regular operation, wherein case of circuit breaker it will isolate and close the circuit without any replacement. So the *circuit breaker* is defined as the mechanical device which is designed to isolate and close the contact members and thus electrical circuit under normal and abnormal conditions.

Requirements of the circuit breakers. Following are the main requirements:

(*i*) It must interrupt the normal and short circuit current safely without over-heating and· damage.

(*ii*) It must close and open the circuit on load condition.

(*iii*) It must isolate the circuit under faulty conditions without any delay.

(*iv*) It must not isolate the circuit when momentarily over current flows under healthy conditions of the circuit.

Q. Explain how would you specify the ratings of a circuit breaker?

Ans. A circuit breaker automatically be capable of isolating the circuit under abnormal and faulty conditions. The rating of the circuit breaker is given by the duties it has to perform (for regular standard ratings I.S.-375/1951). The circuit breaker has the following ratings:

(*a*) Rupturing capacity rating.

(*b*) Making capacity rating.

(*c*) Short circuit rating.

(a) **Rupturing capacity rating.** The rupturing or breaking capacity of a circuit breaker is the maximum amount of the current that the circuit breaker is capable of breaking at a predetermined recovery of voltage under the prescribed operating conditions.

It is generally expressed in terms of MVA and is equal to the product of rated current voltage and a factor depending upon the number of phases, *i.e.*, for single phase it is one and for three phase it is $\sqrt{3}$. So the breaking capacity = $\sqrt{3}$. $V \times I \times 10^{-6}$ MVA, where I is the breaking current which can be asymmetrical or symmetrical in amperes and V is the voltage in volts.

(b) **Making capacity.** It is desirable that the circuit breaker must be capable of withstanding the electromagnetic forces developed under faulty conditions, if it is required to close or open the circuit. These forces, obviously are proportional to the square of maximum value of the current.

The making current of the circuit breaker is the peak or maximum value of the maximum current wave (d.c. component) in the first cycle after the circuit is closed by the circuit breaker.

(c) **Short time rating.** Sometimes it is required to carry a short circuit current for a short interval without interrupting the circuit. This may happen momentary because of momentary fault persisting only for one or two seconds. In that case the circuit breaker are calibrated in such a way as not to trip on the predetermined rated value. The value depends upon the type and rating of the circuit breaker.

Q. Write short notes on the followings:

(a) **Air circuit breakers (A.C.B.)**

(b) **Oil circuit breakers (O.C.B.)**

(c) **Air blast circuit breakers (A.B.C.B.).**

Ans. (a) **Air circuit breakers (A.C.B.).** These circuit breakers are designed for low voltage say up to 600 V, for the protection of general lighting power and motor circuits, etc. There are two fixed terminals mounted on the insulated base, one over the other in a vertical plane. These terminals are short circuited by means of a moving contact with heavy pressure by a bridging member by a system of linkage. The arcing contacts are made of carbon alloy arc resistant silver tungsten or copper tungsten alloy to avoid the damage.

These circuit breakers are usually provided with an over load and no load protections. The over load devices are usually set for ranging from 80% to 160% of the ratings. Such circuit breakers are rated as 600 A, 250 to 600 volts A.C. and 1200 A, 250-750 V D.C. mains.

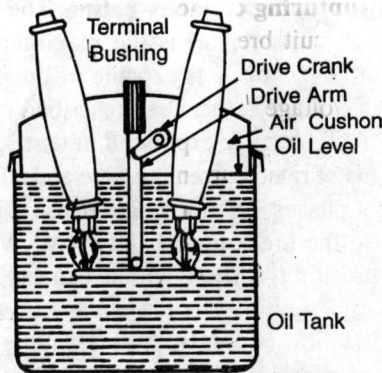

Fig. 27.13. Air circuit breaker. **Fig. 27.14.** O.C.B. Double break O.C.B.

(*b*) **Oil circuit breaker.** The oil is used in these types of the circuit breakers, so these are known as oil circuit breakers. These are high voltage power circuit breakers. The separating contacts of the circuit breaker are immersed within the insulating oil. In case of arcing the oil is vaporised forming bubbles. The bubbles of the gas so formed prevent the restriking of the arc after the current reaches to zero in a particular cycle.

The circuit breakers have large rupturing capacity and generally employed with a solenoid coil or motorised or pneumatic gear for remote controls. A double break oil circuit breaker is shown in Fig. 27.7.

These O.C.B. are, plane break oil circuit breaker – suitable up to 1100 volts 150 MVA. Plain explosion pot type circuit breaker – suitable for low and high capacities and cross Jet type – suitable for interrupting heavy current at high voltages.

Advantages. Following are the advantages of the O.C.B.:

(*i*) The oil used has good dielectric strength, which allows even less clearance sufficiently enough for live conductor and other components.

(*ii*) The oil acts as an insulator.

(*iii*) It (oil) absorbs heat energy of the arc and dissipates to cool down the contacts.

(*iv*) The oil quenches the spark if any.

(*v*) The gases so formed by decomposition of the oil due to arc, have good cooling properties.

Disadvantages. Following are the disadvantages:

(*i*) It is inflammable.

(*ii*) It forms an explosive mixture with air.

(*iii*) It requires periodic maintenance to keep oil clean and clear.

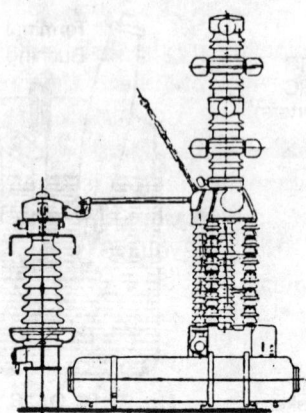

Fig. 27.15. Side view 132 kV 3 phase air-blast circuit breaker.

(c) **Air blast circuit breaker.** These are the air circuit breaker but the compressed air at a pressure of approximately 16-20 kg/cm^2 is used for the arc extinction. The use of compressed air for arc extinction is very suitable and increase its suitability for higher outputs *i.e.,* upto 400 kV with breaking capacity 7500 MVA, (during short circuit).

The Fig. 27.15 shows the construction of the *ABCB* in which to extinct the arc, the compressed air flows along the axis of the arc. In case when the fault comes, the current exceeds the predetermined value and the *ABCB* is opened, hence causes an arc. To extincts the arc compressed air is used. With the use of compressed air even 1.75 cm of short gap can eliminate the high current arcing.

Following are the types of *ABCB*:
(i) Axial Blast air circuit breaker.
(ii) Axial Blast air circuit breaker with side moving contact.
(iii) Cross Blast air circuit breaker.

Advantages. Following are the advantages:
(i) No risk of fire or explosion.
(ii) The Air Blast circuit breaker operates at a high speed.
(iii) Duration of arc is very small, because the compressed air extincts the arc quickly.
(iv) Less possibility of damaging the contacts, because of less duration of arc.
(v) It is frequent in operation as the fresh air is available for every successive operation.

(*vi*) Requires less maintenance.

Disadvantages. Following are the disadvantages:

(*i*) In order of frequent operations of the circuit breaker, it is necessary that compressor should have sufficient capacity of high pressure air.

(*ii*) The maintenance of the compressor and other allied equipment such as pipe and automatic controls is required.

(*iii*) Leakage in the high pressure pipe may result.

(*iv*) Sensitivity to restriking voltage.

(*v*) Current chopping.

Q. What do you understand by MCB? Describe its construction, operations, utility and applications.

Miniature circuit breakers (Moulded case circuit breaker):

Ans. These circuit breakers are minitype and are known as M.C.B. These are generally used in domestic and industrial applications. These are preferred in low voltage system. With the use of these MCB's the replacement of fusing element problem is solved. These are as sensitive as that of the H.R.C. fuse. These are used in place of kit ket fuses as circuit fuse and as main switch too.

These are easy to operate. During fault it trips off and disconnect the faulty circuit, thus prevent the wiring installation appliance and other equipments from damaging. In general construction it has two terminals, the incoming and outgoing. The main part is the tripping mechanism which has both thermal and magnetic sensing devices. The contacts and sensing devices are assembled in a moulded case. The moulded case has good mechanical and dielectric strengths. The parts in which current flows are prepared by electrolytic copper or silver alloy depending upon the current capacity for which it is designed, say ranging from 6A to 60A. To ensure best suitability all other parts are non-ferrous, non-rusting type. A special provision is done for arc extinguishing. The arc chute chamber is specially designed. The arc is distinguished by magnetic field created by the same arc which diversify because of its unique construction. The arc length increases i.e. it elongates and arc quenches itself.

Arc Extinguishing and Tripping. In these the arc is elongated and split by the arc splitters. These are made of resin bonded fibre glass plate. These are placed perpendicular to the arc, the is pulled by electromagnetic force resulted because of the magnetic field applied in proper direction, so as to pull the arc upwards. The arc is pulled, as a result it is elongated and splited and cooled resulting the extinguishing. Special provision is made on the upper side to release the hot air so as hot air so as not to effect the other working parts of the MCB.

The low range circuit breakers provide these protective release, **thermal tripping** it can be achieved by adjustable bimetallic thermal release, the other is **Super Rapid trip** it is of the order of 13 to 15 msec. which is achieved by collapsing from the fulcrum point instead of actuating the trip bar.

These are in single unit, double unit and triple units suitables for circuits, main switches and three phase main switch also for distribution boxes, a number of units depending upon the current ratings and capacity are assembled there in.

The current ratings of single units are as 0.5 A, 1 A, 2 A, 3 A, 6 A, 10 A, 20 A, 32 A, 40 A, 50 A, 60 A. etc. Their rupturing capacity on a.c. are 3 kA 415 V and on *d.c.* 3 kA at 50 V (Non inducting) and 1 kA 110 V (non inductive).

These are extensively used in domestic installation, commercial complexes and in industrial applications as they are easy to operate and autometrically ensure the safety of the circuit; it will not operate unless the fault is clear.

Q. What are the sulphur Hexafluoride circuit breakers and why are these preferred?

Sulphur Hexafluoride (SF$_6$) Circuit Breakers

Now the turning of this century has comprehensively switching on to the use of sulphur Hexafluoride (SF$_6$) circuit breakers instead of OCB and ABCBs, etc.

The sulphur hexafluoride gas is an inert, heavy good dielectric strength and better are quenching properties. Its dielectric strength can be increased by increasing the pressure of the gas. Nowadays this gas is extensively used in electrical equipments, like high voltage metal enclosed cables, H.V. metal clad switch gears, capacitors, circuit breakers, C.T. and terminal bushings etc. SF$_6$, gas is in liquid shape and is generally supplied by chemicals firms. The production cost can be low when manufactured in bulk quantity.

The gas is colourless, odourless, non-toxic, non-inflammable, it is heavy say 5 times the air at 20°C. The heat transfer properties are 2.5 times than that of air (thus the current carrying capacity of the conductor is increased by 250% than that of air. It has remarkable property of low arc time constant i.e., better arc extinguishing property. The gas is inert and electronegative.

These circuit breakers are designed as double pressure type (nowadays obsolete) and single pressure puffer type. It is a popular designed ranging from 3.6 KV to 760 KV circuit breakers.

Q. What do you understand by arc-phenomena? Give the methods of arc interruption.

Ans. Whenever, the contacts working of a circuit breaker begins to isolate, several thousands ampere current passes through gap causing a voltage drop and this gap does not interrupt the current but the heat is produced. The heat causes the gap air to be ionised. This ionized column acts as a conductor and that small voltage is sufficient to maintain the arc. Thus the initiating electrons (charged particles) are produced due to field emission and thermal emission, caused due to voltage drop and heat produced.

Following are the methods of interrupting the arc.

(a) **High resistance interruption.** It is also called as rheostatic interruption. In this method, the effective resistance is increased with time. This method is used in d.c. breakers. This resistance may be increased by lengthening the arc, cooling the arc, the splitting the arc and constraining the arc.

(b) **Low resistance interruption.** It is also known as the current zero interruption. This method is usually applied in all high power a.c. breakers. In this, the resistance is kept low until the current zero where arc automatically extinguishes and is prevented from restriking.

Line Insulators

The insulators used for the over-head line should have the following properties:

(i) High resistance i.e. high insulation strength.
(ii) Good dielectric strength.
(iii) Should be fre from the atmospheric effects *viz* moisture, chemical changes etc.

Generally these insulators are made from glass and porcelain.

Porcelain. It has good mechanical strength.

(ii) It is costly than glass.
(iii) Cracks etc. can not be seen easily.
(iv) Not easily breakable.

Glass. A glass insulator has the following properties.

(i) Less mechanical strength than porcelain.
(ii) Some what cheaper than porcelain.
(iii) Cracks can be seen easily.
(iv) Breaking strength is less.

Type of insulators

The insulators, according to their shapes are of the following types:

(1) **Pin insulators.** The pin insulator supports the wire at the top or at the one side. It is supported with a steel bolt or pin which is fixed on the cross arm. The conductor is first placed in the groove at the top of the insulator and then attached to the insulator by tying it with the help of a wire generally of the same material as that of conductor. The pin insulators have some times one, two or three sheds or petticoats for the suitability on telephone line, L.T. lines and H.T. lines respectively.

Uses. Pin type insulators are used for telephone lines, distribution and transmission lines of electrical power up to 33 kV.

Limitation:

(*i*) These are less flexible.

(*ii*) The size become heavy at higher voltages so these are not economical to use above 33 kV line.

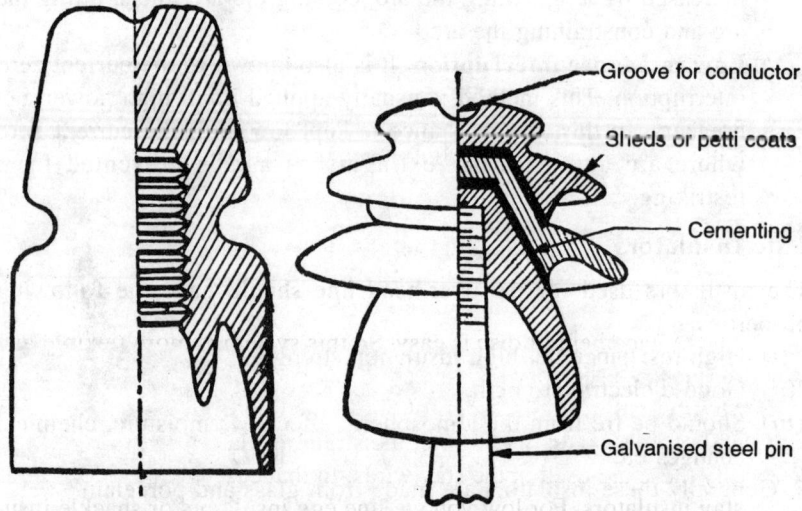

Fig. 27.16. Pin insulators.

2. **Shackle insulator:** The shackle insulators are used to support the wire at dead end, or sharp turns or where the wires take bend at a particular angle. It has hole in its centre and is supported by means of nuts and bolts and a supporting strip.

   Uses: These are used at dead end or sharp turns of L.T. lines.

3. **Suspension insulator:** These are also known as disc insulators. These are used for H.T. transmission line conductors. It consists of a number of porcelain discs connected in series by metal links in the form of a string. A number of discs can be added to meet the demand of increased voltage of H.T. line. The conductors are sus-

pended at the bottom end of the string as shown in Fig. 27.11 while the upper end of the string is secured to the tower or pole cross-arm. *Uses.* These are generally used for H.T. lines above 33 kV.

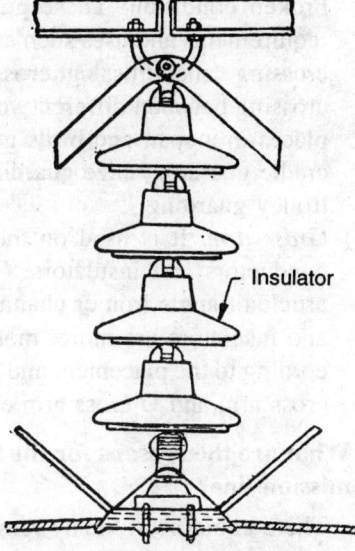

Fig 27.17. Shackle Insulator.   Fig. 27.18. Suspension type insulators.

These insulators are cheap for more than 33 kV voltage range. Each disc is designed for approximately 11 kV so number of units can be increased in series to meet any voltage requirement above 33 kV. The replacement of disc is easy. So this system is more flexible and frequently used for H.T. lines.

4. **Strain insulators.** In order to relieve the line from the strings at the end of line, or sharp ends etc, the strain insulators are used. Generally these are used in the stay wires. In that case these are termed as stay insulators. For low voltage line egg insulators or shackle insulator can be used as a strain insulator. However for H.T. lines disc or suspension type insulators are also used. If the strain is much higher in that case two parallel stays are also used. These are used to avoid flow of any leakage current to earth. These are provided at height about 3 m. above the ground level.

The other components of the overhead line are;

(a) *Stay wires.* At the ends or bends the poles are provided with stay wires. These are galvanised steel stranded wires and taken up the strain of the line.

(b) *Pole clamps.* Each pole is provided with clamp to fix up the street light bracket.

(c) *Guard wire.* These are the small pieces of G.I. wire and generally connected between the neutral wire and earth wires. The main function is to earth the phase conductor in case of accidentally broken condition. These guards are classified according to the requirements and uses such as road crossing, railway crossing, river crossing canal crossing, crossing of telegraph and telephone lines, crossing between different voltage lines, both side of pole and two places in a span and birds guard too. Thus these are classified as cradle guarding, cage guarding, bird guarding, bead guarding and trolley guarding.

(d) *Gross arm.* It is used on the top side of the pole to support the conductors and insulators. Generally these are made of wooden article or angle iron or channels. Some-times when the conductors and insulators are more, more than one cross arms are used. According to the placement and use these are known as cross arm, side cross arm and U cross arm etc.

Q. What are the reasons for the failure of an insulators in the transmission line?

Ans. There could be so many reason's for the failure of an insulator in the transmission or distribution line.

(a) Cracking of an insulator

The thermal expansion of steel, porcelain and cement is different, so as the climatic condition changes, the temperature rises, the expansion is different and uneven which will crack the insulator. So the thermal stresses developed due to seasons i.e. heat, cold or dryness, dampness, rain etc. are the causes of cracking of an insulator.

(b) Defective raw material

If the raw material is defective, on the actual performance the insulator will fail so the defective raw material is also a cause of failure.

(c) Porosity in the material

If during manufacturing some porous are left (low temperature casting) the insulator will puncture in dampness or rain etc. or in long use.

(d) Glazing

If the insulation is not properly glazed the water will stick to its outer surface which will effect its dielectric strength and the insulator will fail.

(e) Flashover

If by any reason there is a flash between the metal parts the continuous slow heating will spoil the insulator and it will fail.

(f) Mechanical stresses

When the insulator is installed and the whole assembly or any one of it is not in equilibrium state, some mechanical stresses will develop because of drawing of line conductors. These stresses will result the mechanical failure of insulator.

(g) Sometimes Short Circuit

Sometimes short circuit due to birds etc. near the insulator happens, it will cause a flashover and failure of the insulator.

(Comparison between OH and UG system)

Comparison between overhead and underground systems is given below:

Comparison Between Overhead and Underground Systems

S. No.	Particular	Overhead System	Underground System
1.	Safety	It is less safe.	It is more safe.
2.	Cost	It is less expensive.	It is more expensive.
3.	Appearance	Not good.	It gives good look.
4.	Flexibility	It is more flexible.	It is not flexible.
5.	Working voltage	It can be worked up to 400 kV.	It cannot be worked above 66 kV.
6.	Chances of faults	Faults occur frequently.	Very little chances of faults
7.	Location of faults	Faulty point can be located easily.	Faulty point cannot be located easily.
8.	Repairing	Overhead lines can be easily repaired.	Underground lines cannot be easily repaired.
9.	Maintenance cost	It has high maintenance cost.	It has low maintenance cost.
10.	Interruptions	It has more chances of supply interruptions.	It has very little chances of interruptions.
11.	Interference with communication system	It interferes.	No interference.
12.	Lightning	More chances of being subjected to lightning.	Very little chances of being subjected to lightning.
13.	Frequency of accidents	It has more chances of accidents.	Very little chances of accidents.
14.	Erection	Difficult in zigzag routes.	Easy to erect.

Q. What are the most important characteristics of the transmissions line?

Ans. The most important characteristics of the transmission line are as follows

(a) Reliability (b) Regulation and (c) efficiency.

The *reliability* of the transmission services are greatly affected by the designing of the live. The conductors and supporting structure must be strong enough to support the heaviest loads which may be imposed upon them under the most adverse weather conditions, sufficient distance or clearance must be provided between the conductor to conductor, conductor to earth to limit the danger of short circuiting or grounding the line. Utmost care must be given for the selection of insulator (shape, size and dielectric strength). So that only the most severe lighting conditions can cause the line to flash over.

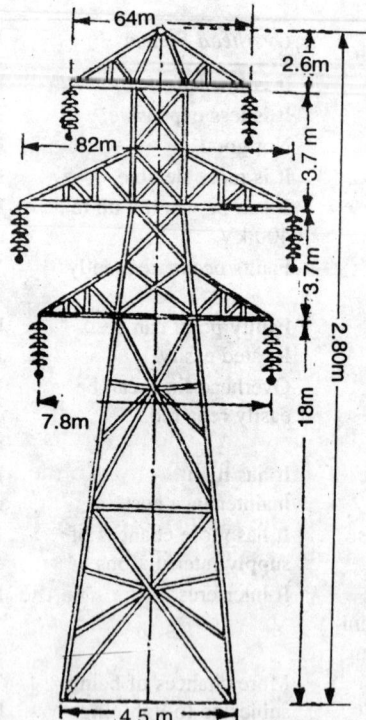

Fig. 27.19. Double circuit tower.

Obviously the voltage *regulation* of the system is defined as the rise in receiver voltage when full load is taken off shown in the percentage of full load voltage with the generator voltage held constant.

The voltage regulation of the transmission system depends on the resistance, reactance and the capacity of the line and the power factor of the load. To have a good voltage regulation the resistance, reactance should be low and the load power factor be high.

The power losses in the line are actually $I_L^2 R_L$ watts. (I_L line current, R_L - line resistance) and comparatively small losses due to leakage over the insulators and corona effects. To reduce these losses the resistance of the line must be as low as possible. It can be achieved by increasing the cross sectional area of the line conductors. The losses can be reduced by decreasing the line load current by increaseing the transmission voltages or by installing synchronous capacitors or static capacitors also.

Q. What are the advantages of high transmission voltage?

Ans. The high voltage transmission line have the following advantages:

 (*i*) By increasing the voltage for a given power we can reduce the line current hence power losses.

 (*ii*) As current is less weight of the line will also reduce.

 (*iii*) Reduction in weight reduces Pole/tower strength, supporting structure and number of structure.

 (*iv*) Losses are reduced hence increasing transmission efficiency.

 (*v*) The losses are less so the efficiency of transmission is improved and the voltage regulation too.

Q. What are the factors limits the increased voltage?

Ans. Although the increased transmission voltage have several uses still there are certain limitations which affects it.

The cost of the line support will increase as the cost of insulators, length of tower, spacing between the conductors, more ground clearance, insulation between ground and line conductors is also increased, long excess arm, all effects the cost and limitation to the higher voltage.

Q. What is corona? Name the factors effecting it. What are the advantages and disadvantages of corona?

Ans. It is an inherent and unique phenomenon in the high voltage transmission lines. When an alternating current of high and very high potential difference is applied to the conductors whose spacing is large as compared to their diameter than the atmospheric air surrounding the wires is subjected to electro-static stresses. These stresses are proportional to the potential gradient at low voltage these are less and at higher voltage are higher. The stresses of air because of hissing sound and sometimes the bluish colour glow around the conductor. This effect of the higher or very higher voltage is termed as *corona*.

So we can say that corona is the phenomenon of hissing sound or noise, in the over head line conductors. It also produces the bluishing glow and production of ozone gas around the conductor.

It is an important factor for designing the transmission lines which actually decides the spacing between the conductors, otherwise at higher potential difference small flash over may take place between the conductors of transmission line.

: *Factor:* There are the following factors that affects the corona.

The shape of the conductor i.e., an stranded conductors will have more corona than the solid single conductor. The *atmosphere* also plays an important role, as the ionisation of air is the main cause so the physical condition and state of the atmosphere for example stormy weather, moisted atmosphere, will carry more ions resulting corona at even much less voltage compared to fair weather conditions. Spacing between the conductors also reduces voltage gradient reducing corona to occur. Finally the voltage of transmission line will affect the intensity of corona, less voltage less corona and high voltage more corona.

Advantages

The virtual diameter of the conductor due to corona (conductive airana around the conductor) is increased which decreases the electrostatic stresses between the conductors.

As the electrostatic stresses are reduced the probability of flash over is reduced, this improves the performance of the system. It also reduces the effects of transients produced by lightning.

Disadvantages:

The loss of energy in electrostatic charge. It is dissipated in the forms of heat, light, sound and chemical reactions. The corona causes the corrosion of the line conductors. It affects the sinusoidal wave of the voltage this effects as an interference in the near by communication lines.

Q. How will you reduce the corona effects?

Ans. The corona effect which may cause flash over instead of hissing sound, should be carefully studies and its effects be reduced. It can be reduced by increasing the conductor size and by increasing the spacing between the conductors.

Poles and Sub-station

Q. Name and draw free hand sketch of the different types of poles used for O.H. distribution purpose.

Ans. The following poles generally used for the distribution lines (LT & HT)

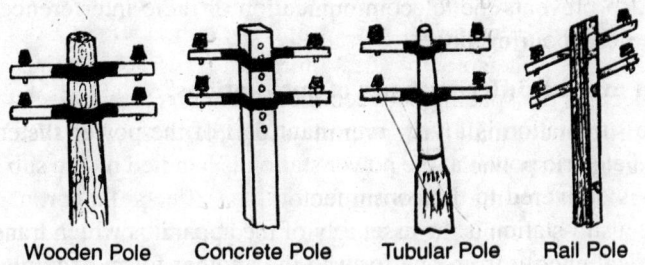

Wooden Pole   Concrete Pole   Tubular Pole   Rail Pole

Fig. 27.20. Poles.

(*a*) Wooden poles.   (*c*) Tubular steel pole.
(*b*) Concrete pole.   (*d*) Rail poles.

Q. What do you mean by the transposition of conductors in O.H. transmission line? What are its advantages?

Ans. When in case of an *OH* line, all the three phases are supposed to be spaced symmetrically keeping the line constants, identical but due to some mechanical considerations it in not so, these are placed irregularly. It disturb the line constants, as a result the voltage drop in the three lines will be different and unequal line voltage is achieved at the receiving end. This difficulty is overcome by the use of transposition of conductors. The conductors of the *OH* transmission line are kept at a regular distance and the system adopted is known as *transposition of conductors* as shown in Fig. 27.21.

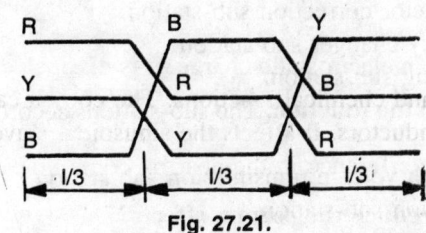

Fig. 27.21.

In this case let the sequence by *RYB*, then after *l/3* distance it is made by transposition as *BRY* and after again *l/3* distance as *YBR*, by interchanging the line conductors, so the arrangement of changing the positions of overhead line conductors of the transmission line at regular distance in order to make line constants symmetrical is called *transposition of conductors*.

Advantages:

(*i*) The line constants are similar for all the three phases.

(*ii*) It also prevents the telecommunication or radio interferences from the neighbouring line.

Q. What are the different types of sub-stations?

Ans. The sub-station is a very important part of the power system. The generated electric power at the power station is handled by the sub station before it is delivered to the consumers.

Hence a sub-station is the assembly of the apparatus which transforms the electrical energy from one form to the another form, example *AC* to *AC* or *AC* to *DC* etc.

The characteristics and operations of the sub-stations may be elaborated as follows:

(*a*) **Switching operation.** Used to switch on and off the power line.

(*b*) **Voltage transformation operation.** To transform the voltage from high voltage to low voltage, low voltage to high voltage.

(*c*) **Converting operation.** Used to convert a.c. into d.c. and vice versa.

(*d*) **Frequency converting.** To convert frequency of the a.c. mains as desired or for the specific purpose.

The sub-stations are classified according to the service, control, mounting, function and type of apparatus used. These are as follows:

(*a*) **According to the service:**

(*i*) Transformer sub-stations.

(*ii*) Industrial sub-stations.

(*iii*) Switching sub-stations.

(*iv*) Power factor correction sub-station.

(*v*) Frequency changer sub-station.

(*vi*) Converting sub-station.

(*b*) **According to the function.** The sub-stations according to the function are:

(*i*) Extra high voltage transmission sub-station.

(*ii*) Distribution sub-station.

(*iii*) Industrial sub-station.

(*iv*) Power factor correction sub-station.

(*v*) Frequency changer sub-station.

(*vi*) D.C. for lighting sub-station.

(*c*) **According to the types of apparatus:**

(*i*) Transformer sub-station.

(*ii*) Rotory converter sub-station.

(*iii*) Rectifier sub-station.

(*iv*) Motor generator sub-station.

(*v*) Frequency changer sub-station.

(*d*) **According to the control:**

(*i*) Manual control sub-station.

(*ii*) Automatic control sub-station.

(*iii*) Supervisory control sub-station.

(*e*) **According to the mounting:**

(*i*) Indoor sub-station.

(*ii*) Outdoor sub-station.

(1) Pole mounting sub-station.

(2) Foundation mounting sub-station.

Q. Describe the sub-station which are classified according to the service.

Ans. The sub-stations according to the service are as under:

(*a*) **Transformer sub station.** The sub-stations which transforms the power from one voltage to the another voltage with the help of the transformers are known as the transformer sub-station. These may be further classified as and are shown in Fig. 27.1.

(*i*) *Transmission or primary sub-stations.* Such stations receive power from the generating station and step up to high voltage 132 kV, 220 kV, 400 kV. etc. for further transmission.

(*ii*) *Sub-transmission or secondary sub-stations.* Such sub-stations are installed after the primary sub-station. It receive power from the primary sub-station through the transmission line and step down to 11 kV, 33 kV, 66 kV for secondary transmission.

(*iii*) *Distribution sub-station.* These receive power from the secondary sub-stations or directly from the power stations and step down for secondary distribution.

(*b*) **Industrial sub-station.** These sub-stations are for industrial individual consumers and are known as the industrial sub-stations.

(*c*) **Switching sub-station.** These are for performing switching operations of power lines without transformation of voltage and are known as the switching sub-stations.

(*d*) **Power factor correction or synchronous sub-station.** The sub-stations are erected for the improvement of the power factor. Generally the bank of capacitors or the synchronous machines are run in over-excited condition to improve the power factor, such sub-stations are called power factor correction sub-station.

(*e*) **Frequency changer sub-station.** Sometimes other than the supply frequency is needed for certain operation, so the frequency changers are used to change the frequency and frequency changers are installed in the sub-station. These sub-stations are called frequency changer sub-station.

(f) **Converting sub-station.** Generation of supply is 90% of A.C. in nature. There are certain applications which are exclusively for D.C., so converting machines are installed in the power stations to convert the A.C. into D.C. These power stations are called converting sub-stations.

Q. Describe the mounting type sub-stations.

Ans. According to the design and mounting, the sub-stations are classified as under:

(a) **Indoor sub-station.** In this case the apparatus are installed within the sub-station building. Such sub-stations are usually up to 11 kV voltage, but can be extended to 33 kV and 66 kV, if the pollution, surrounding pollution because of metal corroding, fumes and conducting gasses and dust etc. permits. The primary side, generally H.T. carries only the oil circuit breaker. From the bus bar various feeders are taken to the different areas. Every feeder is provided with its circuit breaker and isolators. The auxiliaries of the indoor sub-stations are (i) the storage battery (ii) fire fighting equipments, such as water buckets, sand bucket and fire extinguishers etc. The batteries are used for the operation of protective relays and gears, and for emergency lighting in the sub-station.

(b) **Outdoor sub-station.** The outdoor sub-station, as the name implies are not confined to the building. The outdoor sub-station are of two types:

(i) *Pole mounted type.* Such sub-station are erected for mounting distribution transformer of capacity upto 300 kVA. These are cheapest and simple in construction. It does not require much attention for its operation. Thus the maintenance cost is less. Generally, the transformer upto 100 kVA are mounted on the double pole structure, and above 100 kVA to 300 kVA are on the four pole steel structure with a suitable platform at a suitable height. The equipments are generally used:

(1) G.O. – Gang Operating Switch.
(2) T.P.M.O. – Triple Pole Mechanically operated Switch for switching on and off of H.T.
(3) H.T. fuse unit.
(4) Switches and fuses on secondary side.
(5) Lightning arrester.
(6) Proper earthing at two or more than two places.

(*ii*) **Foundation mounted type.** These are build in the open areas. The main sub-station for primary and secondary transmission and for primary distribution are of this type. These type of sub-stations are associated with O.H. lines. The equipment required for much sub-stations are very heavy therefore site selection is done accordingly.

The switch gear consists of circuit breaker of suitable type on both the side but with the increased reliability of the modern transformer.

Merits:

Following are the merits of outdoor sub-stations:

(*i*) The equipments are within the view therefore the fault location is easy.

(*ii*) Extension of installation is easy and less time consuming.

(*iii*) The construction work required is comparatively small and cost is less.

Demerits:

These are certain demerits:

(*i*) The supervision, maintenance and fault repairing is to be performed in open area during all the weathers.

(*ii*) More space is required for the sub-station.

More over while designing the main conditions, fluctuation in ambient temperature, dust, open space, weather changings, are to be kept in mind and hence the cost of the equipments is more.

REVIEW QUESTIONS

1. What do you mean by a circuit breaker?
2. What are the basic requirements of a relay and circuit breaker?
3. Write short not on the followings:
   (*i*) Simple isolator.
   (*ii*) Oil circuit breaker.
   (*iii*) Air blast circuit breaker.
4. What is the phenomenon of arc?
5. What are the sub-stations, briefly explain their characteristics?
6. What are the different types of sub-station?
7. Explain the pole mounting sub-stations.
8. What are the uses of lightning arrester?
9. Explain the horn gap type lightning arrester.
10. Explain the construction of a simple underground cables.

28

Electrical Appliances

HEATING APPLIANCES

The principle of heating, *i.e.* "whenever an electrical current flows through a conductor heat is produced, which is directly proportional to the square of current, time and resistance", is used in many heating appliances. Generally for domestic heating appliance the nicrome wire or ribbon of different sizes is used. The length and diameter of the wire or ribbon decide the heating capacity *i.e.* wattage of the appliance.

In all heating appliance the heating element is wound over an insulated and fire proof material for example, mica, china clay, porcelain etc.

Q. Describe some heating appliances with neat sketches.

Ans. Heaters. The heaters are used for general heating purpose. These are of the following types:

(a) **Ordinary heater.** The heating element is made of spiral shape wound with nicrome wire. The element is placed in the grooves of heater china clay plate as shown in Fig. 28.1. The plates are available in different diameters. The two ends of the element are connected to the two screws from where two wires (pin) are connected to the porcelain connector. The wires between the terminal and heater element are insulated by means of insulated porcelain beats. The beats are also of different diameters (generally known by numbers) used for different heaters depending upon the capacity. Generally 21 SwG, 26 SWG and 22 SWG, nicrome wire are used. The wattage range from 500 W, 750 W, 1000 W, 1500 W and 2000 W etc.

Fig. 28.1. Ordinary heater.

The supply is connected to the heater by means of a suitable lead having earth continuity conductor.

(b) **Room heater.** These are also known as the radiant heater or bar type heater. These are used to heat a room, or a particular place.

In general construction of the heater, it is wound with spiral nicrome wire over a china clay bar. Two ends of the element are brought out to the terminals on both the ends of the rod. The rod is fixed in the connector by means of screws, supply is given to these two ends. To increase the efficiency the polished reflector is used behind the rod, shown in Fig. 28.2, which radiate the heat efficiently.

Fig. 28.2. Room Heater.

These heaters are having sometimes the bowl of china clay and a round polished reflector.

The position of the reflector can be adjusted according to the need in any direction. The number of the bar may be one or more than one. Heating element in both the cases is fixed possibly at the focus of the reflector.

(c) **Immersion heater.** As the name implies, these heaters are used in immersed conditions. These are used to heat the water or other liquid directly. The heat produced is directly dissipated to the water and thus the water is heated. In general construction the heating element is made of spiral shape wound with nicrome wire. The element is placed in the copper tube and insulated from the walls etc. by means of the insulated and fire proof powder or sand all around. The ends of the tube are sealed with thermal and water proof compound. Two ends are taken to the connector for main supply. The tube is bend into the spiral shape to concentrate in a less space, as shown in Fig. 28.3. There is a level indicator marked to avoid the damage of the tube. For the safety the earth wire is connected to the tube. These heaters are available in different shape and wattage, for example 1000 W, 1500 W and 2000 W etc.

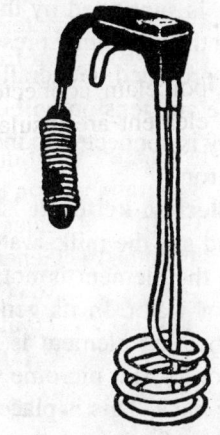

Fig. 28.3. Immersion heater.

Precaution. Any immersion heater should not be connected to the supply without immersing into water or liquid First the heater should be immersed in water (liquid) then it should be connected and switched on to the supply.

Electric iron. It is the domestic heating appliance which is used to iron the clothes. There are the following main parts:

Fig. 28.4. Electric iron.

(*i*) *Sole plate.* It is made of cast iron which in most of the cases is chromium plated. Sometimes the sole plate has a heel rest and the plate is said 'the sole plate with heel rest'. The sole plate should have fine and plane surface.

(*ii*) *Pressure plate.* It is also made of cast iron and is kept over the heating element. It offers a pressure on the element and keep the element in position. It also solves the purpose of weight for iron the clothes.

(*iii*) *Heating element.* It is also known as the press element. The press element is made of nicrome ribbon, which is wound over the mica sheets. The element is covered with two mica layers so that the nicrome wire may not come in contact with the metallic part of the iron. The two ends are connected with two strips of brass or copper which are connected to the terminals of the electric iron.

(*iv*) *Iron case.* It is used to cover the element and the pressure plate etc. It has a connector to connect the leads from the element and mains also. It may be chrome plated or painted.

(*v*) *Handle.* It is used to handle the iron. The wooden or bakelite handle is supported by the screws or strips.

In the *automatic* presses, an arrangement for the temperature control is also provided. Generally a bimetallic strip is controlled by a knob. The space and tension will decide the temperature range. A violet lamp generally is connected to indicate the operation *i.e.* 'on' and 'off' positions of the iron.

Electric kettle. It is used to heat the liquid say the milk, water etc., but in this case the element is not immersed in liquid or water. In its general construction, the heating element is made of nicrome ribbon or flat nicrome wire. It is wound over mica. This is placed in between two covers of mica, so that the nicrome wire

Fig. 28.5. Electric Kettle.

may not come in contact with any metallic part of the kettle. The two ends of the element are connected to the terminals of the kettle through two brass strips.

The element is placed in the position by means of cast iron pressure plate. These are available in different wattages.

In same electric kettles the immersion type heater element are also used. In some kettles safety device is incorporated to prevent damage to it, if it is switched dry.

Soldering iron. A soldering iron is used for soldering the joints etc. In its simple construction the heating element is made by winding the nicrome wire or ribbon over the mica sheats. Generally there are two elements in an iron which are connected in parallel or in series. These elements are pressed in a case of iron sheat which is tightly placed in a solid iron. The two main leads of the element are connected to the supply in a connector. The body of the iron is connected with earth for the purpose of safety from shock. These are available in different wattages 10 W, 15 W, 20 W, 35 W, 40 W, 65 W, 125 W, 250 W, etc.

Boiling plate or hot plate. In this type of heater the element is not opened. The nicrome wire heating element is kept in a fire proof china clay cement. The terminals of the element are brought out to the plug connector. There is a cast iron plate over the heating element. The whole arrangement is assembled in a body of iron sheet. These are available in different wattage and the heat can be controlled by means of a controlling switch, connecting the elements in series and in parallel as shown in Fig. 28.6.

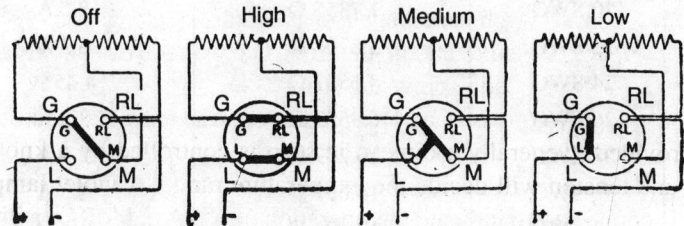

Fig. 28.6. Different connections of hot plate.

Q. Why nicrome wire is used for the heating elements? How will you design a heater element?

Ans. Generally, all the electrical heating appliances having the heating element made of nicrome wire or ribbon. The principle of heat conversion of electrical energy into thermal energy is used in most of the appliances. The nicrome which is a alloy of 80% Nickel and 20% Chromium is used, because of the following properties:

(*i*) It has a high specific resistance *i.e.*, 9×10^{-8} Ωm. Therefore less length of wire will be sufficient to produce the required resistance.

(*ii*) It can withstand the thermal stress.

(*iii*) Its mechanical strength is good.

(*iv*) It has high melting temperature *i.e.* approximately 1380°C.

(*v*) It has low temperature coefficient *i.e.*, 0.00017 per degree centigrade.

In most of the cases, all the heating elements are designed to produce a temperature approximately 500°C, so far the designing of heating element these points are taken into account:

(*i*) The wire of proper size and capacity, according to the wattage of the heater.

(*ii*) Sufficient length of the wire required according to the wattage of the heater.

In most of the cases, for general heater element 20 SWG nicrome wire is used for 2000 W heater 22 SWG for 1500 W, 24 SWG for 1000 W heater and 26 SWG wire for 750 W heater.

Table 28.1 shows the safe current capacity for the different size of wire and resistance per metre.

Table 28.1

S.No.	SWG of wire	Resistance in ohm per metre at 500°C	Ampere to produce 500°C
1.	18 SWG	0.9744 Ω	12.6 A
2.	20 SWG	1.7355 Ω	8.6 A
3.	22 SWG	2.8707 Ω	6.3 A
4.	24 SWG	4.6587 Ω	4.45 A
5.	26 SWG	6.9553 Ω	3.5 A
6.	28 SWG	10.269 Ω	2.8 A
7.	30 SWG	14.665 Ω	2.3 A
8.	32 SWG	19.291 Ω	1.99 A

Example. *Calculate the length of wire and size of nicrome wire for a heater of 1000 W for 250 V.*

Solution. The wattage of the heater is 1000W

So the current drawn $= \dfrac{1000}{250} = 4$ A.

The wire suitable for this current is 24 SWG, which can carry a current of 4.45 A for 500°C of temperature, which is quite on the safer side. This wire has a resistance of 4.6587 Ω/meter.

how the resistance required $R = \dfrac{V^2}{W}$

$$= \dfrac{250 \times 250}{1000} = 62.5\,\Omega$$

The length required $= \dfrac{\text{Total resistance}}{\text{Resistance per metre}}$

$$= \dfrac{62.5}{4.6587} = 13.415 \text{ metre}$$

$$\approx 13 \text{ metre and } 42 \text{ cm.}$$

so for a 1000 W 250 V heater 24 SWG nicrome wire of 13.421 metre is needed.

Air-conditioner. It is a device which can create artificial controlled atmosphere in a room or big office, according to its capacity. Air-conditioning takes various forms depending upon the requirements, but essentially it is system of delivering cooled or hot air to a given space and maintaining the area at a given temperature and sometimes at a certain humidity level also. Thus it can create a pleasant atmosphere both in summer and winter. It works on the principle of replacing the air of a place by the clean and purified air at a controlled temperature and humidity. Air is circulated through the coils and temperature is maintained by the thermostat in the room which cuts off the source, as soon as the required and predetermined conditions are achieved.

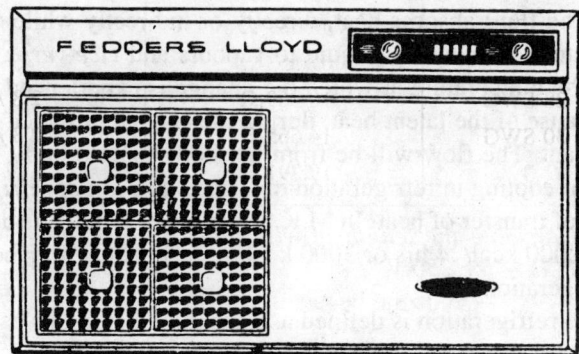

Fig, 28.7. Air-conditioner.

Automatic Voltage Stabilizer

The automatic stabilizer is used to keep the secondary load voltage as constant say 220 ± 9%, or so, whatsoever the input voltage may be. (Generally ranging from 180V to 250V). It employs a voltage sensing circuit comprising of zenor diode and BC-148. This circuit is used for SK-100 and the relay. The zener diode does not function during normal voltage. This keeps BC-148B off while second BC-148, SK-100 and relay is functioning in ON state.

While the supply voltage is high the zener diode operates and off the relay. Now A.C. can be obtained from the upper relay point which is actually 200V point. It can be operated to function with the help of preset 100Ω. If the supply voltage is always on high voltage side the middle and top connections of the transformer may be interchanged to obtain the constant voltage on secondary side.

Refrigeration

Q. What do you understand by refrigeration?

Ans. The *refrigeration* is defined as the branch of science which deals with the process of reducing and maintaining the temperature of a space or material below the surrounding temperature. This phenomenon is widely used in storage, ice manufacturing, medicines, preservation, in hospitals, in hotels, in houses and so many other places.

Q. On what principle the refrigeration depends? Also state a ton of refrigeration.

Ans. The refrigeration is based on the principle of Joule Thomson effect, which states that when a cooled and compressed gas is allowed to pass through a narrow passage it falls in temperature. In particular it can be stated as, "The fluid absorbs heat, directly or indirectly while changing from one state of another *i.e.,* liquid to vapours and *vice versa.* In addition to this the temperature at which the change is experienced, remains constant because of the latent heat, during the change provided the pressure is constant. The flow will be from hot body to cold body."

The rate of cooling in refrigeration in termed as *a ton of refrigeration.* It is the unit of transfer of heat. In M.K.S. system of units, it is defined as the rate of 72000 kcal/24 hrs or 3000 kcal/hr or 50 K cal/min. cooling as one ton refrigeration.

The one to refrigeration is defined as a machine having its capacity to produce cooling effect of 200 B.T.U./min or 50 K cal/min.

Q. Describe in brief the vapour compression cycle of refrigeration.

Ans. The simple arrangement and a schematic diagram of the vapour compression cycle of refrigeration is shown in Fig. 28.8.

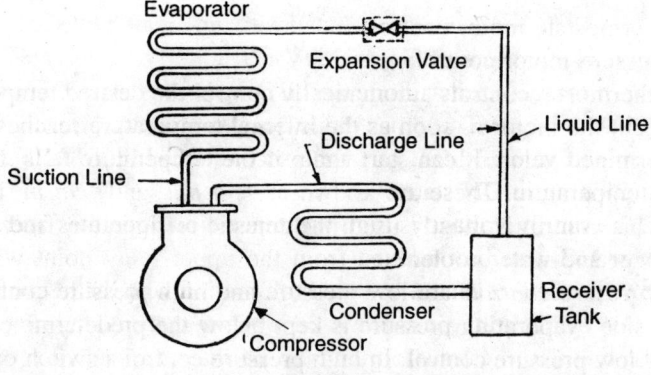

Fig 28.8.

The evaporator provides a heat transfer through which heat can pass from the refrigerated space into the vaporised refrigerant. Then it passes through the suction line inlet to compressor at low pressure though suction line. The compressor raises its pressure and temperature to such a value, so that it may be easily condensed. The flow is also maintained by the compressor. This is then discharged to condenser at high pressure. The condenser provides a heat transfer. The receiver tank stores the refrigerant from condenser and supplies to the evaporator through to the liquid line and expansion valve, where actually the pressure reduces considerably, so that the refrigerant may take heat from the load or the refrigerating space.

Q. What are the important factors to be considered for the selection of a motor for a particular plant?

Ans. In addition to the supply A.C. or D.C. and voltage available, there are certain important factors to be kept in mind while selecting the proper type of motor:
   (*i*) Capacity and starting torque.
  (*ii*) Working conditions, temperature, ambient temperature dust, moisture or explosive materials.
 (*iii*) Starting current limitation.
  (*iv*) Speed, single or multi speed operations.
   (*v*) Continuous or intermittent ratings.
  (*vi*) Efficiency and power factor.

Q. Why electric controls are required for the motor compressor?

Ans. The controls are required for starting, stopping the unit and to protect against over loading. The over loading may be caused by high refrigerant pressure, or excessive current drawn by the motor. The motor controls are of two types:

(*a*) Thermostate motor control.

(*b*) Pressure motor control.

The thermostat controls automatically control the desired temperature and cut out the motor as soon as the internal temperature reaches to the pre-determined value. It can start again if the temperature falls than the desired temperature. These are known as '*cut out*' and '*cut in*' respectively. This control is mostly used in domestic refrigerator, freezer, air-conditioner and water cooler etc.

The *pressure controls* are low pressure and high pressure controls. If the low side evaporating pressure is kept below the predetermined level it is said low pressure control. In high pressure control a switch operates automatically by the high pressure side, when pressure reaches to too high level.

Q. Explain the different types of motor starting relays.

Ans. The motor starting relays are as under:

(*a*) Current relay.

(*b*) Voltage or potential relay.

(*c*) Hot wire or thermal relay.

(*d*) Solid state electronics relay.

(*a*) **Current relay.** Generally it is used with small motors. It is a magnetic relay and is connected in series with the running winding as shown in Fig. 28.9. It is wound with thick conductor of few turns. When the motor is energised and high current because of locked rotor passes through it, the heavy current produces more flux resulting the contacts, which were normally opened, short circuited. Thus the starting winding is energised. Now the torque developed is because of both the starting and running windings. Now when the rotor is free or current reduces the starting contacts are open circuited and motor continue to run only on the running winding.

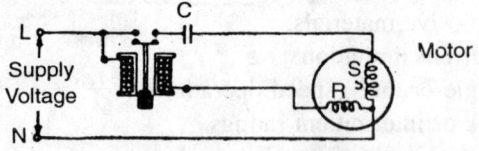

Fig. 28.9. Current relay.

(*b*) **Voltage or potential relay.** This type of relays are used with the motors of high torque *viz.* capacitor start, capacitor start and run motors as shown a Fig. 28.10. It is wound with thin conductors having more number of turns to withstand the line voltage. The relay are connected in series with the starting capacitor and are closed when the motor is not running.

When both the starting and running windings are working and motor is running, the voltage across starting winding increases considerably. The winding of relay produce more magnetic flux pulling the contact points and opening the starting circuit. The relay coil is connected across the starting winding.

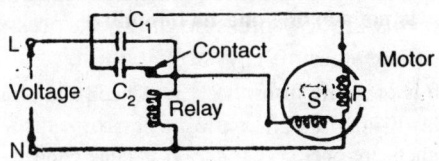

Fig. 28.10. Voltage relay.

(*c*) **Thermal or bimetallic relay.** The principle of bimetallic strips is used in this type of relay. These are of two types, the bimetallic strip control and resistance wire control relay.

(*d*) **Solid state electronics relay.** The relay circuit having electronics components like diode, triode and transistors are being used in controlling the hermatic motors.

Q. Draw the circuit diagram of a refrigerator motor connection with overload thermostatic switch.

Ans. Connection diagram. Figs. 28.10 and 28.11 show the different electrical connections.

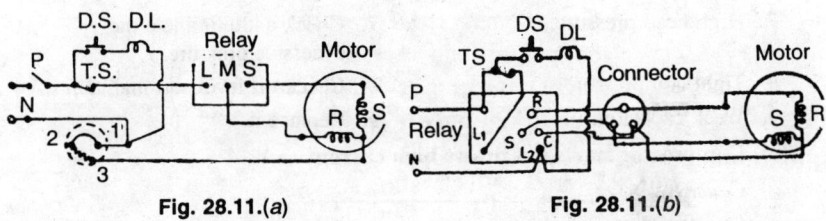

Fig. 28.11.(*a*)   **Fig. 28.11.(*b*)**

D.S. (Door Switch), D.L. (Door Light), OLR (over load relay), T.S. (thermostatic switch).

Q. What is meant by hermatic and semi-hermatic units?

Ans. When an electric motor (compressor drive) and compressor are mounted on the same shaft and are enclosed in the same case or sealed unit, the compressor is called hermatic compressor.

In case of semi-hermatic compressors, the body itself serves as the casing. The crank is extend to hold the motor, and whole assembly is tightened with nuts and bolts. These are generally used in domestic as well as commercial fields.

Q. What are the common faults and their remedies in hermatic compressor?

Ans. Following are the common electrical faults and their remedies:

S. No.	Faults and their possible causes	Remedies
(a)	**Compressor is not starting and no humming**	
	Causes:	
	1. Main switch is off or no fuses in the main switch.	1. Check and arrange for supply.
	2. Control contacts are open.	2. Check controls.
	3. Open protector.	3. Check and reset.
	4. Open circuit in starter	4. Check and if damaged replace the starter.
(b)	**Compressor only hums**	
	Causes:	
	1. Wiring connection are not proper.	1. Check wiring connections and remove the faults.
	2. Low voltage.	2. Check the voltage.
	3. Capacitor circuit in open condition.	3. Check and remove fault or replace condenser if needed.
	4. Relay contact not closing.	4. Check and remove the fault.
	5. Open circuit in starting winding.	5. Check and remove the fault.
	6. Starter winding earthed.	6. Check and remove the fault.
	7. High head pressure.	7. Check and eliminate the excessive pressure.
	8. Tight compressor.	8. Check oil level and maintain it.
	9. Weak capacitor.	9. Replace it.
(c)	**Compressor starts but draws high current**	
	Causes:	
	1. Low line voltage.	1. Check and arrange if can be.
	2. Connections are not proper.	2. Check and reconnect.
	3. Defective relay.	3. Check and replace.
	4. Running capacitor short circuited.	4. Check and replace if faulty.
	5. Short circuit in running winding.	5. Check and remove the fault.
	6. Weak starting capacitor.	6. Replace the capacitor.
	7. High discharge pressure.	7. Check and properly maintain it.
	8. Tight compressor.	8. Check oil level and bushes etc.

(Contd.)

S. No.	Faults and their possible causes	Remedies

(d) Compressor starts runs but cycles on the protector

Causes:

1.	Low line voltage.	1. Check it.
2.	Additional current passing through protects.	2. Check for additional load and remove it.
3.	Suction pressure is too high.	3. Check and maintain the pressure.
4.	Discharge pressure is too high.	4. Check ventilation etc.
5.	Running capacitor defective.	5. Check current and if necessary change over load.
6.	Weak protector.	Check the winding and capacitor and remove fault.
7.	Partial short circuit and tight compressor.	7. Check oil level.

(e) Starting capacitor burnt out

Causes:

1.	Short cycling.	1. Produce head pressure if possible and reduce number of starts.
2.	Starting winding operation are too much.	2. Reduce starting load.
3.	Sticking relay contact.	3. Clean contacts or replace relay.
4.	Improper relay setting.	4. Replace relay or set properly.
5.	Improper capacitor type and voltage.	5. Check and replace if necessary.
6.	Capacitor terminals are short circuited because of moisture or water.	6. Check and change capacitor if necessary.

(f) Running capacitor burnt out

Causes:

1.	Excessive line voltage.	1. Check and arrange for correct voltage.
2.	More load.	2. Reduce load.
3.	Capacitor rating is low.	3. Check and arrange correct rating.

(g) Relay burnt out

Causes:

1.	Excessive line voltage.	1. Reduce line voltage if possible.
2.	Incorrect rating.	2. Check and correct.
3.	Incorrect relay.	3. Use proper relay as recommended by the manufacturer.

REVIEW QUESTIONS

1. Write down an essay on the use of electricity in home and in agriculture.
2. What do you understand by the heating appliance, explain any one of them?
3. Explain the construction of a room heater.
4. Write short note on the following:
    1. Hot plate.
    2. Electric press iron.
    3. Electric kettle.
    4. Air-conditioner.
5. What is refrigerator? Explain the refrigeration cycle.
6. What are the electrical faults and their remedies in a refrigerator?
7. Draw the circuit diagram of the electrical wiring in a refrigerator.
8. Why relays are used in the refrigerator? Explain current relay.

Part II

Workshop Calculation and Science

1. MENSURATION

Area and volume

Area. The area of the following figures is an given under:

   (*i*) **Rectangle.** The plane figure enclosed by four sides having its opposite sides equal and all angles as right angles, is known as the rectangle.

        Area of the rectangle = Length × breadth

$$A = l \times b$$

Here in Fig. *ABCD, AB, DC* and CB, DA are the four sides, *AC* and *BD* are two diagonals. The sum of the all four sides is called perimeter.

$$\text{Perimeter} = 2 \times (\text{length} + \text{breadth})$$
$$= 2\,(l + b).$$

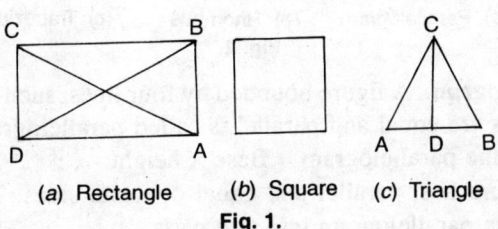

(*a*) Rectangle   (*b*) Square   (*c*) Triangle

Fig. 1.

   (*ii*) **Square.** The square is a particular form of a rectangle having all the four sides equal as shown in Fig. 1(*b*).

        Area of the square = side × side

$$A = (\text{side})^2 = (a)^2 \qquad (\text{where } a = \text{side})$$

The *diagonal* $= a\sqrt{2}$

and   Perimeter $= 4a.$

(iii) **Triangle.** Any figure having three closed sides or bounded by three sides, is called triangle as shown in Fig. 1(c).

$$\text{Area of the triangle} = \frac{1}{2} \times \text{base} \times \text{height} \quad \text{(Right angled triangle)}$$

$$= \sqrt{s(s-a)(s-b)(s-c)} \quad \text{(any triangle)}$$

where $\qquad s = \dfrac{a+b+c}{2}$ and *a*, *b* and *c* are sides.

(iv) **Circle.** A circle is a plane figure bounded by a curved line, every point of which is at the same distance from the centre.

The line joining centre to the circumference is known as radius and joining two points on the circumference through centre is called diameter. If *r* is the radius then the dia = 2*r* = *D*.

$$\text{Area of the circle} = \pi r^2 = \frac{\pi D^2}{4}$$

The circumference $\quad = 2\,\pi r = \pi D$.
The area of a circular ring having *R* and *r* radius (if *R* > *r*).
Area of the ring = $\pi(R^2 - r^2)$.

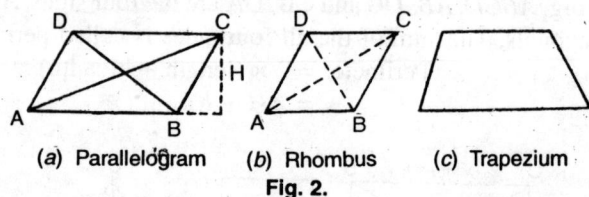

(*a*) Parallelogram   (*b*) Rhombus   (*c*) Trapezium
Fig. 2.

(v) **Parallelogram.** A figure bounded by four lines, such that the opposite sides are equal and parallel is called parallelogram.
Area of the parallelogram = Base × height.
This figure has, parallel and equal opposite sides. The diagonals divide the parallelogram into two parts.

(vi) **Rhombus.** If all the four sides of the figure are parallel and equal it is called rhombus. It is a tilted square.

(vii) **Trapezium.** A plane figure having all four sides such that at least one pair of opposite side is parallel, is called trapezium.

(viii) **Quadrilateral.** A figure bounded by any four sides so that none of them is equal.

$$\text{Area of quadrilateral} = \frac{1}{2} \times \text{product of the two diagonal.}$$

Volume. It is the amount of the space enclosed by a body.

(*i*) **Cube.** It has all the three length, breadth and height equal and the volume is equal to

V = length × breadth × height

= (side)^3.

(*ii*) **Cuboid.** It has all enclosed space by six rectangular planes. The volume = length × breadth × height

The diagonals = $\sqrt{\text{length}^2 + \text{breath}^2 + \text{height}^2}$.

(*iii*) **Right circular cylinder.** A right angled prism having its top and base as circular and made just by rotating a rectangle above one of its arm.

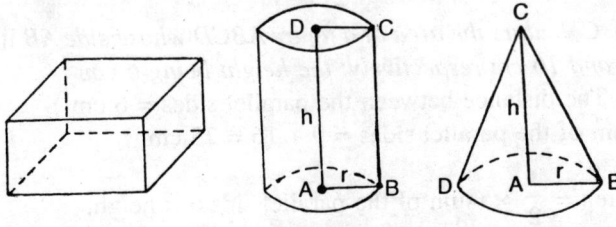

Fig. 3.

Volume of the right circular cylinder = $\pi r^2 h$

= area of the base × height.

(*iv*) **Cone.** The figure or volume obtained just by revolving the triangle around its one side. It is otherwise a pyramid having base as circle.

Volume = $\dfrac{1}{3}\pi r^2 h$.

The distance from the vertex to the circumference of the base is called *slant height*.

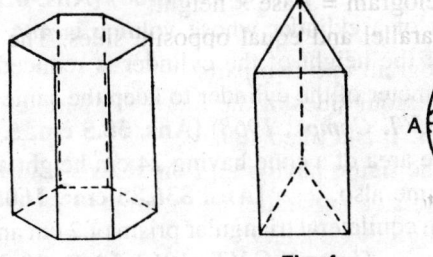

Fig. 4

(*v*) **Prism.** It is figure obtained just by joining two parallel congruent polygons and the sides joining the vertex.

Volume = Area of base × height.

(vi) **Sphere.** It is figure obtained just by joining two parallel congruent polygons and the sides joining the vertex.

$$\text{Volume} = \frac{4}{3} \pi (\text{radius})^2.$$

Example. *Find the area of a triangle whose base is 10 cm and height is 6 cm.*

Solution. Area of the triangle $= \frac{1}{2} \times \text{base} \times \text{height}$

$$= \frac{1}{2} \times 10 \times 6 = \textbf{30 cm}^2 \textbf{ Ans.}$$

Example. *Calculate the area of a figure ABCD whose side AB || CD and are 9 cm and 16 cm respectively; the height being 6 cm.*

Solution. The distance between the parallel sides = 6 cm
The sum of the parallel sides = 9 + 16 = 25 cm

Now area $= \frac{1}{2} \times (\text{sum of the parallel sides}) \times \text{height}$

$$= \frac{1}{2} \times 25 \times 6 = \textbf{75 cm}^2 \textbf{ Ans.}$$

UNSOLVED EXERCISES

1. A conduit pipe weights 0.6 kg per metre. Find the weight of 9 conduit pipes each measuring 4.3 m.
   (N.C.V.T., 1974) [Ans. **23.22 kg**]
2. A steel washer has the following dimensions, outside diameter 14 mm, inside diameter 7 mm and thickness 1 mm. Find the volume of the steel in cm^3. *(N.C.V.T., 1968)* [Ans. **0.115 cm**^{3}]
3. Find out the height of a cylinder whose volume is 484 cm^3 and diameter is 4 cm. If the height of the cylinder is reduced to half, what will be the diameter of the cylinder to keep the same volume.
   (Inter N.C.V.T. Compt., 1968) [Ans. **38.5 cm, 5.656 cm**]
4. Calculate the surface area of a cone having 24 cm height and 8 cm radius. Find its volume also.   [Ans. **836.88 cm**^{2}, **1608.5 cm**^{3}]
5. find the volume of an equilateral triangular prism of 2 cm and height 6 cms.   *(Inter N.C.V.T., 1969)* [Ans. **10.392 cm**^{3}]
6. Find the angle between two adjacent sides of a regular pentagon.
   (N.C.V.T., 1970) [Ans. **108°**]

7. Find the weight of a hexagonal steel rod, one side of which is 10 mm, and length of rod is 1.75 m, weight of one cubic cm rod is 7.75 gm.   *(N.C.V.T., 1972)* **[Ans. 3.547 kg]**

8. Find the cost of painting of a trapezium whose sum of parallel sides is 1.7 m and height is 1 metre. Rate of painting is Rs. 5.00 per square metre.   *(N.C.V.T., 1984)* **[Ans. Rs. 13.50]**

9. The perimeter of a right angle triangle is 60 cm. If the hypotenuse is 26 cm, find the other two sides and area of the triangle.

     [Ans. 10, 24, 120 cm^2]

10. The area of a rectangle is 520 cm^2. If its one side is 26 cm calculate the other side.   **[Ans. 20 cm]**

2. TRIGONOMETRY AND LOGARITHM

The branch of mathematics which deals with the angles, measurement and their applications is known as trigonometry.

Let in a $\triangle ABC$, there are three sides a, b and c and three angles $\angle A$, $\angle B$ and $\angle C$. These three sides and three angles have their six relationships. In any right angled triangle, here $\triangle ABC$, there are one base, one perpendicular and one hypotenuse, as BC, AC and AB respectively.

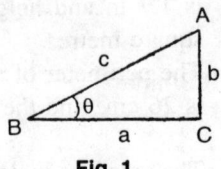

Fig. 1.

Here considering $\angle ABC = \theta$, there are six relations, which are as follows:

1. $\sin\theta = \dfrac{\text{Perpendicular}}{\text{Hypotenuse}} = \dfrac{AC}{AB} = \dfrac{b}{c}$

2. $\cos\theta = \dfrac{\text{Base}}{\text{Hypotenuse}} = \dfrac{BC}{AB} = \dfrac{a}{c}$

3. $\tan\theta = \dfrac{\text{Perpendicular}}{\text{Base}} = \dfrac{AC}{BC} = \dfrac{b}{a}$

4. $\cot\theta = \dfrac{\text{Base}}{\text{Perpendicular}} = \dfrac{BC}{AC} = \dfrac{a}{b}$

5. $\sec\theta = \dfrac{\text{Hypotenuse}}{\text{Base}} = \dfrac{AB}{BC} = \dfrac{c}{a}$

6. $\operatorname{cosec}\theta = \dfrac{\text{Hypotenuse}}{\text{Perpendicular}} = \dfrac{AB}{AC} = \dfrac{c}{b}$

Q. Prove that $\operatorname{cosec}\theta = \dfrac{1}{\sin\theta}$, $\tan\theta = \dfrac{1}{\cot\theta}$, $\sec\theta = \dfrac{1}{\cos\theta}$.

Ans. In $\triangle ABC$,

(i) $\operatorname{cosec}\theta = \dfrac{\text{Hypotenuse}}{\text{Perpendicular}} = \dfrac{AB}{AC}$.

Now dividing the numerator and denominator by AB,

$$= \dfrac{\dfrac{AB}{AB}}{\dfrac{AC}{AB}} = \dfrac{1}{\dfrac{AC}{AB}}$$

Here *AC/AB* in the trigonometrical ratio is equal to the perpendicular/hypotenuse *i.e.* sin θ.

So $$\operatorname{cosec}\theta = \dfrac{1}{\dfrac{AC}{AB}} = \dfrac{1}{\sin\theta}$$ **Proved**

(ii) Similarly

$$\tan\theta = \dfrac{AC}{BC}$$

$$= \dfrac{\dfrac{AC}{AC}}{\dfrac{BC}{AC}} = \dfrac{1}{\dfrac{BC}{AC}} = \dfrac{1}{\cot\theta}$$ **Proved**

(iii) $$\sec\theta = \dfrac{\text{Hypotenuse}}{\text{Base}} = \dfrac{AB}{BC}.$$

$$= \dfrac{\dfrac{AB}{AB}}{\dfrac{BC}{AB}} = \dfrac{1}{\dfrac{BC}{AB}} = \dfrac{1}{\cos\theta}$$ **Proved**

Square Relations

Example. *Prove that*
(i) $\sin^2\theta + \cos^2\theta = 1$,
(ii) $1 + \cot^2\theta = \operatorname{cosec}^2\theta$,
(iii) $1 + \tan^2\theta = \sec^2\theta$.

Solution. *(i)* In a right angle triangle *ABC*, $AB^2 = AC^2 + BC^2$ ...(*i*)
Now dividing by AB^2, the equation will be

$$\dfrac{AB^2}{AB^2} = \dfrac{AC^2}{AB^2} + \dfrac{BC^2}{AB^2}$$

Now changing these functions into the trigonometrical functions.

$$1 = \left(\sin\theta\right)^2 + \left(\cos\theta\right)^2$$

or $$1 = \sin^2\theta + \cos^2\theta$$ **Proved**

(ii) Now dividing equation *(i)* by AC^2

$$\frac{AB^2}{AC^2} = \frac{AC^2}{AC^2} + \frac{BC^2}{AC^2}$$

Changing these into trigonometrical function

$$\left(\frac{AB}{AC}\right)^2 = \left(\frac{AC}{AC}\right)^2 + \left(\frac{BC}{AC}\right)^2$$

$$\mathbf{cosec^2\theta = 1 + cot^2\theta} \qquad\qquad \textbf{Proved}$$

(iii) Again dividing equation *(i)* by BC^2

$$\frac{AB^2}{BC^2} = \frac{AC^2}{BC^2} + \frac{BC^2}{BC^2}$$

or

$$\left(\frac{AB}{BC}\right)^2 = \left(\frac{AC}{BC}\right)^2 + 1$$

or

$$\mathbf{sec^2\theta = 1 + tan^2\,\theta} \qquad\qquad \textbf{Proved}$$

Important Formulae

1. $\sin(-\theta) = -\sin\theta,$ $\qquad\qquad \sin(180 - \theta) = \sin\theta$
   $\cos(-\theta) = \cos\theta,$ $\qquad\qquad \cos(180 - \theta) = -\cos\theta$
   $\tan(-\theta) = -\tan\theta,$ $\qquad\qquad \tan(180 - \theta) = -\tan\theta.$

2. $\sin(90 - \theta) = \cos\theta,$ $\qquad\qquad \sin(90 + \theta) = \cos\theta$
   $\cos(90 - \theta) = \sin\theta,$ $\qquad\qquad \cos(90 + \theta) = -\sin\theta$

3. $\sin(180 - \theta) = \sin\theta,$ $\qquad\qquad \sin(180 + \theta) = -\sin\theta$
   $\cos(180 - \theta) = -\cos\theta,$ $\qquad\qquad \cos(180 + \theta) = -\cos\theta.$

4. $\sin(A + B) = \sin A\cos B + \cos A\sin B$
   $\sin(A - B) = \sin A\cos B - \cos A\sin B$
   $\cos(A + B) = \cos A\cos B - \sin A\sin B$
   $\cos(A - B) = \cos A\cos B + \sin A\sin B$

   $$\tan(A + B) = \frac{\tan A + \tan B}{1 - \tan A\tan B}$$

   $$\tan\left(\frac{\pi}{4} + \theta\right) = \frac{1 + \tan\theta}{1 - \tan\theta}$$

$$\tan(A - B) = \frac{\tan A - \tan B}{1 + \tan A \tan B}$$

$$\tan\left(\frac{\pi}{4} - \theta\right) = \frac{1 - \tan\theta}{1 + \tan\theta}$$

$$\cot(A + B) = \frac{\cot A \cot B - 1}{\cot B + \cot A}$$

$$\cot(A - B) = \frac{\cot A \cot B + 1}{\cot B - \cot A}$$

$$\tan(A + B + C) = \frac{\tan A + \tan B + \tan C - \tan A \tan B \tan C}{1 - \tan A \tan B - \tan B \tan C - \tan C \tan A}.$$

5. $\sin(A + B)\sin(A - B) = \sin^2 A - \sin^2 B$
    $\cos(A + B)\cos(A - B) = \cos^2 A - \sin^2 B$

6. $2\sin A \cos B = \sin(A + B) + \sin(A - B)$
    $2\cos A \sin B = \sin(A + B) - \sin(A - B)$
    $2\cos A \cos B = \cos(A + B) + \cos(A - B)$
    $2\sin A \sin B = \cos(A + B) - \cos(A - B).$

7. $\sin 2A = 2\sin A \cos A = \dfrac{2\tan A}{1 + \tan^2 A}$

    $\cos 2A = \cos^2 A - \sin^2 A = 1 - 2\sin^2 A$

    $\qquad = 2\cos^2 A - 1 = \dfrac{1 - \tan^2 A}{1 + \tan^2 A}$

    $\tan 2A = \dfrac{2\tan A}{1 - \tan^2 A}$

    $\sin 3A = 3\sin A - 4\sin^3 A$

    $\cos 3A = 4\cos^3 A - 3\cos A.$

    $\tan 3A = \dfrac{3\tan A - \tan^3 A}{1 - \tan^2 A}$

8. $\dfrac{\sin A}{a} = \dfrac{\sin B}{b} = \dfrac{\sin C}{c}$

    $a = b\cos C + c\cos B$

$$b = c \cos A + a \cos C$$
$$c = a \cos B + b \cos A$$
$$a^2 = b^2 + c^2 - 2bc \cos A$$
$$b^2 = c^2 + a^2 - 2ca \cos B$$
$$c^2 = a^2 + b^2 - 2ab \cos C.$$

9. $S = \sqrt{s(s-a)(s-b)(s-c)}.$

10. $(m+n)\cot\theta = m\cot C - n\cot B$

$(m+n)\cot\theta = n\cot\beta - n\cot\alpha.$

Example. *Represent 100° in radiation.*

Solution. 180 = π radian *i.e.* 180 × 1° = π radian.

Now $\qquad 100° = 100 \times 1° = 100 \times \dfrac{\pi}{180} = \dfrac{5\pi}{9}$ **radian. Ans.**

Example. *Represent 50° 20′ 30″ into radian.*

Solution. This angle $50° \, 20' \, 30'' = 50 + \left(\dfrac{20}{60}\right)^{\circ} + \left(\dfrac{36}{60 \times 60}\right)$

$$= 50 + 0.333 + 0.01 = 50.343°$$

∴ radians = $50.343 \times \dfrac{\pi}{180}$ = **0.2797 π radian. Ans.**

Example. *Prove that* $\dfrac{\sec\theta}{\cos\theta} - 1 = \tan^2\theta,$

Solution. Taking $\qquad$ L.H.S. $= \dfrac{\sec\theta}{\cos\theta} - 1$

$$= \sec\theta \dfrac{1}{\cos\theta} - 1$$

$$= \sec\theta \, \sec\theta - 1 = \sec^2\theta - 1$$

and $\qquad \sec^2\theta - 1 = \tan^2\theta$ so it is equal to $\tan^2\theta.$

∴ $\qquad$ L.H.S. = R.H.S. **Proved**

Example. *Prove that, sec A − cos A = sin A tan A.*

Solution. Taking L.H.S. = sec A − cos A

$$= \frac{1}{\cos A} - \cos A = \frac{1 - \cos^2 A}{\sec A} = \frac{\sin^2 A}{\sec A}.$$

$$= \frac{\sin A}{\cos A} = \sin A \tan A. \qquad \textbf{Proved}$$

Example. *An object is placed at 45 m. from the base of a minar. The angle of elevation of the top to the object is 30°, calculate the height of the minar.*

Solution. Here as given $\angle APQ = 30°$.

In the right angle triangle $\triangle PQO$.

$$\frac{PQ}{OQ} = \tan Q.$$

Here $\quad AP \parallel QO$

so $\qquad \angle APO = PQO = 30°$

so $\qquad \tan 30 = \dfrac{PQ}{OQ} = \dfrac{PO}{45}$

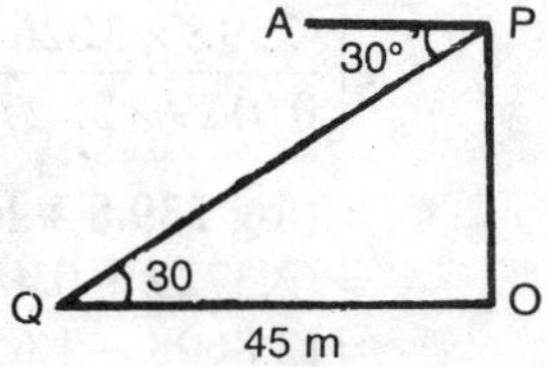

$$PO = 45 \tan 30° = 45 \times \frac{1}{\sqrt{3}} = 25.98$$

Fig. 2.

so the height of the minar = **25.98 m. Ans.**

Example. *In a right angle triangle ABC, angle $\angle A = 90°$, $\angle B = 28°$ and side AC = 8 cm, with help of the trigonometrical tables find side AB and BC.*

Solution. In the triangle $\triangle ABC$ as shown in Fig. 3.

$$\sin 28° = 0.4695 = \frac{AC}{BC} = \frac{8}{BC}$$

$\therefore \qquad BC = \dfrac{8}{0.4695} = \textbf{17 cm. Ans.}$

Again $\quad \tan 28 = 0.5317 = \dfrac{AC}{AB} = \dfrac{8}{AB}$

$\therefore \qquad AB = \dfrac{8}{0.5317} = \textbf{15 cm. Ans.}$

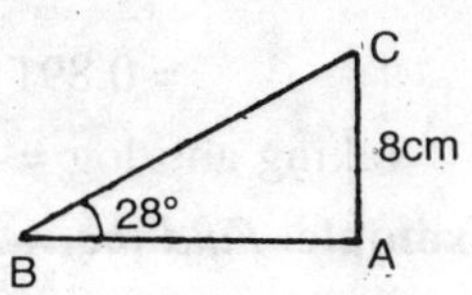

Fig. 3.

Logarithms:

There are these three formulae of the logarithm calculations.

If there are two quantities '*m*' and '*n* then:
  (*i*) **If these are to be multiplied then**
     log(m × n) = log m + log n.
  (*ii*) **If these are to be divided then**

$$\log\left(\frac{m}{n}\right) = \log m - \log n.$$

(*iii*) **If there is the power then**
     log(m)^n = n log m.

Example. *Evaluate* $\dfrac{239.8 \times 3.026}{0.0129 \times 35.27}$ *by* log.

(*N.C.V.T., 1968*)

Solution. It is a combination of multiplication and division, so taking log

$$\log\left[\frac{239.8 \times 3.026}{0.0129 \times 35.27}\right]$$

$$= (\log 239.8 + \log 3.026) - (\log 0.0129 + \log 35.27)$$
$$= (2.3729 + 0.4809) - (2.1106 + 1.5474)$$
$$= (2.8608 - 1.6580 = 3.2028$$

Taking anti log of this = **1595. Ans.**

Example. *Solve by using Logarithms*

$$\sqrt[5]{2.709} \times \sqrt[7]{1.2379}$$
(*N.C.V.T., 1985*)

Solution. Hence $y = \sqrt[5]{2.709} \times \sqrt[7]{1.2379}$

$$\log y = \left[\log \sqrt[5]{2.709} + \log \sqrt[7]{1.2379}\right]$$

$$= \log(2.709)^{1/5} + \log(1.2379)^{1/7}$$

$$= \frac{1}{5}(0.4556) + \frac{1}{7}(0.0927).$$

$$= 0.891 + 0.0137 = 0.1023$$

Taking anti log = **1.266. Ans.**

Example. *Find the value of given expression by log*

$$= \frac{\sqrt{3.12} \times 5.68 \times 0.825}{4.872 \times 0.21}$$

Solution. Let $\qquad x = \dfrac{\sqrt{3.21} \times 5.68 \times 0.825}{4.872 \times 0.21}$

Taking log of both sides

$$\log x = \log \sqrt{3.12} + \log 5.68 + \log 0.825 - (\log 4.872 + \log 0.21)$$

$$= \frac{1}{2}(0.4942) + 0.7543 + \overline{1} \cdot 9165 - (0.6877 - \overline{1}.3222)$$

$$= 0.2471 + 0.7543 + \overline{1} \cdot 9165 - 0.0099 = 0.9080.$$

Taking anti log of 0.9080 = **8.091. Ans.**

UNSOLVED EXERCISE

1. Change into radians : 1°, 90°, 10°, 180° and 150°

$$\left[\text{Ans. } \frac{\pi}{180}, \frac{\pi}{2}, \frac{\pi}{18}, \pi, \frac{5\pi}{6} \right]$$

2. In a right angled triangle *ABC,* sin *B* = 3/8 and the hypotenuse is 24 cm. Calculate the side *AC.*
   [Ans. 9 cm]

3. Find the values of cos θ, sec θ and cosec θ if sin θ = 0.6.

   [Ans. 0.8, 1.25, 1.67]

4. Prove that
   (*i*) sec *A* − cos *A* = sin *A* tan *A.*
   (*ii*) cosec *A* tan *A* cos *A* = 1.
   (*iii*) $\tan^2\theta - \cot^2\theta = \sec^2\theta - \csc^2\theta$.
   (*iv*) $(\sin\theta + \cos\theta)^2 + (\sin\theta - \cos\theta)^2 = 2$.

5. In right angled triangle *ABC,* base is 4 cm, and angle between hypotenuse and base is 55° with the help of trigonometrical table, find the value of hypotenuse and perpendicular.
   (*N.C.V.T., 1974*) **[Ans. 6.972 cm, 5.7124 cm]**

6. In a triangle *ABC,* given that *BC* = 2 cm, *AB* = 5 cm and angle *ACB* = 52°, find, the side *AC* and the area of the triangle.
   (*N.C.V.T., 1985*) **[Ans. 5.96 cm, 4.702 cm²]**

7. A line man who is working on a road place, his ladder which is 12 m in length rests at a point on the road such that it makes an angle of 60° with the ground when it is placed against a lamp post on one side of the road and makes 30° when placed against another lamp post directly on the opposite side of the road. Find the distance between the two lamp posts.
   (*N.C.V.T., 1981*) **[Ans. 16.392 m]**

Solve by logarithms

8. $\dfrac{47.32 \times 12.56^3}{0.0678 \times 56.89^3}$ **[Ans. 7.51]**

9. $\sqrt[7]{981.376}$ **[Ans. 267]**

10. $\dfrac{115.2 \times (0.113)^2}{55.11}$ (*N.C.V.T., 1978*) **[Ans. 0.02669]**

11. Find square root of 966.5 (*N.C.V.T., 1978*) **[Ans. 1.4926]**

12. $\dfrac{17.32 \times (1.34)^2}{0.0592 \times (21.34)^2}$ (*N.C.V.T., 1980*) **[Ans. 1.149]**

3. WORKSHOP SCIENCE

Science Definitions

Q. Define the followings: density, relative density, specific heat, latent heat, principle of Archimedes, law of parallelogram of forces, equilibrium principle of moments, elastic limit, stress, strain, elasticity, malleability, mechanical advantage and velocity ratio.

Ans. Density: *(N.C.V.T. 1970, 66)*

It is defined as the mass per unit volume. It's unit is gm/c.c., or kg/m^2

$$\therefore \qquad \text{Density} = \frac{\text{Mass}}{\text{Volume}} = \frac{M}{V}.$$

Relative density: *(N.C.V.T., 1970, 72)*

It is also termed as specific gravity. It is defined as the ratio of the weight of a given volume of a substance to the weight of the same volume of water. It is a ratio, hence no unit.

$$\text{Specific gravity} = \frac{\text{Weight of given volume of substance}}{\text{Weight of equal volume of water}}.$$

Specific heat: *(N.C.V.T., 1974)*

It is defined as the ratio of heat required to raise the temperature of a certain mass of the substance through a certain degree rise of temperature to the water of same mass through same degree rise of temperature.

Latent heat. It is amount of heat absorbed or released per unit mass of a substance to change its state e.g., latent heat of the ice is 80 cal/gm and latent heat of vaporisation 536 cal/gm.

Principle of Archimedes: *(N.C.V.T. 1968, 73)*

According to the principle of Archimedes, whenever a body is dipped into the water or liquid it suffers a loss in weight, the loss is weight will be equal to the water or liquid displaced.

Laws of parallelogram of forces. If two forces acting on a body are represented by the two vectors shown in magnitude and direction, by the adjacent sides of a parallelogram, their resultant will be shown by the diagonal of the parallelogram passing through the point of intersection of acting forces.

It can be given, by R, if the forces are A and B at an angle θ.

$$R = \sqrt{A^2 + B^2 + 2AB\cos\theta}$$

Equilibrium. It is the state of body if it is kept in rest position or if the resultant of the number of forces acting on a body is zero, the position is said the equilibrium state.

Principle of moment. It is nothing but the product of the force acting and the perpendicular distance from the line of action. In the equilibrium state, the moment of the forces in anti-clockwise direction is equal to the moment of the forces in clockwise direction. It is known as the principle of moment.

Elastic limit. It is a limit of change of shape *i.e.*, it can be defined as the magnitude of the applied force beyond which the body cannot regain its original shape or size.

Strain. When a force acting upon a body produces a relative displacement in its parts or shape or in both, the body is said strained. This can be longitudinal, volumetric and shearing strains.

$$\text{The longitudinal strain } = \frac{\text{Change in original length}}{\text{Original length}} = \frac{\delta l}{L}$$

Stress. Whenever the external force is applied on a body some internal forces are developed which tends to bring back the original state of the body *i.e.*, restoring tendency.

So the force per unit area is defined as the stress.

Elasticity. It is the property of a body by virtue of which it offers a resistive force to the force tending to change its volume, shape or both.

The body will return to its originality if it is within the elastic limits.

The ratio of stress and strain is called modulus of elasticity.

$$\textit{i.e. Modulus of elasticity } = \frac{\text{Stress}}{\text{Strain}} = \gamma.$$

Malleability. It is the property of the material so that it can be hammered to sheets or the sheets can be prepared.

Gold, silver, copper etc. are good examples of having malleability.

Mechanical advantages. It is the ratio of load to effort of a machine. If P is the force applied and W is the load to be lifted then

$$M.A. = \frac{\text{Load}}{\text{Efforts}} = \frac{W}{P}$$

Velocity ratio. It is the ratio of distance moved by the effort to the distance moved by the load.

$$\textit{i.e.} \qquad V.R. = \frac{\text{Distance moved by the effort}}{\text{Distance moved by the load}}.$$

Part III

Engineering Drawing

Q. *(a)* Draw a free hand sketch of a grid used for lead acid battery plate.

(b) Draw a free hand sketch of shackle insulator with clamp.

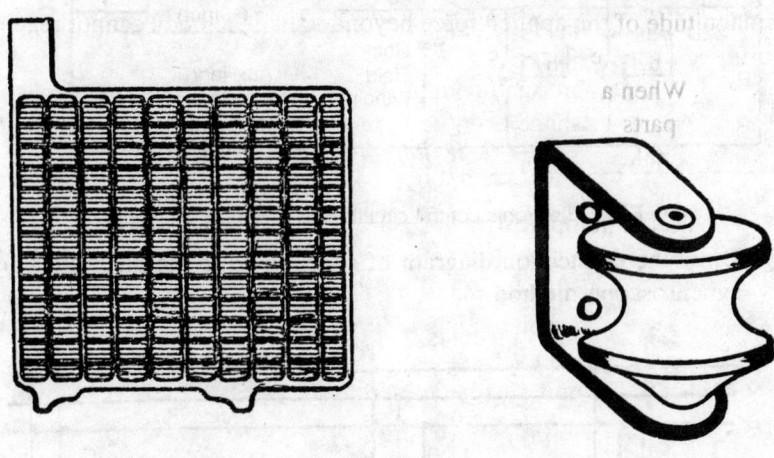

(a) L.A. Cell plate grid.   **Fig. ED-1**   *(b)* Shackle insulator.

Q. Draw a section view of a pin insulator.

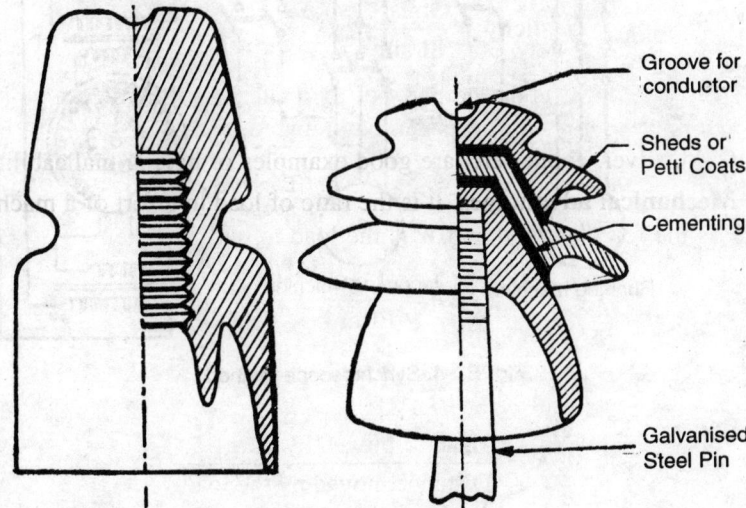

Groove for conductor

Sheds or Petti Coats

Cementing

Galvanised Steel Pin

Fig. ED-2. Pin insulators.

Q. Draw a connection diagram of battery charging while using an alternator with electronics control circuit.

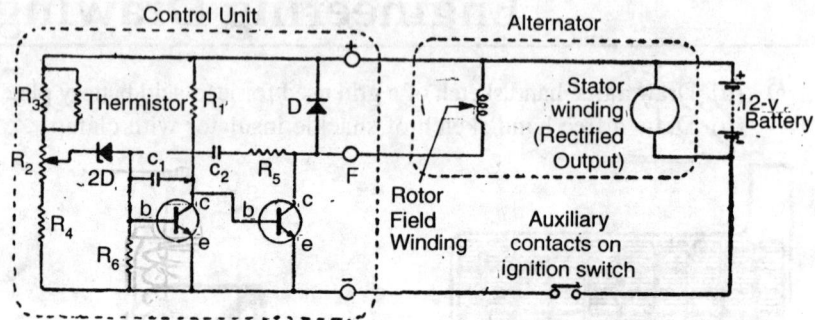

Fig. ED-3. Electronic control circuit used in battery charging.

Q. Draw the connection diagram of synchronizing the alternators by synchroscope method.

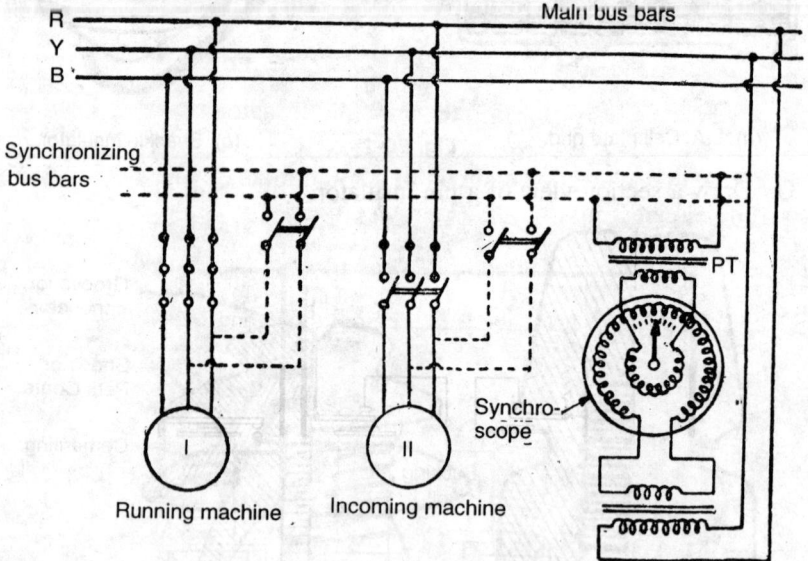

Fig. ED-4. Synchroscope method.

Q. How the conductors are supported in case of H.T. line using suspension type insulators.

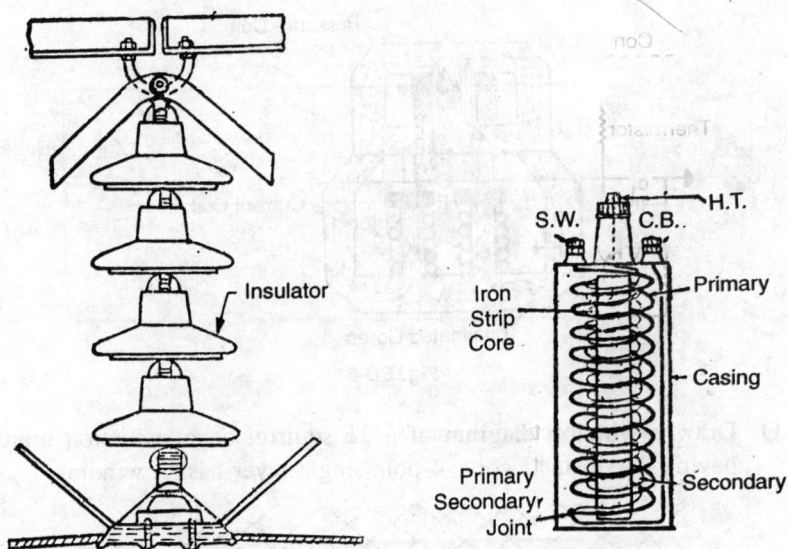

Fig. ED-5. Suspension type insulators.   **Fig. ED-6. Ignition coil.**

Q. Draw the use of transformer principle in automobile engineering (Ignition coil).

Q. Draw the connection diagram of a 3 unit regulator used in automobile engineering.

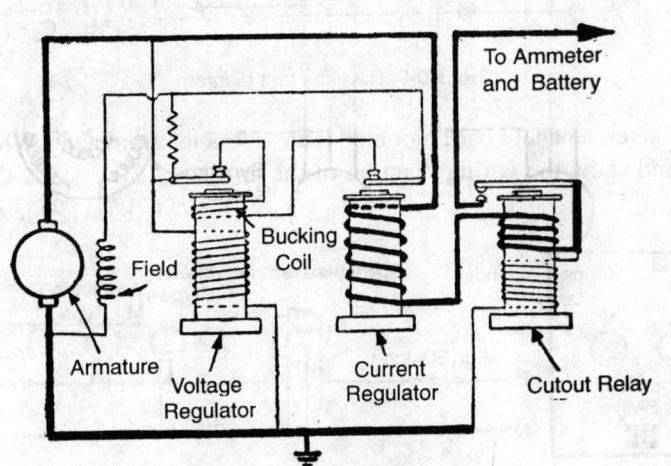

Fig. ED-7. Unit regulator.

Q. Draw a free hand sketch of a modern set of combined stampings of a single phase energy meter.

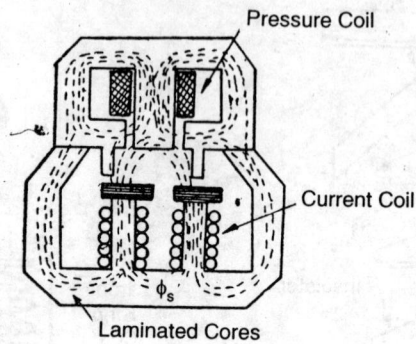

Fig. ED-8

Q. Draw a winding diagram of a 3φ squirrel cage induction motor having 24 slots, 12 coils, 4 pole single layer basket winding.

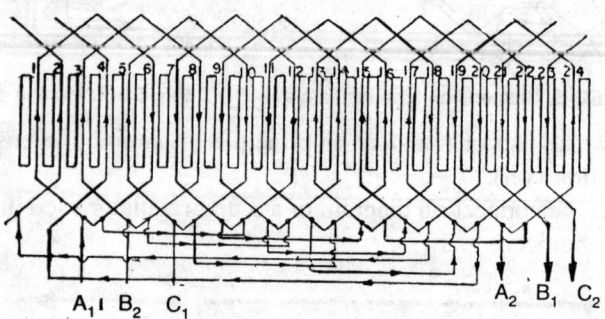

Fig. ED-9. Development diagram

Q. Sketch a panel board mounted with voltmeter ammeter, kWh meter and show the wiring diagram of the panel.

(N.C.V.T., 1968, 71, 86)

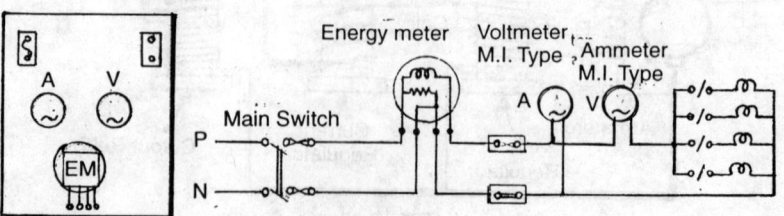

Fig. ED-10

Q. Draw the front view diagram of a panel board for 3-phase alternator, also show the connection diagrams.

(*N.C.V.T., 1971, 74, 80, App.*)

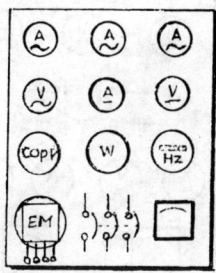

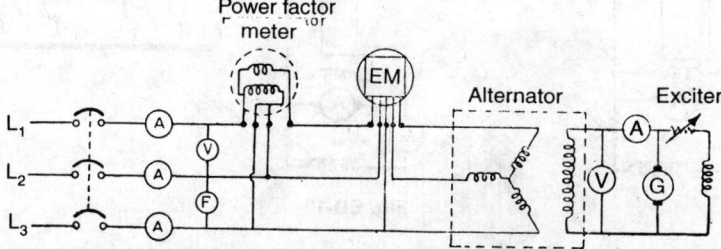

Fig. ED-11

Q. Draw the circuit diagram of a panel board layout of an electroplating generator with I.S.I. symbols. (*I.T.I., 1969, 70, 72, 73*)

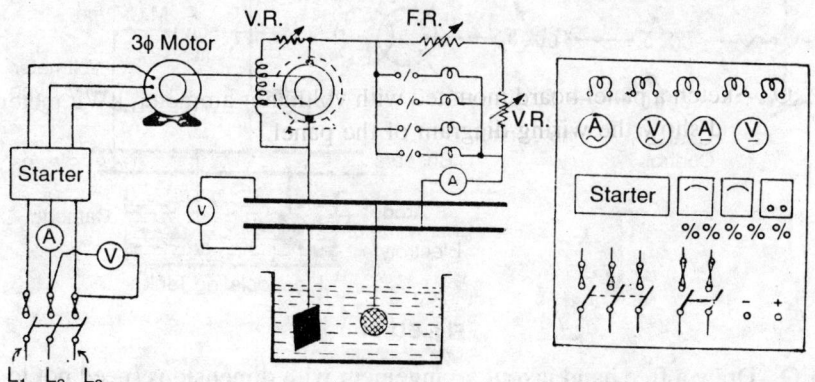

Fig. ED-12

Q. Make panels for two compound 50 kW 440 V.D.C. generators

suitable for working in parallel. All necessary accessories and measuring instruments should be provided, use standard symbols.

(Inter I.T.I. Comp., 1972)

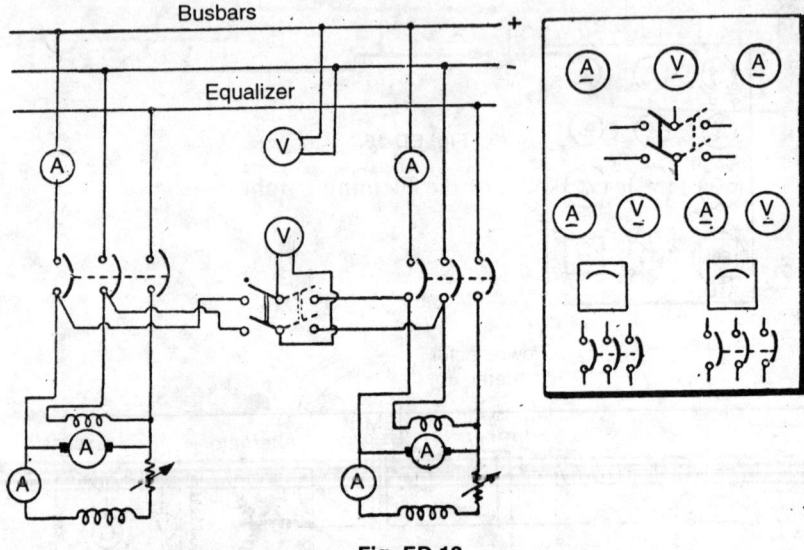

Fig. ED-13

Q. Draw a free hand diagram of an electroplating set being operated on A.C. 3φ. 440 V with coarse and fine controls.

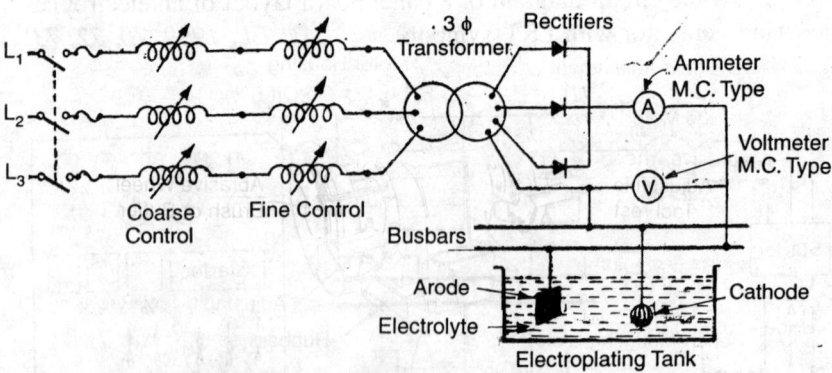

Fig. ED-14

Q. Draw a free hand layout arrangement with dimensions (need not to be drown to scale) of a 5 H.P. D.C. motor panel board with switches, starters, ammeter, voltmeter, controlling and protective devices.

(A.I. Comp. 1968)

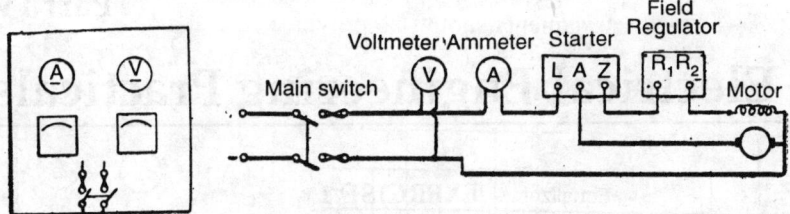

Fig. ED-15.

Q. Draw free hand sketch of the Fleming's right hand rule.

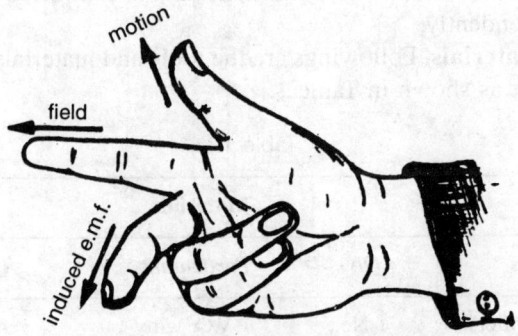

Fig. ED-16.

Q. Draw free hand sketch of a bench grinder and label its parts.

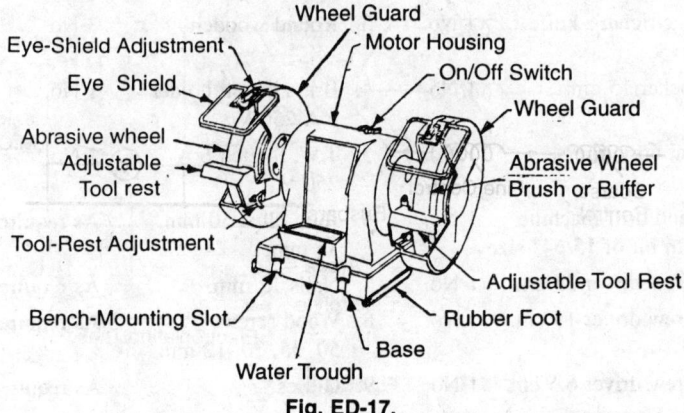

Fig. ED-17.

Part IV

Electrical Engineering Practicals

EXERCISE 1

Q. Control one lamp from two locations in batten wiring system.

Introduction. This circuit is used where a lamp point is to be controlled from two places, as in case of stairs. The lamp is controlled from both the places independently.

Tools and materials. Followings are the tools and materials used for this wiring circuit as shown in Table 1.

Table 1.

Tool		Materials	
Particulars	*Qty.*	*Particulars*	*Qty.*
1. Insulated pliers 15 cm.	1 No.	1. P.V.C. wire/ C.T.S. wire 3/22 SWG single core	As required
2. Side cutter 15 cm. mm.	1 No.	2. Batten 12 × 12 mm	As required
3. Electrician's knife block.	1 No.	3. Round wooden	3 No.
4. Pocker 15 cm.	1 No.	4. Brass batten holder 6 A 250 V.	1 No.
5. Flat nose pliers,	1 No.	5. T.W. Switch 6 A 250 V.	2 No.
6. Hand drill machine with bit of 15/64″ size	1 No.	6. Joint clips 50 mm, 35 mm.	As required
7. Screw driver 25 cm.	1 No.	7. Nails 12 mm.	As required
8. Screw driver 15 cm.	1 No.	8. Wood screws 50, 35, 20, 12 mm.	As required
9. Screw driver 6.5 cm.	1 No.	9. Gitties.	As required
10. Hand saw 40 cm.	1 No.	10. Round porcelain cleats.	As required
11. Rasp cut file 15 cm.	1 No.		
12. Firmer chisel.	1 No.		

Procedure:

(*i*) Make the layout and connection diagrams as shown in fig. P-1, P-2.

(*ii*) Select and cut the batten of required size.

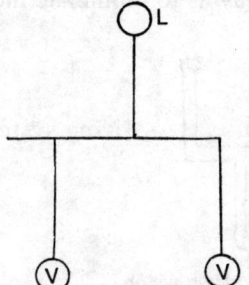

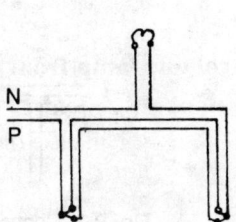

Fig. P-1. Lay out diagram.   **Fig. P-2.** Connection diagram.
(stair cage wiring)

(*iii*) Make proper joints and fix them.

(*iv*) Fix the joint clips with the help of nails, keeping in mind the number of wires drawn.

(*v*) Now fix the batten on the wiring board or fix the batten on the place of wiring with the help of gitties and porcelain round cleats.

(*vi*) Run the wires is the clips. Tight the clips so that wires are not loose in them.

(*vii*) Now fix the blocks and take out wires after making proper holes in the blocks.

(*viii*) Fix the accessories and make connections.

(*ix*) Check the circuit again and if correct energise it.

Precautions:

(*i*) Make a suitable and accessible layout and connection diagram.

(*ii*) The joints should be right and tight.

(*iii*) The block should be cut carefully. There should not be any gap left.

(*iv*) The clips should be fixed carefully and properly.

(*v*) Accessories should be tight and properly fixed.

(*vi*) Screw must be counter shunk.

(*vii*) No joint should be used in the running wiring except at the terminals of the accessories.

(*viii*) The bend should not be sharp.

EXERCISE 2

Carry out the godown wiring circuit having three lamps. Wiring is to

be done in P.V.C. wire on T.W. batten. **Give connection diagram and list of materials required.**

(N.C.V.T. 1970, A.I.C. 1971, 73)

Introduction. This wiring circuit is used in the places of godown etc. where a lamp is to be controlled in such a way as to illuminate the successive post.

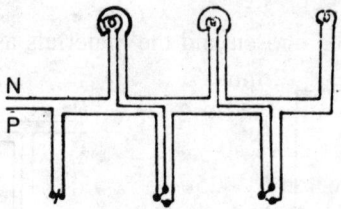

Fig. P-3. Connection diagram of godown wiring.

List of the material required:

1. T.W. Batten 12 × 12 mm ...4.5 m.
2. Round wooden books 87 × 25 mm ...6 No.
3. Bakelite or Brass batten holder 6 A 250 V ...3 No.
4. Single way switch 6 A 250 V ...1 No.
5. Two way switch 6 A 250 V ...2 No.
6. P.V.C. wire 1.5 mm^2 ...13 m.
7. Joint clips 25 mm ...one packet.
8. Nails 12 mm ...app. 50 gm.
9. Wood screws 12, 20, 30, 50 mm. ...as required.

EXERCISE 3

Carry out the intermediate switch wiring in batten wiring system. OR Control one lamp from three locations in batten wiring system. Write the precautions used.

Precautions:

(i) Prepare the layout diagram and make necessary joints. The joints should be well tight and right.

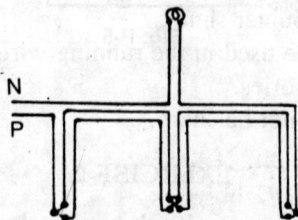

Fig. P-4. Intermediate wiring.

(ii) Cut the batten to the size and fix the joint clips keeping in mind the number of wires. In no case the horizontal distance be more than 10 cm and vertical distance 15 cm.

(iii) In case of moisture/round porcelain cleats should be used.

(iv) The connections should be right and tight.

Material Required:

Let the dimensions be chosen and the materials as under:

1.	T.W. Batten 12 × 12 mm	...2.20 M.
2.	Round wooden block 87 × 25 mm	...4 No.
3.	Two way switch 6 A 250 V.	...2 No.
4.	Intermediate switch 6 A 250 V.	...1 No.
5.	Batten holder 6 A 250 V.	...1 No.
6.	P.V.C. wire 1.5 mm²...	...5 m.
7.	Nails 12 mm	...50 gm.
8.	Joint clips 25, 37 and 50 mm	...as required.
9.	Wood screws 12, 20, 30, 50 mm	...as required.

EXERCISE 4

The two lamps are to be so connected that both either burn bright or dim, or both go off, not more than two switches may be used for this part of wiring.

(N.C.V.T., 1971, Compt. 1971, Inter I.T.I 1972)

Connection diagram

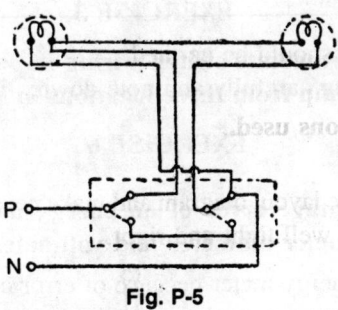

Fig. P-5

S_1 up S_2 up L_1, L_2 on

S_1 up S_2 down L_1 on

S_1 down S_2 up both off

S_1 down, S_2 down $L_1 L_2$ in series.

EXERCISE 5

Connect a heater and measure the energy consumption in half an hour. Draw connection diagram.

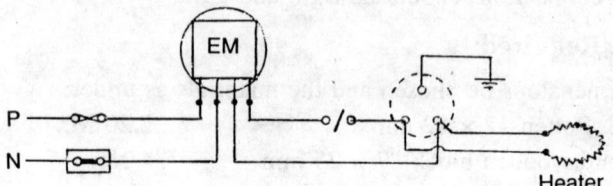

Fig. P-6.

Precautions.
1. Check individual equipment.
2. Check the connection diagram and see it should be tight.
   Make the table and get the energy consumed.

S.No.	Time	Reading		$E_2 - E_1 = E$	Remarks
		E_1	F_2		
1.					
2.					
3.					
.					
.					
.					

3. The plug point should be earthed.
4. Take the reading carefully and note down.

EXERCISE 6

Find out the percentage error of an energy meter at unity power factor load by voltmeter ammeter and voltmeter method.

Introduction. The energy meter because of error can read wrong which will effect the consumption and finally the monthly bill of the house/industry, so it should be checked properly.

Tools and Materials. Following tools and materials are required:

Tools			Materials and meter		
S.No.	Particulars	Qty.	S.No.	Name	Qty.
1.	Insulated pliers 6″	1 No.	1.	P.V.C. wire 1/18 SWG	as required
2.	Screw Driver 6″	1 No.	2.	Cotton waste	do
3.	Screw Driver 1½″	1 No.	3.	Single phase resistive load	1 No.
4.	Electrician's knife	1 No.	4.	Ammeter 0-5 A	1 No.
			5.	Voltmeter 0-250 V	1 No.
			6.	Energy meter 5 A 250 V 1 kW	1 No.
			7.	Wattmeter 1 kW 5 A 250 V	1 No.

Connection diagram

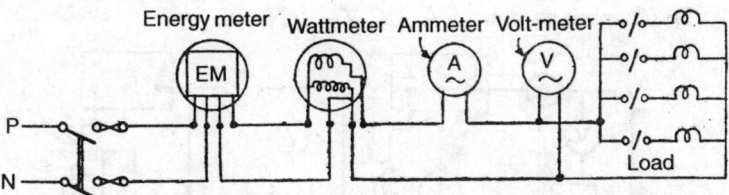

Fig. P-7. Energy meter Percentage error.

Procedure:

(i) Test each appliance and instrument.

(ii) Make the connection diagram as shown in Fig. P-7.

(iii) On the main switch and connect appropriate load.

(iv) Take down the readings of ammeter, voltmeter and wattmeter and see the energy consumption *i.e.* $\dfrac{VIt}{3600000}$ kWH observed energy).

(v) Calculate the energy shown by the energy meter by just connecting the number of revolutions made in a particular time period (recorded energy) *i.e.* n/k. When nth revolution made, k = rcv/kWH.

(vi) Find the percentage error.

$$\% \text{ error} = \frac{\text{Recorded energy} - \text{Observed energy}}{\text{Observed energy}}$$

Make the table. (Let the meter has $k = 1200$ Rev/kWh 5A 250 V Energymeter).

Volt meter reading, V	Ammeter reading, I	No. of rev. made	Time in sec.	Observed energy W	E Recorded energy by energy meter	% error
220 V	2 A	15	62 sec.	0.01137	0.9125	9.94%
220 V	4 A	20	62 sec.	0.01516	0.01667	–
220 V	5 A	30	60 sec.	0.08133	0.025	–

EXERCISE 7

Connect a d.c. compound motor or a shunt motor with a suitable starter. Test the connections and run the motor. Reverse the direction of the motor.

Draw the wiring diagram of connections.

(*N.C.V.T., 1968, 70; A.I.T., 1965, 68; Inter I.T.I. 1969, 74*)

Material required. Let the D.C. motor be of 1 HP. The following material is required:

1. DPIC main switch 15 A 250 V – 1 No.

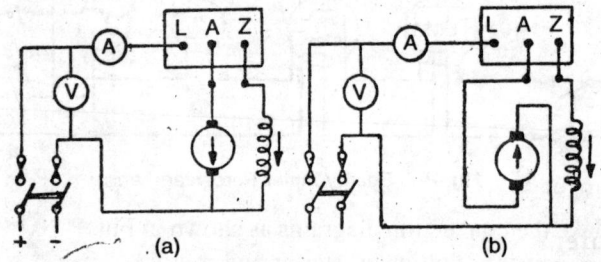

Fig. P-8. Shunt motor with starter.

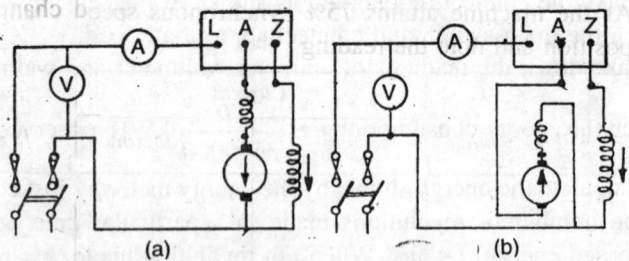

Fig. P-9. Compound motor with starter.

2. D.C. three point starter 1 HP 230 V   – 1 No.
3. D.C. shunt or compound motor 1 HP 250 V – 1 No.
4. Ammeter MC type 0-10 A   – 1 No.

5. Voltmeter MC type 0-250 V   – 1 No.
6. Connecting wires 25 mm   – as required.
7. Bare copper or aluminium wire for earthing as required.
8. Fuse wire 10 A as required.
9. Series testing board–   1 No.

EXERCISE 8

Carry out the proper connections for a 3 f squirrel cage induction motor (3 or 5 HP) with star delta starter and record starting current. After it, change the d.o.r. of the motor.

(N.C.V.T., 1966, 67, 69, Inter I.T.I., 1966, 68, 69, 71, 73, 75)

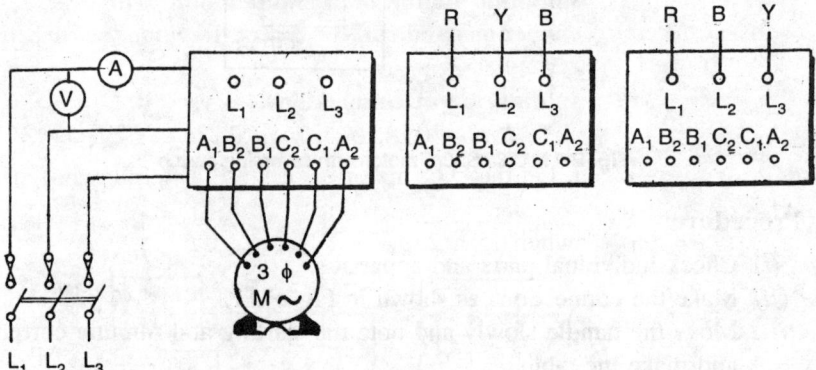

Fig. P-10

Procedure:

  (i) Make the connection diagrams as shown in Fig. P-10 by connecting the ammeter, voltmeter, starter and motor.

  (ii) Run the motor is star and take down the reading for starting current. As the machine attains 75% synchronous speed change to delta position and note the reading.

S.No.	Line voltage	Current		Remark
		Starting	*Running*	
1.				
2.				
3.				
·				
·				
·				

EXERCISE 9

Connect a D.C. shunt motor with starter and reversing switch. Make the connection diagram and note the starting and running current of the motor.

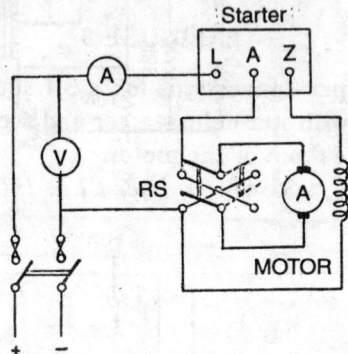

Fig. P-11. D.C. Shunt motor with reversing switch

Procedure:

(i) Check individual parts and apparatus.

(ii) Make the connections as shown in Fig. P-11.

(iii) Move the handle slowly and note the starting and running current and make the table.

S.No.	Voltage	Starting current	Running current	Remarks
1.				
2.				
3.				

EXERCISE 10

Wire up a singing bird door bell circuit. Give the list of the material used and the precautions observed.

Introduction. It is a circuit used for bell purposes in the houses etc.

Electrical connection.

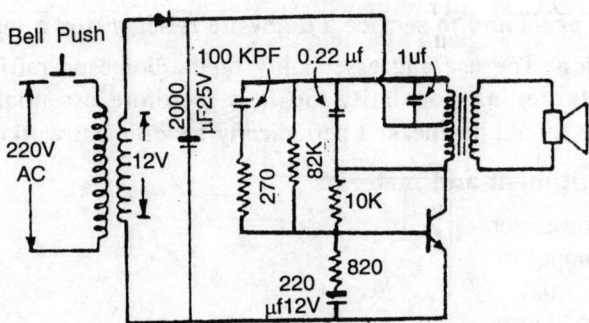

Fig. P-12. Singing bird door bell circuit diagram.

Materials:

Transistor	S.L. 100	...1 No.
Capacitors	2000 μF 25 V	...1 No.
	220 μF 25 V	...1 No.
	1 μF 12 V	...1 No.
	0.25 μF 12 V	...1 No.
	0.1 μF 12 V	...1 No.
Resistors	82 kΩ 0.25 W	...1 No.
	10 kΩ 0.25 W	...1 No.
	820 Ω 0.25 W	...1 No.
	270 Ω	...1 No.
Diode	IN 4001	...1 No.
Transformer 220 V/12 V A.C.		...1 No.
Out put push pull circuit		...1 No.
Load speaker 2″ −8 Ω		...1 No.
Bell push		...1 No.
Connecting wires		...as required.

Precautions:

(*i*) Every component should be tested before use.

(*ii*) The circuit should be checked very carefully.

(*iii*) The soldering should be done very fine. There should not be any dry soldering.

(*iv*) Clean every soldering by the brush after soldering.

(*v*) Check the circuit on motor if possible and see the response.

EXERCISE 11

Study the parts and to service a domestic refrigerator a sealed unit.

Introduction. The use and essentiality of the domestic refrigerator is now widely getting popularity. So it is therefore essential that the refrigerator should be checked periodically for efficient working.

Tools equipment and material:

1. Thermometer.
2. Clean cloth.
3. Wire brush.
4. Small blower.
5. Screw driver connector.
6. Connecting wires.
7. voltmeter.
8. Spirit level.
9. Torch.
10. Match box.

Procedure:

1. Check the installation.
2. All connections should be well tight and checked properly for their insulation and breakage.
3. Noise and rise of temperature should be checked.
4. The voltmeter should be used for voltage measurement.
5. The current drawn should be checked by an ammeter.
6. The insulation resistance should be checked by means of megger.
7. For any leakage the torch should be used.
8. The thermostat switch, door gasket, door switch, lamp holder and lamp should be checked properly.
9. The cabinet and cooling coil temperature should be checked properly.
10. Cleaning the condenser, the suction line, discharge line should be done.
11. The operation should be checked for its smooth and noiseless operation.
12. Give supply and start the refrigerator, check its performance.

Precautions:

1. Connections should be right and tight.
2. The mechanical and electrical parts of the refrigerator should be properly cleaned.
3. Properly rating and capacity transformer should be used.

Appendix

Table 1: Current Ratings
Single Circuit Twin and Multi-Core
P.V.C. Insulated Sheathed Aluminium Conductor Cables
(Cables Provided With Close Excess-Current Protection)

These current ratings are based on the following assumptions:
- (*a*) Ambient air temperature 40°C.
- (*b*) Maximum conductor temperature 70°C.
- (*c*) Thermal resistivity of P.V.C. 650° m/w.

Nominal Cross-Sectional Area	Number and Diameter of Wire	Enclosed in Conduit for Trunking Twin Cables Single-phase A.C. or D.C.	Clipped Direct to a Surface or on a Cable Tray and Enclosed Twin cables Single-phase A.C. or D.C.
(1) In mm²	(2) In mm²	(3) (Amps.)	(4) (Amps.)
1.5	1/1.40	12	15
2.5	1/1.80	16	19
4.0	1/2.24	21	26
6.0	1/2.80	27	32
10	1/3.55	37	42
16	7/1.70	50	58
25	7/2.24	57	66
35	7/2.50	69	82
50	7/3.00	87	98
70	19/2.24	–	120
95	19/2.50	–	145
120	37/2.06	–	170
150	37/2.24	–	195
185	37/2.50	–	225
240	37/3.00	–	265
300	61/2.50	–	305

Table 2: Current Carrying Capacity of P.V.C. and V.I.R.
Cable (Aluminium)

Nearest M.M. Size and No. of Wire	Equivalent Copper Conductor	Max. Current Taken by	
		V.I.R. Cable	P.V.C. Cable
1/1.40	1/.044 (1/18)	3 amp.	12
1/1.80	3/.029 (3/22)	5 amp.	16
	3/.036 (3/20)	10 amp.	–
1/2.24	7/.029 (7/22)	15 amp.	21
1/2.80	7/.036 (7/20)	24 amp.	27
1/3.55	7/.044 (7/18)	31 amp.	37
7/1.70	7/.052 (7/17)	37 amp.	50
7/2.24	7/.064 (7/16)	46 amp.	57
7/2.50	19/.044 (19/18)	53 amp.	69
7/3.00	19/.052 (19/17)	65 amp.	87
19/2.50	19/.083 (19/14)	100 amp.	120
19/2.24	19/.064 (19/16)	63 amp.	144
37/2.06	37/.072 (37/15)	120 amp.	170
37/2.24	37/.103 (37/12)	250 amp.	195

Table 3: Copper Wire
British Legal Standard or S.W.G.

Gauge No.	Dia. inch	Area	
		Circ. Mills.	Square inch
10	0.128	16284	0.01287
11	0.116	13456	0.01067
12	0.114	10816	0.00849
13	0.092	8460	.00665
14	0.082	6400	.00503
15	0.072	5184	0.00407
16	0.064	4096	0.00322
17	0.056	3136	0.00246
18	0.048	2304	0.001810
19	0.040	1600	0.001257

(Contd.)

Gauge No.	Dia. inch	Area	
		Circ. Mills.	Square inch
20	0.035	1296	0.001018
21	0.032	1024	0.000004
22	0.028	784	0.000616
23	0.024	576	0.000452
24	0.022	484	0.000380
25	0.020	400	0.000314
26	0.0180	324	0.0002545
27	0.0164	269	0.0002120
28	0.0184	219	0.0001720
29	0.0136	185	0.0001453
30	0.0124	145	0.0001208
31	0.0116	135	0.0001057
32	0.0108	117	0.0000916
33	0.0100	100	0.0000785
34	0.0092	85	0.0000665
35	0.0084	71	.0000554
36	0.0076	58	0.0000455
37	0.0068	46	0.0000363
38	0.0060	36	0.0000283
39	0.0052	27	0.0000212
40	0.0048	23	0.0000181

Table 4: Data of Super Enamelled and Resistance Wire

Size S.W.G.	Diam. Inch	M/M	Super Enamelled Medium Covering Minimum increase in diameter		Nickel-Chromium Heating Purposes Resistance per Foot at 20°C approx Ohms	Resistance Wire For Starter Regulator etc, Resistance per Foot at 20°C approx Ohms.
			units	min.		
14	.080	2.03	2.9	.074	.102	.046
15	.072	1.02	2.8	.071	.127	.057
16	.064	1.62	2.7	.069	.160	.072

(Contd.)

Size S.W.G.	Diam. Inch	M/M	Super Enamelled Medium Covering Minimum increase in diameter		Nickel-Chromium Heating Purposes Resistance per Foot at 20°C approx Ohms	Resistance Wire For Starter Regulator etc, Resistance per Foot at 20°C approx Ohms.
			units	min.		
17	.056	1.42	2.6	.066	.209	.094
18	.048	1.21	2.5	.063	.284	.128
19	.040	1.01	2.4	.061	.410	.184
20	.036	.91	2.3	.058	.506	.228
21	.032	.81	2.2	.056	.640	.288
22	.028	.71	2.1	.053	.834	.376
23	.021	.60	2.0	.051	1.14	.511
24	.022	.55	1.9	.048	1.35	.609
25	.020	.50	1.9	.048	1.64	.737
26	.018	.45	1.8	.046	2.02	.910
27	.0164	.41	1.7	.043	2.43	1.09
28	.0148	37	1.6	.041	2.99	1.35
29	.0136	.34	1.6	.041	3.54	1.59
30	.0124	.31	1.5	.038	4.26	1.91
31	.0116	.29	1.3	.033	4.87	2.14
32	.0108	.27	1.3	.033	5.62	2.53
33	.0100	.25	1.3	.033	6.56	2.95
34	.0092	.23	1.2	.030	7.45	3.48
35	.0084	.21	1.2	.030	9.29	4.18
36	.0076	.19	1.1	.028	11.4	5.10
37	.0068	.17	1.1	.028	14.2	6.37
38	.0060	.15	.9	.023	18.2	8.17
39	.0052	.13	.9	.023	24.2	10.88
40	.0048	.12	.8	.020	28.5	12.77
41	.0044	.11	.8	.020	34.0	15.22
42	.0040	.10	.7	.018	41.0	18.38
43	.0036	.09	.6	.015	50.7	22.69
44	.0032	.08	.6	.015	64.0	28.79
45	.0028	.07	.5	.0125	83.7	37.60
46	.0024	.06	.4	.0100	114.0	51.17
47	.0020	.05	.3	.0075	164.0	73.67
48	.0016	.04			256.0	115.13

Super enamelled: Synthetic Enamelled Copper wire.

Nickel Chromium: Nichrome wire.

contains 80 per cent Nickel and 20 per cent Chromium.

Resistance Wires: Eureka, Advance.

contains 44 per cent Nickel and 56 per cent Copper.

Table 5: Aluminium Conductor Paper Insulated Lead Covered Cables
Estimated Maximum Continuous Current Rating A.C.
Cables in Ground

Standard Aluminium Conductor		Conductor Current Rating in Ground		
Area mm²	No. and diameter mm	Single core Unarmoured Amps	Twin core Armoured Amps	3/3½ & 4 core Armoured Amps
6	1/2.80	50	57	48
10	1/3.55	70	74	62
16	7/1.70	90	96	81
25	7/2.24	115	122	107
35	7/2.50	138	147	128
50	{ 7/3.00 19/1.80	172	180	158
70	19/2.24	208	219	192
15	19/2.50	244	262	224
120	37/2.06	278	302	257
150	37/2.24	316	346	296
185	37/2.50	359	398	336
225	37/2.80	399	450	380
300	61/2.50	466	536	438
400	61/3.00	553	618	513

Table 6: Current Ratings for Heavy Duty 1100 V. P.V.C. Cables

Current Ratings in Amperes of P.V.C. Insulated Aluminium Conductor Heavy Duty 1100 V. Grade Cables

		In Air			Laid Direct in Ground			
n area	No. and diameter of wires	Three, 1-core cables in 3-phase system, laid side by side 7 cm. apart	2 core	3 core and 4 core	Three 1-core Cable in 3-phase system, laid side by side 7 cm apart	2 core	3 core and 4 core	Nominal area
mm²	mm	amp	amp	amp	amp	amp	amp	mm²
4	1/2.24	...	32	29	...	40	36	4
6	1/2.80	...	42	38	...	52	48	6
10	1/3.55	...	58	50	...	70	64	10
16	7/1.70	92	75	70	115	96	85	16
25	7/2.24	115	100	85	145	125	110	25
35	7/2.50	140	120	105	175	150	130	35
50	7/3.00	170	150	130	215	190	160	50
70	19/2.24	210	180	155	260	225	195	70
95	19/2.50	250	215	190	310	210	235	95
120	37/2.06	285	245	215	355	303	270	120
150	37/2.24	320	280	250	400	350	310	150
185	37/2.50	350	310	285	440	390	355	185
240	37/3.00	400	365	330	500	455	410	240
300	61/2.50	445	410	375	555	510	470	300
400	61/3.00	505	490	450	584	610	560	400

Table 7: Current Ratings for Aluminium Conductor Wires/Cables Subject to Voltage drop vulcanised Rubber, P.V.C. or Polythene Insulated Cables Sheathed with tough Rubber, P.V.C. Lead Twin. 3-Core and 4-Core. Conditions of Insulation: One cable is run in conduit troughing or casing or in free air or open trench. Ambient temperature: 30°C (86°F)

Size of conductor			One twin-core cable D.C. or single-phase A.C.		One 3-core or 4-core cable balanced three-phase	
Nominal area	Number and diameter of wires	Current rating	Approximate length of run for 1-volt drop		Current rating	Approximate length of run for 1 volt drop
			D.C.	A.C.		
mm^2	mm	amp.	metres	metres	amp.	metres
1.5	1/1.40	10	2.3	2.3	7	3.7
2.5	1/1.80	15	2.5	2.5	11	3.9
4	1/2.24	20	2.9	2.9	14	4.8
6	1/2.80	27	3.4	3.4	19	5.5
10	1/3.55	34	4.2	4.2	24	6.8
16	7/1.70	43	5.3	5.3	30	8.7
25	7/2.24	59	6.6	6.6	42	10.8
35	7/2.50	69	7.1	7.1	48	11.7
40	7/3.00	91	7.7	7.7	62	31.1
	19/1.80					
70	19/2.24	118	9.0	8.8	82	14.7
95	19/2.50	135	9.8	9.5	94	15.7
120	37/2.06	162	10.8	10.3	114	16.8
150	37/2.24	181	11.4	10.7	127	17.5
185	37/2.50	209	12.3	11.2	146	18.4
225	37/2.80	240	13.5	11.7	169	19.1
240	37/3.00	263	14.0	11.9	185	19.5
300	61/2.50	279	14.6	12.1	202	20.0

These ratings apply for P.V.C. insulated and polythene insulated cables the maximum conductor size of 50 mm². At rated current.

Ambient Temperature. For ambient temperatures other than 30°C (86°F) multiply the current ratings and divide the lengths for 1 volt drop by the appropriate factor below:

Ambient air temperature	25°C	35°C	40°C	45°C
Rating factor	(77°F)	(95°F)	(114°F)	113°F)
Grouping factor	1.13	0.86	0.69	0.47
Number of cables	2	3	4	5
Rating factor	0.8	0.7	0.65	0.6

Table 8: Copper Conductor Flexible Wires

Size of Conductor		Number and diameter of wires		Approximate overall diameter		Maximum allowable resistance at 20°C (68°F) for tinned wires		Current rating twin; 3-core 4-core (subject to voltage drop)	Approximate voltage drop per 10 metre run	Maximum Permissible weight supported by twin flexible	
Nominal area											
1	2	3	4	5	6	7	8	9	10	11	12
inch²	mm²	inch	mm²	inch	mm	ohm/100 yds	ohm/km	amp.	volts	lb.	kg.
0.0006	–	14/.0076	14/.193	0.034	0.86	42.03	45.95	2	2.0	3	1.360
–	0.5	16/.00787	16/.200	0.037	0.95	34.25	37.46	3.3	2.7	3.3	1.5
0.001	–	23/.0076	23/.193	0.064	1.17	25.57	27.97	5	3.0	5	2.268
–	0.75	24/.00787	24/.200	0.049	1.24	22.83	24.97	6.2	3.3	5.5	2.25
–	1.00	32/0.0787	32/.200	0.052	1.31	17.13	18.73	8.9	3.6	10	4.5
0.0017	–	40/.0076	40/.193	0.057	1.45	14.71	16.08	10	3.5	10	4.536
–	0.5	48/.00787	48/.200	0.065	1.64	11.42	12.49	11.2	3.0	10	4.5
0.003	–	70/.0076	70/.193	0.076	1.93	8.41	9.19	15	3.0	10	4.536
–	2.5	80/.00787	80/.200	0.082	2.02	6.85	7.49	17.4	2.8	10	4.5
0.0048	–	110/.0076	110/.193	0.092	2.34	5.35	5.85	20	2.5	10	4.536
–	4.0	128/.00787	127/.200	0.104	2.64	4.31	4.72	24.2	2.5	10	4.5
0.007	–	162/.0076	162/.200	0.120	3.05	3.63	3.97	25	2.1	10	4.536

The resistances given are for straight single cores. For twisted and multicore flexible cords, an allowance to exceeding 5 per cent must be added for the extra due to the lay of the cores.

(1) Conductor sizes and resistance from I.S. 434-1953.

(2) Ratings are based on Table 27 of I.E.E. Regulation.

Table 9: Current ratings for Copper Single Core Rubber, P.V.C. or Polythene Insulated Wires Cables Included Tough Rubber P.V.C. Lead or Aluminium-Sheath Cable Run/Bunched, and Enclosed in one Conduit, Troughing.

Size of Conductor			Two cables D.C. or single phase A.C.		Three or four cables balanced three-phase A.C.		For cables D.C. or single phase A.C.	
Nominal cross-sectional area in²	Number & dia. (in.) of wires	S.W.G. No.	Current Rating — A.C. Single phase or D.C. amps.	Approx. length of run for 1 volt. drop at rated current — A.C. Single phase or D.C. metres.	Current Rating — Balanced 3-phase A.C. amp.	Approx. length of run for 1 volt drop at rated current — Balanced 3-phase A.C. metres	Current Rating — D.C. or single phase A.C. amps.	Approx. length of run for 1 volt drop at rated current — D.C. or single phase A.C. metres
0.0015	1/.044	1/18	5	4.9	5	5.8	5	4.9
0.002	3/.029	3/22	10	3.0	10	3.7	10	3.0
0.003	3/.036	3/20	15	3.4	13	4.3	13	3.7
0.0045	7/.029	7/22	20	3.7	15	5.8	15	4.9
0.007	7/.036	7/20	28	4.0	25	5.2	22	5.2
0.01	7/.044	7/18	36	4.9	32	6.1	29	6.1
0.0145	7/.052	7/17	43	5.5	39	7.0	34	7.0
0.0225	7/.064	7/16	53	7.0	48	8.8	42	8.8
0.03	19/.044	19/18	62	7.6	56	9.8	50	9.4
0.04	19/.052	19/17	74	8.8	67	11.8	58	11.0
0.06	19/.064	19/16	97	10.0	88	12.8	78	12.5

Table 10: Maximum Capacity of Conduits for V.I.R., PVC and Aluminium Insulated Cables

Conductor of cables		Approximate overall dia. of cables				Size of conduit — Maximum Number of Cables											
		250 volts		660 volts		¾ inch		1 inch		1¼ inch		1½ inch		2 inch		2½ inch	
Nominal area inch²	Number and diameter of wires	inch	mm	inch	mm	250 ν	660 ν	250 ν	660 ν	250 ν	660 ν	250 ν	660 ν	250 ν	660 ν	250 ν	660 ν
.0015	1/.044	0.129	—	.200	—	6	4	10	9	14	12	—	—	—	—	—	—
.002	3/.029	0.165	—	.215	—	6	4	10	9	14	10	—	—	—	—	—	—
.003	3/.036	0.180	—	.230	—	5	4	10	8	14	9	—	—	—	—	—	—
Nominal area mm²		inch	mm	inch	mm												
1.5	1/1.40	0.165	4.20	0.213	5.40	6	4	10	9	14	10	—	—	—	—	—	—
2.5	1/1.80	0.181	4.60	0.236	6.00	5	3	10	6	14	8	—	—	—	—	—	—
4.0	1/2.24	0.207	5.25	0.268	6.80	4	2	6	5	10	7	—	—	—	—	—	—
6.0	1/2.80	0.236	6.00	0.289	7.35	4	—	4	4	10	6	—	7	—	7	—	—
10.0	1/3.55	0.280	7.10	0.319	8.10	2	—	4	3	5	5	—	6	—	5	—	—
16.0	7/1.70	0.348	8.85	0.380	9.65	—	—	2	2	4	3	—	3	—	5	6	—
25.0	7/2.24	0.425	10.80	0.453	11.50	—	—	1	1	2	1	5	2	5	3	7	4
35.0	7/2.50	0.462	11.75	0.482	12.25	—	—	1	—	2	—	3	1	3	2	5	2
—	7/3.00	—	—	—	—	—	—	—	—	—	—	—	—	—	—	—	—
50.0	19/1.80	0.527	13.40	0.547	13.90	—	—	—	—	—	—	2	—	1	—	3	—
70.0	19/2.24	—	—	0.657	16.70	—	—	—	—	—	—	—	—	3	3	—	5
95.0	19/2.50	—	—	0.752	19.70	—	—	—	—	—	—	—	—	1	2	—	2

Note: The above table applies to all types of conduits. Irrespective of whether they are light or heavy gauge.

Table 11: Estimated Current for A.C. Motors

H.P. of motor	Single Phase		Three-phase 440 Volts
	230 Volts	440 Volts	
¼	2.0	1.1	0.63
½	3.7	2.0	1.05
¾	5.0	3.0	1.50
1	6.5	4.0	1.80
2	11.5	7.0	3.30
3	17.5	9.5	4.60
5	24	14.0	7.30
8.5	34	21.0	10.50
10	48	27.0	13.60
15	70	40.0	20.00
20	92	53.0	27.00
25	102	63.0	33.60
30	130	78.0	40.00
40	–	–	53.50
50	–	–	66.00
60	–	–	79.00
75	–	–	98.00
105	–	–	129.00
150	–	–	139.00
200	–	–	250.00

Table 12: Resistivity and Temperature Coefficients

Metals	Resistivity at 20°C in $\mu\Omega/cm^3$	Temperature Coefficient of resistance (α) per°C
Silver	1.63	0.0038
Copper Soft	1.72	0.0043
Copper Hard	1.77	0.0041
Aluminium	2.83	0.0038
Iron	10.00	0.0062
Steel	18.00	0.003
Lead	22.00	0.0043
Mercury	95.8	0.0009
Nickel	7.8	0.006
Platinum	11.00	0.0035
Tin	11.5	0.0045
Tungsten	5.51	0.0051
Zinc	6.1	0.0037

(Contd.)

Metals	Resistivity at 20°C in $\mu\Omega/cm^3$	Temperature Coefficient of resistance (α) per°C
Resistance Alloys		
Eureka	49.00	Negligible
German Silver	16.40	0.23 to 0.6
Platinoid	34.00	0.00025
Manganin	44.5	Negligible
Nicrome	110.00	0.00017
Kanthal	140.00	
Carbon (Graphite)	750–2500	−0.0002 to −0.0005

Table 13: Illumination
Spacing Mounting Height in Metres

Direct, Semi-direct and General Lighting Fittings			Semi and Totally Indirect Lighting Fittings			
Mounting height of units		Max. Spacing distance between points	Ceiling height		Max. spacing distance between points	Suspension distance ceiling to top of reflector
Above plane of work	Above floor		Above plane of work	Above floor		
1.21	1.98	1.82	1.52	2.28	2.28	0.38
1.52	2.28	2.28	1.83	2.59	2.74	0.45
1.83	2.59	2.74	2.13	2.89	3.20	0.53
2.13	2.89	3.20	2.43	3.2	3.66	0.60
2.43	3.2	3.66	2.74	3.5	4.11	0.68
2.74	3.5	4.11	3.05	3.81	4.57	0.76
3.05	3.81	4.57	3.35	4.11	5.03	0.84
3.35	4.11	5.03	3.66	4.27	5.48	0.91
3.66	4.27	5.48	3.96	4.72	5.94	0.99
3.96	4.72	5.94	4.27	5.03	6.40	1.06
4.27	5.03	6.40	4.57	5.33	6.86	1.14
4.57	5.33	6.86	4.87	5.64	7.31	1.22
4.87	5.64	7.31	5.48	6.25	8.23	1.30
5.48	6.25	8.23	6.1	6.86	9.14	1.60
6.1	6.86	9.14	6.4	7.16	9.6	1.83

Table 14: Current Carrying capacity of wires.

Wire Gauge	Current in A.	Wire Insulation
14	7.5	Double silk covering
15	6.1	"
16	4.8	"
17	3.7	"
18	2.7	"
19	1.9	Enamel covering
20	1.5	"
21	1.2	"
22	0.92	"
23	0.68	"
24	0.57	"
25	0.4	"
26	0.38	"
27	0.32	"
28	0.26	"
29	0.22	"
30	0.18	"
31	0.158	"
32	0.138	"
33	0.118	"
34	0.100	"
35	0.083	"
36	0.068	"
37	3.054	"
38	0.012	"
39	0.032	"
40	0.027	"

Table 15: Best possible values of core sizes and output of small transformers upto 10-1000 VA
(These values are based on experience and designed datas)

S.No.	Sectional area of core cm²	Output V × A	Approximate winding area cm²	Turn/Volt	Actual size of core	Width of tongue	Approx. Quantity of 0.035 cm core	Remarks
(1)	(2)	(3)	(4)	(5)	(6)	(7)	(8)	
1.	1.94	10	4.03	23.3	1.59 × 1.43	1.59	36	
2.	2.58	15	4.84	17.5	1.59 × 1.9	1.59	48	
3.	3.23	20	5.65	14.0	1.9 × 1.9	1.9	48	
4.	3.87	25	5.65	11.7	1.9 × 2.22	1.9	56	
5.	4.52	30	5.8	10.0	2.22 × 2.22	2.22	60	
6.	5.16	37.5	5.98	8.78	2.38 × 2.38	2.38	60	
7.	5.80	43	6.15	7.8	2.54 × 2.54	2.54	64	
8.	6.45	50	6.45	7.0	2.54 × 2.86	2.54	71	
9.	8.06	75	8.06	5.6	3.175 × 2.86	3.175	71	
10.	9.68	100	8.87	4.65	3.175 ×3.49	3.175	88	
11.	11.3	125	9.68	4.00	3.175 × 3.49	3.175	100	
12.	11.3	150	10.48	4.00	3.175 × 3.49	3.175	100	
13.	12.9	175	11.29	3.5	3.81 × 3.81	3.81	96	
14.	12.9	200	12.096	3.5	3.81 × 3.81	3.81	96	
15.	14.51	225	12.5	3.1	4.13 × 4.13	4.13	104	

(Contd.)

S.No.	Sectional area of core cm^2	Output V × A	Approximate winding area cm^2	Turn/Volt	Actual size of core	Width of tongue	Approx. Quantity of 0.035 cm core	Remarks
(1)	(2)	(3)	(4)	(5)	(6)	(7)	(8)	
16.	16.13	250	12.90	2.8	4.13 × 4.45	4.13	112	
17.	16.13	300	14.52	2.8	4.45 × 4.45	4.45	112	
18.	17.75	325	14.85	2.5	4.45 × 4.45	4.45	112	
19.	19.35	350	15.32	2.3	4.76 × 4.76	4.76	120	
20.	19.35	400	19.35	2.3	5.08 × 4.76	5.08	120	
21.	20.97	450	20.16	2.16	5.08 × 5.08	5.08	120	
22.	22.58	500	25.81	2.00	5.08 × 5.4	5.08	128	
23.	24.19	750	32.26	1.86	5.08 × 5.71	5.08	136	
24.	25.8	850	35.3	1.75	5.08 × 5.71	5.08	144	
25.	29.03	1000	37.1	1.55	5.71 × 5.71	5.71	144	

Table 16: Multiplying Factor for Calculating the Sizes of Capacitor for Power Factor Improvement

Power factor of load before using capacitors	Size of capacitors in KV Ar per KW of load for raising the power factor to												
	0.80	0.85	0.90	0.91	0.92	0.93	0.94	0.95	0.96	0.97	0.98	0.99	Unity
1	2	3	4	5	6	7	8	9	10	11	12	13	14
0.45	1.230	1.360	1.501	1.532	1.561	1.592	1.626	1.659	1.695	1.737	1.784	1.846	1.988
0.46	1.179	1.309	1.446	1.473	1.502	1.533	1.567	1.600	1.636	1.677	1.725	1.786	1.929
0.47	1.130	1.260	1.397	1.425	1.454	1.485	1.519	1.552	1.588	1.629	1.677	1.758	1.881
0.48	1.076	1.206	1.343	1.370	1.400	1.430	1.464	1.497	1.534	1.575	1.623	1.684	1.826
0.49	1.030	1.160	1.297	1.326	1.355	1.386	1.420	1.453	1.489	1.530	1.578	1.639	1.782
0.50	0.982	1.112	1.248	1.276	1.303	1.337	1.369	1.403	1.441	1.481	1.529	1.590	1.732
0.51	0.936	1.066	1.202	1.230	1.257	1.291	1.323	1.357	1.395	1.435	1.483	1.544	1.686
0.52	0.894	1.024	1.160	1.188	1.215	1.249	1.281	1.315	1.353	1.393	1.441	1.502	1.644
9.53	0.850	0.980	1.116	1.144	1.171	1.205	1.237	1.271	1.309	1.349	1.397	1.458	1.600
0.54	0.809	0.939	1.075	1.103	1.132	1.164	1.196	1.230	1.268	1.308	1.356	1.417	1.559
0.55	0.769	0.899	1.035	1.063	1.090	1.124	1.156	1.190	1.228	1.268	1.316	1.377	1.519
0.56	0.730	0.860	0.996	1.024	1.051	1.085	1.117	1.151	1.189	1.229	1.277	1.338	1.480
0.57	0.692	0.822	0.958	0.986	1.013	1.047	1.079	1.113	1.151	1.191	1.239	1.300	1.442
0.58	0.655	0.785	0.921	0.949	0.976	1.010	1.042	1.076	1.114	1.154	1.202	1.263	1.405
0.59	0.618	0.748			0.939	0.973	1.005	1.039	1.077	1.117	1.165	1.226	1.368
0.60	0.584	0.714	0.849	0.878	0.905	0.939	0.971	1.005	1.043	1.083	1.131	1.192	1.334
0.61	0.549	0.679	0.815	0.843	0.870	0.904	0.936	0.970	1.008	1.048	1.096	1.157	1.299
0.62	0.515	0.645	0.781	0.809	0.836	0.870	0.902	0.936	0.974	1.014	1.062	1.123	1.265

(Contd.)

1	2	3	4	5	6	7	8	9	10	11	12	13	14
0.63	0.483	0.613	0.749	0.777	0.804	0.838	0.870	0.904	0.942	0.982	1.030	1.091	1.233
0.64	0.450	0.580	0.716	0.744	0.771	0.805	0.837	0.871	0.909	0.949	0.997	1.058	1.200
0.65	0.419	0.549	0.685	0.713	0.740	0.774	0.806	0.840	0.878	0.918	0.966	1.027	1.169
0.66	0.388	0.518	0.654	0.682	0.709	0.743	0.775	0.809	0.847	0.887	0.935	0.996	1.138
0.67	0.358	0.488	0.624	0.652	0.679	0.713	0.745	0.779	0.817	0.857	0.905	0.966	1.108
0.68	0.329	0.459	0.595	0.623	0.650	0.684	0.716	0.750	0.788	0.828	0.876	0.937	1.079
0.69	0.299	0.429	0.565	0.593	0.620	0.654	0.686	0.720	0.758	0.798	0.840	0.907	1.049
0.70	0.270	0.400	0.536	0.564	0.591	0.625	0.657	0.691	0.729	0.769	0.811	0.878	1.020
0.71	0.242	0.372	0.508	0.536	0.563	0.597	0.629	0.663	0.701	0.741	0.783	0.850	0.992
0.72	0.218	0.343	0.479	0.507	0.534	0.568	0.600	0.634	0.672	0.712	0.754	0.821	0.963
0.73	0.186	0.316	0.452	0.480	0.507	0.541	0.573	0.607	0.645	0.685	0.727	0.794	0.936
0.74	0.159	0.289	0.425	0.453	0.480	0.514	0.546	0.580	0.618	0.658	0.767	0.700	0.909
0.75	0.132	0.262	0.398	0.426	0.453	0.487	0.519	0.553	0.591	0.631	0.673	0.740	0.882
0.76	0.106	0.235	0.371	0.399	0.426	0.460	0.492	0.526	0.564	0.604	0.652	0.713	0.855
0.77	0.079	0.209	0.345	0.373	0.400	0.434	0.466	0.500	0.538	0.578	0.620	0.687	0.829
0.78	0.053	0.183	0.319	0.347	0.374	0.408	0.440	0.474	0.512	0.552	0.594	0.661	0.803
0.79	0.026	0.156	0.293	0.320	0.347	0.381	0.413	0.447	0.485	0.525	0.567	0.634	0.776
0.80	—	0.130	0.266	0.294	0.321	0.355	0.387	0.421	0.459	0.499	0.641	0.608	0.750
0.81	—	1.104	0.240	0.288	0.295	0.329	0.361	0.395	0.433	0.473	0.515	0.582	0.724
0.82	—	0.078	0.214	0.242	0.269	0.303	0.335	0.369	0.407	0.447	0.489	0.556	0.698
0.83	—	0.052	0.188	0.216	0.243	0.277	0.309	0.343	0.381	0.421	0.463	0.530	0.672
0.84	—	0.026	0.162	0.190	0.217	0.251	0.283	0.317	0.355	0.395	0.437	0.504	0.645
4.85	—	—	0.136	0.164	0.191	0.225	0.257	0.291	0.329	0.369	0.417	0.478	0.620

(Contd.)

1	2	3	4	5	6	7	8	9	10	11	12	13	14
0.86	–	–	0.109	0.140	0.167	0.198	0.230	0.264	0.301	0.343	0.390	0.450	3.593
0.87	–	–	0.083	0.114	0.141	0.172	0.204	0.238	0.275	0.517	0.364	0.424	0.567
0.88	–	–	0.054	0.085	0.112	0.143	0.175	0.209	0.246	0.288	9.335	0.395	0.538
0.89	–	–	0.028	0.059	0.084	0.117	0.149	0.183	0.230	0.262	0.309	0.369	0.512
0.90	–	–	–	0.031	0.058	8.089	0.121	0.155	0.192	0.234	0.281	0.341	0.484
9.91	–	–	–	–	0.027	0.058	.090	0.124	0.161	0.203	0.250	0.310	0.453
0.92	–	–	–	–	–	0.031	0.063	0.097	0.134	0.176	0.223	0.283	0.426
0.93	–	–	–	–	–	–	0.032	0.066	0.103	0.145	0.192	0.252	0.395
0.94	–	–	–	–	–	–	–	0.034	0.071	0.113	0.160	0.220	0.363
0.95	–	–	–	–	–	–	–	–	0.037	0.079	0.126	0.186	0.329
0.96	–	–	–	–	–	–	–	–	–	0.042	0.089	0.149	0.292
0.97	–	–	–	–	–	–	–	–	–	–	0.047	0.107	0.250
0.98	–	–	–	–	–	–	–	–	–	–	–	0.060	0.203
0.99	–	–	–	–	–	–	–	–	–	–	–	–	0.143

Example: A load of 200 kW at power factor 0.70 is to be improved to 0.95 power factor, so from the above table the capacitor kVAr rating will be = 200 × 0.691 = 138.2 kVAr.

Extracts from the Indian Electricity Rules

1. INTRODUCTION

Before actually studying Indian Electricity Rules (I.E. Rules) and other precautions, we should realise, why these rules and regulations have been framed.

These are framed by Institution of Electrical Engineers to:

(*i*) safe guard consumers (users) of electrical energy from shock,

(*ii*) minimise fire risk and

(*iii*) ensure, as far as possible, satisfactory operation of equipment and apparatus used.

2. DEFINITIONS

1. **"Ampere"** means a unit of electric current and is the unvarying electric current which when passed through a solution of silver nitrate in water in accordance with the specifications set out deposits silver at the rate of 0.001118 gramme per sec;

2. **"Apparatus"** means electrical apparatus and includes, all machines, fittings, accessories and appliances in which conductors are used;

3. **"Cable"** means a length of insulated single conductor (solid or standard) or of two or more such conductors, each provided with its own insulation, which are laid up together. Such insulated conductor or conductors may or may not be provided with and over all mechanical protective covering;

4. **"Flexible Cable"** means a cable consisting of one or more cores each formed of a group of wires, the diameter and the physical properties of the wires and the insulating material being such that afford flexibility;

5. **"Circuit Breaker"** means a device, capable of making and breaking the circuit under all conditions, and unless otherwise specified so designed as to break the current automatically under abnormal conditions;

6. **"Conductor"** means any wire, cable, bar, tube, rail or plate used for conducting energy and so arranged as to be electrically connected to system;

7. **"Conduit"** means rigid or flexible metallic tubing or mechanically strong and fire resisting non-metallic tubing in which a cable or cables may be drawn for the purpose of affording it or them mechanical protection;

8. **"Electrician"** means a person over 21 years of age who is competent for the purposes of the rule in which the term is used and who has been appointed in writing by the lessee-owner, agent or manager of installation;

9. **"Guarded"** means covered, shielded, fenced or otherwise protected by means of suitable casings, barrier, rails or metal screens to remove the possibility of dangerous contact of approach by persons or objects to a point of danger;

10. **"Inspector"** means an Electrical Inspector appointed under section 36;

11. **"Lightning Arrestor"** means a device which has the property of diverting to earth any electrical surge of excessively high amplitude applied to its terminals and is capable of interrupting follow current if present and restoring itself three after to its original operating conditions;

12. **"Live"** means electrically charged;

13. **"Neutral conductor"** means that conductor of a multi-wire system, the voltage of which is normally midway between the voltages of other conductors of the system;

14. **"Ohm"** means a unit of electrical resistance and is the resistance offered to an unvarying electric current by a column of mercury at the temperature of melting ice 14.4521 grammes in mass of an uniform cross-sectional area and of length of 106.3 centimetres;
    The aforesaid unit is represented by the resistance between the terminals of the instrument marked "Government of India Ohm Standard Verified" to the passage of an electric current when the coil of wire, forming part of the aforesaid instrument and connected to the aforesaid terminals is in all parts at a temperature of 30°C.

15. **"Overhead Line"** means any electric supply line which is placed above ground and in the open space but including live rails of a traction system;

16. **"Span"** means the horizontal distance between two adjacent supporting points of an overhead conductor;

17. **"Switch"** means a manually operated device for opening and closing or changing the connection of a circuit;

18. **"Switch Gear"** shall denote switches, breakers, cut-outs and other apparatus used for operation, regulation and control of circuits;

19. **"Volt"** means a unit of electromotive force and is the electric pressure which, when steadily applied to a conductor, the resistance of which is one ohm will, produce a current of one ampere;

20. **"Voltage"** means the difference of electrical potential measured in volts between any two conductors or between any part of either conductor and the earth as measured by suitable voltmeter and is said to be;

    "Low" where the voltage does not exceed 250 volts under normal conditions however, subject to the percentage variations allowed by these rules;

    "Medium" where the voltage does not exceed 650 volts under normal conditions however, subject to the percentage variation allowed by these rules;

    "High" where the voltage does not exceed 33,000 volts under normal conditions however, subject to the percentage variation allowed by these rules;

    "Extra High" where the voltage exceeds 33,000 volts under normal conditions however, subject to the percentage variation allowed by these rules.

GENERAL SAFETY PRECAUTIONS

Rule 10. Construction, Installation, Protection, Operation and maintenance of electric supply lines and apparatus.

All electric supply lines and apparatus shall be sufficient in power and size and of sufficient mechanical strength for the work they may be required to do, and so far as practicable, shall be constructed, installed, protected, worked and maintained in accordance with standards for the Indian Standards Institution so as to prevent danger.

Rule 30. Service lines and apparatus on consumer's premises.

(1) The supplier shall ensure that all electric supply lines, wires fittings and apparatus belonging to him or under his control which are on a consumer's premises are in a safe condition and in all respects fit for supplying energy, and the supplier shall take due precautions to avoid danger arising on such premises from such supply lines, wires, fittings and apparatus.

(2) The consumer shall also ensure that the installation under his control is maintained in a safe condition.

Rule 31. Cut-out consumer's premises. The supplier shall provide a suitable cut-out in each conductor of every line other than an earthed or earthed neutral conductor or the earthed external conductor of a concentric cables within a consumer's premises in an accessible position Such cut-out shall be contained within adequately enclosed fire-proof receptacle:

Where more than one consumer is supplied through a common service-line, each such consumer shall be provided with an independent cut-out at the point of junction to the common service.

Rule 33. Earthed terminal on consumer's Premises. The supplier shall provide and maintain on the consumer's premises for the consumer's use of suitable earthed terminal in an accessible position at or near the point of commencement of supply as defined under Rule 58.

Provided that in the case of medium, high or extra high voltage installation the consumer shall, in addition to the afore-mentioned arrangement provide his own earthing system with an independent electrode.

Rule 35. Caution Notices. The owner of every medium, high and extra voltage installation shall affix permanently in a conspicuous position a caution notice in Hindi and the local language of the district and of a type approved by the Inspector on:

  (a) every motor, generator, transformer and other electrical plant and equipment together with apparatus used for controlling or regulating the same;

  (b) all supports of high extra high voltage overhead lines;

  (c) luminous tube signs requiring high voltage supply X-ray and similar high-frequency installation.

Rule 48. Precautions against leakage before collecting:

  (1) The supplier shall not contact with his works the installation or apparatus on the premises of any applicant for supply unless he is reasonably satisfy that the connection will not at the time of making the connection cause a leakage from that installation or the apparatus exceeding one thousandth part of the maximum current supplied to the premises.

  (2) If the supplier declines to make connection under the provisions of sub-rule (1) he shall serve upon the applicant a notice in writing stating his reason for so declining.

Rule 52. Appeal to Inspector in regard to defects.

  (1) If any applicant for supply for a consumer is dissatisfied with the action of the supplier in deciding to commence or to recommence the supply to energy his premises on the ground that the installation is defective or is likely to constitute danger, he may appeal to the Inspector to test the installation and the supplier shall not, if the

Inspector or under his orders, any other officer appointed to assist the Inspector, is satisfied that the installation is free from the defect and danger complained of, be entitled to refuse supply to the consumer on the grounds aforesaid, and shall within twenty-four hours after the receipt of such intimation from the inspector, commence, continue or recommence the supply of energy.

(2) Any test for which application has been made under the provisions of sub-rule (1) shall be carried out within seven days after the receipt of such application.

(3) This rule shall be endorsed on every notice given under the provisions of rule 47, 48 and 49.

Rule 53. The cost of any inspection and test made by the Inspector, at the request of the consumer or other interested party, shall be borne by the consumer or other interested party unless the Inspector directs otherwise.

Rule 54. Declared Voltage of supply to Consumer. Except with the written consent of the consumer or the previous sanction of the State Government a supplier shall not permit the voltage at the point of commencement of supply as defined under rule 58 to vary from the declared voltage by more than 5 percent in the case of low or medium voltage or by more than 12.5 percent in the case of high or extra high voltage.

Rule 55. Declared Frequency of supply to consumer. Except with the written consent of the consumer or with the previous sanction of the State Government a supplier shall not permit the frequency of an alternating current supply to vary from the declared frequency by more than 5 per cent.

Overhead Lines

Rule 74. Material and strength.

(1) All conductors of overhead lines other than those specified in sub-rule (1) of rule 86 shall have a breaking strength of not less than 317.51 kg. (700 lbs.)

(2) Where the voltage is low and the span is of less than 50 ft. (15.24 metres) and is on the owner's or consumer premises, a conductor having an actual breaking strength of not less than 300 lbs. (136.08 kg.) may be used.

Rule 75. Joints. Joints of conductors overhead lines shall be mechanically and electrically secure under the conditions of operation. The ultimate strength of joint shall not be less than 95 per cent of that of the conductor and the electrical conductivity not less than that of the conductor.

Rule 76. Maximum stresses: Factor of safety. (1) (*a*) The owner of every overhead line shall ensure that it has the following minimum factors of safety for supports based on crippling load shall be as follows:

(*i*)	for metal supports	2.0
(*ii*)	for mechanically processed concrete supports	2.5
(*iii*)	for hand moulded concrete supports	3.0
(*iv*)	for wood supports	4.5

The said owner shall also ensure that the strength of the supports in the direction of the line is not less than one fourth of the strength required in the direction transverse to the line.

Rule 77. Clearances above ground of the lowest Conductor.

(1) No conductor of an overhead line, including service lines erected across a street shall at any part thereof be at height less than:

    (*a*) for low and medium voltage lines   5.791 m

    (*b*) for high voltage lines   6.096 m

(2) No conductor of an overhead line including service lines erected along any street shall at any part there of be at a height less than:

    (*a*) for low and medium voltage lines   5.486 m

    (*b*) for high voltage lines   5.791 m

(3) No conductor of an overhead line including service lines, erected elsewhere than along or across any street shall be at a height less than:

    (*a*) for low, medium and high voltage lines upto and including 11,000 volts, if bare   4.572 m

    (*b*) for low medium and high voltage lines up to and including 11,000 volts if insulated   3.963 m

    (*c*) for high voltage lines above 11,000 volts   5.182 m

(4) For extra-high voltage lines the clearance above ground shall not be less than 5.182 m plus 0.3048 m for every 33,000 volts.

Provided that the minimum clearance along or across any street shall not be less than 6.096 metres.

Rule 78. Clearance between conductors and trolley wires.

No conductor of an overhead line crossing a tramway or trolley bus route using trolley wires shall have less than the following clearances above any trolley wire.

    (*a*) low and medium voltage lines   1.219 m

    Provided that where an insulated conductor suspended from a bare wire crosses over a trolley wire the minimum clearance for such insulated conductor shall be 0.6096 m.

    (*b*) high voltage lines upto and including 11,000 volts   1.829 m

    (*c*) high voltage lines above 11000 volts   2.439 m

    (*d*) extra-high voltage lines   3.048 m

Rule 79. Clearances from building of low and medium voltage lines and service lines. (1) where a low or medium voltage overhead line passes above or adjacent to or terminates on any building, the following minimum clearances from any accessible point, on the basis of maximum sag, shall be observed:
- (*a*) For any flat roof, open balcony, verandah roof and lean-to-roof–
  - (*i*) When the line passes above the building, a vertical clearance of 2.439 m from the highest point, and
  - (*ii*) When the line passes adjacent to the building, a horizontal clearance of 1.219 m from the nearest point, and
- (*b*) For pitched roof.
  - (*i*) When the line passes above the building, a vertical clearance of 1.219 m immediately under tee lines; and
  - (*ii*) When the line passes adjacent to the building, a horizontal clearance of 1.219 m.
- (2) Any conductor so situated as to have a clearance less than that specified in sub-rule (*i*) shall be adequately insulated and shall be attached by means of metal clips at suitable intervals to a bare earthed bearer wire having a breaking strength of not less than 517.51 kg.
- (3) The horizontal clearance shall be measured when the line is at maximum deflection from the vertical due to wind pressure.

Rule 81. Conductors at different voltages on same supports. Where conductors forming parts of systems at different voltages are erected on the same supports the owner shall make adequate provision to guard against danger to linesmen and others from the lower voltage system being charged above its normal working voltage by leakage from or contact with the higher voltage system and the methods of construction and the clearance between the conductors of the two systems shall be subject to the prior approval of the Inspector.

Rule 84. Proximity to aerodrome. Overhead lines shall not be erected in the vicinity of aerodrome until the aerodrome authorities have approved in writing the route of the proposed lines.

Rule 88. Guarding. (1) Where guarding is required under these rules, the provisions of sub-rule (2) to (4) shall apply.
- (2) Every guard-wire shall be connected with earth at each point at which its electrical continuity is broken.
- (3) Every guard-wire shall have an actual breaking strength of not less than 632.02 kg and if made of iron or steel, shall be galvanised.
- (4) Every guard-wire of cross-connected system or guard-wires, shall have sufficient current-carrying capacity to ensure the rendering

dead, without risk of the guard-wire or wire till the contact of any line wire has been removed.

Rule 92. Protection against lightning. (1) The owner of every overhead line which is so exposed as to be liable to injury from lightning shall adopt efficient means for diverting to earth any electrical surges due to lightning.

Rule 93. Unused Overhead lines. Where an overhead line used to be ceased as an electric supply line, the owner shall maintain in a safe mechanical condition in accordance with rule 75 or shall remove it.

Log Tables
Logarithms

	0	1	2	3	4	5	6	7	8	9	1	2	3	4	5	6	7	8	9
10	0000	0043	0086	0128	0170						5	9	13	17	21	26	30	34	38
						0212	0253	0294	0334	0374	4	8	12	16	20	24	28	32	36
11	0414	0453	0492	0531	0569						4	8	12	16	20	23	27	31	35
						0607	0645	0682	0719	0755	4	7	11	15	18	22	26	29	33
12	0792	0828	0864	0899	0934						3	7	11	14	18	21	25	28	32
						0969	1004	1038	1072	1106	3	7	10	14	17	20	24	27	31
13	1139	1173	1206	1239	1271						3	6	10	13	16	19	23	26	29
						1303	1335	1367	1399	1430	3	7	10	13	16	19	22	25	29
14	1461	1492	1523	1553	1584						3	6	9	12	15	19	22	25	28
						1614	1644	1673	1703	1732	3	6	9	12	14	17	20	23	26
15	1761	1790	1818	1847	1875						3	6	9	11	14	17	20	23	26
						1903	1931	1959	1987	2014	3	6	8	11	14	17	19	22	25
16	2041	2068	2095	2122	2148						3	6	8	11	14	16	19	22	24
						2175	2201	2227	2253	2279	3	5	8	10	13	16	18	21	23
17	2304	2330	2355	2380	2405						3	5	8	10	13	15	18	20	23
						2430	2455	2480	2504	2529	3	5	8	10	12	15	i7	20	22
18	2553	2577	2601	2625	2648						2	5	7	9	12	14	17	19	21
						2672	2695	2718	2742	2765	2	4	7	9	11	14	16	18	21
19	2788	2810	2833	2856	2878						2	4	7	9	11	13	16	18	20
						2900	2923	2945	2967	2989	2	4	6	8	11	13	15	17	19
20	3010	3032	3054	3075	3096	3118	3139	3160	3181	3201	2	4	6	8	11	13	15	17	19
21	3222	3243	3263	3284	3304	3324	3345	3365	3385	3404	2	4	6	8	10	12	14	16	18
22	3424	3444	3464	3483	3502	3522	3541	3560	3579	3598	2	4	6	8	10	12	14	15	17
23	3617	3636	3655	3674	3692	3711	3729	3747	3766	3784	2	4	6	7	9	11	13	15	17
24	3802	3820	3838	3856	3874	3892	3909	3927	3945	3962	2	4	5	7	9	11	12	14	16
25	3979	3997	4014	4031	4048	4065	4082	4099	4116	4133	2	3	5	7	9	10	12	14	15
26	4150	4166	4183	4200	4216	4232	4249	4265	4281	4298	2	3	5	7	8	10	11	13	15
27	4314	4330	4346	4362	4378	4393	4409	4425	4440	4456	2	3	5	6	8	9	11	13	14
28	4472	4487	4502	4518	4533	4548	4564	4579	4594	4609	2	3	5	6	8	9	11	12	14
29	4624	4639	4654	4669	4683	4698	4713	4728	4742	4757	1	3	4	6	7	9	10	12	13
30	4771	4786	4800	4814	4829	4843	4857	4871	4886	4900	1	3	4	6	7	9	10	11	13
31	4914	4928	4942	4955	4969	4983	4997	5011	5024	5038	1	3	4	6	7	8	10	11	12
32	5051	5065	5079	5092	5105	5119	5132	5145	5159	5172	1	3	4	5	7	8	9	11	12
33	5185	5198	5211	5224	5237	5250	5263	5276	5289	5302	1	3	4	5	6	8	9	10	12
34	5315	5328	5340	5353	5366	5378	5391	5403	5416	5428	1	3	4	5	6	8	9	10	11
35	5441	5453	5465	5478	5490	5502	5514	5527	5539	5551	1	2	4	5	6	7	9	10	11
36	5563	5575	5587	5599	5611	5623	5635	5647	5658	5670	1	2	4	5	6	7	8	10	11
37	5682	5694	5705	5717	5729	5740	5752	5763	5775	5786	1	2	3	5	6	7	8	9	10
38	5798	5809	5821	5832	5843	5855	5866	5877	5888	5899	1	2	3	5	6	7	8	9	10
39	5911	5922	5933	5944	5955	5966	5977	5988	5999	6010	1	2	3	4	5	7	8	9	10
40	6021	6031	6042	6053	6064	6075	6085	6096	6107	6117	1	2	3	4	5	6	8	9	10
41	6128	6138	6149	6160	6170	6180	6191	6201	6212	6222	1	2	3	4	5	6	7	8	9
42	6232	6243	6253	6263	6274	6284	6294	6304	6314	6325	1	2	3	4	5	6	7	8	9
43	6335	6345	6355	6365	6375	6385	6395	6405	6415	6425	1	2	3	4	5	6	7	8	9
44	6435	6444	6454	6464	6474	6484	6493	6503	6513	6522	1	2	3	4	5	6	7	8	9

Logarithms

	0	1	2	3	4	5	6	7	8	9	1	2	3	4	5	6	7	8	9
45	6532	6542	6551	6561	6571	6580	6590	6599	6609	6618	1	2	3	4	5	6	7	8	9
46	6628	6637	6646	6656	6665	6675	6684	6693	6702	6712	1	2	3	4	5	6	7	7	8
47	6721	6730	6739	6749	6758	6767	6776	6785	6794	6803	1	2	3	4	5	5	6	7	8
48	6812	6821	6830	6839	6848	6857	6866	6875	6884	6893	1	2	3	4	4	5	6	7	8
49	6902	6911	6920	6928	6937	6946	6955	6964	6972	6981	1	2	3	4	4	5	6	7	8
50	6990	6998	7007	7016	7024	7033	7042	7050	7059	7067	1	2	3	3	4	5	6	7	8
51	7076	7084	7093	7101	7110	7118	7126	7135	7143	7152	1	2	3	3	4	5	6	7	8
52	7160	7168	7177	7185	7193	7202	7210	7218	7226	7235	1	2	2	3	4	5	6	7	7
53	7243	7251	7259	7267	7275	7284	7292	7300	7308	7316	1	2	2	3	4	5	6	6	7
54	7324	7332	7340	7348	7356	7364	7372	7380	7388	7396	1	2	2	3	4	5	6	6	7
55	7404	7412	7419	7427	7435	7443	7451	7459	7466	7474	1	2	2	3	4	5	5	6	7
56	7482	7490	7497	7505	7513	7520	7528	7536	7543	7551	1	2	2	3	4	5	5	6	7
57	7559	7566	7574	7582	7589	7597	7604	7612	7619	7627	1	2	2	3	4	5	5	6	7
58	7634	7642	7649	7657	7664	7672	7679	7686	7694	7701	1	1	2	3	4	4	5	6	7
59	7709	7716	7723	7731	7738	7745	7752	7760	7767	7774	1	1	2	3	4	4	5	6	7
60	7782	7789	7796	7803	7810	7818	7825	7832	7839	7846	1	1	2	3	4	4	5	6	6
61	7853	7860	7868	7875	7882	7889	7896	7903	7910	7917	1	1	2	3	4	4	5	6	6
62	7924	7931	7938	7945	7952	7959	7966	7973	7980	7987	1	1	2	3	3	4	5	6	6
63	7993	8000	8007	8014	8021	8028	8035	8041	8048	8055	1	2	2	3	3	4	5	5	6
64	8062	8069	8075	8082	8089	8096	8102	8109	8116	8122	1	1	2	3	3	4	5	5	6
65	8129	8136	8142	8149	8156	8162	8169	8176	8182	8189	1	1	2	3	3	4	5	5	6
66	8195	8202	8209	8215	8222	8228	8235	8241	8248	8254	1	1	2	3	3	4	5	5	6
67	8261	8267	8274	8280	8287	8293	8299	8306	8312	8319	1	1	2	3	3	4	5	5	6
68	8325	8331	8338	8344	8351	8357	8363	8370	8376	8382	1	1	2	3	3	4	4	5	6
69	8388	8395	8401	8407	8414	8420	8426	8432	8439	8445	1	1	2	2	3	4	4	5	6
70	8451	8457	8463	8470	8476	8482	8488	8494	8500	8506	1	1	2	2	3	4	4	5	6
71	8513	8519	8525	8531	8537	8543	8549	8555	8561	8567	1	1	2	2	3	4	4	5	5
72	8573	8579	8585	8591	8597	8603	8609	8615	8621	8627	1	1	2	2	3	4	4	5	5
73	8633	8639	8645	8651	8657	8663	8669	8675	8681	8686	1	1	2	2	3	4	4	5	5
74	8692	8698	8704	8710	8716	8722	8727	8733	8739	8745	1	1	2	2	3	4	4	5	5
75	8751	8756	8762	8768	8774	8779	8785	8791	8797	8802	1	1	2	2	3	3	4	5	5
76	8808	8814	8820	8825	8831	8837	8842	8848	8854	8859	1	1	2	2	3	3	4	5	5
77	8865	8871	8876	8882	8887	8893	8899	8904	8910	8915	1	1	2	2	3	3	4	4	5
78	8921	8927	8932	8938	8943	8949	8954	8960	8965	8971	1	1	2	2	3	3	4	4	5
79	8976	8982	8987	8993	8998	9004	9009	9015	9020	9025	1	1	2	2	3	3	4	4	5
80	9031	9036	9042	9047	9053	9058	9063	9069	9074	9079	1	1	2	2	3	3	4	4	5
81	9085	9090	9096	9101	9106	9112	9117	9122	9128	9133	1	1	2	2	3	3	4	4	5
82	9138	9143	9149	9154	9159	9165	9170	9175	9180	9186	1	1	2	2	3	3	4	4	5
83	9191	9196	9201	9206	9212	9217	9222	9227	9232	9238	1	1	2	2	3	3	4	4	5
84	9243	9248	9253	9258	9263	9269	9274	9279	9284	9289	1	1	2	2	3	3	4	4	5
85	9294	9299	9304	9309	9315	9320	9325	9330	9335	9340	1	1	2	2	3	3	4	4	5
86	9345	9350	9355	9360	9365	9370	9375	9380	9385	9390	1	1	2	2	3	3	4	4	5
87	9395	9400	9405	9410	9415	9420	9425	9430	9435	9440	0	1	1	2	2	3	3	4	4
88	9445	9450	9455	9460	9465	9469	9474	9479	9484	9489	0	1	1	2	2	3	3	4	4
89	9494	9499	9504	9509	9513	9518	9523	9528	9533	9538	0	1	1	2	2	3	3	4	4
90	9542	9547	9552	9557	9562	9566	9571	9576	9581	9586	0	1	1	2	2	3	3	4	4
91	9590	9595	9600	9605	9609	9614	9619	9624	9628	9633	0	1	1	2	2	3	3	4	4
92	9638	9643	9647	9652	9657	9661	9666	9671	9675	9680	0	1	1	2	2	3	3	4	4
93	9685	9689	9694	9699	9703	9708	9713	9717	9722	9727	0	1	1	2	2	3	3	4	4
94	9731	9736	9741	9745	9750	9754	9759	9763	9768	9773	0	1	1	2	2	3	3	4	4
95	9777	9782	9786	9791	9795	9800	9805	9809	9814	9818	0	1	1	2	2	3	3	4	4
96	9823	9827	9832	9836	9841	9845	9850	9854	9859	9863	0	1	1	2	2	3	3	4	4
97	9868	9872	9877	9881	9886	9890	9894	9899	9903	9908	0	1	1	2	2	3	3	4	4
98	9912	9917	9921	9926	9930	9934	9939	9943	9948	9952	0	1	1	2	2	3	3	3	4
99	9956	9961	9965	9969	9974	9978	9983	9987	9991	9996	0	1	1	2	2	3	3	3	4

Antilogarithms

	0	1	2	3	4	5	6	7	8	9	1	2	3	4	5	6	7	8	9
.00	1000	1002	1005	1007	1009	1012	1014	1016	1019	1021	0	0	1	1	1	1	2	2	2
.01	1023	1026	1028	1030	1033	1035	1038	1040	1042	1045	0	0	1	1	1	1	2	2	2
.02	1047	1050	1052	1054	1057	1059	1062	1064	1067	1069	0	0	1	1	1	1	2	2	2
.03	1072	1074	1076	1079	1081	1084	1086	1089	1091	1094	0	0	1	1	1	1	2	2	2
.04	1096	1099	1102	1104	1107	1109	1112	1114	1117	1119	0	1	1	1	1	2	2	2	2
.05	1122	1125	1127	1130	1132	1135	1138	1140	1143	1146	0	1	1	1	1	2	2	2	2
.06	1148	1151	1153	1156	1159	1161	1164	1167	1169	1172	0	1	1	1	1	2	2	2	2
.07	1175	1178	1180	1183	1186	1189	1191	1194	1197	1199	0	1	1	1	1	2	2	2	2
.08	1202	1205	1208	1211	1213	1216	1219	1222	1225	1227	0	1	1	1	1	2	2	2	3
.09	1230	1233	1236	1239	1242	1245	1247	1250	1253	1256	0	1	1	1	1	2	2	2	3
.01	1259	1262	1265	1268	1271	1274	1276	1279	1282	1285	0	1	1	1	1	2	2	2	3
.11	1288	1291	1294	1297	1300	1303	1306	1309	1312	1315	0	1	1	1	2	2	2	2	3
.12	1318	1321	1324	1327	1330	1334	1337	1340	1343	1346	0	1	1	1	2	2	2	2	3
.13	1349	1352	1355	1358	1361	1365	1368	1371	1374	1377	0	1	1	1	2	2	2	3	3
.14	1380	1384	1387	1390	1393	1396	1400	1403	1406	1409	0	1	1	1	2	2	2	3	3
.15	1413	1416	1419	1422	1426	1429	1432	1435	1439	1442	0	1	1	1	2	2	2	3	3
.16	1445	1449	1452	1455	1459	1462	1466	1469	1472	1476	0	1	1	1	2	2	2	3	3
.17	1479	1483	1486	1489	1493	1496	1500	1503	1507	1510	0	1	1	1	2	2	2	3	3
.18	1514	1517	1521	1524	1528	1531	1535	1538	1542	1545	0	1	1	1	2	2	2	3	3
.19	1549	1552	1556	1560	1563	1567	1570	1574	1578	1581	0	1	1	1	2	2	2	3	3
.20	1585	1589	1592	1596	1600	1603	1607	1611	1614	1618	0	1	1	1	2	2	3	3	3
.21	1622	1626	1629	1633	1637	1641	1644	1648	1652	1656	0	1	1	2	2	2	3	3	3
.22	1660	1663	1667	1671	1675	1679	1683	1687	1690	1694	0	1	1	2	2	2	3	3	3
.23	1698	1702	1706	1710	1714	1718	1722	1726	1730	1734	0	1	1	2	2	2	3	3	4
.24	1738	1742	1746	1750	1754	1758	1762	1766	1770	1774	0	1	1	2	2	2	3	3	4
.25	1778	1782	1786	1791	1795	1799	1803	1807	1811	1816	0	1	1	2	2	2	3	3	4
.26	1820	1824	1828	1832	1837	1841	1845	1849	1854	1858	0	1	1	2	2	3	3	3	4
.27	1862	1866	1871	1875	1879	1884	1888	1892	1897	1901	0	1	1	2	2	3	3	3	4
.28	1905	1910	1914	1919	1923	1928	1932	1936	1941	1945	0	1	1	2	2	3	3	4	4
.29	1950	1954	1959	1963	1968	1972	1977	1982	1986	1991	0	1	1	2	2	3	3	4	4
.30	1995	2000	2004	2009	2014	2018	2023	2028	2032	2037	0	1	1	2	2	3	3	4	4
.31	2024	2046	2051	2056	2061	2065	2070	2075	2080	2084	0	1	1	2	2	3	3	4	4
.32	2089	2094	2099	2104	2109	2113	2118	2123	2128	2133	0	1	1	2	2	3	3	4	4
.33	2138	2143	2148	2153	2158	2063	2168	2173	2178	2183	0	1	1	2	2	3	3	4	4
.34	2188	2193	2198	2203	2208	2213	2218	2223	2228	2234	1	1	2	2	3	3	4	4	5
.35	2239	2244	2249	2254	2259	2265	2270	2275	2280	2286	1	1	2	2	3	3	4	4	5
.36	2291	2296	2301	2307	2312	2317	2323	2328	2333	2339	1	1	2	2	3	3	4	4	5
.37	2344	2350	2355	2360	2366	2371	2377	2382	2388	2393	1	1	2	2	3	3	4	4	5
.38	2399	2404	2410	2415	2421	2427	2432	2438	2443	2449	1	1	2	2	3	3	4	4	5
.39	2455	2460	2466	2472	2477	2483	2489	2495	2500	2506	1	1	2	2	3	3	4	5	5
.40	2512	2518	2523	2529	2535	2541	2547	2553	2559	2564	1	1	2	2	3	4	4	5	5
.41	2570	2576	2582	2588	2594	2600	2606	2612	2618	2624	1	1	2	2	3	4	4	5	5
.42	2630	2636	2642	2649	2655	2661	2667	2673	2679	2685	1	1	2	2	3	4	4	5	6
.43	2692	2698	2704	2710	2716	2723	2729	2735	2742	2748	1	1	2	3	3	4	4	5	6
.44	2754	2761	2767	2773	2780	2786	2793	2799	2805	2812	1	1	2	3	3	4	4	5	6
.45	2818	2825	2831	2838	2844	2851	2858	2864	2871	2877	1	1	2	3	3	4	5	5	6
.46	2884	2891	2897	2904	2911	2917	2924	2931	2938	2944	1	1	2	3	3	4	5	5	6
.47	2951	2958	2965	2972	2979	2985	2992	2999	3006	3013	1	1	2	3	3	4	5	5	6
.48	3020	3027	3034	3041	3048	3055	3062	3069	3076	3083	1	1	2	3	4	4	5	6	6
.49	3090	3097	3105	3112	3119	3126	3133	3141	3148	3155	1	1	2	3	4	4	5	6	6

Antilogarithms

	0	1	2	3	4	5	6	7	8	9	1	2	3	4	5	6	7	8	9
.50	3162	3170	3177	3184	3192	3199	3206	3214	3221	3228	1	1	2	3	4	4	5	6	7
.51	3236	3243	3251	3258	3266	3273	3281	3289	3296	3304	1	2	2	3	4	5	5	6	7
.52	3311	3319	3327	3334	3342	3350	3357	3365	3373	3381	1	2	2	3	4	5	5	6	7
.53	3388	3396	3404	3412	3420	3428	3436	3443	3451	3459	1	2	2	3	4	5	6	6	7
.54	3467	3475	3483	3491	3499	3508	3516	3524	3532	3540	1	2	2	3	4	5	6	6	7
.55	3548	3556	3565	3573	3581	3589	3597	3606	3614	3622	1	2	2	3	4	5	6	7	7
.56	3631	3639	3648	3656	3664	3673	3681	3690	3698	3707	1	2	3	3	4	5	6	7	8
.57	3715	3724	3733	3741	3750	3758	3767	3776	3784	3793	1	2	3	3	4	5	6	7	8
.58	3802	3811	3819	3828	3837	3846	3855	3864	3873	3882	1	2	3	4	4	5	6	7	8
.59	3890	3899	3908	3917	3926	3936	3945	3954	3963	3972	1	2	3	4	5	5	6	7	8
.60	3981	3990	3999	4009	4018	4027	4036	4046	4055	4064	1	2	3	4	5	6	6	7	8
.61	4074	4083	4093	4102	4111	4121	4130	4140	4150	4159	1	2	3	4	5	6	7	8	9
.62	4169	4178	4188	4198	4207	4217	4227	4236	4246	4256	1	2	3	4	5	6	7	8	9
.63	4266	4276	4285	4295	4305	4315	4325	4335	4345	4355	1	2	3	4	5	6	7	8	9
.64	4365	4375	4385	4395	4406	4416	4426	4436	4446	4457	1	2	3	4	5	6	7	8	9
.65	4467	4477	4487	4498	4508	4519	4529	4539	4550	4560	1	2	3	4	5	6	7	8	9
.66	4571	4581	4592	4603	4613	4624	4634	4645	4656	4667	1	2	3	4	5	6	7	9	10
.67	4677	4688	4699	4710	4721	4732	4742	4753	4764	4775	1	2	3	4	5	7	8	9	10
.68	4786	4797	4808	4819	4831	4842	4853	4864	4875	4887	1	2	3	4	6	7	8	9	10
.69	4898	4909	4920	4932	4943	4955	4966	4977	4989	5000	1	2	3	5	6	7	8	9	10
.70	5012	5023	5035	5047	5058	5070	5082	5093	5105	5117	1	2	4	5	6	7	8	9	11
.71	5129	5140	5152	5164	5176	5188	5200	5212	5224	5236	1	2	4	5	6	7	8	10	11
.72	5248	5260	5272	5284	5297	5309	5321	5333	5346	5358	1	2	4	5	6	7	8	10	11
.73	5370	5383	5395	5408	5420	5433	5445	5458	5470	5483	1	3	4	5	6	8	9	10	11
.74	5495	5508	5521	5534	5546	5559	5572	5585	5598	5610	1	3	4	5	6	8	9	10	12
.75	5623	5636	5649	5662	5675	5689	5702	5715	5728	5741	1	3	4	5	7	8	9	10	12
.76	5754	5768	5781	5794	5808	5821	5834	5848	5861	5875	1	3	4	5	7	8	9	11	12
.77	5888	5902	5916	5929	5943	5957	5970	5984	5998	6012	1	3	4	5	7	8	10	11	12
.78	6026	6039	6053	6067	6081	6095	6109	6124	6138	6152	1	3	4	6	7	8	10	11	13
.79	6166	6180	6194	6209	6223	6237	6252	6266	6281	6295	1	3	4	6	7	9	10	11	13
.80	6310	6324	6339	6353	6368	6383	6397	6412	6427	6442	1	3	4	6	7	9	10	12	13
.81	6457	6471	6486	6501	6516	6531	6546	6561	6577	6592	2	3	5	6	8	9	11	12	14
.82	6607	6622	6637	6653	6668	6683	6699	6714	6730	6745	2	3	5	6	8	9	11	12	14
.83	6761	6776	6792	6808	6823	6839	6855	6871	6887	6902	2	3	5	6	8	9	11	13	14
.84	6918	6934	6950	6966	6982	6998	7015	7031	7047	7063	2	3	5	6	8	10	11	13	15
.85	7079	7096	7112	7129	7145	7161	7178	7194	7211	7228	2	3	5	7	8	10	12	13	15
.86	7244	7261	7278	7295	7311	7328	7345	7362	7379	7396	2	3	5	7	8	10	12	13	15
.87	7413	7430	7447	7464	7482	7499	7516	7534	7551	7568	2	3	5	7	9	10	12	14	16
.88	7586	7603	7621	7638	7656	7074	7691	7709	7727	7745	2	4	5	7	9	11	12	14	16
.89	7762	7780	7798	7816	7834	7852	7870	7889	7907	7925	2	4	5	7	9	11	13	14	16
.90	7943	7962	7980	7998	8017	8035	8054	8072	8091	8110	2	4	6	7	9	11	13	15	17
.91	8128	8147	8166	8185	8204	8222	8241	8260	8279	8299	2	4	6	8	9	11	13	15	17
.92	8318	8337	8356	8375	8395	8414	8433	8453	8472	8492	2	4	6	8	10	12	14	15	17
.93	8511	8531	8551	8570	8590	8610	8630	8650	8670	8690	2	4	6	8	10	12	14	16	18
.94	8710	8730	8750	8770	8790	8810	8831	8851	8872	8892	2	4	6	8	10	12	14	16	18
.95	8913	8933	8954	8974	8995	9016	9036	9057	9078	9099	2	4	6	8	10	12	15	17	19
.96	9120	9141	9162	9183	9204	9226	9247	9268	9290	9311	2	4	6	8	11	13	15	17	19
.97	9333	9354	9376	9397	9419	9441	9462	9484	9506	9528	2	4	7	9	11	13	15	17	20
.98	9550	9572	9594	9616	9638	9661	9683	9705	9727	9750	2	4	7	9	11	13	16	18	20
.99	9772	9795	9817	9840	9863	9886	9908	9931	9954	9977	2	5	7	9	11	14	16	18	20

Natural Sines

Deg	0' 0°.0	6' 0°.1	12' 0°.2	18' 0°.3	24' 0°.4	30' 0°.5	36' 0°.6	42' 0°.7	48' 0°.8	54' 0°.9	Mean Differences 1	2	3	4	5
0	.0000	0017	0035	0052	0070	0087	0105	0122	0140	0157	3	6	9	12	15
1	.0175	0192	0209	0227	0244	0262	0279	0297	0314	0332	3	6	9	12	15
2	.0349	0366	0384	0401	0419	0436	0454	0471	0488	0506	3	6	9	12	15
3	.0523	0541	0558	0576	0593	0610	0628	0645	0663	0680	3	6	9	12	15
4	.0698	0715	0732	0750	0767	0785	0802	0819	0837	0854	3	6	9	12	15
5	.0872	0889	0906	0924	0941	0958	0976	0993	1011	1028	3	6	9	12	14
6	.1045	1063	1080	1097	1115	1132	1149	1167	1184	1201	3	6	9	12	14
7	.1219	1236	1253	1271	1288	1305	1323	1340	1357	1374	3	6	9	12	14
8	.1392	1409	1426	1444	1461	1478	1495	1513	1530	1547	3	6	9	12	14
9	.1564	1582	1599	1616	1633	1650	1668	1685	1702	1719	3	6	9	12	14
10	.1736	1754	1771	1788	1805	1822	1840	1857	1874	1891	3	6	9	12	14
11	.1908	1925	1942	1959	1977	1994	2011	2028	2045	2062	3	6	9	11	14
12	.2079	2096	2113	2130	2147	2164	2181	2198	2215	2232	3	6	9	11	14
13	.2250	2267	2284	2300	2317	2334	2351	2368	2385	2402	3	6	8	11	14
14	.2419	2436	2453	2470	2487	2504	2521	2538	2554	2571	3	6	8	11	14
15	.2588	2605	2622	2639	2656	2672	2689	2706	2723	2740	3	6	8	11	14
16	.2756	2773	2790	2807	2823	2840	2857	2874	2890	2907	3	6	8	11	14
17	.2924	2940	2957	2974	2990	3007	3024	3040	3057	3074	3	6	8	11	14
18	.3090	3107	3123	3140	3156	3173	3190	3206	3223	3239	3	6	8	11	14
19	.3256	3272	3289	3305	3322	3338	3355	3371	3387	3404	3	5	8	11	14
20	.3420	3437	3453	3469	3486	3502	3518	3535	3551	3567	3	5	8	11	14
21	.3584	3600	3616	3633	3649	3665	3681	3697	3714	3730	3	5	8	11	14
22	.3746	3762	3778	3795	3811	3827	3843	3859	3875	3891	3	5	8	11	14
23	.3907	3923	3939	3955	3971	3987	4003	4019	4035	4051	3	5	8	11	14
24	.4067	4083	4099	4115	4131	4147	4163	4179	4195	4210	3	5	8	11	13
25	.4226	4242	4258	4274	4289	4305	4321	4337	4352	4368	3	5	8	11	13
26	.4384	4399	4415	4431	4446	4462	4478	4493	4509	4524	3	5	8	10	13
27	.4540	4555	4571	4586	4602	4617	4633	4648	4664	4679	3	5	8	10	13
28	.4695	4710	4726	4741	4756	4772	4787	4802	4818	4833	3	5	8	10	13
29	.4848	4863	4879	4894	4909	4942	4939	4955	4970	4985	3	5	8	10	13
30	.5000	5015	5030	5045	5060	5075	5090	5105	5120	5135	3	5	8	10	13
31	.5150	5165	5180	5195	5210	5225	5240	5255	5270	5284	2	5	7	10	12
32	.5299	5314	5329	5344	5358	5373	5388	5402	5417	5432	2	5	7	10	12
33	.5446	5561	5476	5490	5505	5519	5534	5548	5563	5577	2	5	7	10	12
34	.5592	5606	5621	5635	5650	5664	5678	5693	5707	5721	2	5	7	10	12
35	.5736	5750	5764	5779	5793	5807	5821	5835	5850	5864	2	5	7	10	12
36	.5878	5892	5906	5920	5934	5948	5962	5976	5990	6004	2	5	7	9	12
37	.6018	6032	6046	6060	6074	6088	6101	6115	6129	6143	2	5	7	9	12
38	.6157	6170	6184	6198	6211	6225	6239	6252	6266	6280	2	5	7	9	11
39	.6293	6307	6320	6334	6347	6361	6374	6388	6401	6414	2	4	7	9	11
40	.6428	6441	6455	6468	6481	6494	6508	6521	6534	6547	2	4	7	9	11
41	.6561	6574	6587	6600	6613	6626	6639	6652	6665	6678	2	4	7	9	11
42	.6691	6704	6717	6730	6743	6756	6769	6782	6794	6807	2	4	6	9	11
43	.6820	6833	6845	6858	6871	6884	6896	6909	6921	6934	2	4	6	8	11
44	.6947	6959	6972	6984	6997	7009	7022	7034	7046	7059	2	4	6	8	10

Natural Sines

Deg 0' 0°.0	6' 0°.1	12' 0°.2	18' 0°.3	24' 0°.4	30' 0°.5	36' 0°.6	42' 0°.7	48' 0°.8	54' 0°.9	Mean Differences 1	2	3	4	5
45 .7071	7083	7096	7108	7120	7133	7145	7157	7169	7181	2	4	6	8	10
46 .7193	7206	7218	7230	7242	7254	7266	7278	7290	7302	2	4	6	8	10
47 .7314	7325	7337	7349	7361	7373	7385	7396	7408	7420	2	4	6	8	10
48 .7431	7443	7455	7466	7478	7490	7501	7513	7524	7536	2	4	6	8	10
49 .7547	7558	7570	7581	7593	7604	7615	7627	7638	7649	2	4	6	8	9
50 .7660	7672	7683	7694	7705	7716	7727	7738	7749	7760	2	4	6	7	9
51 .7771	7782	7793	7804	7815	7826	7837	7848	7859	7869	2	4	5	7	9
52 .7880	7891	7902	7912	7923	7934	7944	7955	7965	7976	2	4	5	7	9
53 .7986	7997	8007	8018	8028	8039	8049	8059	8070	8080	2	3	5	7	9
54 .8090	8100	8111	8121	8131	8141	8151	8161	8171	8181	2	3	5	7	8
55 .8192	8202	8211	8221	8231	8241	8251	8261	8271	8281	2	3	5	7	8
56 .8290	8300	8310	8320	8329	8339	8348	8358	8368	8377	2	3	5	6	8
57 .8387	8396	8406	8415	8425	8434	8443	8453	8462	8471	2	3	5	6	8
58 .8480	8490	8499	8508	8517	8526	8536	8545	8554	8563	2	3	5	6	8
59 .8572	8581	8590	8599	8607	8616	8625	8634	8643	8652	1	3	4	6	7
60 .8660	8669	8678	8686	8695	8704	8712	8721	8729	8738	1	3	4	6	7
61 .8746	8755	8763	8771	8780	8788	8796	8805	8813	8821	1	3	4	6	7
62 .8829	8838	8846	8854	8862	8870	8878	8886	8894	8902	1	3	4	5	7
63 .8910	8918	8926	8934	8942	8949	8957	8965	8973	8980	1	3	4	5	6
64 .8988	8996	9003	9011	9018	9026	9033	9041	9048	9056	1	3	4	5	6
65 .9063	9070	9078	9085	9092	9100	9107	9114	9121	9128	1	2	4	5	6
66 .9135	9143	9150	9157	9164	9171	9178	9184	9191	9198	1	2	3	5	6
67 .9205	9212	9219	9225	9232	9239	9245	9252	9259	9265	1	2	3	4	6
68 .9272	9278	9285	9291	9298	9304	9311	9317	9323	9330	1	2	3	4	5
69 .9336	9342	9348	9354	9361	9367	9373	9379	9385	9391	1	2	3	4	5
70 .9397	9403	9409	9415	9421	9426	9432	9438	9444	9449	1	2	3	4	5
71 .9455	9461	9466	9472	9478	9483	9494	9494	9500	9505	1	2	3	4	5
72 .9511	9516	9521	9527	9532	9537	9542	9548	9553	9558	1	2	3	3	4
73 .9563	9568	9573	9578	9583	9588	9593	9598	9603	9608	1	2	2	3	4
74 .9613	9617	9622	9627	9632	9636	9641	9646	9650	9655	1	2	2	3	4
75 .9659	9664	9668	9673	9677	9681	9686	9690	9694	9699	1	1	2	3	4
76 .9703	9707	9711	9715	9720	9724	9728	9732	9736	9740	1	1	2	3	3
77 .9744	9748	9751	9755	9759	9763	9767	9770	9774	9778	1	1	2	3	3
78 .9781	9785	9789	9792	9796	9799	9803	9806	9810	9813	1	1	2	2	3
79 .9816	9820	9823	9826	9829	9833	9836	9839	9842	9845	1	1	2	2	3
80 .9848	9851	9854	9857	9860	9863	9866	9869	9871	9874	0	1	1	2	2
81 .9877	9880	9882	9885	9888	9890	9893	9895	9898	9900	0	1	1	2	2
82 .9903	9905	9907	9910	9912	9914	9917	9919	9921	9923	0	1	1	2	2
83 .9925	9928	9930	9932	9934	9636	9638	9940	9942	9943	0	1	1	1	2
84 .9945	9947	9949	9951	9952	9954	9956	9957	9959	9960	0	1	1	1	2
85 .9962	9963	9965	9966	9968	9969	9971	9972	9973	9974	0	0	1	1	1
86 .9976	9977	9978	9979	9980	9981	9982	9983	9984	9985	0	0	1	1	1
87 .9986	9987	9988	9989	9990	9990	9991	9992	9993	9993	0	0	0	1	1
88 .9994	9995	9995	9996	9996	9997	9997	9997	9998	9998	0	0	0	0	0
89 .9998	9999	9999	9999	9999	1.000	1.000	1.000	.000	1.000	0	0	0	0	0
90 1.000														

Natural Cosines
[Numbers in difference columns to be subtracted, not added.]

Deg	0' 0°.0	6' 0°.1	12' 0°.2	18' 0°.3	24' 0°.4	30' 0°.5	36' 0°.6	42' 0°.7	48' 0°.8	54' 0°.9	1	2	Mean Differences 3	4	5
0	1.000	1.000	1.000	1.000	1.000	1.000	9999	9999	9999	9999	0	0	0	0	0
1	.9998	9998	9998	9997	9997	9997	9996	9996	9995	9995	0	0	0	0	0
2	.9994	9993	9993	9992	9991	9990	9990	9989	9988	9987	0	0	0	1	1
3	.9986	9985	9984	9983	9982	9981	9980	9979	9978	9977	0	0	1	1	1
4	.9976	9974	9973	9972	9971	9969	9968	9966	9965	9963	0	0	1	1	1
5	.9962	9960	9959	9957	9956	9954	9952	9951	9949	9947	0	1	1	1	2
6	.9945	9943	9942	9940	9938	9936	9934	9932	9930	9928	0	1	1	1	2
7	.9925	9923	9921	9919	9917	9914	9912	9910	9907	9905	0	1	1	2	2
8	.9903	9900	9898	9895	9893	9890	9888	9885	9882	9880	0	1	1	2	2
9	.9877	9874	9871	9869	9866	9862	9860	9857	9854	9851	0	1	1	2	2
10	.9848	9845	9842	9839	9836	9833	9829	9826	9823	9820	1	1	2	2	3
11	.9816	9813	9810	9806	9803	9799	9796	9792	9789	9785	1	1	2	2	3
12	.9781	9778	9774	9770	9767	9763	9759	9755	9751	9748	1	1	2	3	3
13	.9744	9740	9736	9732	9728	9724	9720	9715	9711	9707	1	1	2	3	3
14	.9703	9699	9694	9690	9686	9681	9677	9673	9668	9664	1	1	2	3	4
15	.9659	9655	9650	9646	9641	9636	9632	9627	9622	9617	1	2	2	3	4
16	.9613	9608	9603	9598	9593	9588	9583	9578	9573	9568	1	2	2	3	4
17	.9563	9558	9553	9548	9542	9537	9532	9527	9521	9516	1	2	3	3	4
18	.9511	9505	9500	9494	9489	9483	9478	9472	9466	9461	1	2	3	4	5
19	.9455	9449	9444	9438	9432	9426	9421	9415	9409	9403	1	2	3	4	5
20	.9397	9391	9385	9379	9373	9367	9361	9354	9348	9342	1	2	3	4	5
21	.9336	9330	9323	9317	9311	9304	9298	9291	9285	9278	1	2	3	4	5
22	.9272	9265	9259	9252	9245	9239	9232	9225	9219	9212	1	2	3	4	6
23	.9205	9198	9191	9184	9178	9171	9164	9157	9150	9143	1	2	3	5	6
24	.9135	9128	9121	9114	9107	9100	9092	9085	9078	9070	1	2	4	5	6
25	.9063	9056	9048	9041	9033	9026	9018	9011	9003	8996	1	3	4	5	6
26	.8988	8980	8973	8665	8957	8949	8942	8934	8926	8918	1	3	4	5	6
27	.8910	8902	8894	8886	8878	8870	8862	8854	8846	8838	1	3	4	5	7
28	.8829	8821	8813	8805	8796	8788	8780	8771	8763	8755	1	3	4	6	7
29	.8746	8738	8729	8721	8712	8704	8695	8686	8678	8669	1	3	4	6	7
30	.8660	8652	8643	8634	8625	8616	8607	8599	8590	8581	1	3	4	6	7
31	.8572	8563	8554	8545	8536	8526	8517	8508	8499	8490	2	3	5	6	8
32	.8480	8471	8462	8453	8443	8434	8425	8415	8406	8396	2	3	5	6	8
33	.8387	8377	8368	8358	8348	8339	8329	8320	8310	8300	2	3	5	6	8
34	.8290	8281	8271	8261	8251	8241	8231	8221	8211	8202	2	3	5	7	8
35	.8192	8181	8171	8161	8151	8141	8131	8121	8111	8100	2	3	5	7	8
36	.8090	8080	8070	8059	8049	8039	8028	8018	8007	7997	2	3	5	7	9
37	.7986	7976	7965	7955	7944	7934	7923	7912	7902	7891	2	4	5	7	9
38	.7880	7869	7859	7848	7837	7826	7815	7804	7793	7782	2	4	5	7	9
39	.7771	7760	7749	7738	7727	7716	7705	7694	7683	7672	2	4	6	7	9
40	.7660	7649	7638	7627	7615	7604	7593	7581	7570	7559	2	4	6	8	9
41	.7547	7536	7524	7513	7501	7490	7478	7466	7455	7443	2	4	6	8	10
42	.7431	7420	7408	7396	7385	7373	7361	7349	7337	7325	2	4	6	8	10
44	.7193	7181	7169	7157	7145	7133	7120	7108	7096	7083	2	4	6	8	10

Natural Sines

Deg	0' 0°.0	6' 0°.1	12' 0°.2	18' 0°.3	24' 0°.4	30' 0°.5	36' 0°.6	42' 0°.7	48' 0°.8	54' 0°.9	1	2	Mean 3	Differences 4	5
45	.7071	7059	7046	7034	7022	7009	6997	6984	6972	6959	2	4	6	8	10
46	.6947	6934	6921	6909	6896	6884	6871	6858	6845	6833	2	4	6	8	11
47	.6820	6807	6794	6782	6769	6756	6743	6730	6717	6704	2	4	6	9	11
48	.6691	6678	6665	6652	6639	6626	6613	6600	6587	6574	2	4	7	9	11
49	.6561	6547	6534	6521	6508	6494	6481	6468	6455	6441	2	4	7	9	11
50	.6428	6414	6401	6388	6374	6361	6347	6334	6320	6307	2	4	7	9	11
51	.6293	6280	6266	6252	6239	6225	6211	6198	6184	6170	2	5	7	9	11
52	.6157	6143	6129	6115	6101	6088	6074	6060	6046	6032	2	5	7	9	12
53	.6018	6004	5990	5976	5962	5948	5934	5920	5906	5892	2	5	7	9	12
54	.5878	5864	5850	5835	5821	5807	5793	5779	5764	5750	2	5	7	9	12
55	.5736	5721	5707	5693	5678	5664	5650	5635	5621	5606	2	5	7	10	12
56	.5592	5577	5563	5548	5534	5519	5505	5490	5476	5461	2	5	7	10	12
57	.5446	5432	5417	5402	5388	5373	5358	5344	5329	5314	2	5	7	10	12
58	.5299	5284	5270	5255	5240	5225	5210	5195	5180	5165	2	5	7	10	12
59	.5150	5135	5120	5105	5090	5075	5060	5045	5030	5015	3	5	8	10	13
60	.5000	4985	4970	4955	4939	4924	4909	4894	4879	4863	3	5	8	10	13
61	.4848	4833	4818	4802	4787	4772	4756	4741	4726	4710	3	5	8	10	13
62	.4695	4679	4664	4648	4633	4617	4602	4586	4571	4555	3	5	8	10	13
63	.4540	4524	4509	4493	4478	4462	4446	4431	4415	4309	3	5	8	10	13
64	.4384	4368	4352	4337	4321	4305	4289	4274	4258	4242	3	5	8	11	13
65	.4226	4210	4195	4179	4163	4147	4131	4115	4099	4083	3	5	8	11	13
66	.4067	4051	4035	4019	4003	3987	3971	3955	3939	3923	3	5	8	11	14
67	.3907	3891	3875	3859	3843	3827	3811	3795	3778	3762	3	5	8	11	14
68	.3746	3730	3714	3697	3681	3665	3649	3633	3616	3600	3	5	8	11	14
69	.3584	3567	3551	3535	3518	3502	3486	3469	3453	3437	3	5	8	11	14
70	.3420	3404	3387	3371	3355	3338	3322	3305	3289	3272	3	5	8	11	14
71	.3256	3239	3223	3206	3190	3173	3156	3140	3123	3107	3	6	8	11	14
72	.3090	3074	3057	3040	3024	3007	2990	2974	2957	2940	3	6	8	11	14
73	.2924	2907	2890	2874	2857	2840	2823	2807	2790	2773	3	6	8	11	14
74	.2756	2740	2723	2706	2689	2672	2656	2639	2622	2605	3	6	8	11	14
75	.2588	2571	2554	2538	2521	2504	2487	2470	2453	2436	3	6	8	11	14
76	.2419	2402	2385	2368	2351	2334	2317	2300	2284	2267	3	6	8	11	14
77	.2250	2233	2215	2198	2181	2164	2147	2130	2113	2096	3	6	9	11	14
78	.2079	2062	2045	2028	2011	1994	1977	1959	1942	1925	3	6	9	11	14
79	.1908	1891	1874	1857	1840	1822	1805	1788	1771	1754	3	6	9	11	14
80	.1736	1719	1702	1685	1668	1650	1633	1616	1599	1582	3	6	9	12	14
81	.1564	1547	1530	1513	1495	1478	1461	1444	1426	1409	3	6	9	12	14
82	.1392	1374	1357	1340	1323	1305	1288	1271	1253	1236	3	6	9	12	14
83	.1219	1201	1184	1167	1149	1132	1115	1097	1080	1063	3	6	9	12	14
84	.1045	1028	1011	0993	0976	0958	0941	0924	0906	0889	3	6	9	12	14
85	.0872	0854	0837	0819	0802	0785	0767	0750	0732	0715	3	6	9	12	15
86	.0698	0680	0663	0645	0628	0610	0593	0576	0558	0541	3	6	9	12	15
87	.0523	0506	0488	0471	0454	0436	0419	0401	0384	0366	3	6	9	12	15
88	.0349	0332	0314	0297	0279	0262	0244	0227	0209	0192	3	6	9	12	15
89	.0175	0157	0140	0122	0105	0087	0070	0052	0035	0017	3	6	9	12	15
90	.0000														